ELEMENTS OF MECHANICAL ENGINEERING

ELEMENTS OF
MECHANICAL ENGINEERING

Mahesh Kumar

B.Tech., M.E., Ph.D.

Department of Mechanical Engineering
Guru Jambheshwar University of Science & Technology
Hisar

TECHSAR™

NEW DELHI

Published and Printed by
TechSar Pvt. Ltd.
4435-36/7, Ansari Road, Daryaganj,
New Delhi – 110 002 (India)
E-mail: tech_info@techsarworld.com
Website: www.techsarworld.com

ISBN 978-93-82332-40-4

Reprint 2025

Preface

This book is written to meet the requirements of first year engineering students. The present book contains eight units. In each unit various examples are solved systematically in S.I. units without omitting any step. Unsolved problems and multiple choice questions with answers have been given at the end of each unit to enable the students to test their understanding of this subject.

I have no befitting words to express my deep sentiments to my parents to whom this book has been dedicated. I would like to express my appreciation to my wife Promila Devi and my children Reena and Tushar for their wholehearted support, understanding, and patience throughout the preparation of this textbook.

I extend gratitude to my departmental colleagues who were a source of encouragement through the course of this writing work. I express sincere thanks to my friend Pankaj Khatak for his valuable suggestions. I am extremely thankful to my student Rakesh Kumar for his help in preparing the drawings. I express my appreciation and thankfulness to the publisher for bringing out the book in a very short period.

At last I would like to thank God (*Bhagwan*) for giving me the knowledge, strength, courage, and patience for successfully accomplishing this book. I pray that He continues the same for the rest of my life.

In this book, the syllabus material has been arranged in simple and understandable form from the various textbooks and other reliable sources. The author does not claim any originality in the topics of the subject. Care has been taken at every stage of proofreading and checking but perfection cannot be claimed. I shall be very grateful to the readers of this book for pointing out any mistake that might have crept in despite my sincere efforts. Constructive and valuable suggestions for further improvement of this book are warmly welcome. I hope that this book will be very useful to the students and the teachers.

Mahesh Kumar

Contents

1

Properties of Steam and Boilers

1.1 INTRODUCTION

Steam, the gaseous phase of water, is the most common working substance in heat engines and turbines due to its large heat carrying capacity, easy and cheap availability, and non-toxic nature. It is a pure substance, i.e., it has homogeneous and fixed chemical composition irrespective of the phase or phases in which it exists. Steam may be wet, saturated, or superheated. The change in the properties of steam during a process fulfils the required objectives. Steam is used for the purposes of electrical power generation, industrial heating, etc.

The device used for generating steam is called a boiler (or steam generator). A boiler is a closed vessel generally made of steel in which steam is generated from water at a desired pressure and temperature by the application of heat. The hot gases formed by the burning of fuel such as coal, oil, or gas transfer heat to water through the heating surface of the boiler. The boilers are equipped with various fittings and devices which are necessary for the safety of the boiler and its efficient operation and control. These fittings and devices are categorized under two headings, namely, boiler mountings and accessories. The performance of a boiler is measured in terms of equivalent evaporation and boiler efficiency. The pressure difference which maintains the flow of gases in the combustion chamber of a boiler is called draught. Draught may be of natural or artificial type.

1.2 FORMATION OF STEAM AT CONSTANT PRESSURE AND ENTHALPY CHANGES

Consider one kg of ice at −10°C contained in a cylinder under 1 atmospheric pressure exerted by a weight W placed on the frictionless piston (Figure 1.1a). Steam is generally formed at constant pressure. Let the heat be supplied gradually to the cylinder by some means so that the temperature increases at constant pressure. The distinct heat addition regimes are shown in $T - Q$ and $T - v$ diagrams in Figure 1.2.

Process 0-1 The temperature of ice increases from –10°C to 0°C. Its volume also increases with the increase in temperature. The state point 1 represents the saturated solid state at which ice starts to melt.

Enthalpy change for the process 0-1 (from –10°C to 0°C) is given by,

$$(dh)_{0-1} = c_i\,(T_1 - T_0) = 2.09 \times [0 - (-10)] = 20.9 \text{ kJ/kg}$$

where c_i is the specific heat of ice whose value is taken as 2.09 kJ/kg.

Process 1-2 The ice begins to melt into water at 0°C with the addition of heat. At state point 2 the melting process ends during which the temperature and pressure remain constant (Figure 1.1b). During melting of ice at 0°C, its volume decreases. The amount of heat (or energy) absorbed during melting is called the latent heat of fusion and it is equivalent to the amount of energy released during freezing. The latent heat of fusion of water at 1 atmospheric pressure is 337.7 kJ/kg.

Enthalpy change for the process 1-2 (at 0°C) is given by,

$$(dh)_{1-2} = \text{Latent heat of fusion of ice} = 337.7 \text{ kJ/kg.}$$

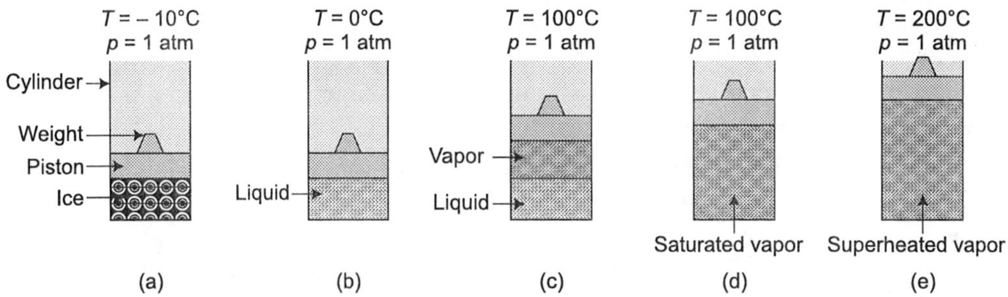

Fig. 1.1

Process 2-3 The water at the temperatures above than 0°C remains in liquid form and is called compressed liquid or sub-cooled liquid, which means it will not vaporize. The addition of more heat beyond state point 2 causes rise in temperature till the water reaches its boiling point which is 100°C at one atmospheric pressure (i.e., 1.01325 bar). At a given pressure, the temperature at which a pure substance changes phase is called saturation temperature. At one atmospheric pressure saturation temperature of water is 100°C. The saturation temperature is a function of saturation pressure. So, a substance at higher pressures boils at higher temperatures and vice versa. The amount of heat added during the process 2-3 is called sensible heat and is denoted by h_f. Sensible heat increases as the pressure increases. When water reaches at state point 3 still it is in liquid form but any further heat addition will vaporize some of its quantity. A liquid that is about to vaporize is called a saturated liquid. So, state 3 is a saturated liquid state. During the process 2-3 with the increase in temperature the volume of water decreases and reaches to its minimum at 4°C and again it increases due to thermal expansion (Figure 1.2).

Enthalpy change for the process 2-3 (from 0°C to 100°C) is given by,

$$(dh)_{2-3} = c_{pw}\,(T_3 - T_2) = 4.186 \times [100 - 0] = 418.6 \text{ kJ/kg}$$

0-1: Solid state; 1-2: Melting (Fusion); 2-3: Liquid state; 3-4: Vaporization; 4-5: Vapor state (Superheating)

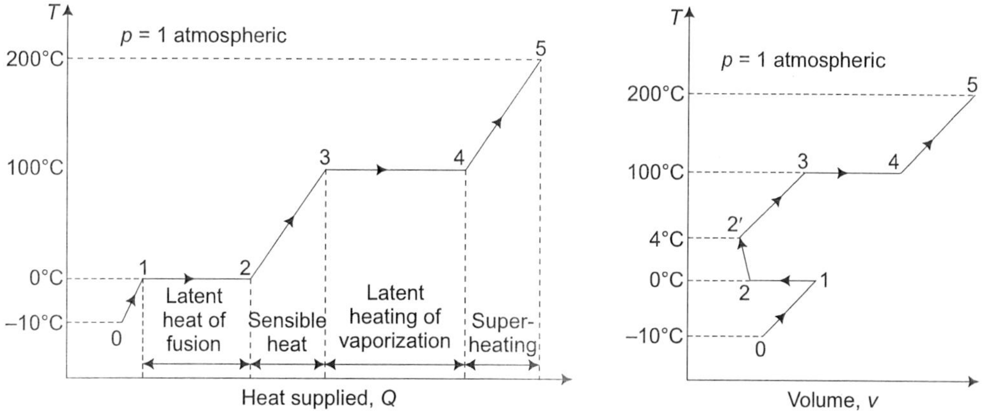

Fig. 1.2

Process 3-4 The water starts boiling at state point 3 and its temperature remains constant until it vaporizes completely at state point 4. Therefore, the boiling or phase change process gets completed at state point 4. The line joining the state points 3 and 4 is called vaporization line. In the midway of the vaporization line, there will be equal amounts of liquid and vapor (Figure 1.1c). The substance between state points 3 and 4 is called saturated liquid-vapor mixture (or wet steam). At the state point 4 the cylinder will be completely filled with vapors (Figure 1.1d). A vapor which is about to condense is called a saturated vapor. At state point 4 any heat loss will lead to condensation of some of the vapors. So, it is called saturated vapor state and the vapors or steam at this stage is called saturated steam (or dry saturated steam).

In wet steam the relative amounts of vapor phase and saturated liquid phase determine the dryness fraction of steam. The dryness fraction of steam is defined as the ratio of the mass of the dry vapor to the total mass of mixture. It is designated by x and is given by,

$$x = \frac{\text{Mass of dry saturated vapor}}{\text{Mass of mixture}} = \frac{m_g}{m_t} = \frac{m_g}{m_f + m_g} \tag{1.1}$$

where m_t = Mass of mixture, m_g = Mass of dry saturated vapor, and m_f = Mass of water vapor (or saturated liquid).

When the dryness fraction is expressed in the percentage it is termed quality of steam and is given by,

$$\text{Quality of steam} = 100x$$

For dry saturated steam, $m_f = 0$

$\therefore$ $x = 1$.

So, quality of dry saturated steam is 100%.

Wetness fraction is denoted by y and it is given by,

$$y = 1 - x = \frac{m_f}{m_f + m_g} \tag{1.2}$$

When wetness fraction is expressed in terms of percentage it is called priming and is given by, $100(1 - x)$. Thus,

$$\text{Quality} + \text{Priming} = 100\%.$$

The amount of heat added during process 3-4 is called latent heat of vaporization (or enthalpy of vaporization) which is denoted by h_{fg} and is equivalent to the energy released during condensation. The latent heat of vaporization for water at 1 atmospheric pressure is $2256.9 \approx 2257$ kJ/kg. During the process 3-4 there will be a large increase in the volume of water due to vapor (or steam) formation.

Enthalpy change for the process 3-4 (at 100°C) is given by,

$$(dh)_{3-4} = \text{Latent heat of vaporization} = 2257 \text{ kJ/kg}.$$

Depending upon the condition of steam the total heat of wet steam which represents the quantity of heat required to convert 1 kg of water at 0°C into wet steam at constant pressure can be obtained as given below:

Let enthalpy of saturated liquid = H_f, and enthalpy of saturated vapor = H_g, then enthalpy of saturated liquid-vapor mixture is given by,

$$H = H_f + H_g$$

$$H = mh$$

$$m_t h = m_f h_f + m_g h_g \qquad \text{(i)}$$

Now substituting $m_f = m_t - m_g$ in equation (i), we have,

$$m_t h = (m_t - m_g)h_f + m_g h_g$$

$$h = \left(1 - \frac{m_g}{m_t}\right) h_f + \frac{m_g}{m_t} h_g$$

So, the total heat of wet steam is given by,

$$h = (1 - x)h_f + xh_g \quad (\because m_g/m_t = x) \qquad (1.3)$$

$$h = h_f - xh_f + xh_g$$

$$h = h_f + x(h_g - h_f)$$

$$= h_f + xh_{fg} \quad (\because h_{fg} = h_g - h_f) \qquad (1.4)$$

The total heat of dry saturated steam represents the quantity of heat required to convert 1 kg of water at 0°C into dry saturated steam at constant pressure. The dryness fraction of dry saturated steam is unity. So, the total heat of dry saturated steam can be obtained by substituting $x = 1$ in equation (1.4), we have,

$$h = h_f + 1 \times (h_g - h_f) = h_g \qquad (1.5)$$

The enthalpy of dry saturated steam increases with the increase in pressure.

Process 4-5 The heat addition process at constant pressure beyond state point 4 is called superheating and the steam (or vapor) so obtained is called superheated steam (Figure 1.1e) which behaves as a perfect gas. During the process 4-5 only vapor (single phase) is there, so further heat addition will result in an increase in both the temperature and volume. At state point 5 the temperature of vapor is very high than the saturation temperature. So, a small amount of heat transfer from the vapor may cause a drop in its temperature but there will not be any condensation as long as its temperature will remain above 100°C (i.e., saturation temperature). Therefore, the vapor at state point 5 is a superheated vapor, means it is not about to condense. The difference between the superheated steam temperature (T_{sup}) and saturated steam temperature (T_{sat}) is called degree of superheating. Thus, degree of superheating is equal to ($T_{sup} - T_{sat}$). The steam occupies more volume of the container when degree of superheat increases.

Enthalpy change for the process 4-5 (from 100°C to say 200°C) is given by,

$$(dh)_{4-5} = c_{ps} (T_5 - T_4) = 2.1 \times [200 - 100] = 210 \text{ kJ/kg.}$$

where c_{ps} is the specific heat of steam which varies from 2.0 kJ/kg to 2.3 kJ/kg.

Depending upon the condition of steam the total heat of superheated steam which represents the quantity of heat required to convert 1 kg of water at 0°C into superheated steam at constant pressure is given by,

$$\begin{aligned} h &= h_f + h_{fg} + c_{ps} (T_{sup} - T_s) \\ &= h_g + c_{ps} (T_{sup} - T_s) \end{aligned} \tag{1.6}$$

Use of superheated steam in thermal power plants has the following advantages:

1. Due to superheating of steam its work capacity increases without increase in its pressure. Superheated steam gives higher thermal efficiency.
2. Superheated steam can be expanded considerably which reduces the heat loss due to condensation to a large extent. At the same time problem of corrosion and erosion of the turbine blades are minimized.
3. Superheated steam has greater heat contents. So, specific fuel consumption is low and thus smaller size of the plant for a given output. Reduction in fuel consumption results in saving the fuel cost.
4. Superheating is done in a superheater by utilizing the heat from the waste gases which otherwise would have passed uselessly to the atmosphere.

1.3 PROPERTY DIAGRAMS FOR PHASE CHANGE PROCESSES

The variations of properties during phase change processes can be studied and understood with the help of properties diagrams such as $T - v$, $p - v$ and $p - T$.

The phase change process for water at one atmospheric pressure is shown in Figure 1.2. Now the same process is repeated by changing the value of pressure by adding different weights on the piston. Assuming that temperature of water in the cylinder-piston arrangement is 0°C. When the heat is transferred to the water at new pressure values the process follows a path like the process at 1 atmospheric pressure but there are some major differences (Figure 1.3) which are given below:

(i) With the increase in pressure the water starts boiling at a higher temperature.

(ii) The specific volume of the saturated liquid increases while the specific volume of saturated vapor decreases. Thus, the line joining the saturated liquid state (i.e., state point 3) and the saturated vapor state (i.e., state point 4) gets shorter and shorter with the increase in pressure. This line becomes a point when the pressure value reaches to 220.09 bar. It means the saturated liquid and saturated vapor states become identical. This point at which the distance between saturated liquid and saturated vapor states becomes zero is called critical point. The temperature, pressure, and volume at the critical point are called critical temperature (T_{cr}), critical pressure (p_{cr}), and critical volume (v_{cr}). The critical point properties of water are T_{cr} = 324.14°C, p_{cr} = 220.09 bar, and v_{cr} = 0.003155 m^3/kg. There will be only a single phase at the pressure above the critical pressure.

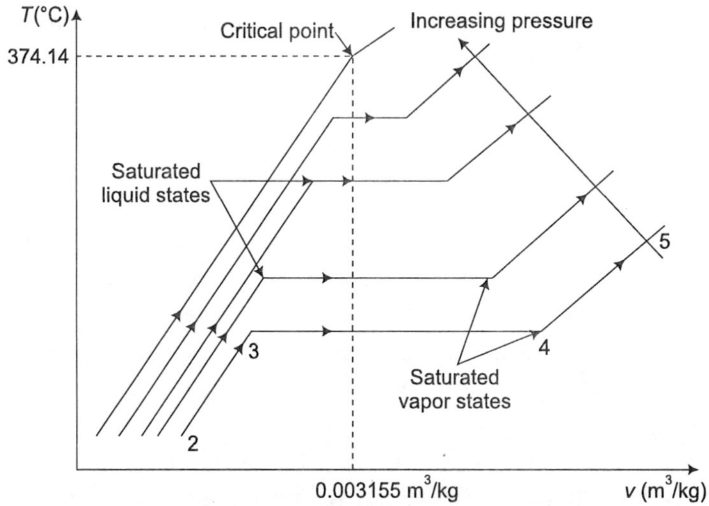

Fig. 1.3

The saturated liquid states in Figure 1.3 can be joined by a line called the saturated liquid line, and the saturated vapor states can be joined by another line, called the saturated vapor line (or dry saturated vapor line). These two lines meet at the critical point and thus form a dome shaped curve as shown in Figure 1.4 (a).

The region left to the saturated liquid line has the compressed liquid states thus it is called compressed liquid region. In the compressed liquid region only liquid phase (water) exists. The region right to the saturated vapor line has the superheated vapor states thus it is called superheated vapor region. In the superheated region only vapor phase (superheated steam) exists. The dome shaped curve region has both the vapor and liquid phase under equilibrium, so this region is called saturated liquid-vapor region or the wet region.

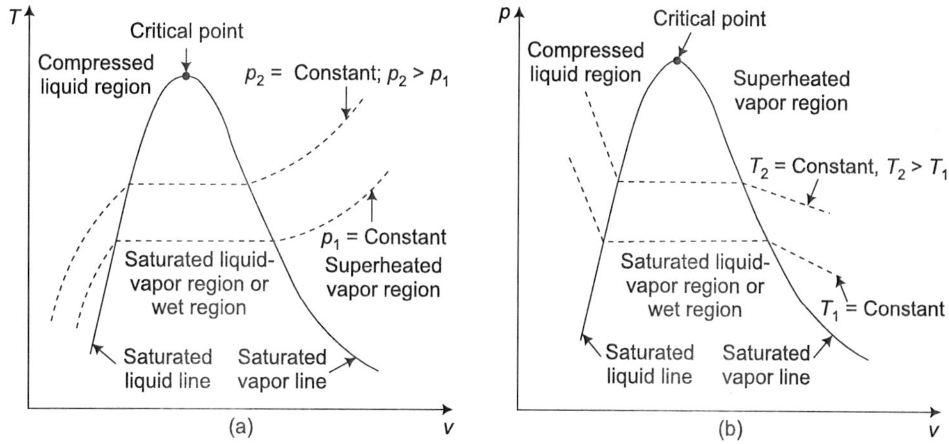

Fig. 1.4

The shape of the $p - v$ diagram is similar to the $T - v$ diagram, but the constant temperature lines have a downward trend as shown in Figure 1.4 (b).

The $p - T$ diagram for water is shown in Figure 1.5 in which solid, liquid, and vapor regions have been depicted.

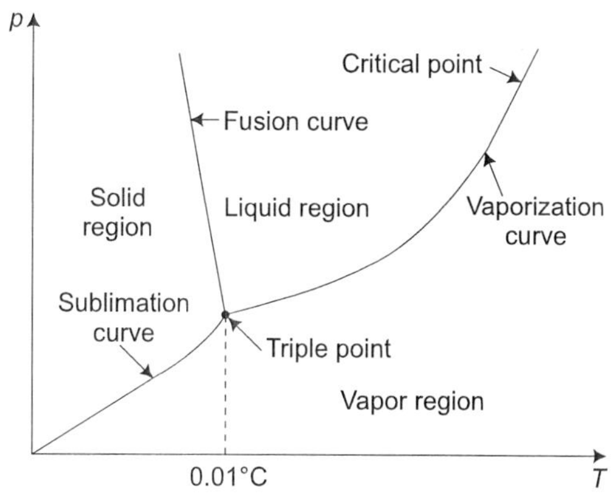

Fig. 1.5

The single-phase regions have been separated from each other by saturation lines. The sublimation, fusion, and vaporization curves meet at triple point at which all the three phases solid, liquid, and vapor coexist in equilibrium. The triple point values for water are: triple point temperature = 0.01°C and triple point pressure = 0.006113 bar. The vaporization curve ends at the critical point where the liquid and vapor phases cannot be distinguished. The fusion curve for water has negative slope, so melting of ice decreases with the increase in pressure. In the sublimation process the solid water or ice directly transforms into saturated vapor without its melting into water.

1.4 THERMODYNAMIC PROPERTIES OF STEAM

The main thermodynamic properties of steam are: pressure (p), temperature (T), specific volume (v), enthalpy (h), internal energy (u), and entropy (s).

(i) Specific Volume The volume per kg of steam is called specific volume of steam. It is denoted by v and is measured in m^3/kg. The specific volume of dry saturated steam is denoted by v_g and the specific volume of saturated liquid (water) is denoted by v_f. The subscript f stands for saturated liquid state and g stands for saturated vapor state.

By applying similar analysis by which enthalpy equation (1.4) is obtained the expression for specific volume can also be derived. Thus, the volume of wet steam is given by,

$$v = v_f + x \times (v_g - v_f)$$

$$= v_f + xv_{fg} \quad (v_{fg} = v_g - v_f) \tag{1.7}$$

At low pressure $v_f << v_g$, so neglecting v_f,

$$\therefore \quad v = xv_g \tag{1.8}$$

The superheated steam behaves like a perfect gas from and at the dry saturated state, so its specific volume with reasonable accuracy at low pressure can be determined by applying Charle's law. Thus

$$\frac{p_g\, v_g}{T_g} = \frac{p_{sup}\, v_{sup}}{T_{sup}}$$

But superheating is done at constant pressure, so $p_g = p_{sup}$. Therefore, specific volume of superheated steam is given by,

$$\frac{v_g}{T_g} = \frac{v_{sup}}{T_{sup}}$$

$$\therefore \quad v_{sup} = v_g \times \frac{T_{sup}}{T_g} \tag{1.9}$$

The specific volume of superheated steam up to a pressure of 20 atmospheric can be calculated by the correlation given by Professor Linds as,

$$v_{sup} = \frac{0.471\, T_{sup}}{p_{sup}} - 0.016 \tag{1.10}$$

where the pressure of superheating (p_{sup}) is in kN/m^2.

(ii) Internal Energy of Steam By applying similar analysis by which enthalpy equation (1.4) is obtained the expression for specific internal energy can also be derived. Thus, the specific internal energy for wet steam is given by,

$$u = u_f + x \times (u_g - u_f)$$

$$= u_f + xu_{fg} \quad (u_{fg} = u_g - u_f) \tag{1.11}$$

Enthalpy (h) is the sum of internal energy (u) and flow energy or flow work (pv). Enthalpy per unit mass is given by,

$$h = u + pv \tag{1.12}$$

$$u = h - pv \tag{1.13}$$

$\therefore$ $$u_f = h_f - pv_f$$

and $$u_g = h_g - pv_g$$

Substituting the values of u_f and u_g in equation (1.11), we have,

$$u = (h_f - pv_f) + x[(h_g - pv_g) - (h_f - pv_f)]$$

$$= (h_f - pv_f) + x[(h_g - h_f) - p(v_g - v_f)]$$

$$= (h_f - pv_f) + x[h_{fg} - pv_{fg}]$$

So, internal energy of wet steam is given by,

$$u = (h_f + xh_{fg}) - p(v_f + xv_{fg}) \tag{1.14}$$

When neglecting v_f,

$$u = (h_f + xh_{fg}) - pxv_g \tag{1.15}$$

Internal energy of dry saturated steam can be obtained by substituting $x = 1$ in equation (1.15), we have,

$$u = (h_f + h_{fg}) - pv_g$$

$$= h_g - pv_g \tag{1.16}$$

Internal energy of superheated steam is given by,

$$u = h_f + h_{fg} + c_{ps}(T_{sup} - T_s) - pv_{sup} \tag{1.17}$$

Internal energy of saturated liquid is given by,

$$u = h_f - p_f v_f \tag{1.18}$$

(iii) Entropy of Steam The entropy change (ds) of steam heating at constant pressure is given by,

$$ds = \frac{\delta Q}{T} = \frac{c_p\, dT}{T} \tag{i}$$

Entropy change from initial state 1 to final state 2 can be obtained by integration of equation (i)

$\therefore$ $$\Delta s = s_2 - s_1 = \int_{T_1}^{T_2} \frac{c_p\, dT}{T} = c_p \ln \frac{T_2}{T_1}$$

If 0°C is taken as the datum then, $s_1 = 0$ and $T_1 = 273$ K, so entropy of steam at any temperature T above the freezing point of water is given by,

$$s = c_p \ln \frac{T}{273} \tag{1.19}$$

Entropy of water at saturation temperature (T_s) is given by,

$$s_f = c_{pw} \ln \frac{T_s}{273} \tag{1.20}$$

The heat addition at constant pressure during evaporation of water at saturation temperature is equal to the latent heat of vaporization. So, change of entropy (or entropy of vaporization) during evaporation is given by,

$$s_{fg} = \frac{h_{fg}}{T_s} \tag{ii}$$

By applying similar analysis by which enthalpy equation (1.4) is obtained the expression for specific entropy can also be derived. Thus, entropy for wet steam is given by,

$$s = s_f + x\,(s_g - s_f)$$

$$= s_f + x s_{fg} \quad (s_{fg} = s_g - s_f) \tag{1.21}$$

Entropy for wet steam can be obtained by substituting equation (ii) in equation (1.21), we have,

$$s = s_f + x\,\frac{h_{fg}}{T_s} \tag{1.22}$$

Entropy of dry saturated steam can be obtained by substituting $x = 1$ in equation (1.22), we have,

$$s = s_f + (1)\,\frac{h_{fg}}{T_s}$$

$$= s_f + s_{fg}$$

$$= s_f + (s_g - s_f) = s_g \tag{1.23}$$

Entropy of superheated steam for water having temperature above 0°C can be given by,

$$s = s_f + s_{fg} + c_{ps} \ln \frac{T_{\text{sup}}}{T_s}$$

$$= s_g + c_{ps} \ln \frac{T_{\text{sup}}}{T_s} \tag{1.24}$$

(iv) **External Work Done during Evaporation and Internal Latent Heat** The heat addition at constant pressure to water causes its evaporation, means its phase changes from saturated water to saturated steam. During this evaporation process its volume increases from v_f to v_g which results in the displacement of the piston, thus external work is obtained. The external work done during evaporation is given by,

$$\text{External work} = p(v_g - v_f) \tag{1.25}$$

At low pressure $v_f << v_g$, so neglecting v_f, we have,

$$\text{External work} = p.v_g \tag{1.26}$$

In the case of wet steam with dryness fraction x, the external work of evaporation is given by,

$$\text{External work} = p(xv_g) \tag{1.27}$$

During the phase change process certain amount of latent heat supplied to saturated water is absorbed in the weakening of the molecular forces and thus causes the expansion. The difference between the latent heat of steam supplied and the external work of evaporation gives the internal latent heat. So, in the case of dry saturated steam it is given by,

$$\text{Internal latent heat} = h_{fg} - pv_g \tag{1.28}$$

1.5 STEAM TABLES

The values of thermodynamic properties of steam such as pressure (p), temperature (T), specific volume (v), enthalpy (h), internal energy (u), and entropy (s) are required to study the thermodynamic cycles and to determine the performance of work producing devices. These properties are determined experimentally or theoretically and are presented in tabular form known as steam tables. In the solution of engineering problems the values of these properties are taken from the steam tables directly. There is no need to calculate them by using many equations each time which is a complex task.

In preparing the tables of properties for water, the internal energy of liquid water at triple point (i.e., 0.01°C) is taken as zero. The enthalpy of saturated water (h_f) at its triple point is negligible. The entropy of saturated water at its triple point is also zero.

The steam tables are presented in two different forms:

1. Saturated water and steam table which may be based on pressure or temperature
   (a) Saturated water and steam table based on temperature (Appendix-Table A-1.1)
   (b) Saturated water and steam table based on pressure (Appendix-Table A-1.2)

   During the phase change process in the liquid-vapor mixture region (wet region) at a constant pressure value the temperature will have a fixed value called saturation temperature and vice versa. So, the state of system can be fixed by a single property either the pressure or temperature. The saturated water and steam tables list the properties at two different states in the mixture region, namely, the saturated liquid state (the properties will have subscript f) and saturated vapor state (the properties will have subscript g).
2. Superheated steam tables (Appendix-Table A-1.3)

   Steam can exist at any temperature above the saturation temperature for a given pressure in the superheated region. Each temperature corresponds to a different state. In the superheated steam tables various properties are listed for various temperatures for a given pressure. In the superheated region, it is necessary to specify two independent variables such as temperature and pressure to describe its state.

The thermodynamic properties of steam which are tabulated in the form of tables are given below:

p = Absolute pressure (bar),
T = Saturation temperature (°C),
v_f = Specific volume of saturated liquid (m^3/kg),
v_g = Specific volume of saturated vapor (m^3/kg),
h_f = Enthalpy of saturated liquid (kJ/kg),
h_g = Enthalpy of saturated vapor (kJ/kg),
s_f = Entropy of saturated liquid (kJ/kgK),
s_g = Entropy of saturated vapor (kJ/kgK)

The steam tables are given in the Appendix Tables A-1.1 to A-1.3 at the end of this book.

The following relations are also applicable:

Change of enthalpy during evaporation,

$$h_{fg} = h_g - h_f \tag{1.29}$$

Change of entropy during evaporation,

$$s_{fg} = s_g - s_f \tag{1.30}$$

Change of volume during evaporation,

$$v_{fg} = v_g - v_f \tag{1.31}$$

Internal energy for all the states of the fluid can be calculated by using the following relationship,

$$u = h - pv \tag{1.32}$$

The properties which are not given in the tables can be determined by the method of interpolation. For example, if we have to determine the values of properties s, h, v, and T for dry saturated steam at 12.6 bar pressure. Then using interpolation (Figure 1.6) by the governing linear equation for the required property say z can be determined by,

$$z = z_1 + \frac{p - p_1}{p_2 - p_1}(z_2 - z_1) \tag{1.33}$$

For saturated enthalpy (h_g) at 12.6 bar:

$$h_g = \text{saturated enthalpy at 12 bar} + \frac{(12.6 - 12)}{(13 - 12)} \times$$
$$(\text{saturation enthalpy at 13 bar} - \text{saturation enthalpy at 12 bar})$$

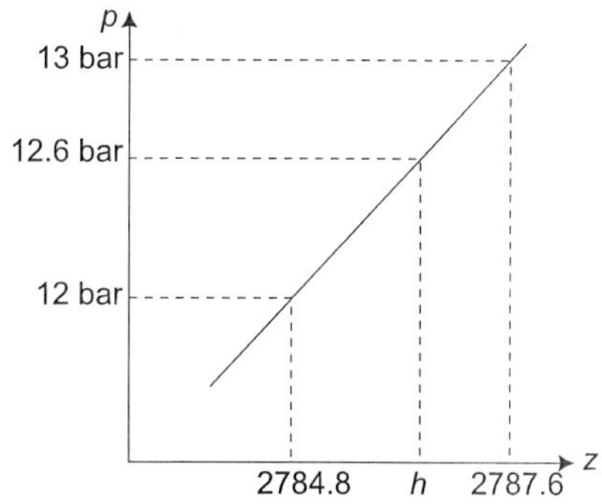

Fig. 1.6

$$\therefore \qquad h_g = 2784.8 + \frac{12.6 - 12}{13 - 12} \times (2787.6 - 2784.8)$$

$$= 2786.48 \text{ kJ/kg}$$

Similarly, entropy and specific volume can be determined as given below:

For entropy (s_g) at 12.6 bar:

$$s_g = 6.5233 + \frac{12.6 - 12}{13 - 12} \times (6.4953 - 6.5233)$$

$$= 6.4811 \text{ kJ/kgK}$$

For specific volume (v_g) at 12.6 bar:

$$v_g = 0.1633 + \frac{12.6 - 12}{13 - 12} \times (0.15125 - 0.1633)$$

$$= 0.15607 \text{ m}^3\text{/kg}$$

For temperature (T_g) at 12.6 bar:

$$T_g = 187.99 + \frac{12.6 - 12}{13 - 12} \times (191.64 - 187.99)$$

$$= 190.18°\text{C}$$

1.6 STEAM PROPERTY DIAGRAMS OR CHARTS

The property values of steam can be presented in graphical forms known as charts or diagrams which are very useful in solving the engineering problems. The two most common diagrams are (i) temperature-entropy ($T - s$) diagram, and (ii) enthalpy-entropy ($h - s$) diagram which is popularly known as Mollier diagram suggested by Dr. R. Mollier in 1904. The salient features of these two diagrams are given below:

(i) Temperature-Entropy (T-s) Diagram Generally, liquid-vapor transformations are of interest, so in Figure 1.7 only liquid and vapor regions for water are shown.

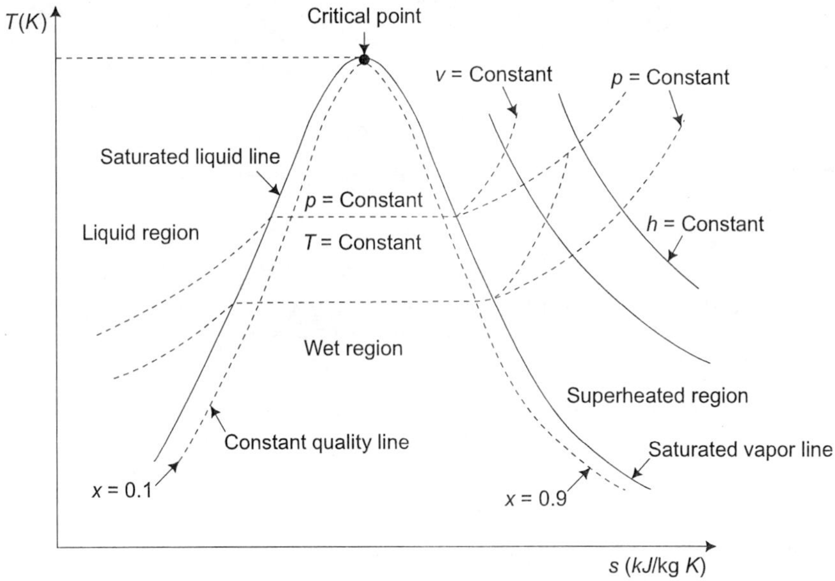

Fig. 1.7 Schematics of $T-s$ Diagram for Water

The absolute temperature is plotted on the axis of ordinate and entropy on the abscissa. The saturated liquid line and saturated vapor line divide the diagram into three regions. To the left of saturated liquid line there is liquid region. Between the saturated liquid line and saturated vapor line there is wet region. To the right of saturated vapor line there is superheated region. The saturated liquid and saturated vapor lines meet at critical point.

In the wet region, the constant pressure lines and constant temperature lines are horizontal and parallel to each other. The nature of constant pressure lines in liquid, wet, and superheated regions is not same and their slopes are governed by the following relations:

$$\left(\frac{\partial T}{\partial s}\right)_v = \frac{T}{c_v} \quad \text{and} \quad \left(\frac{\partial T}{\partial s}\right)_p = \frac{T}{c_p} \tag{1.34}$$

In superheated region, the entropy increases with increase of temperature, so the constant pressure lines are inclined. In this region, the constant volume lines are steeper than constant pressure lines. In any process on $T-s$ diagram, the area under the process line represents the heat addition or rejection.

∴ Heat added or rejected,

$$Q = \int T\,ds \tag{1.35}$$

The isothermal and isentropic (or reversible adiabatic) processes are represented by horizontal and vertical lines respectively. A throttling process is an irreversible process, so it cannot be truly represented in $T-s$ diagram. It is shown by dotted line and the area under this curve does not represent the heat addition. All these three processes are shown in Figure 1.8. The constant quality lines and constant specific volume lines are shown by dotted lines. The constant enthalpy lines are also shown in $T-s$ diagram.

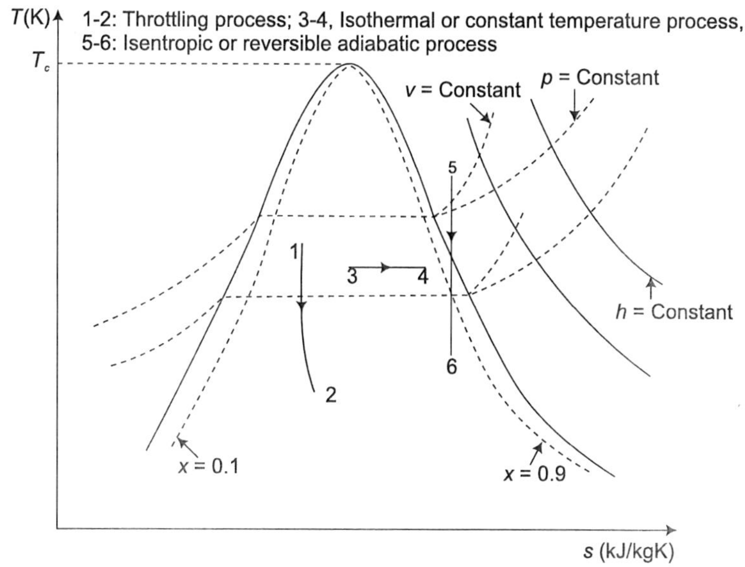

Fig. 1.8

(ii) Enthalpy-Entropy (h-s) Diagram The enthalpy-entropy ($h - s$) diagram is known as Mollier diagram. It is very useful in solving the engineering problems where change of enthalpy occurs such as in steam turbines and nozzles. In this diagram enthalpy is plotted along the axis of ordinate and entropy along the abscissa. Figure 1.9 shows complete Mollier diagram showing saturated liquid and saturated vapor lines which meet at critical point.

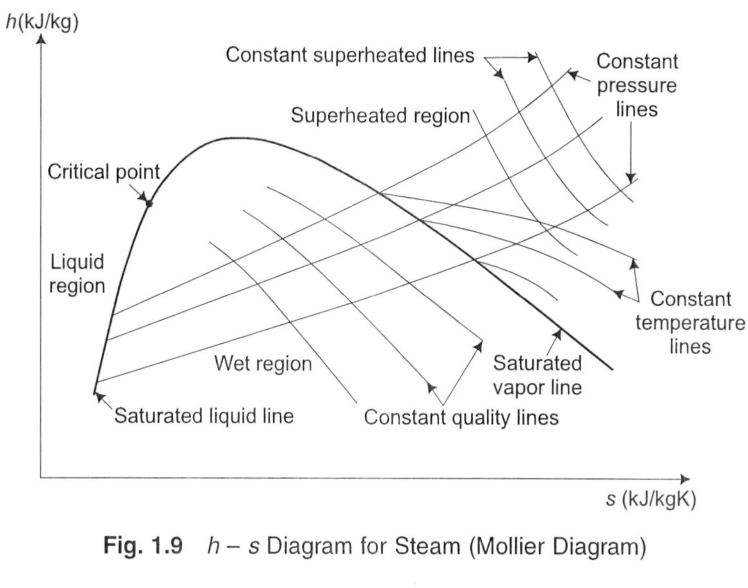

Fig. 1.9 $h - s$ Diagram for Steam (Mollier Diagram)

This diagram has three regions, namely, liquid, wet, and superheated regions. The intercepts between saturated liquid and saturated vapor lines at different pressures are divided into equal parts and these are connected to obtain the constant quality lines. The lines of constant quality are drawn in the diagram and in some diagrams constant specific volume lines may also be drawn. The constant volume lines are steeper than constant pressure lines. Since, $(\partial T/\partial s)_p = T$, so the slope of constant pressure lines will be equal to absolute temperature lines in the $h - s$ diagram. In the wet region, the temperature of steam remains constant thus the constant pressure lines are straight, coincide with constant temperature lines, and have converging tendency towards origin.

In the superheated region, the constant pressure lines bend slightly upward, this could be due to the fact that at high temperature superheated steam behaves like an ideal gas. At the points on the saturation vapor line the constant temperature lines diverge from the constant pressure lines and at low values of pressure and temperature these lines become almost horizontal.

Generally, the saturated liquid region is not required for solving engineering problems. Thus, only a part of this diagram near the saturated vapor region and superheated region is plotted in the commercial Mollier diagram as shown in Figure 1.10.

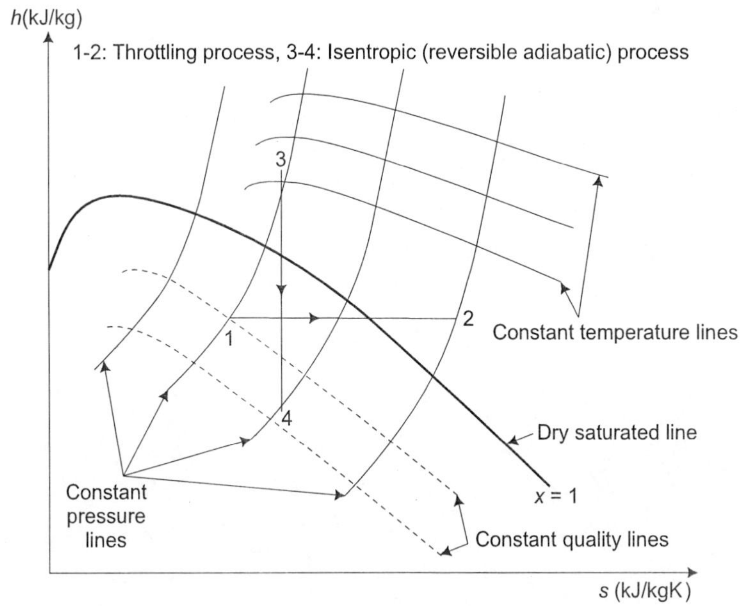

Fig. 1.10 Commercial Mollier Diagram

Throttling process, being a constant enthalpy process, is shown by a horizontal line 1-2 on Mollier chart. And an isentropic (reversible adiabatic) process being a constant entropy process is shown by a vertical line 3-4. During a process all the changes in properties can be known very easily with the help of the Mollier diagram by marking the initial and final conditions of steam on it.

1.7 THERMODYNAMIC PROCESSES FOR STEAM

Steam does not behave like ideal gas and thus ideal gas equation $pv = RT$ is not valid for steam. However, the basic equations derived from the first and second law of thermodynamics can be used for steam. The basic energy equations for non-flow and flow processes are:

$$dq = du + pdv \quad \text{(For non-flow processes)} \tag{1.36}$$

and

$$dq = dh + vdp \quad \text{(For flow processes)} \tag{1.37}$$

The different non-flow processes using vapor (or steam) are given below:

(i) Constant Pressure Process Let the steam undergoe a constant pressure process, for example, steam generation in boilers. Its initial condition is represented by point 1 in the wet region and the final condition by point 2 in the superheated region as shown in Figure 1.11 on $p - v$, $T - s$, and $h - s$ diagrams.

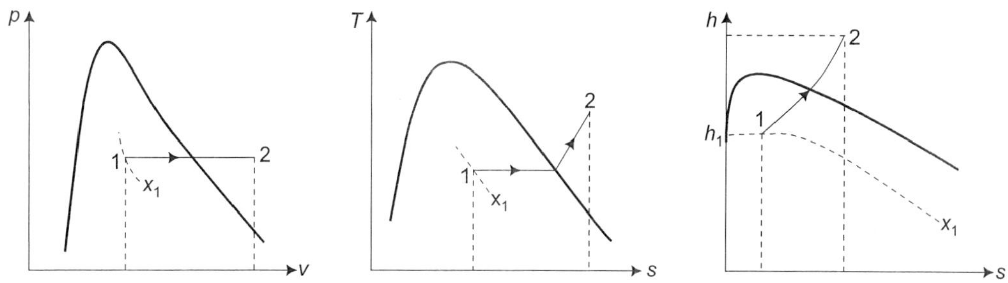

Fig. 1.11

The work done during reversible non-flow process at constant pressure is given by,

$$w_{1-2} = \int_1^2 p\, dv = p(v_2 - v_1) \text{ kJ/kg} \tag{1.38}$$

where $v_1 = v_{f1} + x(v_{g1} - v_{f1})$,

if v_{f1} is neglected then, $v_1 = xv_{g1}$, and

$$v_2 = v_{sup} = v_{g2} \times \frac{T_{sup}}{T_{sat}}$$

The heat transferred during reversible non- flow process at constant pressure is given by,

$$dq = du + pdv$$

or

$$q_{1-2} = (u_2 - u_1) + \int_1^2 pdv$$

$$q_{1-2} = (u_2 - u_1) + p_2 v_2 - p_1 v_1$$

$$= (u_2 + p_2 v_2) - (u_1 + p_1 v_1)$$

$$= (h_2 - h_1) \text{ kJ/kg} \tag{1.39}$$

where $$h_1 = h_{f1} + x(h_{g1} - h_{f1})$$

(ii) Constant Volume Process The constant volume heating process is represented on $p - v$, $T - s$, and $h - s$ diagrams as shown in Figure 1.12. The steam is heated from state point 1 (wet region) at pressure p_1 to state point 2 (superheated region) at pressure p_2.

$$v_1 = v_{f1} + x\,(v_{g1} - v_{f1}),$$

and $$v_2 = v_{sup} = v_{g2} \times \frac{T_{sup}}{T_{sat}}$$

As $v_1 = v_2$, so, $v_{f1} + x\,(v_{g1} - v_{f1}) = v_2$

$\therefore$ $$x = \frac{v_2 - v_{f1}}{v_{g1} - v_{f1}}$$

Work done, $$w_{1-2} = \int_1^2 p\,dv = p(v_2 - v_1)$$

$$= p(0) = 0 \quad (\because v_2 = v_1) \tag{1.40}$$

The heat transferred is given by,

$$q_{1-2} = (u_2 - u_1) + \int_1^2 p\,dv$$

$$= (u_2 - u_1) + 0$$

$$= (h_2 - p_2\,v_2) - (h_1 - p_1\,v_1) \text{ kJ/kg} \tag{1.41}$$

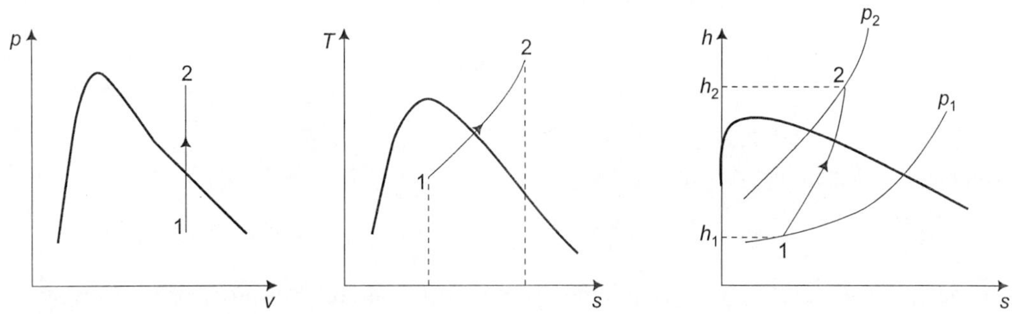

Fig. 1.12

(iii) Constant Temperature or Isothermal Process The constant temperature heating process is represented on $p - v$, $T - s$, and $h - s$ diagrams as shown in Figure 1.13.

Let the initial condition of steam be wet (state point 1) and final condition be superheated (state point 2).

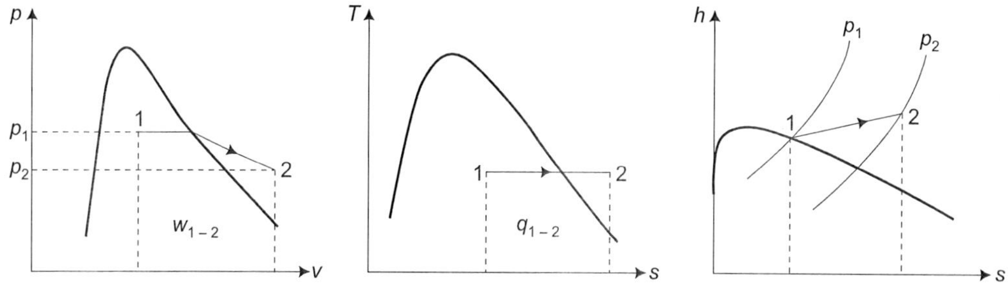

Fig. 1.13

Heat transferred is given by,

$$q_{1-2} = \int_1^2 T ds = T \int_1^2 ds$$

$$= T(s_2 - s_1) \text{ kJ/kg} \quad (1.42)$$

Work done for non-flow process is given by,

$$w_{1-2} = q_{1-2} - (u_2 - u_1) \quad (1.43)$$

Work done for steady flow process is given by,

$$w_{1-2} = q_{1-2} - (h_2 - h_1) \quad (1.44)$$

(iv) Reversible Adiabatic Process (or Isentropic Process) Let the steam undergoes an adiabatic expansion process. If the process is reversible adiabatic (i.e., isentropic process) then the process is represented by 1-2 on $p - v$, $T - s$, and $h - s$ diagrams as shown in Figure 1.14. If the process is irreversible adiabatic, the process is represented by $1 - 2'$ showing increase in entropy due to irreversibility.

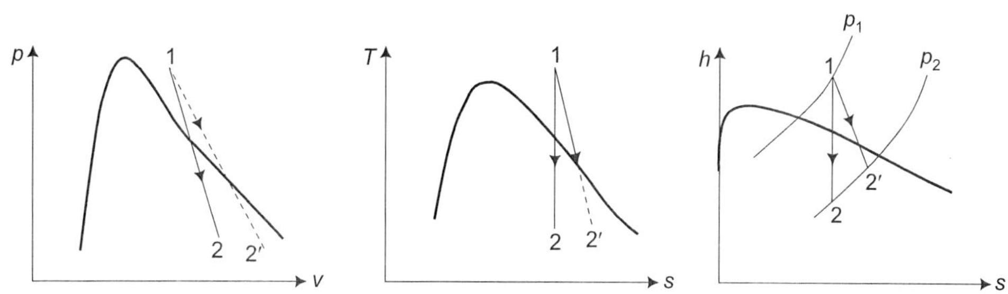

Fig. 1.14

Process is isentropic,

$$\therefore \quad s_1 = s_2$$

$$s_1 = s_{sup} = s_{g1} + c_{ps} \ln \frac{T_{sup}}{T_s}$$

and $$s_2 = s_{\text{wet}} = s_{f2} + xs_{fg2}$$

$$s_{f2} + xs_{fg2} = s_{g1} + c_{ps} \ln \frac{T_{\text{sup}}}{T_s}$$

$$\therefore \quad x = \frac{s_{g1} + c_{ps} \ln (T_{\text{sup}}/T_s) - s_{f2}}{s_{fg2}} \tag{1.45}$$

The heat transferred for non-flow reversible adiabatic process is given by,

$$q_{1-2} = (u_2 - u_1) + \int_1^2 p dv$$

$$= (u_2 - u_1) + w_{1-2}$$

The process is adiabatic,

$$\therefore \quad q_{1-2} = 0$$

$$\therefore \quad w_{1-2} = (u_1 - u_2) \quad \text{(For reversible process)} \tag{1.46}$$

$$w_{1-2} = (u_1 - u_{2'}) \quad \text{(For irreversible process)} \tag{1.47}$$

Work lost for non-flow reversible adiabatic due to irreversibility is given by,

$$w_{\text{lost}} = \text{Eq. (1.46)} - \text{Eq. (1.47)}$$

$$= (u_1 - u_2) - (u_1 - u_{2'}) = (u_{2'} - u_2) \tag{1.48}$$

The heat balance for the steady flow reversible adiabatic process can be written as,

$$u_1 + p_1 v_1 + q - w = u_2 + p_2 v_2$$

So, $$h_1 + 0 - w = h_2 \quad [\because q = 0,\ u + pv = h]$$

$$\therefore \quad w = (h_1 - h_2) \quad \text{(For reversible process)} \tag{1.49}$$

$$w = (h_1 - h_{2'}) \quad \text{(For irreversible process)} \tag{1.50}$$

Work lost for steady-flow reversible adiabatic due to irreversibility is given by,

$$w_{\text{lost}} = \text{Eq. (1.49)} - \text{Eq. (1.50)}$$

$$= (h_1 - h_2) - (h_1 - h_{2'}) = (h_{2'} - h_2) \tag{1.51}$$

Isentropic efficiency, η_{isen} can be given by,

$$\eta_{\text{isen}} = \frac{h_1 - h_{2'}}{h_1 - h_2} \tag{1.52}$$

(v) Polytropic Process Let the steam undergoes polytropic expansion and it follows the law $pv^n = C$. The polytropic process 1-2 is shown on $p - v$ diagram in Figure 1.15 in which state point 1 lies in superheated region and state 2 is in wet region.

The work done for non-flow process during polytropic expansion (changes in kinetic energy and potential energy have been neglected) is given by,

$$w_{1-2} = \int_1^2 p dv = \frac{p_1 v_1 - p_2 v_2}{n-1} \text{ kJ/kg} \tag{1.53}$$

Work done for steady flow process during polytropic expansion is given by,

$$(w_{1-2})_s = -\int_1^2 v dp = \frac{n}{n-1}(p_1 v_1 - p_2 v_2) \text{ kJ/kg} \tag{1.54}$$

Heat transfer for non-flow process during polytropic expansion is given by,

$$q_{1-2} = (u_2 - u_1) + \int_1^2 p dv$$

$$= (u_2 - u_1) + \frac{p_1 v_1 - p_2 v_2}{n-1} \text{ kJ/kg} \tag{1.55}$$

Heat transfer for steady flow process during polytropic expansion is given by,

$$(q_{1-2})_s = dh - \int_1^2 v dp$$

$$= (h_2 - h_1) + \frac{n}{n-1}(p_1 v_1 - p_2 v_2) \text{ kJ/kg} \tag{1.56}$$

$$(q_{1-2})_s = (u_2 + p_2 v_2) - (u_1 + p_1 v_1) + \frac{n}{n-1}(p_1 v_1 - p_2 v_2) \text{ kJ/kg}$$

$$(q_{1-2})_s = (u_2 - u_1) + (p_1 v_1 - p_2 v_2)\left(\frac{n}{n-1} - 1\right) \text{ kJ/kg}$$

$$\therefore \quad (q_{1-2})_s = (u_2 - u_1) + \frac{p_1 v_1 - p_2 v_2}{n-1} \text{ kJ/kg} \tag{1.57}$$

From equations (1.55) and (1.57) it is observed that heat transferred during flow and non-flow polytropic process is same.

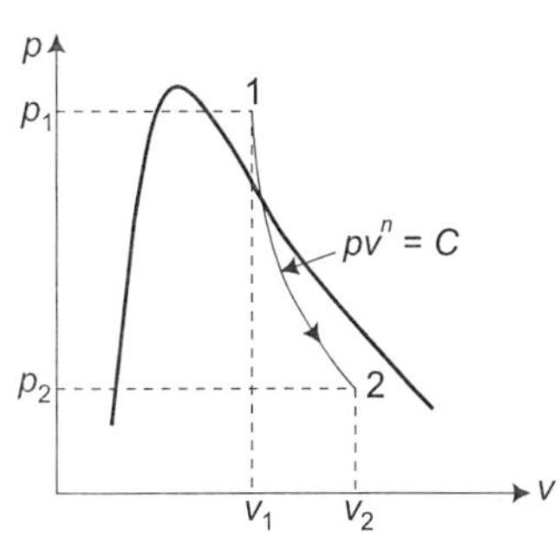

Fig. 1.15

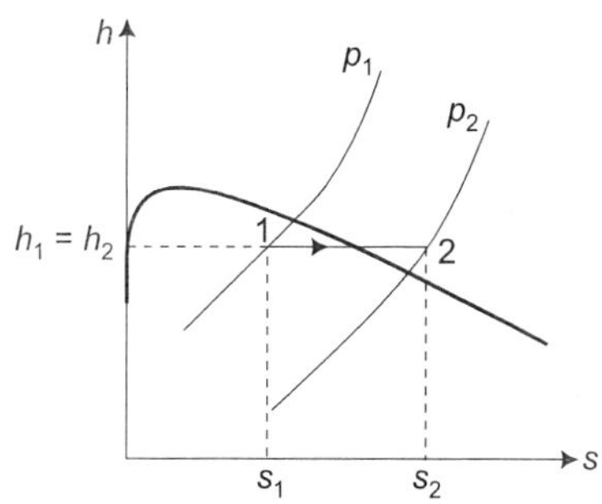

Fig. 1.16

(vi) Throttling Process In the throttling process enthalpy remains constant. The throttling process is an adiabatic irreversible process. This process finds applications in (i) determining the dryness fraction of steam, (ii) controlling the speed of primemovers such as engines and turbines, and (iii) in the refrigeration systems to reduce the pressure and temperature of liquid refrigerant.

Let the initial condition of steam be in wet region represented by point 1 on $h - s$ diagram in Figure 1.16. The final condition of steam lies in superheated region which is denoted by state point 2. In the throttling process,

$$h_1 = h_2, \qquad p_1 > p_2,$$

$$T_1 > T_2, \qquad s_2 > s_1.$$

In the throttling process,

$$h_1 = h_2$$

$$\therefore \quad h_{f1} + x_1\, h_{fg1} = h_{g2} + c_{ps}\,(T_{sup} - T_{g2}) \tag{1.58}$$

If the steam remains only in the wet region,

$$h_{f1} + x_1\, h_{fg1} = h_{f2} + x_2\, h_{fg2}$$

Example 1.1 Determine the dryness fraction of steam which has 1.5 kg of water in suspension with 50 kg of dry saturated steam.

Solution Dryness fraction,

$$x = \frac{m_g}{m_f + m_g} = \frac{50}{50 + 1.5} = 0.9709$$

Example 1.2 Determine the condition of steam, i.e., whether it is wet, dry, or superheated for the following cases: (i) steam has a pressure of 12 bar and specific volume 0.14 m^3/kg, (ii) steam has a pressure of 17.5 bar and a temperature of 250°C, and (iii) steam has a pressure of 25 bar and enthalpy is 2750 kJ/kg.

Solution

(i) From steam table (Table A-1.2), specific volume of dry saturated steam at 12 bar is $v_g = 0.1633$ m^3/kg. The specific volume of the given steam is less than the specific volume of the dry saturated steam. So, the steam is wet and its dryness fraction is given by,

$$v = (1 - x)v_f + x.v_g$$

or

$$0.14 = (1 - x) \times 0.001139 + x \times 0.1633$$

$$\therefore \quad x = \frac{0.138861}{0.162161} = 0.8563$$

(ii) From steam table (Table A-1.2), saturation temperature corresponding to 17.5 bar is 205.76°C. But the temperature of the given steam is 250°C, so the steam is superheated. The degree of superheat = 250 – 205.76 = 44.24°C.

(iii) At 25 bar the dry saturated steam has enthalpy of 2803.1 kJ/kg (from steam table, Table A-1.2), but the enthalpy of given steam is 2750 kJ/kg which is less than the dry saturated steam at that pressure. Thus, the steam is wet.

For determining the dryness fraction applying the relation,

$$h = (1 - x)v_f + xh_g$$

or $$2750 = (1 - x) \times 962.09 + x \times 2803.1$$

$\therefore$ $$x = \frac{1787.91}{1841.01} = 0.9711$$

Example 1.3 A vessel of 0.32 m^3 contains 1.6 kg mixture of water and steam in equilibrium at a pressure of 5.5 bar. Determine the volume and mass of liquid, the volume and mass of water, the enthalpy, and the entropy of steam.

Solution The specific volume of the given mixture,

$$v = (0.32/1.6) = 0.2 \text{ m}^3/\text{kg}$$

The specific volume of saturated liquid and saturated vapor at 5.5 bar from the steam table (Table A-1.2),

$$v_f = 0.001097 \text{ m}^3/\text{kg},$$

and $$v_g = 0.3427 \text{ m}^3/\text{kg}$$

$$v = (1 - x)v_f + xv_g$$

or $$0.2 = (1 - x) \times 0.001097 + x \times 0.3427$$

$\therefore$ $$x = 0.5319$$

$$x = \frac{\text{Mass of dry saturated vapor}}{\text{Mass of mixture}} = \frac{m_g}{m_t} = \frac{m_g}{m_f + m_g}$$

$$m_g = x \times (m_f + m_g)$$

$\therefore$ Mass of vapor, $$m_g = 0.5319 \times 1.6 = 0.851 \text{ kg}$$

$\therefore$ Mass of liquid, $$m_f = 1.6 - 0.851 = 0.749 \text{ kg}$$

$$\text{Volume of vapor} = m_g \times v_g$$

$$= 0.851 \times 0.3427 = 0.2916 \text{ m}^3, \text{ and}$$

$$\text{Volume of liquid} = m_f \times v_f$$

$$= 0.749 \times 0.001097 = 0.00082 \text{ m}^3$$

The enthalpy is given by,

$$h = (1 - x)h_f + xh_g$$

$$= (1 - 0.5319) \times 655.91 + 0.5319 \times 2752.9$$

$$= 1771.29 \text{ kJ/kg}$$

The entropy is given by,

$$s = (1 - x)s_f + xs_g$$

$$= (1 - 0.5319) \times 1.8972 + 0.5319 \times 6.7892$$

$$= 4.4992 \text{ kJ/kgK}$$

Example 1.4 2 kg of saturated liquid water is kept in a container at constant pressure of 5 bar. It is heated until it becomes 70% dry. Determine the volume and enthalpy changes.

Solution The volume and enthalpy at the initial state are obtained from steam table (Table A-1.2) corresponding to 5 bar pressure:

$$v_1 = v_f = 0.001093 \text{ m}^3\text{/kg},$$

and $$h_1 = h_f = 640.21 \text{ kJ/kg}$$

The volume and enthalpy at the final state can be obtained on the basis of 70% dry are:

$$v_2 = (1 - x)\ v_f + xv_g$$

∴ $$v_2 = (1 - 0.70) \times 0.001093 + 0.70 \times 0.3749$$

$$= 0.26276 \text{ m}^3\text{/kg}$$

$$h_2 = (1 - x)\ h_f + xh_g$$

∴ $$h_2 = (1 - 0.70) \times 640.21 + 0.70 \times 2748.7$$

$$= 2116.15 \text{ kJ/kg}$$

Change in volume and enthalpy can be given by,

$$\Delta V = m(v_2 - v_1)$$

$$= 2 \times (0.26276 - 0.001093) = 0.5233 \text{ m}^3$$

$$\Delta H = m(h_2 - h_1)$$

$$= 2 \times (2116.15 - 640.21) = 2951.88 \text{ kJ}$$

Example 1.5 A cylindrical vessel of radius 20 cm and length 100 cm contains saturated steam and water mixture at 250°C. Determine the mass of each if their volumes are equal. Also determine the dryness fraction of steam.

Solution The volume of the vessel $= \pi \times r^2 \times h$

$$= \pi \times (0.2)^2 \times 0.1 = 0.012566 \text{ m}^3$$

Therefore, volume of vapor and water

$$\frac{0.012566}{2} = 0.00628 \text{ m}^3$$

From steam table (Table A-1.1), the specific volume corresponding to saturation temperature 250°C,

$$v_f = 0.001251 \text{ m}^3/\text{kg},$$

and $$v_g = 0.0501 \text{ m}^3/\text{kg}$$

∴ Mass of liquid,

$$m_f = \frac{0.00628}{0.001251} = 5.022 \text{ kg}$$

∴ Mass of vapor,

$$m_g = \frac{0.00628}{0.0501} = 0.1253 \text{ kg}$$

$$v = (1 - x)\, v_f + x v_g$$

or $$0.012566 = (1 - x) \times 0.001251 + x \times 0.0501$$

∴ $$x = 0.2316$$

Example 1.6 Determine the amount of heat required to generate 4.5 kg of steam at a pressure of 10 bar and temperature of 230°C from water at 20°C. Take specific heat for superheated steam equal to 2.1 kJ/kg K.

Solution The steam properties at 10 bar pressure from steam table (Table A-1.2),

$$t_s = 179.91°\text{C}, \qquad h_f = 762.79 \text{ kJ/kg},$$

and $$h_g = 2778.1 \text{ kJ/kg}$$

Total heat of superheated steam is given by,

$$h = h_f + h_{fg} + c_{ps}\,(T_{\text{sup}} - T_s)$$

$$= h_g + c_{ps}\,(T_{\text{sup}} - T_s)$$

∴ $$h = 2778.1 + 2.1 \times (230 - 179.91)$$

$$= 2882.29 \text{ kJ/kg}$$

Heat contents of one kg of water at 20°C is given by,

$$h_w = 1 \times 4.186 \times (20 - 2)$$

$$= 83.72 \text{ kJ/kg}$$

Amount of heat required per kg of steam,

$$q = 2882.29 - 83.72$$

$$= 2798.57 \text{ kJ/kg}$$

Therefore, total amount of heat required,

$$Q = 4.5 \times 2798.57$$
$$= 12593.56 \text{ kJ}$$

Example 1.7 Determine the enthalpy and entropy of steam when the pressure is 20 bar and the specific volume is 0.0795 m^3/kg.

Solution The steam properties at 20 bar pressure from steam tables (Table A-1.2),

$$v_f = 00.001177 \text{ m}^3\text{/kg}, \qquad v_g = 0.09963 \text{ m}^3\text{/kg},$$
$$h_f = 908.77 \text{ kJ/kg}, \qquad h_g = 22799.5 \text{ kJ/kg},$$
$$s_f = 22.4473 \text{ kJ/kgK}, \quad \text{and} \qquad s_g = 6.3408 \text{ kJ/kgK}$$

Since $v_f < v < v_g$

$\therefore$ We have mixture of liquid and vapor

$$v = (1 - x)\, v_f + xv_g$$
$$0.0795 = (1 - x) \times 0.001177 + x \times 0.09963$$
$$\therefore \qquad x = 0.7955$$

The enthalpy is given by,

$$h = (1 - x)\, h_f + xh_g$$
$$= (1 - 0.7955) \times 908.77 + 0.7955 \times 2799.5$$
$$= 2412.84 \text{ kJ/kg}$$

The entropy is given by,

$$s = (1 - x)\, s_f + xs_g$$
$$= (1 - 0.7955) \times 2.4473 + 0.7955 \times 6.3408$$
$$= 5.5446 \text{ kJ/kgK}$$

Example 1.8 Find the mass of 0.30 m^3 of wet steam at 6 bar pressure and 0.90 dryness fraction. Also determine, (i) entropy of one kg of steam, and (ii) the total heat content of one m^3 of this steam.

Solution The steam properties at 6 bar pressure from steam table (Table A-1.2),

$$v_f = 0.001101 \text{ m}^3\text{/kg}, \qquad v_g = 0.3157 \text{ m}^3\text{/kg},$$
$$h_f = 670.54 \text{ kJ/kg}, \qquad h_g = 2756.8 \text{ kJ/kg},$$
$$s_f = 1.9311 \text{ kJ/kgK}, \quad \text{and} \qquad s_g = 6.7600 \text{ kJ/kgK}$$

$$v = (1 - x)\, v_f + xv_g$$
$$v = (1 - 0.90) \times 0.001101 + 0.90 \times 0.3157$$
$$= 0.2842 \text{ m}^3\text{/kg}$$

$$\text{Mass} = \frac{0.30}{0.2842}$$

$$= 1.055 \text{ kg}$$

$$\text{Mass of 1 m}^3 \text{ of steam} = \frac{1}{0.9 \times 0.2846}$$

$$= 3.909 \text{ kg}$$

The entropy is given by,

$$s = (1 - x)\, s_f + x s_g$$

$$= (1 - 0.90) \times 1.9311 + 0.90 \times 6.7600$$

$$= 6.2771 \text{ kJ/kgK}$$

The enthalpy of 1 kg of steam is given by,

$$h = (1 - x)\, h_f + x h_g$$

$$= (1 - 0.90) \times 670.54 + 0.90 \times 2756.8$$

$$= 2548.17 \text{ kJ/kg}$$

∴ Total heat content of steam

$$= 3.909 \times 2548.17 = 9960.79 \text{ kJ}$$

Example 1.9 The pressure of steam in a condenser is 0.15 bar and the dryness fraction is 0.88. Determine the amount of heat that must be abstracted from steam to condense (i) 1 kg of steam, and (ii) 1 m^3 of steam.

Solution The steam properties at 0.15 bar pressure from steam tables (Table A-1.2),

$v_f = 0.001014$ m^3/kg, $v_g = 10.022$ m^3/kg,

$h_f = 225.91$ kJ/kg, and $h_g = 2599.1$ kJ/kg

Heat to be removed per kg of steam is given by,

$$h = x(h_g - h_f)$$

$$= 0.88 \times (2599.1 - 225.91) = 2088.40 \text{ kJ/kg}$$

Specific volume of steam of 0.88 dryness fraction (neglecting volume of water),

$$v = x v_g$$

$$v = 0.88 \times 10.022 = 8.8194 \text{ m}^3\text{/kg}$$

∴ Heat to be removed per m^3 of steam

$$= \frac{2088.41}{8.8194} = 236.79 \text{ kJ}$$

Example 1.10 Determine the enthalpy, volume, internal energy, and entropy of superheated steam at 14 bar pressure and 200°C. Neglect the volume of water and take specific heat of superheated steam equal to 2.1 kJ/kg.

Solution The steam properties at 14 bar pressure from steam tables (Table A-1.2),

$v_g = 0.14084$ m^3/kg, $h_f = 830.29$ kJ/kg,

$h_g = 2790.0$ kJ/kg, $s_f = 2.2842$ kJ/kgK,

$s_g = 66.4692$ kJ/kgK, $h_{fg} = 1959.71$ kJ/kg,

$s_{fg} = 4.185$ kJ/kgK, and $T_{sat} = 195.07$°C

Total heat of 1 kg of steam is given by,

$$h = h_f + h_{fg} + c_{ps}\,(T_{sup} - T_s)$$

$$= h_g + c_{ps}\,(T_{sup} - T_s)$$

$$\therefore \quad h = 830.29 + 1959.71 + 2.1 \times (200 - 195.07)$$

$$= 2800.35 \text{ kJ/kg}$$

Volume of superheated steam is given by,

$$v_{sup} = v_g \times \frac{T_{sup}}{T_g}$$

$$= 0.14084 \times \frac{(200 + 273)}{(195.07 + 273)}$$

$$= 0.1423 \text{ m}^3\text{/kg}$$

Internal energy of superheated steam is given by,

$$u = [h_f + h_{fg} + c_{ps}\,(T_{sup} - T_s)] - pv_{sup}$$

$$= [2800.35] - \frac{14 \times 10^5 \times 0.1423}{10^3}$$

$$= 2800.35 - 199.22 = 2601.13 \text{ kJ/kg}$$

The entropy of superheated steam is given by,

$$s = s_f + s_{fg} + c_{ps} \ln \frac{T_{sup}}{T_s}$$

$$\therefore \quad s = 2.2846 + 4.185 + 2.1 \times \ln \frac{(200 + 273)}{(195.07 + 273)}$$

$$= 6.4912 \text{ kJ/kgK}$$

Example 1.11 Determine the change in internal energy when 1 kg of steam expands from 12 bar and 300°C to 0.5 bar and 0.85 dry. Take specific heat of superheated steam equal to 2.2 kJ/kg.

Solution The steam properties at 12 bar pressure from steam table (Table A-1.2),

$$v_g = 0.1633 \text{ m}^3/\text{kg}, \quad \text{and} \quad h_g = 2784.8 \text{ kJ/kg}$$

Internal energy of superheated steam is given by,

$$u_1 = h_f + h_{fg} + c_{ps}(T_{sup} - T_s) - pv_{sup}$$

$$= h_g + c_{ps}(T_{sup} - T_s) - pv_g \times \frac{T_{sup}}{T_g}$$

$$u_1 = 2784.8 + 2.2 \times (300 - 187.99) - \frac{12 \times 10^5}{10^3} \times 0.1633 \times \frac{573}{460.99}$$

$$= 2787.65 \text{ kJ/kg}$$

The steam properties at 0.5 bar from steam table (Table A-1.2),

$$v_g = 3.240 \text{ m}^3/\text{kg}, \quad h_f = 340.47 \text{ kJ/kg},$$

$$h_g = 2645.9 \text{ kJ/kg}, \quad \text{and} \quad h_{fg} = 2305.43 \text{ kJ/kg}$$

Internal energy of wet steam is given by (neglecting volume of water),

$$u_2 = (h_f + xh_{fg}) - px_{vg}$$

$$u_2 = (340.47 + 0.85 \times 2305.43) - 0.5 \times 10^2 \times 0.85 \times 3.240$$

$$= 2162.38 \text{ kJ/kg}$$

$$\therefore \quad \Delta u = u_2 - u_1$$

$$= 2162.38 - 2787.65 = -625.27 \text{ kJ/kg}$$

The negative sign indicates that the internal energy decreases during the process.

Example 1.12 Steam at 14 bar and 0.88 dry is cooled at constant volume at 141.32°C. Find the final dryness fraction of steam.

Solution From steam table (Table A-1.2) the value of saturation temperature at 14 bar pressure is 195.07°C.

The temperature falls to 141.32°C, so the pressure will also fall. The saturation pressure value corresponding to 141.32°C from steam table (Table A-1.2) is 3.75 bar.

For constant volume cooling, we can write (neglecting volume of water),

$$x_1 v_{g1} = x_2 v_{g2}$$

where v_{g1} is the specific volume corresponding to 14 bar pressure and v_{g2} the specific volume corresponding to 3.75 bar.

$$\therefore \quad x_2 = x_1 \times \frac{v_{g1}}{v_{g2}}$$

$$= 0.88 \times \frac{0.114084}{0.4914} = 0.2043$$

Example 1.13 Steam at 12 bar and 0.9 dry is available. Find the final dryness fraction of steam in each of the cases: (i) 150 kJ of heat is removed per kg of steam at constant pressure, (ii) steam is cooled at constant volume and its temperature falls to 124°C, and (iii) it expands isentropically in the turbine developing 210 kJ/kg of steam and the pressure becomes 0.5 bar.

Solution

(i) Initial steam enthalpy at 12 bar pressure is given by,

$$h = h_f + xh_{fg} = h_f + x(h_g - h_f)$$

$$= 798.64 + 0.9 \times (2784.8 - 798.64)$$

$$= 2586.18 \text{ kJ}$$

When 150 kJ of heat is removed, energy equation becomes,

$$h_f + x_1 (h_g - h_f) = 2586.18 - 150$$

$$= 2436.18 \text{ kJ}$$

$$\therefore \quad x_1 = \frac{2436.18 - h_f}{h_g - h_f}$$

$$= \frac{2436.18 - 798.64}{2784.8 - 798.64} = 0.8245$$

(ii) From steam table (Table A-1.2) the value of saturation temperature at 12 bar pressure is 187.99°C.

The temperature falls to 124°C, so the pressure will also fall. The saturation pressure value corresponding to 124°C from steam table (Table A-1.1) is 2.25 bar.

For constant volume cooling, we can write,

$$x_1 v_{g1} = x_2 v_{g2}$$

where v_{g1} is the specific volume corresponding to 12 bar pressure and v_{g2} the specific volume corresponding to 2.25 bar.

$$\therefore \quad x_2 = x_1 \times \frac{v_{g1}}{v_{g2}}$$

$$= 0.9 \times \frac{0.1633}{0.7933} = 0.1853$$

(iii) For the isentropic process, we can write (neglecting kinetic and potential energy changes),

$$h_1 - W = h_2$$

where h_1 is the enthalpy corresponding to 12 bar pressure and h_2 the enthalpy corresponding to 2.25 bar.

$$h_2 = h_{f2} + x_2 h_{fg2}$$

$$= h_{f2} + x_2 (h_{g2} - h_{f2})$$
$$= 340.47 + x_2 \times (2645.9 - 340.47)$$
$$h_1 - W = 2586.18 - 210$$
$$\therefore \quad 340.47 + x_2 \times (2645.9 - 340.47) = 2586.18 - 210 \quad [\because h_1 - W = h_2]$$
$$\therefore \quad x_2 = 0.8830$$

Example 1.14 A pressure cooker contains 2.2 kg of steam at 5 bar pressure and 0.88 dry. Find the quantity of heat to be rejected so that the quality of steam becomes 0.55 dry.

Solution From steam table (Table A-1.2) corresponding to 5 bar pressure the various properties are:

$$v_g = 0.3749 \text{ m}^3/\text{kg}, \qquad h_f = 640.21 \text{ kJ/kg},$$
$$h_g = 2748.7 \text{ kJ/kg}, \quad \text{and} \quad h_{fg} = 2108.49 \text{ kJ/kg}$$

Specific internal energy at initial state 1 is given by,

$$u_1 = (h_{f1} + xh_{fg1}) - p_1 x_1 v_{g1}$$
$$= (640.21 + 0.88 \times 2108.49) - \frac{5 \times 10^5}{10^3} \times 0.88 \times 0.3749$$
$$= 2495.68 - 164.96 = 2330.72 \text{ kJ/kg}$$

For constant volume process (neglecting volume of water),

$$x_1 v_{g1} = x_2 v_{g2}$$
$$\therefore \quad v_{g2} = v_{g1} \times \frac{x_1}{x_2} = 0.3749 \times \frac{0.88}{0.55} = 0.5998 \text{ m}^3/\text{kg}$$

The corresponding steam properties to specific volume = 0.5998 m^3/kg from steam table (Table A-1.2) by interpolation:

$$p_2 = 3.03 \text{ bar}, \qquad h_{f2} = 563.06 \text{ kJ/kg},$$
$$h_{g2} = 2725.81 \text{ kJ/kg}, \quad \text{and} \quad h_{fg2} = 2162.75 \text{ kJ/kg}$$

Specific internal energy at final state 2 is given by,

$$u_2 = (h_{f2} + xh_{fg2}) - p_2 x_2 v_{g2}$$
$$= (563.06 + 0.55 \times 2162.75) - \frac{3.03 \times 10^5}{10^3} \times 0.55 \times 0.5998$$
$$= 1752.57 - 99.96 = 1652.61 \text{ kJ/kg}$$

$$\text{Heat required} = u_1 - u_2 = \Delta u$$
$$= 2330.72 - 1652.61 = 678.11 \text{ kJ/kg}$$
$$\therefore \quad \text{Total heat rejected} = m \times \Delta u$$
$$= 2.2 \times 678.11 = 1491.84 \text{ kJ}$$

Example 1.15 A pressure cooker contains 2 kg of dry and saturated steam at 4 bar while in operation. Find the quantity of heat that must be rejected so that the quality of steam becomes 0.6 dry. Also calculate the pressure and temperature of the steam that exists in the cooker after the heat rejection.

Solution From steam table (Table A-1.2) corresponding to 4 bar pressure,

$$v_{g1} = 0.4625 \text{ m}^3/\text{kg}, \qquad h_{f1} = 604.73 \text{ kJ/kg},$$

$$h_{g1} = 2738.5 \text{ kJ/kg}, \quad \text{and} \quad h_{fg1} = 2133.77 \text{ kJ/kg}$$

Specific internal energy at initial state 1,

$$u_1 = (h_{f1} + xh_{fg1}) - p_1\, x_1\, v_{g1}$$

$$\therefore \quad u_1 = (604.73 + 1 \times 2133.77) \times \frac{4 \times 10^5}{10^3} \times 1 \times 0.4625$$

$$= 2553.5 \text{ kJ/kg}$$

For constant volume process (neglecting volume of water),

$$x_1\, v_{g1} = x_2\, v_{g2}$$

$$\therefore \quad v_{g2} = v_{g1} \times \frac{x_1}{x_2} = 0.4625 \times \frac{1}{0.6} = 0.7708 \text{ m}^3/\text{kg}$$

The corresponding steam properties to specific volume = 0.7708 m^3/kg from steam tables by interpolation are:

$$p_2 = 2.30 \text{ bar}, \qquad h_{f2} = 530.92 \text{ kJ/kg},$$

$$h_{g2} = 2715.42 \text{ kJ/kg}, \quad \text{and} \quad h_{fg2} = 2184.5 \text{ kJ/kg}$$

Specific internal energy at final state 2,

$$u_2 = (h_{f2} + xh_{fg2}) - p_2\, x_2\, v_{g2}$$

$$\therefore \quad u_2 = (530.92 + 0.6 \times 2184.5) - \frac{2.3 \times 10^5}{10^3} \times 0.6 \times 0.7708$$

$$= 1735.25 \text{ kJ/kg}$$

$$\text{Heat rejected} = u_1 - u_2 = \Delta u$$

$$= 2553.5 - 1735.25 = 818.25 \text{ kJ/kg}$$

$$\therefore \quad \text{Total heat rejected} = m \times \Delta u$$

$$= 2 \times 818.25 = 1636.5 \text{ kJ}$$

Example 1.16 Steam at 12 bar and 200°C expands up to the pressure 2 bar. The dryness fraction at the end of expansion is 0.9. Determine the change in the internal energy.

Solution From steam table (Table A-1.3) at pressure 12 bar and 200°C,

$$h_1 = 2815.9 \text{ kJ/kg}, \quad \text{and} \quad v_1 = 0.1793 \text{ m}^3/\text{kg}$$

$$u_1 = h_1 - p_1 v_1$$
$$= 2815.9 - \frac{12 \times 10^5}{10^3} \times 0.1793$$
$$= 2600.74 \text{ kJ/kg}$$

From steam table (Table A-1.2) at pressure 2 bar and 0.9 dry,

$$h_2 = h_{f2} + x_2 h_{fg2}$$
$$= 504.68 + 0.9 \times 2201.92 = 2486.47 \text{ kJ/kg}$$
$$v_2 = x_2 v_{g2}$$
$$= 0.9 \times 0.8857 = 0.7971$$
$$u_2 = h_2 - p_2 v_2$$
$$= 2486.41 - \frac{2 \times 10^5}{10^3} \times 0.7971$$
$$= 2326.99 \text{ kJ/kg}$$

$\therefore$ $$\Delta u = 2326.99 - 2600.74$$
$$= -273.75 \text{ kJ/kg}$$

Example 1.17 Two boilers *X* and *Y* discharge equal amount of steam into a common main steam pipe. The steam from boiler *X* is at 14 bar and 300°C and from *Y* at 14 bar and 0.9 dry. Determine (i) the equilibrium condition after mixing in the main pipe, (ii) the loss of entropy by high temperature steam, (iii) gain of entropy of the low pressure steam, and (iv) net increase or decrease of entropy.

Solution

(i) From steam table (Table A-1.3) for boiler *X* at 14 bar and 300°C,

$$h_{sup} = h_X = 3040.4 \text{ kJ/kg,}$$

and $$s_{sup} = s_X = 6.9533 \text{ kJ/kgK}$$

And for boiler *Y* at 14 bar for dry saturated steam from Table A-1.2,

$h_f = 830.29$ kJ/kg, $h_g = 2790.0$ kJ/kg,

$s_f = 2.2842$ kJ/kgK, $s_g = 6.4692$ kJ/kgK,

$h_{fg} = 1959.71$ kJ/kg, and $s_{fg} = 4.185$ kJ/kgK

$$h = h_Y = h_f + x \times h_{fg}$$
$$= 830.29 + 0.9 \times 1959.71 = 2594.03 \text{ kJ/kg}$$
$$s = s_Y = s_f + x \times s_{fg}$$
$$= 2.2842 + 0.9 \times 4.185 = 6.0507 \text{ kJ/kg}$$

Assuming that 1 kg of steam has been delivered by both the boilers, the enthalpy of mixture is given by,

$$h_{mix} = \frac{h_X + h_Y}{2}$$

$$= \frac{3040.4 + 2894.03}{2} = 2817.22 \text{ kJ/kg}$$

The enthalpy of dry saturated steam at 14 bar = 2790.0 kJ/kg which is less than the enthalpy of mixture. This indicates that mixture steam in the main pipe is superheated.

Corresponding to mixture enthalpy h_{mix} = 2817.22 kJ/kg and pressure 14 bar from steam table (Table A-1.3) for superheated condition by interpolation, we have,

$$(T_{sup})_{mix} = 2205.62°\text{C}, \qquad (s_{sup})_{mix} = 6.5255 \text{ kJ/kgK},$$

and $$(v_{sup})_{mix} = 0.18077 \text{ m}^3\text{/kg}$$

(ii) The loss of entropy by high temperature steam is given by,

$$s_{loss} = (s_{sup})_X - (s_{sup})_{mix}$$

$$= 6.9533 - 6.5255 = 0.4278 \text{ kJ/kgK}$$

(iii) Gain of entropy of the low pressure steam is given by,

$$s_{gain} = (s_{sup})_{mix} - (s_Y)$$

$$= 6.5255 - 6.0507 = 0.4788 \text{ kJ/kgK}$$

(iv) The net increase of entropy is given by,

$$(s_{gain})_{net} = 0.4788 - 0.4278$$

$$= 0.047 \text{ kJ/kgK}$$

Example 1.18 A container of volume 2.5 m^3 is to be completely filled with dry and saturated steam at 10 bar. For this the container is emptied and the necessary amount of water is filled. Determine (i) the necessary amount of water to be filled in the drum and the resulting temperature of steam, and (ii) the amount of water if the steam generated is 0.85 dry.

Solution

(i) From steam table (Table A-1.2) corresponding to 10 bar pressure,

$$T_{sat} = 179.91°\text{C}, \qquad \text{and} \qquad v_g = 0.1944 \text{ m}^3\text{/kg}$$

∴ Mass of water = Mass of steam

$$= (2.5/0.1944) = 12.86 \text{ kg}$$

(ii) The specific volume of wet steam is given by (neglecting volume of water)

$$v = xv_g$$

$$= 0.85 \times 0.1944 = 0.16524 \text{ m}^3\text{/kg}$$

$$\therefore \quad \text{Mass of water} = (2.5/0.16524)$$
$$= 15.13 \text{ kg}$$

Example 1.19 Steam at 10 bar pressure and 0.9 dry enters a superheater and leaves at the same pressure at 300°C. Determine the change in specific entropy of steam. Take specific heat of steam equal to 2.20 kJ/kg K.

Solution Entropy of steam at the entry of superheater is given by,

$$s_1 = s_{f1} + x_1 s_{fg1}$$
$$= 2.1386 + 0.9 \times (6.5864 - 2.1386)$$
$$= 6.1416 \text{ kJ/kgK}$$

Entropy of steam at the exit of superheater is given by,

$$s = s_g + c_{ps} \ln \frac{T_{sup}}{T_s}$$
$$= 6.5864 + 2.20 \times \ln\left(\frac{300 + 273}{179.9 + 273}\right)$$
$$= 7.1038 \text{ kJ/kgK}$$

Change in entropy is given by,

$$\Delta s = s_2 - s_1 = 7.1038 - 6.1416$$
$$= 0.6922 \text{ kJ/kgK}$$

Example 1.20 Find the change in internal energy of 1 kg of superheated steam at a pressure of 12 bar and 300°C expanded to a pressure of 1.5 bar and 0.8 dry. Take specific heat for superheated steam equal to 2.1 kJ/kg K.

Solution Internal energy of superheated steam at 12 bar and 300°C is given by,

$$u = h_f + h_{fg} + c_{ps} (T_{sup} - T_s) - pv_g$$
$$= h_g - pv_g + c_{ps} (T_{sup} - T_s)$$
$$\therefore \quad u_1 = h_g - pv_g + c_{ps} (T_{sup} - T_s)$$
$$\therefore \quad u_1 = 2784.8 - \frac{12 \times 10^5}{10^3} \times 0.1633 + 2.1 \times (300 - 187.99)$$
$$= 2824.06 \text{ kJ/kg}$$

Internal energy of steam at 1.5 bar and 0.8 dry (neglecting volume of water),

$$u_2 = (h_{f2} + x_2 h_{fg2}) - p_2 x_2 v_{g2}$$
$$\therefore \quad u_2 = (467.08 + 0.8 \times 2226.42) - \frac{1.5 \times 10^5}{10^3} \times 0.8 \times 1.1593$$
$$= 2109.1 \text{ kJ/kg}$$

Change in internal energy is given by,

$$\Delta u = u_2 - u_1$$
$$= 2109.1 - 2824.06 = -714.96 \text{ kJ/kg}$$

Negative sign indicates decrease in internal energy.

Example 1.21 Steam expands in a turbine isentropically from pressure 175 bar and temperature 550°C to 0.1 bar. Determine the work done per kg of steam. Take specific heat for superheated steam equal to 2.1 kJ/kg K.

Solution From superheated steam table (Table 1.3) corresponding to 175 bar and 550°C, the saturation temperature, $T_{sat} = 354.75°C$ which is less than 550°C. Therefore, steam is superheated. Thus, steam tables can be used directly, we have,

$$h_{sup} = h_1 = 3421.4 \text{ kJ/kg}, \quad \text{and} \quad s_{sup} = s_1 = 6.4229 \text{ kJ/kgK}$$

At 0.1 bar pressure, entropy is given by,

$$s_2 = s_f + x\,(s_g - s_f)$$
$$= 0.6492 + x \times (8.1501 - 0.6492) = 0.6492 + 7.5013x$$

But $s_1 = s_2$

$$\therefore \quad 6.4229 = 0.6492 + 7.5013\,x$$
$$\therefore \quad x = 0.7697$$

At 0.1 bar pressure, enthalpy is given by,

$$h_2 = h_f + x\,(h_g - h_f)$$
$$= 191.81 + 0.7697 \times (2584.6 - 191.81) = 2033.54 \text{ kJ/kg}$$

$$\text{Work done per kg} = h_1 - h_2$$
$$= 3421.4 - 2033.54 = 1387.86 \text{ kJ}$$

Example 1.22 Three kg of steam expands in a turbine adiabatically from pressure 16 bar and temperature 250°C to 0.5 bar. The steam is just dry and saturated at the end of expansion. Determine (i) the actual work done by the steam, (ii) ideal work done (or isentropic work), and (iii) the % of work lost due to irreversibility.

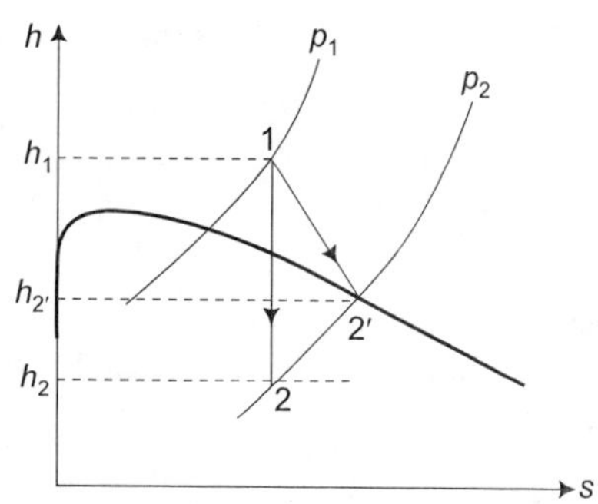

Fig. 1.17

Solution Refer Figure 1.17. Process 1-2 is the isentropic (reversible adiabatic) and process 1 – 2′ is irreversible adiabatic.

(i) From superheated steam table (Table A-1.3) corresponding to 16 bar and 250°C

$$h_{sup} = h_1 = 2919.2 \text{ kJ/kg}, \qquad \text{and} \qquad s_{sup} = s_1 = 6.6732 \text{ kJ/kgK}$$

At 0.5 bar pressure for dry and saturated steam, enthalpy (from Table A-1.2)

$$h_{2'} = h_{g2} = 2645.9 \text{ kJ/kg}$$

So, actual work done by the turbine is given by,

$$w_{1-2'} = h_1 - h_{2'}$$

$$= 2919.2 - 2645.9 = 273.3 \text{ kJ/kg}$$

Total work done is given by,

$$W_{1-2'} = m \times (h_1 - h_{2'})$$

$$= 3 \times (2919.2 - 2645.9) = 819.9 \text{ kJ}$$

(ii) At 0.5 bar pressure, entropy is given by,

$$s_2 = s_f + x\,(s_g - s_f)$$

$$= 1.0910 + x_2 \times (7.5939 - 1.0910)$$

$$= 1.0910 + 6.5029\,x$$

But $s_1 = s_2$

$\therefore$ $$6.6732 = 1.0910 + 6.5029\,x$$

$\therefore$ $$x = 0.8501$$

At 0.5 bar pressure, enthalpy is given by,

$$h_2 = h_f + x\,(h_g - h_f)$$

$$= 340.7 + 0.8501 \times (2645.9 - 340.7)$$

$$= 2300.55 \text{ kJ/kg}$$

Work done due to isentropic expansion is given by,

$$w_{1-2} = h_1 - h_2$$

$$= 2919.2 - 2300.55 = 618.65 \text{ kJ/kg}$$

Total work done is given by,

$$W_{1-2} = m \times (h_1 - h_2)$$

$$= 3 \times (2919.2 - 2300.55) = 1855.95 \text{ kJ}$$

(iii) Work lost due to irreversibility is given by,

$$w_{lost} = (h_1 - h_2) - (h_1 - h_{2'})$$

$$= (h_{2'} - h_2)$$

$$= 2645.9 - 2300.55 = 345.35 \text{ kJ/kg}$$

Total work lost is given by,

$$W_{\text{lost}} = 3 \times 345.35$$

$$= 1036.05 \text{ kJ}$$

$$\text{Percentage of ideal work lost} = \left(\frac{w_{1-2} - w_{1-2'}}{w_{1-2}}\right) \times 100$$

$$\text{Therefore, percentage of ideal work lost} = \left(\frac{618.65 - 273.3}{618.65}\right) \times 100 = 55.82\%$$

Example 1.23 In a boiler 1500 kg of steam is generated per hour at 14 bar and 0.95 dryness fraction. The steam is then heated in a superheater to a temperature of 350°C at constant pressure. The feed water temperature is 25°C. Determine (i) heat supplied per hour in the boiler, (ii) heat supplied per hour in the superheater, (iii) internal energy of superheated steam, and (iv) change in internal energy of steam.

Solution

(i) Heat contained per kg of water is given by,

$$h_1 = c_{pw} \times \Delta T$$

$$= 4.186 \times 25 = 104.65 \text{ kJ/kg}$$

The amount of heat required to generate 0.95 dry steam is given by,

$$h_2 = h_f + x\,(h_g - h_f)$$

$$= 830.29 + 0.95 \times (2790 - 830.29)$$

$$= 2692.01 \text{ kJ/kg}$$

The amount of heat required by 1 kg of water in the boiler is given by,

$$h_3 = h_2 - h_1$$

$$= 2692.01 - 104.65 = 2587.36 \text{ kJ/kg}$$

So, the total amount of heat supplied per hour in the boiler is given by,

$$H_{\text{boiler}} = m \times h_3 = 1500 \times 2587.36$$

$$= 3881040 \text{ kJ/kg} = 3881.04 \text{ MJ/kg}$$

(ii) Enthalpy of superheated steam per kg is given by,

$$h_{\text{sup}} = h_g + c_{ps}\,(T_{\text{sup}} - T_s)$$

$$= 2790.0 + 2.1 \times (350 - 195.07)$$

$$= 3115.35 \text{ kJ/kg}$$

Heat supplied in the superheater per kg is given by,

$$h_4 = h_{sup} - h_3$$
$$= 3115.35 - 2587.36 = 527.99 \text{ kJ/kg}$$

So, the total amount of heat supplied per hour in the superheater is given by,

$$H_{sup} = m \times h_4 = 1500 \times 527.99$$
$$= 791985 \text{ kJ/kg} = 791.985 \text{ MJ/kg}$$

(iii)
$$v_{sup} = v_g \times \frac{T_{sup}}{T_g}$$
$$= 0.14084 \times \left(\frac{350 + 273}{195.07 + 273}\right)$$
$$= 0.1875 \text{ m}^3\text{/kg}$$

Internal energy of superheated steam is given by,

$$u_{sup} = h_{sup} - pv_{sup}$$
$$= 3115.35 - 14 \times 10^5/10^3 \times 0.1875$$
$$= 2852.85 \text{ kJ/kg}$$

(iv) Internal energy of saturated liquid at 25°C is given by,

$$u_1 = h_f - p_f v_f$$
$$= 104.87 - \frac{14 \times 10^5}{10^3} \times 0.001003$$
$$= 103.47 \text{ kJ/kg}$$

Change in internal energy,

$$\Delta u = u_{sup} - u_1$$
$$= 2852.85 - 103.47 = 2749.38 \text{ kJ/kg}$$

Example 1.24 A vessel contains one kg of steam which has 1/3rd liquid and 2/3rd vapor by volume. The temperature of the steam is 179.91°C. Determine the dryness fraction, specific volume, and specific enthalpy of the mixture.

Solution The pressure in the vessel = 10 bar which is corresponding to saturation temperature of steam 179.91°C. Applying given condition, we have,

$$\frac{V_f}{V_g} = \frac{m_f v_f}{m_g v_g} = \frac{1/3}{2/3} = \frac{1}{2}$$

$$\therefore \quad \frac{m_f \times 0.001127}{m_g \times 0.1944} = \frac{1}{2}$$

$$\therefore \quad \frac{m_f}{m_g} = 86.25$$

By definition the dryness fraction is given by,

$$x = \frac{m_g}{m_f + m_g} = \frac{1}{(m_f/m_g) + 1}$$

$$= \frac{1}{87.25} = 0.01146$$

The specific volume of the mixture is given by (From Table A-1.2, corresponding to 10 bar pressure),

$$v = (1 - x)\, v_f + x v_g$$

$$\therefore \quad v = (1 - 0.01146) \times 0.001137 + 0.01146 \times 0.1944$$

$$= 0.003342 \text{ m}^3\text{/kg}$$

The specific enthalpy of the mixture,

$$h = (1 - x)\, h_f + x h_g$$

$$\therefore \quad h = (1 - 0.01146) \times 762.79 + 0.01146 \times 2778.1$$

$$= 785.88 \text{ kJ/kg}$$

Example 1.25 A vessel of 0.5 m^3 capacity contains 2 kg of wet steam at a pressure of 7.5 bar. Determine (i) the volume and mass of liquid, and (ii) volume and mass of vapor.

Solution The specific volume of mixture is given by,

$$v = (0.5/2) = 0.25 \text{ m}^3\text{/kg}$$

From steam table (Table A-1.2) at 7.5 bar,

$$v_{fg} = v_g - v_f$$

$$= 0.2556 - 0.001111 = 0.2545 \text{ m}^3\text{/kg}$$

$$v = (1 - x)\, v_f + x v_g$$

$$0.25 = (1 - x) \times 0.001111 + x \times 0.2556$$

$$\therefore \quad x = 0.978$$

$$x = \frac{\text{Mass of dry saturated vapor}}{\text{Mass of mixture}}$$

$$= \frac{m_g}{m_t} = \frac{m_g}{m_f + m_g}$$

$$m_g = x \times (m_f + m_g)$$

∴ Mass of vapor,

$$m_g = 0.978 \times 2 = 1.956 \text{ kg}$$

∴ Mass of liquid,

$$m_f = 2.0 - 1.956 = 0.044 \text{ kg}$$

$$\text{Volume of vapor} = m_g \times v_g$$
$$= 1.956 \times 0.2556 = 0.4999 \text{ m}^3, \text{ and}$$
$$\text{Volume of liquid} = m_f \times v_f$$
$$= 0.044 \times 0.001111 = 0.0000489 \text{ m}^3$$

Example 1.26 A vessel of 1 m^3 capacity contains steam at a pressure of 8 bar and 0.9 dry. Steam is blown off by opening the valve until the pressure drops to 4 bar. Determine the weight of the steam blown off.

Solution The initial mass of steam in the vessel is given by,

$$m_i = \frac{\text{volume}}{x_1 \times v_g} = \frac{2}{0.9 \times 0.2404} = 9.244 \text{ kg}$$

The steam blowing process has been considered a throttling process.

In the case of throttling process,

$$h_1 = h_2$$

$$\therefore \quad h_{f1} + x_1 h_{fg1} = h_{f2} + x_2 h_{fg2}$$

$$721.10 + 0.9 \times (2769.1 - 721.10) = 604.73 + x_2 (2738.5 - 604.73)$$

$$\therefore \quad x_2 = 0.9184$$

The final mass of steam in the vessel,

$$m_f = \frac{\text{Volume}}{x_2 \times v_g}$$
$$= \frac{2}{0.9184 \times 0.4625} = 4.709 \text{ kg}$$

∴ Mass of steam blown off = 9.244 – 4.709

= 4.535 kg

Example 1.27 A spherical vessel of 1.2 m^3 capacity contains steam at 9 bar and 0.9 dryness fraction. Steam is blown off until the pressure drops to 4.5 bar. The valve is then closed and steam is allowed to cool until the pressure drops to 3.5 bar. Assume that enthalpy of steam in the vessel remains constant during blowing off period. Find (i) the mass of steam blown off, (ii) the dryness fraction of steam in the vessel after cooling, and (iii) the heat lost per kg of steam during cooling.

Solution

(i) The initial mass of steam in the vessel is given by,

$$m_i = \frac{\text{Volume}}{x_2 \times v_g}$$

$$= \frac{1.2}{0.9 \times 0.2150} = 6.202 \text{ kg}$$

The steam blowing process has been considered a throttling process.

In the case of throttling process,

$$h_1 = h_2$$

$$\therefore \quad h_{f1} + x_1 h_{fg1} = h_{f2} + x_2 h_{fg2}$$

$$742.82 + 0.9 \times (2773.9 - 742.82) = 623.24 + x_2 (2743.9 - 623.24)$$

$$2570.79 = 623.24 + 2120.33 x_2$$

$$\therefore \quad x_2 = 0.9184$$

The final mass of steam in the vessel is given by,

$$m_f = \frac{\text{Volume}}{x_2 \times v_g}$$

$$= \frac{1.2}{0.9184 \times 0.4140} = 3.456 \text{ kg}$$

∴ Mass of steam blown off

$$= 6.202 - 3.156 = 3.046 \text{ kg}$$

Internal energy,

$$u_2 = h_2 - p_2 x_2 v_{g2}$$

$$= 2570.79 - \frac{4.5 \times 10^5}{10^3} \times 0.7252 \times 0.4141$$

$$= 2399.69 \text{ kJ/kg}$$

(ii) At 3.5 bar pressure from steam table (Table A-1.2), we have,

$h_f = 584.31$ kJ/kg, $h_g = 2732.4$ kJ/kg,

$v_g = 0.5243$ m^3/kg, and $h_{fg} = 2148.09$ kJ/kg,

$$\therefore \quad x_3 = \frac{1.2}{3.156 \times 0.5243} = 0.7252$$

(iii)

$$h_3 = h_{f3} + x_3 h_{fg3}$$

$$= 584.31 + 0.7252 \times (2432.4 - 584.31)$$

$$= 2142.10 \text{ kJ/kg}$$

$$u_3 = h_3 - p_3\, x_3\, v_{g3}$$

$$= 2142.10 - \frac{3.5 \times 10^5}{10^3} \times 0.7252 \times 0.5243$$

$$= 2009.02 \text{ kJ/kg}$$

∴ Heat lost during cooling/kg of steam,

$$u_2 - u_3 = 2399.69 - 2009.02$$

$$= 390.67 \text{ kJ}$$

Example 1.28 A rigid vessel of 1.5 m^3 capacity contains superheated steam at 18 bar and 350°C. The vessel is cooled at constant volume until the steam is dry saturated. Determine (i) the mass of steam in the vessel, (ii) the final pressure, and (iii) the amount of heat transferred.

Solution

(i) The mass of steam in the vessel,

$$m = \frac{\text{Volume}}{v_1} = \frac{1.5}{0.15457} = 9.704 \text{ kg}$$

(ii) The steam is cooled at constant volume and at the end it is dry saturated

∴ $$v_1 = v_2 = v_{g2} = 0.15457 \text{ m}^3\text{/kg}$$

From steam table (Table A-1.2), the pressure of dry saturated steam corresponding to specific volume 0.15457 m^3/kg (by interpolation) is 12.72 bar.

(iii) For a constant volume pressure the heat transferred is given by,

$$q_{1-2} = (u_2 - u_1) + \int_1^2 p\,dv$$

$$= (u_2 - u_1) + 0$$

$$= (h_2 - p_2\, v_2) - (h_1 - p_1\, v_1) \text{ kJ/kg}$$

$$q_{1-2} = (h_2 - h_1) - v_1\,(p_2 - p_1) \text{ kJ/kg}, \quad [\because v_1 = v_2]$$

At 18 bar and 350°C from steam table (Table A-1.3), we have,

$$h_1 = 3141.2 \text{ kJ/kg, and}$$

At 16 bar for dry saturated steam from Table A-1.2, we have,

$$h_2 = h_g = 2786.82 \text{ kJ/kg}$$

∴ $$q_{1-2} = (2786.82 - 3141.2) - 0.15457 \times \frac{(18 - 12.72) \times 10^5}{10^3}$$

$$= -\,435.99 \text{ kJ/kg}, \quad [\because v_1 = v_2]$$

∴ Total heat transferred = 9.704 × (– 435.99)

$$= -\,4230.85 \text{ kJ}$$

Example 1.29 4.5 kg of superheated steam at 16 bar and 350°C loses 3500 kJ of heat to the surrounding at constant pressure during a steady flow process. Determine the final dryness fraction of steam.

Solution At 16 bar and 350°C from steam table (Table A-1.3),

$$h_1 = 3145.4 \text{ kJ/kg}$$

Let h_2 be the enthalpy of final state of steam.

Applying the given condition,

$$4.5 \times (h_2 - 3145.4) = -3500$$

$$h_2 = 2367.62 \text{ kJ/kg}$$

Also $$h_2 = h_{f2} + x_2 (h_{g2} - h_{f2})$$

At 16 bar from Table A-1.3 by interpolation,

$$h_{f2} = 858.31 \text{ kJ/kg}, \quad \text{and} \quad h_{g2} = 2793.82 \text{ kJ/kg}$$

$$2367.62 = 858.31 + x_2 (2793.82 - 858.31),$$

$$\therefore \quad x_2 = 0.7798$$

Example 1.30 2.5 kg of steam at 12 bar and 250°C undergoes a constant pressure process. The so obtained steam is wet having a dryness fraction of 0.6. Determine work done, change in enthalpy, and heat transferred during this non-flow process.

Solution At 12 bar and 250°C (from steam table A-1.3)

$$v_1 = v_{sup1} = 0.19235 \text{ m}^3\text{/kg, and}$$

$$v_2 = x \times v_{g2} = 0.6 \times 0.1633$$

$$= 0.09796 \text{ m}^3\text{/kg}$$

$$h_1 = h_{sup1} = 2935 \text{ kJ/kg}$$

$$h_2 = h_{f2} + x_2 (h_{g2} - h_{f2})$$

$$= 798.64 + 0.6 \times (2784.8 - 798.64)$$

$$= 1990.34 \text{ kJ/kg}$$

$$\therefore \quad w_{1-2} = m \times [p_1 (v_2 - v_1)]$$

$$= 2.5 \times \frac{12 \times 10^5}{10^3} \times (0.09798 - 0.19235)$$

$$= -283.1 \text{ kJ}$$

$$\therefore \quad q_{1-2} = m \times (h_2 - h_1)$$

$$= 2.5 \times (1990.34 - 2935)$$

$$= -2361.65 \text{ kJ}$$

Example 1.31 Dry saturated steam at 10 bar expands in a cylinder to 1.5 bar having dryness fraction as 0.75. The expansion takes place as per the law $pv^n = C$. Find the index of expansion and work done during the expansion process.

Solution At 10 bar from steam table (Table A-1.2),

$$v_1 = v_{g1} = 0.1944 \text{ m}^3/\text{kg}$$

At 1.5 bar from steam table (Table A-1.2),

$$v_{g2} = 1.1593 \text{ m}^3/\text{kg}$$

$$\therefore \quad v_2 = x_2 \times v_{g2}$$

$$= 0.75 \times 1.1593 = 0.8695 \text{ m}^3/\text{kg}$$

The expansion takes place as per the law

$$pv^n = C.$$

$$\therefore \quad p_1 v_1^n = p_2 v_2^n$$

Taking the logarithm both sides, we have,

$$\ln p_1 + n \ln v_1 = \ln p_2 + n \ln v_2$$

$$\therefore \quad n = \frac{\ln p_1 - \ln p_2}{\ln v_2 - \ln v_1}$$

$$= \frac{\ln (p_1/p_2)}{\ln (v_2/v_1)}$$

$$= \frac{\ln (10/1.5)}{\ln (0.8695/0.1944)} = 1.27$$

$$w_{1-2} = \int_1^2 pdv = \frac{p_1 v_1 - p_2 v_2}{n-1} \text{ kJ/kg}$$

$$\therefore \quad w_{1-2} = \frac{10 \times 10^5 \times 0.1944 - 1.5 \times 10^5 \times 0.8695}{10^3 \times (1.27 - 1)}$$

$$= 236.94 \text{ kJ/kg}$$

Example 1.32 Superheated steam at a pressure of 10 bar and temperature of 250°C expands adiabatically and reversibly in a cylinder to 1.5 bar. Find (i) the work done per kg of steam during the expansion process, and (ii) mean adiabatic index.

Solution At 10 bar and 250°C from steam table (Table A-1.3),

$$v_1 = v_{\text{sup}1} = 0.23268 \text{ m}^3/\text{kg}, \qquad s_1 = s_{\text{sup}1} = 6.9246 \text{ kJ/kg},$$

and $h_1 = h_{\text{sup}1} = 2942$ kJ/kg

At 1.5 bar from steam table (Table A-1.2),

$$v_{g2} = 1.1593 \text{ m}^3\text{/kg}, \qquad h_{f2} = 467.08 \text{ kJ/kg},$$

$$h_{g2} = 2693.5 \text{ kJ/kg}, \qquad h_{fg2} = 2226.42 \text{ kJ/kg},$$

$$s_{f2} = 1.4335 \text{ kJ/kg}, \qquad s_{g2} = 7.2232 \text{ kJ/kg},$$

and $\quad s_{fg2} = 5.7985 \text{ kJ/kgK}$

The process is reversibly adiabatic,

$$\therefore \quad s_1 = s_2 = s_{f2} + x_2 (s_{g2} - s_{f2})$$

$$6.9246 = 1.4335 + x_2 \times (7.2232 - 1.4335),$$

$$\therefore \quad x_2 = 0.9469$$

$$h_2 = h_{f2} + x_2 (h_{g2} - h_{f2})$$

$$= 467.08 + 0.9469 \times 2226.43 = 2575.28 \text{ kJ/kg}$$

$$v_2 = x_2 \times v_{g2}$$

$$= 0.9469 \times 1.1593 = 1.098 \text{ m}^3\text{/kg}$$

The heat transferred for non-flow reversible adiabatic process is given by,

$$q_{1-2} = (u_2 - u_1) + \int_1^2 pdv$$

$$= (u_2 - u_1) + w_{1-2}$$

The process is adiabatic,

$$\therefore \quad q_{1-2} = 0$$

$$w_{1-2} = (u_1 - u_2)$$

$$= (h_1 - p_1 v_1) - (h_2 - p_2 v_2)$$

$$w_{1-2} = \left(2942 - \frac{10 \times 10^5}{10^3} \times 0.23268\right) - \left(2575.28 - \frac{1.5 \times 10^5}{10^3} \times 1.098\right)$$

$$= 298.74 \text{ kJ/kg}$$

Also, $\quad w_{1-2} = \int_1^2 pdv = \dfrac{p_1 v_1 - p_2 v_2}{n-1} \text{ kJ/kg}$

$$\therefore \quad 298.74 = \frac{10 \times 10^5 \times 0.23268 - 1.5 \times 10^5 \times 1.098}{(n-1) \times 10^3}$$

$$\therefore \quad n = 1.23$$

Example 1.33 Dry and saturated steam expands isothermally from a pressure of 20 bar to 3 bar in a steam engine cylinder. Assume the expansion as non-flow process, determine (i) enthalpy change, (ii) internal energy change, (iii) entropy change, (iv) heat transferred, and (v) work done.

Solution At 20 bar from steam table (Table A-1.2), we have,

$$v_{g1} = 0.09963 \text{ m}^3/\text{kg}, \qquad h_{f1} = 908.77 \text{ kJ/kg},$$

$$h_{g1} = 2799.5 \text{ kJ/kg}, \qquad h_{fg1} = 1890.73 \text{ kJ/kg},$$

$$s_{f1} = 2.4473 \text{ kJ/kgK}, \qquad s_{g1} = 6.3408 \text{ kJ/kgK},$$

$$s_{fg1} = 3.8936 \text{ kJ/kgK}, \quad \text{and} \quad T_{g1} = 212.42°\text{C}$$

At 3 bar and 212.42°C from steam table (Table 1.3) by interpolation, we have,

$$v_2 = v_{\text{sup}2} = 0.7362 \text{ m}^3/\text{kg}, \qquad s_2 = s_{\text{sup}2} = 7.362 \text{ kJ/kgK},$$

$$h_2 = h_{\text{sup}2} = 2890.86 \text{ kJ/kg}$$

Change in enthalpy is given by,

$$\Delta h = h_2 - h_1$$

$$= 2890.86 - 2799.5 = 91.36 \text{ kJ/kg}$$

Change in internal energy is given by,

$$\Delta u = (u_2 - u_1) = (h_2 - p_2\, v_2) - (h_1 - p_1\, v_1)$$

$$= \left(2890.86 - \frac{3 \times 10^5}{10^3} \times 0.7362\right) - \left(2799.5 - \frac{20 \times 10^5}{10^3} \times 0.09963\right)$$

$$= 143.38 \text{ kJ/kg}$$

Change in entropy is given by,

$$\Delta s = s_2 - s_1 = 7.362 - 6.3408$$

$$= 1.014 \text{ kJ/kgK}$$

Heat transferred is given by,

$$q_{1-2} = T_{g1}\,(s_2 - s_1)$$

$$= (212.42 + 273) \times 1.014 = 492.22 \text{ kJ/kgK}$$

Work done is given by,

$$w_{1-2} = (u_1 - u_2) + q_{1-2}$$

$$= -\,143.38 + 492.22 = 348.84 \text{ kJ/kg}$$

Example 1.34 4000 kg/h of steam enters a surface condenser at 0.075 bar. The cooling water is passed through the tubes of the condenser at a rate of 2×10^5 litres/hr. The temperature of water increases from 20°C to 31.5°C. The condensate leaves the condenser at a temperature of 27.5°C. Determine the condition of steam entering the condenser.

Solution Volume of cooling water $= 2 \times 10^5 \times 10^{-3} = 200 \text{ m}^3$ $(\because 1 \text{ litre} = 10^{-3} \text{ m}^3)$

Mass of cooling water = density × volume

$$= 200 \times 1000 \text{ kg}$$

∴ Mass of cooling water/kg of steam

$$= \frac{200 \times 1000}{4000} = 50 \text{ kg}$$

At 0.075 bar from steam table (Table A-1.2), we have,

$$h_{f1} = 168.77 \text{ kJ/kg}, \qquad h_{g1} = 2574.8 \text{ kJ/kg},$$

and $h_{fg1} = 2406.03$ kJ/kg

Enthalpy of steam is given by,

$$h_1 = h_{f1} + x_1\, h_{fg1}$$

$$= 168.77 + x_1 \times 2406.03$$

Enthalpy of condensate at 27.5°C by interpolation from Table A-1.1,

$$h_{f2} = 115.32 \text{ kJ/kg}$$

Heat lost by steam = Heat gained by cooling water

$$h_1 - h_{f2} = m \times c_{pw}\,(T_{w2} - T_{w1})$$

$$168.77 + x_1 \times 2406.03 - 115.32 = 50 \times 4.186 \times (31.5 - 21)$$

∴ $x_1 = 0.9806$

Example 1.35 Steam at 14 bar and 300°C expands isentropically in a steam turbine till the temperature drops to 85°C. Find (i) the quality of steam at the end of expansion process, (ii) the work done per kg of steam, and (iii) power produced by the turbine if the steam flow rate is 7.5 kg/s.

Solution At 14 bar and 300°C from steam table (Table A-1.3),

$$s_1 = s_{sup1} = 6.9533 \text{ kJ/kgK}, \quad \text{and} \quad h_1 = h_{sup1} = 3040.4 \text{ kJ/kg}$$

At 85°C from steam table (Table A-1.1),

$$v_{g2} = 1.1593 \text{ m}^3\text{/kg}, \qquad h_{f2} = 355.88 \text{ kJ/kg},$$

$$h_{g2} = 2651.9 \text{ kJ/kg}, \qquad h_{fg2} = 2296.02 \text{ kJ/kg},$$

$$s_{f2} = 1.1342 \text{ kJ/kg}, \qquad s_{g2} = 7.5444 \text{ kJ/kg},$$

and $s_{fg2} = 6.102$ kJ/kgK

Since the process is isentropic,

$$s_1 = s_2 = s_{f2} + x_2\,(s_{g2} - s_{f2})$$

$$6.9533 = 1.1342 + x_2 \times (7.5666 - 1.1342)$$

$\therefore \quad x_2 = 0.9078$

$$h_2 = h_{f2} + x_2 (h_{g2} - h_{f2})$$
$$= 355.88 + 0.9078 \times 2296.01$$
$$= 2440.21 \text{ kJ/kg}$$

The work done per kg of steam is given by,

$$w_{1-2} = h_1 - h_2$$
$$= 3040.4 - 2440.21 = 600.19 \text{ kJ/kg}$$

The power developed is given by,

$$\text{Power} = m \times w_{1-2}$$
$$= 8.5 \times 600.19$$
$$= 5101.62 \text{ kJ/s} = 5.102 \text{ MW}$$

Example 1.36 Steam at 5 bar and 0.85 dry expands in a cylinder to 1.5 bar as per the law $pv^n = C$, where index of expansion is 1.25. Determine the interchange of heat between steam and the cylinder per kg of steam.

Solution At 5 bar from steam table (Table A-1.2), we have,

$v_{g1} = 0.3749 \text{ m}^3/\text{kg}$, $\quad s_{f1} = 1.8606 \text{ kJ/kg}$,

$s_{g1} = 6.8212 \text{ kJ/kg}$, $\quad s_{fg1} = 4.9606 \text{ kJ/kgK}$,

$h_{f1} = 640.21 \text{ kJ/kg}$, $\quad h_{g1} = 2748.7 \text{ kJ/kg}$,

and $\quad h_{fg1} = 2108.49 \text{ kJ/kg}$

$$v_1 = x \times v_{g1}$$
$$= 0.85 \times 0.3749 = 0.318665 \text{ m}^3/\text{kg}$$
$$h_1 = h_{f1} + x \times h_{fg1}$$
$$= 640.21 + 0.85 \times (2108.49)$$
$$= 2432.43 \text{ kJ/kg}$$

And at 1.5 bar from steam table (Table A-1.2), we have,

$T_{g2} = 111.37°\text{C}$, $\quad h_{f2} = 467.08 \text{ kJ/kg}$,

$h_{g2} = 2693.5 \text{ kJ/kg}$, $\quad h_{fg2} = 2226.42 \text{ kJ/kg}$,

and $\quad v_{g2} = 1.1593 \text{ m}^3/\text{kg}$

$$h_2 = h_{f2} + x_2 \times h_{fg2}$$
$$= 467.08 + 0.7202 \times (2226.42)$$
$$= 2070.55 \text{ kJ/kg}$$

The expansion takes place as per the law

$$pv^n = C.$$

$$\therefore \quad p_1 v_1^{1.25} = p_2 v_2^{1.25}$$

$$v_2 = v_1 \times (p_1/p_2)^{1/1.25}$$

$$\therefore \quad v_2 = 0.318665 \times (5/1.5)^{1/1.25}$$

$$= 0.8349 \text{ m}^3\text{/kg}$$

$$v_2 = x_2 v_{g2}$$

$$0.8349 = x_2 \times 1.1593$$

$$\therefore \quad x_2 = 0.7202$$

The heat transferred,

$$q_{1-2} = (h_2 - p_2 v_2) - (h_1 - p_1 v_2) + \frac{p_1 v_1 - p_2 v_2}{n-1} \text{ kJ/kg}$$

$$\therefore \quad q_{1-2} = (2070.55 - 1.5 \times 10^2 \times 0.83491) - (2432.43 - 5 \times 10^2 \times 0.318665) + (5 \times 10^2 \times 0.318665 - 1.5 \times 10^2 \times 0.83491)/(1.25 - 1)$$

$$\therefore \quad q_{1-2} = 1945.32 - 2273.09 + 136.38$$

$$= -191.39 \text{ kJ/kg}$$

The negative sign indicates that heat is rejected from the steam.

1.8 MEASUREMENT OF DRYNESS FRACTION OF STEAM

The dryness fraction of a sample of steam can be determined experimentally by the following methods: 1. Bucket or barrel calorimeter, 2. Separating calorimeter, 3. Throttling calorimeter, and 4. Combined separating and throttling calorimeter.

1. Bucket or Barrel Calorimeter Schematic view of a bucket calorimeter is shown in Figure 1.18. It consists of a copper calorimeter which contains a measured quantity of water. The calorimeter is surrounded by an insulating vessel and some air gap is provided between the two vessels to prevent any heat loss from the calorimeter. The value of water equivalent is determined experimentally and it is marked on the vessel. The complete unit is placed on the weighing balance. The sample of steam enters through the sampling tube via the control valve and is passed through the fine exit holes provided at the end of the pipe into the cold water. The steam condenses as it comes in contact with cold water and as a result the temperature of water increases. The mass of condensed steam is measured from the difference of readings from the weighing balance before and after the condensation of steam. The rise in temperature is measured by a thermometer and the pressure of steam is measured by the pressure gauge.

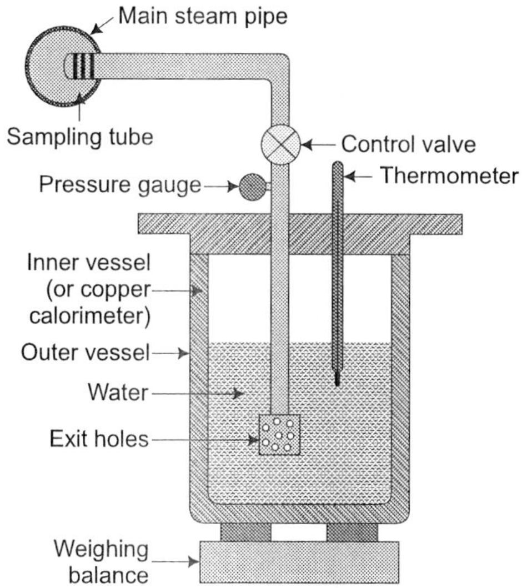

Fig. 1.18 Bucket Calorimeter

Let T_1 = Initial temperature of cold water, T_2 = Final temperature of the mixture after steam has condensed, M_e = Water equivalent of the calorimeter, m_s = Mass of steam condensed, m_w = Mass of cold water, p = Pressure of steam, and T_s = Saturation temperature corresponding to pressure p.

According to the law of conservation the heat lost by steam is equal to heat gained by the water, we have,

$$m_s \times [x \times h_{fg} + c_{ps} (T_s - T_2)] = (M_e + m_w) \times c_{pw} \times (T_2 - T_1)$$

So, dryness fraction is given by,

$$x = \frac{(M_e + m_w) \times c_{pw} \times (T_2 - T_1) - m_s \times c_{ps} \times (T_s - T_2)}{m_s \times h_{fg}} \quad (1.59)$$

The bucket calorimeter method gives approximate results and it is used for determining the dryness fraction of fairly wet steam.

2. Separating Calorimeter A schematic diagram of separating calorimeter is shown in Figure 1.19. The sample of wet steam enters the inner vessel of separating calorimeter through a sampling tube via a controlling valve. The incoming steam strikes the baffle plates (or separators) and its direction gets a sudden reversal. As a result, the heavier water particles are separated and get collected at the bottom of the inner vessel. The quantity of collected water is measured from the gauge having a graduated scale connected to the inner vessel. The dry steam moves upwards and passes down through the annular space between the inner and outer vessels. The dry and saturated steam passes through a calibrated orifice into the bucket calorimeter which is placed on the weighing balance. Thus, the condensed steam is weighed.

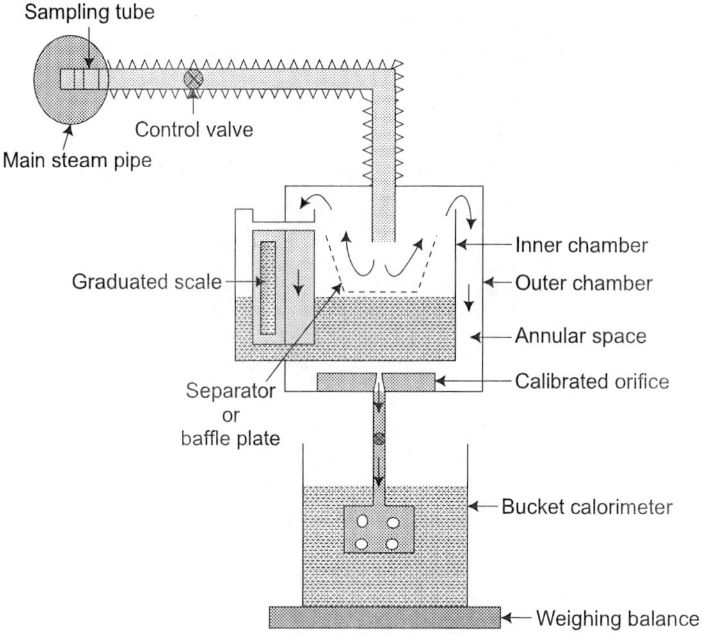

Fig. 1.19 Separating Calorimeter

Let m_s = Mass of steam condensed in bucket calorimeter, m_w = Mass of water separated from the steam in the inner vessel, and x = Dryness fraction of wet steam condensed.

$$\therefore \qquad x = \frac{m_s}{(m_w + m_s)} \qquad (1.60)$$

As complete separation of water is not possible, so this method also gives approximate results.

3. Throttling Calorimeter A schematic diagram of throttling calorimeter is shown in Figure 1.20. It works on the principle of throttling of wet steam so that it becomes superheated by at least 5°C. The steam sample from the steam mains enters into a sampling tube and then it is passed through a throttle valve. The steam is throttled down to a lower pressure (generally, atmospheric pressure) and it gets superheated. The steam then flows down into the inner chamber of the calorimeter and rises up again to enter into the annular space between the inner and outer chambers. Subsequently, the throttled steam is condensed in the condenser and is collected as condensate, or it is allowed to escape to the atmosphere. In order to prevent any heat loss from the calorimeter the outer chamber is insulated. The pressure and temperature of steam after throttling are measured with the help of a manometer and a thermometer.

Let p_1 = Initial pressure of wet steam, x_1 = Dryness fraction of wet steam, h_1 = Enthalpy of steam before throttling, h_{fg1} = Latent heat of vaporization, h_{sup2} = Enthalpy of steam after throttling, p_2 = Pressure of throttled steam, T_{g2} = Saturated temperature of throttled steam corresponding

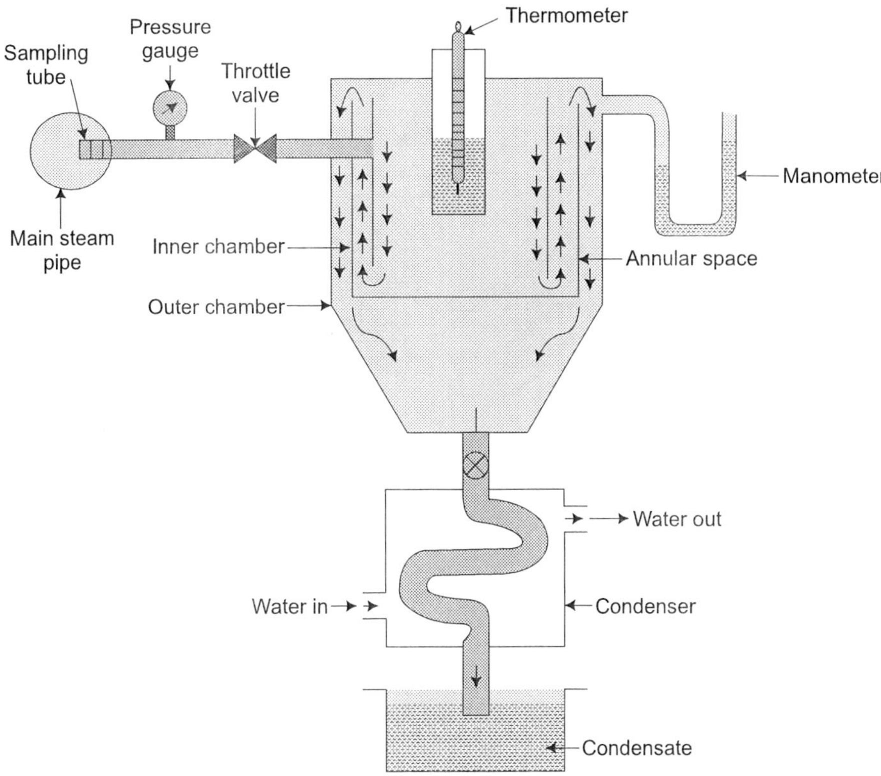

Fig. 1.20 Throttling Calorimeter

to pressure p_2, c_{ps} = Specific heat of superheated steam, and T_{sup2} = Temperature of superheated steam after throttling of steam.

During throttling process, the enthalpy remains constant, i.e.,

Enthalpy before throttling at pressure p_1 = Enthalpy after throttling at pressure p_2

$$h_1 = h_{sup2}$$

$$h_{f1} + x_1 h_{fg1} = h_{g2} + c_{ps} (T_{sup2} - T_{g2})$$

$$\therefore \quad x_1 = \frac{h_{g2} + c_{ps} (T_{sup2} - T_{g2}) - h_{f1}}{h_{fg1}} \tag{1.61}$$

4. Combined Separating and Throttling Calorimeter In the separating calorimeter due to incomplete separation of moisture, the measured value of dryness fraction of the given sample will always be higher than the actual value. Throttling calorimeter cannot measure dryness fraction value of highly wet steam (0.6 to 0.9 dry) as it cannot be superheated up to the required value by its throttling. But the dryness fraction of highly wet steam can be measured accurately by combining the separating and throttling calorimeters in series as shown in Figure 1.21.

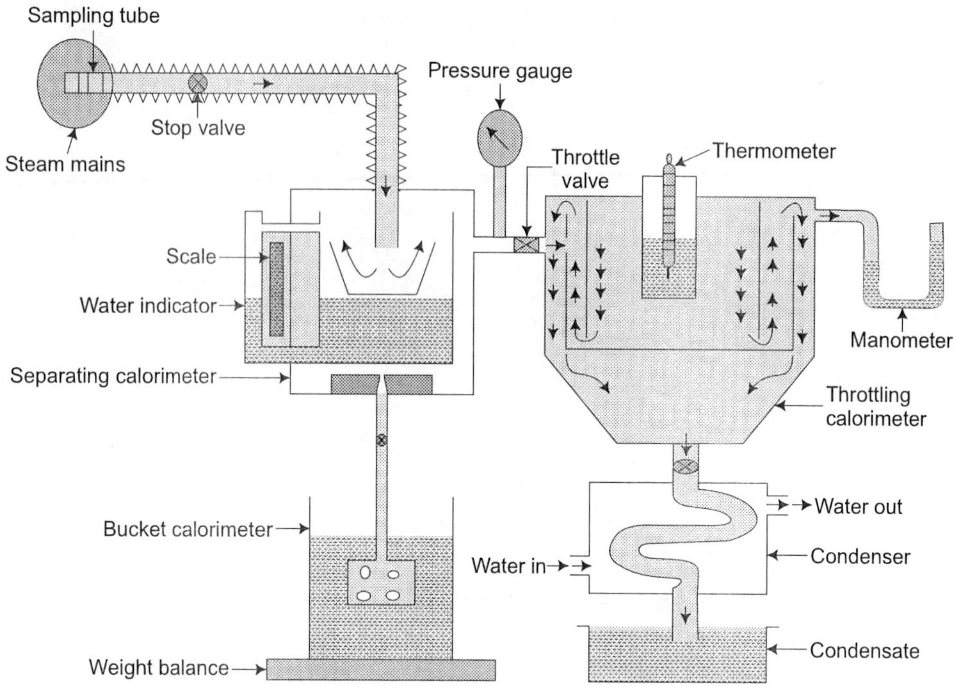

Fig. 1.21 Separating and Throttling Calorimeter

Steam first enters into the separating calorimeter where major part of the moisture (water particles) is separated and gets collected at its bottom. The mass of the separated water in the separating calorimeter is measured. Then the steam with considerable dryness fraction leaves the separating calorimeter and enters the throttling calorimeter. The steam after its throttling gets superheated. The steam leaving the throttling calorimeter is condensed in a condenser and its mass is measured. The temperature and pressure of steam leaving the throttle valve is measured with the help of thermometer and manometer.

Let m = Mass of steam passing through sampling tube, and

m_{s2} = Mass of dry steam after throttling

$\therefore$ Dryness fraction of wet steam in steam main pipe,

$$x = \frac{m_{s2}}{m} \quad \text{(i)}$$

m_1 = Mass of water (moisture) separated in separating calorimeter,

$m_{s1} = (m - m_1)$ = Mass of steam coming out from separating calorimeter,

Also $m_{s1} = m_{s2} + m_2$

where part m_{s2} represents dry steam and part m_2 represents moisture

The dryness fraction of steam after separating calorimeter,

$$x_1 = \frac{m_{s1}}{m} \quad \text{(ii)}$$

The dryness fraction of steam after its throttling,

$$x_2 = \frac{m_{s2}}{m_{s1}} \quad \text{(iii)}$$

From equations (i), (ii), and (iii),

$$x = \frac{m_{s2}}{m} = \frac{m_{s2}}{m_{s1}} \times \frac{m_{s1}}{m} = x_1 \times x_2 \quad (1.62)$$

Thus, the dryness fraction of steam in the steam main is the product of dryness fraction given by separating calorimeter and dryness fraction given by the throttling calorimeter.

Example 1.37 A sample of steam at 12 bar is passed through a throttling calorimeter. After the throttling process the temperature and pressure were measured as 1.25 bar and 125°C. Determine the dryness fraction of steam sample if specific heat of superheated steam is 2.1 kJ/kg K. Also determine the minimum value of dryness fraction that can be measured by this calorimeter if the pressure values remain constant and give your comments.

Solution Enthalpy before and after the throttling process remains same, i.e.,

$$h_{f1} + x_1 h_{fg1} = h_{g2} + c_{ps} (T_{sup2} - T_{g2})$$

At 12 bar from steam table (Table A-1.2),

$$h_{f1} = 798.67 \text{ kJ/kg}, \qquad h_{g1} = 2784.8 \text{ kJ/kg},$$

and $$h_{fg1} = 1986.16 \text{ kJ/kg}$$

At 1.25 bar from steam table (Table A-1.2),

$$h_{g2} = 2685.3 \text{ kJ/kg}$$

$$\therefore \quad x_1 = \frac{h_{g2} + c_{ps} (T_{sup2} - T_{g2}) - h_{f1}}{h_{fg1}}$$

$$= \frac{2685.3 + 2.1 \times (125 - 105.99) - 798.64}{1986.16}$$

$$= 0.97$$

For determining minimum value of dryness fraction, we know that the temperature of steam after throttling should be equal to saturated temperature.

$$\therefore \quad h_{f1} + x_1 h_{fg1} = h_{g2}$$

$$\therefore \quad 798.64 + x_1 \times 1986.16 = 2685.3$$

$$\therefore \quad x_1 = 0.9499$$

If the dryness fraction is lower than 0.9499, then combined separating and throttling calorimeter is to be used.

Example 1.38 Determine the dryness fraction of steam when the following data have been observed for a combined separating and throttling calorimeter: water separated = 2.5 kg, steam discharged from the throttling calorimeter = 22.5 kg, temperature of steam after throttling = 115°C, initial pressure = 11.98675 bar (gauge pressure), barometric pressure = 760 mm of Hg, and final pressure = 50 mm of Hg.

Solution We know that the dryness fraction of steam for a combined separating and throttling calorimeter is the product of dryness fraction given by separating calorimeter and dryness fraction, i.e.,

$$x = \frac{m_{s2}}{m} = \frac{m_{s2}}{m_{s1}} \times \frac{m_{s1}}{m} = x_1 \times x_2$$

The dryness of fraction given by separating calorimeter,

$$x = \frac{m_{s2}}{m} = \frac{22.5}{(22.5 + 2.5)}$$

$$= 0.90$$

Pressure $p_1 = 11.98675 + 1.01325 = 13$ bar

Pressure $p_2 = (760 + 50) = 810$ mm of Hg

$$= \frac{810}{760} \times 1.01325 = 1.08 \text{ bar}$$

For throttling calorimeter,

Enthalpy before throttling at pressure p_1 = Enthalpy after throttling at pressure p_2

$$h_{f1} + x_2\, h_{fg1} = h_{g2} + c_{ps}\,(T_{sup2} - T_{g2})$$

At 13 bar from steam table (Table A-1.2), we have,

$h_{f1} = 814.91$ kJ/kg, $h_{g1} = 2787.6$ kJ/kg,

and $h_{fg1} = 1972.69$ kJ/kg

At 1.08 bar from steam table (Table A-1.2) by interpolation, we have,

$h_{g2} = 2678.41$ kJ/kg, and $T_{g2} = 101.69$°C

$$\therefore \quad x_2 = \frac{h_{g2} + c_{ps}\,(T_{sup2} - T_{g2}) - h_{f1}}{h_{fg1}}$$

$$= \frac{2678.41 + 2.1 \times (115 - 101.69) - 814.91}{1972.69}$$

$$= 0.9588$$

Thus,

$$x = x_1 \times x_2$$

$$= 0.90 \times 0.9588 = 0.863$$

Example 1.39 The following observations have been obtained for a combined separating and throttling calorimeter: water separated in the separator = 0.6 kg, steam condensed after throttling = 3.5 kg, temperature of steam after throttling = 125°C, initial steam pressure before throttling = 10 bar, barometric pressure = 76 cm of Hg, final pressure = 20 cm of Hg. Determine the dryness fraction of steam if specific heat of superheated steam is 2.15 kJ/kg K.

Solution We know that the dryness fraction of steam for a combined separating and throttling calorimeter is the product of dryness fraction given by separating calorimeter and dryness fraction, i.e.,

$$x = \frac{m_{s2}}{m} = \frac{m_{s2}}{m_{s1}} \times \frac{m_{s1}}{m} = x_1 \times x_2$$

The dryness of fraction given by separating calorimeter,

$$x_1 = \frac{m_{s1}}{m} = \frac{3.5}{(0.6 + 3.5)}$$
$$= 0.854$$

Pressure $p_1 = 10$ bar

Pressure $p_2 = (76 + 20) = 96$ cm of Hg

$$= \frac{96}{76} \times 1.01325 = 1.28 \text{ bar}$$

For throttling calorimeter,

Enthalpy before throttling at pressure p_1 = Enthalpy after throttling at pressure p_2

$$h_{f1} + x_2 h_{fg1} = h_{g2} + c_{ps} (T_{\text{sup2}} - T_{g2})$$

At 10 bar pressure from steam table (Table -1.2), we have,

$h_{f1} = 762.79$ kJ/kg, $h_{g1} = 2778.1$ kJ/kg,

and $h_{fg1} = 2015.31$ kJ/kg

At 1.28 bar from steam table (Table A-1.2) by interpolation, we have,

$h_{g2} = 2686.28$ kJ/kg, and $T_{g2} = 106.64$°C

$$\therefore \quad x_2 = \frac{h_{g2} + c_{ps} (T_{\text{sup2}} - T_{g2}) - h_{f1}}{h_{fg1}}$$

$$= \frac{2686.28 + 2.15 \times (125 - 106.64) - 762.79}{2015.31}$$

$$= 0.974$$

Thus, $x = x_1 \times x_2$

$$= 0.854 \times 0.974 = 0.832$$

Example 1.40 The steam at pressure of 11 bar having a flow rate of 3.5 kg/min is passed through a combined separating and throttling calorimeter. The moisture is separated at a rate of 0.5 kg/min. The steam is throttled to a pressure of 2.5 bar and passed through a cooling coil which removes heat at a rate of 600 kJ/min. The dryness fraction of steam coming out from the coil is 0.925. Determine the temperature of steam after throttling and its dryness fraction before and after the separator if specific heat of superheated steam is 2.15 kJ/kg K.

Solution We know that the dryness fraction of steam for a combined separating and throttling calorimeter is given by,

$$x = \frac{m_{s2}}{m} = \frac{m_{s2}}{m_{s1}} \times \frac{m_{s1}}{m} = x_1 \times x_2$$

The dryness fraction by separating calorimeter,

$$x_1 = \frac{m_{s1}}{m} = \frac{3.5}{(0.5 + 3.5)}$$

$$= 0.857$$

The heat removed at cooling coil is given by,

$$q_1 = \frac{600}{3} = 200 \text{ kJ/kg}$$

Enthalpy of steam leaving the coil is given by,

$$h = h_f + xh_{fg}$$

$$= 535.34 + 0.925 \times 2181.56$$

$$= 2553.28 \text{ kJ/kg}$$

The total heat content in steam at 2.5 bar when it enters the cooling coil,

$$h_3 = q_1 + h$$

$$= 200 + 2553.28 = 2753.28 \text{ kJ/kg}$$

At pressure p_1 = 11 bar from steam table (Table A-1.2), we have,

$$h_{f1} = 781.32 \text{ kJ/kg}, \qquad h_{g1} = 2781.7 \text{ kJ/kg},$$

and $$h_{fg1} = 2000.38 \text{ kJ/kg}$$

$$h_2 = h_{f1} + x_2\, h_{fg1}$$

$$= 781.32 + x_2 \times 2000.38$$

But $h_2 = h_3$

$$\therefore \quad 2753.28 = 781.32 + x_2 \times 2000.38$$

$$\therefore \quad x_2 = 0.9858$$

Thus dryness fraction of steam,

$$x = x_1 \times x_2$$

$$= 0.857 \times 0.9858 = 0.845$$

At pressure $p_3 = 2.5$ bar, we have,

$$h_{f3} = 535.34 \text{ kJ/kg}, \qquad h_{g3} = 2716.9 \text{ kJ/kg},$$

$$h_{fg3} = 2181.56 \text{ kJ/kg}, \quad \text{and} \quad T_{g3} = 127.43°\text{C}$$

Since $h_3 > h_{g3}$, so steam leaving the throttling calorimeter is superheated.

The enthalpy after throttling is given by,

$$h_3 = h_{g3} + c_{ps} (T_{sup} - T_{g3})$$

or

$$2753.28 = 2716.9 + 2.15 \times (T_{sup} - 127.43)$$

$\therefore$

$$T_{sup} = 144.35°\text{C}$$

1.9 STEAM BOILERS

The device used for generating steam is called a boiler or steam generator. Boiler is a closed vessel, generally made of steel in which steam is generated from water at a desired pressure and temperature by the application of heat. The hot gases formed by the burning of fuel (coal, oil, or gas) transfer heat to water through the heating surface of the boiler. The steam thus generated may be used for the following purposes:

(i) **Power Generation** Electrical power or mechanical work is obtained by the expansion of steam in steam turbines or engines.

(ii) **Process Heating** Steam is utilized for industrial heating process such as in textile, jute, chemical, paper, sugar, and many other industries.

(iii) **Space Heating** In the winter, steam is used for heating the buildings and water for hot water supply.

1.10 REQUIREMENTS OF A GOOD BOILER

Few of the primary requirements of a good boiler are given below:

1. **Safety** Boiler should conform to the safety regulations laid down in the Boiler Act. Water should be safely contained and steam must be delivered safely in the required condition.
2. **Accessibility** Boiler should be easily accessible for cleaning, inspection, and repair.
3. **Cost** Initial, installation, and maintenance costs should be low.
4. **Fuel and Refractory Material Consumption** Fuel and refractory material consumption should be low.

5. **Starting and Steam Generation Capability** Starting should be quick and should be capable to meet the fluctuating demands of steam supply.
6. **Constructional Design, Weight, and Space Requirements** Its design should be simple so that its components can be easily dismantled and transported. It should be light in weight and occupy less space.
7. **Combustion Efficiency and Heat Transfer** It should have better fuel combustion efficiency and high rate of heat transfer. The heating surface should be free from any deposition of mud.

1.11. CLASSIFICATION OF BOILERS

Boilers may be classified in various ways. The important ones are given below:

1. According to the relative position of water and hot gases, the boilers are classified as water tube boiler and fire tube boiler.

   Water Tube Boiler The water circulates through the tubes and hot gases surround them. Examples of water tube boilers are Babcock and Wilcox boiler, Stirling boiler, and Benson boiler.

   Fire Tube Boiler The hot gases pass through the boiler tubes which are surrounded by water. Examples of fire tube boilers are Cochran boiler, Lancashire boiler, Cornish boiler, and Locomotive boiler.
2. According to the direction of the axis of the boiler shell, the boilers are classified as vertical boilers and horizontal boilers. However, in some of the cases, the axis may be inclined also.

   Vertical Boilers In vertical boilers, the axis of the shell is vertical, for example, Cochran boiler.

   Horizontal Boilers In horizontal boilers, the axis of the shell remains horizontal, for example, Lancashire boiler.
3. According to the method of firing, the boilers are classified as internally fired and externally fired boilers.

   Internally Fired Boiler In internally fired boilers, the furnace is provided inside the boiler shell. Examples of internally fired boilers are Cochran boiler, Lancashire boiler, and Scotch boiler.

   Externally Fired Boilers In externally fired boilers, the furnace is located outside the boiler shell. Examples of externally fired boilers are, Babcock and Wilcox boiler, and Locomotive boiler.
4. According to the method of water circulation, boilers are classified as natural circulation and forced circulation boilers.

   Natural Circulation Boiler In natural circulation boiler, water is circulated by natural convection currents set up due to difference in temperature. Examples of natural circulation are Babcock and Wilcox boiler, Locomotive boiler, and Lancashire boiler.

Forced Circulation Boiler In the forced circulation boiler, water is forced to circulate in the boiler by pump. Examples of forced circulation are high pressure boilers such as Velox boiler and Lamont boiler.

5. According to working pressure, boilers are classified as low pressure and high pressure boilers.

   Low Pressure Boilers In low pressure boilers, the pressure of generated steam will be below 80 bar. Examples of low pressure boilers are Babcock and Wilcox boiler, Cochran boiler, Lancashire boiler, Cornish boiler, and locomotive boiler.

   High Pressure Boilers In high pressure boiler, the pressure of generated steam will be more than 80 bar. Examples of high pressure boilers are Velox boiler, Benson boiler, and Lamont boiler.

6. According to the use of boiler, boilers are classified stationary, portable, and mobile boilers.

   Stationary Boilers Boilers which are used for electric power generation in stationary plants are called as stationary boilers.

   Portable Boilers Boilers which can be easily dismantled and transported from one place to the other are called portable boilers.

   Mobile Boilers Boilers which are used to move some carriers like locomotive and ship are called mobile boilers.

7. According to the nature of draught, boilers are classified as natural draught and forced draught boilers.

   Natural Draught Boilers Boiler in which the fuel is burnt with the natural circulation of air is called natural draught boiler.

   Forced Draught Boilers Boiler in which air is forced by a blower to burn the fuel is called forced draught boiler.

1.12 BOILER PARTS: DETAILS AND TERMINOLOGY

1. **Boiler Drum** The boiler drum consists of a cylindrical-shaped shell with closed ends called the boiler heads. It contains water and steam.
2. **Furnace** Furnace is the space provided above the grate and below the boiler shell, where the fuel is burnt efficiently to produce hot gases.
3. **Grate** Grate is the cast iron bars structure on which fuel is burnt. It is placed at the bottom of the furnace. The air passes through the space between the iron bars and thus helps in proper burning of the fuel.
4. **Setting** Setting is a kind of bricks formation which forms the wall of furnace, combustion chamber, and passage of hot gases.
5. **Boiler Mountings** The devices which are necessary for the safety and control are known as boiler mountings. For example, safety valve, fusible plug, pressure gauge, water level indicator, and blow off cock.

6. **Boiler Accessories** The devices which are installed to increase the efficiency of the boiler and help in the proper working of the steam power plant are known as boiler accessories. For example, economizer, superheater, air preheater, and feed pump.
7. **Flue Gases** Flue gases means the hot gases generated as a result of combustion of fuel in the furnace.
8. **Steam and Water Space** The space occupied by the water inside the shell is called water space, whereas the space left above the water is called steam space.
9. **Heating Space** Heating space means the surface of boiler exposed to flue gases.
10. **Draught** Draught means the small pressure difference which causes flow of hot gases and air through a boiler.

1.13 FIRE TUBE BOILERS

1.13.1 Cochran Boiler

Cochran boiler is a vertical, multi-tubular, fire tube boiler. It is internally fired and natural circulation boiler. These boilers are used for small capacity steam generation varying from 20 kg per hour to 3000 kg per hour. The steam can be generated up to a pressure of 20 bar. The efficiency of these boilers ranges from 70% to 75%. The schematic view of Cochran boiler is shown in Figure 1.22.

These boilers are available in different sizes. The shell diameter varies from 1 m to 3 m and height varies from 2 m to 6 m. The typical data of a Cochran boiler of 2.75 m shell diameter are: Height of shell = 5.78 m, Heating surface = 120 m^2, Steam pressure = 6.7 bar, Steam capacity = about 3500 kg per hour, Efficiency = 70% (when coal fired) and 75% (when oil fired)

Constructional and Working Details It consists of a vertical cylindrical shell surrounded by a hemispherical crown at its top. The hemispherical shaped shell provides maximum strength, maximum volume of space, and permits maximum absorption of radiant heat. The fire box (or furnace) is also hemispherical in shape which is constructed in single piece thus it has no joints.

A manhole is provided at the top of the crown through which a man can enter for cleaning the boiler. The connections for steam stop valve, safety valve, and pressure gauge are also provided at the top of the crown.

The grate is provided at the bottom of the firebox on which the coal is fed through the fire door for burning. The ash pit is located below the grate to collect the ash formed which is removed manually after regular intervals. In the firebox a damper is also provided to control the flow of flue gases. The firebox and the combustion chamber are connected through a short flue pipe. The combustion chamber is lined with fire bricks (refractory material) which prevent the damage of shell due to overheating caused by flue gases. Various horizontal fire tubes (about 165 to 170 in numbers and of 62.5 mm external diameter) provided near the combustion chamber are attached

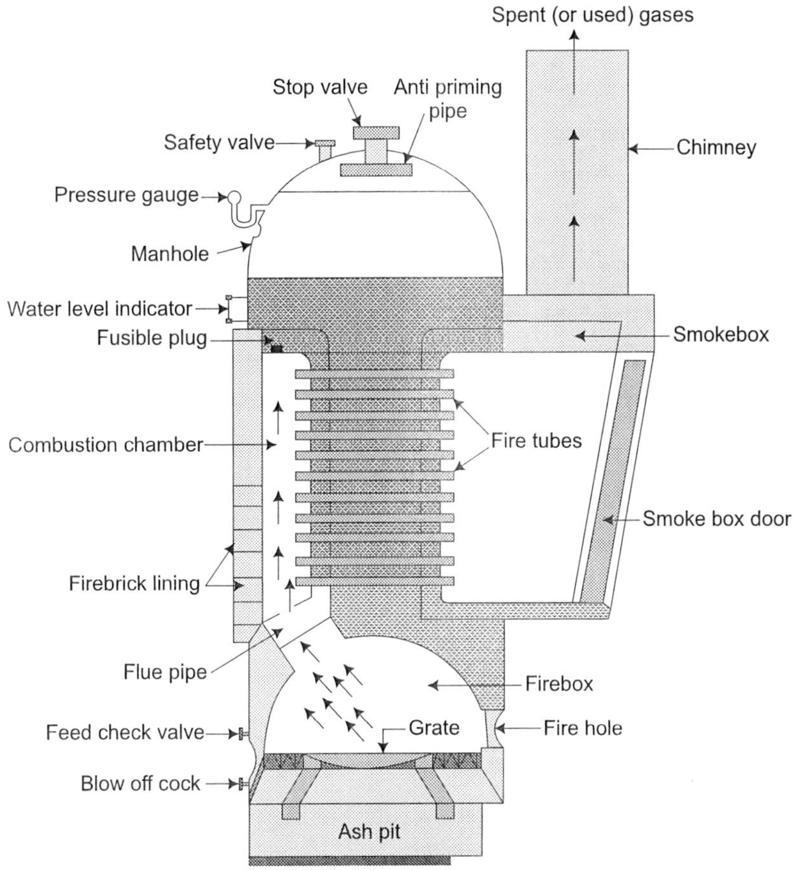

Fig. 1.22 Cochran Boiler

to the smoke box. The smokebox has hinged door which provides access to fire tubes for their cleaning and inspection. A chimney (or stack) is provided at the top of smokebox for the exit of the spent (or used) gases to the atmosphere.

A fusible plug is mounted over the grate to stop the burning of fuel on the grate when the water level falls below a predetermined level. A blow off cock is fitted at the lowest water level to remove the impurities such as mud, salt, etc. In the case of oil fired boiler, an oil burner is provided at the fire hole instead of a grate and thus the bottom of the firebox is lined with firebricks.

When the coal (fuel) is burnt on the grate in the firebox, the hot gases produce which flow through the flue pipe into the combustion chamber. Then the flue gases pass through a large number of horizontal fire tubes to smokebox. The fire tubes and firebox are surrounded by water. Thus, the water converts into steam which collects over the space in the boiler drum. The steam can be taken out through the connection provided for steam stop valve. Finally, these gases are discharged to the atmosphere through the chimney.

1.13.2 Locomotive Boiler

Locomotive boiler is an internally fired, multi-tubular, fire tube boiler. It is an ideal portable boiler which can be used for small power plants. Its steaming capacity varies from 4500 kg/hr to 7000 kg/hr and its working pressure is limited up to 17.5 bar. Its steaming rate is 20 kg/m^2 of the heating surface. The boiler shell (or barrel) is 1.5 m in diameter and 4 m in length. It has about 157 number of thin fire tubes. The schematic view of this boiler is shown in Figure 1.23.

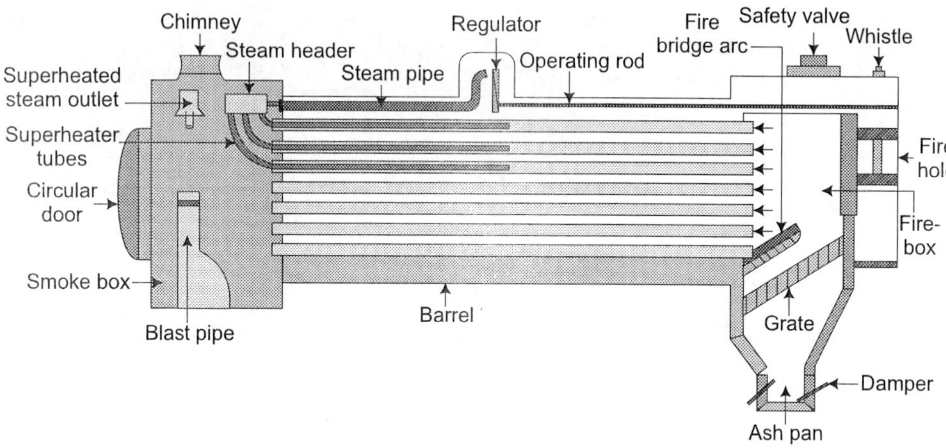

Fig. 1.23 Locomotive Boiler

The firebox is a separate unit which is attached with the one end of the horizontal boiler shell. A smokebox is attached to the other end of the shell. The hot gases of combustion produced in the firebox pass to smokebox through the fire tubes surrounded with water. The steam is collected in steam drum fitted on the top of the shell. To access the smokebox and the fire tubes for its cleaning a door is provided.

In the firebox, the grate is sloped downwards towards the front which deflects the hot gases. As a result, the whole heating surface of the firebox comes in contact with the hot gases. Finally, the hot gases pass through the fire tubes.

When the regulator (stop valve) is opened, the wet steam enters the inlet headers of the superheater. After passing through the superheater tubes these gases return to the outlet header of the superheater from which it is supplied to the primemover.

The boiler draught is provided by the steam exhausted from the primemover which is discharged through the blast pipe fitted in the smokebox to the chimney. To exhaust the hot gases, a diverging section is provided at the top of the smokebox to increase its pressure more than the atmospheric pressure. A steam blower is also provided for use when the steam supply to the engine is shut-off.

1.14 WATER TUBE BOILERS

1.14.1 Babcock and Wilcox Boiler

Babcock and Wilcox boiler is one of the good designs of water tube boiler which is used in steam power plants for generating electricity. It is horizontal, externally fired, natural circulation type of stationary boiler. It is externally fired boiler, so all types of fuel can be burnt. Its evaporative capacity ranges from 20,000 kg/h to 40,000 kg/h and its efficiency varies from 60% to 80%. The normal operating pressure range is 11.5 bar to 17.5 bar, but it can be increased up to 42 bar.

Constructional and Working Details The schematic view of Babcock and Wilcox boiler is shown in Figure 1.24. It consists of high pressure horizontal drum mounted at the top. One half of the drum is filled with water and the steam remains in the other half. The length of drum generally varies from 6 m to 9 m and its diameter varies from 1.2 m to 1.8 m. All mountings are provided on the drum. A large number of water tubes are arranged below the drum. These water tubes are connected to the drum ends with the uptake and downtake headers. The water tubes are kept inclined at an angle of 5° to 15° to the horizontal which promotes the water circulation. The headers are provided with hand holes in front of the tubes which help in cleaning and inspection of water tubes. A mud box having a blow off cock for blowing out the mud or sediments is attached

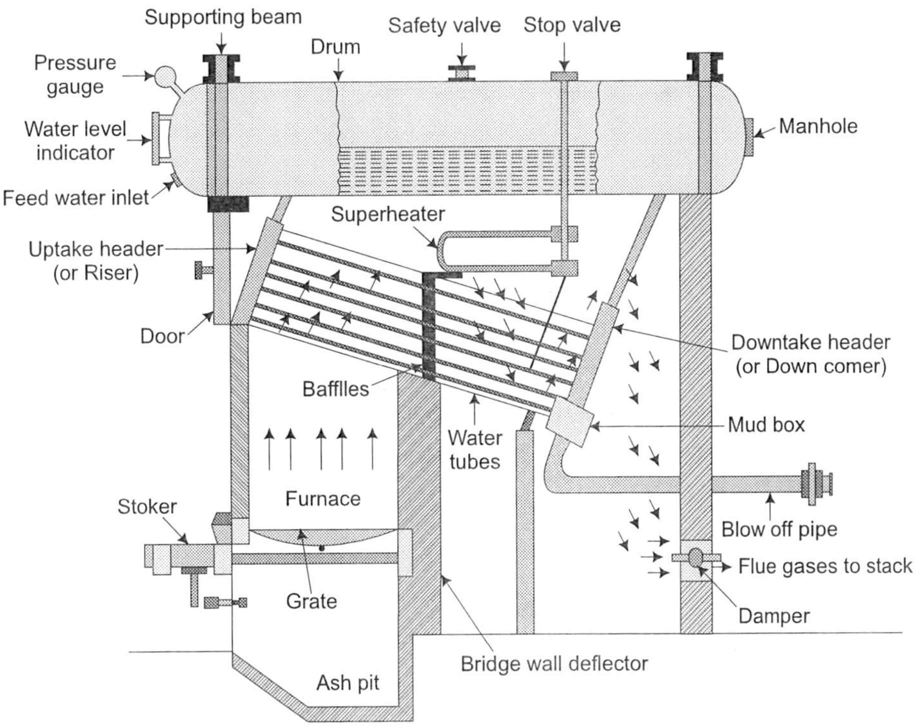

Fig.1.24 Babcock and Wilcox Boiler

at the bottom of the down-take header. Superheater is horizontally arranged between the drum and the water tubes. The entire assembly of water tubes, drum, and all fittings are suspended by steel slings from steel girders resting on steel columns to avoid any strains on the setting (masonry) due to the expansion and contraction of the boiler caused by change in temperature.

The furnace is arranged below the uptake header. The coal is fed to the chain grate stoker through the fire door. The heating surface of this boiler includes the outer surface of the water tubes and half of the cylindrical surface of drum exposed to hot gases. The bridge wall deflector deflects the hot gases upwards. Two firebrick baffles are arranged which provide three passes of the hot gases means the gases first rise up then move down and again rise up. The smokebox having a chimney at its top is provided at the back of the furnace. A damper is placed at the inlet of chimney to regulate draught.

The hot gases produced by the burning of fuel rise upwards and pass across the front portion of the water tubes and the drum. The baffles deflect the gases and thus the gases flow in zigzag fashion over the water tubes and along the superheater tubes. Finally, the hot gases escape to the atmosphere through the chimney. During this travel of hot gases heat is transferred to the water which gets converted into steam. The front portion of the water tubes which comes in contact with the hot gases is relatively heated at higher temperature than the remaining part of it. Thus, the density of water decreases which results in the rise of this hot water and steam through the uptake header (or riser) into the drum where steam and water get separated. The steam being lighter is collected in the upper part of the drum. Then the water from the back portion of the drum enters into the water tubes through the down comer (or downtake header). The movement of water that results from the continual replacement of the hot water in the water tubes is called natural convective currents. Thus, a continuous circulation of water is maintained in the boiler and thus it is also called natural circulation type boiler.

The steam accumulated in the steam space of the drum is taken into the superheater through a pipe. In the superheater, the steam receives additional amount of heat as a result it gets superheated and then it is taken out through the steam stop valve.

1.14.2 Stirling Boiler (or Bent Tube Boiler)

Stirling boiler is a multi-tube, multi-drum, water tube boiler in which bent tubes are used.

Figure 1.25 shows a Stirling boiler with two water drums (mud drums) and three steam drums connected in series.

The water drums are 10 cm to 25 cm larger in diameter than the steam drums. The steam drums are suspended by the steel structures and are connected in series by the steam and water equalizing tubes. The water drums are connected to the steam drums by the bent tubes. The steam and water equalizing tubes maintain the pressure and water level in the steam drums.

Feed water first enters the steam drum 1 from which it passes down to the water drum through the tubes. The mud particles collect in this drum. Then water circulates to the upper steam drums 2 and 3. The hot combustion gases from the fire grate rise and pass over the bent tubes. The baffles

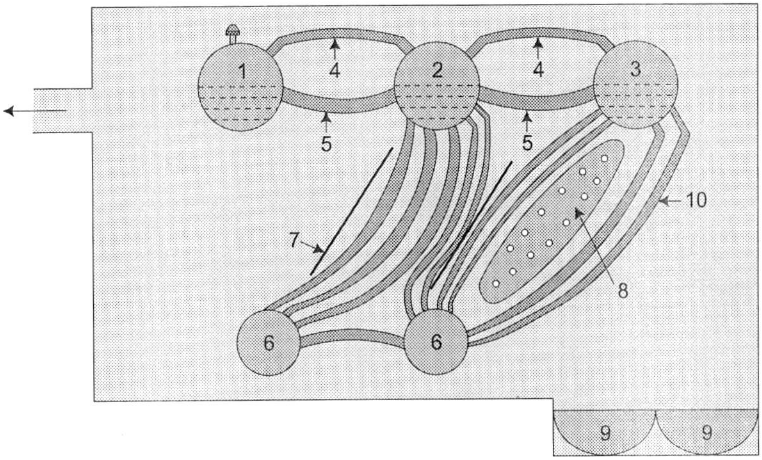

1,2,3: Steam drums; 4: Steam equalizing tubes; 5: Water equalizing tubes; 6: Mud drums; 7: Buffles; 8: Superheater; 9: Grate; 10: Water tubes

Fig. 1.25 Stirling Boiler (or Bent Tube Boiler)

deflect and distribute the gases properly. Finally, the hot gases leave the boiler through the gas outlet. The wet steam is passed through the superheater to superheat it.

These boilers can generate steam up to a pressure of 60 bar and temperature 450°C with a capacity of about 50,000 kg/hr. These boilers are lighter in construction and are more flexible in operation than straight tube boilers. But the bent tubes are difficult to clean and inspect.

1.15 COMPARISON BETWEEN FIRE TUBE AND WATER TUBE BOILERS

The important differences between water tube and fire tube boilers are tabulated below:

Water tube boilers	*Fire tube boilers*
1. The water flows inside the tubes and hot flue gases surround the tubes.	1. The hot flue gases flow inside the tubes and water surrounds the tubes.
2. Due to larger heating surface its steam generation rate is high. Thus, it is suitable for power plants.	2. Due to smaller heating surface its steam generation rate is low. Thus, it is not suitable for power plants.
3. These boilers can work up to pressure as high as 200 bar.	3. The working pressure of these boilers is limited to about 20 bar.
4. For a given power it occupies less floor area but requires more height.	4. For a given power it requires more floor area.
5. These boilers are generally externally fired, thus the furnace can be altered as per the types of fuel.	5. These boilers are generally internally fired, thus the furnace cannot be altered as per the types of fuel.
6. The efficiency of these boilers ranges from 80% to 95%.	6. The efficiency of these boilers ranges from 60% to 80%.

7. Design of these boilers is complex, thus their initial cost is high.	7. Design of these boilers is simple, thus their initial cost is low.
8. For operating these boilers more skill is required.	8. For operating these boilers less skill is required.
9. All parts of these boilers are easily accessible for cleaning, inspection, and repair.	9. All parts of these boilers are not easily accessible for cleaning, inspection, and repair.
10. Large load fluctuations can be handled effectively.	10. Large load fluctuations cannot be handled easily.
11. The chances of explosion are more but the risk of explosion is less serious and the process of steam generation will not stop. If any water tube bursts it can be plugged or replaced.	11. The chances of explosion are less but the risk of explosion is very serious due to large quantity of steam.
12. These boilers can be transported and erected easily as its different parts can be dismantled.	12. Transportation of these boilers is not so easy due to large size of the boiler shell.
13. The water circulation in these boilers is better, thus there are less chances of deposit of impurities on the heated surface.	13. The water circulation in these boilers is poor, thus there are chances of deposit of impurities on the heated surface.
14. Presence of impurities in water cause scale deposits inside the tubes which may lead to overheating and bursting of tubes. Thus, treatment of water is essentially required which is a costly affair.	14. The feed water does not require any treatment.
15. Steam generation capacity of water tube boiler is much larger.	15. Maximum steam generating capacity of fire tube boiler is about 9000 kg/hr.

1.16 BOILER MOUNTINGS AND ACCESSORIES

The boilers are equipped with various fittings and devices which are necessary for the safety of the boiler and its efficient operation and control. These components (fittings and devices) are categorized under two headings, namely, boiler mountings and accessories.

1.16.1 Boiler Mountings

Boiler mountings are the fittings and devices which are necessary for the safety of the boiler and control of steam generation process. The safety fittings are safety valves, fusible plug, water level indicator, and high steam low water safety valve. The control fittings are pressure gauge, steam stop valve, feed check valve, blow off cock, manhole, and mud box. The mountings form an integral part of the boiler.

(i) Safety Valve When the steam pressure in a boiler exceeds the safe working pressure limit the safety valve discharges some of the steam automatically from the boiler to the atmosphere. Thus, it prevents the building up of excessive pressure in the boiler. The safety valves are placed on the top of the boiler shell. As a precautionary measure two safety valves are fitted so that at least one of them works at a time. The commonly used safety valves are dead weight safety valve, spring loaded safety valve, lever safety valve, and high steam and low water safety valve which are given below:

(a) Dead Weight Safety Valve A schematic view of a dead weight safety valve is shown in Figure 1.26.

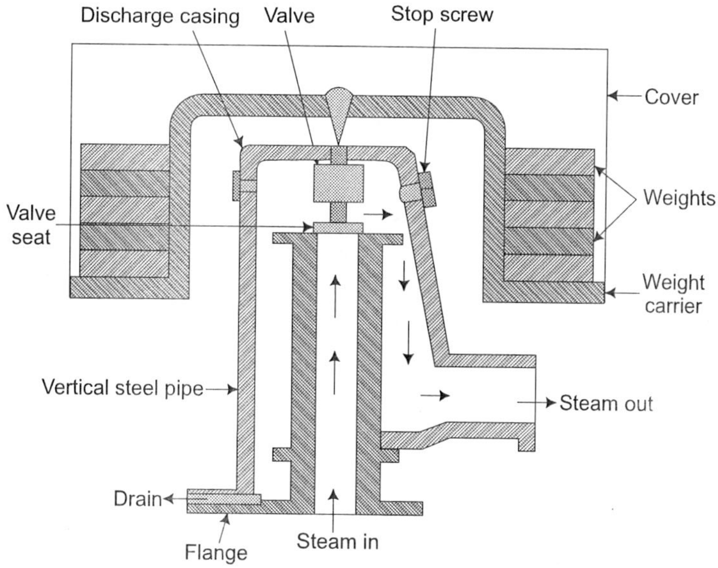

Fig. 1.26 Dead Weight Safety Valve

It consists of a vertical pipe with a flange at its bottom which is bolted to the mounting block on the boiler shell. On the top of the other end of the vertical pipe, a gun metal seat is secured on which gun metal valve rests. The dead weight of cast iron rings are placed on the weight carrier which is suspended from the top of the valve. The steam pressure acting in the upward direction is balanced by the downward force equivalent to the load on the valve. The total load on the valve includes weight of the carrier, weight of cast iron rings, weight of the cover, and weight of the valve itself. Dead weight safety valves are simple in construction and are used for stationary boilers.

When the force due to steam pressure inside the boiler is less than the total dead weight on the valve, the valve remains closed. But when the steam pressure exceeds the total weight, the valve is lifted off and thus the steam escapes through the discharge pipe. Thus, steam pressure in the boiler shell becomes normal and then the valve again closes due to dead weight. The lift of valve is controlled by the stop screws fitted on the discharge casing. The steam condensed in the discharge casing is removed by a drain pipe fitted at the bottom of the casing.

(b) **Spring Loaded Safety Valve** A spring loaded safety valve (Ramsbottom safety valve) is shown in Figure 1.27. It consists of a cast iron body provided with two branch pipes for the steam flow which are connected to a common block. The spring holds the gun metal valves of the same size on their seats provided over the branch pipes by pulling the lever downward. The lever has two conical pivots, one of which is integrally forged with the lever and the other is pin connected to its end. Both pivots rest on the centers of the valves. One end of the spring is hooked to the lever at its middle while the other end is hooked to

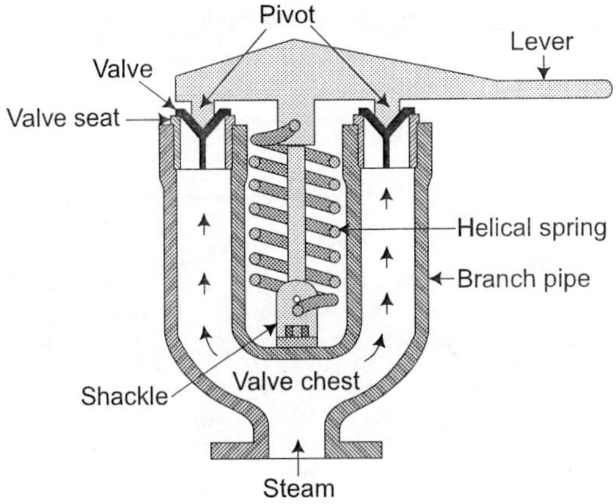

Fig. 1.27 Spring Loaded Safety Valve

the shackle fixed to the valve chest by studs and nuts. The shackle and the lever are also connected by two links.

When the steam pressure in the boiler is normal the valves rest on their seats due to spring force. But when the steam pressure exceeds the working pressure, the valves open and the steam escapes to the atmosphere till the pressure is reduced to the working pressure.

As the working operation of spring loaded safety valve is free from vibrations and jerks thus it is suitable for locomotive and marine boilers.

(c) **Lever Safety Valve** Figure 1.28 shows schematic view of a lever safety valve. The working operation of this valve depends on the second system of levers. The cast iron valve body is bolted to the boiler shell. The gun metal valve rests on the valve seat which is screwed

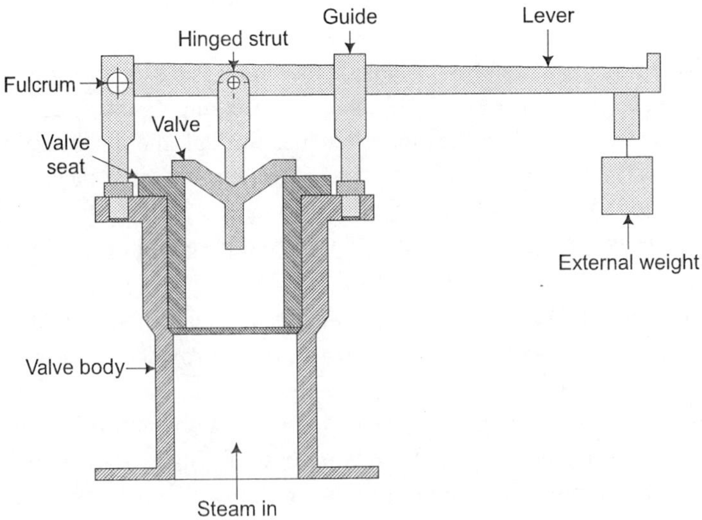

Fig. 1.28 Lever Safety Valve

or fixed to the valve body. The valve is held in its position by a lever made of mild steel or wrought iron. One end of the lever is pivoted at the fulcrum and the other end carries the external weight. The thrust is provided to the valve through a hinged strut. A guide is provided to prevent the lateral movement of the lever and to limit the lift of the valve.

The position of the weight on the lever can be adjusted to keep the valve on its seat in closed position under normal working condition. But when the steam pressure exceeds the working pressure the valve lifts up and some quantity of steam escapes to the atmosphere. Thus, the steam pressure in the boiler falls down and the valve again rests on its seat. The lever safety valve is suitable for stationary boilers.

(d) **High Steam and Low Water Safety Valve** The high steam and low water safety valve blows off the seam when the steam pressure in the boiler becomes higher than the normal working pressure or when the water level in the boiler falls below the predetermined level. Thus, this valve not only safeguards the boiler against too high pressure but also against too low water level in the boiler. Figure 1.29 shows a high steam and low water safety valve.

It consists of two valves V_1 and V_2. The main valve V_2 rests upon the valve seat and it provides a seat for hemispherical valve V_1. The valve V_2 is loaded by a lever which is hinged at its one end and carries a weight at the other end. The arrangement for the valve V_2 is similar to that of lever safety valve. A short pivot attached to the lever by a strut keeps the valve V_2 at its seat under normal working condition. If the pressure inside the boiler exceeds

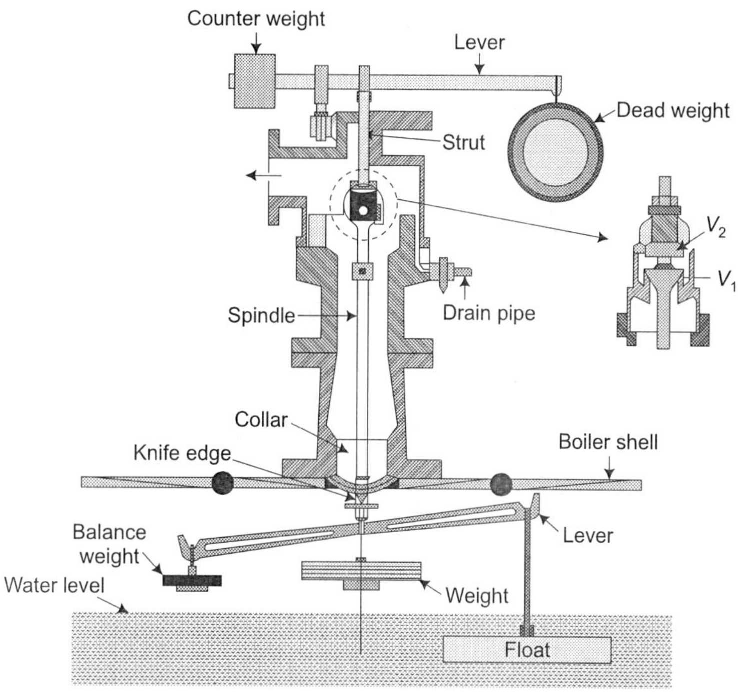

Fig. 1.29 High Steam and Low Water Safety Valve

the working pressure limit, the valve V_2 along with the valve V_1 lifts up and thus the steam is allowed to escape through the passage between the valve seat and the main valve V_2.

The arrangement for the valve V_1 is inside the boiler. It has a lever hinged from the boiler shell. A float is attached at the end of the lever and a balance weight is attached at the other end. The valve V_1 is connected with a spindle which carries a weight at its end. Under normal working pressure the knife edge fitted on the lever touches the collar on the spindle of the valve and the float remains submerged inside the water. When the water level falls below a certain level, the float is uncovered, the whole system becomes unbalanced which causes swing in the lever. Due to this swing of lever the knife edge pushes the collar and the spindle in the upward direction and the valve V_1 opens. Thus, the steam escapes through the narrow passage between the valves V_1 and V_2. This produces a loud noise which acts as a warning to the boiler attendant and thus the corrective measure has to be taken. A drain connection is provided to drain off the condensed steam.

This valve is suitable for stationary boilers such as Lancashire and Cornish boilers.

(ii) Water Level Indicator Water level indicator indicates the level of water in the boiler. It is also called water gauge. Generally, two water gauges are fitted in front of the boiler from where it is easily visible to the observer. Figure 1.30 shows a water level indicator which is attached to the front end of the boiler. It consists of a strong gauge glass tube and three cocks say '*A*', '*B*', and '*C*'. The ends of the glass tube pass through the stuffing boxes fitted in the hollow gun metal castings which are connected to the boiler plate. Generally, the gauge glass tube is marked to show the safe level of water inside the boiler. The steam cock '*A*' connects steam space to glass tube

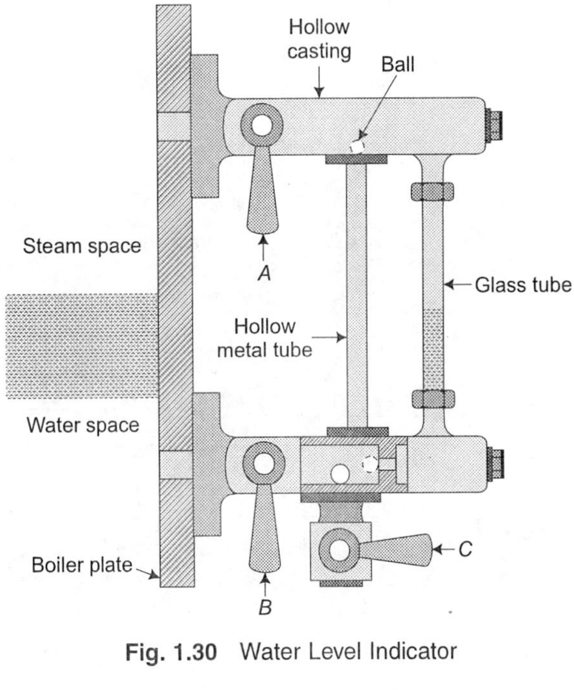

Fig. 1.30 Water Level Indicator

and the water cock 'B' connects water space to the glass tube. The drain cock '*C*' is used to blow off water from the glass tube at regular intervals to avoid accumulation of sediments or foreign particles. The upper and lower gun metal castings are connected by a hollow metal tube having two balls. In the normal working condition these balls close the metal tube openings. If the glass tube breaks, the rush of water through the bottom casting and steam through the upper casting carries the balls to close the glass tube openings. Thus, steam and water supply stop to the glass tube. Now the attendant can safely close the cocks and can replace the glass tube.

(iii) Pressure Gauge A pressure gauge indicates the pressure of steam inside the boiler. It is fitted in front of the boiler such that the operator can read it easily. The constructional details of most commonly used Bourdon pressure gauge is shown in Figure 1.31.

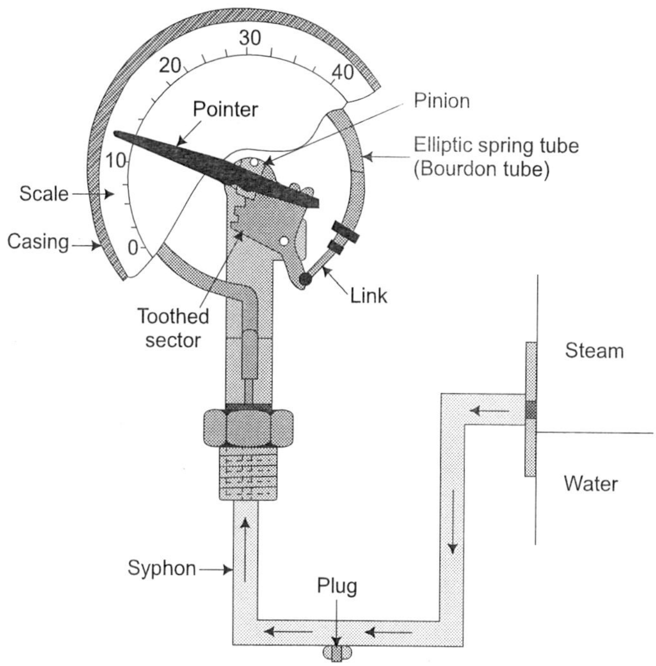

Fig. 1.31 Pressure Gauge

It consists of an elastic metallic tube called Bourdon tube which senses the pressure. The Bourdon tube is made of phosphor bronze and it is of elliptical cross section bent in the form of a circular arc. One end of the tube is fixed which is connected to the steam space of the boiler through a syphon filled with water. The other closed end of the tube is connected to a toothed sector wheel through a link. The toothed sector meshes with a pinion fixed on a spindle which carries the pointer to read the pressure on a dial gauge.

The steam pressure forces the water inside the elliptical tube. The tube tries to become straight thus the closed end of Bourdon tube moves outward. This slight movement of the tube actuates the sector and thus the pinion rotates on which the pointer is mounted. Hence, the slight movement of the Bourdon tube is considerably magnified and the pointer moves on the graduated scale which gives the steam pressure in the boiler.

(iv) Fusible Plug The fusible plug helps in extinguishing the fire in the boiler furnace when the water level in the boiler shell falls below the safe limit. Thus, it prevents the explosion which may occur due to overheating of the fire tubes or the firebox crown plate. Generally, it is fitted at the crown plate of the furnace or over the combustion chamber. It consists of a hollow gun metal body with hexagonal flange which is screwed to the firebox crown plate as shown in Figure 1.32. A hollow gun metal plug is screwed into the gun metal body. Another solid copper plug with conical top and rounded bottom is fitted in the hole of hollow gun metal plug by filling fusible metals in the depressions between the mating surfaces. The fusible metals may be lead or tin which have a low fusion point. Under normal working condition, the plug remains submerged in the water, thus the temperature of the fusible metal will be below its melting point. But as the water level falls below the safe limit, the fusible metal is exposed to the steam by which it gets overheated and melts. The copper plug falls down and held within the gun metal body by the ribs: It opens a way between steam space and the furnace. Thus, water-steam mixture rushes into the furnace through the hole and extinguishes the fire before any damage happens to the boiler due to overheating.

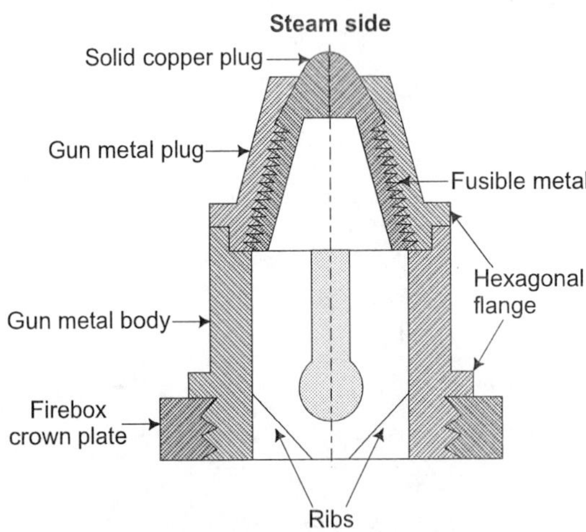

Fig. 1.32 Fusible Plug

(v) Steam Stop Valve or Junction Valve The function of the steam stop valve or junction valve is to regulate the flow of steam from the boiler to the prime mover (turbine or engine) as per the requirement and to stop the steam flow when not required. When a valve is mounted directly on the boiler drum it is called junction valve and when it is placed near the prime mover or between the steam pipes it is called stop valve. Generally, the smaller size valve is called stop valve and larger size refers to junction valve. A commonly used steam stop valve is shown in Figure 1.33. It consists of a valve chest made of cast iron or cast steel. It has two flanges at right angles. One

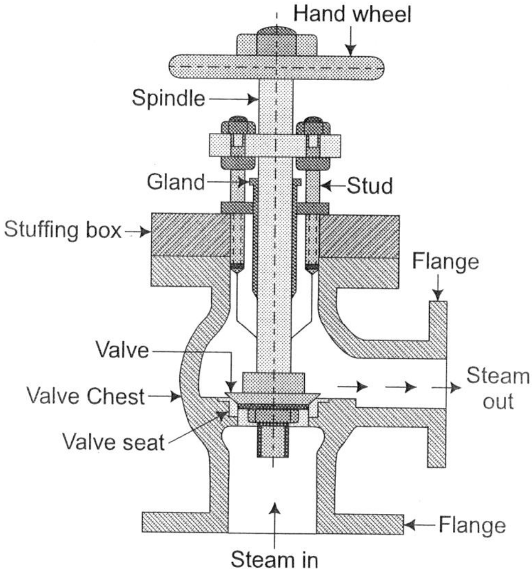

Fig. 1.33 Steam Stop Valve or Junction Valve

flange is bolted to the mounting block of the boiler for the inlet of the steam. The other flange is bolted to the steam pipe for the outlet of the steam. A hand wheel is mounted on the valve spindle to which a gun metal valve disc is screwed by a nut. The valve disc rests on the valve seat which is fixed on the valve body. The spindle is made to pass through glands of a screwed body part and then through a stuffing box. When the wheel is rotated the valve moves up or down and thus it opens or closes the passage for steam flow.

(vi) Feed Check Valve The function of the feed check valve is to allow the feed water under pressure to the boiler from the feed pump. It also prevents the back flow of water from the boiler to the pump when the pump pressure is less than the boiler pressure, or when the feed pump fails to work. So, feed check valve is a non-return valve fitted at the end of the delivery pipe from the feed pump. It is mounted on the top of the boiler shell slightly below the working level of water. Figure 1.34 shows a schematic view of a commonly used feed check valve.

It consists of two valves, namely, feed valve and check valve. The check valve allows the water to flow from the feed pump to the boiler side only. The feed valve is operated by a hand wheel and the check valve operates automatically due to the existence of pressure difference of water.

Under normal working condition, the pressure on the feed pump side is more than the boiler side which causes lifting up of the check valve from its seat and water enters the check valve. The feed valve is to be lifted up manually by operating the hand wheel so that water enters the boiler. The supply of feed water can be controlled by adjusting the position of the feed valve.

(vii) Blow off Cock The function of blow off cock is to discharge periodically the sediments collected in the bottom most part of the water space in the boiler while the boiler is working. It is also used to lower the water level rapidly when the level becomes too high and to empty the

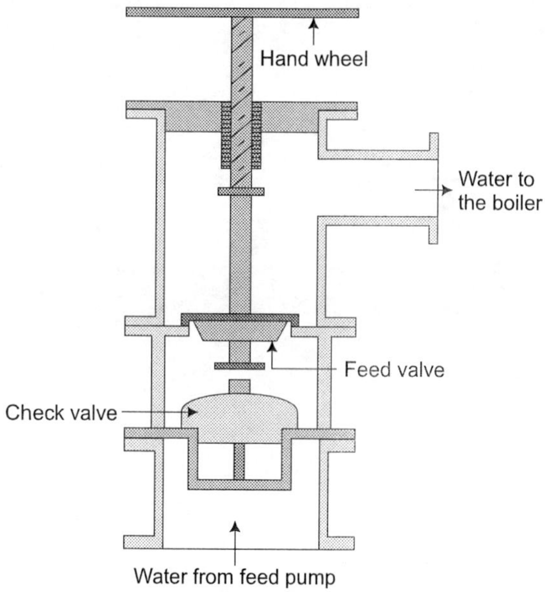

Fig. 1.34 Feed Check Valve

boiler when it is to be cleaned or inspected. It is mounted to the lowest part of the boiler either directly with the boiler shell or to an elbow pipe connected with the boiler shell. A common type of blow off cock is shown in Figure 1.35.

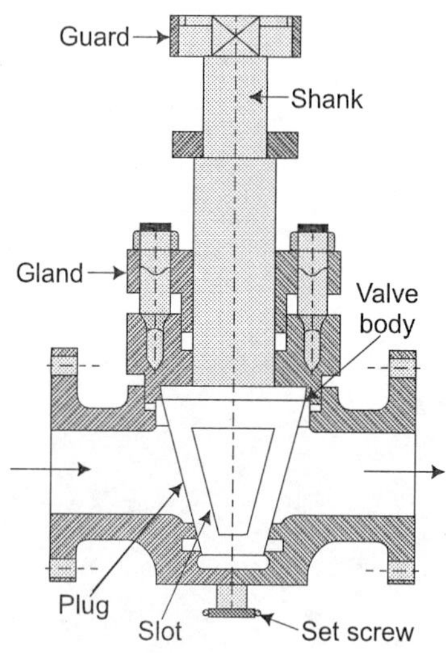

Fig. 1.35 Blow Off Cock

It consists of a conical gun metal plug having a hole or slot in it. The plug is fitted accurately into a similar casing which is provided with two flanges. One of the flanges is bolted to the boiler and the other is connected to a water outlet pipe. The shank of the plug passes through glands and stuffing boxes to avoid any leakage. The hole or slot in the plug is brought in line with the passage of the casing by rotation of the plug. This results in the discharging of water from the boiler. The discharge of water can be stopped by rotating the plug again. In order to remove the sediments quickly, the blow off cock should be operated when the boiler is working. A set screw provided at the bottom of the plug is used to force the plug off its seat when it gets jammed.

(viii) Manhole and Mud Box Manhole is the opening to allow men to enter inside the boiler for cleaning and inspection of the interior boiler parts. It is provided at any suitable location on the boiler shell so that a man can enter through the opening (or hole) inside the boiler. Mud box is placed at the bottom of the boiler to collect the mud.

1.16.2 Boiler Accessories

Boiler accessories are the auxiliary fittings and devices which are provided to increase the efficiency and to help smooth functioning of the plant. The main accessories are: superheater, air preheater, economizer, feed pump, steam separator, injector, and steam trap.

(i) Superheater Superheater is a surface heat exchanger in the form of tubes which is placed in the path of flue gases to extract some of their heat. The function of a superheater is to increase the temperature of saturated steam. In superheaters, steam pressure remains constant but its volume increases due to increase in temperature.

Sudgen's Hair Pin Type Superheater Sudgen's hair pin type superheater is shown in Figure 1.36 which is mostly used in low pressure boilers such as Lancashire and Cornish boilers. It consists of headers and '*U*' shaped steel tubes. When the superheater is working the valve V_1 remains closed and steam passes through '*U*' tubes via valve V_3 and the header where it gets superheated and then goes to the steam mains via valve V_2. When superheated steam is not required then the steam directly passes to the steam mains via valve V_1 and hence valve V_3 remains closed. The temperature of the superheated steam is controlled by adjusting the position of damper which alters the amount of flue gases.

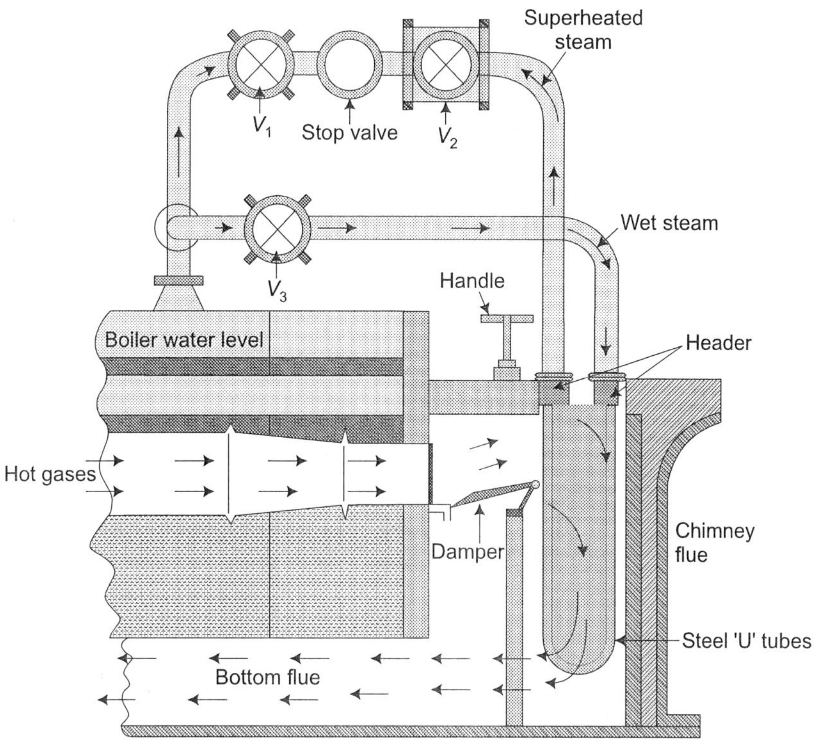

Fig. 1.36 Sudgen's Hair Pin Type Superheater

Classification of Superheaters According to the mode of heat received, superheaters can be classified as: convection superheater, radiant superheater, and combined superheater. Convection superheaters are placed in the path of flue gases and it receives heat by convection, whereas radiant superheaters are placed in one or more walls of the furnace and it receives heat by direct radiation. In the combination superheaters, the steam first enters the radiant superheaters and then in the convective superheaters.

In large installation, the superheater may be located in an independent fire furnace and are called portable superheaters, or separately fired superheaters.

On the basis of flow of steam and flue gases the superheaters are classified as counter flow, parallel flow, and the combined flow. Among these, counter flow superheaters give maximum efficiency and also these are lighter in weight and smaller in size.

According to the position in the furnace with respect to the water tubes superheaters can be classified as interdeck, overdeck, and inter-bank superheaters. Interdeck superheaters are placed in the space between the water tubes which are located near the furnace. Overdeck superheaters are placed in the space over the water tubes as in the case of Babcock and Wilcox boilers. Generally, both the interdeck and overdeck superheaters are convective type. Inter-bank superheaters are placed between the bank, or row of water tubes.

(ii) Economizer Economizer is a device which extracts some of the heat from the spent flue gases by preheating the water being fed to the boiler. It is placed in the path of flue gases between the boiler and the chimney. Generally, economizers are of independent and integral types. The independent type of economizer is installed apart from the boiler setting, whereas the integral type of economizer is installed within the boiler setting.

The main advantages of preheating the boiler feed water by installing economizers are given below:

(i) The feed water supplied to the boiler is at a high temperature, so it requires less amount of heat for its conversion into steam. This results in saving of fuel.

(ii) Thermal efficiency of the plant improves. It increases by about 1% for each 5°C to 6°C rise in temperature of feed water by an economizer.

(iii) Steaming capacity and life of boiler increases.

Green's economizer shown in Figure 1.37 is an independent type of economizer which is used for boilers of medium pressure range up to about 25 bar. It consists of large numbers of vertical cast iron pipes or tubes which are hydraulically pressed into the bottom and top headers. The feed water from the feed pump enters the bottom header. From the bottom header it passes through vertical tubes and reaches to the top header. From the top header it goes to the boiler through the feed check valve. The hot gases pass over the external surface of the water tubes and transfer its heat to the feed water through the tubes surface. The flow of flue gases results in soot formation on the water tubes surfaces which reduces the heat flow rate from the hot gases to the feed water. In order to keep the tubes surfaces clean, each tube is provided with a scrapper to move up and down. A pair of scrapper for adjacent tubes is coupled together by a chain passing over the pulley which is driven with the help of a chain and gear arrangement. The movements of scrappers are

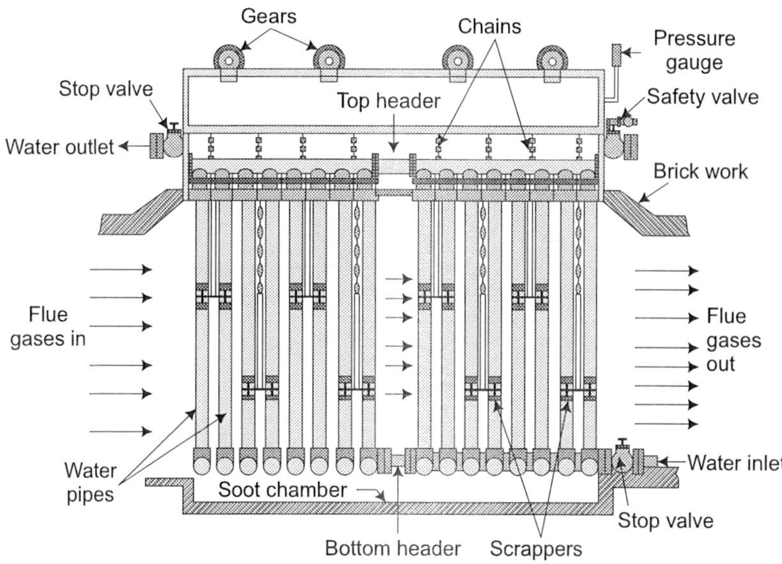

Fig. 1.37 Green's Economizer

adjusted in such a way that when one scrapper moves up the other comes down. The soot removed from the tubes surfaces get collected in the soot chamber provided below the bottom header. A safety valve is also provided on the top header for the safety of pipes against high pressure of water developed due to any reason.

When the economizer is to be repaired or inspected, it can be isolated by passing the flue gases directly to the chimney with the help of the dampers.

(iii) Air Preheater (or Air Heater) Air preheater is like an economizer which extracts some portion of heat of the spent flue gases and transfers it to the air before it is fed to the combustion chamber of the boiler for the burning of fuel. Air preheater is placed after the economizer and before the chimney as shown in Figure 1.38.

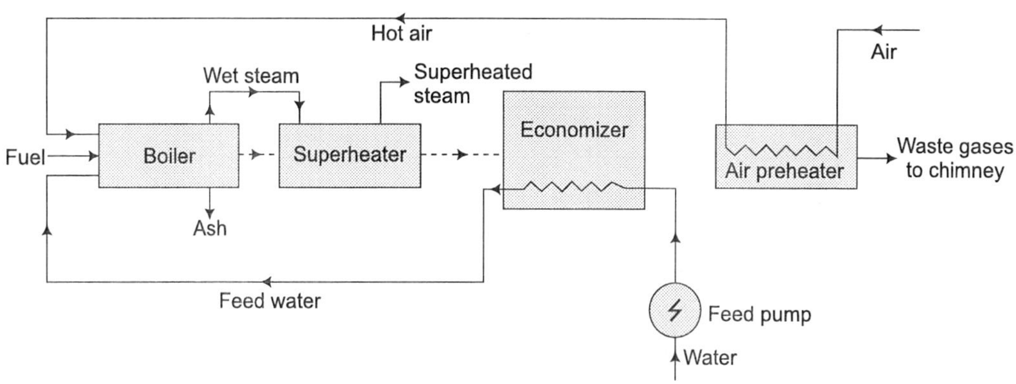

Fig. 1.38 Schematics for the Location of Different Accessories

Generally, air preheaters are of two types, namely, recuperative type and regenerative type. In recuperative air preheaters, heat is directly transferred from the hot gases to the air through a metal wall. Recuperative air preheaters are usually tubular type (Figure 1.39), but some plate types are still in use.

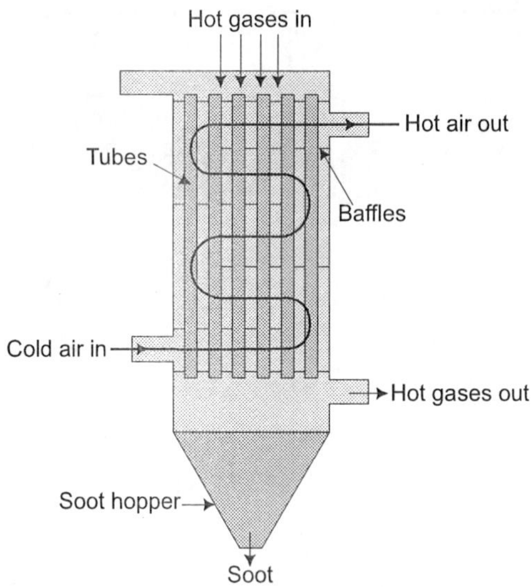

Fig. 1.39 Tubular Type Air Preheater

Tubular types are counter flow in which the hot gases flow inside the tubes and air flows outside over the tubes surface. Baffles are provided to flow the air in a zigzag fashion by which the period of air contact with the hot tubes increases. The plate type consists of a number of parallel plates in which air and hot gases pass in the alternate passages.

Regenerative air preheaters are also known as storage type heat exchangers which have an energy storage medium called matrix. In these preheaters, air and flue gases are made to pass alternatively through the matrix. When the hot flue gases flow through the matrix in the first half of the cycle it gets heat from the hot gases and transfers it to the cold air flowing through the matrix in the next half of the cycle. Thus, this cycle gets repeated. The Ljungstrom air preheater is the most common type of regenerative air preheater.

The advantages of using air preheater are given below:

(i) The boiler efficiency improves by about 1% for a drop of 20°C to 22°C in the gas temperature. The evaporative rate also increases.

(ii) The furnace temperature increases due to preheating of air which results in better combustion of fuel with less soot, smoke, and ash.

(iii) Low grade and inferior quality of fuels can also be used.

(iv) Feed Pumps Feed pump is a device which is required to pump water into the boiler shell. As the pressure inside the boiler is high thus the feed pump has to raise the pressure of water about

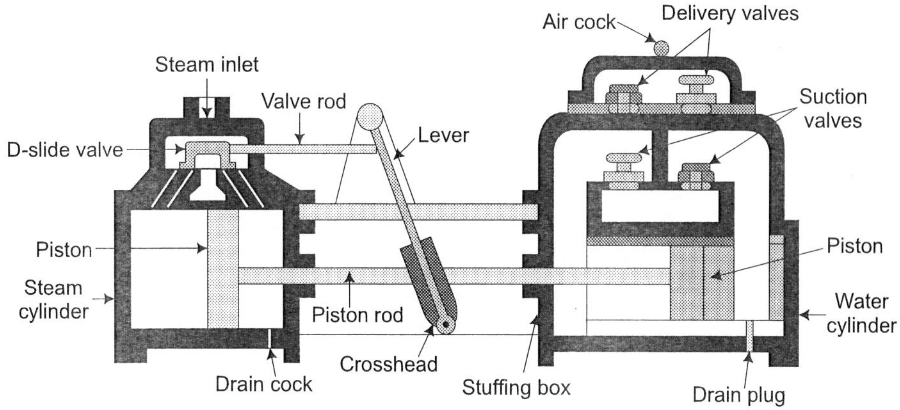

Fig. 1.40 Duplex Feed Pump

10% to 15% higher than that of the boiler. The reciprocating and rotary pumps are commonly used as feed pumps.

The rotary pumps are high-speed centrifugal type and are used to deliver large quantity of water into the boiler. These consist of casing and a rotating element called impeller which is fitted over a shaft. The impeller is made up of disc over which vanes are fitted. The impeller shaft is run by an electric motor, or by a steam turbine. This pump utilizes the centrifugal force of the rotating impeller for pumping water from one place to the other.

The reciprocating pumps are run by the steam from the same boiler in which water is to be fed. On the basis of the number of pump cylinders these may be classified as simplex, duplex, and triplex feed pumps. Duplex feed pump is the common type in which two pumps are mounted on the same block side by side as shown in Figure 1.40. Each pump unit consists of one water cylinder and one steam cylinder. D-slide valve controls the steam to the cylinder. It is operated by the crosshead on the piston rod of opposite cylinders through links and rocker arms. The pumps work alternately by which feed water is continuously supplied to the boiler.

(v) Injector The injector performs the same function as that of feed pump, means it injects feed water to the boiler with the help of a steam jet. The steam used for this purpose may be live steam from the same boiler, or exhaust steam from an engine. The injector is commonly used in vertical and locomotive boilers but it is not suitable for large capacity high pressure boilers.

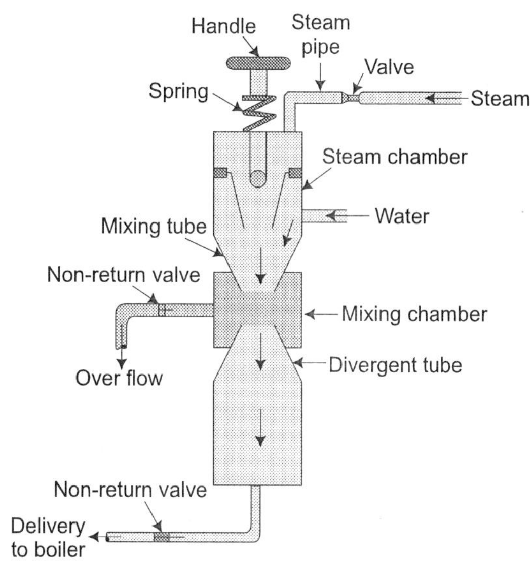

Fig. 1.41 Injector

It consists of a group of nozzles arranged in such a way that the steam flowing through these nozzles forces the water into the boiler. Figure 1.41 shows a schematic view of a steam injector.

The steam coming through the steam pipe expands in the converging nozzle where its pressure drops and consequently, its velocity increases. Due to decrease in pressure of steam at the exit of converging nozzle, the feed water enters from the feed water tank and gets mixed with steam in the mixing tube and mixing chamber. In the mixing chamber, steam condenses and a strong vacuum is created due to which more water flows to the mixing chamber from the feed water tank. The velocity at the end of mixing tube is very high and thus the velocity of water jet in the mixing chamber is also high. This high velocity jet of water enters the divergent tube where its kinetic energy gets converted into the pressure energy. This increased pressure forces the water into the boiler through the feed check valve.

In the beginning, when the injector is started the pressure in the mixing chamber is more than the atmosphere thus steam and water come out through the overflow valve. There will be no overflow when the steam and water are mixed in the proper ratio which is obtained by the adjustment of the gap between nozzle and the mixing tube with the help of hand wheel.

The use of injector results in the following advantages: (i) it is simple and requires minimum space, (ii) it is thermally more efficient and thus results in decreased thermal stresses in the boiler, and (iii) its initial and maintenance costs are low.

Due to these advantages sometimes injector is kept as standby or reserve feed water pump.

(vi) Anti-Priming Pipe Anti-priming pipe is a pipe of length 1 m to 2 m with closed ends having perforations on its upper half end. It is fitted in the steam space inside the boiler through which steam is taken out. Its function is to reduce the quantity of water particles moving along with the steam. When the steam containing water particles passes through it, the heavier particles separate out and get collected at the bottom of the pipe from where it is drained to the boiler.

(vii) Steam Separator Steam separator is also known as steam drier. It performs the function of anti priming pipe. It separates the water particles carried by the steam on its way from the boiler to the steam turbine or engine and thus any damage to the turbine blades by the moisture present in the steam can be avoided. It is installed on the main steam pipe from the boiler near the turbine or engine. Steam separators are of different types such as baffle plate type, reverse current type, and centrifugal type.

A baffle plate type separator is shown in Figure 1.42 which consists of a cylindrical vessel fitted with baffle plates. When the steam containing water particles passes through the chamber it strikes the baffle plates which deflect it and thus the flow direction is changed. Consequently, heavier water particles get collected at the bottom of the vessel from which it is drained out at regular intervals. A water gauge is also fitted in the separator to see the water level.

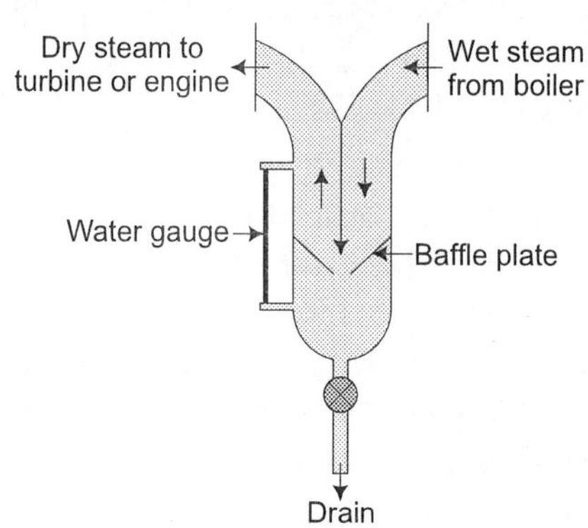

Fig. 1.42 A Baffle Plate Type Separator

(viii) Steam Trap Steam traps are used to collect and automatically drain off water resulting from partial condensation of steam from steam pipes and jackets without allowing any steam to escape with it. It is located in the steam pipe near the steam turbine or engine. The condensed water in the steam pipe flows to the steam trap under gravity. The steam traps are of two types: (i) Bucket or float type, and (ii) Thermal expansion type. Figure 1.43 shows a bucket type steam trap.

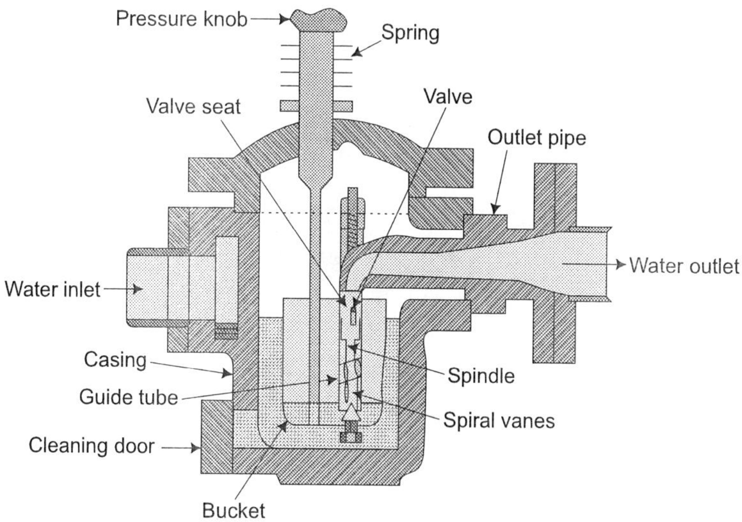

Fig. 1.43 A Bucket Type Steam Trap

The condensed steam enters the casing which always contains sufficient amount of water so that bucket floats. When the water level in the casing reaches above the top edge of the bucket it overflows into the bucket. So, the bucket sinks into the water due to which the valve gets opened and water is forced out through the guide tube. The flow of water through the guide tube strikes the spiral vanes due to which the spindle and the bucket rotate. This rotation is further assisted by the condensed water entering into the casing tangentially. The rotating spindle grinds the valve and seat surfaces by which it remains clean from any deposition. When all the water from the bucket is forced out, it will float up and close the valve. In order to test the sinking of the bucket, a knob is fitted at the top of the trap which is manually operated. A door is provided for removing the sediments collected in the casing.

The expansion type steam trap works on the principle of expansion or contraction of metals under heating or cooling. The difference in temperature causes a change in the length of tube to open a valve which allows the water to escape and to close it when the water is discharged.

(ix) Pressure Reducing Valve Figure 1.44 shows a typical pressure reducing valve. Pressure reducing valve maintains constant pressure of the steam on its delivery side of the valve under fluctuating conditions. It is fitted in the steam supply pipe line. It works on the principle of variable degree of throttling of the steam passing through the valve. It is difficult to deliver constant

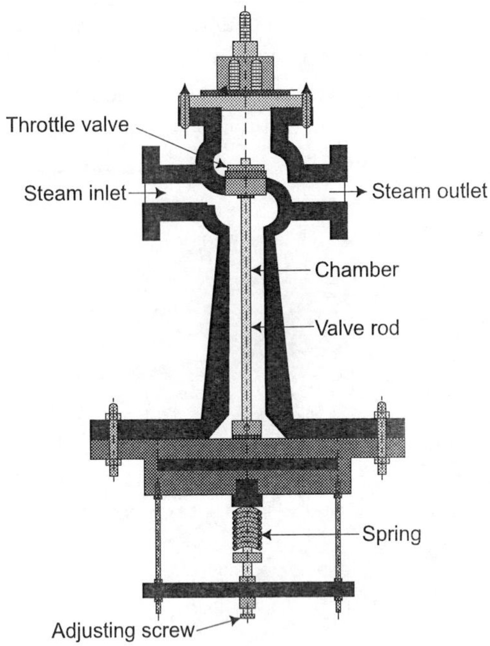

Fig. 1.44 Pressure Reducing Valve

pressure steam under fluctuating demands by the low steaming capacity boilers. Thus, steam is generated at a high pressure than required by the turbine or engine. Under these conditions pressure reducing valve is used to maintain the pressure of steam supplied to the prime mover.

The high pressure steam from the boiler enters through the steam inlet flange and passes through the throttle valve via steam outlet flange to the primemover. When the high pressure steam passes through the throttle valve it gets throttled, consequently, its pressure drops. The throttle valve is actuated by a spring and valve mechanism. The required exit pressure of the steam can be maintained by adjusting the opening of throttle valve with the help of the force exerted on the spring by the adjusting screw.

1.17. DIFFERENCES BETWEEN BOILER MOUNTINGS AND ACCESSORIES

Boiler Mountings	*Boiler Accessories*
1. Boiler mountings are the fittings and devices which are necessary for the safety of the boiler and control of steam generation process.	1. Boiler accessories are the auxiliary fittings and devices which are provided to increase the efficiency and to help in the smooth functioning of the plant.
2. Boiler mountings form the integral part of the boiler.	2. Boiler accessories do not form integral part of the boiler.

Boiler mountings	Boiler accessories
3. Boilers cannot be operated without mountings.	3. Boilers can be operated without accessories.
4. Boiler mountings are generally mounted on the boiler shell.	4. Boiler accessories are generally mounted outside the boiler shell.
5. Examples of boiler mountings: safety valves, fusible plug, water level indicator, high steam low water safety valve, pressure gauge, steam stop valve, feed check valve, blow off cock, manhole, and mud box.	5. Examples of boiler accessories: superheater, air preheater, economizer, feed pump, steam separator, injector, and steam trap.

1.18 BOILER PERFORMANCE

The boiler generates steam at a certain temperature and pressure by utilizing the heat of combustion of fuel. The performance (or capacity) of a boiler is expressed in terms of equivalent evaporation and boiler efficiency.

Evaporative Capacity Evaporative capacity of the boiler means the amount of water evaporated by the boiler per unit time (kg/hr). It may be expressed on the basis of (i) Fuel burnt (kg/kg of fuel), (ii) Grate area (kg/m^2 hr), and (iii) Furnace volume (kg/m^3 hr).

Equivalent Evaporation Generally, the term equivalent evaporation is used to compare the performance of different boilers with respect to evaporative capacities. It is defined as the amount of water evaporated from water at 100°C to dry saturated steam at 100°C. Equivalent evaporation of 1 kg of water requires 2256.9 kJ ≈ 2257 kJ of latent heat at atmospheric pressure.

If m_s is the mass of water actually evaporated into steam (kg/kg of fuel), h_f is the enthalpy of feed water (kJ/kg), and h_s is the enthalpy of steam (kJ/kg) then the equivalent evaporation m_e can be given by,

$$m_e = \frac{m_s \times (h_s - h_f)}{h_{fg}} = m_s \times F \tag{1.63}$$

where $F = \left(\dfrac{h_s - h_f}{h_{fg}}\right)$ is the factor of evaporation which is always greater than unity.

Boiler Efficiency It is the ratio of the heat utilized in the generation of steam to the heat released by the combustion of fuel in the same time period. Mathematically, it is given by,

$$\eta_{\text{boiler}} = \frac{m_s \times (h_s - h_f)}{m_f \times CV} \tag{1.64}$$

where m_f is the amount of fuel consumed in kg/hr and CV is the calorific value of fuel in kJ/kg.

Boiler efficiency can also be expressed as,

$$\eta_{\text{boiler}} = \frac{m_f \times CV - \text{Losses}}{m_f \times CV} \tag{1.65}$$

A boiler consists of an economizer and a superheater is considered a single unit and the efficiency of such a unit is called overall efficiency of the plant.

Boiler efficiency depends on the fixed factors and variable factors which are given below:

(a) Fixed factors: (i) Boiler design: shape and size of furnace, arrangement of water circulation and flue gases, (ii) Heat recovery equipment: superheater, economizer, and air preheater, (iii) Rate of firing, and (iv) Properties of fuel.

(b) Variable factors: (i) Fuel condition at the time of firing, (ii) Humidity and temperature of atmospheric air, (iii) Firing rate, (iv) Excess air fluctuations, and (v) Draught.

Boiler Power It is used to measure the capacity of a boiler. It can be calculated as,

$$\text{Boiler power,} \quad P_{\text{boiler}} = \frac{\text{Equivalent evaporation from and at } 100°\text{C/hr}}{21.296} \tag{1.66}$$

Losses in a Boiler The difference between the heat released by the fuel in the furnace and the heat used to generate steam is called losses in a boiler. The main heat losses are given below:

(i) Heat Loss due to Dry Flue Gases This is the largest loss which is associated with flue gases leaving the boiler. It is given by,

$$Q_{dfg} = m_g \times c_{pg} \times (T_g - T_a) \tag{1.67}$$

where m_g is the mass of flue gases generated, c_{pg} is the specific heat of flue gases (kJ/kg), T_g is the temperature of flue gases, and T_a is the temperature of the air.

(ii) Heat Loss due to Unburnt Fuel This is the heat lost due to unburnt fuel which is collected on the ash pit. It is given by,

$$Q_{uf} = m_{uf} \times CV \tag{1.68}$$

where m_{uf} is the mass of the unburnt fuel per kg of fuel and CV is its calorific value.

(iii) Heat Loss due to Incomplete Combustion of Fuel. This loss is due to incomplete combustion of carbon to carbon monoxide instead of carbon dioxide. One kg of carbon releases 33800 kJ/kg of heat when it burns to CO_2, whereas it releases 10120 kJ/kg when it burns to CO.

Thus, heat loss due to incomplete combustion of one kg of carbon is given by,

$$Q_{ic} = 33800 - 10120 = 23680 \text{ kJ}$$

If the volume percentage of CO and CO_2 in the flue gases is known then the heat loss due to incomplete combustion per kg of fuel can be given as,

$$Q_{ic} = \left(\frac{\text{CO} \times \text{C}}{\text{CO}_2 + \text{CO}}\right) \times 23680 \text{ kJ/kg of fuel} \tag{1.69}$$

where 'C' is the fraction of carbon in one kg of fuel.

(iv) Heat Loss due to Radiation and Convection This loss is due to the exposure of high temperature surfaces to the atmosphere which can be reduced by providing better thermal insulation over the boiler surface. This loss is given by,

$$Q_{rc} = Q - [Q_{dfg} + Q_{uf} + Q_{ic}] \tag{1.70}$$

where Q is the amount of heat liberated by the burning of one kg of fuel.

Example 1.41 1600 kg of fuel is consumed in 24 hours and 12000 kg of water is evaporated during a trial operation of a boiler. The steam pressure is 7 bar and steam condition is dry saturated. The fuel contained 3% moisture and 4% ash and it has a calorific value of 30,000 kJ/kg. The feed water temperature is observed 30°C. Determine (i) the boiler efficiency, (ii) the equivalent evaporation per kg of dry coal, and (iii) equivalent evaporation per kg of combustibles.

Solution

Steam produced per kg of coal = (12000/1600) = 7.5 kg

(i) Boiler efficiency, $\eta_{\text{boiler}} = \dfrac{m_s \times (h_s - h_f)}{m_f \times CV}$

$$\therefore \quad \eta_{\text{boiler}} = \frac{7.5 \times (2763.5 - 4.186 \times 30)}{30000}$$

$$= 0.6595 \text{ or } 65.95\%$$

(ii) Equivalent evaporation per kg of dry coal,

$$m_e = \frac{m_s \times (h_s - h_f)}{h_{fg}}$$

$$= \frac{7.5 \times (2763.5 - 4.186 \times 30)}{2257 \times (1 - 0.03)}$$

$$= 9.04 \text{ kg}$$

(iii) Equivalent evaporation per kg of combustibles

$$= \frac{9.04 \times 0.97}{0.93} = 9.43 \text{ kg}$$

Example 1.42 The data observed during a trial of two marine boilers are: trial duration = 10 hrs, total heating surface area = 2500 m^2, total grate area = 4 m^2, mean pressure of steam = 10 bar, total coal burnt = 6000 kg, total mass of water evaporated = 60,000 kg, temperature feed water = 30°C, condition of steam = 0.96, and calorific value of fuel = 30,000 kJ/kg. Determine (i) the mass of coal burnt/m^2 of grate area per hr, (ii) the mass of water evaporated per kg of coal under actual conditions, (iii) the equivalent evaporation from and at 100°C, (iv) the equivalent evaporation per m^2 of coal heating surface per hour, and (v) the boiler efficiency.

Solution

(i) Steam formed per hour = (60000/10) = 6000 kg

Coal burnt per hour = (6000/10) = 600 kg

Coal burnt per/m^2 of grate per hour = (600/4) = 150 kg

(ii) Mass of water evaporated per kg of coal = (6000/600) = 10 kg

(iii) Equivalent evaporation from and at 100°C per kg of coal,

$$m_e = \frac{m_s \times (h_s - h_f)}{h_{fg}}$$

$$\therefore \quad m_e = \frac{10 \times [(762.79 + 0.96 \times 2015.31) - (4.186 \times 30)]}{2257}$$

$$= 11.39 \text{ kg}$$

(iv) Equivalent evaporation per m^2 of heating surface per hr

$$= \frac{11.39 \times 600}{2500}$$

$$= 2.734 \text{ kg}$$

(v) Boiler efficiency, $$\eta_{\text{boiler}} = \frac{m_s \times (h_s - h_f)}{m_f \times CV}$$

$$\therefore \quad \eta_{\text{boiler}} = \frac{10 \times (2778.1 - 4.186 \times 30)}{30000}$$

$$= 0.8842 \text{ or } 88.42\%$$

Example 1.43 The equivalent evaporation of a boiler from and at 100°C is 10.5 kg of steam per kg of fuel. If the boiler generates 15000 kg/hr of steam at 20 bar pressure from feed water at 30°C and the fuel consumption is 1600 kg/hr, determine the boiler efficiency. Also determine the condition of steam if the calorific value of fuel is 29,500 kJ/kg.

Solution Equivalent evaporation from and at 100°C per kg of coal,

$$m_e = \frac{m_s \times (h_s - h_f)}{h_{fg}}$$

or $$10.5 = \frac{15000 \times [h_s - (4.186 \times 30)]}{1600 \times 2257}$$

$$\therefore \quad h_s = 2653.42 \text{ kJ/kg}$$

Boiler efficiency, $$\eta_{\text{boiler}} = \frac{m_s \times (h_s - h_f)}{m_f \times CV}$$

$$\therefore \quad \eta_{\text{boiler}} = \frac{15000 \times (2653.42 - 4.186 \times 30)}{1600 \times 29500}$$

$$= 0.8033 \text{ or } 80.33\%$$

From steam table at 20 bar,

$$h_f = 908.77 \text{ kJ/kg}, \quad \text{and} \quad h_{fg} = 1890.73 \text{ kJ/kg}$$

Therefore, $908.77 + x \times 1890.73 = 2653.42$

$\therefore$ Condition of steam, $x = 0.923$

Example 1.44 A boiler plant delivers steam at 17.5 bar and 300°C to a steam turbine which develops 1500 kW at the rate of 10 kg/kWh. The grate area is to be designed to burn 500 kg of coal per m^2 per hour. Determine the grate area if the following data are provided: feed water temperature 60°C, combustion efficiency = 0.92, boiler efficiency with superheater = 0.72, and the calorific value of fuel is 27,000 kJ/kg. Take specific heat value equal to 2.1 kJ/kgK.

Solution Let the grate area be A m^2, therefore, coal burnt per hour = (500 × A) kg/hr

Steam supplied to the turbine per hour = 1500 × 10 = 15000 kg/hr

Heat supplied in the boiler and superheater per hour,

$$= 15000 \times [h_g \times c_{ps} \times (T_{\text{sup}} - T_s) - h_f]$$

$$= 500 \times A \times CV \times \eta_{\text{combustion}} \times \eta_b$$

$$15000 \times [2796.4 + 2.15 \times (300 - 205.76) - 4.186 \times 60]$$

$$= 500 \times A \times 27000 \times 0.92 \times 0.72$$

Therefore, grate area $A = \dfrac{41.22 \times 10^6}{8942400} = 4.61\ \text{m}^2$

Example 1.45 A boiler delivers steam at the rate of 5500 kg/hr with an efficiency of 80%. Pressure of steam generated is 10 bar and dryness fraction is 0.95. The feed water is supplied at 35°C and the calorific value of fuel is 30,500 kJ/kg. Determine (i) the rate of fuel consumption, (ii) the equivalent evaporation, (iii) efficiency of the boiler when the temperature of the steam leaving the superheater is 250°C. Take specific heat of superheated steam as 2.25 kJ/kg K.

Solution

(i) Boiler efficiency,

$$\eta_{\text{boiler}} = \frac{m_s \times (h_s - h_f)}{m_f \times CV}$$

$$0.80 = \frac{5500 \times [(762.79 + 0.95 \times 2015.31) - 4.186 \times 35)}{m_f \times 30500}$$

$$\therefore \quad m_f = 570.47 \text{ kg/hr}$$

(ii) Equivalent evaporation from and at 100°C per kg of coal,

$$m_e = \frac{m_s \times (h_s - h_f)}{h_{fg}}$$

$$\therefore \quad m_e = \frac{5500 \times [(762.79 + 0.95 \times 2015.31) - 4.186 \times 35)}{570.47 \times 2257}$$

$$= 10.81 \text{ kg/kg of fuel}$$

(iii) Boiler efficiency,

$$\eta_{\text{boiler}} = \frac{m_s \times (h_s - h_f)}{m_f \times CV}$$

$$\therefore \qquad \eta_{boiler} = \frac{5500 \times [2778.10 + 2.25 \times (250 - 179.91) - (4.186 \times 35)]}{570.47 \times 30500}$$

$$= 88.17\%$$

Example 1.46 The data observed during a trial of a boiler are: steam pressure = 10 bar, coal burnt = 700 kg, feed water supplied per hour = 7000 kg, temperature feed water entering the economizer = 20°C, temperature feed water leaving the economizer = 75°C, condition of steam leaving the boiler = 0.96, and calorific value of coal used = 33,000 kJ/kg. Determine (i) heat absorbed by feed water in various components as a percentage of the total heat absorbed, and (ii) the thermal efficiency of the plant.

Solution

(i) Heat absorbed by one kg of feed water in the economizer,

$$q_{economizer} = 44.186 \times (75 - 20)$$

$$= 230.23 \text{ kJ/kg}$$

Heat absorbed by one kg of water in the boiler,

$$q_{boiler} = (762.79 + 0.96 \times 2015.31) - (4.186 \times 75)$$

$$= 2383.54 \text{ kJ/kg}$$

Heat absorbed in superheater,

$$q_{superheater} = 2942.6 - (762.79 + 0.96 \times 2015.31)$$

$$= 245.11 \text{ kJ/kg}$$

Therefore, total heat absorbed by one kg of water,

$$q = q_{economizer} + q_{boiler} + q_{superheater}$$

$$= 230.23 + 2383.54 + 245.11$$

$$= 2858.88 \text{ kJ/kg}$$

Therefore, percentage of heat absorbed in the economizer

$$= \frac{230.23}{2858.88} \times 100 = 8.05\%$$

Therefore, percentage of heat absorbed in the boiler

$$= \frac{2383.54}{2858.88} \times 100 = 83.37\%$$

Therefore, percentage of heat absorbed in the superheater

$$= \frac{245.11}{2858.88} \times 100 = 8.57\%$$

(ii) ∴ Thermal efficiency of the plant,

$$\eta_{plant} = \frac{7000 \times 2858.88}{700 \times 33000} = 86.63\%$$

Example 1.47 The data observed during a trial of a boiler for one hour are: steam pressure = 12 bar, coal burnt = 700 kg, steam generated = 5500 kg, temperature feed water entering the economizer = 20°C, temperature of hot well = 40°C, condition of steam entering the superheater = 0.94, temperature of steam leaving the superheater = 250°C, and calorific value of coal used = 31,000 kJ/kg. Determine (i) equivalent evaporation without superheater, (ii) equivalent evaporation with superheater, (iii) thermal efficiency of the boiler without superheater, (iv) thermal efficiency of the boiler with superheater, and (v) amount of heat supplied by the superheater per hour. Take specific heat of flue gases as 2.15 kJ/kg K.

Solution Mass of steam generated per kg of fuel,

$$m_s = (5500/700) = 7.86 \text{ kg}$$

(i) Equivalent evaporation without superheater,

$$m_e = \frac{m_s \times (h_s - h_f)}{h_{fg}}$$

$$\therefore \quad m_e = \frac{7.86 \times [(798.64 + 0.94 \times 1986.16) - (4.186 \times 40)]}{2257}$$

$$= 8.64 \text{ kg}$$

(ii) Equivalent evaporation with superheater,

$$m_e = \frac{7.86 \times [(2784.8 + 2.15 \times (250 - 187.99) - (4.186 \times 40)]}{2257}$$

$$= 9.58 \text{ kg}$$

(iii) Thermal efficiency of the boiler without superheater,

$$\eta_{\text{boiler}} = \frac{m_s \times (h_s - h_f)}{m_f \times CV}$$

$$\therefore \quad \eta_{\text{boiler}} = \frac{7.86 \times [(798.64 + 0.94 \times 1986.16) - (4.186 \times 40)]}{31000 \times 100}$$

$$= 63.34\%$$

(iv) Thermal efficiency of the boiler with superheater,

$$\eta_{\text{boiler}} = \frac{7.86 \times [(2784.8 + 2.15 \times (250 - 187.99) - (4.186 \times 40)]}{31000 \times 100}$$

$$= 69.74\%$$

(v) Heat supplied by superheater per hour,

$$= 5500 \times [(1 - 0.94) \times 1986.16 + 2.15 \times (250 - 187.99)]$$

$$= 13.89 \times 10^5 \text{ kJ/hr}$$

Example 1.48 The data observed for a steam power plant consisting of a boiler, an economizer, and a superheater: steam pressure = 12 bar, coal burnt = 500 kg/hr, mass of steam produced = 3500 kg/hr, temperature feed water entering the economizer = 50°C, temperature feed water leaving the economizer = 90°C, condition of steam leaving the boiler = 0.96, temperature of superheated steam leaving the superheater = 250°C, and calorific value of coal used = 25,000 kJ/kg. Determine (i) equivalent evaporation, (ii) boiler efficiency, (iii) boiler efficiency with economizer and superheater.

Solution

(i) Equivalent evaporation,

$$m_e = \frac{3500 \times (798.64 + 0.96 \times 1986.16) - (4.186 \times 90)}{500 \times 2257}$$

$$= 7.22 \text{ kg/kg of fuel}$$

(ii) Boiler efficiency,

$$\eta_{\text{boiler}} = \frac{3500 \times (798.64 + 0.96 \times 1986.16) - (4.186 \times 90)}{500 \times 25000}$$

$$= 65.2\%$$

(iii) Boiler efficiency with economizer,

$$\eta_{\text{boiler}} = \frac{3500 \times (798.64 + 0.96 \times 1986.16) - 50 \times 4.186)}{500 \times 25000}$$

$$= 69.89\%$$

(iv) Boiler efficiency with superheater,

$$\eta_{\text{boiler}} = \frac{3500 \times (2935 - 90 \times 4.186]}{500 \times 25000}$$

$$= 71.63\%$$

1.19 BOILER DRAUGHT (DRAFT)

For efficient combustion of fuel it is necessary to maintain continuous flow of fresh air and to exhaust the spent gases. The pressure difference which maintains the flow of gases in the combustion chamber of a boiler is called draught. This pressure difference is very small and generally measured in terms of mm of water (1 mm of water column is equivalent to 9.80665 N/m^2). The draught obtained by a chimney is called natural draught, whereas draught obtained with the help of fans, steam jets, etc., is called artificial draught. The classification of draught is shown in Figure 1.45.

1.19.1 Natural Draught (Chimney Draught)

The difference in static pressure by the chimney is produced due to the difference in the weight of the column of hot gases inside the chimney and the weight of equal column of cold air outside the

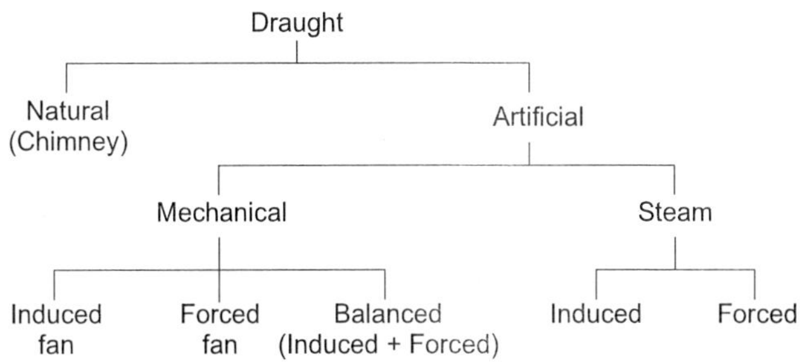

Fig. 1.45

chimney. The density of cold air outside the chimney is more than the density of hot gases inside the chimney. Thus, the pressure outside the level of chimney base is more than the pressure inside the chimney at its base due to which air flows from the surroundings to the furnace grate. The schematic view of a chimney of height 'H' m above the grate level is shown in Figure 1.46.

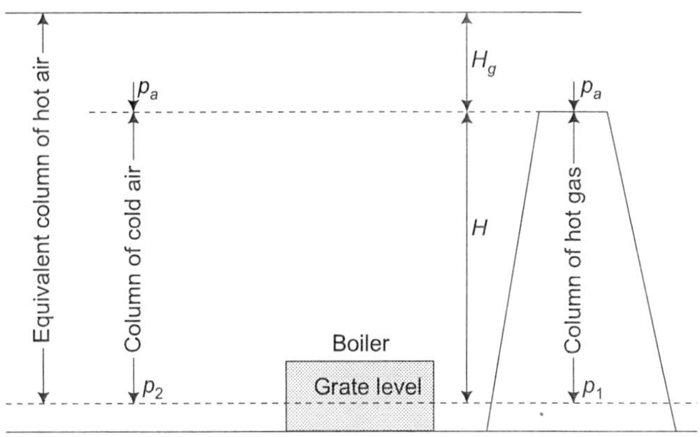

Fig. 1.46

Let p_a = Atmospheric pressure in N/m^2, $w_g = \rho_g g$ = Average specific weight of hot gases in N/m^3, ρ_g = Average mass density of hot gases, H = Height of chimney measured above from the grate level in m, ρ_a = Mass density of air, $w_a = \rho_a g$ = Specific weight of air in N/m^3, $p_1 = p_a + \rho_g gH$ = Pressure at the grate level on the chimney side, and $p_2 = p_a + \rho_g gH$ = Pressure at the grate level on outside the chimney.

The net pressure difference or the natural draught causing the flow of gases through the combustion chamber is given by,

$$\Delta p = p_2 - p_1 = (\rho_a - \rho_g)\, gH$$

$$= (w_a - w_g)\, H \tag{1.71}$$

This pressure difference is very small which is known as static pressure. The value of static pressure in chimney is generally less than 12 mm of water column which is measured with the help of a water manometer.

Determination of Height and Diameter of Chimney Let us assume that volume of flue gases in the chimney is equal to the volume of air supplied, and both volumes are measured at the same temperature.

At N.T.P., one kg of air occupies volume,

$$v = \frac{R \times T}{p} = \frac{287 \times 283}{1.01325 \times 10^5}$$

$$= 0.7734 \text{ m}^3/\text{kg}$$

The pressure difference is very small therefore, it is assumed constant. The volume of m kg of air outside the chimney,

$$v_a = v \times \frac{T_a}{T} = 0.7734 \times m_a \times \frac{T_a}{273},$$

and density

$$\rho_a = \frac{m_a}{v_a} = \frac{273 \times m_a}{0.7734 \times T_a \times m_a}$$

$$= \frac{353}{T_a}$$

Same volume of hot gases be inside the chimney,

$$v_g = 0.7734 \times m_a \times \frac{T_g}{273}$$

Density

$$\rho_g = \frac{(m_a + 1)}{v_g} = \frac{(m_a + 1)}{0.7734 \times (T_g/273) \times m_a}$$

$$= \left(\frac{m_a + 1}{m_a}\right) \times \frac{353}{T_g}$$

Natural draught

$$\Delta p = (\rho_a - \rho_g)\, gH$$

$$= gH \times \left[\frac{353}{T_a} - \left(\frac{m_a + 1}{m_a}\right) \times \frac{353}{T_g}\right]$$

$$\Delta p = 353\, gH \times \left[\frac{1}{T_a} - \left(\frac{m_a + 1}{m_a}\right) \times \frac{1}{T_g}\right] \text{ N/m}^2 \qquad (1.72)$$

The natural draught can be expressed in terms of water column h_w as,

$$h_w = \frac{\Delta p}{\rho_w g} = 353\, gH \times \left[\frac{1}{T_a} - \left(\frac{m_a + 1}{m_a}\right) \times \frac{1}{T_g}\right] \times 1000 \times \left(\frac{1}{1000 \times g}\right)$$

$$\therefore \quad h_w = 353\, H \times \left[\frac{1}{T_a} - \left(\frac{m_a + 1}{m_a}\right) \times \frac{1}{T_g}\right] \text{ mm of water column} \qquad (1.73)$$

Equations (1.72) and (1.73) represent the theoretical value of the natural draught which cannot be achieved in a chimney. The actual value of draught is less than the theoretical value due to frictional resistance to the passage of air flow, head loss due to bends, etc.

The natural draught can be expressed in terms of meter of hot gases H_g as,

$$H_g = \frac{\Delta p}{\rho_w g}$$

$$= 353\ gH \times \left[\frac{1}{T_a} - \left(\frac{m_a + 1}{m_a}\right) \times \frac{1}{T_g}\right] \times \left(\frac{m_a}{m_a + 1}\right) \times \frac{T_g}{353} \times \frac{1}{g}$$

$$\therefore \quad H_g = H \times \left[\frac{m_a}{m_a + 1} \times \frac{T_g}{T_a} - 1\right] \text{ m of hot gas column} \tag{1.74}$$

Theoretical velocity of hot gas through the chimney,

$$V_g = \sqrt{2g\ (H_g - \Delta H_g)} \tag{1.75}$$

where ΔH_g is the head loss (m of hot gas) due to friction in the chimney.

$$V_g = \sqrt{2g}\ \sqrt{H_g}\ \sqrt{1 - \frac{\Delta H_g}{H_g}}$$

$$= k\ \sqrt{H_g},$$

where $k = \sqrt{2g \times \left(1 - \frac{\Delta H_g}{H_g}\right)}$,

which depends on the nature of the chimney. For brick chimney, $k = 0.825$, and for steel chimney, $k = 1.1$.

The mass of gases flowing through any cross section of the chimney,

$$m_g = \rho_g \times A \times V_g.$$

If D is the diameter of the chimney, then

$$m_g = \rho_g \times \frac{\pi}{4} \times D^2 \times V_g$$

$$\therefore \quad D = 1.128 \times \sqrt{\frac{m_g}{\rho_g\ V_g}} \tag{1.76}$$

Condition for Maximum Discharge through the Chimney Let us assume the losses are negligible then velocity of gas through the chimney is given by,

$$V_g = \sqrt{2g\ H_g}$$

Now substituting the value of H_g, we have,

$$V_g = \sqrt{2g \times H \times \left[\frac{m_a}{m_a + 1} \times \frac{T_g}{T_a} - 1\right]}$$

The density of hot gases

$$\rho_g = \frac{p}{RT_g},$$

and mass flow rate $m_g = \rho_g \times A \times V_g$

So,
$$m_g = \frac{p}{RT_g} \times A \times \sqrt{2g \times H \times \left[\frac{m_a}{m_a + 1} \times \frac{T_g}{T_a} - 1\right]}$$

$$= \frac{K}{T_g} \times \sqrt{\left[\frac{m_a}{m_a + 1} \times \frac{T_g}{T_a} - 1\right]}$$

where
$$K = \frac{p \times A \times \sqrt{2gh}}{R} = \text{Constant}$$

For a given value of H, T_a, and m_a, m_g will be maximum when

$$\frac{dm_g}{dT_g} = 0$$

$$\therefore \quad \frac{d}{dT_g}\left[\frac{K}{T_g} \times \sqrt{\left[\frac{m_a}{m_a + 1} \times \frac{T_g}{T_a} - 1\right]}\right] = 0, \text{ after solving, we get,}$$

$$\frac{T_g}{T_a} = 2 \times \frac{m_a + 1}{m_a}$$

$$T_g = 2 \times \left(\frac{m_a + 1}{m_a}\right) \times T_a \qquad (1.77)$$

From the expression (1.77) it is observed that for the maximum discharge the temperature of hot gases is slightly greater than twice the atmospheric temperature.

The draught in mm of water column for maximum discharge can be obtained by substituting the value of T_g in equation (1.73), we have,

$$h_w = 353\, H \times \left[\frac{1}{T_a} - \left(\frac{m_a + 1}{m_a}\right) \times \frac{m_a}{2 \times (m_a + 1)} \times \frac{1}{T_a}\right] \text{ mm of water colume}$$

$$\therefore \quad h_w = 176.5 \frac{H}{T_a} \text{ mm of water column} \qquad (1.78)$$

The height of hot gas column for the maximum discharge can be obtained by substituting the value of T_g from equation (1.77) in equation (1.74), we have,

$$H_g = H \times \left[\frac{m_a}{m_a + 1} \times 2 \times \left(\frac{m_a + 1}{m_a}\right) \times T_a \times \frac{1}{T_a} - 1\right]$$

$$= \text{H m of hot gas column} \qquad (1.79)$$

It indicates that for maximum discharge the draught produced in terms of the hot gas column in the chimney is equal to its height.

Efficiency of a Chimney The temperature of flue gases leaving a chimney is higher in the case of natural draught system than in artificial system because certain minimum temperature is required to produce a given draught with the given height of a chimney. This results in loss of heat and thus lowers the overall efficiency of a plant.

Let T_{gn} = Absolute temperature of the flue gases leaving the chimney in the natural draught system in K, T_{ga} = Absolute temperature of the flue gases leaving the chimney in the artificial draught system in K, c_{pg} = Average specific heat of flue gases in kJ/kg K.

Extra heat carried away by 1 kg of flue gases = $c_{pg} \times (T_{gn} - T_{ga})$ kJ/kg

The draught pressure produced by the natural draught system in meter of hot gases is given by,

$$H_g = H \times \left[\frac{m_a}{m_a + 1} \times \frac{T_g}{T_a} - 1\right] \text{ m of hot gas column}$$

The maximum energy this head would give to 1 kg of flue gases at the cost of extra heat,

$$m \times g \times H_g = 1 \times g \times H \times \left[\frac{m_a}{m_a + 1} \times \frac{T_g}{T_a} - 1\right]$$

$$= \frac{9.81 \times H}{1000} \times \left[\frac{m_a}{m_a + 1} \times \frac{T_g}{T_a} - 1\right] \text{ kJ}$$

Efficiency of chimney,

$$\eta = \frac{\text{Maximum energy which would have given}}{\text{Extra heat carried away by flue gases}}$$

$$\therefore \quad \eta = \frac{9.81 \times H}{1000} \times \left[\frac{m_a}{m_a + 1} \times \frac{T_g}{T_a} - 1\right] \times \frac{1}{c_{pg} \times (T_{gn} - T_{ga})} \tag{1.80}$$

From equation (1.80) it is observed that the efficiency of a chimney is proportional to its height. But the efficiency is less than 1% even for a very tall chimney. It means the chimney is a very inefficient device for producing draught. The loss in draught in a chimney is about 20% of the total draught produced by it. This means the available draught is about 80% of the total static draught calculated theoretically.

1.19.2 Artificial Draught

In actual power plant, the required draught is about 300 mm of water which can be achieved only by artificial draught system. The artificial draught system may be of mechanical type or of steam type. These systems are more economical when the required draught is more than 40 mm of water. The steam jet draught system is used for small installations such as in locomotive, while mechanical draught system is used in central power stations.

Mechanical Draught It may be classified as forced draught, induced draught, and balanced draught depending on the system used to produce it.

(i) **Forced Draught** In a forced draught system, a fan or blower is installed near or at the base of the boiler to force the air to pass through the furnace, flues, economizer, air-pre-heater, and to the stack or chimney. The pressure generated by the fan or blower ranges from 25 mm to 75 mm of water. This draught system is known as positive draught system because the pressure of air throughout the system is above the atmospheric pressure.

(ii) Induced Draught In the induced draught system, a fan or blower is installed near, or at the base of the chimney. The fan or blower sucks the gases from the furnace as a result of which the pressure inside the furnace reduces below the atmospheric pressure. This induces the surrounding air from the atmosphere to flow through the grate. The action of this system is similar to that of chimney draught system. Induced fan is generally employed when economizer and air preheaters are used in the system.

The main advantages of forced draught system over induced draught system are: (i) Forced system handles cold air, so for the same draught, the fan size and the power requirement is low which is about 1/3rd to half that of induced system, (ii) It does not require water cooled bearing, (iii) The air leak tendency is less, and (iv) There is no heat loss.

(iii) Balanced Draught Balanced draught is a combination of the forced and induced draught system. If only forced system is used, the furnace cannot be opened because the high pressure air inside the system blows out suddenly. If only induced draught system is used, again the furnace cannot be opened because the cold air rushes into the system due to low pressure inside the system. So, in the forced as well as in the induced draught system the opening of furnace either for firing or inspection purposes leads to the reduction of effective draught and dilution of the combustion gases due to air leakage. To overcome these difficulties balanced draught system is used.

Advantages of Mechanical Draught Generally, mechanical draught is being used in modern power plant. Its main advantages are: (i) It allows to burn low grade fuel, (ii) The efficiency of mechanical draught system is high (about 7%), (iii) It is better in control, (iv) The air flow is uniform and it can be regulated as per the requirement, (v) The fuel consumption is about 15% less than in natural draught system, so it is more economical, (vi) The fuel burning capacity of the grate is more thus evaporative capacity of the boiler is higher, (vii) The chimney height reduces because the chimney is used only to exhaust the gases high in the atmosphere to avoid pollution, and (viii) The smoke formation is less.

Disadvantages of Mechanical Draught Main disadvantages of mechanical draught systems are: (i) High installation cost, (ii) High running cost, and (iii) High maintenance cost.

Power Required to Drive a Draught Fan

Let T_a = Temperature of atmospheric air, m_a = Mass of air supplied per kg of fuel, T_g = Average temperature of hot gas, h_w = Draught in mm of water = gh_w N/m^2, $v = \frac{RT}{p} = \frac{287 \times 273}{1.01325 \times 10^5} = 0.7734$ m^3/kg = Specific volume of air at N.T.P., m_f = Mass of fuel burnt, kg/s, η_m = Mechanical efficiency of the fan, and V = Total volume of gas handled by the fan in m^3/s.

Therefore, power required,

$$P = \frac{gh_w \times V}{\eta_m} \tag{1.81}$$

(i) Power required for forced draught fan (F.D. Fan)

Mass of air handled = $m_a \times m_f$

Specific volume of atmospheric air,

$$v_a = v \times \frac{T_a}{T} = 0.7743 \times \frac{T_a}{273} \quad (p = \text{constant})$$

Total volume of gas handled by fan,

$$V = (m_a \times m_f) \times 0.7734 \times \frac{T_a}{273} \text{ m}^3\text{/s}$$

$\therefore$ Power required,

$$P = \frac{gh_w \times V}{\eta_m} = \frac{gh_w}{\eta_m} \times (m_a \times m_f) \times 0.7734 \times \frac{T_a}{273} \tag{1.82}$$

(ii) Power required for induced draught fan (I.D. Fan)

Mass of air handled = $(m_a + 1) \times m_f$

Specific volume of hot gases,

$$v_g = 0.7743 \times \frac{T_a}{273} \times \left(\frac{m_a}{m_a + 1}\right) \text{ m}^3\text{/kg}$$

Total volume of gas handled,

$$V = (m_a + 1) \times m_f \times 0.7734 \times \frac{T_a}{273} \times \left(\frac{m_a}{m_a + 1}\right) \text{ m}^3\text{/s}$$

$\therefore$

$$P = \frac{gh_w \times V}{\eta_m}$$

$$= \frac{gh_w}{\eta_m} \times (m_a + 1) \times m_f \times 0.7734 \times \frac{T_a}{273} \times \frac{m_a}{m_a + 1} \tag{1.83}$$

$$\text{Power ratio} = \frac{\text{Power of I.D. fan}}{\text{Power of F.D. fan}} = \frac{T_g}{T_a} \tag{1.84}$$

The flue gas temperature T_g is slightly more than twice the atmospheric temperature T_a. Therefore, for the same amount of draught, the power required by an induced draught fan is slightly more than twice the power required by a forced draught fan.

Steam Jet Draught In this draught system, steam is passed through a jet to produce the required draught. It may be induced draught, or forced draught depending on the location of steam jet. When the steam jet is directed into the chimney it is called induced draught in which the air is induced through the flues, the grate, and ash pit to the smokebox. But when the jet is installed in the ash pit it is called forced draught system in which air is forced to the fuel bed, furnace, and flues to the chimney.

The induced draught system is mostly used in locomotive boilers, where the exhaust steam from the steam engine is discharged through a blast pipe placed at the smokebox and below the chimney. In forced steam draught system, the steam from the boiler is throttled by a valve to about 2.5 bar gauge pressure and passed through the nozzles projected in a diffuser. Thus, the steam moves with a greater velocity and drags the air along with it. The forced draught is also called turbine draught.

The advantages of steam jet draught system are: (i) It requires very little attention, (ii) It is economical and occupies less space, (iii) Cheap and low grade fuel can also be used, (iv) Draught automatically adjusts to suit the requirements, and (v) Its maintenance cost is almost negligible.

The disadvantages of steam jet draught system are: (i) It cannot be started until steam pressure is available, and (ii) In the forced system, steam passes through the furnace carry away some heat.

Example 1.49 Determine the minimum temperature of the flue gases required to produce a draught of 10 mm of water by a chimney of 25 m height if the air-fuel ratio required for combustion is 20:1 and atmospheric temperature is 30°C.

Solution The draught is given by,

$$h_w = 353\, H \times \left[\frac{1}{T_a} - \left(\frac{m_a + 1}{m_a}\right) \times \frac{1}{T_g}\right] \text{ mm of water column}$$

$$10 = 353 \times 25 \times \left[\frac{1}{303} - \left(\frac{20 + 1}{20}\right) \times \frac{1}{T_g}\right]$$

$$\therefore \quad T_g = 484.51 \text{ K} \quad \text{or} \quad 211.51°\text{C}$$

Example 1.50 Determine the amount of air used for burning coal in a boiler having a chimney 30 m high to produce a draught of 10 mm of water, when the temperature of flue gas in the chimney is 350°C and the boiler house temperature is 35°C. State does this chimney satisfy the condition of maximum discharge.

Solution The draught is given by,

$$h_w = 353\, H \times \left[\frac{1}{T_a} - \left(\frac{m_a + 1}{m_a}\right) \times \frac{1}{T_g}\right] \text{ mm of water column}$$

$$15 = 353 \times 35 \times \left[\frac{1}{308} - \left(\frac{m_a + 1}{m_a}\right) \times \frac{1}{623}\right]$$

$$\therefore \quad m_a = 7.13 \text{ kg/kg of coal}$$

For maximum discharge,

$$T_g = 2 \times \left(\frac{m_a + 1}{m_a}\right) \times T_a$$

$$= 2 \times \left(\frac{7.13 + 1}{7.13}\right) \times 308$$

$$= 702.39 \text{ K} \quad \text{or} \quad 429.39°\text{C}$$

The actual flue gas temperature,

$$T_g = 350°C < 429.39°C,$$

therefore, condition is satisfied.

Example 1.51 If the flue gas temperature is 275°C, ambient air temperature is 27°C, and the mass of flue gas is 20 kg/kg of fuel burnt in the furnace, determine the draught produced by a chimney of 35 m height. Also determine the mass of the flue gases passing through the chimney per second if the diameter of chimney is 1400 mm and 25% of the theoretical draught is lost in friction.

Solution The draught is given by,

$$h_w = 353\, H \times \left[\frac{1}{T_a} - \left(\frac{m_a + 1}{m_a}\right) \times \frac{1}{T_g}\right] \text{ mm of water column}$$

$$\therefore \quad h_w = 353 \times 35 \times \left[\frac{1}{300} - \left(\frac{19 + 1}{19}\right) \times \frac{1}{548}\right]$$

$$= 18.01 \text{ mm of water column}$$

$$H_g = H \times \left[\frac{m_a}{m_a + 1} \times \frac{T_g}{T_a} - 1\right]$$

$$= 35 \times \left[\frac{19}{20} \times \frac{548}{300} - 1\right] = 25.74 \text{ m}$$

$$\text{Actual available draught} = 0.75 \times H_g$$

$$= 0.25 \times 25.74 = 19.305 \quad (\because \text{Draught lost} = 25\%)$$

$$V_g = \sqrt{2gH_g} = \sqrt{2 \times 9.81 \times 19.305} = 19.462 \text{ m/s}$$

$$\rho_g = \frac{(m_a + 1)}{m_a} \times \frac{353}{T_g} = \frac{20}{19} \times \frac{353}{548} = 0.678 \text{ kg/m}^3$$

$$A = \frac{\pi}{4}(D^2) = \frac{\pi}{4}(1.4)^2 = 1.539 \text{ m}^2$$

$$\therefore \quad m_g = \rho_g \times A \times V_g$$

$$= 0.678 \times 1.539 \times 19.462 = 20.31 \text{ kg/s}$$

Example 1.52 If the height of a chimney is 25 m, temperature of flue gases is 300°C, ambient air temperature is 25°C, mass of air is 15 kg/kg of fuel burnt in the furnace, and atmospheric pressure is 730 mm of Hg is given, determine the value of draught produced.

Solution The draught is given by,

$$h_w = 353\, H \times \left[\frac{1}{T_a} - \left(\frac{m_a + 1}{m_a}\right) \times \frac{1}{T_g}\right] \text{ mm of water column}$$

$$\therefore \qquad h_w = 353 \times \frac{730}{760} \times 25 \times \left[\frac{1}{298} - \frac{16}{15} \times \frac{1}{573}\right]$$

$$= 12.66 \text{ mm of water}$$

Example 1.53 Find the height of a chimney required for a boiler having an average coal consumption of 1525 kg/hr and producing 15.5 kg of dry flue gases per kg of coal fired. The pressure drop in fuel bed is 4.9 mm, in boiler flue is 7.5 mm, in boiler tube passages is 2.5 mm, and in bends is 0.5 mm. Actual draught may be taken as 30% greater than the theoretical draught value. If the flue gases have a mean temperature of 275°C and ambient air temperature is 27°C, state whether the chimney fulfils the condition of maximum discharge.

Solution Actual draught required to overcome the losses

$$= 4.5 + 7.5 + 2.5 + 0.5 = 15 \text{ mm of water}$$

$$\therefore \qquad \text{Theoretical draught} = \frac{\text{Actual draught}}{1.3} = \frac{15}{1.3}$$

$$= 11.54 \text{ mm of water}$$

$$h_w = 353\, H \times \left[\frac{1}{T_a} - \left(\frac{m_a + 1}{m_a}\right) \times \frac{1}{T_g}\right] \text{ mm of water column}$$

$$11.54 = 353\, H \times \left[\frac{1}{300} - \frac{15.5}{14.5} \times \frac{1}{548}\right]$$

$\therefore$ Chimney height, $\qquad H = 23.65$ m

For maximum discharge,

$$T_g = 2 \times \left(\frac{m_a + 1}{m_a}\right) \times T_a$$

$$= 2 \times \frac{15.5}{14.5} \times 300$$

$$= 642.85 \text{ K} \quad \text{or} \quad 369.85°\text{C}$$

The actual flue gas temperature,

$$T_g = 275°\text{C} < 369.85°\text{C},$$

therefore, condition is satisfied.

Also $\qquad h_w = 176.5 \dfrac{H}{T_a} = 176.5 \times \dfrac{23.65}{300}$

$$= 13.91 \text{ mm of water column}$$

Therefore, theoretical draught 11.54 < 13.91 mm of water. Both conditions show that chimney fulfills the condition.

Example 1.54 If the flue gas mean temperature is 260°C, ambient air temperature is 25°C, find the height of a chimney to produce a static draught of 25 mm of water. Also determine the mass of flue gases formed per kg of fuel if barometer reads 760 mm of Hg. Take the density of air and hot gases as 1.293 kg/m^3 and 1.35 kg/m^3 respectively.

Solution The density of air at temperature T_a,

$$\rho_a = \rho \times \frac{T}{T_a} = 1.293 \times \frac{273}{298}$$

$$= 1.184 \text{ kg/m}^3$$

The density of hot gases at temperature T_g,

$$\rho_g = 1.35 \times \frac{273}{533}$$

$$= 0.691 \text{ kg/m}^3$$

$$\text{Draught produced} = \rho_w \times g \times h_w \times 10^{-3}$$

$$= (\rho_a - \rho_g)\, gH$$

$$1000 \times 25 \times 10^{-3} = (1.184 - 0.691) \times H$$

$$\therefore \quad H = 50.71 \text{ m}$$

$$h_w = 353\, H \times \left[\frac{1}{T_a} - \left(\frac{m_a + 1}{m_a}\right) \times \frac{1}{T_g}\right] \text{ mm of water column}$$

$$25 = 353 \times 50.71 \times \left[\frac{1}{298} - \left(\frac{m_a + 1}{m_a}\right) \times \frac{1}{533}\right]$$

$$\therefore \quad m_a = 22.62 \text{ kg/kg of fuel}$$

$$\therefore \quad \text{Mass of gas formed} = (m_a + 1) = 22.62 + 1$$

$$= 23.62 \text{ kg/kg of fuel}$$

Example 1.55 If the flue gas mean temperature is 220°C, ambient air temperature is 27°C, and air fuel ratio is 15:1, find the height of a chimney to produce a draught of 20 mm of water column. Also determine the power required if forced fan is used for producing the given draught when fuel consumption is 1250 kg/hr and mechanical efficiency of fan is 80%.

Solution

$$h_w = 353\, H \times \left[\frac{1}{T_a} - \left(\frac{m_a + 1}{m_a}\right) \times \frac{1}{T_g}\right] \text{ mm of water column}$$

$$20 = 353 \times H \times \left[\frac{1}{300} - \frac{15}{14} \times \frac{1}{493}\right]$$

$$\therefore \quad H = 48.84 \text{ m}$$

Power, $$P = \frac{gh_w}{\eta_m \times 1000} \times (m_a \times m_f) \times 0.7734 \times \frac{T_a}{273} \text{ kW}$$

$$\therefore \quad P = \frac{9.81 \times 20}{0.80 \times 1000} \times (15 \times 1) \times 0.7734 \times \frac{300}{273}$$

$$= 3.126 \text{ kW}$$

Example 1.56 If the flue gas mean temperature is 275°C, ambient air temperature is 25°C, air supplied per kg of fuel is 17.5, fuel consumption is 1600 kg/hr, draught required is 75 mm of water, mechanical efficiency is 82%, and leakage factor is 1.12, determine the power of forced and induced draught fans. Specific volume of air at N.T.P. is 0.7734 m^3/kg. Also find the % increase in power for induced draught fan.

Solution Power required by forced fan,

$$P = \frac{gh_w}{\eta_m \times 1000} \times (m_a \times m_f) \times 0.7734 \times \frac{T_a}{273} \text{ kW}$$

$$\therefore \quad P = \frac{9.81 \times 75}{0.82 \times 1000} \times 17.5 \times (1600/3600) \times 0.7734 \times \frac{298}{273} \times 1.12$$

$$= 6.598 \text{ kW}$$

Power required by induced fan,

$$P = \frac{gh_w}{\eta_m \times 1000} \times (m_a + 1) \times m_f \times 0.7734 \times \frac{T_a}{273} \times \left(\frac{m_a}{m_a + 1}\right) \times k \text{ kW}$$

$$\therefore \quad P = \frac{9.81 \times 75}{0.82 \times 1000} \times 18.5 \times \frac{1600}{3600} \times 0.7734 \times \frac{548}{273} \times \left(\frac{17.5}{18.5}\right) \times 1.12$$

$$= 12.134 \text{ kW}$$

% increase in power for induced draught fan

$$= \frac{12.134 - 6.598}{6.598} \times 100$$

$$= 83.90\%$$

Example 1.57 Draught required is 65 mm of water which is obtained by a forced fan. If the flue gas mean temperature is 325°C, ambient air temperature is 27°C, air supplied per kg of fuel is 15, fuel consumption is 1500 kg/hr, mechanical efficiency of fan is 86%, and transmission efficiency is 0.92, determine the power of the forced fan. Also determine the percentage increase in power if induced fan is used instead of forced fan to produce the same draught if the efficiency of induced fan is 85% and same efficiency of driving gear.

Solution Power required by a forced fan is given by,

$$P = \frac{gh_w}{\eta_m \times \eta_t \times 1000} \times (m_a \times m_f) \times 0.7734 \times \frac{T_a}{273} \text{ kW}$$

$$\therefore \quad P = \frac{9.81 \times 65}{0.86 \times 0.92 \times 1000} \times 15 \times (1500/3600) \times 0.7734 \times \frac{300}{273}$$

$$= 4.281 \text{ kW}$$

Power required by induced fan,

$$P = \frac{gh_w}{\eta_m \times \eta_t \times 1000} \times (m_a + 1) \times m_f \times 0.7734 \times \frac{T_a}{273} \times \left(\frac{m_a}{m_a + 1}\right) \text{kW}$$

$$\therefore \quad P = \frac{9.81 \times 65}{0.86 \times 0.92 \times 1000} \times (16) \times \frac{1500}{3600} \times 0.7734 \times \frac{598}{273} \times \left(\frac{15}{15 + 1}\right)$$

$$= 8.099 \text{ kW}$$

% increase in power for induced draught fan

$$= \frac{8.099 - 4.281}{4.281} \times 100$$

$$= 98.18\%$$

Example 1.58 The draught lost through the grate is 25 mm of water of a boiler plant using forced draught fan supplying air at 15 m/s. The fuel consumption is 8500 kg/hr, air supplied is 16 kg/kg of fuel, ambient air temperature is 27°C, pressure is 1 bar, and mechanical efficiency of the fan is 80%. Determine the shaft power required to drive the fan. Take the density of air at N.T.P. equal to 1.293 kg/m^3.

Solution Velocity head of air developed by fan,

$$h = V^2/2g = (15^2)/2 \times 9.81$$

$$= 11.47 \text{ m}$$

So, pressure head at N.T.P

$$= \rho gh = 1.293 \times 9.81 \times 11.47$$

$$= 145.49 \text{ N/m}^2$$

Draught lost in the grate = 25 mm of water

$$= 25 \times 9.81 = 245.25 \text{ N/m}^2$$

Therefore, total draught required,

$$h_w = 145.49 + 245.25$$

$$= 390.74 \text{ N/m}^2$$

Now applying,

$$pV = mRT$$

$$1 \times 10^5 \times V = (8500/3600) \times 287 \times 300$$

∴ Volume, $V = 32.53\ m^3/s$

∴ Shaft power, $$P = \frac{h_w \times V}{\eta_m} = \frac{390.74 \times 32.53}{0.80 \times 1000}$$

$$= 15.98\ kW$$

Example 1.59 A chimney has a height of 105 m. For the maximum discharge condition, calculate the temperature of the chimney gases and the draught produced if the air supplied per kg of fuel is 20 kg. Also determine the efficiency of the chimney if the temperature of the chimney gases in artificial draught system is limited to 125°C. Take the boiler house temperature as 30°C and specific heat of flue gases as 1.005 kJ/kg K.

Solution For maximum discharge, temperature of chimney gases,

$$T_g = 2 \times \left(\frac{m_a + 1}{m_a}\right) \times T_a$$

$$= 2 \times \left(\frac{20 + 1}{20}\right) \times 303 = 636.3\ K$$

Draught, $$h_w = 353\ H \times \left[\frac{1}{T_a} - \left(\frac{m_a + 1}{m_a}\right) \times \frac{1}{T_g}\right] \text{ mm of water column}$$

$$h_w = 353 \times 105 \times \left[\frac{1}{303} - \left(\frac{20 + 1}{20}\right) \times \frac{1}{636.3}\right]$$

$$= 61.16 \text{ mm of water column}$$

Extra heat carried away by the flue gases per kg due to temperature being higher than in artificial draught system,

$$= 1 \times 1.005 \times (363.3 - 398)$$

$$= 239.49\ kJ/kg$$

The static head of 105 m will give energy

$$= 1 \times 9.81 \times 105 \times 10^{-3}$$

$$= 0.981\ kJ/kg$$

∴ Efficiency of chimney = 0.981/239.49 × 100

$$= 0.409\%$$

Example 1.60 The minimum temperature apart from the draught of the waste gases leaving a boiler is 125°C. A chimney of 50 m height is used to produce natural draught in which the gases temperature is 300°C. If the air supplied to the furnace is 20 kg/kg of coal burnt, the temperature of air is 10°C and the mean specific heat of the gases is 1.005 kJ/kg K, determine the draught produced and the efficiency of the chimney. Express the extra expenditure of heat on the production of draught in the chimney as a percentage of the heat of the fuel when calorific value of the fuel is 32000 kJ/kg.

Solution

Draught, $$h_w = 353\ H \times \left[\frac{1}{T_a} - \left(\frac{m_a + 1}{m_a}\right) \times \frac{1}{T_g}\right] \text{ mm of water column}$$

$$h_w = 353 \times 50 \times \left[\frac{1}{283} - \left(\frac{20 + 1}{20}\right) - \frac{1}{573}\right]$$

$$= 30.02 \text{ mm of water column}$$

The draught in m of hot gases,

$$H_g = H \times \left[\frac{m_a}{m_a + 1} \times \frac{T_g}{T_a} - 1\right]$$

$$= 50 \times \left[\frac{20}{20} \times \frac{573}{283} - 1\right]$$

$$= 46.41 \text{ m of hot gas column}$$

Static head of 46.41 m will give energy

$$= 1 \times 9.81 \times 46.41 \times 10^{-3}$$

$$= 0.455 \text{ kJ/kg}$$

Extra heat carried away by the flue gases per kg due to temperature being higher than in artificial draught system,

$$= 1 \times 1.005 \times (573 - 398)$$

$$= 175.87 \text{ kJ/kg}$$

∴ Efficiency of chimney $$= \frac{0.455}{175.87} \times 100$$

$$= 0.259\%$$

Extra heat carried away by the flue gases

$$= \frac{21 \times 175.87}{32000} \times 100$$

$$= 11.54\%$$

EXERCISE

(i) Objective Type Questions

1. If m_f is the mass of water vapor in suspension, m_g is the mass of dry steam, then dryness fraction of steam x is

(a) $x = \dfrac{m_g}{m_g + m_f}$ (b) $x = \dfrac{m_g}{m_f}$ (c) $x = \dfrac{m_f}{m_g + m_f}$ (d) $x = \dfrac{m_f}{m_g}$

2. If the dryness fraction of a sample of steam is 0.9, the mass of water particles contain in 5 kg of steam will be:
   (a) 0.4 kg (b) 0.7 kg (c) 0.5 kg (d) 0.3 kg
3. By increasing the pressure, the specific volume of steam:
   (a) decreases (b) increases
   (c) remains same (d) decreases and then increases
4. Mollier diagram is a plot of:
   (a) pressure and volume (b) pressure and enthalpy
   (c) temperature and entropy (d) enthalpy and entropy
5. Slope of constant pressure line on $h - s$ diagram of steam with increasing pressure:
   (a) remains same (b) decreases
   (c) increases (d) none of these
6. At triple point:
   (a) liquid water and dry saturated steam coexist
   (b) ice has maximum specific volume
   (c) ice becomes superheated vapor on heating
   (d) solid, liquid, and vapor coexist
7. Only throttling calorimeter is used for measuring:
   (a) dryness fraction of steam up to 0.6
   (b) dryness fraction of low pressure steam
   (c) dryness fraction of high pressure steam
   (d) very high dryness fraction 0.98
8. Sublimation is the process of:
   (a) changing from solid phase to liquid phase
   (b) changing from liquid phase to solid phase
   (c) changing from solid phase to vapor phase directly
   (d) none
9. Sublimation occurs in:
   (a) air (b) liquid nitrogen
   (c) solid carbon dioxide (d) water
10. Steam coming out of the whistle of a pressure cooker is:
    (a) wet vapor (b) dry saturated vapor
    (c) superheated vapor (d) ideal gas
11. At critical pressure (220.09 bar):
    (a) latent heat of vaporization is zero
    (b) entropy is zero

(c) sublimation occurs from a solid to vapor
(d) saturated vapor converts to gas

12. In water tube boiler hot gases flow:
(a) outside the tube (b) inside the tube
(c) outside only in supercritical boilers (d) outside the tube in subcritical boilers

13. In supercritical boiler, the operating pressure is:
(a) between 5-100 bar (b) between 100-150 bar
(c) between 150-221 bar (d) more than 221 bar

14. The following device is not present in supercritical boilers:
(a) economizer (b) superheater (c) reheater (d) evaporator

15. In the natural draft boiler, draft in terms of water column (mm of water) is:
(a) $353\, H \times \left[\frac{1}{T_a} - \left(\frac{m_a + 1}{m_a}\right) \times \frac{1}{T_g}\right]$ (b) $353\, H \times \left[\frac{1}{T_g} - \left(\frac{m_a + 1}{m_a}\right) \times \frac{1}{T_a}\right]$
(c) $H \times \left[\frac{1}{T_g} - \left(\frac{m_a + 1}{m_a}\right) \times \frac{1}{T_a}\right]$ (d) $353\, H \times \left[\frac{1}{T_g} + \left(\frac{m_a + 1}{m_a}\right) \times \frac{1}{T_a}\right]$

16. Critical temperature and pressure for steam are:
(a) 313.14°C, 220 bar (b) 374.14°C, 220.09 bar
(c) 375°C, 221 bar (d) 309.14°C, 421 bar

17. The triple point of water in °C is:
(a) 0.05 (b) 0.01 (c) 0.10 (d) 10

18. At critical point the enthalpy of vaporization is:
(a) minimum (b) maximum
(c) dependent on temperature only (d) zero

19. The number of degrees of freedom for a pure substance at its triple point is:
(a) 1 (b) 2 (c) 0 (d) none

20. Steam throttling is a process of:
(a) constant entropy (b) constant enthalpy
(c) constant volume (d) constant pressure

21. In throttling process pressure of steam:
(a) constant (b) increases (c) does not change (d) decreases

22. Following are the boiler accessories except:
(a) fusible plug (b) air preheater (c) economizer (d) superheater

23. Following are the boiler mountings except:
(a) fusible plug (b) blow-off cock (c) safety valve (d) injector

24. The saturated liquid line and saturated vapor line meet at:
(a) ice point (b) triple point (c) critical point (d) boiling point

25. At the critical point specific volume of water in m^3/kg is:

(a) 0.003155 (b) 0.00563 (c) 0 (d) 0.001

26. Following are the water tube boilers except:

(a) Lancashire (b) Babcock and Wilcox

(c) Stirling (d) Benson

27. Following are the fire tube boilers except:

(a) Lancashire (b) Cochran (c) Cornish (d) Stirling

28. Heat transferred during flow and non-flow process are:

(a) not equal (b) equal

(c) during flow greater than non-flow (d) none

29. Irreversibility is given by:

(a) w_{actual} (b) $w_{reversible}$

(c) $w_{actual} - w_{reversible}$ (d) $w_{reversible} - w_{actual}$

30. Lancashire boiler has fire tubes equal to:

(a) 4 (b) 3 (c) 2 (d) 1

31. Fusible plug body is made of:

(a) gun metal (b) copper (c) aluminium (d) brass

32. If x_1 and x_2 are the dryness fraction of steam sample given by separating calorimeter, and throttling calorimeter respectively, then the actual dryness fraction is:

(a) x_1/x_2 (b) x_2/x_1 (c) $x_1\, x_2$ (d) $1/x_1\, x_2$

33. The major loss in the boiler occurs in:

(a) moisture (b) excess air (c) unbrunt carbon (d) exhaust gas

34. The overall efficiency of modern boiler with complete accessories ranges from

(a) 70-75% (b) 75-80% (c) 80-85% (d) 85-90%

35. Internal latent heat is given by:

(a) $h_{fg} - pv_g$ (b) $h_f - pv_g$ (c) $h_g - pv_g$ (d) $h_g - xpv_g$

36. Entropy of wet steam is given by:

(a) $s = s_{fg} - pv_g$ (b) $s = s_f + x(h_{fg}/T_s)$

(c) $s = s_g + x(h_{fg}/T_s)$ (d) $s = s_{fg} + x(h_{fg}/T_s)$

37. The chimney efficiency is about:

(a) 20% (b) 10% (c) 25% (d) 0.25%

38. For maximum discharge of hot gases through the chimney the height of hot gas column producing draught is:

(a) twice the height of chimney (b) equal to the height of the chimney

(c) half the height of chimney (d) more than twice the height of chimney

39. For the same draught produced the power of forced draught fan in comparison to induced draught fan is:

(a) more (b) equal (c) less (d) double

40. The pressure at the furnace is minimum in the case of:

(a) balanced draught (b) forced draught

(c) natural draught (d) induced draught

ANSWERS

1. (a)	2. (c)	3. (a)	4. (d)	5. (c)	6. (d)	7. (d)	8. (c)	9. (c)	10. (a)
11. (a)	12. (a)	13. (d)	14. (d)	15 (a)	16. (b)	17. (b)	18. (d)	19. (c)	20. (b)
21. (d)	22. (a)	23. (d)	24. (c)	25. (a)	26. (a)	27. (d)	28. (b)	29. (d)	30. (c)
31. (a)	32. (c)	33. (d)	34. (d)	35. (a)	36. (b)	37. (d)	38. (b)	39. (c)	40. (a)

(ii) Review Questions

1. Explain the change of phase of water at constant pressure with the help of $T-v$ and $T-s$ diagrams.
2. Define (i) Internal latent heat, (ii) External work of evaporation, (iii) Sensible heat of water, (iv) Latent heat of vaporization, and (v) Total heat of steam.
3. Define dryness fraction of steam. Name the various methods for measuring it.
4. Define triple point and critical point of water and show these points on different diagrams.
5. Explain the process of heat generation and enthalpy changes during steam generation at constant pressure with the help of $p-v$ and $T-s$ diagrams.
6. Draw a schematic plot for temperature-entropy for water and show constant pressure, constant temperature, constant volume, and constant dryness fraction lines on it.
7. Explain the Mollier chart for steam with a suitable diagram. Also draw the commercial Mollier chart.
8. Differentiate between wet, dry saturated, and superheated steam.
9. Explain the method of finding the dryness fraction of steam by bucket calorimeter.
10. Explain the principle of throttling calorimeter used for determining the dryness fraction of steam. Also give its limitations.
11. Describe the separating and throttling calorimeter with a neat sketch for measuring the quality of steam. Also derive an expression for dryness fraction of steam.
12. Draw the following process on $p-v$, $T-s$, and $h-s$ diagrams: (i) Constant volume process, (ii) Constant pressure process, (iii) Constant temperature process, and (iv) Isentropic process.

13. What is a steam boiler? How the boilers can be classified?
14. Sketch and describe the constructional features of a Cochran boiler.
15. Describe the constructional and working features of Babcock and Wilcox boiler and also state where water tube boiler is exclusively used.
16. Explain the differences between boiler mountings and accessories.
17. Name the various accessories used in steam generating plant.
18. Explain with neat sketches the following boiler mountings: (i) Pressure gauge, (ii) Water level indicator, (iii) Blow off cock, (iv) Feed check valve, (v) Junction or steam stop valve, and (vi) High steam and low water safety valve.
19. Explain with neat sketches the following boiler accessories: (i) Superheater, (ii) Air preheater, (iii) Economizer, (iv) Injector, and (v) Steam separator.
20. Explain with neat sketches the following boiler mountings: (i) Dead weight safety valve, (ii) Spring loaded safety valve, (iii) Lever safety valve, (iv) Feed check valve, (v) Fusible plug, and (vi) Blow off cock.
21. Explain with neat sketches the following boiler accessories: (i) Feed pump, (ii) Anti-priming pipe, (iii) Steam separator, (iv) Steam trap, and (v) Pressure reducing valve.
22. Define (i) Equivalent evaporation from and at 100°C, and (ii) Boiler efficiency.
23. What are the sources of heat loss in boiler plants and the methods to reduce them?
24. What do you understand by boiler draught?
25. Derive an expression connecting the height of a chimney and the draught it produces in terms of temperature of outside air and the mean temperature of flue gases. Also give the assumptions made.
26. Derive the condition for maximum discharge of flue gases through a chimney of a given height.
27. Derive an expression for draught in mm of water column for maximum discharge.
28. Write the advantages of artificial draught system over natural draught system.
29. Explain the various types of mechanical draught systems.
30. Derive the expressions for the power required in forced and induced draught fans.

(iii) Numerical Problems

1. A tank contains 50 kg of liquid water and 2 kg of water vapor under saturation conditions at 30°C. If the specific volume of saturated vapor at that temperature is given as 32.893 m^3/kg, determine the volume of tank and the moisture contents of the mixture.

   [**Ans.** 65.8362 m^3, 0.9615]

2. A vessel of volume 0.05 m^3 contains saturated steam and water mixture at 300°C. If the mass of liquid is 10 kg then determine the pressure, the mass, the specific volume, internal energy, and the entropy of steam.

   [**Ans.** 86 bar, 1.6648 kg, 11.6648 kg, 0.004286 m^3/kg, 18033 kJ, 17724 kJ, 42.06 kJ/kg K]

3. Evaluate enthalpy, internal energy, volume, and entropy of one kg of steam having a dryness fraction 0.9 and pressure 20 bar. Also evaluate the same properties at the same pressure when steam is superheated by 50°C. Take specific heat for superheated steam equal to 2.1 kJ/kg K.

[**Ans.** 2610.58 kJ/kg, 0.08978 m^3/kg, 2431.02 kJ/kg, 5.9515 kJ/kg K, 2904.5 kJ/kg, 0.1099 m^3/kg, 2882.52 kJ/kg, 6.5467 kJ/kg K]

4. Determine the state of the steam in the following cases: (a) pressure 8 bar and specific volume 0.196 m^3/kg, (b) pressure 14 bar and temperature 225°C, and (c) pressure 25 bar and 2700 kJ/kg of heat is required to generate steam from water at 0°C.

[**Ans.** x = 0.815, degree of superheat = 30°C, x = 0.945]

5. A spherical shell of 100 cm in diameter contains saturated steam and water at 300°C. Determine the mass of saturated steam and saturated water in the shell if their volumes are equal. [**Ans.** 12.074 kg, 186.61 kg]

6. Find the specific volume, enthalpy, internal energy, and entropy of wet steam at 15 bar pressure and dryness fraction 0.85.

[**Ans.** 0.1120 m^3/kg, 2500.01 kJ/kg, 2332.01 kJ/kg, 5.8253 kJ/kg K]

7. Steam at 8 bar and 0.85 dry is available. Find the final dryness fraction of steam in each of the cases: (i) 125 kJ of heat is removed per kg of steam at constant pressure, (ii) steam is cooled at constant volume and its temperature falls to 151.86°C, and (iii) it expands isentropically in the turbine developing 200 kJ/kg of steam and the pressure becomes 0.75 bar.

[**Ans.** 0.7889, 0.5451, 0.7691]

8. Two boilers *X* and *Y* discharge equal amount of steam into a common main steam pipe. The steam from boiler *X* is at 12 bar and 350°C and from *Y* at 12 bar and 0.92 dry. Determine (i) the equilibrium condition after mixing in the main pipe, (ii) the loss of entropy by high temperature steam, (iii) gain of entropy of the low pressure steam, and (iv) net increase or decrease of entropy.

[**Ans.** (i) 231.01°C, 0.1874 m^3/kg, 6.7383 kJ/kg K, 0.4737, (ii) 0.4737 kJ/kg K, (iii)1.0332 kJ/kg K, (iv) 0.5595]

9. A container of volume 3.0 m^3 is to be completely filled with dry and saturated steam at 12 bar. For this, the container is emptied and the necessary amount of water is filled. Determine (i) the necessary amount of water to be filled in the drum and the resulting temperature of steam, and (ii) the amount of water if the steam generated is 0.95 dry.

[**Ans.** 18.31 kg, 19.34 kg]

10. 3 kg of steam at 200°C is kept in a container of volume 0.30 m^3. Find the pressure, enthalpy, and mass of vapor. [**Ans.** 15.538 bar, 2372.05 kJ/kg, 2.349 kg]

11. Determine the internal energy of steam at 15 bar and 0.9 dry in a vessel of 0.04 m^3 capacity. Due to heat loss the pressure of steam falls 5 bar. Find the dryness fraction of steam in the final stage and heat lost. [**Ans.** 2419.49 kJ/kg, 1247.83 kJ/kg, 0.3163, –1171.66 kJ/kg]

12. Steam at 10 bar and 0.9 dryness fraction is available. Determine the final dryness fraction of steam in each of the two cases: (i) 170 kJ of heat is removed per kg of steam at constant pressure, and (ii) steam expands isentropically to a pressure 0.5 bar in a turbine in a flow process. The turbine develops 300 kJ of work per kg of steam. [**Ans.** 0.815, 0.83]

13. A pressure cooker contains 1.5 kg of saturated steam at 5 bar. Find the quantity of heat that must be rejected so as to reduce quality 60% dry. Find the pressure, temperature at new state, and amount of heat transferred. [**Ans.** 2.91 bar, 133.1°C, –1223.01 kJ]

14. A pressure cooker contains 2 kg of saturated steam at 3 bar. Find the quantity of heat that must be rejected so as to reduce quality 60% dry. Find the pressure, temperature at new state, and amount of heat transferred. [**Ans.** 1.75 bar, 116°C, –1668.17 kJ]

15. In a boiler 1000 kg of steam is generated per hour at 10 bar and 0.95 dryness fraction. The steam is then heated in a superheater to a temperature of 300°C at constant pressure. The feed water temperature is 30°C. Determine (i) heat supplied per hour in the boiler, (ii) heat supplied per hour in the superheater, (iii) internal energy of superheated steam, and (iv) change in internal energy of steam.

[**Ans.** 2551.75 MJ/hour, 478.54 MJ/hour, 2784.39 kJ/kg, 2659.62 kJ/kg]

16. Determine the entropy of 5 kg of steam at 20 bar and 300°C. Take specific heat for superheated steam equal to 2.1 kJ/kg K. [**Ans.** 33.445 kJ/kg K]

17. Determine the specific entropy and enthalpy of steam when pressure is 10 bar and specific volume is 0.16 m^3/kg. If 5 kg of steam at the given state is heated at constant pressure to a temperature of 250°C, find the heat transfer during the process.

[**Ans.** 5.79 kJ/kg K, 2.419 MJ/kg, 2.616 MJ]

18. During a steady flow process 4 kg of steam at 14 bar and 250°C loses 3600 kJ of heat to the surrounding at constant pressure. Determine the final condition of steam. [**Ans.** 0.611]

19. Steam at a pressure of 15 bar and 0.95 dryness fraction is generated in a boiler is made to pass through its superheater where additional quantity of heat is supplied to heat at constant pressure. As a result, the temperature of steam rises to 350°C. Determine the heat supplied in the superheater and change in internal energy. [**Ans.** 452.71 kJ/kg, 333.67 kJ/kg]

20. In a perfectly insulated cylinder 2.5 kg/s of steam expands isentropically from an initial pressure of 20 bar and 350°C to a final temperature of 85°C. Determine the quality of steam at the end of the process and non-flow work done.

[**Ans.** 0.9082, 572.59 kJ/kg, 1431.47 kJ]

21. Water and steam are contained in a boiler at 1 bar pressure. The air is removed by opening the stop valve and then it is closed. Determine (i) the heat required converting one kg of water from this condition to dry saturated steam at 5 bar, and (ii) when the stop valve is opened and boiler supplies steam at 5 bar pressure then determine the heat required per kg of steam formed from feed water at 45°C. [**Ans.** 2143.91 kJ/kg, 2560.28 kJ/kg]

22. 2.5 kg of steam expands in a turbine isentropically from pressure 20 bar and temperature 350°C to a temperature of 95°C. Determine (i) the quality of the steam at the end of expansion process, and (ii) steady flow work done.

[**Ans.** 0.9348, 625.32 kJ/kg, 1563.3 kJ]

23. A combined separating and throttling calorimeter has been used to find the dryness fraction of steam at a pressure of 9 bar. The pressure and temperature of steam after throttling are 1.25 bar and 115°C. The mass of steam condensed after throttling is 2.25 kg and mass of water collected in the separator is 0.25 kg. Determine the dryness fraction of steam if specific heat for superheated steam is 2.15 kJ/kg K. [**Ans.** 0.869]

24. During a test on throttling calorimeter the following observations were made: pressure in the main steam = 10 bar, pressure after throttling = 1.01325 bar, and temperature after throttling = 120°C. Take specific heat for superheated steam as 2.1 kJ/kg K and determine the dryness fraction of steam. Also find the maximum degree of wetness which can be measured by this calorimeter. [**Ans.** 0.9492]

25. The data observed during a trial of a boiler are: trial duration = 24 hrs, heating surface area = 3200 m^2, grate area = 3.5 m^2, pressure of steam = 15 bar, total coal burnt = 6000 kg, mass of water evaporated = 20,000 kg, temperature feed water = 40°C, condition of steam = 0.94, and calorific value of fuel = 20,000 kJ/kg. Determine (i) the mass of coal burnt/m^2 of grate area per hr, (ii) the equivalent evaporation from and at 100°C, (iii) the equivalent evaporation per m^2 of coal heating surface per hour, and (iv) the boiler efficiency.

[**Ans.** 71.43 kg, 3.70, 41.79%, 0.2893]

26. The following data were recorded for a boiler: feed water temperature = 50°C, boiler pressure = 10 bar, condition of steam = 0.95, coal consumption = 500 kg/hr, feed water supplied = 4000 kg/hr, and CV of fuel = 30,500 kJ/kg. Determine (i) the evaporation factor, and (ii) the equivalent evaporation from and at 100°C. [**Ans.** 1.093, 8.745]

27. The data observed during a trial of a boiler are: steam pressure = 10 bar, coal burnt = 720 kg, feed water supplied per hour = 7000 kg, temperature feed water entering the economizer = 25°C, temperature feed water leaving the economizer = 80°C, condition of steam leaving the boiler = 0.95, and calorific value of coal used = 34,000 kJ/kg. Determine the thermal efficiency of the plant, and heat absorbed by feed water in various components as a percentage of the total heat absorbed. [**Ans.** 8.11%, 9.42%, 82.46%, 77.9%]

28. The data observed for a steam power plant consisting of a boiler, an economizer, and a superheater: steam pressure = 15 bar, coal burnt = 700 kg/hr, mass of steam produced = 5000 kg/hr, temperature feed water entering the economizer = 40°C, temperature feed water leaving the economizer = 100°C, condition of steam leaving the boiler = 0.97, temperature of superheated steam leaving the superheater = 300°C, and calorific value of coal used = 30,000 kJ/kg. Determine (i) equivalent evaporation, (ii) boiler efficiency, and (iii) boiler efficiency with economizer and superheater. [**Ans.** 7.32, 55%, 61%, 62.4%]

29. A chimney has a height of 100 m. For the maximum discharge condition, calculate the temperature of the chimney gases and the draught produced if the air supplied per kg of fuel is 18 kg. Also determine the efficiency of the chimney if the temperature of the chimney gases in artificial draught system is limited to 120°C. Take the boiler house temperature as 40°C and specific heat of flue gases 1.005 kJ/kg K. [**Ans.** 388°C, 56.4, 0.409%]

30. Determine the minimum temperature of the flue gases required to produce a draught of 15 mm of water by a chimney of 30 m height if the air-fuel ratio required for combustion is 20:1 and atmospheric temperature is 27°C. [**Ans.** 274.75°C]

2

Steam Turbines and Condensers

2.1 INTRODUCTION

Steam turbine is a type of turbomachine which is an assembly of nozzles and blades. It converts a part of the energy of high temperature and high pressure steam into mechanical energy (or shaft work). The operation of steam turbine completely depends on the dynamic action of the steam expanding in nozzles. The steam turbines are used for the generation of electricity in steam power plants varying from 1 MW to 1500 MW capacity. These are also used for marine propulsion. The steam turbines operate at very high speed (up to about 40,000 rpm) and are able to give efficiency about 40% which is higher than the other power producing devices.

The steam expands in a turbine from high pressure to low pressure (below the atmospheric pressure). Steam cannot be exhausted to atmosphere at a pressure lower than the atmospheric pressure. This is made possible by using an additional unit called condenser. A steam condenser is a closed vessel in which vacuum is maintained and the exhausted steam is condensed by extraction of heat. About 50% to 60% of the heat energy associated with steam is lost in a steam condenser. Cooling water supplied to the condenser for the condensation of steam becomes hot. The cooling towers and the cooling ponds are used to cool this hot water coming out of the condenser so that the cooled water can be reutilized in the condenser again.

2.2 STEAM NOZZLE AND ITS GENERAL FLOW ANALYSIS

A nozzle is a duct of smoothly varying cross section by means of which pressure energy of steadily flowing fluid is converted into kinetic energy. The fluid enters the nozzle with a relatively small velocity and high pressure. As it flows through the nozzle, its pressure falls and a part of the enthalpy of steam gets converted into kinetic energy. The amount of energy which converts into kinetic energy depends on the pressure ratio and the type of expansion. Generally, nozzles are designed to obtain isentropic expansion which imparts maximum conversion. The velocity increases from entrance to the exit of the nozzle.

The applications of nozzles are: (i) In steam and gas turbines for power generation, (ii) In aviation purposes such as propulsion of jet engines and rocket, (iii) In injectors and ejectors, and (v) For flow measurements, etc.

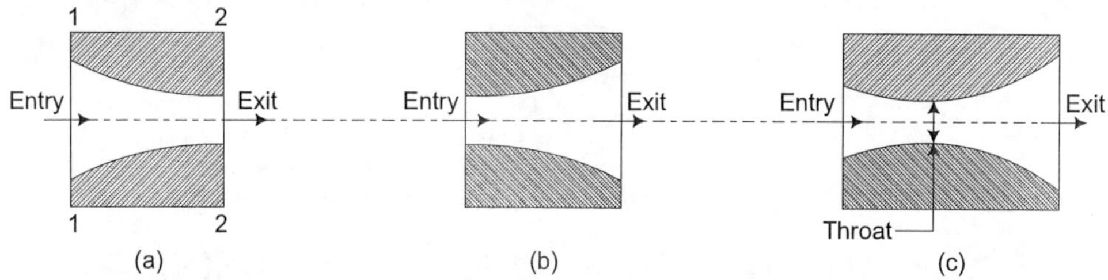

Fig. 2.1 Types of Nozzles

The nozzles are of three types:

(i) **Convergent Nozzle** If the cross section of the nozzle decreases continuously from the entrance to the exit, it is called convergent nozzle shown in Figure 2.1 (a).

(ii) **Divergent Nozzle** If the cross section of the nozzle increases continuously from the entrance to the exit, it is called divergent nozzle shown in Figure 2.1 (b).

(iii) **Convergent-divergent Nozzle** If the cross section of the nozzle first decreases and then increases, it is called convergent-divergent nozzle shown in Figure 2.1 (c).

In nozzles the flow is considered adiabatic (i.e., $q = 0$) and no work is done on or by the fluid (i.e., $w = 0$). Consider a nozzle (Figure 2.1a) in which section 1-1 is the entrance of the nozzle and section 2-2 is the exit of the nozzle. Let the flow of fluid be initially at pressure p_1, velocity V_1, and enthalpy h_1 expands through the nozzle. Applying steady flow energy equation, we have,

$$h_1 + \frac{V_1^2}{2} = h_2 + \frac{V_2^2}{2} \quad (\because q = 0 \text{ and } w = 0)$$

The entering velocity of steam V_1 is small as compared with velocity V_2. Thus, neglecting V_1, we have,

$$h_1 = h_2 + \frac{V_2^2}{2}$$

$$V_2 = \sqrt{2 \times (h_1 - h_2)}$$

Generally, enthalpy is expressed in kJ/kg, then the exit velocity in m/s is given by,

$$V_2 = \sqrt{2 \times 1000 \times (h_1 - h_2)}$$

$$= 44.72 \times \sqrt{(h_1 - h_2)} \text{ m/s} \quad (2.1)$$

When the nozzle efficiency (η_{nozzle}) is also considered, then the exit velocity is given by,

$$V_2 = 44.72 \times \sqrt{\eta_{nozzle} \times (h_1 - h_2)} \text{ m/s} \quad (2.2)$$

The general thermodynamic relation for flow of gas through a duct is given by,

$$\frac{dA}{A} = \frac{dV}{V}(M^2 - 1) \quad (2.3)$$

where A is the area of the duct, V is the velocity of fluid flow, and M is the Mach number. The Mach number is defined as the ratio of flow velocity to sound (or sonic) velocity in the flowing medium, i.e.,

$$M = V/C.$$

The following conclusion can be drawn from Eq. (2.3):

Case 1: Accelerated Flow This type of flow takes place in nozzles. Along the flow direction pressure decreases and velocity increases. Based on the velocity of the fluid two conditions arise for change in cross section.

(i) When $M < 1$, (i.e., $V < C$), $(dA/A) = -\text{ve}$, so for subsonic flow the nozzle must be convergent nozzle. These nozzles are called subsonic nozzles. But when $M = 1$, (i.e., $V = C$), $(dA/A) = 0$, the throat of the nozzle is reached and the flow is sonic.

(ii) When $M > 1$, (i.e., $V > C$), $(dA/A) = +\text{ve}$, so for supersonic flow the nozzle must be a divergent nozzle. These nozzles are called supersonic nozzles or diverging nozzles.

Case 2. Decelerated (or retarded) flow. This type of flow takes place in diffusers. Diffuser converts kinetic energy into pressure energy. Along the flow direction pressure increases and velocity decreases. Based on the velocity of the fluid again two conditions arise for change in the cross section.

(i) When $M < 1$, (i.e., $V < C$), $(dA/A) = +\text{ve}$, so for subsonic flow the diffuser must be divergent. But when $M = 1$, (i.e., $V = C$), $(dA/A) = 0$, the throat of the diffuser is reached and the flow is sonic.

(ii) When $M > 1$, (i.e., $V > C$), $(dA/A) = -\text{ve}$, so for supersonic flow the diffuser must be convergent.

2.3 PRINCIPLE OF OPERATION OF STEAM TURBINES

The principle of operation of turbines is based on Newton's second law of motion. In steam turbines, steam first expands in nozzles where due to fall in pressure of steam its thermal energy converts into kinetic energy, resulting in the emission of a high velocity jet of steam. The high velocity steam impinges on the curved blades mounted on a wheel gets deflected by an angle. The change in flow direction causes the change in momentum by which a force is exerted on the blades. Thus, the wheel rotates and produces power (Figure 2.2).

When the blade is locked, the jet enters and leaves with equal velocity ($V_1 = V_2$), and thus develops maximum force if we neglect friction in the blade (Figure 2.3 a). In this case no mechanical work is done, as the blade is stationary means its velocity is zero ($u = 0$). When the blade is allowed to move, the leaving velocity of steam jet from the blade reduces and thus force also reduces. Due to the blade velocity work will be done and maximum work will be obtained when the blade velocity is half of the steam inlet velocity (i.e., $u = V_1/2$). In this case the steam velocity at the blade outlet will be zero, since all the kinetic energy of steam is converted into work (Figure 2.3 b). The force and work done will be zero when the blade velocity is equal to steam velocity.

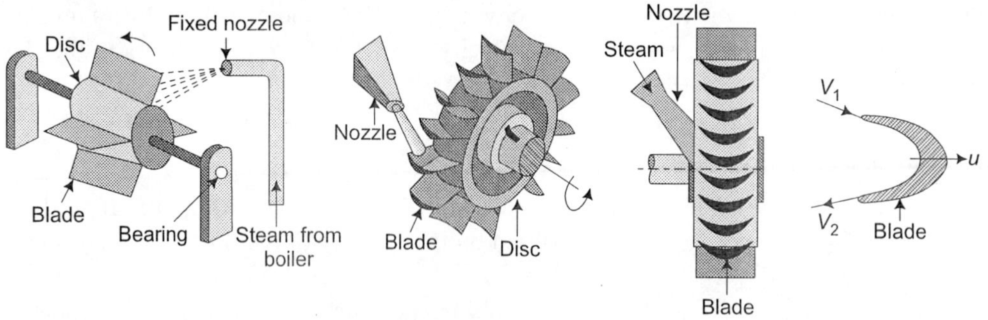

Fig. 2.2 Principle of Impulse Turbine

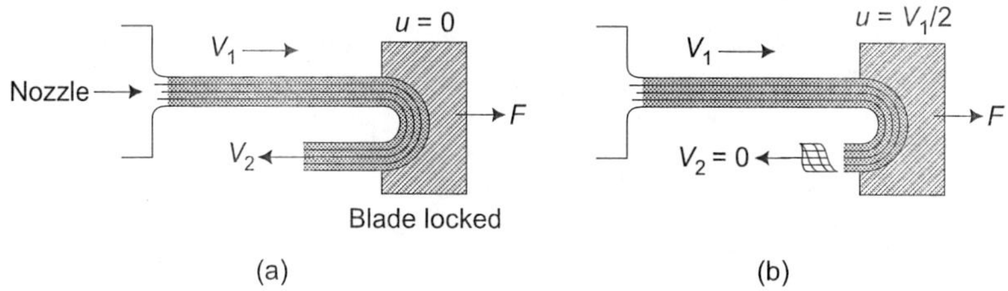

Fig. 2.3 Jet Action on Blade

About 2000 years ago, the first steam turbine was made by Hero of Alexandria which worked on pure reaction principle. There was no provision for driving any device by it, but it was a type of toy (Figure 2.4). It consisted of a hollow ball mounted on between two pivots. The ball was provided with two converging tubes (nozzles). The steam generated in a cauldron placed below the ball was supplied to the ball through one pivot. The steam was expanded through the nozzles to atmospheric pressure which results in a reactive force on the ball and made it to rotate between the pivots.

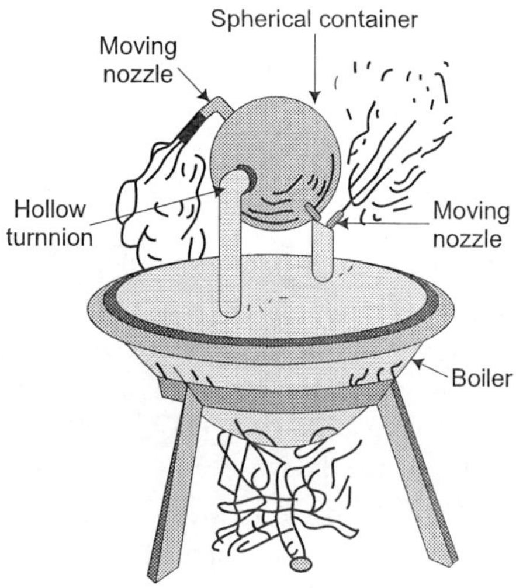

Fig. 2.4 Hero's Turbine (Principle of Reaction Turbine)

2.4 CLASSIFICATION OF STEAM TURBINES

On the basis of the principles of operation (i.e., mode of steam action) the steam turbines can be divided into two types: (i) Impulse turbine, and (ii) Reaction turbine (or impulse-reaction turbine).

(i) Impulse Turbine In impulse turbines the steam expands only in the nozzles. It means the pressure drop (enthalpy drop) takes place only in nozzles and not in moving blades. The moving blades only deflect the steam through an angle. An impulse turbine works on the principle of impulse, means the kinetic energy of steam is used to exert a force on the moving blades. This is achieved by having the symmetrical blades, means the cross sectional area of blades is constant as shown in Figure 2.5. Thus, steam pressure remains constant while it flows through the moving blades.

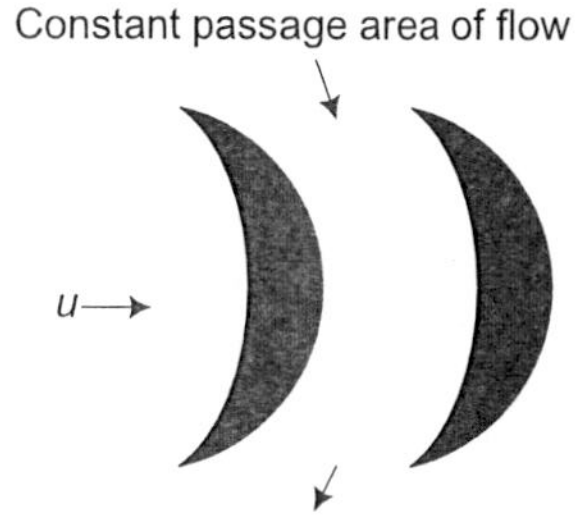

Fig. 2.5 Impulse Blade

A simple impulse turbine consists of a nozzle or a set of nozzles, a rotor (or runner) mounted on a shaft, one set of moving blades fixed to the runner, and a casing. A set (or row) of nozzles and moving blades makes a stage. The schematic view and the flow of steam through a simple impulse turbine are shown in Figure 2.6 in which the pressure and velocity variation have also been illustrated. It can be seen from the figure that the pressure drop takes place only in nozzles. So, the complete expansion of steam from the steam chest pressure to the condenser pressure takes place only in one set of nozzles and it leaves with a high

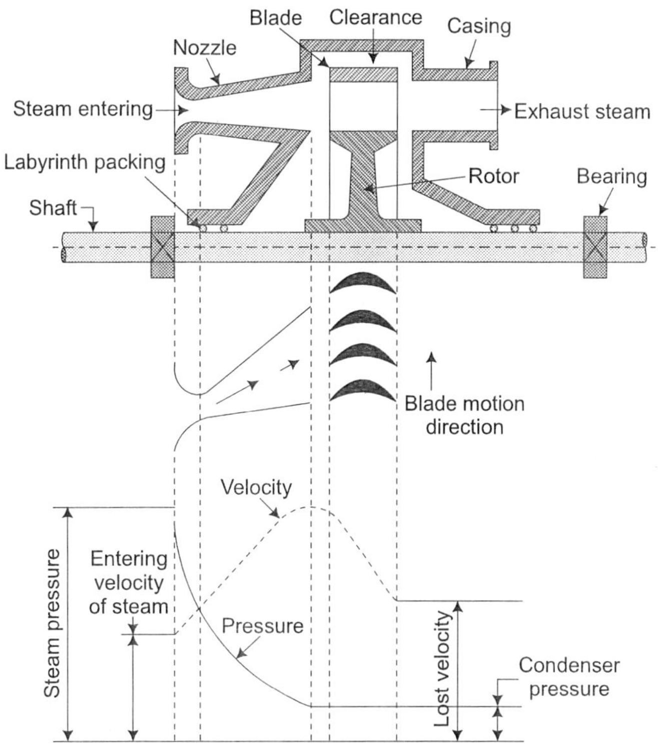

Fig. 2.6 Schematic View of Simple Impulse Turbine

velocity. The pressure in the recess between nozzles and blades is assumed same. The velocity of steam continuously decreases in moving blades. The steam leaving velocity is very high which causes energy loss, called carry-over loss. The carry-over loss may be about 10% to 12% of the initial kinetic energy. The principal example of this type of turbine is De-Laval turbine. Since the expansion of steam occurs in single stage means all the kinetic energy is absorbed by one ring of moving blades, so the velocity of turbine rotor will be very high (about 30,000 rpm). Such high speeds of the turbine rotor are not useful for practical purposes. At the same time, centrifugal stresses set up in the rotor are very high, so there is a risk of structural failure. Thus, a speed reduction gearing is required in between the turbine and the generator which reduces the plant efficiency.

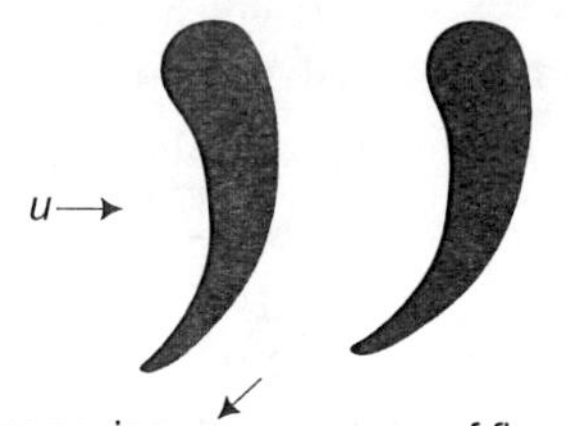

Fig. 2.7 Reaction Blade

(ii) Impulse-Reaction Turbine In practice an impulse-reaction turbine is known as reaction turbine. In an impulse-reaction turbine the steam expands in nozzles (fixed blades) as well as in moving blades. Therefore, the pressure drop occurs gradually and continuously over fixed and moving blades in succession. The pressure drop of steam in moving blades results in further generation of kinetic energy within the blades which gives reaction force. This reaction force (due to pressure drop) adds to the other propelling force (due to change in momentum) through the rotor to the turbine shaft. So, an impulse-reaction turbine works on the principle of impulse and reaction both. This is achieved by varying the blade passage cross sectional area. Its blades are not symmetrical and are thicker at one end which provides converging shaped blade passage (like that of a nozzle) for the expansion of steam as shown in Figure 2.7.

In an impulse-reaction turbine a number of rows of fixed blades are attached to the casing with an equal number of moving blades mounted on the rotor. The blade arrangement for an impulse-reaction turbine is shown in Figure 2.8 in which the variation of pressure and velocity of steam passing through the turbine blades is also depicted. It can be seen in the figure that there is a gradual drop in pressure in both the moving blades and the fixed blades.

In reaction turbine the steam is admitted for the whole blade annulus and hence there is an all round or complete admission. The steam path in a reaction turbine is divided into stages. A row of fixed blades followed by a row of moving blades constitutes a stage. The steam passing through the first set of fixed blades undergoes a small drop in pressure and thus its velocity increases. It then enters the first row of moving blades where it suffers a change in its direction and therefore, change in momentum. This change in momentum causes an impulse on the blades. But in this type of turbine, the blades passage varies due to which there will be drop in steam pressure in moving blades which causes increase in the kinetic energy of steam. This kinetic energy produces a reaction in the direction opposite to that of added velocity. Thus, the propelling or driving force is the vector sum of impulse force and reaction force. This type of turbine is known as an impulse-reaction turbine or a reaction turbine. The carry-over loss in these turbines is about 1% to 2% of

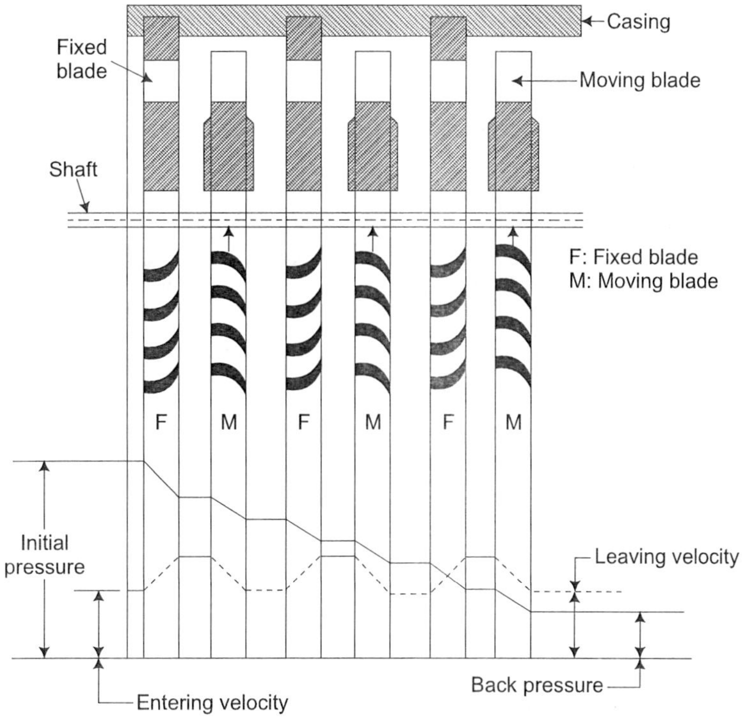

Fig. 2.8 Schematic View of a Reaction Turbine

the total initial available kinetic energy. The example of an impulse reaction turbine is Parson's reaction turbine.

The steam turbine can also be classified on the following basis:

1. On the basis of direction of flow: (i) Radial flow turbine, and (ii) Axial flow turbine.
2. On the basis of method of compounding: (i) Velocity compounded turbine, (ii) Pressure compounded turbine, and (iii) Pressure-velocity compounded turbine.
3. On the basis of number of stages: (i) Single-stage turbine, and (ii) Multistage turbine.
4. On the basis of position of shaft: (i) Horizontal shaft turbine, and (ii) Vertical shaft turbine.
5. On the basis of pressure of steam at the inlet: (i) Low pressure steam turbine, and (ii) High pressure steam turbine.
6. On the basis of exhaust condition of steam: (i) Condensing turbine, and (ii) Non-condensing turbine.
7. On the basis of their use or service.

2.5 COMPARISON BETWEEN IMPULSE AND REACTION TURBINES

Impulse Turbine	*Reaction Turbine*
1. The pressure drop occurs in the nozzles only and pressure remains constant in moving blades.	1. The pressure drops in fixed blades (nozzles) as well as in moving blades.
2. The relative velocity of steam passing over the moving blades remains same, or it reduces slightly due to friction.	2. The relative velocity of steam increases due to the continuous expansion of steam.
3. The shape of the blade is of profile type and symmetrical. Its manufacturing is not so difficult thus it is not costly.	3. The shape of the blades is of aerofoil type and asymmetrical. Its manufacturing is difficult thus it is costly.
4. It has constant blade channel area.	4. It has varying blade channel area (like a nozzle).
5. The pressure on both ends of the moving blade remains same.	5. The pressure on both ends of the moving blade varies.
6. The admission of steam is at fixed locations of nozzle or incomplete.	6. The admission of steam is all round or complete.
7. Diaphragm fixed in the casing contains the nozzle and the rotor construction is of disc or wheel type.	7. Fixed blades similar to moving blades attached to casing serve as nozzles which guide the steam and the rotor construction is of drum type.
8. For the same power developed the numbers of stages are less due to large pressure drop in nozzles.	8. For the same power developed the number of stages are more due to small pressure drop in each stage.
9. The blade speed and velocity are high.	9. The blade speed and velocity are low.
10. Efficiency of the blades is low.	10. Efficiency of the blades is high.
11. It is suitable for small power requirements.	11. It is suitable for medium and high power requirements.
12. For the same power output space requirement is less.	12. For the same power output space requirement is more.
13. It is not economical under part load conditions.	13. It is economical under part load conditions.
14. Frictional losses as compared to leakage losses are more.	14. Leakage losses as compared to frictional losses are more.
15. It works on the principle of impulse.	15. It works on the principle of impulse and reaction both.

2.6 COMPARISON BETWEEN STEAM ENGINE AND STEAM TURBINE

Steam Engine	*Steam Turbine*
1. It works on static action of steam.	1. It works on dynamic action of steam.
2. Presence of reciprocating parts makes it a poorly balanced machine and thus heavy foundation is required.	2. Absence of reciprocating parts makes it a perfectly balanced machine and thus heavy foundation is not required.

3. Higher speed above 3000 rpm is not advisable. Also higher range of speed is not possible.	3. Speed up to 40,000 rpm may be used by which power output per unit volume as well as per unit weight can be increased. Greater range of speed is possible.
4. The thermal efficiency is low which varies from 15% to 20%.	4. The thermal efficiency is high which varies from 35% to 40%.
5. It works on modified Rankine cycle. Expansion of steam is possible up to 0.2 bar to 0.3 bar.	5. It works on Rankine cycle. Expansion of steam is possible up to 0.035 bar.
6. Mechanical efficiency is low and power is not produced at uniform rate, so flywheel is required.	6. Mechanical efficiency is high and it produces power at uniform rate, so no flywheel is required.
7. Wear, tear, and maintenance cost per unit of power developed is high.	7. Wear, tear, and maintenance cost per unit of power developed is low.
8. Steam consumption increases with the years of service.	8. Steam consumption does not increase with the years of service.
9. Large power unit is difficult to design, build, and operate.	9. Large power unit is possible and thus suitable for large thermal stations.
10. Internal lubrication is required.	10. No internal lubrication is required.
11. Overloading results in great reductions in its efficiency.	11. Overloading results in slight reductions in its efficiency.
12. Electrical generators are difficult to couple with the steam engines.	12. Electrical generators are directly coupled with the steam turbines.

2.7 COMPOUNDING OF IMPULSE TURBINE

As already discussed that the rotor speed in the case of simple impulse turbine is very high and the carry-over losses are also high. Therefore, it becomes necessary to overcome these difficulties associated with the use of single-stage impulse turbines. Compounding is the method employed for reducing the rotational speed of the impulse turbine to practical limits. This is achieved by making use of more than one set of nozzles, blades, rotors in series keyed to a common shaft so that either the steam pressure or steam jet velocity can be absorbed by the turbine in stages. Thus, the performance of the turbines can be improved by its compounding, i.e., by the use of more than one stage. There are three methods for compounding of impulse turbines, namely, velocity compounding, pressure compounding, and pressure-velocity compounding.

2.7.1 Velocity Compounding (The Curtis Turbine)

In this type of turbine the pressure drop occurs only in a single row of nozzles and the velocity drop is achieved in many small drops through two, or more than two rows of moving blades. It consists of one set of nozzles, two or more than two rows of moving blades attached to the rotor or wheel, and one row of fixed blades attached to the casing in a reversed manner in between two rows of moving blades as shown in Figure 2.9. The fixed blades only direct the steam from the first row of moving blades to the next row of moving blades, so these are also called guide blades. The pressure drop (enthalpy drop) takes place only in the nozzle at the first stage and it

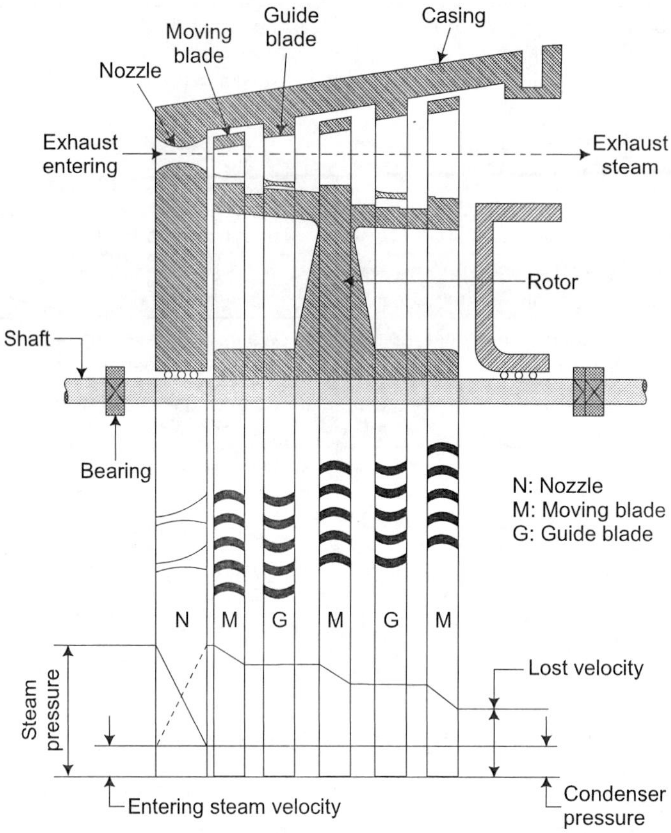

Fig. 2.9 Velocity Compounding (The Curtis Turbine)

gets converted into kinetic energy. There is no pressure drop either in moving blades or the guide blades, means it remains constant as in the case of simple impulse turbine.

The high velocity steam from the nozzles enters the first row of moving blades where its velocity reduces slightly by giving up its kinetic energy to the rotor. It then enters the first row of guide blades which direct it to the second row of moving blades. A slight drop in velocity occurs in the guide blades due to friction. When the steam passes through the second row of moving blades again there will be slight drop in its velocity due to giving up some more kinetic energy to the rotor. Then it is again directed by the second row of guide blades to the third row of moving blades where further velocity drop takes place, and finally, it leaves the rotor axially with a reduced velocity. The pressure and velocity variation is shown in Figure 2.9.

Therefore, the kinetic energy of the steam gained in the nozzles is successively absorbed in stages by the rows of moving blades and gets converted into mechanical energy. Finally, the steam is exhausted from the last row of the moving blades. The example of velocity compounded impulse turbine is the Curtis turbine named after its designer. The velocity compounded turbines are used as drives for centrifugal compressors, centrifugal pumps, small generators, and for driving feed pumps in large power plants.

Advantages of Velocity Compounded Turbines Due to relatively fewer numbers of stages (2 to 3 only) its initial cost is low. It is easy to start, requires less space, and is reliable. The turbine and its casing are not subjected to high pressure as all the expansion of steam takes place in the nozzles, so these need not be manufactured of high strength. Due to low pressure inside the turbine the leakage losses are less.

Disadvantages of Velocity Compounded Turbines Due to very high steam initial velocity the frictional losses are more which causes low efficiency. The ratio of blade velocity to steam velocity is not optimum for all the wheels which also results in low efficiency of the turbine. Therefore, the efficiency of the turbine decreases with the increase in the number of stages. The power developed in each successive blade row decreases with the increase in the number of rows and all the rows require same space, material, and initial cost. So, all stages are not used with equal economy.

2.7.2 Pressure Compounding (The Rateau Turbine)

In this type of turbine the whole pressure drop from the steam chest pressure to the condenser pressure divides into a series of smaller pressure drops across many stages of impulse turbine, thus it is called pressure compounded impulse turbine.

It has a number of simple impulse turbines in a series on the same shaft. Each of the simple impulse turbines consists of one row of nozzles and one row of moving blades which is termed a stage of the turbine. The rows of moving blades are mounted on the rotor or wheel which is keyed to the turbine shaft in series. The rows of fixed blades (nozzles) are fitted into the diaphragm which separates the two stages and is fixed to the turbine casing. The pressure drop is divided up equally between all the nozzles rows. The pressure and velocity variations are shown in Figure 2.10.

The steam from the boiler enters the first row of nozzles where its pressure reduces partially and its velocity increases. Then it passes over the first row of moving blades where nearly all its velocity is absorbed and its pressure remains constant. The exhaust steam from the first row of moving blades enters the next nozzles row. The steam again partially expands and its velocity again increases which gets absorbed in the second row of moving blades, and so on.

This turbine was invented by Professor Rateau, so it is known as Rateau turbine. In this turbine, the ratio of blade velocity to steam velocity remains constant, so this is the most efficient type of impulse turbine. But it is very costly as it has large number of stages. As the steam pressure decreases from one stage to the next stage, the specific volume of the steam increases, therefore, the blade height increases towards the low pressure side of the turbine.

2.7.3 Pressure-Velocity Compounding

It is a combination of pressure and velocity compounding. The total drop in steam pressure is carried out in two stages and the velocity obtained in each stage is also compounded. This type of arrangement for two rotors is shown in Figure 2.11.

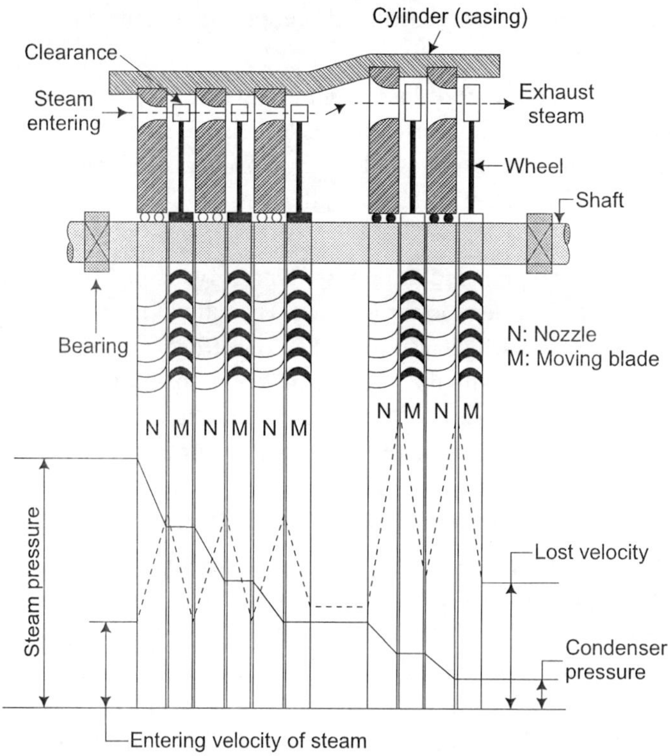

Fig. 2.10 Pressure Compounding (The Rateau Turbine)

In the first row of nozzles, there is a drop in pressure of the steam but its velocity increases. The pressure remains constant in the two rows of moving blades of the first rotor and in the first row of guide blades (fixed blades). But there is a velocity drop in the moving blades row. In the second set of nozzles the remaining pressure drop occurs and velocity increases. The pressure remains constant in the two rows of moving blades of the second rotor and in the second row of guide blades. But the velocity drops in the moving blades of the second rotor.

So, in each rotor, velocity drop is achieved by the many rows of moving blades and hence its velocity is compounded. And the total pressure drop occurs into two small pressure drops in two nozzles rows, hence, it is pressure compounded. This method of compounding is used in Moore and the Curtis turbines. This arrangement results in a more compact turbine than a pressure compounded turbine, but its efficiency is relatively lower. The specific volume of steam is higher in the second stage, so, the blade height is greater than the first stage.

2.8 VELOCITY DIAGRAMS FOR IMPULSE TURBINE

To evaluate the force on the blades and the power developed by a turbine, it is necessary to determine the rate of change of momentum of steam across the moving blades. The steam should

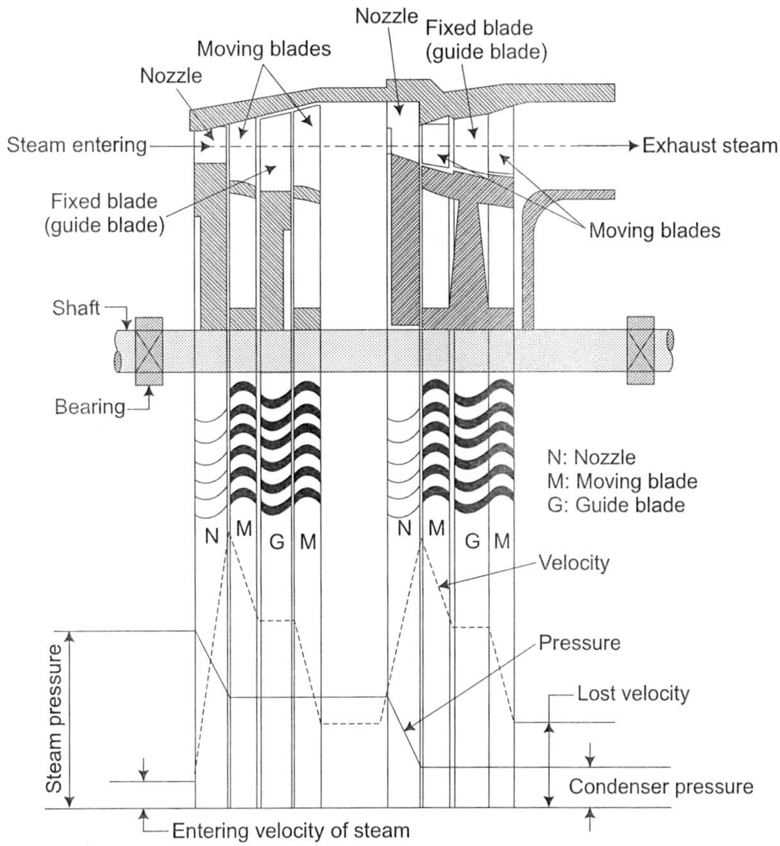

Fig. 2.11 Pressure-Velocity Compounding

enter and leave the blade without any shock for which the inlet and outlet angles of the moving blades should be evaluated. In order to fulfil these objectives, it is essential to draw the vector diagrams at the inlet and outlet of the moving blades showing the variations of velocity of steam during its flow through the blade passage.

The following notations are used for the velocity diagrams:

Suffixes 1 and 2 denote the inlet and outlet conditions respectively, for moving blades as shown in Figure 2.12 and Figure 2.13.

u = Tangential or circumferential velocity of blades which remains constant at inlet and outlet of moving blades because of small blades height, i.e., $u_1 = u_2 = u$

V_1 and V_2 = Absolute velocity of steam at the inlet and outlet respectively

V_{w1} and V_{w2} = Velocity of whirl at the inlet and outlet respectively, (i.e., tangential component of V_1)

V_{f1} and V_{f2} = Velocity of flow at the inlet and outlet respectively, (i.e., axial component of V_1 and V_2 respectively)

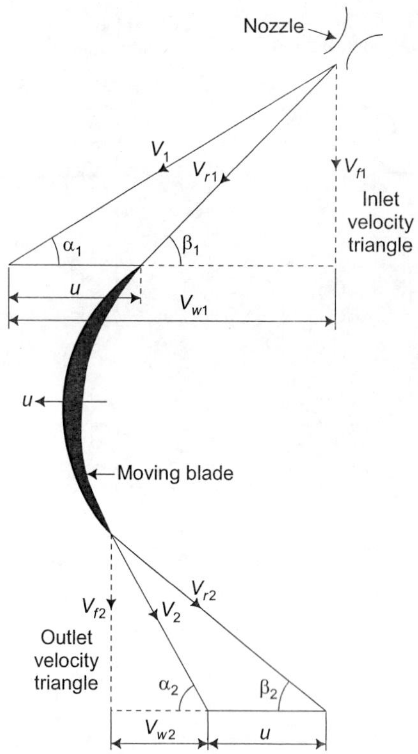

Fig. 2.12 Velocity Diagram for Impulse Turbine

V_{r1} and V_{r2} = Relative velocity of steam with respect to the blade at inlet and outlet,
m = Mass flow rate of steam in kg/s,
α_1 = Nozzle angle or angle of absolute velocity at inlet,
α_2 = Angle of absolute velocity at outlet (or inlet angle of fixed blade),
β_1 = Blade angle at inlet or angle of relative velocity at inlet,
β_2 = Blade angle at outlet or angle of relative velocity at outlet,
d = Mean diameter of blade drum, and
h = Blade height

Speed ratio,

$$\rho = \frac{\text{Tangential velocity of blade}}{\text{Absolute velocity of steam at inlet}} = \frac{u}{V_1} \tag{2.4}$$

Blade velocity coefficient,

$$K = \frac{V_{r2}}{V_{r1}} \quad \text{where} \quad V_{r2} < V_{r1} \tag{2.5}$$

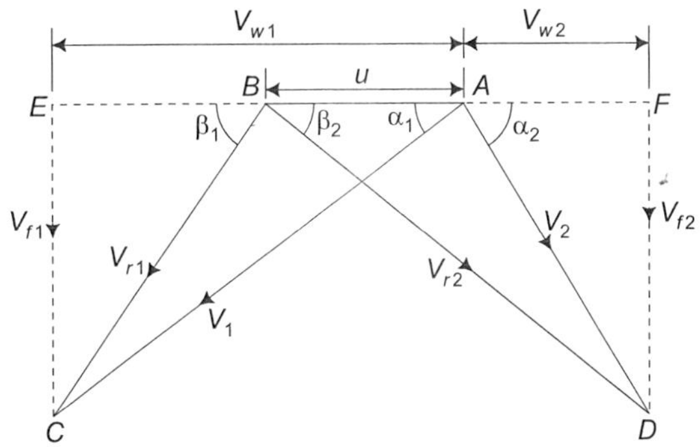

Fig. 2.13 Combined Velocity Diagram for Impulse Turbine

The value of K varies from 0.85 to 0.9 and for smooth blades $K = 1$, means $V_{r1} = V_{r2}$.

The procedure for drawing the combined velocity diagram is given below:

1. Draw horizontal line AB equal to blade velocity u to some suitable scale.
2. Draw a line AC at an angle α_1 with AB. Cut $AC = V_1$.
3. Join B and C which represents the relative velocity V_{r1} at inlet. The blade inlet angle β_1 is measured and its value is evaluated.
4. From point C draw a perpendicular CE on AB produced. CE represents flow velocity (axial velocity) at inlet and AE represents whirl velocity (tangential velocity) at inlet.
5. From point B draw a line BD at an angle β_2 (blade outlet angle).

   Cut $BD = V_{r2} = KV_{r1}$. Join A and D which represents the absolute velocity (V_2) at outlet. The angle α_2 is measured.
6. From point D draw a perpendicular DF on BA produced. DF represents flow velocity (axial velocity) at outlet and AF represents whirl velocity (tangential velocity) at outlet.

   Thus, velocity triangles get completed.

2.8.1 Force and Work Done on Turbine Blades and Efficiency

1. Force in the tangential direction (F_t).

$$F_t = \text{mass} \times \text{acceleration in tangential direction}$$

$$= \text{mass per second} \times \text{change in velocity in tangential direction}$$

$$F_t = m \times (V_{w1} - V_{w2}) \text{ N}$$

(when V_{w1} and V_{w2} are in opposite direction)

$$F_t = m \times (V_{w1} + V_{w2}) \text{ N}$$

(when V_{w1} and V_{w2} are in same direction)

Thus general expression for tangential force becomes,

$$F_t = m \times (V_{w1} \pm V_{w2}) \text{ N} \tag{2.6}$$

2. Force in the axial direction or axial thrust (F_a).

$$F_a = \text{mass per second} \times \text{change in velocity in axial direction}$$

$$F_a = m \times (V_{f1} - V_{f2}) \text{ N} \tag{2.7}$$

3. Work done on the blades (W).

$$W = F_t.u = m \times (V_{w1} \pm V_{w2}) \,.\, u \text{ Nm/s} \tag{2.8}$$

4. Power developed by the turbine (P).

$$P = \frac{m \times (V_{w1} \pm V_{w2}) \,.\, u}{1000} \text{ kW} \tag{2.9}$$

5. Diagram or blade efficiency (η_b).

   It is defined as the ratio of the work done by blade per second to the energy entering the blade per second.

$$\therefore \qquad \eta_b = \frac{m(V_{w1} \pm V_{w2}) \,.\, u}{(1/2) \times m \times V_1^2}$$

$$= \frac{2u(V_{w1} \pm V_{w2})}{V_1^2} \tag{2.10}$$

6. Energy lost due to blade friction (E_f).

$$E_f = \frac{1}{2} m \left(V_{r1}^2 - V_{r2}^2\right) \text{ Nm/s} \tag{2.11}$$

7. Stage efficiency or gross efficiency (η_s).

   It is defined as the ratio of the work done by blades per kg of steam flowing through the stage (i.e., nozzles and moving blades) to the isentropic enthalpy drop per kg of steam in that stage.

Thus,

$$\eta_s = \frac{(V_{w1} \pm V_{w2}) \,.\, u}{\Delta h} \tag{2.12}$$

where Δh = Isentropic enthalpy drop per kg of steam in the nozzle of the impulse turbine.

The stage efficiency also takes into account the losses in nozzle,

$$\therefore \qquad \eta_s = \eta_b \times \eta_n \tag{2.13}$$

where η_b = Blade efficiency, and η_n is the nozzle efficiency.

Nozzle efficiency,

$$\eta_n = \frac{V_1^2}{2(h_0 - h_1)} \tag{2.14}$$

where h_0 = Enthalpy of steam at the entrance of nozzle in J/kg, and

h_1 = Enthalpy of steam at the nozzle exit in J/kg.

2.8.2 Condition for Maximum Efficiency of Impulse Turbine

From the combined velocity diagram (Figure 2.13)

$$V_{w1} = V_1 \cos \alpha_1 = V_{r1} \cos \beta_1 + u,$$

and

$$V_{w2} = V_2 \cos \alpha_2 = V_{r2} \cos \beta_2 + u$$

$$\therefore \quad V_{w1} + V_{w2} = V_{r1} \cos \beta_1 + V_{r2} \cos \beta_2$$

$$= V_{r1} \cos \beta_1 \left[1 + \frac{V_{r2}}{V_{r1}} \cdot \frac{\cos \beta_2}{\cos \beta_1}\right]$$

$$= V_{r1} \cos \beta_1 \, [1 + KC]$$

where $K = \dfrac{V_{r2}}{V_{r1}}$, and $C = \dfrac{\cos \beta_2}{\cos \beta_1}$

$$V_{w1} + V_{w2} = (V_1 \cos \alpha_1 - u) \, [1 + KC]$$

Rate of work done per kg of steam per second

$$= (V_1 \cos \alpha_1 - u) \, [1 + KC] \, . \, u$$

The blade efficiency is given by,

$$\eta_b = \frac{(V_1 \cos \alpha_1 - u) \, [1 + KC] \, . \, u}{(V_1^2/2)}$$

$$= \frac{(V_1 \cos \alpha_1 - u) \, [1 + KC] \, . \, 2u}{V_1^2}$$

Speed ratio is given by,

$$\rho = (u/V_1)$$

Then

$$\eta_b = 2 \, (\rho \cos \alpha_1 - \rho^2) \, [1 + KC] \tag{2.15}$$

If α_1, K, and C are assumed constant then blade efficiency depends on the ρ.

For maximum blade efficiency,

$$\frac{d\eta_b}{d\rho} = 0$$

$$\therefore \quad 2(1 + KC) \, (\cos \alpha_1 - 2\rho) = 0$$

$$\therefore \quad \rho = \frac{\cos \alpha_1}{2}, \quad \text{because} \quad (1 + KC) \neq 0$$

$$\therefore \quad (\eta_b)_{max} = 2[1 + KC] \left[\frac{\cos \alpha_1}{2} \cos \alpha_1 - \frac{\cos^2 \alpha_1}{4}\right]$$

$$= [1 + KC] \, \frac{\cos^2 \alpha_1}{2} \tag{2.16}$$

Assuming that blades are symmetrical and smooth (friction is zero).

Then $$\beta_1 = \beta_2$$

$$\therefore \quad C = 1, \quad \text{and} \quad K = 1$$

$$\therefore \quad (\eta_b)_{max} = \cos^2\alpha_1 \tag{2.17}$$

Figure 2.14 shows the variations of blade (or diagram) efficiency with blade speed ratio. Generally, blade velocity of 250 m/s is maintained but it may also be used to about 370 m/s.

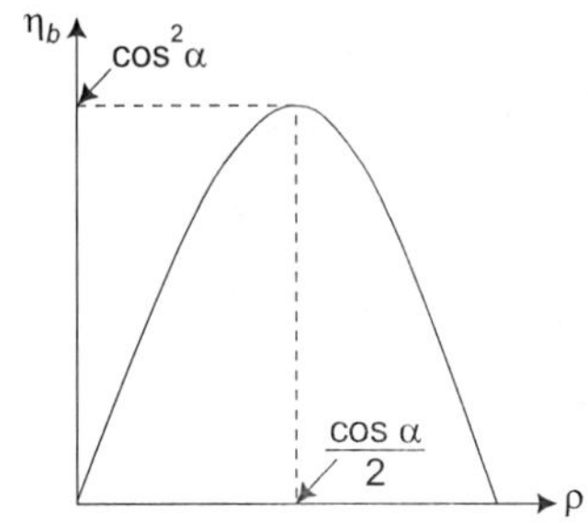

Fig. 2.14

Rate of work done per kg of steam per second, $(V_1 \cos \alpha_1 - u)[1 + KC] \,.\, u$

$$\rho = \frac{u}{V_1} = \frac{\cos \alpha_1}{2}$$

$$\therefore \quad V_1 = \frac{2u}{\cos \alpha_1}$$

Then maximum rate of work done per kg of steam per second,

$$= \left[\frac{2u}{\cos \alpha_1} \,.\, \cos \alpha_1 - u\right][1 + 1 \times 1] \,.\, u = 2u^2 \tag{2.18}$$

2.9. VELOCITY DIAGRAMS FOR IMPULSE-REACTION TURBINE

The combined velocity diagram for impulse-reaction turbine is shown in Figure 2.15. In reaction turbines the steam expands continuously while passing over the fixed and moving blades rows. This is made possible by the use of tapered rotor with progressively increasing height of the blades. When the steam flows through the moving blades the relative velocity at the exit (V_{r2}) will always be higher than the relative velocity at the inlet (V_{r1}).

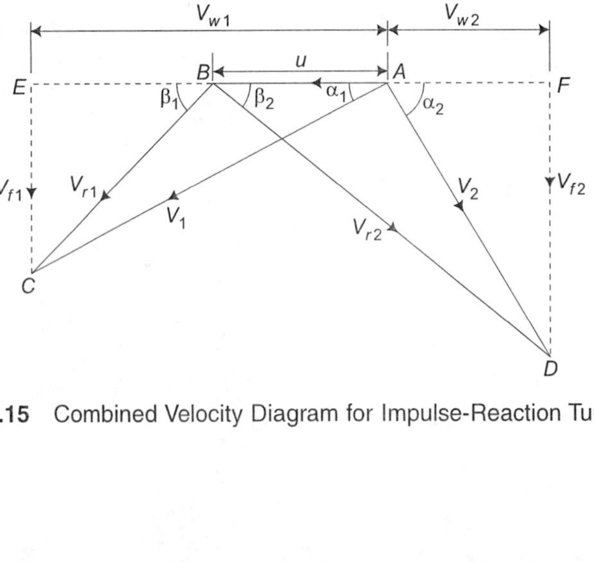

Fig. 2.15 Combined Velocity Diagram for Impulse-Reaction Turbine

2.9.1 Degree of Reaction (R)

The degree of reaction is defined as the ratio of enthalpy drop in the moving blades to the sum of the enthalpy drops in the fixed and the moving blades, i.e., in a stage. Figure 2.16 shows the mounting of fixed and moving blades with $h - s$ diagram in which points 1, 2, and 3 indicate the actual conditions of steam.

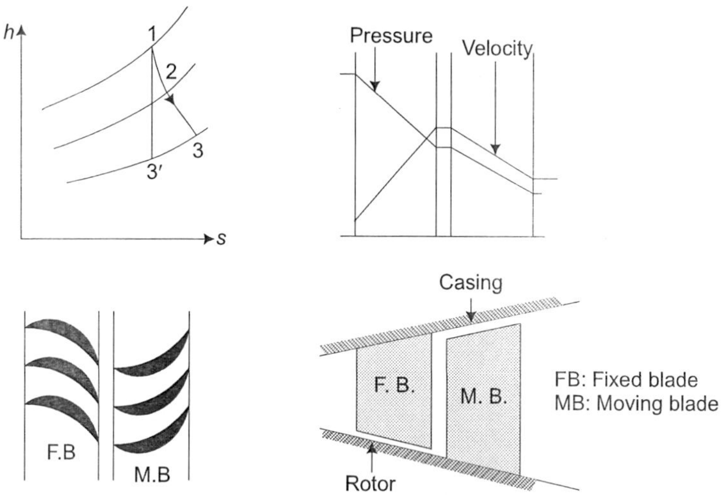

Fig. 2.16 Isentropic Expansion with Pressure-Velocity Distribution in a Reaction Turbine Stage

The degree of reaction can be given by,

$$R = \frac{(\Delta h)_{\text{moving blades}}}{(\Delta h)_{\text{fixed blades}} + (\Delta h)_{\text{moving blades}}}$$

$$= \frac{(\Delta h)_m}{(\Delta h)_f + (\Delta h)_m} = \frac{h_2 - h_3}{h_1 - h_3} \tag{2.19}$$

In pure impulse turbine, degree of reaction is zero and in pure reaction turbine it is one.

The enthalpy drop in moving blades is given by,

$$\Delta h_m = \frac{V_{r2}^2 - V_{r1}^2}{2} \tag{2.20}$$

And enthalpy drop in fixed blades is given by,

$$\Delta h_f = \frac{V_1^2 - V_2^2}{2} \tag{2.21}$$

The total enthalpy drop in a stage = work done by the steam in the stage, i.e.,

$$\Delta h_m + \Delta h_f = u(V_{w1} + V_{w2}) \tag{2.22}$$

$$R = \frac{V_{r2}^2 - V_{r1}^2}{2u(V_{w1} + V_{w2})} \tag{2.23}$$

From Figure 2.15, we have,

$$V_{r2} = V_{f2} \operatorname{cosec} \beta_2,$$

$$V_{r1} = V_{f1} \operatorname{cosec} \beta_1,$$

and $$(V_{w1} + V_{w2}) = V_{f1} \cot \beta_1 + V_{f2} \cot \beta_2$$

The velocity of flow remains constant, i.e.,

$$V_{f1} = V_{f2} = V_f$$

So, degree of reaction can be given by,

$$R = \frac{(V_{f2} \operatorname{cosec} \beta_2)^2 - (V_{f1} \operatorname{cosec} \beta_1)^2}{2uV_f (\cot \beta_1 + \cot \beta_2)}$$

$$R = \frac{V_f (\operatorname{cosec} \beta_2 - \operatorname{cosec} \beta_1)^2}{2u(\cot \beta_1 + \cot \beta_2)}$$

$$R = \frac{V_f}{2u}\left[\frac{(\cot^2\beta_2 + 1) - (\cot^2\beta_1 + 1)}{(\cot \beta_1 + \cot \beta_2)}\right]$$

$$R = \frac{V_f}{2u}\left[\frac{(\cot^2\beta_2 - \cot^2\beta_1)}{(\cot \beta_1 + \cot \beta_2)}\right]$$

$$\therefore \quad R = \frac{V_f}{2u}(\cot \beta_1 + \cot \beta_2) \tag{2.24}$$

If the reaction turbine is designed for 50% reaction then Eq. (2.24) can be written as,

$$\frac{1}{2} = \frac{V_f}{2u}(\cot \beta_1 + \cot \beta_2)$$

$$u = V_f (\cot \beta_1 + \cot \beta_2)$$

The blade velocity u can also be given by,

$$u = V_f (\cot \alpha_1 + \cot \beta_1) = V_f (\cot \beta_2 + \cot \alpha_2)$$

Therefore, $\beta_1 = \alpha_2$, and $\beta_2 = \alpha_1$

It means that the moving blades and fixed blades must have the same shape if the degree of reaction is 50%. This condition gives symmetrical velocity diagrams. This type of turbine is known as Parson's reaction turbine.

2.9.2 Blade or Diagram Efficiency of a Reaction Turbine

For determining the efficiency the following assumptions are made:

(i) The degree of reaction is 50%,

(ii) The fixed and moving blades are of the same shape, and

(iii) The velocity of steam at exit from the preceding stage is same as the velocity of steam at the entrance to the succeeding stage.

The work done per kg of steam,

$$W = u(V_{w1} + V_{w2})$$
$$= u[V_1 \cos \alpha_1 + (V_{r2} \cos \beta_2 - u)]$$

Now as per the assumptions,

$$\beta_2 = \alpha_1, \quad \text{and} \quad V_{r2} = V_1$$
$$W = u[2V_1 \cos \alpha_1 - u)$$
$$= V_1^2 \left[\frac{2uV_1 \cos \alpha_1}{V_1^2} - \frac{u^2}{V_1^2}\right]$$

But, $(u/V_1) = \rho$,

so, $$W = V_1^2 (2\rho \cos \alpha_1 - \rho^2) \tag{2.25}$$

Kinetic energy supplied to fixed blade = $(V_1^2/2)$, and

Kinetic energy supplied to moving blade = $\dfrac{V_{r2}^2 - V_{r1}^2}{2}$

∴ Total energy supplied to the stage,

$$\Delta h = \frac{V_1^2}{2} + V_{r2}^2 - \frac{V_{r1}^2}{2}$$
$$= V_1^2 - \frac{V_{r1}^2}{2} \quad (\because V_{r2} = V_1) \tag{2.26}$$

Now substituting,

$$V_{r1}^2 = V_1^2 + u^2 - 2V_1 u \cos \alpha_1 \text{ in Eq. (2.26)}$$

We have, $$\Delta h = V_1^2 - \frac{1}{2}\left[V_1^2 + u^2 - 2V_1 u \cos \alpha_1\right]$$
$$= \frac{1}{2}\left[V_1^2 + 2V_1 u \cos \alpha_1 - u^2\right]$$
$$\Delta h = \frac{V_1^2}{2}\left[1 + \frac{2u}{V_1} \cos \alpha_1 - \frac{u^2}{V_1^2}\right]$$
$$= \frac{V_1^2}{2}(1 + 2\rho \cos \alpha_1 - \rho^2) \tag{2.27}$$

We know that the blade efficiency of the reaction turbine,

$$\eta_b = \frac{W}{\Delta h}$$

Substituting the values of W and Δh in blade efficiency expression, we have,

$$\eta_b = \frac{W}{\Delta h}$$

$$= \frac{V_1^2 (2\rho \cos \alpha_1 - \rho^2)}{(V_1^2/2)(1 + 2\rho\cos \alpha_1 - \rho^2)}$$

$$\eta_b = \frac{2\rho (2 \cos \alpha_1 - \rho)}{1 + 2\rho \cos \alpha_1 - \rho^2}$$

$$= \frac{(4\rho \cos \alpha_1 - 2\rho^2)}{(1 + 2\rho \cos \alpha_1 - \rho^2)} \qquad (2.28)$$

For maximum efficiency,

$$\frac{d\eta_b}{d\rho} = 0,$$

i.e.,

$$\frac{d}{d\rho}\left[\frac{(4\rho \cos \alpha_1 - 2\rho^2)}{(1 + 2\rho \cos \alpha_1 - \rho^2)}\right] = 0$$

$$\frac{(1 + 2\rho \cos \alpha_1 - \rho^2)(4 \cos \alpha_1 - 4\rho) - (4\rho \cos \alpha_1 - 2\rho^2) . (2 \cos \alpha_1 - 2\rho)}{(1 + 2\rho \cos \alpha_1 - \rho^2)} = 0$$

$$4 (\cos \alpha_1 - \rho)(1 + 2\rho \cos \alpha_1 - \rho^2) - 4\rho (\cos \alpha_1 - \rho)(2 \cos \alpha_1 - \rho) = 0$$

$$4 (\cos \alpha_1 - \rho) [(1 + 2\rho \cos \alpha_1 - \rho^2) - \rho (2 \cos \alpha_1 - \rho)] = 0$$

$$4 (\cos \alpha_1 - \rho) [1 + 2\rho \cos \alpha_1 - \rho^2 - 2\rho \cos \alpha_1 + \rho] = 0$$

$$4 (\cos \alpha_1 - \rho) . [1] = 0$$

$$\therefore \qquad \rho = \cos \alpha_1 \qquad (2.29)$$

Substituting the value of ρ from equation (2.29) in equation (2.28), the value of maximum efficiency is given by,

$$(\eta_b)_{max} = \frac{(4 \cos^2\alpha_1 - 2 \cos^2\alpha_1)}{(1 + 2 \cos^2\alpha_1 - \cos^2\alpha_1)}$$

$$\therefore \qquad (\eta_b)_{max} = \frac{2 \cos^2\alpha_1}{(1 + \cos^2\alpha_1)} \qquad (2.30)$$

2.10 LOSSES IN STEAM TURBINES

The steam turbine losses tend to decrease the efficiency and work output of a turbine. The various losses in steam turbines are given below.

Residual Velocity Loss The steam leaves the turbine with certain absolute velocity (V_1). The energy loss due to absolute exit velocity of steam is equivalent to ($V_1^2/2$). In a single-stage impulse turbine, the residual velocity loss may be about 10% to 12% which can be reduced by using multistage turbine.

Loss Due to Friction and Turbulence Friction loss occurs in nozzles, turbine blades, and between steam and rotating disc. The friction loss in the nozzle is taken into account by nozzle efficiency. The loss due to friction and turbulence is about 10%.

Losses Due to Leakage Some amount of steam leaks from the chamber without doing useful work. The total leakage loss is about 1% to 2%. The steam leakage occurs: (i) between the turbine shaft and the bearings, (ii) at the clearance between the diaphragm and the stationary disc in the case of impulse turbines, (iii) at the blade tips in the case of reaction turbines, and (iv) leakage of steam through the labyrinth glands.

Loss Due to Mechanical Friction and Bearing This loss occurs due to friction between shaft and bearing which can be reduced by a proper lubricating system.

Losses in Regulating Valves The steam enters the turbine through the stop and regulating (governor) valves. In these valves, the steam gets throttled and the pressure of steam at the entry to the turbine is less than the boiler pressure. Although throttling is a constant enthalpy process, the enthalpy drop available for work in the turbine is reduced. The magnitude of this loss may be of the order of 5% to 10%.

Losses Due to Wetness of Steam The condition of steam after last stage of the turbine is wet. The velocity of water particles is less than that of steam. Therefore, the water particles have to be dragged along with the steam which results in loss of the kinetic energy of steam.

Radiation Loss Turbines are heavily insulated to avoid any heat loss to the surroundings, so this loss is negligible.

2.11. EFFICIENCIES AND REHEAT FACTOR

Due to various internal losses in the stage of steam turbine the work available is less than that due to isentropic expansion. This is schematically presented on the $h-s$ diagram as shown in Figure 2.17. The process $1-2s$ shows the isentropic expansion in the nozzle and 1-2 shows the actual path of the process which is due to friction loss.

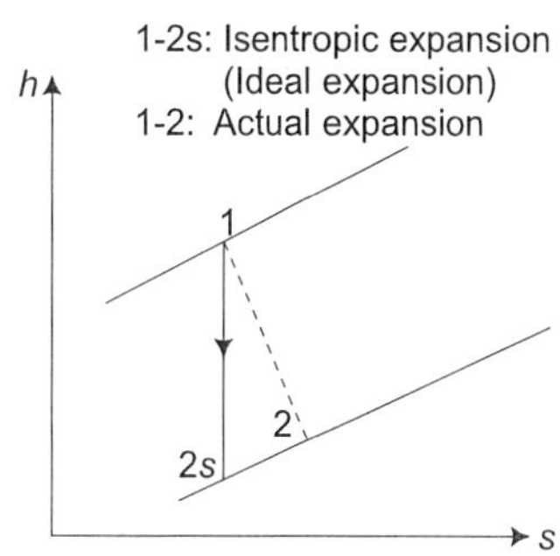

Fig. 2.17 Expansion of Steam in Single Stage on $h-s$ Diagram

Stage Efficiency It is defined as the ratio of actual heat drop to the isentropic heat drop. It is given by,

$$\eta_s = \frac{\text{Actual heat drop}}{\text{Isentropic heat drop}}$$

$$= \frac{h_1 - h_2}{h_1 - h_{2s}} = \frac{\Delta h_{1-2}}{\Delta h_{1-2s}} \quad (2.31)$$

Figure 2.18 shows the expansion of steam through a four-stage turbine.

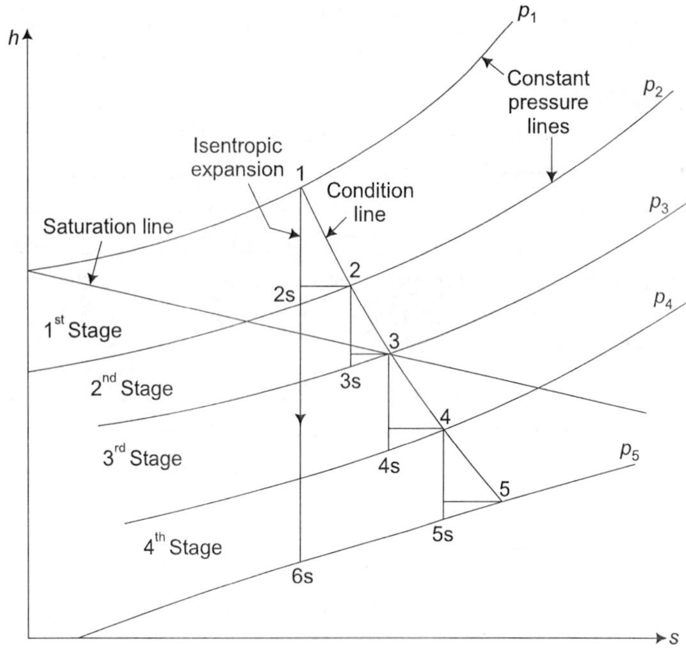

Fig. 2.18 Expansion of Steam in Multistage Turbine on $h - s$ Diagram

The isentropic heat drops are shown by points $1 - 2s$, $2 - 3s$, $3 - 4s$ and $4 - 5s$, whereas actual heat drops are shown by points 1-2, 2-3, 3-4, and 4-5. The curve joining the points 1, 2, 3, 4, and 5 is called the condition line. The heat drops Δh_{1-2}, Δh_{2-3}, Δh_{3-4}, Δh_{4-5}, are the actual heat drops which get converted into useful work. The heat drops Δh_{1-2s}, Δh_{2-3s}, Δh_{3-4s} and Δh_{4-5s} are the isentropic heat drops. The sum of the isentropic heat drops is called cumulative heat drop which is denoted by Δh_c. The heat drop Δh_{1-6s} is called direct isentropic heat drop (or Rankine heat drop) which is denoted by Δh_{ise}. The constant pressure lines on the $h - s$ diagram diverges as we move from left to right, so the cumulative heat drop is always greater than Rankine heat drop.

Reheat Factor (RF) It is defined as the ratio of the cumulative heat drop to the direct isentropic heat drop (or Rankine heat drop). It can be given by,

$$RF = \frac{\text{Cumulative isentropic heat drop}}{\text{Rankine heat drop}} \tag{2.32}$$

Referring to Figure 2.18,

$$RF = \frac{\Delta h_{1-2s} + \Delta h_{2-3s} + \Delta h_{3-4s} + \Delta h_{4-5s}}{\Delta h_{1-6s}}$$

$$= \frac{\Delta h_c}{\Delta h_{ise}} \tag{2.33}$$

As the cumulative heat drop is always greater than Rankine heat drop, so the value of reheat factor is always greater than unity. The reheat factor depends on turbine efficiency, initial pressure, superheat, exit pressure, and number of stages in a given pressure range. The reheat factor is greater if the number of stages are more for a given pressure range and lower the stage efficiency. The lower value of reheat factor is always desirable. The value of reheat factor generally lies in the range of 1.02 to 1.06.

Internal Efficiency It is defined as the ratio of the internal work done (actual work done) in the turbine to the direct Rankine work done. It is given by,

$$\eta_i = \frac{\Delta h_{1-2} + \Delta h_{2-3} + \Delta h_{3-4} + \Delta h_{4-5}}{\Delta h_{1-6s}}$$

$$= \frac{\Delta h_i}{\Delta h_{ise}} \tag{2.34}$$

If the stage efficiency remains same for all the stages then we have,

$$\eta_s = \frac{\Delta h_{1-2}}{\Delta h_{1-2s}} = \frac{\Delta h_{2-3}}{\Delta h_{2-3s}} = \frac{\Delta h_{3-4}}{\Delta h_{3-4s}} = \frac{\Delta h_{4-5}}{\Delta h_{4-5s}}$$

$$\eta_s = \frac{\Delta h_{1-2} + \Delta h_{2-3} + \Delta h_{3-4} + \Delta h_{4-5}}{\Delta h_{1-2s} + \Delta h_{2-3s} + \Delta h_{3-4s} + \Delta h_{4-5s}}$$

$$= \frac{\Delta h_i}{\Delta h_c} \tag{2.35}$$

or

$$\Delta h_i = \eta_s \,.\, \Delta h_c \tag{2.36}$$

But

$$\Delta h_c = RF \times \Delta h_{ise} \quad \text{(From Eq. 2.33)}$$

$\therefore$

$$\Delta h_i = \eta_s \,.\, (RF \times \Delta h_{ise}) \tag{2.37}$$

Substituting equation (2.37) in equation (2.34), we have,

$$\eta_i = \frac{\Delta h_i}{\Delta h_{ise}} = \frac{\eta_s \,.\, (RF \times \Delta h_{ise})}{\Delta h_{ise}}$$

$$= \eta_s \times RF \tag{2.38}$$

2.12 GOVERNING OF STEAM TURBINES

The purpose of governing is to maintain the speed of the turbine fairly constant irrespective of load on the turbine. The methods of governing which are used in steam turbines are: (1) Throttle governing, (2) Nozzle control governing, (3) By-pass governing, (4) Combination of throttle and nozzle governing, and (5) Combination of throttle and bypass governing.

2.12.1 Throttle Governing

The main parts of a simple throttle governing system are shown schematically in Figure 2.19. The turbine stop valve '*A*' is used to start and stop the turbine by its opening or closing with the help

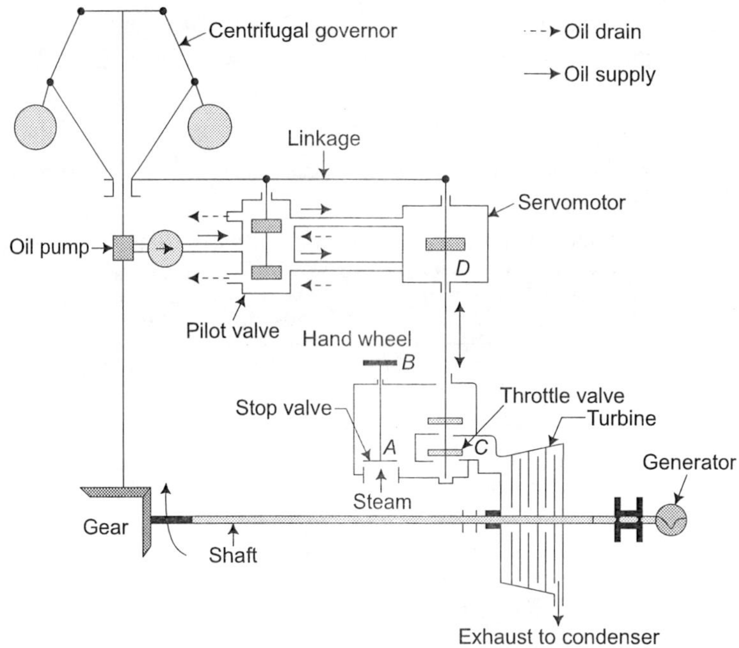

Fig. 2.19 Schematic View of Throttle Governing

of a hand wheel '*B*'. The valve '*A*' remains fully open when the turbine is on full load condition. To sense the change in speed of turbine shaft a centrifugal governor is used which is driven by bevel worm gears from the turbo-generator shaft. The relay system consists of a pilot valve and a servomotor. The throttle valve '*C*' (or governor valve) controls the speed of turbine by throttling of the steam. The opening or closing of the throttle valve is made by the movement of the servomotor piston '*D*' by the high pressure oil entering from the pilot valve into the servomotor which is in turn controlled by the centrifugal speed governor.

Under normal working conditions the servomotor piston remains in the middle position and pilot valve keeps the oil ports closed. When the speed of turbine increases the oil enters the upper half of the servomotor which moves the servomotor piston '*D*' in the downward direction. This causes closing of the throttle valve which results in the reduction of steam flow rate and thus the work output till the speed is maintained to the normal working conditions. Simultaneously, with the downward movement of the servomotor piston, the oil from lower half of the servomotor gets drained out through the pilot valve port.

When the speed of the turbine decreases the oil under pressure enters the lower half of the servomotor, then servomotor piston gets lifted up causing lifting of the throttle valve. This causes increase in steam supply rate and thus the work output till the normal speed is attained by the turbine. Simultaneously, the oil from the upper half of the servomotor gets drained out through the pilot valve port.

2.12.2 Nozzle Control Governing

In the nozzle control governing, the steam supply from the main valve is divided into two, three, or more lines. Each line feeds a set of nozzles controlled by valves. The steam supply to one or more of the nozzle sets is cut-off by the valves operated by the governor when the load on the turbine decreases.

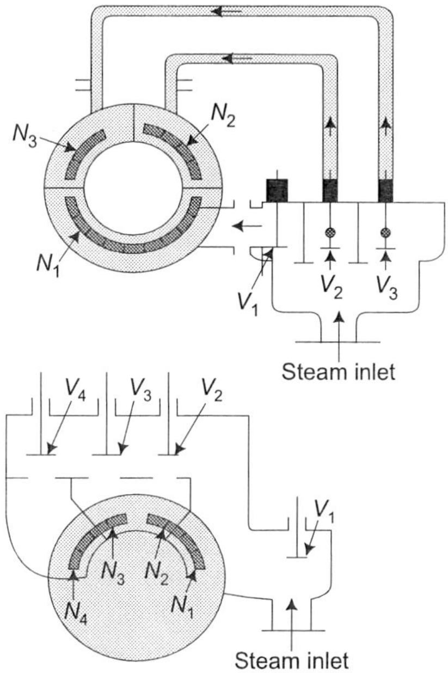

Fig. 2.20 Schematic View of Nozzle Control Governing

Figure 2.20 shows arrangements for nozzle control governing, where nozzles are divided into three sets N_1, N_2, and N_3 each controlled by valves V_1, V_2, and V_3 respectively and four sets N_1, N_2, N_3, and N_4 each controlled by valves V_1, V_2, V_3, and V_4 respectively. This method is suitable for medium and large steam turbines. It is generally employed at the first stage of turbine and is not practical for multistage impulse turbines.

2.12.3 Bypass Governing

In this method of governing, the extra quantity of steam from the first stage nozzle box of the turbine is bypassed into the latter stages as shown in Figure 2.21.

The main valve and the bypass valve are under the control of speed governor. The total amount of steam entering the turbine passes through the main valve. For all loads greater than the economic load, the bypass valve gets opened which allows the steam to pass from the first stage nozzle box into the steam belt and thus into the nozzles of the downstream stages. When the load reduces the bypass valve closes.

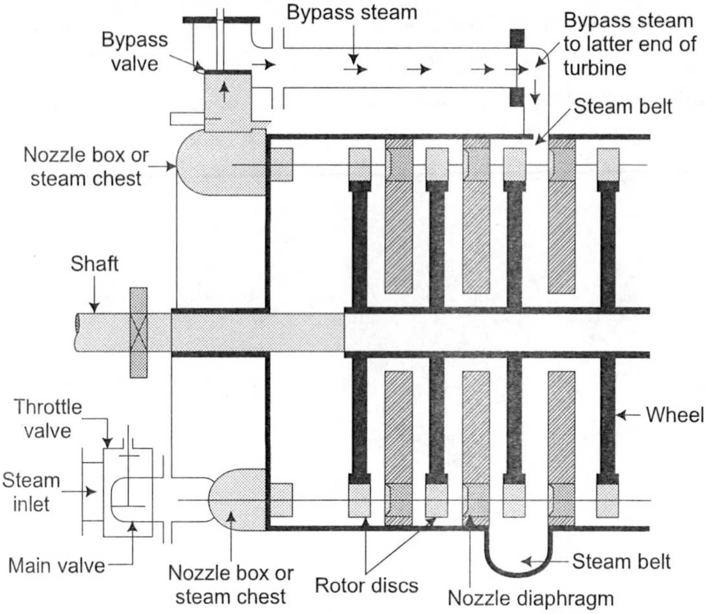

Fig. 2.21 Schematic View of Bypass Governing

Example 2.1 Steam flows from the nozzles of a single row impulse turbine with a velocity of 500 m/s in a direction which is inclined at an angle of 15° to the peripheral velocity. Steam comes out of the moving blades with an absolute velocity of 100 m/s and the direction at 120° with the direction of blade motion. The blades are equiangular and steam flow rate is 7.5 kg/s. Calculate (i) the power developed, (ii) power loss due to friction, and (iii) blade efficiency. Solve the problem analytically.

Solution Refer Figure 2.22.

$$CB = V_{f1} = V_1 \sin \alpha_1$$

$$= 500 \times \sin 15° = 129.41 \text{ m/s}$$

$$CD = V_1 \cos \alpha_1 - u$$

$$= 500 - \cos 15° - u = (482.96 - u) \text{ m/s}$$

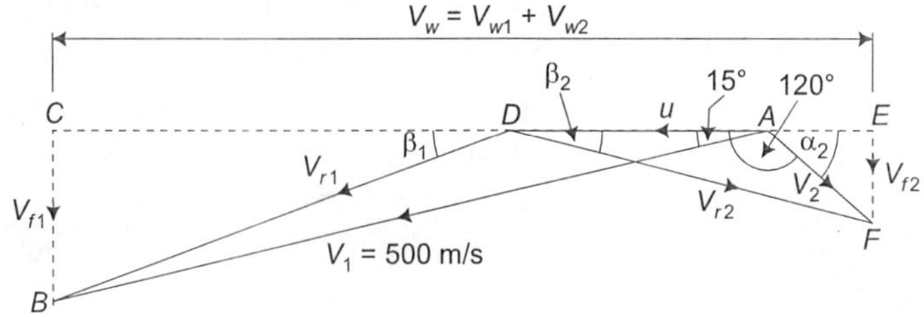

Fig. 2.22

$$\tan \beta_1 = \frac{CB}{CD} = \frac{129.41}{482.96 - u} \quad \text{(i)}$$

$$EF = V_{f2} = AF \sin \alpha_2$$

$$= 100 \times \sin 60° = 86.60 \text{ m/s}$$

$$AE = AF \cos \alpha_2 = 100 \times \cos 60° = 50 \text{ m/s}$$

$$DE = u + AE = u + 50$$

$$\tan \beta_2 = \frac{EF}{DE} = \frac{86.60}{u + 50} \quad \text{(ii)}$$

Since $\beta_1 = \beta_2$,

$\therefore$ $\tan \beta_1 = \tan \beta_2$

$$\frac{129.41}{482.96 - u} = \frac{86.60}{u + 50}$$

or $u = 163.67$ m/s

$\therefore$ $CD = 482.96 - 163.67 = 319.29$ m/s

and $DE = 163.67 + 50 = 213.67$ m/s

$$V_w = CD + DE = 532.96 \text{ m/s}$$

(i) $$\text{Blade power} = \frac{m \times u \times V_w}{1000}$$

$$= \frac{7.5 \times 163.67 \times 532.96}{1000} = 654.22 \text{ kW}$$

(ii) $$\text{Power loss due to friction} = \frac{m \times \left(V_{r1}^2 - V_{r2}^2\right)}{2 \times 1000} \text{ kW}$$

$$V_{r1} = \sqrt{(CB)^2 + (CD)^2}$$

$$= \sqrt{(129.41)^2 + (319.29)^2} = 344.52 \text{ m/s}$$

$$V_{r2} = \sqrt{(EF)^2 + (DE)^2}$$

$$= \sqrt{(86.60)^2 + (213.67)^2} = 230.55 \text{ m/s}$$

$\therefore$ Power loss due to friction

$$= \frac{7.5 \times (344.52^2 - 230.55^2)}{2 \times 1000} = 245.78 \text{ kW}$$

(iii) Blade efficiency, $$\eta_b = \frac{2 \times u \times V_w}{V_1^2}$$

$$= \frac{2 \times 163.67 \times 532.96}{500^2} = 0.6978 \text{ or } 69.78\%$$

Example 2.2 In a De-Laval turbine the steam enters the wheel through a nozzle with a velocity of 550 m/s and at an angle of 20° to the direction of motion of the blade. The blade speed is 250 m/s and the exit angle is 24°. By following the graphical method, find (i) the inlet angle of moving blade, (ii) exit velocity of steam and its direction, (iii) power developed per kg of steam, and (iv) the diagram efficiency.

Solution Following are the steps to draw the velocity diagram:

1. Choosing scale, 1 cm = 50 m/s

$\therefore$ $u = (200/50) = 4$ cm

and $V_1 = (500/50) = 10$ cm

2. Draw $AB = u = 4$ cm. Draw line AC at 20° ($\because \alpha_1 = 20°$), cut this line $V_1 = 10$ cm.
3. Join BC which represents V_{r1}. Measure BC and multiply it by scale to get V_{r1}.
4. Draw line at B at 24° ($\because \beta_2 = 24°$). With B as centre and radius BC swing an arc meeting line through B at D. Join BD which represents V_{r2}.
5. Join AD which represents V_2.
6. From points C and D draw perpendiculars meeting line AB produced at E and F respectively. CE represents V_{f1}, and DF represents V_{f2}. AE represents V_{w1}, and AF represents V_{w2}.

This completes the velocity diagram as shown in Figure 2.23.

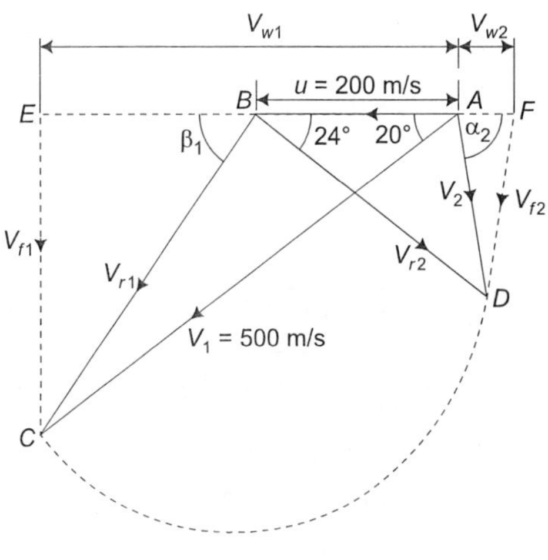

Fig. 2.23 (Not to scale)

(i) Inlet angle of the moving blade (β_1): By measurement from the velocity triangle,

we found, $\angle EBC = \beta_1 = 32°$

(ii) Exit velocity of steam (V_2) and its direction (α_2): By measurement from the velocity triangle,

we get, $AD = 3.2$ cm

$$\therefore \quad V_2 = 3.2 \times 50 = 160 \text{ m/s}$$

and $$\angle FAD = \alpha_2 = 55.5°$$

(iii) Power developed per kg of steam: By measurement from the velocity triangle, we get,

$$AE = 9.3 \text{ cm}$$

$$\therefore \quad V_{w1} = 9.3 \times 50 = 465 \text{ m/s}$$

and $$AF = 1.8 \text{ cm}$$

$$\therefore \quad V_{w2} = 1.8 \times 50 = 90 \text{ m/s}$$

Power developed per kg of steam

$$= \frac{m \times u \times (V_{w1} + V_{w2})}{1000}$$

$$= \frac{1 \times 200 \times (465 + 90)}{1000} = 111 \text{ kW}$$

(iv) Diagram efficiency: By measurement from the velocity triangle, we get,

$$V_w = (V_{w1} + V_{w2})$$

$$= (465 + 90) = 555 \text{ m/s}$$

Diagram efficiency, $$\eta_b = \frac{2 \times u \times V_w}{V_1^2}$$

$$= \frac{2 \times 200 \times 555}{500^2} = 0.888 \text{ or } 88.8\%$$

Example 2.3 In an impulse turbine the steam enters with a velocity of 450 m/s at an angle 20° to the plane of rotation of the blades. The blade mean speed is 200 m/s. The exit angle from the blades is 25°. There is 15% loss in relative velocity due to friction in the blades. Determine (i) the inlet angle of the blades, (ii) exit velocity of steam and its direction, (iii) the power developed per kg of steam, (iii) the diagram efficiency, and (iv) the axial thrust. Solve the problem graphically.

Solution Following are the steps to draw the velocity diagram:

1. Choosing scale

$$1 \text{ cm} = 50 \text{ m/s}$$

$$\therefore \quad u = (200/50) = 4 \text{ cm}$$

and $$V_1 = (450/50) = 9 \text{ cm}$$

2. Draw $AB = u = 4$ cm. Draw line AC at 20° ($\because \alpha_1 = 20°$), cut this line $V_1 = 9$ cm.
3. Join BC which represents V_{r1}. Measure BC and multiply it by scale to get V_{r1}. Cut off length CG which is 15% of BC.

4. Draw line at B at 25° ($\because \beta_2 = 25°$). With B as centre and radius BG swing an arc meeting line through B at D. Join BD which represents V_{r2}.
5. Join AD which represents V_2.
6. From points C and D draw perpendiculars meeting line AB produced at E and F respectively. CE represents V_{f1}, and DF represents V_{f2}. AE represents V_{w1}, and AF represents V_{w2}.

This completes the velocity diagram as shown in Figure 2.24.

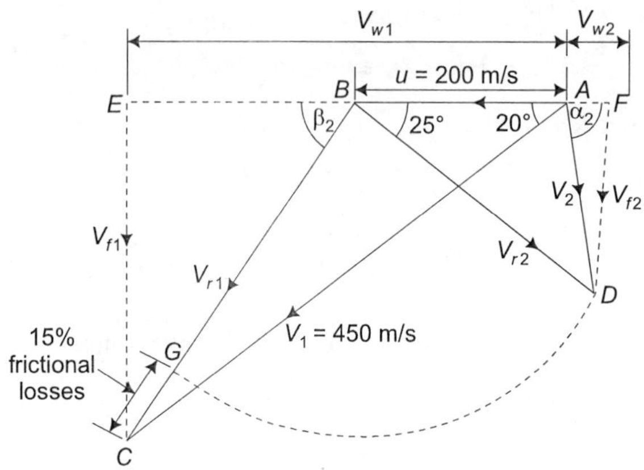

Fig. 2.24 (Not to scale)

(i) Inlet angle of the moving blade (β_1): By measurement from the velocity triangle,

we found $\quad \angle EBC = \beta_1 = 35°$

(ii) Exit velocity of steam (V_2) and its direction (α_2): By measurement from the velocity triangle,

we get, $\quad AD = 1.9$ cm

$\therefore \quad V_2 = 1.9 \times 50 = 95$ m/s

and $\quad \angle FAD = \alpha_2 = 86°$

(iii) Power developed per kg of steam: By measurement from the velocity triangle,

$AE = 8.4$ cm

$\therefore \quad V_{w1} = 8.4 \times 50 = 420$ m/s

and $\quad AF = 0.5$ cm,

$\therefore \quad V_{w2} = 0.5 \times 50 = 25$ m/s

Power developed per kg of steam

$$= \frac{m \times u \times (V_{w1} + V_{w2})}{1000}$$

$$= \frac{1 \times 200 \times (420 + 25)}{1000} = 89 \text{ kW}$$

(iv) Diagram efficiency: By measurement from the velocity triangle,

$$V_w = (V_{w1} + V_{w2})$$

$$= (420 + 25) = 445 \text{ m/s}$$

Diagram efficiency, $$\eta_b = \frac{2 \times u \times V_w}{V_1^2}$$

$$= \frac{2 \times 200 \times 445}{450^2} = 0.8790 \text{ or } 87.90\%$$

(v) Axial thrust per kg of steam (F_a): By measurement from the velocity triangle, we get,

$$EC = 3.1 \text{ cm}$$

$\therefore$ $$V_{f1} = 3.1 \times 50 = 155 \text{ m/s}$$

and $$FD = 2 \text{ cm}$$

$\therefore$ $$V_{f2} = 2 \times 50 = 100 \text{ m/s}$$

$$F_a = m \times (V_{f1} - V_{f2})$$

$$= 1 \times (155 - 100) = 50 \text{ V}$$

Example 2.4 In a stage of an impulse turbine provided with a single row wheel, the mean diameter of blade ring is 0.80 m and the speed of rotation is 3000 rpm. The steam issues from the nozzle with a velocity of 300 m/s and the nozzle angle is 20°. The blades are equiangular and blade coefficient is 0.85. Determine the power developed in the blades when the axial thrust on the blade is 150 N. Also determine the diagram efficiency.

Solution Refer Figure 2.25.

$$u = \pi \times d \times (N/60)$$

$$= \pi \times 0.80 \times (3000/60) = 125.66 \text{ m/s}$$

$$V_{w1} = V_1 \cos \alpha_1$$

$$= 300 \times \cos 20° = 281.91 \text{ m/s}$$

$$BE = V_{w1} - u = 281.91 - 152.66$$

$$= 156.25 \text{ m/s}$$

$$CE = V_{f1} = \sqrt{V_1^2 - V_{w1}^2}$$

$$= \sqrt{(300)^2 - (281.91)^2} = 102.59 \text{ m/s}$$

$$\tan \beta_1 = \frac{CE}{BE} = \frac{102.59}{156.52} = 0.6554$$

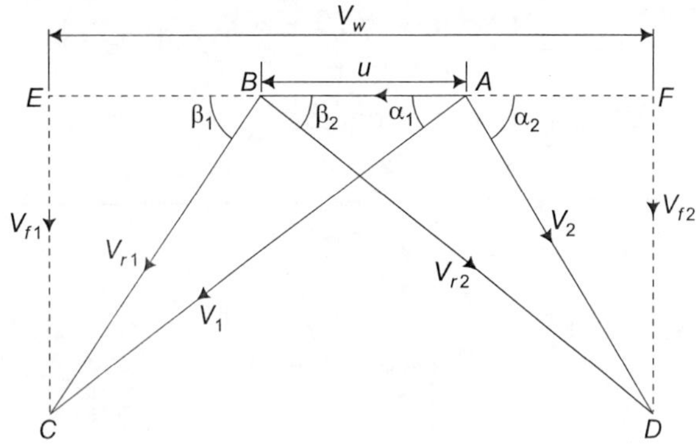

Fig. 2.25

$$\beta_1 = 33.28°$$

$\therefore$ $$\beta_1 = \beta_2 = 33.28°$$

$$V_{r1} = \sqrt{V_{f1}^2 + BE^2}$$

$$= \sqrt{(102.59)^2 + (156.25)^2} = 186.92 \text{ m/s}$$

But $$K = 0.85 = \frac{V_{r2}}{V_{r1}},$$

$\therefore$ $$V_{r2} = 0.85 \times 186.92 = 158.88 \text{ m/s}$$

$\therefore$ $$V_{w2} = V_{r2} \cos \beta_2 - u$$

$$= 158.88 \times \cos (33.28) - 125.66 = 7.16 \text{ m/s}$$

$$V_{f2} = V_{r2} \times \sin \beta_2$$

$$= 158.88 \times \sin (33.28°) = 87.18 \text{ m/s}$$

$$F_a = m \times (V_{f1} - V_{f2})$$

$\therefore$ $$m = \frac{F_a}{(V_{f1} - V_{f2})}$$

$$= \frac{150}{(102.59 - 87.18)} = 9.734 \text{ kg/s}$$

$$\text{Blade power} = \frac{m \times u \times (V_{w1} + V_{w2})}{1000}$$

$$= \frac{9.734 \times 125.66 \times (281.91 + 7.16)}{1000}$$

$$= 353.583 \text{ kW}$$

$$\eta_b = \frac{2 \times u \times V_w}{V_1^2}$$

$$= \frac{2 \times 125.66 \times (281.91 + 7.29)}{300^2}$$

$$= 0.8076 \text{ or } 80.76\%$$

Example 2.5 Steam at 350 m/s is supplied to the single stage impulse turbine through a nozzle. The nozzle angle is 25°. The mean diameter of the blade rotor is 1.0 m and it has a speed of 2000 rpm. Find the blade angles if there is no axial thrust. If the steam flow rate is 5 kg/s and the blade coefficient is 0.85 then determine the power developed. Also determine the blade efficiency and energy lost in the blade friction.

Solution Refer Figure 2.25.

$$u = \pi \times d \times (N/60)$$

$$= \pi \times 1.0 \times (2000/60) = 104.72 \text{ m/s}$$

$$V_{w1} = V_1 \cos \alpha_1 = 350 \times \cos 25°$$

$$= 317.21 \text{ m/s}$$

$$BE = V_{w1} - u = 317.21 - 104.72$$

$$= 212.49 \text{ m/s}$$

$$CE = V_{f1} = V_{f2} = \sqrt{V_1^2 - V_{w1}^2}$$

$$= \sqrt{(350)^2 - (317.21)^2} = 147.91 \text{ m/s}$$

$$V_{r1} = \sqrt{V_{f1}^2 + BE^2}$$

$$= \sqrt{(147.91)^2 + (212.49)^2} = 258.90 \text{ m/s}$$

$$\tan \beta_1 = \frac{CE}{BE} = \frac{147.91}{212.49}$$

$$= 0.6961$$

$$\therefore \quad \beta_1 = 34.84°$$

$$K = 0.85 = \frac{V_{r2}}{V_{r1}}$$

$$\therefore \quad V_{r2} = 0.85 \times 258.90 = 220.06 \text{ m/s}$$

$$\sin \beta_2 = \frac{V_{f2}}{V_{r2}} = \frac{157.91}{220.06} = 0.6721$$

$$\therefore \quad \beta_2 = 42.23°$$

$$\therefore \quad V_{w2} = V_{r2} \cos \beta_2 - u$$

$$= 220.06 \times \cos(42.23) - 104.72 = 58.22 \text{ m/s}$$

$$\text{Power developed} = \frac{m \times u \times (V_{w1} + V_{w2})}{1000}$$

$$= \frac{5 \times 104.72 \times (317.21 + 58.22)}{1000} = 196.58 \text{ kW}$$

Blade efficiency,
$$\eta_b = \frac{2 \times u \times V_w}{V_1^2}$$

$$= \frac{2 \times 104.72 \times (317.21 + 58.22)}{350^2}$$

$$= 0.6419 \text{ or } 64.19\%$$

Power loss due to friction

$$= \frac{m \times \left(V_{r1}^2 - V_{r2}^2\right)}{2 \times 1000}$$

$$= \frac{5 \times (258.90^2 - 220.06^2)}{2 \times 1000}$$

$$= 46.51 \text{ kW}$$

Example 2.6 In a stage of Parson's reaction turbine the mean diameter of the blade ring is 80 cm and its speed is 1500 rpm. The steam velocity at the inlet of moving blades is 165 m/s and the outlet blade angle is 20°. The mass flow rate of steam is 6.5 kg/s and the stage efficiency is 85%. Determine (i) blade inlet angle, (ii) power developed, and (iii) available isentropic enthalpy drop.

Solution Refer Figure 2.26.

$$u = \pi \times d \times (N/60)$$

$$= \pi \times 0.80 \times (1500/60) = 62.83 \text{ m/s}$$

In Parson's reaction turbine,

$$V_{f1} = V_{f2} = V_f,$$

$$\beta_1 = \alpha_2 \quad \text{and} \quad \beta_2 = \alpha_2$$

$$V_{w1} = V_1 \cos \alpha_1$$

$$= 165 \times \cos 20° = 155.05 \text{ m/s}$$

$$V_{f1} = V_{f2} = V_f = V_1 \sin \alpha_1$$

$$= 165 \times \sin 20° = 56.43 \text{ m/s}$$

$$BE = V_{w1} - u$$

$$= 155.05 - 62.83 = 92.22 \text{ m/s}$$

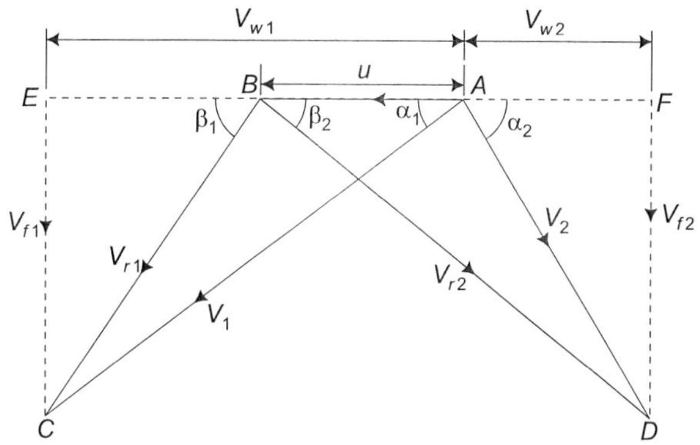

Fig. 2.26

$$\tan \beta_1 = \frac{CE}{BE} = \frac{56.43}{92.22} = 0.6119$$

$$\beta_1 = \alpha_2 = 31.46°$$

$$\tan \alpha_2 = \frac{V_{f2}}{V_{w2}}$$

or

$$V_{w2} = \frac{V_{f2}}{\tan \alpha_2} = \frac{56.43}{\tan (31.46)} = 92.23 \text{ m/s}$$

$$P = \frac{m \times u \times (V_{w1} + V_{w2})}{1000}$$

$$= \frac{6.5 \times 62.83 \times (155.05 + 92.23)}{1000}$$

$$= 100.99 \text{ kW}$$

$$\eta_s = \frac{\text{Work done by turbine}}{\text{Isentropic enthalpic drop}}$$

$$= \frac{P}{\Delta h_{isn}}$$

$\therefore$

$$\Delta h_{isn} = \frac{P}{\eta_s} = \frac{100.99}{0.85}$$

$$= 118.81 \text{ kW}$$

Example 2.7 In a stage of impulse turbine the blade angles are equal and the nozzle angle is 20°. The velocity coefficient of the blade is 0.85. Find the maximum blade efficiency. If the actual blade efficiency is 90 per cent of maximum blade efficiency, determine the blade speed ratio.

Solution $\beta_1 = \beta_2$, $\alpha_1 = 20°$, and $K = 0.85$

$$(\eta_b)_{max} = [1 + KC]\,\frac{\cos^2\alpha_1}{2}$$

$$\therefore \qquad (\eta_b)_{max} = [1 + 0.85 \times 1]\,\frac{\cos^2(20)}{2}$$

$$= 0.8168 \text{ or } 81.68\%$$

$$\text{Actual efficiency} = 0.90 \times 0.8168 = 0.7351 \text{ or } 73.51\%$$

Blade efficiency of a single row impulse turbine in terms of blade speed ratio is given by,

$$\eta_b = 2\,(\rho \cos \alpha_1 - \rho^2)\,[1 + KC]$$

$$0.7351 = 2\,(\rho \cos 20° - \rho^2)\,[1 + 0.85]$$

$$= 2\,(\rho \times 0.9397 - \rho^2) \times 1.85$$

or $\qquad \rho \times 0.9397 - \rho^2 = 0.7315/(2 \times 1.83)$

or $\quad \rho^2 - 0.9397\,\rho + 0.1986 = 0$

$$\therefore \qquad \rho = 0.9397 \pm \sqrt{\frac{(0.9397)^2 - 4 \times 1 \times 0.1986}{2 \times 1}}$$

$$= 1.15 \text{ or } 0.729$$

Example 2.8 In a stage of Pàrson's steam turbine the following data are given: mean diameter of moving blade ring = 0.70 m, speed = 3000 rpm, steam velocity at exit and steam velocity at the inlet = 160 m/s, blade outlet angle = 20°, steam mass flow rate = 7 kg/s. Determine (i) blade inlet angle, (ii) tangential force on moving blades ring, and (iii) power developed in stage. Solve the problem graphically.

Solution

$$u = \pi \times d \times (N/60)$$

$$= \pi \times 0.70 \times (3000/60) = 109.95 \text{ m/s}$$

Choosing scale $\qquad$ 1 cm = 40 m/s

$$\therefore \qquad u = (109.95/40) = 2.74 \text{ cm}$$

$$V_1 = V_{r2} = (160/40) = 4 \text{ cm}$$

Steps to Draw the Velocity Diagram

1. Draw line $AB = u = 2.74$ cm.
2. Draw line AC at 20° and BD at 20° because $\beta_2 = \alpha_1$. Cut these lines for $V_1 = V_{r2} = 4$ cm at points C and D.
3. Construct the inlet and outlet velocity triangles as shown in Figure 2.27.

   Measure $V_w = 4.7 \text{ cm} = 4.7 \times 40 = 188$ m/s

   Measure blade inlet angle, $\beta_1 = 52°$

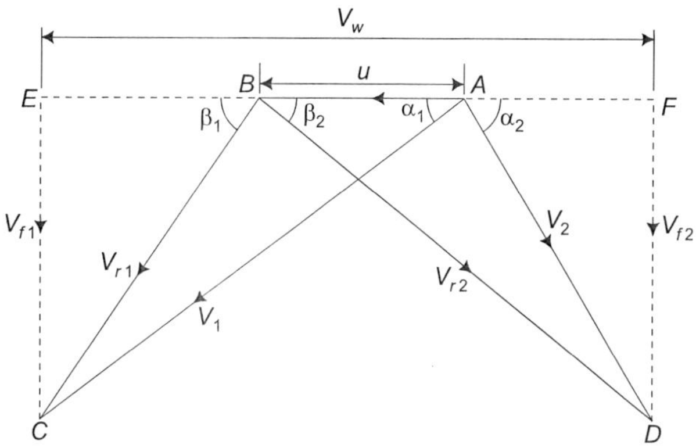

Fig. 2.27 (Not to scale)

Tangential force, $F_t = m \times V_w$

$= 7 \times 188 = 1316$ N

$$\text{Power developed} = \frac{m \times V_w \times u}{1000}$$

$$= \frac{7 \times 188 \times 109.95}{1000} \text{ kW} = 144.69 \text{ kW}$$

2.13 STEAM CONDENSERS

The efficiency of a steam power plant working on Carnot cycle is given by $1 - (T_2/T_1)$, where T_2 is the temperature at which heat is rejected, and T_1 is the temperature at which heat is supplied. From this expression it is clear that efficiency increases with the increase in temperature T_1 and decrease in temperature T_2. The maximum value of temperature T_1 is limited by metallurgical considerations. The value of temperature T_2 can be reduced below the atmospheric temperature if the exhaust of the steam occurs below the atmospheric pressure. The steam cannot be exhausted to the atmosphere if it expands in the turbine (or engine) to a back pressure which is lower than the atmospheric pressure. This is made possible by exhausting the steam into a closed vessel called condenser in which the pressure is maintained below the atmosphere by continuously condensing the steam by means of circulating cold water at atmospheric temperature. Condensation of steam enables expansion of steam to a lower back pressure and the utilization of large enthalpy drop (heat drop) by which both the efficiency and power output of the plant increases.

Therefore, the condenser is a closed vessel heat transfer device in which the exhaust steam of a turbine (or engine) is condensed by means of cooling water at a pressure below than the atmospheric pressure. The condensed steam is called condensate which can be reused in the boiler as feed water.

The hatched area in Figure 2.28(a) and 2.28(b) represents the increase in work output in a steam engine and steam turbine respectively, by exhausting the steam into a condenser.

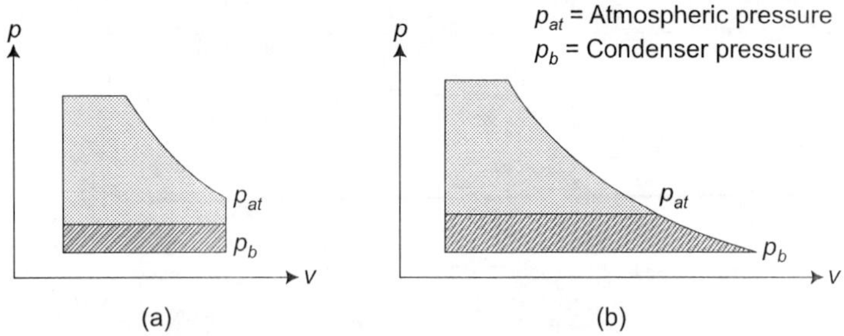

Fig. 2.28 Increase in Work Done on *p-V* Diagram

Advantages of a Steam Condenser

1. The work output and efficiency of the steam power plant increases considerably due to low pressure (vacuum pressure) of condenser.
2. The recovery of condensate which is good quality feed water for reuse in the boiler reduces treated water requirement and thus reduces the cost of power generation. Hot feed water is sent to the boiler thus thermal stresses get reduced.
3. Due to increased work output for the same heat input the steam consumption gets reduced.

2.14 ELEMENTS OF A STEAM CONDENSING PLANT

The essential components of a steam condensing plant in a steam power plant are shown in Figure 2.29.

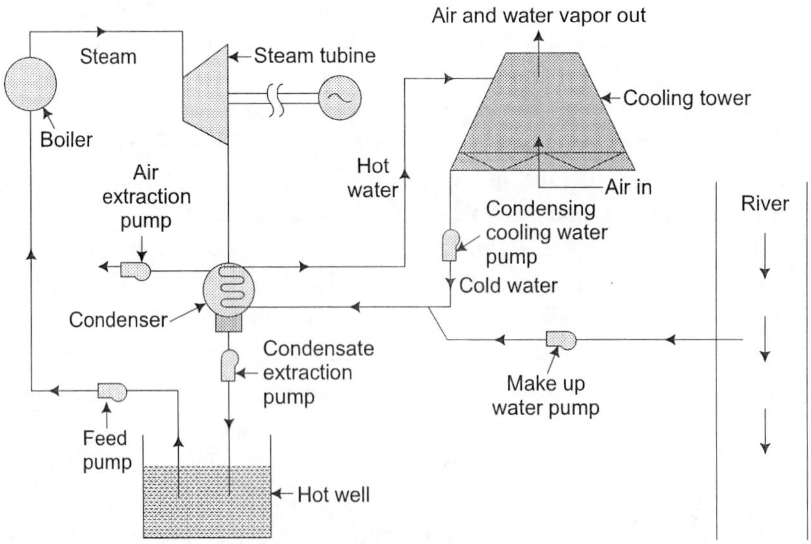

Fig. 2.29 Steam Power Plant Schematics with Condensing Unit Elements

1. **Condenser** It is a closed vessel heat exchanger in which exhaust steam from the turbine is condensed. During the condensation process the steam gives up heat to circulating cooling water.
2. **Condensate Extraction Pump** It is a pump required to extract the condensate from the condenser to the hot well.
3. **Air Extraction Pump** It is a pump to remove air and non-condensable gases. Sometimes a single pump known as wet air pump is used to remove both air and the condensate.
4. **Hot Well** It is a sump provided between the condenser and the boiler. In a hot well the condensate pumped by the condensate extraction pump is collected.
5. **Boiler Feed Pump** It is a pump used to force the condensate from hot well into the boiler. This is done by increasing the pressure of the condensate above the boiler pressure.
6. **Condenser Cooling Water Pump** It is a pump used to maintain the circulation of cooling water through the condenser.
7. **Make Up Water Pump** It is a pump used to compensate the water lost due to evaporation during cooling of hot water in the cooling tower. It circulates the required amount of cooling water into the condenser from the river or some other water source.
8. **Cooling Tower or Spray Pond** A cooling tower is used to re-cool the hot water coming out of the condenser. The hot water is cooled by evaporative cooling in which heat is rejected to the atmospheric air. The use of cooling tower becomes essential when there is shortage of cooling water, and it has to be used over and over again in the condenser.
9. **Relief Valve** It is used to relieve the steam from the condenser when the condenser does not work properly by which the plant becomes non-condensing.

2.15 TYPES OF STEAM CONDENSERS

The steam condensers are broadly classified into two types, namely, jet condenser (or mixing type condenser), and surface condenser (or non-mixing type condenser).

1. Jet Condensers or Mixing Type Condensers In jet condensers there is direct contact between the exhaust steam and cooling water. The temperature of the condensate is same as that of the cooling water leaving the condenser. Heat exchange occurs by direct conduction between the steam and water. If the cooling water is not pure and free from harmful impurities then the condensate cannot be reused as feed water to the boilers. Due to loss of condensate and high power requirement by the pump these condensers are rarely used in modern steam power plants.

Jet condensers are of the following types:

(i) Low-level jet condensers: These are of the following two types
 (a) Parallel flow type: The steam and cooling water flow in the same direction.
 (b) Counter flow type: The steam and cooling water flow in the opposite directions.

(ii) High-level jet condensers, and

(iii) Ejector jet condensers.

2. Surface Condensers or Non-mixing Type Condensers In surface condensers, there is no direct contact between the exhaust steam and the cooling water. The steam surrounds the tubes fitted in the condenser shell and the cooling water circulates through these tubes. The steam gets condensed due to the heat transfer to cooling water by conduction and convection. The condensate collected from these condensers is reused as feed water in the boiler. Thus, these condensers are most suitable for modern steam power plants and chemical industries. These are generally used where a large quantity of inferior water is available and better quality of feed water is to be supplied to the boiler. So, these condensers are universally used in marine engines where seawater is used for cooling purposes. The only drawback of these condensers is its high initial cost but is recovered by the saving in running cost.

The surface condensers may be classified according to:

(i) Direction of condensate flow and tube arrangement:

- (a) Down flow type,
- (b) Central flow type, and
- (c) Inverted flow type.

(ii) The number of water passes: single pass, double pass, or multi pass, and

(iii) Evaporative type.

2.15.1 Low-level Jet Condensers

Low-level jet condenser is placed at low levels such that vacuum inside the condenser draws the cooling water into it from the cooling water source. Parallel flow and counterflow low-level jet condensers are shown in Figure 2.30.

In counterflow jet condenser the water and steam flows in opposite directions but in parallel flow both flow in the same direction. In both the cases the cooling water enters at the top of the condenser and passes through the perforated trays so that it breaks into sprays and increases the heat transfer rate by providing more contact surface area.

When the steam comes into contact with cooling water it gets condensed. The extraction of air is done from the top of the condensers. The vacuum created in the condenser is sufficient to draw the cold water from the cooling pond. The pressure causing the water flow from the cooling pond (or cold tank) to the condenser top is given by $(p_a - p_c)$, where p_a is the atmospheric pressure and p_c is the condenser pressure. The condensate is extracted by the extraction pump and is discharged to the hot well. The excess amount of condensate from hot well flows into the cooling pond by an overflow pipe and the remaining water is pumped to the boiler as feed water. These condensers have the disadvantage of flooding the turbine if water extraction pump fails due to any reason.

2.15.2 High-level Jet Condensers

A high-level jet condenser is also called barometric condenser because the condenser shell is placed above the hot well by more than barometric height of water column of 10.363 m. A long tailpipe, more than 10.363 m in length is attached between the bottom of the condenser and the

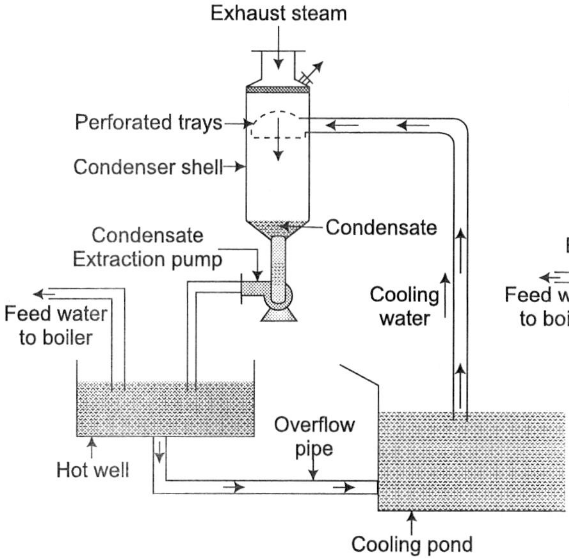

(a) Parallel flow type low-level jet condenser

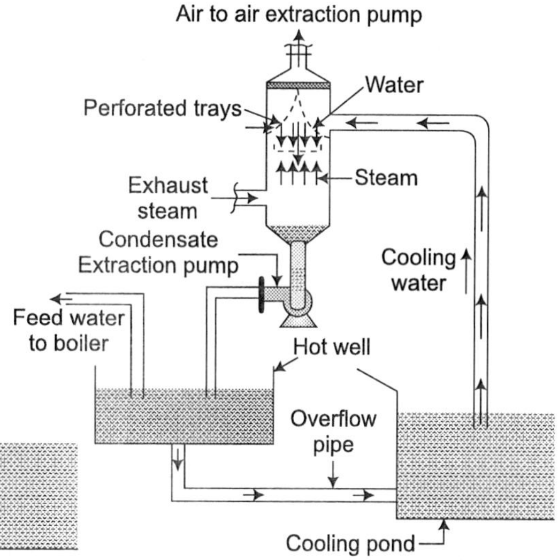

(b) Counter flow type low-level jet condenser

Fig. 2.30 Low-level Jet Condensers

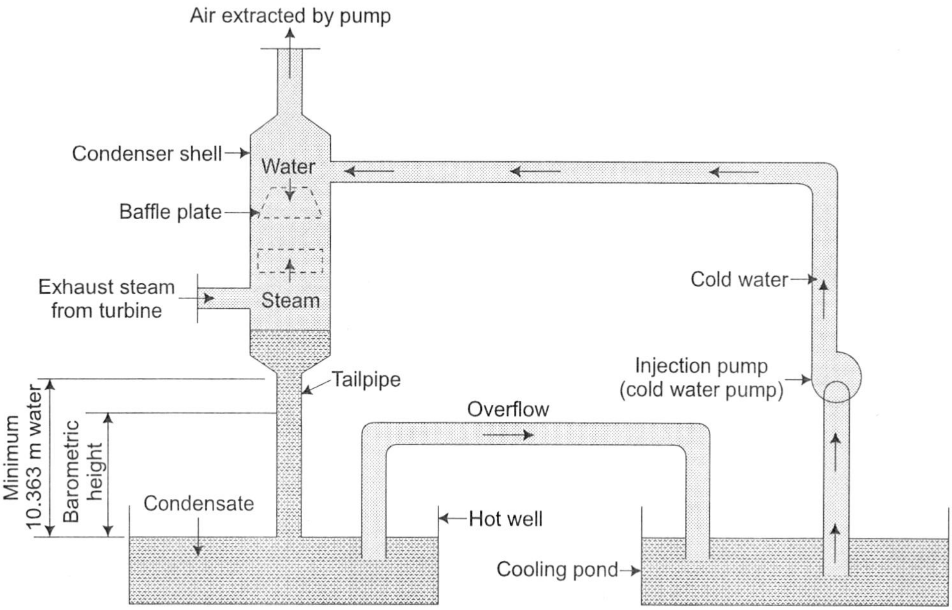

Fig. 2.31 High-level Jet Condenser

hot well. The pressure at the bottom of the pipe is equal to atmospheric pressure, whereas at its top in the condenser shell vacuum pressure is maintained. It avoids the rise of water in the tail pipe and water extraction pump is also not required. The condensate and water from the condenser go down to the hot well under the gravity and maintain a water leg in the tail pipe depending upon the vacuum in the condenser. As the height of the shell is large so, an injection pump is required to pump water to the top of the shell. A schematic view of high jet condenser is shown in Figure 2.31.

The working and other details of this condenser are similar to low-level counter flow jet condenser. Its drawbacks are high costs and loss of vacuum between the turbine and the condenser. It is used where sufficient head required for tailpipe is available.

2.15.3 Ejector Condenser

A schematic view of an ejector condenser is shown in Figure 2.32. It is suitable for moderate vacuum only.

The cooling water under the head of about 5 m to 6 m enters the condenser from its top. It is discharged through a series of convergent nozzles in which the potential energy get converted into the kinetic energy and thus partial vacuum is created. The exhaust steam enters the condenser through a non-return valve. It condenses through the mixing with cooling water and thus vacuum further increases. The condensate after passing through the series of convergent nozzles passes through the divergent nozzle. When it passes through the divergent nozzle the kinetic energy of condensate reconverts into potential energy and thus a higher pressure than the atmospheric pressure is obtained which forces out the condensate to the hot well. Thus, in this condenser no air extraction pump is required. A non-return valve is fitted at the exhaust steam pipe to the condenser to prevent any rush of water from the hot well to the turbine in the case of cooling water failure.

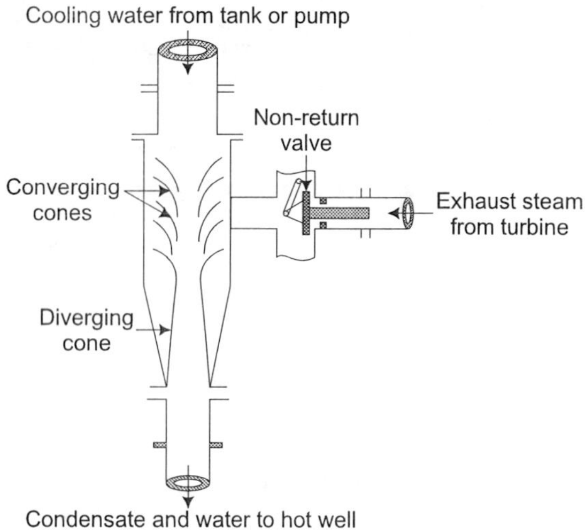

Fig. 2.32 Ejector Condenser

2.15.4 Surface Condenser

A schematic view of the surface condenser is shown in Figure 2.33.

It consists of an airtight cylindrical shell closed at each end. A number of parallel brass water tubes are fitted in the tube plates which are fixed between each cover head and the shell. The space between the tube plates and the cover head is known as water box. A baffle plate partitions the water box into two sections. The cooling water enters the shell at the lower half section and flows in one direction. Then it returns in the opposite direction through the upper half section of

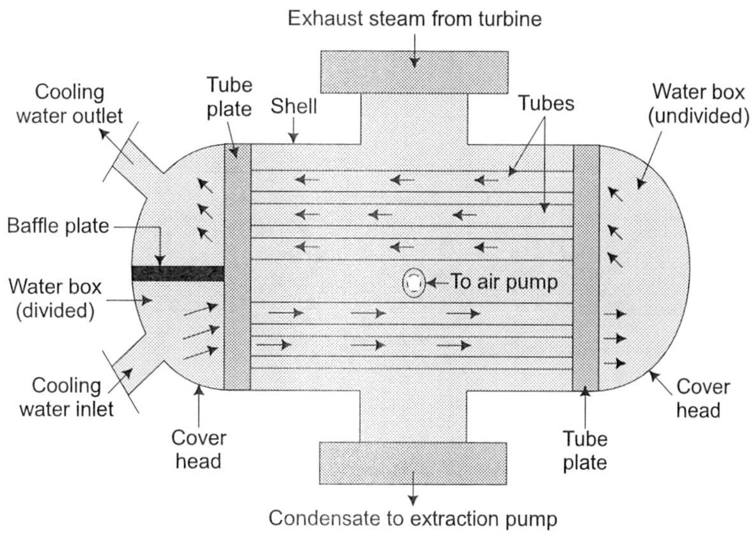

Fig. 2.33 Surface Condenser

the condenser and finally, it leaves through the outlet. In this type of condenser the water traverses two times, so it is called the two-pass condenser. The exhaust steam from the turbine enters at the top of the condenser and surrounds the tubes through which cooling water flows under force. The steam condenses when it comes in contact with the cold surface of the tubes. The water gets warmed in the condenser is discharged into the cooling tower, river, or cooling pond. The condensate is taken out from the condenser by a condensate extraction pump. The air is removed by an air extraction pump.

Down Flow Surface Condenser Figure 2.34 (a) shows a sectional view of a down flow surface condenser which is also known as dry vacuum type condenser. The steam enters at the top and flow downwards over the tubes carrying cooling water. A section of tubes near the air pump suction is screen off by providing a baffle plate. This screen is also called air cooler which helps in extracting air at a lower temperature than the condensate. The low temperature of air reduces its volume and hence the size of the pump gets reduced by as much as 50 per cent.

Central Flow Surface Condenser Figure 2.34 (b) shows a sectional view of a central flow surface condenser in which the air extraction pump is provided at the centre and the air is extracted from the centre. The exhaust steam and air enter from the top and flow radially towards the centre by passing over the entire periphery of the tubes. The condensate is extracted at the bottom by condensate extraction pump. This condenser is an improvement over the down flow type as steam has an access to the whole periphery of the cooling tubes.

Inverted Flow Surface Condenser In the inverted flow type surface condenser the air suction pump is provided at the top. So, the steam entering from the bottom of the condenser flows upwards. The condensed steam collects in the bottom section from where it is extracted with the help of a condensate extraction pump.

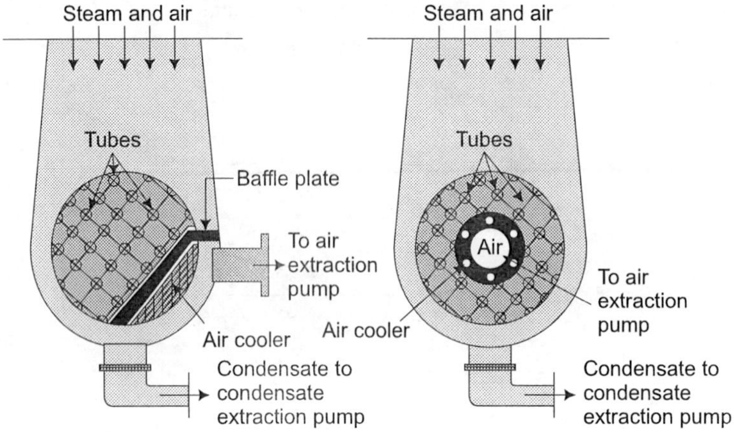

Fig. 2.34 (a) Down Flow Surface Condenser (b) Central Flow Surface Condenser

2.15.5 Evaporative Condenser

A schematic arrangement of the evaporative condenser is shown in Figure 2.35.

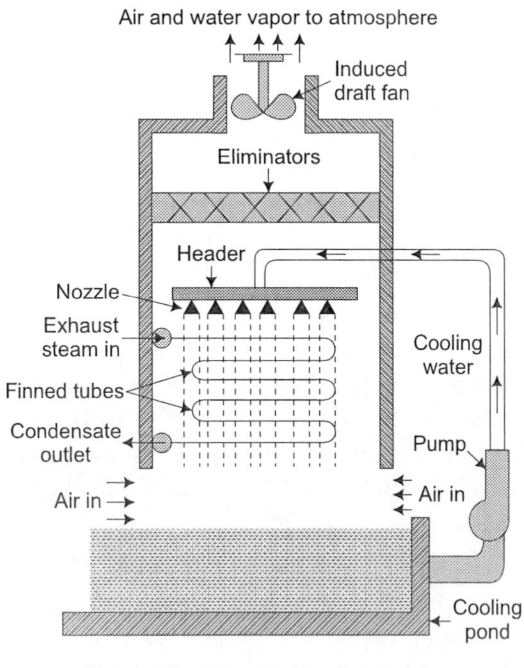

Fig. 2.35 Evaporative Condenser

These are used where there is scarcity of water. The exhaust steam from the turbine enters a coiled finned pipe. The water from the cooling pond is pumped by means of a pump to a horizontal header which is provided with spray nozzles. The sprayed water forms thin film over the pipe surface and gets evaporated in passing over the pipe under a small partial pressure and thus cools the steam inside the pipes. The air is drawn over the surface of finned pipe with the help of induced draft fan to increase the evaporation of cooling water which further increases the condensation of steam in pipes. Eliminators are provided to prevent the exit of water vapors with the leaving heated air. The cooling water gets collected into the cooling pond. In the cooling pond the water lost due to evaporation is replenished by the addition of required amount of cold make up water.

The arrangement of this type of condenser is simple, cheap, and requires small quantity of cooling water thus capacity of water circulating pump is small. The vacuum maintained in this type of condenser is not as high as in the case of surface condensers. These condensers are used in small power plants and are extensively used in refrigeration plant units.

2.16 REQUIREMENTS OF A GOOD SURFACE CONDENSER

The following are the requirements of a good surface condenser:

1. The steam should enter the condenser with least possible resistance for its easy flow.
2. For effective condensation the steam should be well distributed in the vessel and there should be minimum pressure drop.
3. The circulating cooling water should flow through the tubes with least friction. The rise in temperature of cooling water should be limited to 10°C for obtaining better thermal efficiency.
4. There should be no undercooling of the condensate, so the steam should lose only its latent heat to the circulating water. This is made possible by regulating the quantity of circulating water in such a way that its exit temperature is same as the saturation temperature of steam.
5. To obtain maximum heat transfer rate the tubes should be made of high thermal conductivity material. The water should flow through tubes and steam outside so that the outer surface of the tubes does not get deposited with sediments. If the cooling water is dirty then sediments will get deposited inside the tubes. These sediments can be cleaned by motor driven brushes after removing the end cover plates.
6. There should be no leakage of air from the condenser.
7. Minimum energy should be spent to extract the air from the condenser. This is achieved by fitting a baffle plate at the coolest section where air pump is fitted. This arrangement reduces the specific volume of air and thus reduces the size of the pump.

2.17 ADVANTAGES AND DISADVANTAGES OF CONDENSERS

2.17.1 Jet Condensers

The following are the advantages and disadvantages of jet condensers.

Advantages

1. Due to more intimate mixing of steam and cooling water it requires less quantity of circulating water for the condensation of steam.
2. Due to direct mixing it requires less building space.
3. The arrangement of jet condenser is simple in construction and low in cost. Its maintenance cost is also low.
4. Low-level jet condenser does not require cooling water pump. In barometric and ejector condensers there is no need of condensate extraction pump.

Disadvantages

1. There is wastage of condensate.
2. If the condensate is to be used as feed water then the cooling water should be pure and free from any harmful impurities.

3. In the barometric condenser use of long pipe increase the cost of the condenser.
4. In the low-level jet condenser if the condensate extraction pump fails then there is greater possibility of flooding of the engine.
5. In the case of barometric condenser a vacuum loss of about 1 to 1.5 cm of Hg occurs due to leakage in the long exhaust pipe line.
6. Vacuum more than 66 cm of Hg cannot be achieved. This is because the dissolved air in the cooling water gets liberated at low pressures.
7. The air extraction pump needs high power which may be about double the power required by a surface condenser.

2.17.2 Surface Condensers

The following are the advantages and disadvantages of surface condensers.

Advantages

1. A high vacuum can be achieved (as much as 73.5 cm of Hg) and thus gives greater plant efficiency.
2. Since the cooling water and steam do not mix, the condensate is recovered and can be used as feed water to the boiler. Due to this advantage, these condensers are used in all steam power plants.
3. Since the cooling water and steam do not mix, any kind of cooling water can be used. This results in considerable reduction in the cost of water softening plant.
4. The chances of vacuum loss are minimized.
5. It requires much less power to run the air extraction pump and for water pumping.
6. It ideally suits high capacity plants.
7. It requires less quantity of make up water (about 4% to 5%).

Disadvantages

1. The system is bulky and requires large floor area.
2. It requires high capital cost and maintenance cost.
3. It requires more cooling water.

2.18 COMPARISON OF JET AND SURFACE CONDENSERS

Jet Condenser	Surface Condenser
1. Cooling water and steam mix together.	1. Cooling water and steam do not mix together.
2. Condensate contains impurity, so it cannot be reused as feed water.	2. Condensate is pure and free from any harmful impurities, so it can be reused as feed water.
3. It requires less quantity of cooling water for condensation of steam due to direct mixing of water with steam.	3. It requires more quantity of cooling water for condensation of steam since there is no mixing.

4. The cooling water should be free from impurities.	4. The cooling water of any quality can be used.
5. Vacuum efficiency is low. Thus, not suitable for high capacity plants.	5. Vacuum efficiency is high. Thus, more suitable for high capacity plants.
6. The system design is simple, compact, and economical.	6. The system design is complicated, bulky, and costly.
7. It requires small floor area.	7. It requires large floor area.
8. Maintenance is simple and less costly.	8. Maintenance is costly and requires skilled worker.
9. It requires more power for air pump.	9. It requires less power for air pump.
10. It requires more power for water pumping.	10. It requires less power for water pumping.
11. It does not require condensate extraction pump.	11. It requires condensate extraction pump.
12. Vacuum more than 66 cm of Hg cannot be achieved.	12. A high vacuum as much as 73.5 cm of Hg can be achieved.

2.19 CONDENSER VACUUM AND ITS MEASUREMENT

The pressure inside the condenser is below the atmospheric pressure and is called vacuum pressure. The vacuum in the condenser is the difference between the barometric pressure and the absolute pressure in the condenser. The difference between the barometer reading and the vacuum gauge reading gives the absolute pressure in the condenser as shown in Figure 2.36.

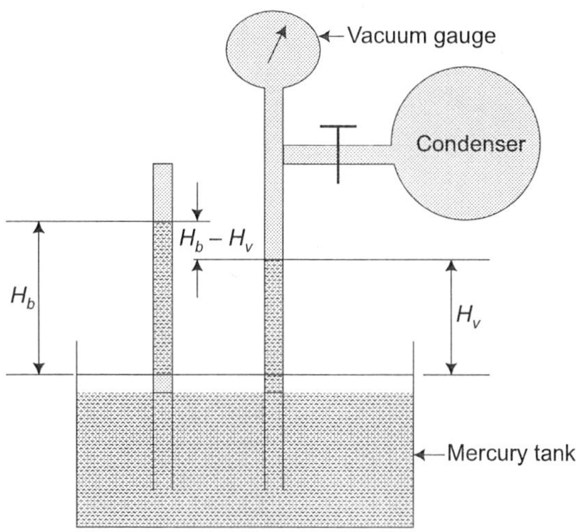

Fig. 2.36 Vacuum Measurement

Let the vacuum in the condenser be equivalent to H_v cm of mercury which is measured by vacuum gauge and the barometric pressure is equivalent to H_b cm of mercury.

Therefore, absolute pressure = $(H_b - H_v)$ cm of Hg

We know that 76 cm of Hg = 1.01325 bar

Therefore, pressure equivalent to 1 cm of Hg = (1.01325/76) = 0.0133322 bar

If it is desired to know the absolute pressure in the condenser then the barometric reading and gauge reading must be known. If barometric reading is 76 cm of Hg and gauge reading is 68 cm of Hg then the absolute pressure in the condenser will be

$$= (76 - 68) = 8 \text{ cm of Hg}$$

$$= 8 \times 0.0133322 = 0.10666 \text{ bar.}$$

The barometric head H_b varies according to atmospheric conditions, and hence the absolute pressure in the condenser is a function of the barometric pressure.

If the barometric reading (atmospheric pressure) is 76 cm of Hg at sea-level then the corrected vacuum in the condenser is given by,

$$= 76 - \text{absolute pressure in the condenser in cm of Hg}$$

$$= 76 - (H_b - H_v) \qquad (2.39)$$

2.20 SOURCES OF AIR IN THE CONDENSER, ITS EFFECTS AND REMOVAL

Sources of Air Leakage into the Condenser Following are the sources of air leakage into the condenser:

1. As the pressure inside the condenser is below atmospheric, so the air tends to leak into it through the joints, through the vents from the atmospheric relief valve and other accessories.
2. The air dissolved in exhaust steam is liberated at low pressure.
3. The dissolved air in the water is also carried with the cooling water in jet condensers which gets liberated at low pressure. The quantity of air carried with cooling water is about 0.005% of the cooling water.

In a properly maintained surface condenser for a steam turbine plant the amount of air leakage is about 5 kg per 10,000 kg of steam condensed, whereas for steam engine plants the air leakage is about 15 kg per 10,000 kg of steam.

Effects of Air on the Performance of Condenser The presence of air in the condenser affects its performance in the following ways:

1. It increases the condenser pressure or back pressure of the turbine (or engine) and reduces the work output.
2. Due to lower thermal conductivity of air the rate of heat transfer also lowers, thus the surface area of the condenser tubes is to be increased for a condenser duty.
3. The presence of air decreases the partial pressure of steam, so at low pressure the steam has more latent heat. To remove this increased amount of heat more amount of cooling water is to be supplied.

4. The air presence causes reduction in the rate of condensation of steam because the extraction of heat by circulating water is partly from steam and partly from air.
5. An air extraction pump is required to maintain vacuum in the condenser. In spite of shielding some quantity of steam escapes with the air which reduces the amount of the condensate. Moreover, the condensate gets under-cooled, so more heat is to be supplied to the feed water in the boiler.
6. Air present in the condenser increases the corrosive action.

Procedure to Check Air Leakage Run the plant till the pressure and temperature in the condenser become steady. Then isolate the condenser from the rest of the plant by shutting off the steam supply and stopping the air extraction and condensate pumps. If the condenser has any leakage then the readings of vacuum gauge and thermometer will fall.

Methods to Check Air Leakage The point of air leakage can be located by any one of the following methods:

(i) **Soap Bubble Method** The condenser is filled with air at pressure and then its effect on soap water is noted at the locations where infiltration is likely to occur. At the point of air leakage bubbles can be seen.

(ii) **Candle Flame Method** When the condenser is under vacuum then the large leakages can be detected by passing a candle flame over suspected locations.

(iii) **Peppermint Oil Method** Peppermint oil is applied over suspected locations when the condenser is working. If any leakage is there then oil fumes will enter into the condenser and there will be oil odour in the discharge from the condenser.

2.21 DALTON'S LAW OF PARTIAL PRESSURES

The law states that the total pressure exerted by a non-reactive mixture of gases or a mixture of gas and vapor is equal to the sum of partial pressure of the individual constituents of the mixture at the same temperature (T).

The partial pressure of each constituent of the mixture is the pressure exerted by the constituent taken separately in the same volume (V) of the vessel as that of mixture and at the same temperature. This is explained by taking the following example as shown in Figure 2.37.

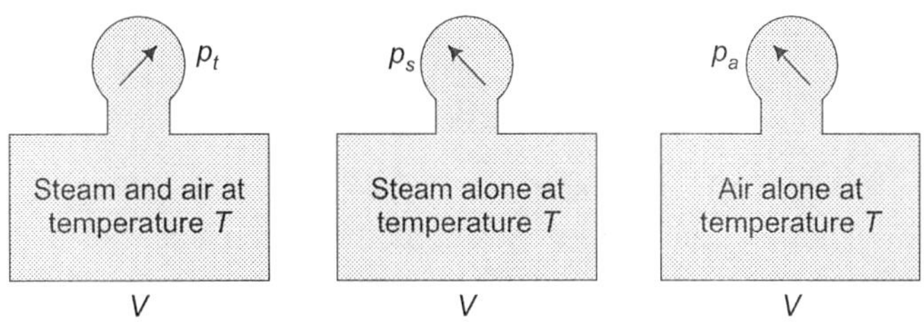

Fig. 2.37 Dalton's Law of Partial Pressure

The total pressure (p_t) in the condenser or vessel is the sum of the partial pressures of steam (p_s) and air (p_a).

According to Dalton's law of partial pressure,

$$p_t = p_s + p_a$$

or

$$p_a = p_t - p_s$$

The vacuum gauge reads the total pressure in the condenser, i.e., p_t. The steam pressure p_s can be known from the steam table corresponding to condenser temperature. As p_t and p_s are known thus p_a can be determined.

2.22 VACUUM EFFICIENCY

The vacuum efficiency is a measure of the degree of perfection of maintaining a desired vacuum in the condenser. The ideal vacuum means the vacuum due to steam alone when air is absent. Under such a condition total pressure in the condenser will reach to the pressure of steam corresponding to the saturation temperature of the steam.

Let

p_s = Saturation pressure of steam in bar corresponding to the temperature of water entering the condenser,

p_t = Total pressure of air and steam

$= p_s + p_a$, and

p_b = Barometric pressure or atmospheric pressure

$$\text{Ideal vacuum} = p_b - p_s \tag{2.40}$$

$$\text{Actual vacuum} = p_b - p_t$$

$$= p_b - (p_a + p_s) \tag{2.41}$$

The vacuum efficiency (η_{vacuum}) is defined as the ratio of actual vacuum recorded by the vacuum gauge to the ideal vacuum.

$$\eta_{\text{vacuum}} = \frac{\text{Actual vacuum recorded by gauge}}{\text{Ideal vacuum}}$$

$$= \frac{p_b - (p_a + p_s)}{p_b - p_s} = \frac{p_b - p_t}{p_b - p_s} \tag{2.42}$$

If there is no air leakage into the condenser then $p_a = 0$ and hence the vacuum efficiency becomes 100%. The vacuum efficiency depends upon the effectiveness of the air cooling and the rate at which the air is removed by the air pump. Generally, the vacuum efficiency is about 98% to 99%.

2.23 CONDENSER EFFICIENCY

In an ideal condenser the steam should give only its latent heat to the circulating water so that temperature of the condensate becomes equal to the saturation temperature corresponding to the condenser pressure. It means there should be no undercooling of the condensate. Maximum temperature to which the cooling water can be raised is the condensate temperature. The condenser efficiency ($\eta_{\text{condenser}}$) is then defined as the ratio of actual rise in the temperature of the cooling water to the maximum possible rise, i.e.,

$$\eta_{\text{condenser}} = \frac{T_2 - T_1}{T_s - T_1} \qquad (2.43)$$

where T_2 = Outlet temperature of cooling water, T_1 = Inlet temperature of cooling water, and T_s = Saturation temperature corresponding to condenser temperature.

Example 2.9 The following readings are recorded during a test on a steam condenser: vacuum in condenser = 71.5 cm of Hg, barometer reading = 76.5 cm of Hg, mean temperature of condenser = 33°C, hot well temperature = 29°C, inlet temperature of cooling water = 9°C, outlet temperature of cooling water = 26.5°C. Calculate (i) corrected vacuum to standard barometer, (ii) vacuum efficiency, (iii) under-cooling of the condensate, and (iv) condenser efficiency.

Solution

(i) $$\text{Corrected vacuum} = 76 - (H_b - H_g)$$
$$= 76 - (76.5 - 71.5) = 71 \text{ cm of Hg}$$

(ii) The absolute pressure of steam corresponding to condenser temperature of 33°C (from steam table)

$$= 0.05029 \text{ bar}$$
$$= (0.05029/0.0133322) = 3.772$$

$$\therefore \qquad \eta_{\text{vacuum}} = \frac{\text{Actual vacuum recorded by gauge}}{\text{Ideal vacuum}}$$

$$= \frac{71.5}{76.5 - 3.772} = 0.9831 \text{ or } 98.31\%$$

(iii) Under cooling of condensate = (condenser temperature – hot well temperature)

Therefore, under-cooling of the condensate = (33 – 29) = 4°C

(iv) Absolute condenser pressure = (barometric reading – vacuum reading)

$$= (76 - 71.5) = 4.5 \text{ cm of Hg}$$
$$= 4.5 \times 0.0133322 = 0.05999 \text{ bar}$$

Saturation temperature corresponding to 0.05999 bar = 35.85°C (from steam table)

Therefore, the maximum temperature to which cooling water can be raised is 35.85°C.

$$\therefore \qquad \eta_{condenser} = \frac{T_2 - T_1}{T_s - T_1} = \frac{26.5 - 9}{35.85 - 9}$$

$$= 0.6518 \text{ or } 65.18\%$$

Example 2.10 In a condenser, vacuum gauge reads 71.5 cm of Hg while barometer reads 75.5 cm of Hg. The temperature of condenser is 25°C. Determine (i) the pressure of steam and air, (ii) mass of air per kg of steam, and (iii) corrected vacuum to standard barometer, and (iv) vacuum efficiency.

Solution

(i) Total absolute pressure (p_t) in the condenser

$$= (H_b - H_v) = (75.5 - 71.5) = 4 \text{ cm of Hg}$$

$$= 4 \times 0.0133322 = 0.05332 \text{ bar}$$

(ii) Partial pressure of steam (p_s) at condensate temperature of 25°C

$$p_s = 0.03166 \text{ bar (From steam table)}$$

Partial pressure of air,

$$p_a = p_t - p_s = 0.05332 - 0.03166$$

$$= 0.02166 \text{ bar}$$

Specific volume of steam = 43.402 m^3/kg. Air will also occupy the same volume.

Applying ideal gas equation,

$$p_a v_a = m_a R_a T_a$$

$$(0.02166 \times 10^5) \times 61.02 = m_a \times 287 \times (25 + 273)$$

$$\therefore \qquad m_a = 1.094 \text{ kg/kg of steam}$$

(iii) $$\text{Corrected vacuum} = 76 - (H_b - H_g) = 76 - (75.5 - 71.5)$$

$$= 72 \text{ cm of Hg}$$

(iv) From steam table, the absolute pressure of steam corresponding to condenser temperature of 25°C

$$= 0.03166 \text{ bar} = (0.03166/0.0133322)$$

$$= 2.375$$

$$\therefore \qquad \eta_{vacuum} = \frac{\text{Actual vacuum recorded by gauge}}{\text{Ideal vacuum}}$$

$$= \frac{71.5}{75.5 - 2.37}$$

$$= 0.9777 \text{ or } 97.77\%$$

Example 2.11 A condenser deals with 900 kg of steam per hour with a dryness fraction of 0.9. The temperature of condenser is 40°C. The air associated with the steam in the condenser is 200 kg/hour. Determine the vacuum reading while barometer reads 75.5 cm of Hg. Correct this vacuum to a standard barometer reading of 76 cm of Hg.

Solution From steam table, partial pressure of steam at 40°C,

$$p_s = 0.07375 \text{ bar, and}$$

Volume of one kg of steam at 40°C,

$$v_s = 19.55 \text{ m}^3$$

$\therefore$ Total volume of steam $= m \times (x \times v_s) = 900 \times (0.9 \times 19.55)$

$$= 15835.5 \text{ m}^3.$$

Air will also occupy the same volume.

$$p_a v_a = m_a R_a T_a$$

$$p_a \times 15835.5 = 200 \times 287 \times (40 + 273)$$

$$\therefore \quad p_a = 1134.55 \text{ N/m}^2 = 0.011345 \text{ bar}$$

Total condenser pressure $= 0.07375 + 0.011345 = 0.085095$ bar

or $= (0.085095/0.0133322) = 6.38$ cm of Hg

$\therefore$ Vacuum reading $= 75.5 - 6.38 = 69.12$ cm of Hg

Corrected vacuum $= 76 - (75.5 - 69.12) = 69.62$ cm of Hg

Example 2.12 The following data are recorded during a test on a steam condenser: vacuum in condenser = 71 cm of Hg, barometer reading = 76.5 cm of Hg, mean temperature of condenser = 35°C, hot well temperature = 28°C, inlet temperature of cooling water = 8.5°C, outlet temperature of cooling water = 25.6°C, condensate collected = 1900 kg/hour, and cooling water quantity = 59,500 kg/hour. Calculate (i) corrected vacuum to standard barometer, (ii) vacuum efficiency, (iii) under-cooling of condensate, (iv) condenser efficiency, (v) quality of steam entering the condenser, and (iv) mass of air present per m^3 of the condenser volume and per kg of uncondensed steam.

Solution

(i) Corrected vacuum $= 76 - (H_b - H_v)$

$$= 76 - (76.5 - 71) = 70.5 \text{ cm of Hg}$$

(ii) The absolute pressure of steam corresponding to the condenser temperature of 35°C (from steam table)

$$= 0.05622 \text{ bar} = (0.05622/0.0133322)$$

$$= 4.22 \text{ cm of Hg}$$

$$\therefore \qquad \eta_{vacuum} = \frac{\text{Actual vacuum recorded by gauge}}{\text{Ideal vacuum}}$$

$$= \frac{71}{76.5 - 4.22} = 0.9823 \text{ or } 98.23\%$$

(iii) Under cooling of condensate = (condenser temperature – hot well temperature)

$$= (35 - 28) = 7°C$$

(iv) Absolute condenser pressure = (barometric reading – vacuum reading)

$$= (76.5 - 71) = 5.5 \text{ cm of Hg}$$

$$= 5.5 \times 0.0133322 = 0.07333 \text{ bar}$$

Saturation temperature corresponding to 0.05999 bar = 39.11°C (from steam table)

Therefore, the maximum temperature to which cooling water can be raised is 35.85°C

$$\therefore \qquad \eta_{condenser} = \frac{T_2 - T_1}{T_s - T_1} = \frac{25.6 - 8.5}{39.11 - 8.5}$$

$$= 0.5586 \text{ or } 55.86\%$$

(v) Heat lost by steam = Heat gained by water (neglecting heat received by air)

$$\therefore \qquad m_s\,(h_1 + x_1\,h_{fg1} - h_c) = m_w\,c_{pw}\,(T_2 - T_1) \qquad (h_c \text{ is } h_f \text{ value of condensate at 28°C})$$

$$1900 \times (166.89 + x_1 \times 2407.21 - 117.3) = 59500 \times 4.2 \times (25.6 - 8.5)$$

$$\therefore \qquad x_1 = 0.914$$

(vi)

$$p_a = p_t - p_s = 0.07333 - 0.05622$$

$$= 0.01711 \text{ bar}$$

To determine the mass of air per m^3 of the condenser at 0.01711 bar pressure and at 35°C applying ideal gas equation,

$$p_a\,v_a = m_a\,R_a\,T_a$$

$$0.01711 \times 10^5 \times 1 = m_a \times 287 \times (35 + 273)$$

$$\therefore \qquad m_a = 0.01936 \text{ kg/m}^3$$

The volume of one kg mass of uncondensed steam is given by v_s which is the specific volume of saturated steam at 35°C

So, volume of one kg of uncondensed steam = 25.25 m^3/kg (from steam table)

The mass of air in one kg of uncondensed steam is calculated by applying ideal gas equation,

$$0.01911 \times 10^5 \times 25.25 = m_a \times 287 \times (35 + 273)$$

$$\therefore \qquad m_a = 0.489 \text{ kg}$$

2.24 MASS OF COOLING WATER REQUIRED IN A CONDENSER

In a jet condenser the exhaust steam and the cold water mix directly, so the resulting temperature of the condensed steam and water is the same at its outlet. But in a surface condenser the temperature of the condensate and the circulating water may differ.

Let m = Quantity of circulating water in kg/h, m_s = Quantity of steam condensed in kg/h, T = Temperature of the wet exhaust steam in °C, T_c = Temperature of the condensate in °C, h_f = Sensible heat or enthalpy of water at T°C, h_{fg} = Latent heat or enthalpy of evaporation of steam at T°C, x = Dryness fraction of exhaust steam, T_1 = Inlet temperature of cooling water in °C, T_2 = Outlet temperature of cooling water in °C, h_c = Sensible heat or enthalpy of condensate, c_w = Specific heat of cooling water.

Now for a condenser,

Heat given up by steam = Heat absorbed by cooling water

For a surface condenser,

$$m_s \times (h_f + xh_{fg} - h_c) = c_{pw} \times m \times (T_2 - T_1)$$

$$\therefore \quad m = \frac{m_s \times (h_f + xh_{fg} - h_c)}{c_{pw} \times (T_2 - T_1)} \tag{2.44}$$

For a jet condenser,

$$T_2 = T_c$$

$$\therefore \quad m = \frac{m_s \times (h_f + xh_{fg} - h_c)}{c_{pw} \times (T_c - T_1)} \tag{2.45}$$

Example 2.13 A 175 kW steam engine consumes 9 kg of steam per kWh. The back pressure of the engine and the condenser pressure are equal to 0.15 bar. The temperature of the cooling water at the inlet and outlet are 15°C and 28°C respectively. The temperature of the condensate is 29°C. Determine the quantity of cooling water required per hour if the steam exhausted to the condenser is dry saturated.

Solution From steam table (Table A-1.2) at 0.15 bar,

$$h_f = 225.91 \text{ kJ/kg}, \qquad h_g = 2599.1 \text{ kJ/kg},$$

$$h_{fg} = (2599.1 - 225.91) = 2373.19 \text{ kJ/kg}$$

Cooling water/kg of steam,

$$m = \frac{(h_f + xh_{fg} - h_c)}{c_{pw} \times (T_2 - T_1)}$$

$$\therefore \quad m = \frac{(225.91 + 1 \times 2373.19 - 4.186 \times 28)}{4.186 \times (28 - 15)}$$

$$= 45.61 \text{ kg/h/kg of steam}$$

Total steam used per hour

$$= 175 \times 9 = 1575 \text{ kg}$$

∴ Total circulating water used per hour

$$= 45.61 \times 1575 = 71835.75 \text{ kg}$$

2.25 AIR PUMP

Air pump maintains the required vacuum in the condenser by removing air or air and the condensate both. Commonly, dry air pumps and wet air pumps are used for this purpose. Dry air pump removes air only, whereas the wet air pump removes the mixture of condensate, water, and other gases. These pumps may be reciprocating type or rotary type. Here Edward's air pump which is of reciprocating type is discussed.

Edward's Air Pump Edward's air pump consists of a piston which reciprocates in the pump barrel. The piston is flat on its upper surface and conical at the bottom. The barrel has a ring of ports near its lower end through which it communicates with the condenser. The top of the barrel has cover in which head or delivery valves are provided as shown in Figure 2.38.

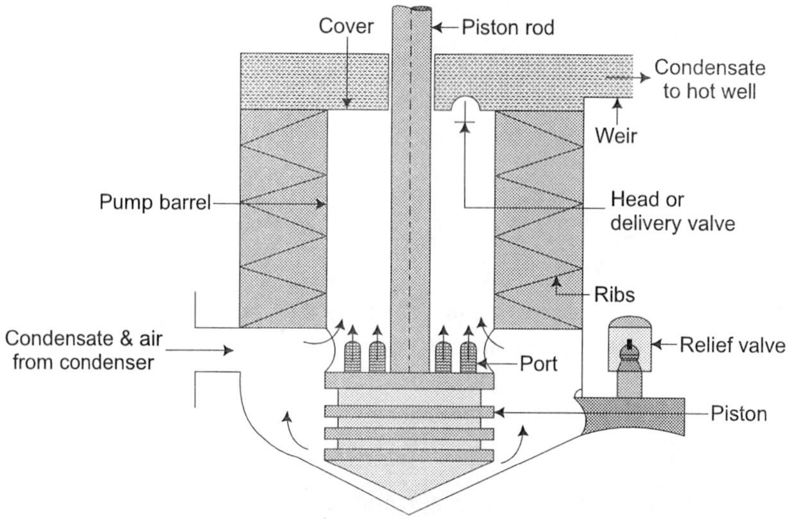

Fig. 2.38 Edward's Air Pump

When the piston moves in the downward direction a partial vacuum is produced in the barrel above the piston due to which the delivery valve closes. The air and water vapors from the condenser rush into the space above the piston through the ports. Further, motion of the piston causes its conical end to displace the condensate rapidly through the ports. Thus, water vapors, air, and the condensate fills the space above the piston.

When the piston moves in the upward direction the charge of water vapors, air, and the condensate get compressed to a pressure slightly above the atmospheric pressure. This opens the head valves through which the charge flows over the weir to the hot well. The weir maintains a sufficient head of water over head valves and thus seals them against any air leakage.

A relief valve is provided at the base of the barrel to release the charge if the pressure exceeds the atmospheric pressure due to any reason.

These pumps are generally used for low vacuum requirements. Its main advantage over other pumps is that it does not have any inaccessible valves.

2.26 COOLING TOWERS AND COOLING PONDS

Cooling water is supplied to the condenser for the condensation of steam. In this process the cooling water becomes hot. To cool this hot water coming out of the condenser the cooling towers and the cooling ponds are used so that the cooled water can be reutilized in the condenser again.

2.26.1 Cooling Towers

The cooling towers are very useful when there is scarcity of both the water and the land. It is an artificial device by which the hot water coming out of the condenser is cooled effectively. By using cooling towers the cooling water requirement is reduced and only makeup water is to be supplied. The principle of cooling the water is similar to that of the evaporative condenser. Some water about 1% goes into air in the form of water vapor by absorbing its latent heat of vaporization from the remaining water and thus causes the reduction in the water temperature. The cooling towers reduce the cooling water demand about 75 times but it is achieved at the expense of large capital, land, and operational costs. The types of cooling towers on the basis of the draught (method of air circulation) are:

(i) Natural draught cooling towers

(ii) Mechanical draught cooling towers: (a) Forced draught cooling towers, and (b) Induced draught cooling towers.

(i) Natural Draught Cooling Towers The schematic view of natural cooling tower is shown in Figure 2.39. The hot water from the condenser is pumped to a height of about 8 m to 12 m which enters the tower and then sprayed over the woodwork and trays. The water in the form of sprays meets the air entering from the bottom of the tower which is open to the atmosphere. Hot water gives up its heat to the air and gets cooled. The hot air along with some water vapor leaves the tower at top and the cooled water falls down in the form of rain and gets collected in the pond at the bottom of the tower. The cooled water from the pond is again supplied to the condenser. No fan is used in natural draught cooling towers. The bottom of the tower is kept open through which air enters into the tower. The airflow is maintained due to pressure difference caused by the difference in density between the hot air inside the tower and the outside atmospheric air. The water vapors leaving the tower along with the air is prevented by using water eliminators. Still some loss of water will be there which will be compensated by adding fresh water called make up water. The reduction in temperature of the water is called range. The disadvantage of this type of tower is that to produce large natural draught the tower will be very high.

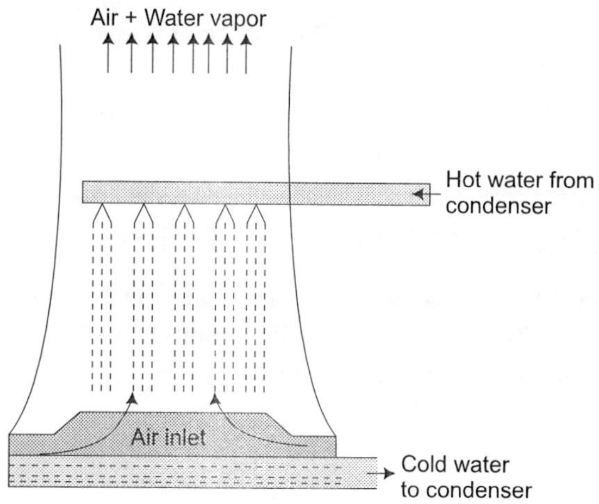

Fig. 2.39 Schematics of Natural Draught Cooling Tower

(ii) Mechanical Draught Cooling Towers In the mechanical draught cooling towers the air is circulated with the help of a mechanical device like a fan or a blower. Depending on the position of the fan or blower these are classified as forced draught or induced draught cooling towers. When the fan is installed at the bottom of the tower it is known as forced draught cooling tower as shown in Figure 2.40 (a). When the fan is installed at the top of the tower it is called induced draught cooling tower as shown in Figure 2.40 (b).

Hot water coming from the condenser enters the tower from its top and is sprayed through the nozzles. The sprayed water meets with the air going upwards. Eliminators are provided at the

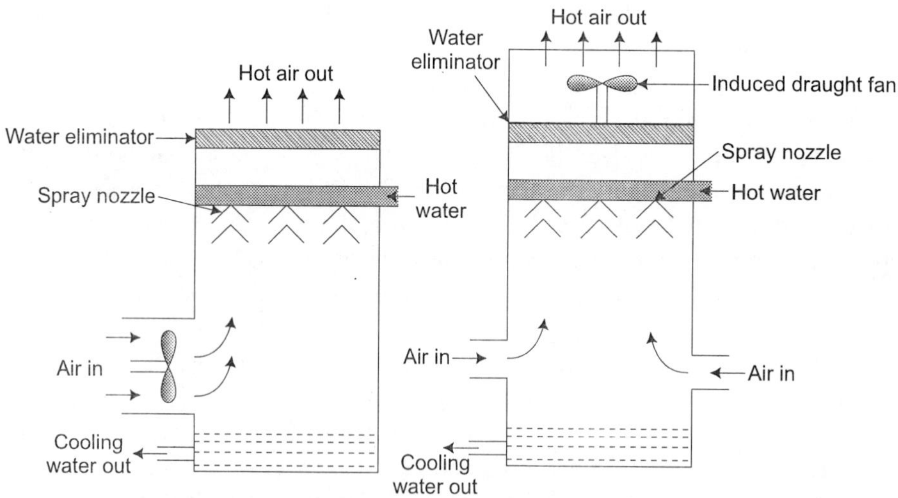

(a) Forced draught cooling tower (b) Induced draught cooling tower

Fig. 2.40 Schematics of Mechanical Draught Cooling Tower

top to prevent the escaping of water droplets with air leaving from the top. Cooling towers are generally hyperbolic in shape and are made of steel, concrete, or timber. The induced draught cooling towers are generally used in large capacity power plants. In the case of forced draught tower power requirement is high and maintenance cost of the fan is high. The induced draught tower occupies less space as the fan drives are installed at the top of the tower. In the induced draught tower the air is drawn by the fan from all the sides of the tower through the openings at low velocity and thus the cooling effect is obtained across the entire cross section of the tower. Also it handles warm air, so there will be no freezing problems during winter season as in the case of forced draught cooling towers.

2.26.2 Cooling Ponds

It is the simplest method of removing heat from the hot water by discharging it through a pipeline into a large open pond exposed to the atmosphere. The water is cooled by blowing air over the surface of the pond and mixing with cold water of the pond. The heat from the hot water will be transferred to the air by evaporation and convection processes. The loss of water by evaporation and wind blowing over the cooling pond is about 2% to 3%. To reduce the area of pond the water is sprayed into the air over the pond surface by water spray nozzles. For effective cooling these nozzles are fitted at a height of about 1 m to 2.5 m above the surface of water. Even after the use of spray nozzles the area required for cooling is large due to the evaporative cooling process. The factors which affect the dissipation of heat from the cooling pond are: area and depth of pond, temperature of water entering the pond, relative humidity, air velocity, atmospheric temperature and pressure, and solar radiation.

The cooling ponds are classified into two categories, namely, non-directed flow type and directed flow type. The non-directed flow type cooling pond is shown in Figure 2.41 (a) and directed flow type is shown in Figure 2.41 (b).

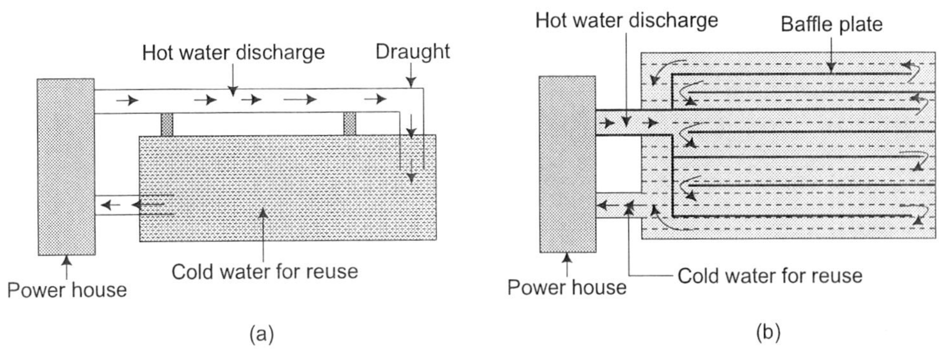

Fig. 2.41 Cooling Ponds

In the case of non-directed flow type ponds, the hot water is discharged into the open pond. But in the case of directed flow type cooling ponds, the pond is divided into a number of channels by providing baffle plates which direct the flow of hot water. The cooling effect in directed flow cooling ponds will be more due to thorough mixing of hot and cold water streams.

On the basis of spray nozzles arrangement, the cooling ponds may also be classified as single deck or double deck ponds. In a single deck cooling pond, the spray nozzles are arranged at the same elevation, whereas in double deck system the nozzles are arranged at different elevations. The double deck cooling ponds are more efficient than single deck system.

The disadvantages of cooling ponds are: (i) Large area is required for cooling, for example, the surface area required in a cooling pond is about 30 times the size of cooling tower for the same duty, (ii) The loss of water due to air blow is large and it is unprotected against dust, and (iii) There is no control over the temperature of cooled water, and its cooling efficiency is low.

The cooling ponds are suitable only for small capacity steam plants where land is easily and cheaply available.

EXERCISE

(i) Objective Type Questions

1. Nozzle is designed for:

(a) maximum pressure and maximum discharge

(b) maximum pressure at outlet

(c) maximum discharge

(d) none of the above

2. For supersonic flow, the nozzle should be:

(a) convergent (b) convergent-divergent

(c) constant area duct (d) divergent

3. For subsonic flow, the nozzle should be:

(a) divergent (b) convergent-divergent

(c) convergent (d) constant area duct

4. Velocity of steam at the exit of nozzle is given by:

(a) $44.72\sqrt{\eta_{nozzle}(h_1 - h_2)}$ (b) $4.472\sqrt{\eta_{nozzle}(h_1 - h_2)}$

(c) $4472\sqrt{\eta_{nozzle}(h_1 - h_2)}$ (d) $0.4472\sqrt{\eta_{nozzle}(h_1 - h_2)}$

5. In the case of nozzle, the whole of friction loss is assumed:

(a) between inlet and throat (b) between throat and exit

(c) between inlet and outlet (d) none of the above

6. Moore and Curtis turbine is:

(a) velocity compounded turbine (b) pressure compounded turbine

(c) reaction turbine (d) pressure-velocity compounded turbine

7. Parson's turbine is:
   (a) impulse turbine (b) impulse-reaction turbine
   (c) velocity compounded turbine (d) pressure compounded turbine
8. Curtis turbine is:
   (a) velocity compounded turbine
   (b) pressure compounded turbine
   (c) pressure-velocity compounded turbine
   (d) reaction turbine
9. Rateau turbine is:
   (a) velocity compounded turbine
   (b) pressure compounded turbine
   (c) pressure-velocity compounded turbine
   (d) reaction turbine
10. De-Laval turbine is:
    (a) simple impulse turbine (b) reaction turbine
    (c) pressure compounded turbine (d) velocity compounded turbine
11. The pressure on the two sides of the moving blades of a reaction turbine is:
    (a) same (b) higher at inlet (c) lower at inlet (d) none of the above
12. In case of impulse turbine:
    (a) enthalpy drops only in moving blades
    (b) enthalpy drops only in fixed blades
    (c) enthalpy drops in moving and fixed blades
    (d) none of the above
13. The steam consumption in reciprocating steam engine in comparison to steam turbine is:
    (a) equal (b) less (c) more (d) none of the above
14. Stage efficiency is also called:
    (a) gross efficiency (b) blade efficiency
    (c) diagram efficiency (d) mechanical efficiency
15. Stage efficiency of steam turbine is:
    (a) $\eta_{nozzle} \times \eta_{blade}$ (b) $\eta_{nozzle}/\eta_{blade}$ (c) $\eta_{blade}/\eta_{nozzle}$ (d) none of the above
16. In case of reaction turbine, the enthalpy drops in:
    (a) moving blades only (b) fixed blades only
    (c) both fixed and moving blades (d) none of the above
17. Degree of reaction is the ratio of:
    (a) enthalpy drop in moving blade to the sum of enthalpy drop in moving and fixed blades
    (b) sum of enthalpy drop in moving and fixed blades to enthalpy drop in moving blade

(c) enthalpy drop in fixed blade to sum of enthalpy drop in moving and fixed blades
(d) none of the above

18. Degree of reaction for Parson's turbine is:
(a) 50% (b) 60% (c) 70% (d) 80%

19. Reheat factor is defined as:
(a) ratio of isentropic enthalpy drop to cumulative enthalpy drop
(b) ratio of cumulative isentropic heat drop to Rankine heat drop
(c) ratio of Rankine heat drop to cumulative isentropic heat drop
(d) none of the above

20. The value of reheat factor generally varies from:
(a) 0.6 to 1.06 (b) 0.02 to 0.06 (c) 0.8 to 1.08 (d) 1.02 to 1.06

21. The value of reheat factor increases if the number of stages are:
(a) less (b) more (c) same (d) none of the above

22. Internal efficiency is the ratio of:
(a) useful heat drop to heat supplied
(b) adiabatic heat drop to heat supplied
(c) total useful heat drop to adiabatic heat drop
(d) none

23. The number of stages in pressure compounded steam turbine in comparison to velocity compounded is:
(a) more (b) less (c) equal (d) none of the above

24. The efficiency of pressure compounded steam turbine in comparison to velocity compounded is:
(a) more (b) less (c) equal (d) none of the above

25. If η_s is constant for all stages then reheat factor is given by:
(a) $RF = \eta_s/\eta_i$ (b) $RF = \eta_i/\eta_s$ (c) $RF = \eta_s \times \eta_i$ (d) none of the above

26. Maximum blade or diagram efficiency of an impulse turbine is:
(a) $[1 + C] \cos \alpha$ (b) $[1 + KC] \cos \alpha$ (c) $\cos \alpha$ (d) $[1 + KC] (\cos \alpha)/2$

27. The maximum rate of work done for a single stage impulse turbine is:
(a) u^2 (b) $2u$ (c) $3u^2$ (d) $2u^2$

28. For maximum blade efficiency of a single stage impulse turbine:
(a) $\rho = \cos \alpha_1/2$ (b) $\rho = \cos^2\alpha_1/2$ (c) $\rho = \cos \alpha_1$ (d) $\rho = \cos^2\alpha_1$

29. The ratio of actual vacuum to the ideal vacuum is termed:
(a) condenser efficiency (b) boiler efficiency
(c) nozzle efficiency (d) vacuum efficiency

30. The ratio of actual temperature rise to the maximum possible rise is termed:
(a) condenser efficiency (b) boiler efficiency
(c) nozzle efficiency (d) vacuum efficiency

31. The cooling system used for big power plants:
(a) spray pond (b) hyperbolic cooling tower
(c) mechanical draught cooling tower (d) natural draught cooling

32. The ratio of sensible and evaporation cooling in wet cooling tower is:
(a) 20:80 (b) 40:60 (c) 40:80 (d) 60:80

33. The total absolute pressure in a condenser is equal to the:
(a) atmospheric pressure – gauge pressure
(b) atmospheric pressure + gauge pressure
(c) barometric pressure – vacuum pressure
(d) barometric pressure + vacuum pressure

34. Actual vacuum in a condenser is equal to the:
(a) barometric pressure – actual pressure
(b) barometric pressure + actual pressure
(c) atmospheric pressure – gauge pressure
(d) atmospheric pressure + gauge pressure

35. Pure cooling water is used in:
(a) surface condenser (b) jet condenser
(c) in both jet and surface condensers (d) none of the above

36. The rotor speed in steam turbines can be reduced by:
(a) velocity compounding (b) pressure compounding
(c) velocity-pressure compounding (d) all of the above

37. The vacuum efficiency of a condenser is about:
(a) 60% (b) 70% (c) 80% (d) 90%

38. Cooling water requirement in a surface condenser is:
(a) about 50 times of steam condensed (b) about 40 times of steam condensed
(c) about 30 times of steam condensed (d) about 20 times of steam condensed

39. Thermal efficiency of steam power plant increases by the employment of surface condenser because:
(a) average temperature of heat addition is increased
(b) average temperature of heat addition is reduced
(c) average temperature of heat rejection is reduced
(d) none of the above

40. Edward's air pump:

(a) removes air, vapor, and the condensed water

(b) removes air only

(c) removes air and vapor

(d) removes vapor only

41. For Parson's reaction turbine, the maximum efficiency is given by:

(a) $\dfrac{\cos^2\alpha_1}{1+\cos^2\alpha_1}$ (b) $\dfrac{2\cos^2\alpha_1}{1+\cos^2\alpha_1}$ (c) $\cos^2\alpha_1$ (d) $\dfrac{1+\cos^2\alpha_1}{\cos^2\alpha_1}$

42. For maximum blade efficiency for Parson's reaction steam turbine the speed ratio is equal to:

(a) $\cos^2\alpha$ (b) $\cos^2\alpha/2$ (c) $2\cos^2\alpha$ (d) $\cos\alpha$

ANSWERS

1. (c)	2. (d)	3. (c)	4. (a)	5. (b)	6. (d)	7. (b)	8. (a)	9. (b)	10. (a)
11. (b)	12. (b)	13. (c)	14. (a)	15. (a)	16. (c)	17. (a)	18 (a)	19. (b)	20. (d)
21. (b)	22. (c)	23. (a)	24. (a)	25. (a)	26. (d)	27. (d)	28. (a)	29. (d)	30. (a)
31. (b)	32. (a)	33. (c)	34. (a)	35. (b)	36. (d)	37. (d)	38. (a)	39. (c)	40. (a)
41. (b)	42. (d)								

(ii) Review Questions

1. Describe with neat sketches the compounding of steam turbine. Why compounding of steam turbine is necessary?
2. What is degree of reaction? Prove that for a turbine with 50% degree of reaction the guide blades and moving blades are identical.
3. Explain the various losses that occur in a steam turbine.
4. Discuss the advantages of steam turbine over steam engine.
5. Define the terms related to steam turbine: (i) Speed ratio, (ii) Blade velocity coefficient, (iii) Diagram efficiency, (iv) Stage efficiency, and (v) Internal efficiency.
6. How are the steam turbines classified?
7. Explain the term reheat factor as applied to steam turbine.
8. Derive the condition for maximum efficiency of an impulse turbine and show that the maximum efficiency is $\cos^2\alpha$ where α is the angle at which the steam enters the blades.
9. Give the comparisons between impulse and reaction turbines.
10. Explain the working principles of impulse and reaction turbines with neat sketches.
11. Derive an expression for maximum diagram efficiency of a reaction turbine.

12. Derive an expression for maximum efficiency of an impulse turbine.
13. Describe the various types of cooling towers with neat sketches.
14. Define (i) Vacuum efficiency, and (ii) Condenser efficiency.
15. Give the requirements of a modern surface condenser.
16. What do you mean by a steam condenser? Explain its functions.
17. Write short notes on

    (i) Cooling ponds, (ii) Cooling towers, (ii) Edward's air pump.
18. What are the main types of condensers? State the advantages of a surface condenser over a jet condenser.
19. Differentiate clearly between the jet and surface condensers.
20. Explain the governing methods of steam turbines with neat sketches.
21. Explain the sources of air in the condenser, its effects, and removal in detail.

(iii) Numerical Problems

1. The rotor of an impulse turbine is 0.26 m diameter and runs at 20,500 rpm. The nozzle angle is 20° and issues a steam jet with a velocity of 910 m/s. The steam mass flow rate through the turbine nozzles is 2.0 kg/s. Draw the velocity diagram and determine (i) tangential force on blades, (ii) axial force on blades, (iii) power developed by the turbine, and (iv) blade efficiency.

   [**Ans.** $F_t = 2228$ N, $F_a = 0$, Power developed = 621.6 kW, and blade efficiency = 75%]
2. Steam issues from the nozzles of a De-Laval turbine with a velocity of 1200 m/s. The nozzle angle is 20°, the mean blade velocity is 400 m/s, and the inlet and outlet angles of blade are equal. The rate of mass flow through the turbine is 900 kg/hr. Determine (i) the blade angles, (ii) the relative velocity of steam entering the blade, (iii) the tangential force on the blade, (iv) the axial force on the blade, (v) the power developed, (vi) the blade efficiency, and (vii) energy lost due to friction in blade per kg of steam. Take velocity of coefficient as 0.8.

   [**Ans.** (i) 29.4°, (ii) 840 m/s, (iii) 328 N, (iv) 20 N, (v) 131.2 kW, (vi) 73%, and (vii) 127 kJ/ kg.]
3. A single row impulse turbine develops 135 kW at a blade speed of 175 m/s using steam mass flow rate of 2 kg/s. Steam leaves the nozzle at 400 m/s. Steam leaves the turbine blade axially. Determine the nozzle angle, blade angle at entry and at exit, assuming no shock and velocity coefficient of the blade is 0.9. [**Ans.** 15.35°, 26.67°, 34.55°]
4. In a simple impulse turbine the nozzles are inclined at 20° to the direction of motion of the moving blades. The steam leaves the nozzles at 375 m/s. The blade speed is 165 m/s. Find suitable inlet and outlet angles for the blades in order that the axial thrust is zero. The relative velocity of steam as it flows over the blade is reduced by 15% by friction. Also determine the power developed for a flow rate of 10 kg/s. [**Ans.** 34.4°, 41.6°, 532 kW]

5. Steam enters the blade row of an impulse turbine with a velocity of 600 m/s at an angle of 25° to the plane of rotation of the blades. The mean blade speed is 255 m/s. The blade angle on the exit side is 30°. The blade friction coefficient is 10%. Determine (i) the angle of the blade on the entry side, (ii) the diagram efficiency, (iii) the power developed per kg of steam, and (iv) axial thrust per of steam per second.

[**Ans.** 41.5°, 83.65, 150.45 kW, – 90 N/kg/s]

6. In a stage reaction turbine the mean diameter of the rotor is 1.4 m. The speed ratio is 0.7. Determine the blade inlet angle if the blade outlet angle is 20°. The rotor speed is 50 rps. Also find the diagram efficiency. Find the percentage increase in diagram efficiency and rotor speed if the rotor is designed to run at the best theoretical speed, the exit angle is 20°. [**Ans.** 20°, 90.5%, 3.65%, 67.08 rps]

7. The vacuum reading of a condenser is 70.5 cm of Hg when the barometer reads 76 cm of Hg and the condensate temperature is 31°C. Calculate vacuum efficiency. [**Ans.** 97.06%]

8. The vacuum in a surface condenser is found to be 70.5 cm of Hg when the barometer reads 76 cm of Hg. The cooling water enters the condenser at 20°C and leaves at 36.5°C. Determine the condenser efficiency. [**Ans.** 87%]

9. A 180 kW steam engine consumes 9.5 kg of steam per kWh. The back pressure of the engine and the condenser pressure are equal to 0.15 bar. The temperature of the cooling water at the inlet and outlet are 18°C and 34°C respectively. The temperature of the condensate is 35°C. Determine the quantity of cooling water required per hour if the steam exhausted to the condenser is dry saturated. [**Ans.** 62568 kg]

10. In a condenser the following data were recorded: vacuum = 70 cm of Hg, barometer = 76 cm of Hg, mean temperature of the condenser = 35°C, hot well temperature = 29°C, mass of cooling water = 45500 kg/hour, inlet temperature = 16.5°C, outlet temperature = 31°C. Determine (i) the mass of air present per unit condenser volume, (ii) dryness fraction of steam entering the condenser, (iii) vacuum efficiency, (iv) the condensate undercooling, and (v) condenser efficiency. [**Ans.** 0.02686 kg/m^3, 94%, 97.51%, 6°C, 78.37%]

3

Internal Combustion Engines and Gas Turbines

3.1 INTRODUCTION

A heat engine is a device which transforms the chemical energy of a fuel into thermal energy and uses a part of this thermal energy to produce mechanical work. Broadly, heat engines can be classified into two categories, namely, internal combustion engines (I.C. engines) and external combustion engines (E.C. engines).

In internal combustion engines the combustion of fuel takes place inside the working cylinder and the working fluid is the product of combustion of fuel-air mixture. For example, petrol, gas, diesel engines, Wankel engine, open cycle gas turbine, jet engines, and rockets are internal combustion engines. In internal combustion engines the thermal energy obtained due to combustion of fuel is directly converted to mechanical power. Due to direct combustion of fuel inside the cylinder the temperature of flue gases is quite high. The high temperature flue gases results in higher thermal efficiency.

In external combustion engines the combustion of fuel takes place outside the working cylinder and the products of combustion of air and fuel transfer heat to a second fluid which then becomes the working fluid for producing power. The most common examples of external combustion engines are the steam engines and steam turbines where steam is used as the working fluid. Other examples of external combustion engines are the hot air engines, closed cycle gas turbine, and Stirling engines.

A gas turbine is a rotary machine similar in principle to a steam turbine. The gas turbine is the most satisfactory power developing machine due to its exceptional reliability, freedom from vibrations, and ability to produce large powers from units of relatively smaller size and weight. The efficiency and work output of gas turbines can be improved by employing regeneration, reheating, and intercooling.

3.2 APPLICATIONS OF INTERNAL COMBUSTION ENGINES

The most important applications of I.C. engines are in transport on land, sea, and air. The other applications of I.C. engines include industrial applications and prime movers for electric generators. Some of the important applications of I.C. engines are summarized as: (i) Road vehicles: mopeds, scooters, motorcycles, cars, and commercial vehicles, (ii) Railway locomotives, (iii) Aircraft, helicopters, and airplanes, (iv) Marine: motor boats and ships, (v) Off road vehicles: agriculture, earth moving, and military, (vi) Industrial, stationary: electric power, and (vii) Lawn movers.

3.3 ADVANTAGES OF I.C. ENGINES OVER E.C. ENGINES

The internal combustion engines have the following advantages over external combustion engines: (i) Higher overall efficiency due to lower heat losses, (ii) Generally, lower weight to power ratio, (iii) Require less space, (iv) Greater mechanical simplicity, (v) Generally, lower first cost, and (vi) Relatively easy starting under cold conditions.

3.4 ADVANTAGES OF E.C. ENGINES OVER I.C. ENGINES

The external combustion engines have the following advantages over internal combustion engines: (i) Due to external combustion of fuel, cheaper and solid fuels can be used, (ii) These engines have higher starting torque, (iii) These engines are self starting except closed cycle gas turbine, (iv) Lubrication problem is negligible, and (v) In large plants overall efficiency is similar to internal combustion engines.

3.5 CLASSIFICATION OF I.C. ENGINES

The internal combustion engines may be classified on the following basis:

1. According to the type of fuel used:
   - (i) **Gas engines** The gas used may be compressed natural gas (CNG), coal gas, and producer gas.
   - (ii) **Petrol engine** Petrol is used as fuel.
   - (iii) **Diesel engine** Diesel is used as fuel.
   - (iv) **Oil engine** Oil used may be crude oil, mineral oil, or biodiesel.
   - (v) **Kerosene engine** Kerosene is used as fuel.
   - (vi) **Bi-fuel engine** In these engines main fuel is gas but in addition small quantity of liquid fuel is also injected at the end of compression stroke for starting purposes.
2. According to the thermodynamic cycle on which the engine works:
   - (i) **Otto cycle or constant volume combustion cycle** Mostly petrol engine and gas engines work on this cycle.
   - (ii) **Diesel cycle or constant pressure combustion cycle** Mostly low speed diesel engine works on this cycle.
   - (iii) **Mixed or limited pressure or dual cycle** Mostly high speed diesel engine works on this cycle.
3. According to the method of ignition of fuel:
   - (i) **Spark ignition engines (S.I. Engines)** Petrol and gas engines are spark ignition engines. The spark is produced by a spark plug with the help of a battery or magneto.
   - (ii) **Compression ignition engines (C.I. Engines)** The diesel engines are the compression ignition engines. The high compression ratio in diesel engines causes high temperature which helps in burning the fuel.

4. According to the number of strokes for the completion of one cycle:
   (i) **Two-stroke cycle engine** When the engine cycle is completed in two strokes (i.e., one revolution of crankshaft) the engine is called two-stroke cycle engine.
   (ii) **Four-stroke cycle engine** When the engine cycle is completed in four strokes (i.e., two revolutions of the crankshaft) the engine is called a four-stroke cycle engine.
5. According to the number of cylinders:
   (i) **Single cylinder engine** The engine has single cylinder.
   (ii) **Multi-cylinder engine** The engine has more than one cylinder.
6. Arrangement of cylinder or cylinders. If the engine has more than one cylinder then the classification can be according to their relative arrangement. The various arrangements are given below:
   (i) **In-line engines** All cylinders are arranged linearly and transmit power to a single crankshaft. This type is very popular with automobile engines. Four and six cylinder in-line engines are popular in automotive applications.
   (ii) **V-engines** An engine with two in line engines inclined at an angle to each other and with one crankshaft. Most of the high powered automobiles use the eight cylinder V- engine, four in-line on each side of the V.
   (iii) **Opposed cylinder engine** An engine with two in-line arrangement and 180° apart forms opposed cylinder engine. This design has been used in small aircraft.
   (iv) **Opposed piston engines** In opposed piston engine a single cylinder houses two pistons, each of which drives a separate crankshaft.
   (v) **Radial engines** In these engines more than two cylinders in each row are equally spaced around the crankshaft. This arrangement is commonly used in conventional air cooled aircraft engines in which 3, 5, 7, or 9 cylinders may be used in one bank and two to four banks of cylinders may be used.
7. According to the fuel supply system:
   (i) **Carburettor engine** Carburettor prepares the required charge and supplies it to the engine during suction stroke.
   (ii) **Airless injection or Solid injection** Injection of fuel directly into the combustion chamber without the aid of compressed air is called solid injection.
   (iii) **Air injection** In this system fuel is forced into the cylinder by means of compressed air.
8. According to the combustion of fuel on one side of the piston or both sides of piston:
   (i) **Single acting engines** Combustion of fuel takes place on one side of the piston. I.C. engines are mostly single acting.
   (ii) **Double acting** Combustion takes place on both sides of the piston.
9. According to the speed of the engine:
   (i) **High-speed engine** Engines having speed above 900 rpm are called high-speed engines.
   (ii) **Low-speed engines** Engines having speed less than 400 rpm are called low-speed engines.

(iii) **Medium-speed engines** Engines having speed in between 400 rpm and 900 rpm are called medium-speed engines.

10. According to the cooling of engine cylinder:
    (i) **Water cooled (or indirectly cooled engines)** Water is used and made to circulate through the water jackets provided around the cylinder, cylinder head, valve ports, and seats where it extracts most of the heat. Medium and large sized engines and the automobile engines are water cooled.
    (ii) **Air cooled engines** By providing fins over the cylinder and cylinder head surface area can be increased considerably which results in increased rate of cooling. Small engines such as scooter and motorcycle are air cooled engines.
11. According to the engine lubrication system:
    (i) **Mist lubrication system** This system is used for 2-stroke cycle engines (light duty engines). Such engines are lubricated by adding 2 to 3 percent lubricating oil in the fuel tank.
    (ii) **Wet sump lubrication system** In this system, the bottom of the crankcase contains an oil sump from which the lubricating oil is pumped to the various engine components by a pump. This system is used for heavy duty engines.
12. According to the method of governing (speed and load control of engines):
    (i) **Quantity governing** The speed and load control of the engine is obtained by the regulation of the throttle valve. The air-fuel ratio is nearly constant for all loads. The S.I. engines are quantity governed engines.
    (ii) **Quality governing** The speed and load control of the engine is obtained by the regulation of the quantity of the fuel injected. The supply of the air remains constant irrespective of load. The air-fuel ratio varies considerably with the load. The C.I. engines are quality governed engines.
13. According to the type of valves used

    Engines may be classified on the basis of the type of valves used such as poppet, sleeve, rotary, etc.
14. According to how an engine is started

    Engines are classified according to how an engine is started, i.e., by compressed air, inertia, electric motor.
15. According to function or field of application
    (i) **Stationary engines** These engines are used in pumping stations, powerhouses, and industries.
    (ii) **Mobile engines** These engines are used in automobiles, aeroplanes, locomotives.

3.6 I.C. ENGINES BASIC TERMINOLOGY

The following terminology is generally used in internal combustion engines as illustrated in Figure 3.1.

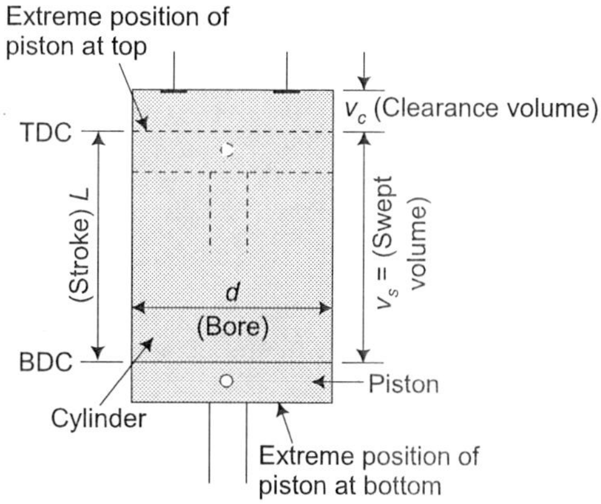

Fig. 3.1 I.C. Engines Terminology

1. **Bore** The inside diameter of the cylinder is called bore. It is designated by d and is given in millimetre (mm).
2. **Dead Centre** The position of the piston at the moment when the direction of the piston motion is reversed at either end point of the stroke is called dead centre. There are two dead centres in the engines, namely, top dead centre and bottom dead centre

   Top Dead Centre (TDC) The extreme position of the working piston at the top of the cylinder (i.e., head side) is known as top dead centre. In the case of horizontal engines, it is known as the inner dead centre (IDC).

   Bottom Dead Centre (BDC) The extreme position of the working piston at the bottom of the cylinder is known as bottom dead centre. In the case of horizontal engines, it is known as outer dead centre (ODC).
3. **Stroke** The nominal distance between the two dead centres is called the stroke or stroke length. It is denoted by L and is usually expressed in millimetre (mm). Stroke length is equal to double the crank radius.
4. **Swept Volume** The volume swept by the working piston in one stroke is called swept volume which is denoted by v_s. It is expressed in terms of cubic centimetres (cc) and is given by,

$$v_s = A \times L = \frac{\pi}{4} d^2L \tag{3.1}$$

   where A is the piston area, d is the piston diameter or cylinder bore, and L is the stroke length.
5. **Clearance Volume** The volume of the space above the piston at the top dead centre is known as clearance volume. It is denoted by v_c. In other words, clearance volume is the volume of compressed gases which is also the volume of the combustion chamber.
6. **Cylinder Volume** It is the sum of the clearance volume and the swept volume. It is denoted by v.

$$\therefore \quad v = v_c + v_s \tag{3.2}$$

7. **Compression Ratio** It is the ratio of the total volume of the cylinder to that of the clearance volume. It is denoted by r.

$$r = \frac{v}{v_c} = \frac{v_c + v_s}{v_c} = 1 + \frac{v_s}{v_c} \quad (3.3)$$

The higher the compression ratio, the better will be the ideal efficiency of an engine. Most of the S.I. engines have compression ratio in the range of 6 to 10, whereas C.I. engines have in the range of 16 to 20.

8. **Piston Speed** It is an important parameter in engine application and is defined as the distance travelled by the piston in one minute. The piston speed value lies in between 8 m/s to 15 m/s. It is denoted by s and is given by,

$$s = 2LN \quad (3.4)$$

where L is the stroke and N is the rotational speed of the crank shaft in rpm.

3.7 ENGINE PERFORMANCE PARAMETERS

1. Mean Effective Pressure. The device which measure the variation of pressure in the cylinder over a part or full cycle is called an indicator and the plot of such information obtained is called an indicator diagram. The indicator diagram provides a very good indication of the process of conversion of heat into mechanical work and thus it becomes a useful tool for studying the performance of engines. Basically, there are two types of indicator diagrams, namely, pressure-volume ($p - v$) plot and pressure-crank angle ($p - \theta$) plot which can be taken from various indicators. Figure 3.2 shows such an indicator diagram of a reciprocating engine.

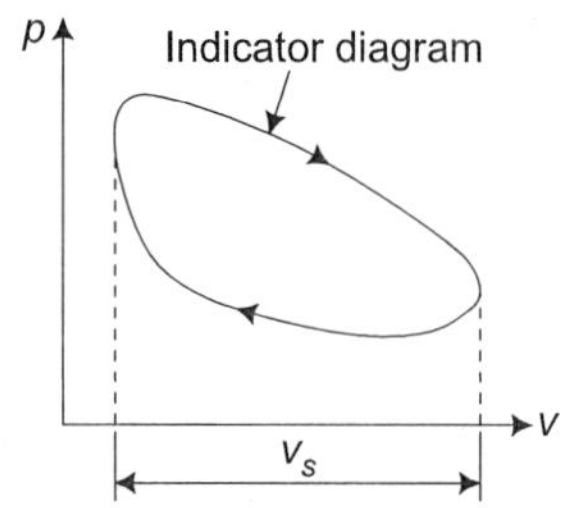

Fig. 3.2 I.C. Engine Indicator Diagram

The area enclosed by the indicator diagram is a measure of work output per cycle from the engine. Mean effective pressure is defined as a hypothetical average pressure which is thought to be acting on the piston throughout the power stroke. It is denoted by p_m.

It is observed that the height on the ($p - v$) diagram is proportional to the pressure on the engine piston. The average height of the diagram in proper pressure units is the indicated mean effective pressure. This average or mean height can be easily determined by measuring the enclosed area with a planimeter and dividing this area by the length of the diagram. Then by the definition of the spring constant, indicated mean effective pressure is given by,

$$i.m.e.p = p_{im} = \left(\frac{a}{l}\right) \times K \quad (3.5)$$

where a the area of indicator diagram, K is the spring constant, and l is the length of indicator diagram.

Length of the indicator diagram is given by the difference between the total volume and the clearance volume.

2. Indicated Power The heat energy liberated inside the cylinder due to combustion of fuel cannot be fully utilized for driving the piston as there are losses to exhaust, to the coolant, and to radiation. The heat energy which is converted to power and is utilized to drive the piston is called indicated power. It is denoted by i.p. The energy distribution in an I.C. engine is shown in Figure 3.3.

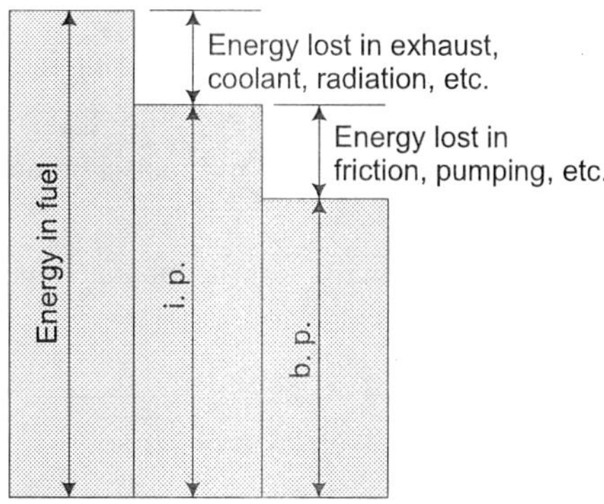

Fig. 3.3 Energy Distribution in an I.C. Engine

Indicated power per cylinder is given by,

$$\text{i.p.} = \frac{p_{im} \times L \times A \times N}{n} \text{ Watts} \tag{3.6}$$

where p_{im} = indicated mean effective pressure in N/m^2, L = stroke length in m, A = area of piston = $(\pi/4)d^2$ in m^2, d = piston diameter, N = rotational speed of the engine in *rps*, n = number of revolutions required to complete one engine cycle, for two-stroke engine n = 1 and for four-stroke engine n = 2.

Thus, for 4-stroke engine, i.p. per cylinder

$$= p_{im} \times L \times A \times (N/2) \tag{3.7}$$

and for 2-stroke engine, i.p. per cylinder

$$= p_{im} \times L \times A \times N \tag{3.8}$$

3. Friction and Brake Power The energy available to drive the piston passes through the connecting rod to the crankshaft. In this transmission there are energy losses due to bearing friction, pumping, etc. The sum of all these losses is termed friction power (f.p). The remaining energy is the useful mechanical energy and is called brake power (b.p). In other words, it is the power available at the crankshaft of the engine. Brake power per cylinder is given by,

$$\text{b.p.} = \frac{p_{im} \times L \times A \times N}{n} \text{ Watts} \tag{3.9}$$

where p_{im} is the brake mean effective pressure.

Brake power is also given by,

$$\text{b.p.} = 2\pi NT \tag{3.10}$$

where T is torque in N-m.

Therefore, from equations (3.9) and (3.10), we have,

$$2\pi NT = \frac{p_{im} \times L \times A \times N}{n}$$

$$\therefore \qquad T = \frac{p_{im} LA}{n} \times \frac{1}{2\pi} \qquad (3.11)$$

Thus, the torque and mean effective pressure are related by the engine size. Higher the mean effective pressure, higher will be the power developed by the engine for a given displacement.

4. Thermal Efficiency It determines how efficiently the fuel is being utilized in the engine. It is of two types:

(i) Indicated Thermal Efficiency It is the ratio of the indicated work done to the energy supplied by the fuel. It is denoted by η_{ith}.

$$\eta_{ith} = \frac{\text{i.p.}}{\text{Energy given by fuel per sec}}$$

$$= \frac{\text{i.p.}}{m_f \times CV} \qquad (3.12)$$

where m_f = mass of fuel in kg/s, CV = Calorific value of fuel in kJ/kg

(ii) Brake Thermal Efficiency It is the ratio of brake or shaft power obtained to the energy supplied by the fuel. It is denoted by η_{bth}.

$$\eta_{bth} = \frac{\text{b.p.}}{\text{Energy given by fuel per sec}}$$

$$= \frac{\text{b.p.}}{m_f \times CV} \qquad (3.13)$$

The value of brake thermal efficiency varies from 25% to 30% in S.I. engines and 25% to 35% in C.I. engines.

5. Mechanical Efficiency It is the ratio of brake power (delivered power) to the indicated power (power provided to the piston). It accounts for the loss due to friction. It is denoted by η_m.

$$\eta_m = \frac{\text{b.p.}}{\text{i.p.}} \qquad (3.14)$$

Also

$$\eta_m = \frac{\text{Brake thermal efficiency}}{\text{Indicated thermal efficiency}}$$

$$= \frac{\eta_{bth}}{\eta_{ith}} \qquad (3.15)$$

6. Relative Efficiency or Efficiency Ratio It is defined as the ratio of indicated thermal efficiency to the corresponding air standard cycle efficiency. It is denoted by η_{rel}.

$$\eta_{rel} = \frac{\text{Indicated thermal efficiency}}{\text{Air standard efficiency}}$$

$$= \frac{\eta_{ith}}{\eta_{air}} \qquad (3.16)$$

7. Volumetric Efficiency It is defined as the ratio of air actually sucked at ambient conditions to the swept volume of the engine. It can be calculated by considering mass or volume of air. However, mass basis is preferred. It is denoted by η_v.

$$\eta_v = \frac{\text{Air or charge actually induced per stroke}}{\text{Swept volume}} \tag{3.17}$$

The volumetric efficiency at full throttle for S.I. engines varies from 80% to 85%, whereas for C.I. engines it varies in between 85% and 90%.

8. Specific Fuel Consumption It is one of the most important parameters for the comparison of the engines. It is defined as the mass of the fuel consumed per kilowatt developed per hour. It is denoted by *sfc*.

When *sfc* is based on brake power, it is called brake specific fuel consumption. It is denoted by *bsfc*.

Thus,
$$bsfc = \frac{m_f}{\text{b.p.}} \text{ kg/kWh} \tag{3.18}$$

where m_f = mass of fuel in kg/h.

Similarly, when *sfc* is based on indicated power, it is called indicated specific fuel consumption. It is denoted by *isfc*.

$\therefore$
$$isfc = \frac{m_f}{\text{i.p.}} \text{ kg/kWh} \tag{3.19}$$

For the gas engines,
$$bsfc = \frac{\text{Gas used in m}^3\text{/h}}{\text{b.p.}} \text{ m}^3\text{/kWh} \tag{3.20}$$

Similarly,
$$isfc = \frac{\text{Gas used in m}^3\text{/h}}{\text{i.p.}} \text{ m}^3\text{/kWh} \tag{3.21}$$

The *sfc* is a criterion of economical power production of the engine or in other words, it tells about how efficiently an engine is converting fuel into work. The value of *bsfc* for diesel engines is low both at full and part loads relative to petrol engines. For diesel engine it is about 0.27 kg/kWh, whereas for petrol engine is 0.34 kg/kWh.

9. Calorific Value Calorific value of a fuel is the energy released per unit quantity of the fuel when the fuel is burned completely and the products of combustion are cooled back to the initial temperature of the combustible mixture. Other terms used for calorific value are heat of combustion and heating value.

When the products of combustion are cooled to 25°C practically all the water vapor resulting from the combustion process will get condensed. The heating value so obtained is called higher calorific value of the fuel. The lower calorific value is heat released when water in the products of combustion does not condense and it remains in the vapor form.

For practical purposes generally lower calorific value of fuel is used in calculations.

10. Fuel-Air (F/A) or Air-Fuel (A/F) Ratio The relative proportions of the fuel and air are very important as far as combustion and engine efficiency are concerned. Either it is expressed as the ratio of the mass of the fuel to that of the air or vice versa.

$$\therefore \quad F/A = (m_f/m_a) \tag{3.22}$$

$$A/F = (m_a/m_f) \tag{3.23}$$

In S.I. engine the fuel-air ratio practically remains constant over a wide range of operation. In the C.I. engine at a particular speed the air flow does not change with load instead it is the fuel flow that varies with the load. Hence, the term fuel-air ratio is generally used instead of air-fuel ratio.

A mixture that contains just enough air for complete combustion of all fuel present in the mixture is called a chemically correct or stoichiometric fuel-air ratio. A mixture having more fuel than that in the chemical correct mixture is called rich mixture, whereas a mixture that contains less fuel than that in the chemical correct mixture is called a lean mixture. The ratio of actual fuel-air ratio to stoichiometric fuel air ratio is called equivalence ratio and is denoted by ϕ .

$$\therefore \quad \phi = \frac{\text{Actual fuel-air ratio}}{\text{Stoichiometric fuel-air ratio}} \tag{3.24}$$

From the above expression it is observed that (i) If $\phi = 1$, the mixture is stoichiometric, (ii) If $\phi > 1$, the mixture will be rich, and (iii) If $\phi < 1$, the mixture will be lean.

Example 3.1 The bore and stroke of a four-cylinder petrol engine are 6.85 cm and 7.5 cm respectively. The compression ratio is 8. Determine the cubic capacity of the engine and the clearance volume of each cylinder.

Solution Swept volume of each cylinder is given by,

$$v_s = (\pi/4) \times d^2 \times L$$

$$= (\pi/4) \times (6.85)^2 \times 7.5 = 276.4 \text{ cm}^3$$

$\therefore$ Cubic capacity of the engine = Swept volume of one cylinder × number of cylinder

$$= 276.4 \times 4 = 1105.6 \text{ cm}^3$$

The compression ratio, $r = \dfrac{v}{v_c} = \dfrac{v_c + v_s}{v_c} = 1 + \dfrac{v_s}{v_c}$

$$8 = 1 + \frac{276.4}{v_c}$$

$$\therefore \quad v_c = \frac{276.4}{7} = 39.51 \text{ cm}^3$$

Example 3.2 A four-cylinder petrol engine has cylinder diameter = 10 cm and stroke = 12 cm. The clearance volume per cylinder is 90 cc. The fuel consumption is 5.5 gm per sec and the developed torque is 175 N-m at a speed of 4200 rpm. Determine (i) compression ratio, (ii) brake power, (iii) brake mean effective pressure, and (iv) brake thermal efficiency if calorific value of fuel is 44000 kJ/kg.

Solution

(i) Swept volume of each cylinder,

$$v_s = (\pi/4) \times d^2 \times L$$

$$= (\pi/4) \times (10^2) \times 12 = 942.48 \text{ cm}^3$$

(ii) Compression ratio,

$$r = \frac{v}{v_c} = \frac{v_c + v_s}{v_c}$$

$$= \frac{90 + 942.48}{90} = 11.47$$

(iii) Brake power,

$$\text{b.p.} = \frac{2\pi NT}{60 \times 1000}$$

$$= \frac{2\pi \times 4200 \times 175}{60 \times 1000} = 76.97 \text{ kW}$$

(iv) Brake mean effective pressure,

$$p_{bm} = \frac{\text{b.p} \times n}{L \times A \times N \times k}$$

$$= \frac{76.97 \times 2}{0.12 \times \pi/4 \times (0.10^2) \times (4200/60) \times 4}$$

$\therefore$ $$p_{bm} = 5.833 \times 10^5 \text{ N/mm}^2 = 5.833 \text{ bar}$$

(v) Brake thermal efficiency,

$$\eta_{bth} = \frac{\text{b.p.}}{m_f \times CV}$$

$$= \frac{76.97}{(5.5/1000) \times 44000} = 0.3180 \text{ or } 31.8\%$$

Example 3.3 A four-stroke S.I. engine develops brake power of 475 kW with mechanical efficiency of 82%. The measured fuel and air consumption are 165 kg/hr and 2500 kg/hr respectively. Calculate the indicated power, friction power, indicated thermal efficiency, and brake thermal efficiency. Take calorific value of the fuel as 43000 kJ/kg.

Solution

Mechanical efficiency, $\text{i.p.} = \frac{\text{b.p.}}{\eta_m} = \frac{475}{0.82} = 579.27 \text{ kW}$

Friction power, $\text{f.p.} = \text{i.p.} - \text{b.p.} = 579.27 - 475 = 104.27 \text{ kW}$

Indicated thermal efficiency,

$$\eta_{ith} = \frac{\text{i.p.}}{m_f \times CV} = \frac{579.27}{(165/3600) \times 43000}$$

$$= 0.2939 \text{ or } 29.39\%$$

Brake thermal efficiency,

$$\eta_{bth} = \frac{\text{b.p.}}{m_f \times CV} = \frac{475}{(165/3600) \times 43000}$$

$$= 0.2410 \text{ or } 24.1\%$$

Example 3.4 A single cylinder four-stroke diesel engine having a swept volume of 900 cm^3 is tested at 300 rpm. When a braking torque 60 Nm is applied, analysis of an indicated diagram gives a mean effective pressure of 10.5 bar. Calculate the break power and mechanical efficiency of the engine.

Solution Brake power,

$$\text{b.p.} = \frac{2\pi NT}{60 \times 1000} = \frac{2\pi \times 300 \times 75}{60 \times 1000} = 1.885 \text{ kW}$$

$$\text{i.p.} = \frac{p_m \times L \times A \times N}{2}$$

$$= \frac{10 \times 10^2 \times 900 \times 10^{-6} \times 300}{60 \times 2} = 2.25 \text{ kW}$$

$$\eta_m = \frac{\text{b.p.}}{\text{i.p.}} = \frac{1.885}{2.25}$$

$$= 0.8378 \text{ or } 83.78\%$$

Example 3.5 A diesel engine develops 4.5 kW. Its indicated thermal efficiency is 29% and mechanical efficiency is 75%. Determine (i) the fuel consumption of engine in kg/hr and in litres per hour, (ii) the indicated specific fuel consumption, and (iii) the brake specific fuel consumption. Take specific gravity of oil 0.86 and calorific value of oil 42 MJ/kg.

Solution

(i)
$$\text{i.p.} = \frac{\text{b.p.}}{\eta_m} = \frac{4.5}{0.75} = 6 \text{ kW}$$

$$\eta_{ith} = \frac{\text{i.p.}}{m_f \times CV}$$

$$m_f = \frac{\text{i.p.}}{\eta_{ith} \times CV} = \frac{6 \times 3600}{0.29 \times 42000} = 1.773. \text{ kg/h}$$

Fuel consumption = (1.773/0.86) = 2.062 litres/h

(ii) Indicated specific fuel consumption,

$$isfc = (1.773/6) = 0.2955 \text{ kg/kWh}$$

(iii) Brake specific fuel consumption,

$$bsfc = (1.773/4.5) = 0.394 \text{ kg/kWh}$$

Example 3.6 Determine the bore and stroke of a four-cylinder, four-stroke petrol engine for the following data: compression ratio = 6, brake power = 300 kW, speed = 300 rpm, stroke = 1.2 times the bore, mechanical efficiency = 0.80, indicated mean effective pressure = 10 bar, relative efficiency = 0.53, and calorific value = 43000 kJ/kg. Also calculate the fuel consumption in kg/s and in kg/kWh. Take γ = 1.4.

Solution

$$\text{i.p.} = \frac{\text{b.p.}}{\eta_m} = \frac{300}{0.80} = 375 \text{ kW}$$

But $$\text{i.p.} = p_{im} \times \frac{LAN}{2} \times k$$

$$375 = 10 \times 10^2 \times (1.2 \times d) \times \left(\frac{\pi}{4} \times d^2\right) \times \frac{300}{2 \times 60} \times 4$$

$\therefore$ $$d = 0.0405 \text{ m} = 4.05 \text{ cm}$$

$\therefore$ $$L = 1.2d = 1.2 \times 4.05 = 4.86 \text{ cm}$$

$$\eta_{air} = 1 - \frac{1}{r^{(\gamma - 1)}} = 1 - \frac{1}{6^{(1.4 - 1)}} = 0.5116$$

$$\eta_{rel} = \frac{\eta_{ith}}{\eta_{air}}$$

$$\eta_{ith} = \eta_{rel} \times \eta_{air}$$

$\therefore$ $$\eta_{ith} = 0.53 \times 0.5116 = 0.2711 \text{ or } 27.11\%$$

$$\eta_{ith} = \frac{\text{i.p.}}{m_f \times CV}$$

$$m_f = \frac{\text{i.p.}}{\eta_{ith} \times CV} = \frac{375}{0.2711 \times 43000}$$

$$= 0.032 \text{ kg/s}$$

$$b.s.f.c. = \frac{m_f \times 3600}{\text{b.p.}} = \frac{0.032 \times 3600}{300}$$

$$= 0.384 \text{ kg/kWh}$$

Example 3.7 Following observations were taken during the testing of a single cylinder four-stroke oil engine: brake wheel diameter = 66 cm, rope diameter = 2.5 cm, speed = 500 rpm, load on rope = 275 N, spring balance reading = 35 N, area and length of indicator diagram = 4.3 cm^2 and 6.4 cm respectively, spring stiffness = 11.5 bar/cm, bore = 10.5 cm, stroke = 15.5 cm, specific fuel consumption = 0.32 kg/kWh, and calorific value of fuel = 42000 kJ/kg. Determine b.p., i.p., η_{ith} and η_{bth}.

Solution The effective brake radius,

$$r = (D + d)/2 = (66 + 2.5)/2 = 34.25 \text{ cm}$$

Torque, $$T = (W - S) \times r = (275 - 35) \times 0.3425 = 82.2 \text{ Nm}$$

$$\text{b.p.} = \frac{2\pi NT}{60 \times 1000}$$

$$= \frac{2\pi \times 500 \times 82.2}{60/1000} = 4.303 \text{ kW}$$

$$p_m = \frac{a}{l} \times k = \frac{4.3}{6.4} \times 11.5 = 7.73 \text{ bar}$$

$$\text{i.p.} = \frac{p_m \times L \times A}{60 \times 1000} \times \frac{N}{2}$$

$$= \frac{7.73 \times 10^5 \times 0.155 \times (\pi/4) \times 0.105^2}{60 \times 1000} \times \frac{500}{2}$$

$$= 4.323 \text{ kW}$$

$$\eta_m = \frac{\text{b.p.}}{\text{i.p.}} \times 100 = \frac{4.303}{4.323} \times 100 = 99.54\%$$

Fuel consumption per hour = 0.32 × 4.303 = 1.377 kg/h

$$\eta_{ith} = \frac{\text{b.p.} \times 60}{m_f \times CV} \times 100$$

$$= \frac{4.303 \times 3600}{1.377 \times 42000} \times 100 = 26.78\%$$

$$\eta_{bth} = \eta_m \times \eta_{ith}$$

$$= 0.9954 \times 0.2678 = 0.2666 \text{ or } 26.66\%$$

Example 3.8 The air-fuel ratio used in S.I. engine of 85 kW capacity is 12:1. Determine the amount of air consumed by the engine at full load when the brake thermal efficiency is 26%. Also determine m^3 of air and m^3 of fuel required per hour if density of air is 1.2 kg/m^3 and density of fuel is 3.45 kg/m^3. If the calorific value of fuel is 43000 kJ/kg, how many m^3 per hour of the mixture is required?

Solution

$$\eta_{bth} = \frac{3600}{m_f \times CV}$$

$$m_f = \frac{3600}{\eta_{bth} \times CV} = \frac{3600}{0.26 \times 43000} = 0.322 \text{ kg/kWh}$$

Fuel consumption per hour = m_f × b.p. = 0.322 × 85 = 27.37 kg/h

Air consumption per hour

$$m_a = \text{fuel consumed} \times \frac{A}{F}$$

$$= 0.322 \times \frac{12}{1} = 328.44 \text{ kg/h}$$

Volume of air,
$$v_a = \frac{\text{Mass of air}}{\text{Density of air}}$$

$$= \frac{328.44}{1.2} = 273.7 \text{ m}^3\text{/h}$$

Volume of fuel,
$$v_f = \frac{\text{Mass of fuel}}{\text{Density of fuel}}$$

$$= \frac{27.37}{3.45} = 7.93 \text{ m}^3\text{/h}$$

∴ Mixture volume = 273.7 + 7.53 = 281.63 m^3/h

Example 3.9 A two-stroke C.I. engine delivers 4950 kW while using 950 kW to overcome friction losses. It consumes 2250 kg of fuel per hour at an air fuel ratio of 18:1. The calorific value of the fuel is 42500 kJ/kg. Determine i.p., η_m, η_{ith} and η_{bth}.

Solution

$$\text{i.p.} = \text{b.p.} + \text{f.p.}$$

$$= 4950 + 950 = 59000 \text{ kW}$$

$$\eta_m = \frac{\text{b.p.}}{\text{i.p.}}$$

$$= \frac{4950}{5900} = 0.839 \text{ or } 83.9\%$$

Air consumption per hour,

$$m_a = \text{fuel consumed} \times \frac{A}{F}$$

$$= 2250 \times \frac{18}{1} = 40500 \text{ kg/h}$$

$$\eta_{ith} = \frac{\text{i.p.} \times 3600}{m_f \times CV} \times 100$$

$$= \frac{5900 \times 3600}{2250 \times 42500} \times 100 = 22.21\%$$

$$\eta_{bth} = \eta_m \times \eta_{ith}$$

$$= 0.839 \times 0.2221 = 0.1863 \text{ or } 18.63\%$$

Example 3.10 A 4-stroke 4-cylinder diesel engine running at 2100 rpm develops 62 kW. Its brake thermal efficiency is 29%, mechanical efficiency is 81%, and calorific value of fuel is 42000 kJ/kg. Engine has a bore of 12 cm and stroke of 10 cm. Take density of air as 1.2 kg/m^3 and air-fuel ratio = 15:1. Determine (i) the fuel consumption of engine in kg/min, (ii) air consumption in m^3/s, (iii) the indicated thermal efficiency, (iv) volumetric efficiency, (v) brake mean effective pressure, and (vi) mean piston speed.

Solution

(i)

$$m_f = \frac{\text{b.p.}}{\eta_{bth} \times CV} = \frac{62 \times 60}{0.29 \times 42000}$$

$$= 0.3054 \text{ kg/min or } 5.09 \times 10^{-3} \text{ kg/s}$$

(ii) Air consumption per second

$$m_a = \frac{\text{Mass of fuel}}{\rho_a} \times \frac{A}{F}$$

$$= \frac{5.09 \times 10^{-3}}{1.2} \times \frac{15}{1} = 6.36 \times 10^{-3} \text{ m}^3/\text{s}$$

(iii)
$$\text{i.p.} = \frac{\text{b.p.}}{\eta_m} = \frac{62}{0.81} = 76.54 \text{ kW}$$

$$\eta_{ith} = \frac{\text{i.p.}}{m_f \times CV} \times 100$$

$$= \frac{76.54}{5.09 \times 10^{-3} \times 42000} \times 100 = 35.8\%$$

3.8 ENGINE PARTS AND THEIR FUNCTIONS

The main parts of an I.C. engine are shown in Figure 3.4 and are explained below:

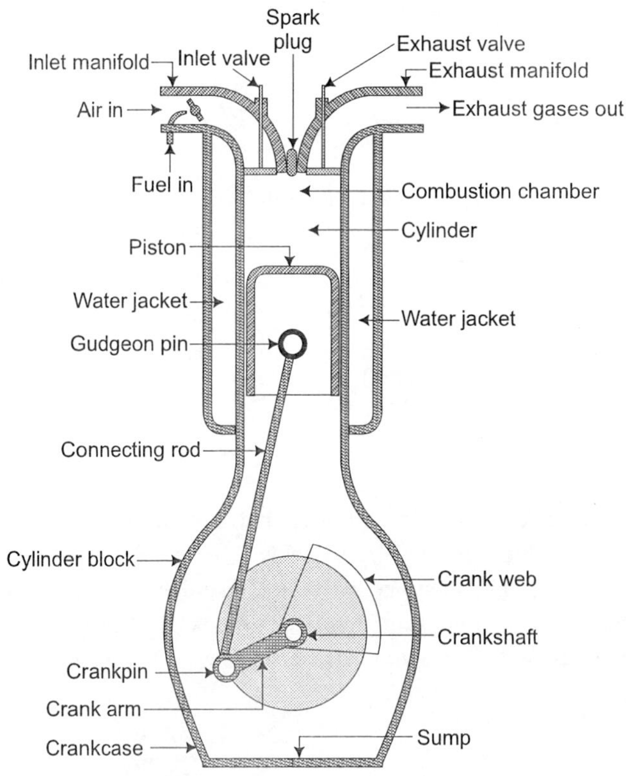

Fig. 3.4 Main Parts of I.C. Engines (S.I. Engine)

1. Cylinder and Cylinder Head The cylinder of an I.C. engine is the main body of the engine in which piston reciprocates to develop power. The space enclosed between the cylinder head and the top of the piston is called combustion chamber. The cylinders have to withstand very high pressures (about 70 bar) and the temperatures (more than 2000°C) because there is direct combustion inside the cylinder. Therefore, its material should be such that (i) it can retain strength at high temperatures, (ii) it should be a good conductor of heat, and (iii) it can sufficiently resistant to rapid wear and tear due to fast moving reciprocating parts. Generally, ordinary cast iron is used for cylinders but in the case of heavy duty engines alloy steels are used. In the case of aeroplanes and automobiles, where lightness is an important factor aluminium alloys are used. Sometimes sleeves or liners are inserted into the cylinders which can be replaced when worn out.

In the case of multiple cylinder engines, the cylinders are normally cast in one block known as cylinder block.

The cylinder head closes one end of the cylinder. It houses the inlet and exhaust valves through which the charge is admitted and exhaust gases discharged to the atmosphere. The cylinder head is usually a one-piece casting bolted to the top of the cylinder block. A copper asbestos gasket is provided in between the block and the head to make a gas tight joint. The cylinder head is also made from cast iron. The cylinder and cylinder head are provided with water jackets in the case of water cooling and fins in the case of air cooling.

2. Piston and Piston Rings Piston is a cylindrical component which fits perfectly into the cylinder. The piston compresses the charge during compression stroke and transmits the gas force developed due to the burning of fuel during expansion stroke to the connecting rod and then to the crankshaft. Internal combustion engine pistons are usually made of aluminum, cast iron, and cast steel. Aluminium has the advantage of higher thermal conductivity and low density.

The piston is fitted with minimum three piston rings which are commonly made of silicon cast iron. The upper piston rings (at least two) are called the compression rings. The compression rings do not allow the leakage of gases in the engine cylinder and thus provide a gas tight seal. The lower piston rings are called oil rings or oil control rings. The oil rings have a groove with several holes so as to transfer the surplus lubricating oil from the cylinder walls to the oil sump.

3. Inlet and Exhaust Manifolds The pipe which connects the intake system to the inlet valve of the engine and through which air or air-fuel mixture is sucked into the cylinder is called the inlet manifold.

The pipe which connects the exhaust system to the exhaust valve of the engine and through which waste gases escape into the atmosphere is called exhaust manifold.

4. Inlet and Exhaust Valves The valves are usually mushroom shaped known as poppet valves with conical seating surfaces. The valves may be provided on the top of the engine or on the side of the engine cylinder. The inlet valve controls the admission of the air or fuel-air mixture during suction stroke of the engine. The exhaust valve controls the removal of waste gases during exhaust stroke. The valves are made of specially alloy steel.

5. Connecting Rod The connecting rod connects the piston and the crankshaft. It transmits the gas forces developed during the expansion stroke to the crankshaft and thus converts the reciprocating motion of the piston into the rotary motion of the crankshaft. The smaller end of the connecting rod is connected to the piston by gudgeon pin and the bigger end is connected to the crankshaft by crankpin. Connecting rod is made of forged steel.

6. Camshaft The camshaft controls the opening and closing of the intake and exhaust valves through the cams, cam followers, push rods, and rocker arms. This shaft also provides the drive to the ignition system. The camshaft is driven by crankshaft through timing gears. In 4-stroke engines the working cycle is completed in two revolutions of the crankshaft and every valve works once only. Thus, camshaft should run at half the crankshaft speed.

7. Crank and Crankshaft Crank and crankshaft converts the reciprocating motion of the piston into rotary motion. Both crank and crankshaft are steel forged and machined to a smooth finish. The crank and crankshaft are held together by means of a key. The crankshaft is supported in main bearings at the two ends. The crankshaft is enclosed in a crankcase. In the crankshaft, crankarms and balance weights are provided for static and dynamic balancing of the rotating system.

8. Flywheel Flywheel is made in the form of a heavy thoroughly balanced disk fitted on the crankshaft. It stores excess energy produced during power stroke thus acts as a reservoir and gives up this stored energy for completing the other idle strokes. When energy is absorbed by a flywheel its speed increases, but its speed decreases when it gives up energy. The alternate absorbing and giving out of energy is continuous. Thus, it tends to keep the speed of the engine within the required limits and ensures uniform torque on the crankshaft during each working cycle.

For lower speeds flywheels are manufactured from cast iron but for higher speeds cast steel is used.

9. Governor Governor is run by a drive from the crankshaft. It acts only when load varies. It regulates the amount of charge in the case of petrol engine and amount of fuel in the case of diesel engine so that speed of the engine can be maintained nearly constant as and when the load varies.

10. Carburettor Carburettor is a device which prepares a combustible fuel-air mixture by mixing the proper amount of fuel with air and delivers it to the cylinder through the intake manifold. The amount of charge entering the cylinder is controlled by a throttle valve.

11. Spark Plug This component is used to initiate the combustion process after the completion of the compression process in spark-ignition engines and is generally mounted on the cylinder head. This is used only in petrol engines.

12. Fuel Pump and Fuel Injector Fuel pump or fuel injection pump delivers accurately measured quantity of fuel under high pressure at the correct instant to the injector fitted to the cylinder.

Fuel injector atomizes the fuel into very fine droplets which cause increase in surface area of the fuel droplets resulting in better mixing and eventually leads to better combustion of the fuel particles. Both the fuel pump and fuel injector are used in diesel engines.

3.9 FOUR-STROKE PETROL ENGINE (S.I. ENGINE)

In order to start the engine the crankshaft is to be rotated either manually or by a starter motor causing the connecting rod to move the piston downwards from top dead centre (TDC) to bottom dead centre (BDC). In a 4-stroke petrol engine the cycle of operations is completed in four strokes of the piston, or two revolutions of the crankshaft and petrol is used as fuel. The four strokes are: (i) Intake or suction stroke, (ii) Compression stroke, (iii) Expansion stroke, and (iv) Exhaust stroke. Each stroke consists of 180° of crankshaft rotation and hence a cycle consists of 720° of crankshaft rotation.

The salient features of the four strokes in a petrol engine are explained below:

1. Intake or Suction Stroke (0-1) Suction stroke (process 0-1, Figure 3.5) starts when the piston is at the top dead centre and about to move downwards. At this time the inlet valve is opened and the exhaust valve is closed (Figure 3.6 a). As the piston moves downwards a fresh charge of fuel-air mixture enters the cylinder through the inlet valve due to the suction created. When the piston reaches the bottom dead centre the suction stroke ends and the inlet valve closes. This completes the suction stroke and the crank is rotated by 180° during this stroke.

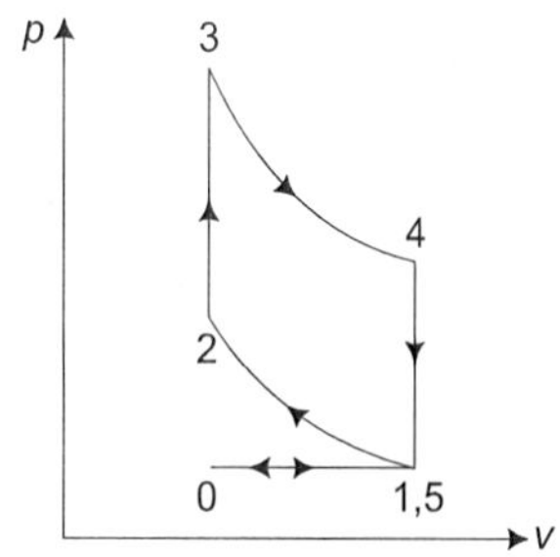

Fig. 3.5 $p - v$ Diagram of a 4-stroke S.I. Engine

2. Compression Stroke (1-2) During this stroke, both the inlet and exhaust valves are closed and the charge is compressed to the clearance volume (Figure 3.6 b). During compression of the charge its temperature and pressure both increase. At the end of compression stroke the air-fuel mixture is ignited with the help of a spark produced by the spark plug. Ideally, it is assumed that the burning takes place instantaneously and thus heat addition takes place at constant volume. During the burning process of the fuel its chemical energy is converted into heat energy and thus produces a temperature rise of above 2000°C (process 2-3, Figure 3.5). Due to heat released from the fuel the pressure at the end of the combustion process increases considerably.

3. Expansion or Power Stroke (3-4) During the expansion stroke both valves remain closed (Figure 3.6 c). The high pressure of the burnt gases forces the piston towards bottom dead centre (process 3-4, Figure 3.5). Thus, power is obtained during this stroke only. Both pressure and temperature decrease during expansion. The exhaust valve opens as the piston reaches the bottom dead centre position and pressure falls suddenly to atmospheric pressure at constant volume (process 4-1, Figure 3.5).

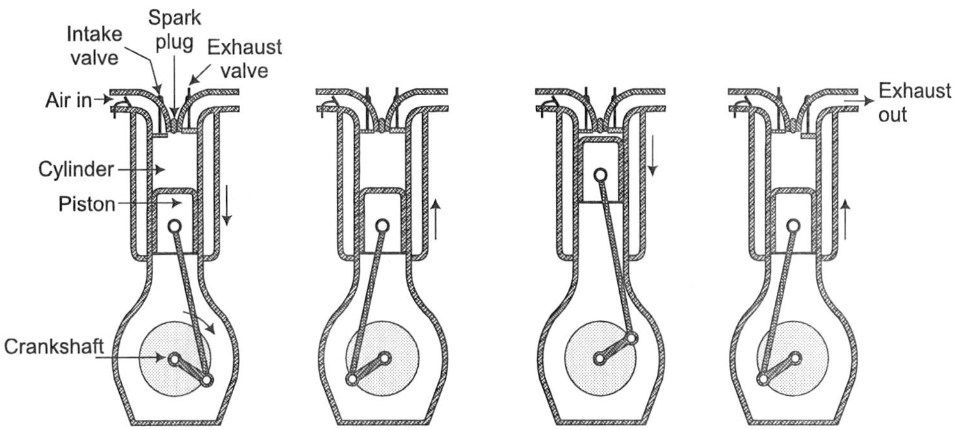

Fig. 3.6 4-Stroke S.I. Engine Cycle Operations

4. Exhaust Stroke At the end of the expansion stroke the exhaust valve opens and the inlet valve remains closed (Figure 3.6 d). The piston starts moving from the bottom dead centre and sweeps out the burnt gases from the cylinder (process 5-0, Figure 3.5). As the piston reaches the top dead centre again the inlet valve opens and fresh charge is taken in during the next downward movement of the piston and the cycle is repeated.

Each cylinder of a four-stroke engine completes the above four operations in two engine revolutions. One revolution of the crankshaft completes during the suction and compression strokes, and the second revolution during the power and exhaust strokes. Thus, for one complete cycle there is only one power stroke while crankshaft turns by two revolutions.

3.10 FOUR-STROKE COMPRESSION IGNITION ENGINE (C.I. ENGINE)

The working cycle of the C.I. engine is completed in four strokes which completes in two revolutions of the crankshaft and diesel is used as the fuel. The four-stroke C.I. engine is similar to the four-stroke S.I. engine but it operates at a much higher compression ratio. The compression ratio of S.I. engines lies in between 6 and 10 while for C.I. engine it lies in between 16 and 20. In C.I. engines a high pressure fuel pump and an injector is provided to inject the fuel into the combustion chamber.

The sequence of operation for the four-stroke C.I. engine is given below:

1. **Suction Stroke** Suction stroke (process 0-1, Figure 3.7) starts when the piston is at TDC position and about to move downwards. At this time the inlet valve is opened and the outlet valve is closed (Figure 3.8 a). As the piston moves downwards fresh air enters the cylinder through the inlet valve due to suction created. This continues until piston reaches BDC thus suction stroke gets completed. It means suction stroke is similar to that in petrol engine except that only air is taken into the cylinder.

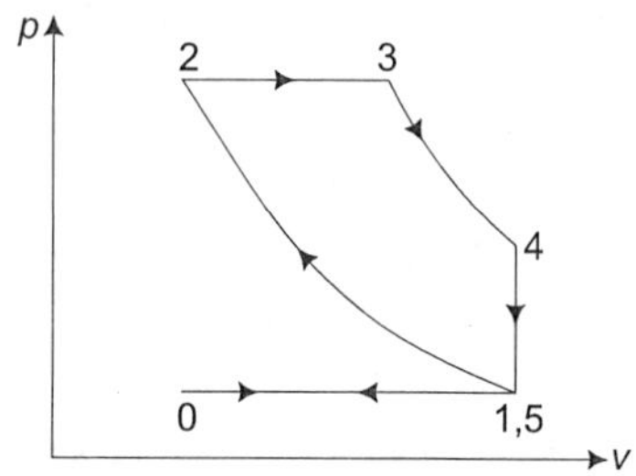

Fig. 3.7 $p - v$ Diagram of a 4-stroke C.I. Engine

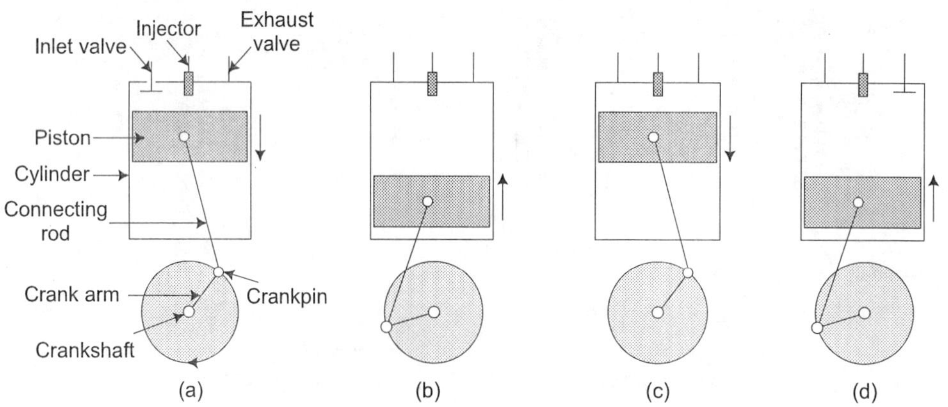

Fig. 3.8 4-Stroke C.I. Engine Cycle Operations

2. **Compression Stroke** Both the valves remain closed during this stroke (Figure 3.8 b). Piston moves from BDC to TDC and thus air inducted during the suction stroke is compressed into the clearance volume. The pressure and temperature of the air increase continuously during this process. Near the end of compression, pressure and temperature of air is about 60 bar and 600°C respectively. This stroke on $p - v$ diagram is shown by process 1-2 (Figure 3.7).
3. **Expansion Stroke or Power Stroke** Both the valves remain closed during the expansion stroke (Figure 3.8 c). Nearly at the end of compression stroke the fuel injection starts. The fuel enters the engine cylinder in the form of a jet and after entering the cylinder, it atomizes, evaporates, and mixes with hot air to form a combustible mixture. After completion of fuel injection (i.e., after cut-off) the products of combustion expand. Since the temperature of air is much above the self-ignition temperature of the fuel thus auto ignition of fuel initiates the combustion process. It is assumed that heat addition during the combustion of fuel occurs at constant pressure (process 2-3, Figure 3.7). The high pressure and high temperature gases produced during the combustion of fuel forces the piston towards BDC (process 3-4). Thus, power is produced during this stroke. This stroke gets completed as the piston reaches its BDC position.
4. **Exhaust Stroke** During the exhaust stroke (process 5-0, Figure 3.7) the exhaust valve is opened and the inlet valve remains closed (Figure 3.8 d). The piston travels from BDC to TDC and thus pushes out the burnt gases. The exhaust stroke completes the cycle and the engine again takes the fresh air inside the cylinder and thus the cycle is repeated.

3.11 COMPARISON BETWEEN PETROL ENGINE (S.I. ENGINE) AND DIESEL ENGINE (C.I. ENGINE)

Both S.I. and C.I. engines have much in common. The detailed comparison is given below:

S.I. Engine	*C.I. Engine*
1. It works on Otto cycle or constant volume heat addition cycle.	1. It works on Diesel cycle or constant pressure cycle.
2. Petrol is used as fuel.	2. Diesel is used as fuel.
3. A homogeneous combustible mixture of fuel and air is supplied during the suction stroke. A carburettor and an ignition system are essential.	3. Only air is sucked during the suction stroke. At the end of compression stroke fuel is directly injected at high pressure into the combustion chamber. A fuel pump and injector are essential.
4. Generally, compression ratio lies in between 6 and 10. The upper limits of compression ratio are fixed by antiknock quality of fuel.	4. The compression ratio in diesel engine lies in between 16 and 20. The upper limit of compression ratio is fixed by weight increase of the engine.
5. Throttle valve controls the quantity of fuel-air mixture. Thus petrol engines are quantity control engines.	5. In the case of diesel engine, the fuel pump regulates the quantity of fuel (diesel). Air quantity is not controlled. Thus, diesel engines are quality control engines.
6. Petrol engines are high-speed engines due to light weight and homogeneous combustion.	6. Diesel engines are low-speed engines due to heavy weight and heterogeneous combustion.

7. Petrol engines require an ignition system with a spark plug in the combustion chamber to initiate combustion.	7. Diesel engines do not require any ignition system and spark plug. High temperature of air causes self-ignition of fuel.
8. Maximum thermal efficiency is lower due to low compression ratio.	8. Maximum thermal efficiency is higher due to higher compression ratio.
9. Petrol engines are lighter in weight due to lower maximum pressure.	9. Diesel engines are heavier due to higher maximum pressure.
10. Initial cost is lower.	10. Initial cost is higher.
11. Petrol engines are used for light duty vehicles.	11. Diesel engines are used for heavy duty vehicles.
12. Noise, pollution, and vibration are lower.	12. Noise, pollution, and vibration are higher.

3.12 TWO-STROKE ENGINES

In two-stroke engines the working cycle is completed in two strokes or one revolution of the crankshaft. In two-stroke engines no piston strokes are required for suction and exhaust operations. This can be achieved as follows:

1. By using the closed crankcase compression. Here the crankcase acts as an air pump due to the up and down movement of the piston. Due to the downward movement of the piston the charge or air gets compressed in the crankcase by its pumping action of the underside of the piston as shown in Figure 3.9. This system is known as three-port system and is commonly used for single cylinder small power engines such as scooter.

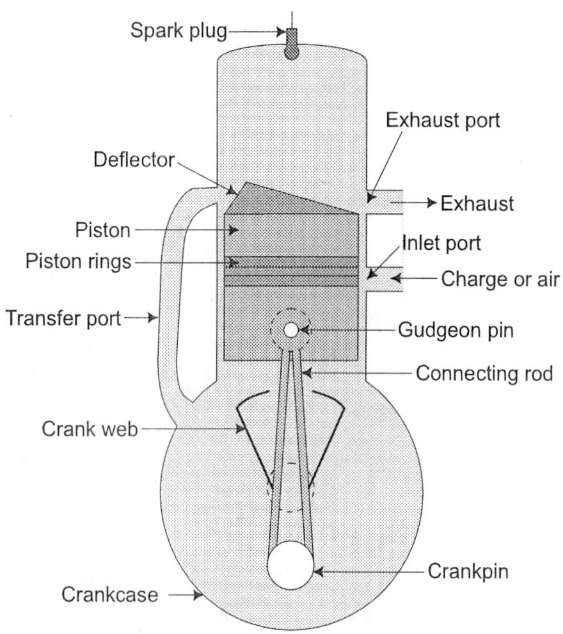

Fig. 3.9 2-Stroke Engine

2. Out side the cylinder a pump is used to compress the charge or air and supply it to the cylinder. This pump is driven by the engine itself. This method is used for large capacity multi-cylinder engines.

Working of Two-Stroke Petrol Engine In the beginning it is considered that the piston is moving towards TDC. At the end of the compression, spark has been produced by the spark plug (Figure 3.10a) thus the compressed charge is ignited. Due to the combustion of fuel high pressure and temperature gases produce which pushes the piston in the downward direction thus produces power. During the downward stroke the piston first uncovers the exhaust port, so the burnt gases leave the cylinder through the exhaust port (Figure 3.10b). Further movement of the piston uncovers the transfer port. The downward movement of the piston simultaneously compresses the charge in the crankcase due to the pumping action of the underside of the piston and thus the compressed charge is supplied to the cylinder through the transfer port. The fresh charge is deflected towards the top of the cylinder by the crown provided on the top of the piston thus the exhaust gases are swept out with the help of the fresh charge (Figure 3.10c). This action of sweeping out the burnt gases with the help of fresh charge is known as scavenging. The crown on the top of the piston serves double purpose of scavenging the upper part of the cylinder and preventing the fresh charge flowing directly to the exhaust port. This will continue till the piston reaches the BDC position. Now the piston starts moving towards TDC. First, it uncovers the inlet port through which charge is filled in the crankcase by the carburettor. Further, movement of the piston covers the transfer port and thus stopping the supply of fresh charge to the cylinder. Then a little later, it covers the exhaust port and actual compression of the charge begins (Figure 3.10d). And at the end of the compression, spark occurs and the cycle is repeated. By this, cycle of the engine is completed within two strokes of the piston or one revolution of the crankshaft

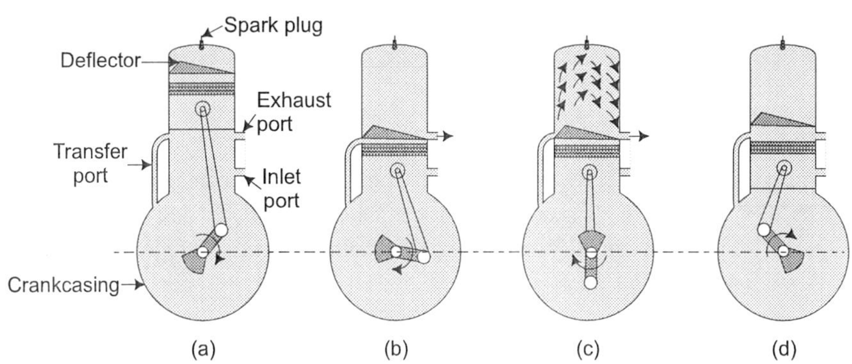

Fig. 3.10 2-Stroke Petrol Engine Cylinder Events

Working of Two-Stroke Diesel Engine The working of two-stroke diesel engine is similar to that of petrol engine. The basic differences are: (i) In the case of diesel engine only air is supplied to the cylinder through the transfer port due to the pumping action of the underside of the piston, (ii) The fuel (diesel) is injected with the help of fuel injector at the end of compression stroke, and (iii) It has high compression ratio means at the end of compression stroke air has high temperature due to which fuel gets self-ignited. Thus, there is no need of spark plug in the case of diesel engine.

3.13 COMPARISON BETWEEN 2-STROKE AND 4- STROKE ENGINES

4-Stroke Engine	*2-Stroke Engine*
1. The working cycle is completed in four strokes of the piston or in two revolutions of the crankshaft.	1. The working cycle is completed in two strokes of the piston or in one revolution of the crankshaft.
2. One power stroke is obtained in every two revolutions of the crankshaft.	2. One power stroke is obtained in each revolution of the crankshaft.
3. Turning moment is not so uniform thus heavier flywheel is required.	3. Turning moment is more uniform hence lighter flywheel is required.
4. Lesser cooling and lubrication is needed.	4. Greater cooling and lubrication is needed.
5. Power produced for the same size of engine is small or for the same power the engine is heavy and bulky.	5. Theoretically it develops double the power for the same size of the engine hence it develops more power or for the same power the engine is light and compact.
6. Lesser rate of wear and tear.	6. Greater rate of wear and tear.
7. The four-stroke engines have valves and valve mechanism.	7. The two-stroke engines have no valves or valve mechanism. It has ports.
8. Initial cost is high.	8. Initial cost is low.
9. Volumetric efficiency of four-stroke engine is high.	9. Volumetric efficiency of two-stroke engine is low.
10. Thermal efficiency of four-stroke engine is high.	10. Thermal efficiency of two-stroke engine is low.
11. Part load efficiency of four-stroke engine is high.	11. Part load efficiency of two-stroke engine is low.
12. Four-stroke engines are used where efficiency is important as in cars, buses, trucks, tractors, airplanes, etc.	12. Two-stroke petrol engines are used in small size only like lawn movers, scooters, mopeds. Two-stroke diesel engines are used in large sizes, e.g., in ships.

3.14. OTTO CYCLE

In 1876, Nicolaus Otto a German scientist proposed a constant volume heat addition cycle which forms the basis for the working of today's spark ignition engines. The indictor diagram given in Figure 3.5 shows the sequence of processes in the operation of S.I. engines. Figure 3.11 shows the ideal Otto cycle on $p - v$ and $T - s$ diagram respectively.

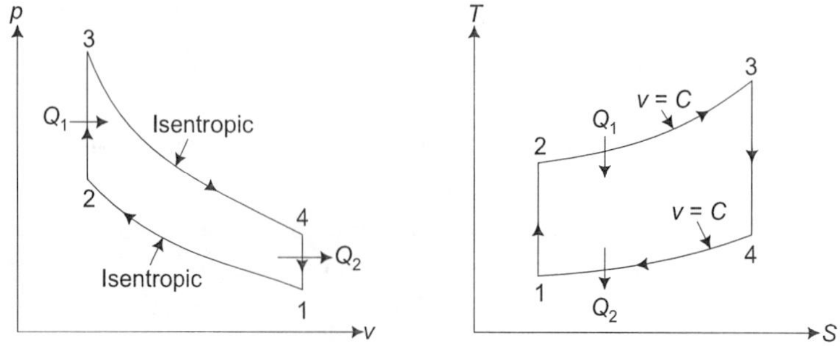

Fig. 3.11 $p - v$ and $T - s$ Diagrams of Otto Cycle

It consists of four internally reversible processes:

1-2: Isentropic compression, 2-3: Constant volume heat addition,

3-4: Isentropic expansion, and 4-1: Constant volume heat rejection.

Expression for Thermal Efficiency

Let m = mass of air supplied to the engine.

Heat supplied, $Q_1 = Q_{2-3} = mc_v\,(T_3 - T_2)$

Heat rejected, $Q_2 = Q_{4-1} = mc_v\,(T_4 - T_1)$

Efficiency,
$$\eta = \frac{\text{Heat supplied} - \text{Heat rejected}}{\text{Heat supplied}}$$

$$\eta = \frac{Q_1 - Q_2}{Q_1} = 1 - \frac{Q_2}{Q_1}$$

$$= 1 - \frac{mc_v\,(T_4 - T_1)}{mc_v\,(T_3 - T_2)}$$

$$\eta = 1 - \frac{(T_4 - T_1)}{(T_3 - T_2)} \qquad (3.25)$$

Process 1-2:
$$\frac{T_2}{T_1} = \left(\frac{v_1}{v_2}\right)^{\gamma - 1},$$

Process 3-4:
$$\frac{T_3}{T_4} = \left(\frac{v_4}{v_3}\right)^{\gamma - 1} = \left(\frac{v_1}{v_2}\right)^{\gamma - 1} = (r)^{\gamma - 1}$$

where r is the compression ratio and $\gamma = (c_p/c_v)$ is the specific heat ratio.

$$\therefore \qquad \frac{T_2}{T_1} = \frac{T_3}{T_4}$$

or
$$\frac{T_3}{T_2} = \frac{T_4}{T_1}$$

or
$$\frac{T_3}{T_2} - 1 = \frac{T_4}{T_1} - 1$$

$$\therefore \qquad \frac{T_4 - T_1}{T_3 - T_2} = \frac{T_1}{T_2} = \left(\frac{v_2}{v_1}\right)^{\gamma - 1} \qquad \text{(i)}$$

Substituting the expression (i) in equation (3.25), we have,

$$\eta = 1 - (v_2/v_1)^{\gamma - 1}$$

$$\therefore \qquad \eta_{\text{Otto}} = 1 - \frac{1}{(r)^{\gamma - 1}} \qquad (3.26)$$

Equation (3.26) shows that the thermal efficiency of an ideal Otto cycle depends on the compression ratio of the engine and the specific heat ratio of the working fluid. The thermal efficiency of the

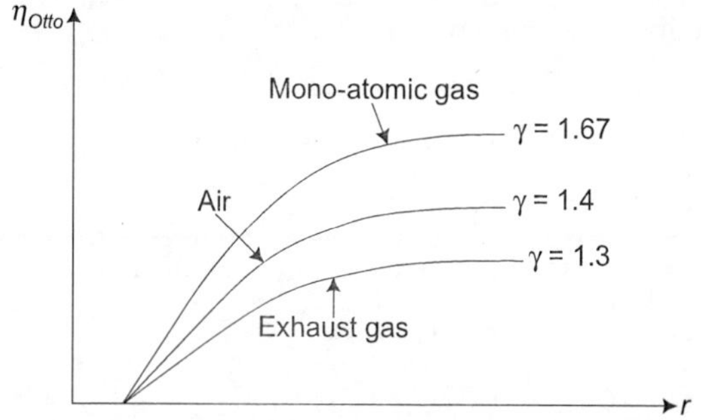

Fig. 3.12 Otto Cycle Efficiency at different r and γ

ideal Otto cycle increases with both the compression ratio and the specific heat ratio as shown in Figure 3.12.

It is independent of the temperature levels at which the cycle operates. Thermal efficiency of S.I. engines can be increased by utilizing higher compression ratio up to about 12. But it cannot be increased beyond a certain limit because of detonation (a noisy and destructive combustion phenomenon) and other problems such as structural strength, bearing reliability, etc.

Expression for Work Output

The net work output for Otto cycle can be given as,

$$W = \text{Area of } p - v \text{ diagram}$$

$$W = \frac{p_3 v_3 - p_4 v_4}{\gamma - 1} - \frac{p_2 v_2 - p_1 v_1}{\gamma - 1} \quad (3.27)$$

Also $$\frac{p_2}{p_1} = \frac{p_3}{p_4} = r^{\gamma},$$

and $$\frac{p_3}{p_2} = \frac{p_4}{p_1} = \text{pressure ratio} = r_p$$

$$v_1 = r v_2$$

and $$v_4 = r v_3$$

$\therefore$ $$W = \frac{p_1 v_1}{\gamma - 1}\left[\frac{p_3 v_3}{p_1 v_1} - \frac{p_4 v_4}{p_1 v_1} - \frac{p_2 v_2}{p_1 v_1} + 1\right]$$

$$= \frac{p_1 v_1}{\gamma - 1}\left[\frac{p_3}{p_2} \cdot \frac{p_2}{p_1} \cdot \frac{1}{(v_1/v_2)} - \frac{p_4}{p_1} \cdot \frac{v_4}{v_4} - \frac{p_2}{p_1} \cdot \frac{1}{(v_1/v_2)} + 1\right]$$

$(\therefore v_1 = v_4 \ \& \ v_2 = v_3)$

$$= \frac{p_1 v_1}{\gamma - 1}\left[\frac{r_p \cdot r^{\gamma}}{r} - r_p - \frac{r^{\gamma}}{r} + 1\right]$$

$$= \frac{p_1 v_1}{\gamma - 1}\left[r_p \cdot r^{\gamma - 1} - r_p - r^{\gamma - 1} + 1\right]$$

$$\therefore \qquad W = \frac{p_1 v_1}{\gamma - 1}(r_p - 1)(r^{\gamma - 1} - 1) \tag{3.28}$$

It is observed from equation (3.28) that work output is directly proportional to pressure ratio r_p.

Expression for Mean Effective Pressure

The mean effective pressure is given by,

$$p_m = \frac{\text{Work output}}{\text{Swept volume}} \tag{i}$$

$$\text{Swept volume} = v_1 - v_2 = v_2\left(\frac{v_1}{v_2} - 1\right)$$

$$= v_2 (r - 1) \tag{ii}$$

Substituting equation (3.28) and the expression (ii) in the expression (i), we have,

$$p_m = \frac{\dfrac{p_1 v_1}{\gamma - 1}(r_p - 1)(r^{\gamma - 1} - 1)}{v_2 (r - 1)}$$

or

$$p_m = \frac{p_1\left(\dfrac{v_1}{v_2}\right)(r_p - 1)(r^{\gamma - 1} - 1)}{(\gamma - 1)(r - 1)}$$

$$\therefore \qquad p_m = \frac{p_1 r (r_p - 1)(r^{\gamma - 1} - 1)}{(\gamma - 1)(r - 1)} \tag{3.29}$$

For Otto cycle, an increase in compression ratio leads to an increase in the mean effective pressure as well as the thermal efficiency. The thermal efficiency of actual S.I. engines varies from 25% to 30%.

Example 3.11 Compare the changes in efficiency of an Otto engine when the compression ratio changes from 5 to 6. Take $\gamma = 1.4$ for air.

Solution

$$(\eta_{\text{Otto}})_1 = 1 - \frac{1}{(r_1)^{\gamma - 1}} = 1 - \frac{1}{(5)^{1.4 - 1}}$$

$$= 0.4747 \text{ or } 47.47\%$$

$$(\eta_{\text{Otto}})_2 = 1 - \frac{1}{(r_2)^{\gamma - 1}} = 1 - \frac{1}{(6)^{1.4 - 1}}$$

$$= 0.5116 \text{ or } 51.16\%$$

Therefore, percentage increase in efficiency,

$$= \frac{(\eta_{\text{Otto}})_2 - (\eta_{\text{Otto}})_1}{(\eta_{\text{Otto}})_2} \times 100$$

$$= \frac{0.5116 - 0.4747}{0.5116} \times 100 = 7.77\%$$

Example 3.12 An engine works on the Otto cycle which is supplied with air at 100 kN/m^2, 30°C. The compression ratio is 7. Heat supplied is 2000 kJ/kg. Calculate (i) cycle efficiency, (ii) the maximum pressure and temperature of the cycle, and (iii) mean effective pressure. Take c_p = 1.005 kJ/kgK, c_v = 0.718 kJ/kgK, and R = 0.287 kJ/kgK for air.

Solution Refer Figure 3.13.

(i)

$$\eta_{Otto} = 1 - \frac{1}{(r_1)^{\gamma - 1}} = 1 - \frac{1}{(7)^{1.4 - 1}}$$

$$= 0.5408 \text{ or } 54.08\%$$

(ii)

$$v_1 = \frac{RT_1}{p_1} = \frac{0.287 \times 303}{100} = 0.87 \text{ m}^3\text{/kg}$$

$$\frac{v_1}{v_2} = 7$$

$$\therefore \quad v_2 = \frac{v_1}{7} = \frac{0.87}{7} = 0.124 \text{ m}^3\text{/kg}$$

$$\frac{T_2}{T_1} = \left(\frac{v_1}{v_2}\right)^{\gamma - 1} = (7)^{1.4 - 1}$$

$$\therefore \quad T_2 = (7)^{1.4 - 1} \times T_1 = (7)^{0.4} \times 303 = 660.54 \text{ K}$$

$$Q_1 = c_v (T_3 - T_2) = 2000 \text{ kJ/kg}$$

$$T_3 - T_2 = \frac{2000}{c_v}$$

$$T_3 - 660.54 = \frac{2000}{0.718}$$

$$\therefore \quad T_3 = T_{max} = 3446.06 \text{ K}$$

$$\frac{p_2}{p_1} = \left(\frac{v_1}{v_2}\right)^{\gamma} = (7)^{1.4} = 15.25$$

$$\therefore \quad p_2 = 15.25 \times p_1 = 15.25 \times 100 = 1525 \text{ kN/m}^2$$

$$\frac{p_3 v_3}{T_3} = \frac{p_2 v_2}{T_2}$$

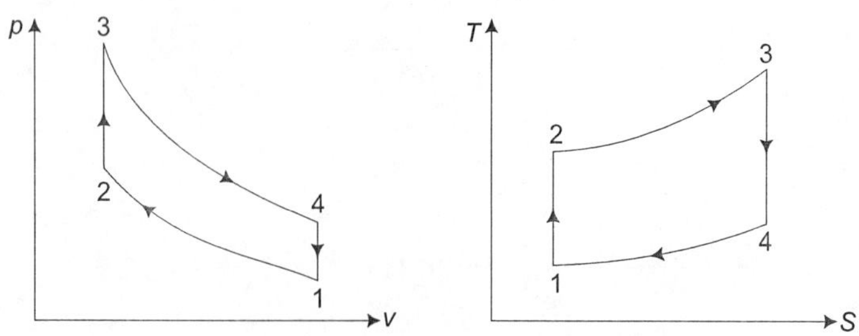

Fig. 3.13

$$\therefore \quad p_3 = p_{max} = T_3 \frac{p_2}{T_2}$$

$$= 3446.06 \times \frac{1525}{660.54} = 7955.98 \text{ kN/m}^2$$

(iii)
$$\eta = \frac{\text{Work done}}{\text{Heat input}} = \frac{W}{Q_1}$$

$$W = \eta \times Q_1 = 0.5408 \times 2000$$

$$= 1031.6 \text{ kJ/kg}$$

Also,
$$W = p_m \times (v_1 - v_2)$$

$$p_m = \frac{W}{(v_1 - v_2)}$$

$$\therefore \quad p_m = \frac{11031.6}{(0.87 - 0.124)} = 1449.87 \text{ kN/m}^2$$

Example 3.13 In an Otto cycle air at 20°C and 1 bar is compressed adiabatically until the pressure is 14 bar. Heat is added at constant volume until the pressure rises to 35 bar. Calculate (i) compression ratio, (ii) air standard efficiency, and (iii) mean effective pressure. Take $c_v = 0.718$ kJ/kgK and $R = 8.314$ kJ/kmol K.

Solution Refer Figure 3.13.

(i)
$$p_1 v_1^\gamma = p_2 v_2^\gamma$$

or
$$\frac{v_1}{v_2} = r = \left(\frac{p_2}{p_1}\right)^{1/\gamma}$$

$$\therefore \quad r = \left(\frac{14}{1}\right)^{1/1.4} = 6.59$$

(ii)
$$\eta_{Otto} = 1 - \frac{1}{(r)^{\gamma - 1}} = 1 - \frac{1}{(6.59)^{1.4 - 1}}$$

$$= 0.5296 \text{ or } 52.96\%$$

(iii)
$$\frac{p_1 v_1}{T_1} = \frac{p_2 v_2}{T_2}$$

$$\therefore \quad T_2 = T_1 \frac{p_2 v_2}{p_1 v_1} = 293 \times \frac{1}{6.59} \times \frac{14}{1}$$

$$= 622.46 \text{ K}$$

$$\frac{p_2 v_2}{T_2} = \frac{p_1 v_1}{T_1}$$

$$\therefore \quad T_3 = T_2 \frac{p_3}{p_2} = 622.46 \times \frac{25}{14}$$

$$= 1556.15 \text{ K}$$

$$Q_1 = c_v (T_3 - T_2)$$

$$= 0.718 \times (1556.15 - 622.46) = 670.39 \text{ kJ/kg}$$

$$v_1 = \frac{R \times T_1}{M \times p_1} = \frac{8314 \times 293}{29 \times 1 \times 10^5}$$

$$= 0.8314 \text{ m}^3\text{/kg} \quad \left[\because R = \frac{R}{M} = \frac{8314}{29}\right]$$

$$W = \eta \times Q_1 = 0.5296 \times 670.39$$

$$= 355.04 \text{ kJ/kg}$$

$$v_1 - v_2 = \frac{v_1 - v_1}{6.59} = \frac{5.59}{6.59} v_1$$

$$= \frac{5.59}{6.59} \times 0.8314 = 0.7052 \text{ m}^3\text{/kg}$$

$$\therefore \quad p_m = \frac{W}{(v_1 - v_2)} = \frac{355.04 \times 10^3}{0.7052 \times 10^5}$$

$$= 5.03 \text{ bar}$$

Example 3.14 Prove that compression ratio for the maximum work per kg of air in Otto cycle between upper and lower limits of absolute temperature T_3 and T_1 is given by, $r = (T_3/T_1)^{1/2(\gamma - 1)}$. Further, prove that the air standard efficiency of the cycle becomes independent of working fluid used, when the above condition is satisfied then prove that $T_2 = T_4 = \sqrt{T_1 T_3}$.

Solution Refer Figure 3.13. The work done per kg of fluid is given by,

$$W = Q_1 - Q_2 = c_v (T_3 - T_2) - c_v (T_4 - T_1)$$

$$\frac{T_2}{T_1} = \left(\frac{v_1}{v_2}\right)^{\gamma - 1} = (r)^{\gamma - 1}$$

$$T_2 = T_1 (r)^{\gamma - 1}$$

$$T_3 = T_4 (r)^{\gamma - 1}$$

$$T_4 = \frac{T_3}{(r)^{\gamma - 1}}$$

$$\therefore \quad W = c_v \left[T_3 - T_1 (r)^{\gamma - 1} - \frac{T_3}{(r)^{\gamma - 1}} + T_1\right]$$

When T_3 and T_1 are constant W is a function of r only.

$$\frac{dW}{dr} = 0$$

$$\frac{d}{dr}\left[T_3 - T_1\,(r)^{\gamma-1} - \frac{T_3}{(r)^{\gamma-1}} + T_1\right] = 0$$

$$-(\gamma-1)\,r^{\gamma-2}\,.\,T_1 - (1-\gamma)\,r^{-\gamma}\,.\,T_3 = 0$$

$$\frac{T_3}{T_1} = r^{2(\gamma-1)}$$

$$\therefore \qquad r = \left(\frac{T_3}{T_1}\right)^{1/2(\gamma-1)}$$

The air standard efficiency is given by,

$$\eta_{\text{Otto}} = 1 - \frac{1}{(r)^{\gamma-1}}$$

Substituting the value of r in the expression of efficiency, we have,

$$\eta_{\text{Otto}} = \frac{1}{[(T_3/T_1)^{1/2(\gamma-1)}]^{\gamma-1}} = 1 - \sqrt{\frac{T_1}{T_3}}$$

This expression shows that efficiency is independent of working fluid used.

Now $T_2 = T_1\,(r)^{\gamma-1}$, substituting the value of r, we have,

$$T_2 = T_1\left[\left(\frac{T_3}{T_1}\right)^{1/2\,(\gamma-1)}\right]^{\gamma-1}$$

$$= T_1\sqrt{\frac{T_3}{T_1}} = \sqrt{T_1T_3}$$

Now $T_4 = T_3/(r)^{\gamma-1)}$, substituting the value of r, we have,

$$T_4 = \frac{T_3}{[(T_3/T_1)^{1/2(\gamma-1)}]^{\gamma-1}} = \sqrt{T_1T_3}$$

Example 3.15 In an S.I. engine working on ideal Otto cycle compression ratio is 6. The pressure and temperature at the beginning of compression are 1 bar and 25°C respectively. The peak pressure is 35 bar. Calculate temperature and pressure at the salient points, the air standard efficiency, and mean effective pressure. Take $\gamma = 1.4$ for air.

Solution. Refer Figure 3.14.

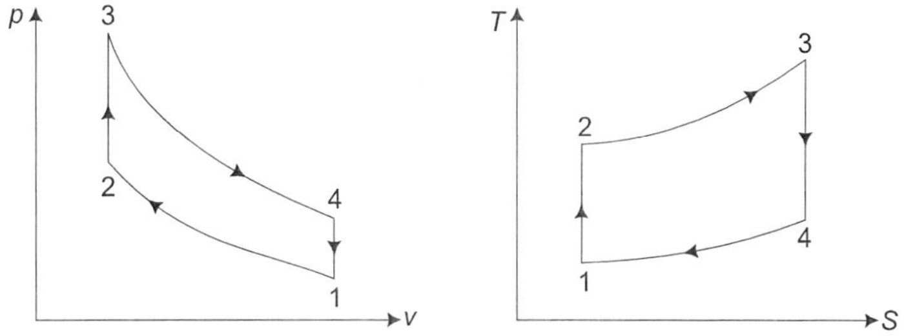

Fig. 3.14

$$v_2 = v_3 = v_c$$

$$v_1/v_2 = r,\ v_1 = r\ v_2 = r\ v_c$$

Process 1-2: $$\frac{p_2}{p_1} = \left(\frac{v_1}{v_2}\right)^{\gamma} = (6)^{1.4} = 12.29$$

$\therefore$ $$p_2 = 12.29 \times p_1 = 12.29 \times 1 \times 10^5$$

$$= 12.29 \times 10^5 \text{ kN/m}^2 = 12.29 \text{ bar}$$

$$\frac{T_2}{T_1} = r^{\gamma - 1} = (6)^{1.4 - 1} = 2.05$$

$\therefore$ $$T_2 = 2.05 \times T_1 = 2.05 \times 298 = 610.9 \text{ K}$$

Process 2-3: $$p_3 = 35 \times 10^5 \text{ kN/m}^2$$

$$\frac{T_3}{T_2} = \frac{p_3}{p_2} = \frac{35 \times 10^5}{12.29 \times 10^5} = 2.85$$

$\therefore$ $$T_3 = 2.85 \times T_2$$

$$= 2.85 \times 610.9 = 1741.07 \text{ K}$$

Process 3-4: $$\frac{p_3}{p_4} = \left(\frac{v_4}{v_3}\right)^{\gamma} = \left(\frac{v_1}{v_2}\right)^{\gamma} = r\gamma = (6)^{1.4} = 12.29$$

$\therefore$ $$p_4 = \frac{p_3}{12.29} = \frac{35 \times 10^5}{12.29} = 2.85 \times 10^5 \text{ N/m}^2$$

$$\frac{T_3}{T_4} = r^{\gamma - 1} = (6)^{1.4 - 1} = 2.05$$

$\therefore$ $$T_4 = \frac{T_3}{2.05} = \frac{174.07}{2.05} = 846.3 \text{ K}$$

$$\eta_{\text{Otto}} = 1 - \frac{1}{(r)^{\gamma - 1}} = 1 - \frac{1}{(6)^{\gamma - 1}} = 0.5116 \text{ or } 51.16\%$$

$$p_m = \frac{\text{Work done}}{\text{Swept volume}} = \frac{\text{Area of } p - v \text{ diagram } 1 - 2 - 3 - 4}{v_s}$$

Area of $p - v$ diagram $1 - 2 - 3 - 4$ = (Area under $3 - 4$) – (Area under $2 - 1$)

$$= \frac{p_3\, v_3 - p_4\, v_4}{\gamma - 1} - \frac{p_2\, v_2 - p_1\, v_1}{\gamma - 1}$$

$$= \frac{3.5 \times 10^5\ v_c - 2.85 \times 10^5 \times 6\ v_c}{1.4 - 1} - \frac{12.29 \times 10^5\ v_c - 1 \times 10^5 \times 6\ v_c}{1.4 - 1}$$

$$= 29 \times 10^5\ v_c$$

$$p_m \times v_s = 29 \times 10^5\ v_c$$

$$\therefore \quad p_m = \frac{29 \times 10^5 \, v_c}{v_s} = \frac{29 \times 10^5 \, v_c}{5 \times v_c}$$

$$= 5.8 \times 10^5 \text{ N/m}^2 \text{ or } 5.8 \text{ bar}$$

Example 3.16 In a constant volume cycle the temperature at the beginning and end of the compression are 40°C and 335°C respectively. Calculate the air standard efficiency and compression ratio. Take $\gamma = 1.4$ for air. A petrol engine with the above compression ratio develops 25 kW indicated power and consumes 8 litres of fuel per hour. The specific gravity of fuel is 0.78 and its calorific value is 44 MJ/kg. Determine the indicated thermal efficiency and the relative efficiency.

Solution Refer Figure 3.14.

$$\frac{T_2}{T_1} = \left(\frac{v_1}{v_2}\right)^{\gamma - 1} = r^{\gamma - 1}$$

$$\therefore \quad r = \left(\frac{T_2}{T_1}\right)^{1/(\gamma - 1)} = \left(\frac{608}{313}\right)^{1/(1.4 - 1)}$$

$$\therefore \quad \eta_{\text{Otto}} = 1 - \frac{1}{(r)^{\gamma - 1}} = 1 - \frac{1}{(5.26)^{1.4 - 1}}$$

$$= 0.4852 \text{ or } 48.52\%$$

$$\text{Fuel consumption} = 8 \times 0.78 \times 1 = 6.24 \text{ kg/h}$$

Indicated thermal efficiency,

$$\eta_{ith} = \frac{\text{i.p.}}{m_f \times CV} = \frac{25 \times 3600}{6.24 \times 44000}$$

$$= 0.3278 \text{ or } 32.78\%$$

Relative efficiency,

$$\eta_{rel} = \frac{\eta_{ith}}{\eta} = \frac{0.3278}{0.4852} \times 100$$

$$= 67.56\%$$

Example 3.17 A gas engine works on Otto cycle has a cylinder of diameter 20 cm and the stroke of 25 cm. The clearance volume is 1500 cm^3. Determine the air standard efficiency. Take $c_p = 1.004$ kJ/kgK and $c_v = 0.717$ kJ/kgK for air.

Solution Swept volume,

$$v_s = (\pi/4)\, d^2 \times L$$

$$= (\pi/4)\,(20)^2 \times 25 = 7853.98 \text{ cm}^3$$

$$r = 1 + \frac{v_s}{v_c} = 1 + \frac{7853.98}{1500} = 6.24$$

$$\gamma = \frac{c_p}{c_v} = \frac{1.004}{0.717} = 1.4$$

$$\therefore \qquad \eta_{Otto} = 1 - \frac{1}{(r)^{\gamma-1}} = 1 - \frac{1}{(6.247)^{1.4-1}}$$

$$= 0.5192 \text{ or } 51.92\%$$

Example 3.18 A petrol engine working on Otto cycle has initial pressure 1 bar and pressure after compression 12 bar. Calculate (i) compression ratio, (ii) percentage clearance, and (iii) thermal efficiency. If the initial temperature is 25°C find (iv) the temperature at the end of compression, (v) mean effective pressure if the maximum pressure in the cycle is 18 bar.

Solution Refer Figure 3.15.

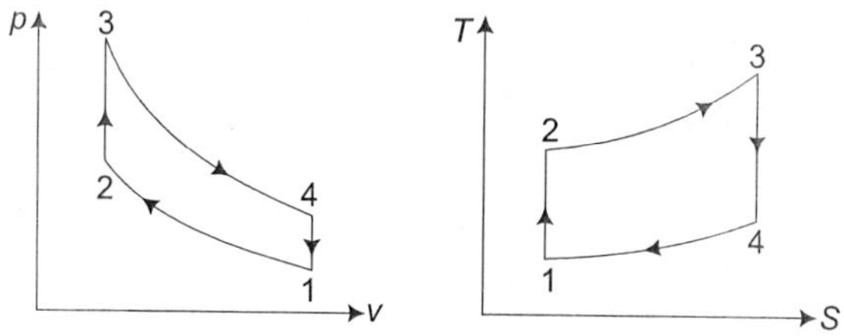

Fig. 3.15

(i) $$r = \frac{v_1}{v_2} = \left(\frac{p_2}{p_1}\right)^{1/\gamma} = \left(\frac{12}{1}\right)^{1/1.4} = 5.9$$

(ii) $$\text{Percentage clearance} = \frac{v_2}{v_1} \times 100 = \frac{100}{5.9} = 16.59\%$$

(iii) $$\eta_{Otto} = 1 - \frac{1}{(r)^{\gamma-1}} = 1 - \frac{1}{(5.9)^{1.4-1}}$$

$$= 0.5083 \text{ or } 5083\%$$

(iv) $$r_p = \frac{p_3}{p_2} = \frac{18}{12} = 1.5$$

$$p_m = \frac{p_1 r (r_p - 1)(r^{\gamma-1} - 1)}{(\gamma - 1)(r - 1)}$$

$$= \frac{1 \times 5.9 \times (1.5 - 1)(5.9^{1.4-1} - 1)}{(1.4 - 1)(5.9 - 1)}$$

$$= 1.56 \text{ bar}$$

3.15 DIESEL CYCLE

The Diesel cycle is an ideal cycle for compressed ignition engines. The difference between Otto and Diesel cycle is that in Diesel cycle heat addition takes place at constant pressure due to which sometimes it is also called constant pressure cycle but this term is generally avoided as it creates confusion with the

Joule cycle. The limitation on compression ratio in the S.I. engines has been overcome by compressing the air alone, instead of the fuel-air mixture, and then injecting the fuel into the cylinder in spray form. After compression the temperature of air is high enough which causes spontaneous burning of fuel after its injection. The sequence of processes in the operation of a C.I. engine is shown in Figure 3.7.

Figure 3.16 shows the ideal Diesel cycle on $p - v$ and $T - s$ diagrams respectively. It consists of four internally reversible processes:

1-2: Isentropic compression, 2-3: Constant pressure heat addition,

3-4: Isentropic expansion, and 4-1: Constant volume heat rejection.

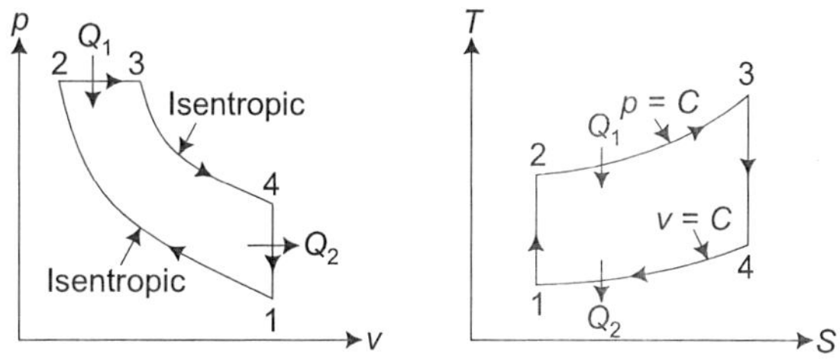

Fig. 3.16 $p - v$ and $T - s$ Diagrams of Diesel Cycle

Expression for Thermal Efficiency

Let m = mass of air supplied to the engine.

Heat supplied, $Q_1 = Q_{2-3} = mc_p (T_3 - T_2)$

Heat rejected, $Q_2 = Q_{4-1} = mc_v (T_4 - T_1)$

Efficiency, $$\eta = \frac{\text{Heat supplied} - \text{Heat rejected}}{\text{Heat supplied}}$$

$$\eta_{\text{diesel}} = \frac{Q_1 - Q_2}{Q_1} = 1 - \frac{Q_2}{Q_1} = 1 - \frac{mc_v (T_4 - T_1)}{mc_p (T_3 - T_2)}$$

$$\eta_{\text{diesel}} = 1 - \frac{(T_4 - T_1)}{\gamma (T_3 - T_2)} \quad (3.30)$$

The efficiency can be represented in terms of the following three ratios:

(i) Compression ratio, $r = (v_1/v_2)$,

(ii) Expansion ratio, $r_e = (v_4/v_3)$, and

(iii) Cut-off ratio, $r_c = (v_3/v_2)$

It is observed that $r = r_e \cdot r_c$

Process 3-4: $$\frac{T_4}{T_3} = \left(\frac{v_3}{v_4}\right)^{\gamma - 1} = \frac{1}{(r_e)^{\gamma - 1}}$$

$\therefore$ $$T_4 = \frac{T_3}{r_e^{\gamma - 1}} = \frac{T_3}{(r/r_c)^{\gamma - 1}} = T_3 \frac{r_c^{\gamma - 1}}{r^{\gamma - 1}}$$

Process 2-3: $$\frac{T_2}{T_3} = \frac{p_2 v_2}{p_3 v_3} = \frac{v_2}{v_3} = \frac{1}{r_c}$$

∴ $$T_2 = \frac{T_3}{r_c}$$

Process 1-2: $$\frac{T_1}{T_2} = \left(\frac{v_2}{v_1}\right)^{\gamma - 1} = \frac{1}{r^{\gamma - 1}}$$

∴ $$T_1 = T_2 \cdot \frac{1}{r^{\gamma - 1}} = \frac{T_3}{r_c} \cdot \frac{1}{r^{\gamma - 1}}$$

Now substituting the values of T_1, T_2, and T_4 in the equation (3.30), we have,

$$\eta_{\text{Diesel}} = 1 - \frac{T_3 \cdot \dfrac{r_c^{\gamma - 1}}{r^{\gamma - 1}} - \dfrac{T_3}{r_c} \cdot \dfrac{1}{r^{\gamma - 1}}}{\gamma (T_3 - T_3/r_c)}$$

$$= 1 - \frac{1}{\gamma} \cdot \frac{1}{r^{\gamma - 1}} \cdot \frac{r_c^{\gamma - 1}}{(r_c - 1)}$$

∴ $$\eta_{\text{Diesel}} = 1 - \frac{1}{r^{\gamma - 1}} \left[\frac{r_c^{\gamma - 1}}{\gamma (r_c - 1)}\right] \quad (3.31)$$

It is observed from equation (3.31) that the efficiency of Diesel cycle differs from that of Otto cycle by the bracketed factor. This factor is always greater than one. Thus, Diesel cycle is less efficient than Otto cycle for the same compression ratio. The higher efficiency of Otto cycle as compared to Diesel cycle for the same compression ratio is of no practical importance. In practice the compression ratio of Diesel engines are much higher as compared to S.I. engines. The normal range of compression ratio for Diesel engine is 16 to 20, whereas for S.I. engine it is 6 to 10. The actual efficiency of Diesel engine is therefore, higher than that of petrol engine. The thermal efficiency of Diesel engines varies from 35% to 40%.

It is also observed that as the cut-off ratio r_c decreases, the efficiency of the Diesel cycle increases as shown in Figure 3.17. For the limiting case when $r_c \approx 1$, the value of bracketed factor in equation (3.31) approaches unity and thus the efficiencies of Otto and Diesel cycles become identical.

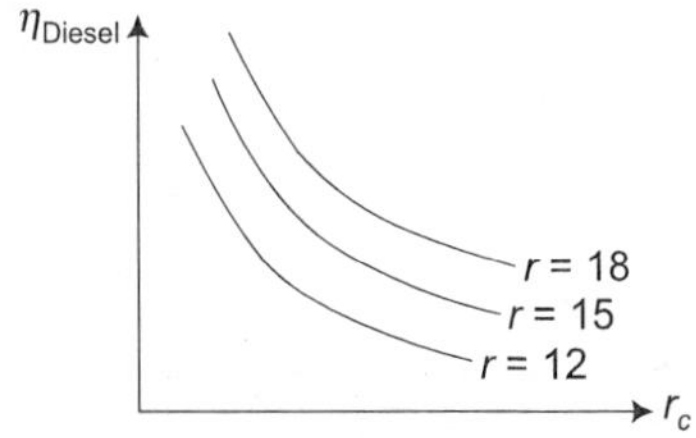

Fig. 3.17 Efficiency of Diesel Cycle for different r_c and r

Expression for Work Output

The net work output for Diesel cycle is given by,

$$W = \text{Area of } p - v \text{ diagram}$$

$$W = \left[p_2 (v_3 - v_2) + \frac{p_3 v_3 - p_4 v_4}{\gamma - 1}\right] - \frac{p_2 v_2 - p_1 v_1}{\gamma - 1} \tag{3.32}$$

$$= p_2 v_2 \left(\frac{v_3}{v_2} - 1\right) + \frac{p_3 r_c v_2 - p_4 r v_2}{\gamma - 1} - \frac{p_2 v_2 - p_1 r v_2}{\gamma - 1}$$

$$= p_2 v_2 (r_c - 1) + \frac{p_3 r_c v_2 - p_4 r v_2}{\gamma - 1} - \frac{p_2 v_2 - p_1 r v_2}{\gamma - 1}$$

$$= v_2 \left[\frac{p_2 (r_c - 1)(\gamma - 1) + p_3 r_c - p_4 r - (p_2 - p_1 r)}{\gamma - 1}\right]$$

$$= v_2 \left[\frac{p_2 (r_c - 1)(\gamma - 1) + p_3 [r_c - (p_4/p_3) \times r] - p_2 [1 - (p_1/p_2) \times r]}{\gamma - 1}\right]$$

$$= p_2 v_2 \left[\frac{(r_c - 1)(\gamma - 1) + (r_c - r_c^{\gamma} . r^{1-\gamma}) - (1 - r^{1-\gamma})}{\gamma - 1}\right]$$

$$\left[\frac{p_4}{p_3} = \frac{p_1}{p_2} = \left(\frac{v_3}{v_4}\right)^{\gamma} = \left(\frac{1}{r_c}\right)^{\gamma} = r_e^{-\gamma} = \left(\frac{r}{r_c}\right)^{-\gamma} = r^{-\gamma} . r_c^{\gamma}\right] \quad \& \quad [p_2 = p_3]$$

$$= p_1 v_1 r^{\gamma - 1} \left[\frac{(r_c - 1)\gamma - r^{1-\gamma} (r_c^{\gamma} - 1)}{\gamma - 1}\right]$$

$$\therefore \quad W = \frac{p_1 v_1 r^{\gamma - 1} \{\gamma (r_c - 1) - r^{1-\gamma} (r_c^{\gamma} - 1)\}}{\gamma - 1} \tag{3.33}$$

Mean Effective Pressure

The mean effective pressure,

$$p_m = \frac{\text{Work output}}{\text{Swept volume}} \tag{i}$$

$$\text{Swept volume} = v_1 - v_2 = v_1 \left(1 - \frac{v_2}{v_1}\right)$$

$$= v_1 \left(1 - \frac{1}{r}\right) = v_1 \left(\frac{r - 1}{r}\right) \tag{ii}$$

Substituting equation (3.33) and the expression (ii) in the expression (i), we have,

$$p_m = \frac{p_1 v_1 r^{\gamma - 1} \left[\frac{\gamma (r_c - 1) - r^{1-\gamma} (r_c^{\gamma} - 1)}{\gamma - 1}\right]}{v_1 [(r - 1)/r]}$$

$$\therefore \qquad p_m = \frac{p_1 \{\gamma . r^\gamma (r_c - 1) - r (r_c^\gamma - 1)\}}{(\gamma - 1)(r - 1)} \qquad (3.34)$$

Example 3.19 A diesel engine has a compression ratio of 16 and cut-off takes place at 6% of the stroke. Determine the air standard efficiency.

Solution Refer Figure 3.16.

Compression ratio, $r = (v_1/v_2) = 16$

$$v_1 = 16 \times v_2$$

$$v_3 = v_2 = 0.06 \times (v_1 - v_2) = 0.06 \times (16v_2 - v_2)$$

$$= 0.9\ v_2,\ v_3 = 1.9\ v_2$$

Cut-off ratio, $r_c = (v_3/v_2) = (0.9v_2/v_2) = 1.9$

$$\eta_{\text{Diesel}} = 1 - 1/r^{\gamma - 1}\left[\frac{r_c^\gamma - 1}{\gamma (r_c - 1)}\right]$$

$$= 1 - \frac{1}{16^{1.4-1}}\left[\frac{16^{1.4} - 1}{1.4 \times (16 - 1)}\right] \times 100 = 72.77\%$$

Example 3.20 A diesel engine operating on a standard Diesel cycle has 200 mm bore and 300 mm stroke. The clearance volume is 420 cm^3. The fuel is injected at constant pressure for 6% of the stroke. Determine the air standard efficiency, if the cut-off is delayed from 6% to 8%. What will be the percentage loss in efficiency? Compression ratio is same in both the cases.

Solution Refer Figure 3.16.

Clearance volume, $v_c = v_2 = 410 \text{ cm}^3$

Swept volume, $v_s = v_1 - v_2 = (\pi/4)\, d^2 \times L$

$$= (\pi/4)\,(20)^2 \times 30 = 9424.78 \text{ cm}^3$$

Cylinder volume, $v_1 = v_s + v_c = 9424.78 + 420 = 9844.78 \text{ cm}^3$

$$v_3 = v_2 + 0.06 \times v_s$$

$$= 420 + 0.06 \times 9424.78 = 985.49 \text{ cm}^3$$

Cut-off ratio, $r_c = (v_3/v_2) = (985.49/420) = 2.35$

Compression ratio, $r = (v_1/v_2) = (9844.78/420) = 23.44$

$$\eta_{\text{Diesel}} = 1 - \frac{1}{r^{\gamma-1}}\left[\frac{r_c^\gamma - 1}{\gamma (r_c - 1)}\right]$$

$$= 1 - \frac{1}{23.44^{1.4-1}}\left[\frac{2.35^{1.4-1}}{1.4 \times (2.35 - 1)}\right] \times 100 = 75.3\%$$

If the fuel is cut-off at 8% of the stroke

$$v_3 = v_2 + 0.08 \times v_s$$

$$= 420 + 0.08 \times 9424.78 = 1173.98 \text{ cm}^3$$

Cut-off ratio, $r_c = (v_3/v_2) = (1173.98/420) = 2.795$

Compression ratio is same, i.e.,

$$r = 23.44$$

$$\therefore \quad \eta_{\text{Diesel2}} = 1 - \frac{1}{23.44^{1.4-1}}\left[\frac{2.795^{1.4} - 1}{1.4 \times (2.795 - 1)}\right]$$

$$= 0.7411 \text{ or } 74.11\%$$

$$\therefore \quad \%\eta \text{ loss} = \frac{\eta_{\text{Diesel}} - \eta_{\text{Diesel2}}}{\eta_{\text{Diesel1}}} \times 100$$

$$= \frac{0.753 - 0.7411}{0.753} \times 100 = 1.58\%$$

Example 3.21 In an air standard diesel cycle, the compression ratio is 17, and at the beginning of compression the temperature is 20°C and the pressure is 0.1 M Pa. Heat is added until the temperature at the end of the constant pressure process is 1485°C. Calculate (i) cut-off ratio, (ii) heat supplied per kg of air, (iii) cycle efficiency, and (iv) mean effective pressure.

Solution Refer Figure 3.18.

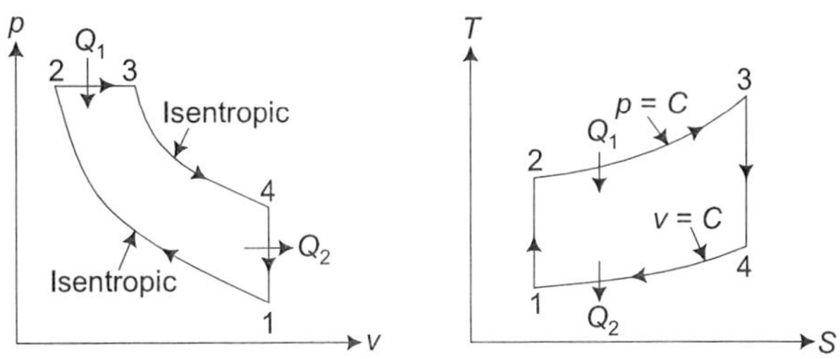

Fig. 3.18

(i) $$\frac{T_2}{T_1} = \left(\frac{v_1}{v_2}\right)^{\gamma-1} = r^{\gamma-1}$$

$$= (17)^{1.4-1} = 3.11$$

$$\therefore \quad T_2 = T_1 \times 3.11 = 293 \times 3.11 = 911.23 \text{ K}$$

$$\frac{p_2 v_2}{T_2} = \frac{p_3 v_3}{T_3}$$

or $$\frac{v_3}{v_2} = \frac{T_3}{T_2}$$

$$\therefore \qquad r_c = \frac{v_3}{v_2} = \frac{T_3}{T_2} = \frac{1758}{911.23} = 1.929$$

(ii) $$Q_1 = c_p\,(T_3 - T_2)$$

$$= 1.005 \times (1758 - 911.23) = 851 \text{ kJ/kg}$$

(iii) $$\frac{T_3}{T_4} = \left(\frac{v_4}{v_3}\right)^{\gamma-1} = \left(\frac{v_1}{v_2} \times \frac{v_2}{v_3}\right)^{\gamma-1}$$

$$= \left(\frac{r}{r_c}\right)^{\gamma-1} = \left(\frac{17}{1.929}\right)^{\gamma-1} = 2.39$$

$$\therefore \qquad T_4 = (T_3/2.39) = (1758/2.39) = 735.56 \text{ K}$$

$$Q_2 = c_p\,(T_4 - T_1)$$

$$= 0.718 \times (735.56 - 293) = 317.76 \text{ kJ/kg}$$

$$\therefore \qquad \eta_{\text{Diesel}} = 1 - \frac{Q_2}{Q_1} = 1 - \frac{317.76}{851}$$

$$= 0.6266 \text{ or } 62.66\%$$

(iv) $$W = \eta \times Q_1$$

$$= 0.6266 \times 851 = 533.24 \text{ kJ/kg}$$

$$v_1 = \frac{RT_1}{p_1} = \frac{0.287 \times 293}{100} = 0.841 \text{ m}^3\text{/kg}$$

$$v_2 = v_1/r = 0.841/17 = 0.0495 \text{ m}^3\text{/kg}$$

$$\therefore \qquad p_m = \frac{\text{Work done}}{\text{Swept volume}} = \frac{W}{v_1 - v_2}$$

$$= \frac{533.24}{0.841 - 0.0495} = 673.71 \text{ kPa}$$

Example 3.22 A diesel engine has a compression ratio of 16. The fuel is cut-off at 0.06 stroke. The relative efficiency is 54%. If the calorific value of fuel is 42000 kJ/kg calculate the mass of fuel required per kW/h.

Solution Refer Figure 3.18.

Compression ratio, $r = (v_1/v_2) = 16,\ v_3 - v_2 = 0.06 \times (v_1 - v_2)$

$$\frac{v_3}{v_2} - 1 = 0.006 \times \left(\frac{v_1}{v_2} - 1\right) = 0.06 \times (r - 1)$$

$$\frac{v_3}{v_2} = r_c = 0.06 \times (16 - 1) + 1 = 1.9$$

$$\eta_{\text{Diesel}} = 1 - \frac{1}{r^{\gamma-1}}\left[\frac{r_c^{\gamma} - 1}{\gamma\,(r_c - 1)}\right]$$

$$= 1 - \frac{1}{16^{1.4-1}}\left[\frac{1.9^{1.4} - 1}{1.4 \times (1.9 - 1)}\right] = 0.7277 \text{ or } 72.77\%$$

But $$\eta_{actual} = \eta_{Diesel} \times \eta_{rel} = 0.7277 \times 0.54$$
$$= 0.3929 \text{ or } 39.29\%$$

Also $$\eta_{actual} = \frac{W}{m_f \times CV}$$
$$\frac{W}{m_f} = \eta_{actual} \times CV = 0.3929 \times 42000 \text{ kJ/kg of fuel}$$

∴ $$m_f = \frac{3600}{0.3929 \times 42000} \text{ kg/kWh}$$
$$= 0.2182 \text{ kg/kWh}$$

Example 3.23 An engine working on air standard Diesel cycle has the pressure and temperature as 1 bar and 30°C at the beginning of the cycle. The compression ratio is 18 and maximum temperature of the cycle is limited to 1525°C. Determine (i) percentage of stroke at which cut-off takes place, (ii) temperature at the end of expansion, and (iii) thermal efficiency of the cycle.

Solution Refer Figure 3.18.

(i) $$T_2 = T_1 \times (v_1/v_2)^{\gamma-1} = T_1 \times r^{\gamma-1}$$
$$= 303 \times (18)^{1.4-1} = 962.83 \text{ K}$$
$$\frac{p_2 v_2}{T_2} = \frac{p_3 v_3}{T_3}$$
$$\frac{v_3}{v_2} = \frac{T_3}{T_2}$$

∴ $$v_3 = \frac{T_3}{T_2} \times v_2 = \frac{1798}{962.83} \times v_2 = 1.87\, v_2$$

∴ $$r_c = (v_3/v_2) = (1.87\, v_2/v_2) = 1.87$$

Percentage of the stroke at which cut-off takes place

$$= \left(\frac{v_3 - v_2}{v_1 - v_2}\right) \times 100$$
$$= \left(\frac{1.87\, v_2 - v_2}{18\, v_2 - v_2}\right) \times 100 = 5.12\%$$

(ii) We know that $$r = r_e \,.\, r_c$$
$$r_e = (r_c/r) = (18/1.87) = 9.63$$
$$\frac{T_4}{T_3} = \left(\frac{v_3}{v_4}\right)^{\gamma-1} = \frac{1}{(r_e)^{\gamma-1}}$$

∴ $$T_4 = \frac{T_3}{r_e^{\gamma-1}} = \frac{1798}{(9.63)^{1.4-1}} = 726.67 \text{ K}$$

(iii)
$$\eta = 1 - \frac{(T_4 - T_1)}{\gamma\,(T_3 - T_2)}$$
$$= 1 - \frac{(726.67 - 303)}{1.4 \times (1798 - 962.83)}$$
$$= 1 - 0.3623 = 0.6377 \text{ or } 63.77\%$$

Example 3.24 A Diesel engine working on Diesel cycle takes air at temperature 25°C and pressure 1 bar. The pressure at the end of compression is 25 bar and the cut-off is 6% of the stroke. Determine (i) compression ratio, (ii) percentage clearance, (iii) heat rejected, (iv) thermal efficiency, and (v) mean effective pressure.

Solution Refer Figure 3.18.

(i)
$$r = (v_1/v_2) = (p_2/p_1)^{1/\gamma}$$
$$= (25)^{1/1.4} = 9.97$$

(ii) Percentage clearance
$$\left(\frac{v_2}{v_1 - v_2}\right) \times 100 = \left[\frac{1}{(v_1/v_2) - 1}\right] \times 100$$
$$= \left(\frac{1}{r - 1}\right) \times 100$$
$$= \left(\frac{1}{9.97 - 1}\right) \times 100 = 11.15\%$$

(iii)
$$T_2 = T_1 \times (v_1/v_2)^{\gamma - 1} = T_1 \times r^{\gamma - 1}$$
$$= 303 \times (9.97)^{1.4 - 1} = 747.64 \text{ K}$$
$$v_1 = 9.97\, v_2,\ v_3 - v_2 = 0.06 \times (v_1 - v_2)$$
$$v_3 = v_2 + 0.06 \times (9.97\, v_2 - v_2) = 1.538\, v_2$$

$\therefore$
$$(v_3/v_2) = r_c = 1.538$$
$$\frac{p_2 v_2}{T_2} = \frac{p_3 v_3}{T_3}$$

or
$$\frac{v_3}{v_2} = \frac{T_3}{T_2}$$

$\therefore$
$$T_3 = \frac{v_3}{v_2} \times T_2$$
$$= 1.538 \times 747.64 = 1149.87 \text{ K}$$
$$Q_1 = c_p\,(T_3 - T_2)$$
$$= 1.005 \times (1149.87 - 747.64) = 404.24 \text{ kJ/kg}$$

$$\frac{T_4}{T_3} = \left(\frac{v_3}{v_4}\right)^{\gamma - 1} = \left[\frac{v_3}{(v_1/v_2) \times v_2}\right]^{\gamma - 1}$$

$$= \left[\frac{v_3}{r \times v_2}\right]^{\gamma - 1} = \left[\frac{1.53.8\ v_2}{9.97 \times v_2}\right]^{1.4 - 1}$$

$\therefore$

$$T_4 = T_3 \times [1.538/9.97]^{0.4}$$

$$= 1149.87 \times [1.538/9.97]^{0.4} = 544.44 \text{ K}$$

$$Q_2 = c_v\,(T_4 - T_1) = 0.717 \times (544.44 - 298)$$

$$= 176.96 \text{ kJ/kg}$$

(iv)

$$\eta = 1 - \frac{(T_4 - T_1)}{\gamma\,(T_3 - T_2)} = 1 - \frac{(544.44 - 298)}{1.4 \times (1149.87 - 747.64)}$$

$$= 1 - 0.4377 = 0.5623 \text{ or } 56.23\%$$

(v)

$$p_m = \frac{p_1\,\{\gamma \,.\, r^{\gamma}\,(r_c - 1) - r\,(r_c^{\gamma} - 1)\}}{(\gamma - 1)\,(r - 1)}$$

$\therefore$

$$p_m = \frac{1 \times \{1.4 \times 9.97^{1.4}\,(1.538 - 1) - 9.97 \times (1.38^{1.4} - 1)\}}{(1.4 - 1)\,(9.97 - 1)}$$

$$= 2.99 \text{ bar}$$

3.16 DUAL CYCLE

The dual cycle is also called mixed cycle or limited pressure cycle. In actual S.I. or C.I. engines it is difficult to simulate the combustion process either a constant volume or a constant pressure heat addition. Therefore, a dual cycle has been proposed which is a compromise between Otto and Diesel cycles. In dual cycle a part of heat is supplied at constant volume and rest at constant pressure. Figure 3.19 shows $p - v$ and $T - s$ diagrams of the dual cycle. It consists of the following internally reversible processes:

1-2: Isentropic compression, 2-3: Constant volume heat addition,

3-4: Constant pressure heat addition, 4-5: Isentropic expansion, and

5-1: Constant volume heat rejection.

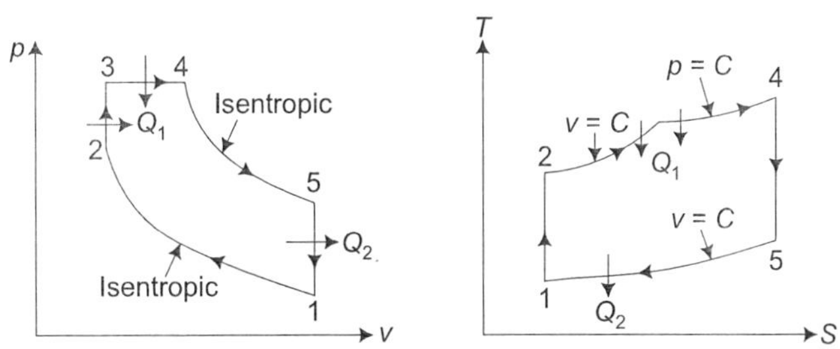

Fig. 3.19 $p - v$ and $T - s$ Diagrams of Dual Cycle

Expression for Thermal Efficiency

Heat is added reversibly partly at constant volume (process 2-3) and partly at constant pressure (process 3-4)

$\therefore$ Heat supplied, $Q_1 = mc_v (T_3 - T_2) + mc_p (T_4 - T_3)$

Heat rejected, $Q_2 = mc_v (T_5 - T_1)$

Efficiency, $$\eta = \frac{\text{Heat supplied} - \text{Heat rejected}}{\text{Heat supplied}}$$

$$\eta_{\text{Dual}} = \frac{Q_1 - Q_2}{Q_1} = 1 - \frac{Q_2}{Q_1}$$

$$= 1 - \frac{mc_v (T_5 - T_1)}{mc_v (T_3 - T_2) + mc_p (T_4 - T_3)}$$

$$\eta_{\text{Dual}} = 1 - \frac{(T_5 - T_1)}{(T_3 - T_2) + \gamma (T_4 - T_3)} \qquad (3.35)$$

The efficiency of the dual cycle can be expressed in terms of the following ratios:

(i) Compression ratio, $r = v_1/v_2$,

(ii) Expansion ratio, $r_e = v_5/v_4$,

(iii) Cut-off ratio, $r_c = v_4/v_3$, and

(iv) Constant volume pressure ratio (or explosion ratio),

$$r_p = p_3/p_2$$

It is observed that $r = r_e \cdot r_c$ or $r_e = r/r_c$

Process 3-4: $$r_c = \frac{v_4}{v_3} = \frac{T_4}{T_3} \cdot \frac{p_3}{p_4} = \frac{T_4}{T_3}$$

$$T_3 = \frac{T_4}{r_c} \qquad \left[\because \frac{p_3 v_3}{T_3} = \frac{p_4 v_4}{T_4}\right]$$

Process 2-3: $$\frac{p_2 v_2}{T_2} = \frac{p_3 v_3}{T_3}$$

$$T_2 = \frac{T_3 p_2}{p_3} = \frac{T_4}{r_p \cdot r_c} \qquad \left[\because \frac{p_2}{p_3} = \frac{1}{r_p} \text{ and } T_3 = \frac{T_4}{r_c}\right]$$

Process 1-2: $$\frac{T_1}{T_2} = \left(\frac{v_2}{v_1}\right)^{\gamma - 1} = \frac{1}{r^{\gamma - 1}}$$

$$T_1 = \frac{T_2}{r^{\gamma - 1}} = \frac{T_4}{r_p \cdot r_c \cdot r^{\gamma - 1}}$$

Process 4-5: $$\frac{T_5}{T_4} = \left(\frac{v_4}{v_5}\right)^{\gamma - 1} = \frac{1}{r_e^{\gamma - 1}}$$

$$T_5 = \frac{T_4}{r_e^{\gamma-1}} = T_4 \cdot \frac{r_c^{\gamma-1}}{r^{\gamma-1}} \quad \left[\because r_e = \frac{r}{r_c}\right]$$

Now substituting the values of T_1, T_2, T_3, and T_5 in equation (3.35)

$$\eta_{\text{Dual}} = 1 - \frac{T_4 \cdot \frac{r_c^{\gamma-1}}{r^{\gamma-1}} - \frac{T_4}{r_p \cdot r_c \cdot r^{\gamma-1}}}{\left(\frac{T_4}{r_c} - \frac{T_4}{r_p \cdot r_c}\right) + \gamma\left(T_4 - \frac{T_4}{r_c}\right)}$$

$$\therefore \qquad \eta_{\text{Dual}} = 1 - \frac{1}{r^{\gamma-1}}\left[\frac{r_p \cdot r_c^{\gamma} - 1}{r_p - 1 + \gamma r_p (r_c - 1)}\right] \qquad (3.36)$$

Figure 3.20 shows the variation of dual cycle efficiency with compression ratio. It is to be observed that at cut-off ratio $r_c = 1$ the efficiency of dual cycle is equal to that of the Otto cycle and at pressure ratio $r_p = 1$ it is equal to Diesel cycle efficiency. It is observed from equation (3.36) that a value of pressure ratio $r_p > 1$ results in an increased efficiency for a given value of r_c and γ. Hence, the dual cycle efficiency always lies in between that of the Otto cycle and Diesel cycle having the same compression ratio.

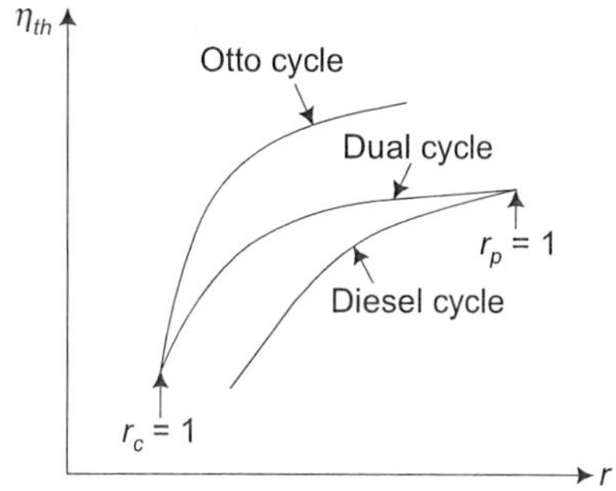

Fig. 3.20 Thermal Efficiency of Dual Cycle versus r

Expression for Work Output

The net work output for Diesel cycle is given by,

$$W = \text{Area of } p - v \text{ diagram}$$

$$W = \left[p_3 (v_4 - v_3) + \frac{p_4 v_4 - p_5 v_5}{\gamma - 1}\right] - [p_2 v_2 - p_1 v_1/\gamma - 1]$$

$$= \frac{p_1 v_1}{\gamma - 1}\left[(\gamma - 1)\left(\frac{p_4 v_4}{p_1 v_1} - \frac{p_3 v_3}{p_1 v_1}\right) + \frac{p_4 v_4}{p_1 v_1} - \frac{p_5 v_5}{p_1 v_1} - \frac{p_2 v_2}{p_1 v_1} + 1\right]$$

$$= \frac{p_1 v_1}{\gamma - 1}\left[(\gamma - 1) r_c r_p r^{\gamma-1} - r_p r^{\gamma-1} + r_c r_p r^{\gamma-1} - r_p r_c^{\gamma} - r^{\gamma-1} + 1\right]$$

$$= \frac{p_1 v_1}{\gamma - 1}\left[\gamma r_c r_p r^{\gamma-1} - \gamma r_p r^{\gamma-1} + r_p r^{\gamma-1} - r_p r_c^{\gamma} - r^{\gamma-1} + 1\right]$$

$$= \frac{p_1 v_1}{\gamma - 1}\left[\gamma r_p r^{\gamma-1} (r_c - 1) + r^{\gamma-1} (r_p - 1) - (r_p r_c^{\gamma} - 1)\right] \qquad (3.37)$$

Expression for Mean Effective Pressure

The mean effective pressure is given by,

$$p_m = \frac{\text{Work output}}{\text{Swept volume}}$$

$$\text{Swept volume} = v_1 - v_2 = v_1 (1 - v_2/v_1)$$

$$= v_1 (1 - 1/r) = v_1 [(r - 1)/r]$$

$$p_m = \frac{p_1 v_1}{\gamma - 1}\left[\frac{\gamma r_p r^{\gamma-1} (r_c - 1) + r^{\gamma-1} (r_p - 1) - (r_p r_c^{\gamma} - 1)}{v_1 (r - 1)/r)}\right]$$

$$\therefore \quad p_m = p_1 \frac{\{\gamma r_p r^{\gamma} (r_c - 1) + r^{\gamma} (r_p - 1) - r (r_p r_c^{\gamma})\}}{(\gamma - 1)(r - 1)} \quad (3.38)$$

Example 3.25 A dual combustion cycle has an adiabatic compression volume ratio of 16:1. The condition at the commencement of the compression are 1 bar, 30°C, and 0.1 m^3. The maximum pressure and maximum temperature of the cycle is 60 bar and 1550°C. If $c_p = 0.71$ kJ/kgK and $\gamma = 1.4$, determine the pressure, volume, and temperature at each point of the cycle and the thermal efficiency.

Solution Refer Figure 3.21.

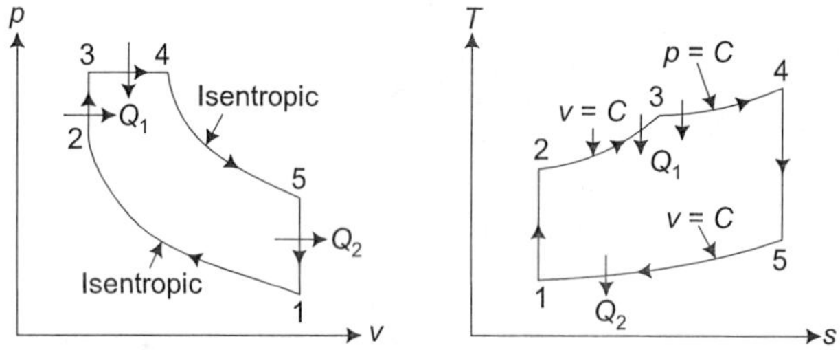

Fig. 3.21

Compression ratio, $r = (v_1/v_2) = 16, v_2 = (0.1/16) = 6.25 \times 10^{-3}$ m^3

$$p_2 = p_1 \times (v_1/v_2)^{\gamma} = 1 \times (16)^{1.4} = 48.5 \text{ bar}$$

$$T_2 = T_1 \times r^{\gamma-1} = 303 \times (16)^{1.4-1} = 918.52 \text{ K}$$

$$T_3 = T_2 \times (p_3/p_2) = 918.52 \times (60/48.5) = 1136.31 \text{ K}$$

$$v_2 = v_3 = 6.25 \times 10^{-3} \text{ m}^3$$

$$v_4 = v_3 \times (T_4/T_3) = 6.25 \times 10^{-3} \times (1823/1136.31) = 0.01 \text{ m}^3$$

$$v_1 = v_5 = 0.1 \text{ m}^3$$

$$p_5 = p_4 \times (v_4/v_5)^{\gamma} = 60 \times (0.01/0.1)^{1.4} = 2.39 \text{ bar}$$

$$T_5 = T_4 \times (v_4/v_5)^{\gamma-1} = 1823 \times (0.01/0.1)^{1.4-1} = 725.75 \text{ K}$$

$$\eta_{\text{Dual}} = 1 - \frac{(T_5 - T_1)}{(T_3 - T_2) + \gamma (T_4 - T_3)}$$

$$\therefore \quad \eta_{\text{Dual}} = 1 - \frac{(725.75 - 303)}{(1136.31 - 918.52) + 1.4 \times (1823 - 1136.31)}$$

$$= 0.6415 \text{ or } 64.15\%$$

Example 3.26 An air standard dual cycle has a compression ratio of 17, and compression begins at 1 bar, 48°C. The maximum pressure is 69 bar. The heat transferred to air at constant pressure is equal to that at constant volume. Calculate (i) the pressure and temperature at all the points of the cycle, (ii) the cycle efficiency, and (iii) mean effective pressure of the cycle. Take $c_v = 0.718$ kJ/kgK and $c_p = 1.005$ kJ/kgK.

Solution Refer Figure 3.21.

(i)
$$T_2 = T_1 \times (v_1/v_2)^{\gamma - 1} = T_1 \times (r)^{\gamma - 1}$$
$$= 321 \times (17)^{1.4 - 1} = 996.98 \text{ K}$$
$$p_2 = p_1 \times (v_1/v_2)^{\gamma} = 1 \times (17)^{1.4} = 52.8 \text{ bar}$$
$$T_3 = T_2 \times (p_3/p_2) = 996.98 \times (69/52.8) = 1302.87 \text{ K}$$

Heat supplied at constant volume,
$$Q_{2-3} = c_v (T_3 - T_2) = 0.718 \times (1302.87 - 996.98)$$
$$= 219.63 \text{ kJ/kg}$$
$$\therefore \quad Q_{2-3} = Q_{3-4} = c_p (T_4 - T_3) = 219.63$$
$$\therefore \quad T_4 = \frac{219.63}{1.005} + 1302.87 = 1521.41 \text{ K}$$
$$\frac{v_4}{v_3} = \frac{T_4}{T_3} = \frac{1521.41}{1302.87} = 1.168$$
$$\therefore \quad \frac{v_5}{v_4} = \frac{v_1}{v_2} \times \frac{v_3}{v_4} = \frac{17}{1.168} = 14.55$$
$$\therefore \quad T_5 = T_4 \times (v_4/v_5)^{\gamma - 1}$$
$$= 1521.41 \times (1/14.55)^{1.4 - 1} = 521.32 \text{ K}$$
$$p_5 = p_1 \times (T_5/T_1) = 1 \times (521.32/321) = 1.624 \text{ bar}$$

(ii)
$$\eta = 1 - \frac{Q_2}{Q_1} = 1 - \frac{c_v (T_5 - T_1)}{c_v (T_3 - T_2) + c_p (T_4 - T_3)}$$
$$= 1 - \frac{c_v (T_5 - T_1)}{2c_v (T_3 - T_2)} = 1 - \frac{(T_5 - T_1)}{2 (T_3 - T_2)}$$
$$\therefore \quad \eta = 1 - \frac{(521.32 - 321)}{2 \times (1302.87 - 996.98)}$$
$$= 0.6726 \text{ or } 67.26\%$$

(iii)
$$v_1 = \frac{RT_1}{p_1} = \frac{0.287 \times 321}{1 \times 10^2} = 0.921 \text{ m}^3\text{/kg}$$
$$v_1 - v_2 = \frac{v_1 - v_1}{17} = \frac{16v_1}{17}$$
$$= 16 \times 0.921/17 = 0.867 \text{ m}^3\text{/kg}$$

$$W = \eta \times Q_1 = 0.6726 \times 611.78 = 411.48 \text{ kJ/kg}$$

$$\therefore \quad p_m = \frac{W}{v_1 - v_2} = 411.48/0.867$$

$$= 474.6 \text{ kN/m}^2 = 4.746 \text{ bar}$$

Example 3.27 An oil engine working on dual cycle has cylinder bore of 25 cm and stroke 45 cm. The compression ratio is 15 and explosion ratio is 1.5. Cut-off occurs at 6% of the stroke. Calculate the air standard efficiency.

Solution

Cut-off ratio,
$$r_c = 1 + k\,(r - 1)$$

$$= 1 + \frac{6}{100} \times (15 - 1) = 1.84$$

$$\eta_{\text{Dual}} = 1 - \frac{1}{r^{\gamma-1}}\left[\frac{r_p \cdot r_c^{\gamma} - 1}{r_p - 1 + \gamma\, r_p\,(r_c - 1)}\right]$$

$$\therefore \quad \eta_{\text{Dual}} = 1 - \frac{1}{15^{1.4-1}}\left[\frac{1.5 \times 1.84^{1.4-1}}{1.5 - 1 + 1.4 \times 1.5 \times (1.84 - 1)}\right]$$

$$= 0.622 \text{ or } 62.2\%$$

Example 3.28 An engine works on limited pressure cycle, the compression ratio is 11. The pressure at the commencement is 1 bar and the temperature is 85°C. If the maximum pressure in the cycle is 48 bar and the constant heat reception continues for 1/20 of the stroke, calculate the work done per kg of air and the ideal thermal efficiency. Take $c_v = 0.718$ kJ/kgK and $c_p = 1.005$ kJ/kgK.

Solution Refer Figure 3.21.

Let $v_3 = v_2 = v_c$, $\quad v_1/v_2 = r = 11$,

$v_1 = 11v_2 = 11v_c$, $\quad v_5 = 11v_c$

$$v_4 = v_3 + \frac{1}{20}(v_1 - v_2)$$

$$= v_c + \frac{1}{20}(11v_c - v_c)$$

$$= v_c + 0.5v_c = 1.5v_c$$

$$\therefore \quad \frac{v_4}{v_c} = \frac{v_4}{v_3} = r_c = 1.5$$

$$\gamma = \frac{c_p}{c_v} = \frac{1.005}{0.718} = 1.4$$

$$T_2 = T_1 \times (v_1/v_2)^{\gamma-1} = T_1 \times (11)^{1.4-1}$$

$$= 358 \times (11)^{1.4-1} = 934.2 \text{ K}$$

$$p_2 = p_1 \times (v_1/v_2)^{\gamma} = 1 \times (11)^{1.4} = 28.7 \text{ bar}$$

$$\frac{T_3}{T_2} = \frac{p_3}{p_2}$$

$$\therefore \quad T_3 = T_2 \times \frac{p_3}{p_2} = 934.2 \times 48/28.7 = 1562.42 \text{ K}$$

$$T_4 = T_3 \times (v_4/v_3) = 1562.4 \times 1.5 = 2343.6 \text{ K}$$

$$T_5 = T_4 \times (v_4/v_3)^{\gamma - 1}$$

$$= 2343.6 \times (1.5v_c/11v_c)^{1.4 - 1} = 1056.24 \text{ K}$$

$$W = Q_1 - Q_2 = c_v (T_3 - T_2) + c_p (T_4 - T_3) - c_v (T_5 - T_1)$$

$$W = 0.718 \times (1562.42 - 934.2) + 1.005 \times (2343.6 - 1562.42) - 0.718 \times (1056.24 - 358)$$

$$\therefore \quad W = 734.81 \text{ kJ/kg}$$

$$Q_1 = c_v (T_3 - T_2) + c_p (T_4 - T_3)$$

$$\therefore \quad Q_1 = 0.718 \times (1562.42 - 934.2) + 1.005 \times (2343.6 - 1562.42)$$

$$= 1236.15 \text{ kJ/kg}$$

$$\therefore \quad \eta_{\text{Dual}} = (W/Q_1) = (615.3/1236.15)$$

$$= 0.5944 \text{ or } 59.44\%$$

Example 3.29 An air standard dual cycle has a compression ratio of 12. The pressure and temperature at the beginning of compression are 1 bar, 25°C. The maximum pressure and the maximum temperature is 45 bar and 1500°C. Calculate the temperature at the end of constant volume heat addition, cut-off ratio, work done per kg of air, and the cycle efficiency. Take $c_v = 0.717$ kJ/kgK and $c_p = 1.004$ kJ/kgK for air.

Solution Refer Figure 3.22.

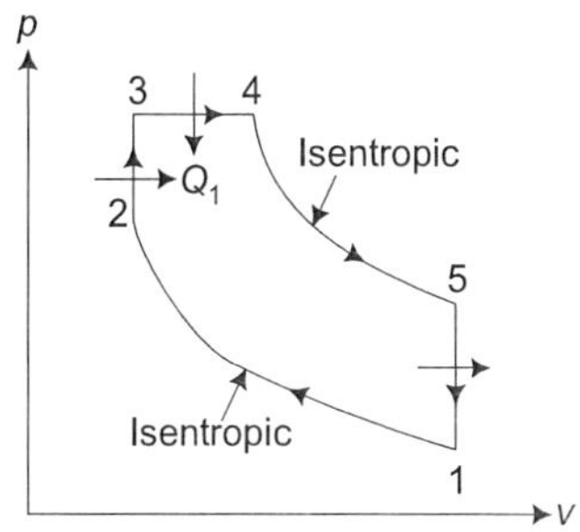

Fig. 3.22

$$\gamma = (c_p/c_v) = (1.005/0.718) = 1.4$$

$$T_2 = T_1 \times (v_1/v_2)^{\gamma - 1} = T_1 \times (r)^{\gamma - 1}$$

$$= 298 \times (12)^{1.4 - 1} = 810.56 \text{ K}$$

$$p_2 = p_1 \times (v_1/v_2)^{\gamma} = 1 \times (12)^{1.4}$$

$$= 32.42 \text{ bar}$$

$$T_3 = T_2 \times (p_3/p_2) = 810.56 \times (45/32.43)$$

$$= 1125.06 \text{ K}$$

$$\therefore \quad r_c = (T_4/T_3) = (1773/1125.06) = 1.576$$

$$\frac{T_4}{T_5} = \left(\frac{v_5}{v_4}\right)^{\gamma - 1} = \left(\frac{r}{r_c}\right)^{\gamma - 1}$$

$$= \left(\frac{12}{1.576}\right)^{1.4 - 1} = 2.2524$$

$\therefore$ $$T_5 = (T_4/2.2524) = (1773/2.2524) = 787.16 \text{ K}$$

$$W = Q_1 - Q_2 = [c_v (T_3 - T_2) + c_p (T_4 - T_3)] - c_v (T_5 - T_1)$$

$$W = [0.717 (1125.06 - 810.56) + 1.004 (1773 - 1125.06)] - 0.717 (787.3 - 298)$$

$\therefore$ $$W = [876.03] - [350.83] = 525.2 \text{ kJ/kg}$$

$\therefore$ $$\eta = (W/Q_1) = (525.2/876.03)$$

$$= 0.5995 \text{ or } 59.95\%$$

Example 3.30 An oil engine works on mixed cycle, the heat liberated at constant pressure being double at constant volume. The compression ratio of the engine is 10 and the expansion ratio is 5.5. The compression and expansion proceses follow the law $pv^{1.2} = C$. The pressure and temperature at the beginning of the compression are 1 bar and 25°C respectively. Take $c_v = 0.717$ kJ/kgK and $c_p = 1.004$ kJ/kgK for air. Find the air standard efficiency and the mean effective pressure.

Solution Refer Figure 3.22.

$$\gamma = (c_p/c_v) = (1.005/0.718) = 1.4$$

$$(v_1/v_2) = r - 1 = 10 - 1 = 9,\ v_1 = 9v_2$$

$$r_c = (r/r_e) = (10/5.5) = 1.82$$

$$\text{Mean effective pressure} = \frac{\text{Area } 1 - 2 - 3 - 4 - 5}{\text{Swept volume}}$$

Area 1 – 2 – 3 – 4 – 5 = Area under 3 – 4 + Area under 4 – 5 – Area under 2 – 1

$$= p_4 (v_4 - v_3) + \frac{p_4 v_4 - p_5 v_5}{n - 1} - \frac{p_2 v_2 - p_1 v_1}{n - 1}$$

$$v_2 = v_3 = v_c$$

$$v_1 = v_5 = rv_c = 10v_c$$

$$v_4 = r_c v_3 = 1.82 v_3$$

$$T_2 = T_1 \times (v_1/v_2)^{n - 1} = T_1 \times (10)^{1.2 - 1}$$

$$= 298 \times (10)^{1.2 - 1} = 472.29 \text{ K}$$

$$p_2 = p_1 \times (v_1/v_2)^n = 1 \times (10)^{1.2} = 15.85 \text{ bar}$$

The heat liberated at constant pressure being double at constant volume,

$$c_p (T_4 - T_3) = 2 \times c_v (T_3 - T_2),\ 1.004 \times (T_4 - T_3)$$

$$= 2 \times 0.717 \times (T_3 - T_2)$$

$$(T_4 - T_3) = 1.428 \times (T_3 - T_2)$$

$$\frac{T_4}{T_3} = \frac{v_4}{v_3} = r_c = 1.82$$

$$T_4 = 1.82 \times T_3$$

$$(1.82T_3 - T_3) = 1.428 \times (T_3 - 472.29)$$

$\therefore$ $$T_3 = 1109.26 \text{ K}$$

$$T_4 = 11.82 \times 1109.26 = 2018.85 \text{ K}$$

$$\frac{p_3}{p_2} = \frac{T_3}{T_2} = \frac{1109.26}{472.29}$$

$\therefore$ $$p_3 = p_2 \times \frac{1109.26}{472.29} = 15.85 \times \frac{1109.26}{472.29} = 37.23 \text{ bar}$$

$$(T_4/T_5) = (r_e)^{n-1} = (5.5)^{1.2-1} = 1.41,\ T_5 = (T_4/1.41)$$

$$= (2018.85/1.41) = 1431.81 \text{ K}$$

$$(p_4/p_5) = (r_e)^n = (5.5)^{1.2},\ p_5 = p_4/(5.5)^{1.2}$$

$$= 37.23/(5.5)^{1.2} = 4.81 \text{ bar}$$

Area 1 – 2 – 3 – 4 – 5 = Area under 3 – 4 + Area under 4 – 5 – Area under 2 – 1

$$= p_4\,(v_4 - v_3) + \frac{p_4 v_4 - p_5 v_5}{n-1} - \frac{p_2 v_2 - p_1 v_1}{n-1}$$

$$= 37.23 \times (1.82\, v_c - v_c) + \frac{37.23 \times 1.82\, v_c - 4.81 \times 10 v_c}{1.2 - 1} - \frac{15.85\, v_c - 1 \times 10 v_c}{1.2 - 1}$$

$$= 30.545\, v_c + 97.975\, v_c - 29.25\, v_c = 99.27\, v_c$$

Also Area 1 – 2 – 3 – 4 – 5 = $p_m \times v_1 = p_m \times 9v_c$

$\therefore$ $$p_m \times 9v_c = 99.27\, v_c$$

$\therefore$ $$p_m = (99.27\, v_c/9v_c) = 11.03 \text{ bar}$$

$$v_1 = \frac{m \times R \times T_1}{p_1} = \frac{1 \times 287 \times 297}{1 \times 10^5}$$

$$= 0.855 \text{ m}^3/\text{kg}$$

$$W = \text{Area } 1 - 2 - 3 - 4 - 5 = 99.27\, v_c$$

$$= 99.27 \times \frac{v_1}{10} = 99.27 \times \frac{0.855}{10} = 848.76 \text{ kJ/kg}$$

Heat added, $$Q_1 = c_v\,(T_3 - T_2) + c_p\,(T_4 - T_3)$$

$$Q_1 = 0.717 \times (1109.26 - 472.29) + 1.004 \times (2018.85 - 1109.26)$$

$$= 1369.84 \text{ kJ/kg}$$

$\therefore \quad \eta = (W/Q_1) = (848.76/1369.84)$

$= 0.6196$ or 61.96%

Example 3.31 A C.I. engine working on the dual cycle takes in two-thirds of its total heat at constant volume and one-third at constant pressure. The compression ratio is 15 and the maximum pressure in the cycle is 50 bar. Air intake is at 20°C and 1 bar. Calculate the temperature at each point of the cycle and thermal efficiency of this cycle. Take $c_v = 0.717$ kJ/kgK and $c_p = 1.004$ kJ/kgK.

Solution Refer Figure 3.22.

$$\gamma = (c_p/c_v) = (1.005/0.718) = 1.4$$

Let $\quad v_3 = v_2 = v_c = 1$ unit

$\therefore \quad v_1 = v_2 \times r = 1 \times 15 = 15$ unit

$$T_2 = T_1 \times (v_1/v_2)^{\gamma - 1} = T_1 \times (r)^{\gamma - 1}$$

$$= 293 \times (15)^{1.4 - 1} = 865.57 \text{ K}$$

$$p_2 = p_1 \times (v_1/v_2)^{\gamma} = 1 \times (15)^{1.4} = 44.31 \text{ K}$$

$$T_3 = T_2 \times (p_3/p_2) = 865.57 \times (50/44.31) = 976.72 \text{ K}$$

Heat added during process 2-3,

$$Q_{2-3} = c_v (T_3 - T_2) = 0.717 \times (976.72 - 865.57)$$

$$= 79.69 \text{ kJ/kg}$$

Let $\quad Q_1 =$ Total heat supplied

$$Q_{2-3} = (2/3)\, Q_1$$

$\therefore \quad Q_1 = (3/2) \times Q_{2-3} = (3/2) \times 79.69$

$$= 119.535 \text{ kJ/kg}$$

$\therefore \quad Q_{3-4} = (1/3) \times Q_1 = (1/3) \times 119.535$

$$= 39.845 \text{ kJ/kg}$$

But $\quad Q_{3-4} = c_p (T_4 - T_3) = 39.845$

$$1.004 \times (T_4 - 976.72) = 39.845$$

$\therefore \quad T_4 = 1016.51$ K

$$v_4 = v_3 \times (T_4/T_3)$$

$$= 1 \times (1016.41/976.72) = 1.041 \text{ m}^3/\text{kg}$$

$$T_5 = T_4 \times (v_4/v_5)^{\gamma - 1}$$

$$= 1016.41 \times (1.040/15)^{1.4 - 1} = 349.63 \text{ K}$$

$$\eta_{Dual} = 1 - \frac{c_v (T_5 - T_1)}{c_v (T_3 - T_2) + c_p (T_4 - T_3)}$$

$$\therefore \quad \eta_{Dual} = 1 - \frac{0.717 \times (349.63 - 293)}{0.717 \times (976.72 - 865.57) + 1.004 \times (1016.41 - 976.72)}$$

$$= 0.6604 \text{ or } 66.04\%$$

3.17 COMPARISON OF OTTO, DIESEL, AND DUAL CYCLES

The relative performance of the Otto, Diesel, and dual cycles can be known by comparing these cycles with regard to thermal efficiency. The important variables which are used as the basis for comparison of the cycles are compression ratio, peak pressure, heat addition, heat rejection, and the net work. The cycles can be analyzed for these parameters.

3.17.1 For Same Compression Ratio and Same Heat Input

Figure 3.23 shows the Otto, dual, and Diesel cycles on $p - v$ and $T - s$ diagrams. As the compression ratio is same so, all the cycles start from the same initial state point 1 and the air is compressed isentropically to the same state 2. In the $T - s$ diagram Area 5 – 2 – 3 – 6 = Area 5 – 2 – 2′ – 3′ – 6′ = Area = 5 – 2 – 3″ – 6″ as this area represents the heat input which is same for all the cycles. The areas under the line 4 – 1, 4′ – 1, and 4″ – 1 represent the heat rejected for Otto, dual, and Diesel cycles respectively.

Thermal efficiency,

$$\eta = 1 - \frac{\text{Heat rejected}}{\text{Heat supplied}} = 1 - \frac{Q_2}{Q_1}$$

For the same compression ratio and for the same heat input, minimum heat is rejected in the Otto cycle and the maximum heat is rejected in the Diesel cycle. Consequently, Otto cycle has the highest work output and efficiency. Diesel cycle has the least efficiency and dual cycle has the efficiency between the two.

$$\therefore \quad \eta_{Otto} > \eta_{Dual} > \eta_{Diesel}$$

It is to be noted that the above condition for same compression ratio is purely theoretical. The actual compression ignition engine is operated at a high compression ratio than S.I. engine which is limited by detonation.

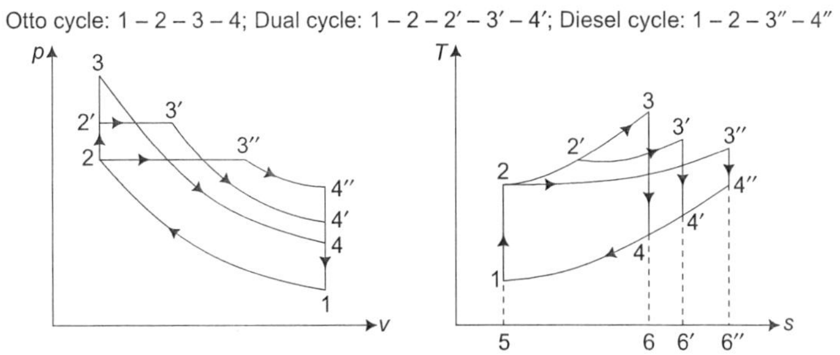

Fig. 3.23 For same Compression Ratio and same Heat Input

3.17.2 For Same Peak Pressure, Peak Temperature, and Heat Rejection

Figure 3.24 shows the Otto and Diesel cycles on $p - v$ and $T - s$ diagrams where the peak pressure, temperature, and the amount of heat rejected are same.

The efficiency of the Otto cycle is given by,

$$\eta_{\text{Otto}} = 1 - \frac{Q_2}{Q_1}$$

where Q_1 = Area 2 – 3 – 6 – 5 in $T - s$ diagram.

The efficiency of the Diesel cycle is given by,

$$\eta_{\text{Diesel}} = 1 - \frac{Q_2}{Q'_1}$$

where Q'_1 = Area 2′ – 3 – 6 – 5 in $T - s$ diagram.

It is clear that $Q'_1 > Q_1$, therefore, Diesel cycle efficiency is greater than the Otto cycle efficiency when both spark ignition and compression ignition engines have to withstand same thermal and mechanical stresses. The dual cycle efficiency will fall between these two.

$$\therefore \quad \eta_{\text{Diesel}} > \eta_{\text{Dual}} > \eta_{\text{Otto}}$$

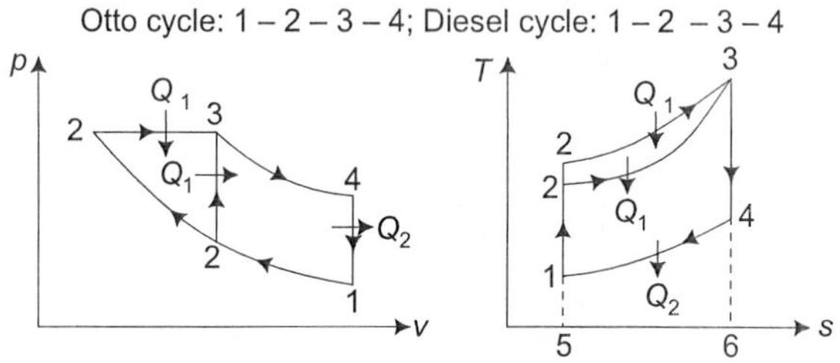

Fig. 3.24 For same Peak Pressure, Temperature, and Heat Rejection

3.17.3 For Constant Maximum Pressure and Same Heat Input

Figure 3.25 compares the Otto and Diesel cycles on the basis of same maximum pressure and same heat input.

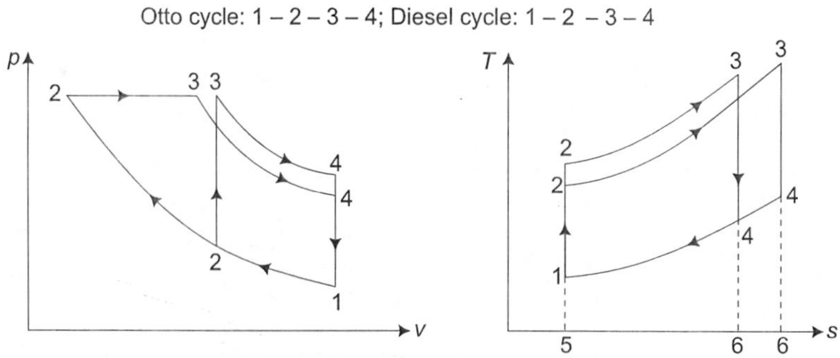

Fig. 3.25 For same Maximum Pressure and Heat Input

For the same maximum pressure the point 3 and 3′ must lie on the same pressure line as shown in $p - v$ diagram and for the same heat input

Area 2 – 3 – 6 – 5 = Area 5 – 2′ – 3′ – 6′ on $T - s$ diagram.

It is seen from the $T - s$ diagram that the heat rejected by the Diesel cycle (Area 4′ – 6′ – 5 – 1) is less than the heat rejected by Otto cycle (Area 4 – 6 – 5 – 1). Hence, the efficiency of Diesel cycle is more than the Otto cycle for the same maximum pressure and heat input. One can also observe that under these conditions the Diesel cycle has higher compression ratio than that of Otto cycle. Also the cycle which will allow the maximum expansion has the highest efficiency. The dual cycle efficiency will be between these two.

$$\therefore \quad \eta_{\text{Diesel}} > \eta_{\text{Dual}} > \eta_{\text{Otto}}$$

3.17.4 For Same Maximum Pressure and Work Output

The efficiency is given by,

$$\eta = \frac{\text{Work done}}{\text{Heat supplied}} = \frac{\text{Work done}}{\text{Work done + Heat rejected}}$$

Refer $T - s$ diagram in Figure 3.25. For the same work output,

Area 1 – 2 – 3 – 4 = Area 1 – 2′ – 3′ – 4′.

These will be equal when point 3 lies on greater entropy than at 3′. It is seen that the heat rejection for Otto cycle is more than that of Diesel cycle. Therefore, for these conditions the Diesel cycle is more efficient than the Otto cycle. The efficiency of dual cycle will lie between the two cycles.

3.17.5 For Same Compression Ratio and Heat Rejection

Figure 3.26 shows the Otto cycle (1 – 2 – 3 – 4) and Diesel cycle (1 – 2 – 3′ – 4) on $p - v$ and $T - s$ diagrams.

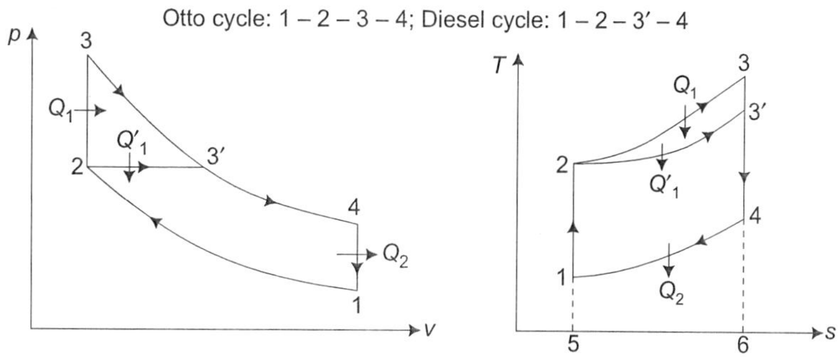

Fig. 3.26 For same Compression Ratio and Heat Rejection

As the compression ratio is same, so the cycles start from the same initial state point 1 and the air is compressed isentropically to same point 2. From the $T - s$ diagram it can be seen that Area 4 – 6 – 5 – 1 is equal which represents the heat rejection.

Efficiency of Otto cycle is given by,

$$\eta_{Otto} = 1 - \frac{Q_2}{Q_1}$$

$$Q_1 = \text{Heat supplied} = \text{Area } (2 - 3 - 6 - 5)$$

$$Q_2 = \text{Heat rejected} = \text{Area } (1 - 4 - 6 - 5).$$

Efficiency of Diesel cycle is given by,

$$\eta_{Diesel} = 1 - \frac{Q_2}{Q'_1}$$

$$Q'_1 = \text{Heat supplied} = \text{Area } (2 - 3' - 6 - 5)$$

$$Q_2 = \text{Heat rejected} = \text{Area } (1 - 4 - 6 - 5).$$

$$\text{Area } (2 - 3 - 6 - 5) > \text{Area } (2 - 3' - 6 - 5)$$

$$\therefore \quad Q_1 > Q'_1$$

That is, heat supplied in Otto cycle is more than that of Diesel cycle but heat rejected is same. Hence, the efficiency of Otto cycle is greater than that of Diesel cycle for a given compression ratio and heat rejection. The efficiency of dual cycle lies between these two cycles.

$$\therefore \quad \eta_{Otto} > \eta_{Dual} > \eta_{Diesel}$$

3.18 GAS TURBINES AND ITS APPLICATIONS

A gas turbine is a rotary machine similar in principle to a steam turbine. The gas turbine is the most satisfactory power developing machine due to its exceptional reliability, freedom from vibrations, and ability to produce large powers from units of relatively smaller size and weight. The main components of a gas turbine are compressor, combustion chamber, and the turbine. The air is compressed and is supplied into a combustion chamber where fuel is burnt. The products of combustion so produced are then made to impinge over rings of turbine blades with high velocity and thus work is produced. About 60% of the work produced is used to drive the compressor and the rest is available as useful work output. Liquid fuels are commonly used in gas turbines. Gaseous fuels are also used if available in natural form as by-products. The gaseous fuels are natural gas, coal gas, producer gas, etc. Solid fuels may also be used after converting them in powder form, e.g., pulverized coal.

The important applications of gas turbines are:

1. Gas turbines have extensive applications in the field of aviation.
2. Gas turbines are used for central electricity generating stations both for base load and for peak load generation. But majority of the gas turbine power plants are used for peak load service.
3. Gas turbines have been used in industry of making pumping sets for oil and gas transmission lines, chemical processing, refineries, etc.
4. Gas turbines are used in combination with conventional steam power plants and are becoming quite popular.

5. In navy all major warships are powered by gas turbines.
6. Gas turbines have also been used on experimental basis in locomotives and in luxury cars.
7. Gas turbines are used for marine propulsion.

3.19 CLASSIFICATION OF GAS TURBINE

Gas turbines are classified in two groups:

1. Constant pressure combustion gas turbine
   (a) Open cycle constant pressure gas turbine.
   (b) Closed cycle constant pressure gas turbine.
2. Constant volume combustion gas turbine.

3.19.1 Constant Pressure Gas Turbine

Figure 3.27 shows a single shaft turbine working on simple open cycle. Its main components are air compressor, combustion chamber, and a gas turbine. Atmospheric air enters the compressor which compresses it to a pressure of 3 or 4 atmosphere and at the same time increasing its temperature. This pressurized air then enters the combustion chamber. Only a part of this air enters the combustor basket to mix and burn with fuel, called primary air. The temperature of the combustion gases at this stage is very high (approximately 1650°C). The temperature of the combustion gases should not exceed 900°C from metallurgical considerations. Modern gas turbines are designed for dependable operation with inlet temperature of 800°C to 850°C. Hence, this high temperature has to be brought down so that temperature stresses do not become excessive and thus breakage of turbine blades may be prevented. This objective

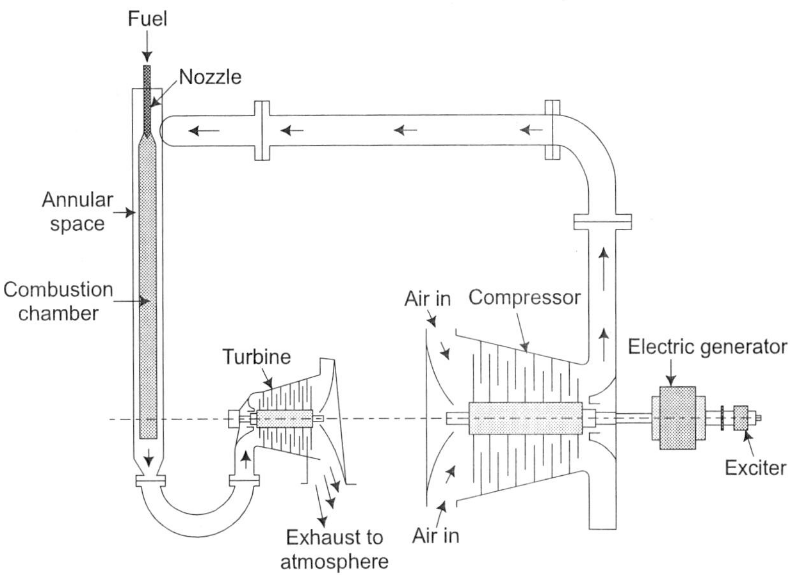

Fig. 3.27 A Constant Pressure Open Cycle Gas Turbine Plant

can be achieved by allowing the remaining air, called secondary air to enter the combustor through the openings farther downstream to mix with and cool down the products of combustion. These high pressure and high temperature gases are then expanded in the gas turbine up to atmospheric pressure and thus produces work. About 60% of the power so developed by the turbine is consumed in driving the compressor and the remaining power available is used to run the electric generator.

3.19.2 Open and Closed Cycle Constant Pressure Gas Turbine

The schematics of open and closed cycle constant pressure gas turbines are shown in Figure 3.28. The ideal constant pressure gas turbine works on Brayton (or Joule) cycle. The $T - s$ diagrams of open and closed turbine cycle are shown in Figure 3.29.

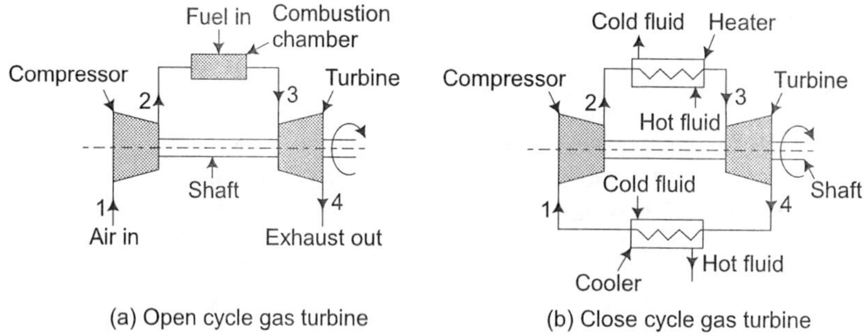

Fig. 3.28 Schematics of Open and Closed Cycle Gas Turbine

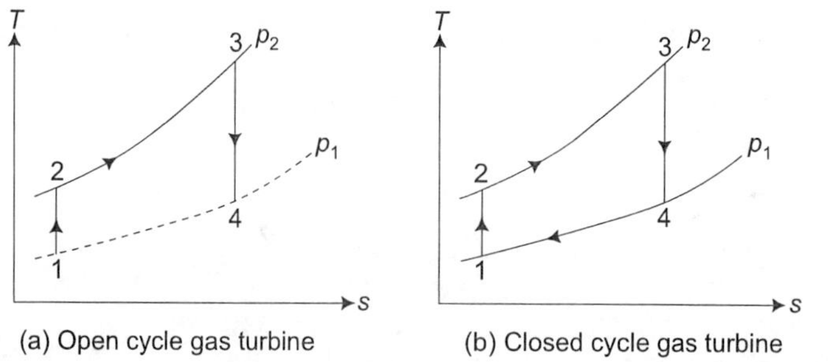

Fig. 3.29 $T - s$ diagrams for Open and Closed Cycle Gas Turbine

In an open cycle constant pressure gas turbine the air is sucked into the compressor from the atmosphere at condition 1 and it is compressed to condition 2. The heat is added at constant pressure during the process 2-3 by direct burning of the fuel into the air in the combustion chamber. The high pressure and high temperature gases so produced due to the combustion of fuel inside the combustion chamber are then passed through the turbine where it expands and thus produces some useful amount of work. Then the products of combustion are exhausted to the atmosphere at condition 4.

In closed cycle gas turbine the working fluid does not mix with the atmospheric air or any other gas and it is recirculated again and again through the machine. The compressed fluid coming out from the compressor is heated in a heat exchanger (heater) by burning the fuel externally. This heat addition takes place at constant pressure. Then the high temperature and high pressure working fluid passes to the turbine where it develops the useful work. The fluid coming out after the expansion through the turbine is cooled to its original temperature at condition 1 in the heat exchanger (cooler) by using an external cooling source. After cooling of the working fluid it is again fed to the compressor and thus the cycle is repeated.

The major difference between the closed cycle and the open cycle is that the working fluid in the case of closed cycle is recirculated repeatedly, whereas it is replaced continuously in an open cycle gas turbine. The process 4-1 represents the cooling of the working fluid in the heat exchanger (cooler) in closed cycle, whereas process 4-1 does not exist in the open cycle instead the cycle ended at point 4 by exhausting the burnt gases to the atmosphere.

3.19.3 Constant Volume Combustion Gas Turbine

A constant volume gas turbine unit works on Atkinson cycle in which the combustion of fuel takes place at constant volume. In the very early stages of development of gas turbine, constant volume gas turbines were manufactured and were known as Holzworth gas turbines. This type of turbine was replaced by constant pressure gas turbine working on Brayton cycle. Figure 3.30 shows the Holzworth constant volume gas turbine. The axial flow compressor sucks the air from the atmosphere and compresses it up to a pressure of 3 bar to 4 bar. The compressed air is then supplied to the combustion chamber through the auto valves V_1 and V_2. The required amount of fuel is also supplied to the combustion chamber with the help of a fuel pump through an automatic valve V_3. Then so formed air fuel mixture is ignited with the help of a spark plug which results in combustion of fuel-air mixture at constant volume. The combustion of fuel-air mixture increases the pressure and temperature of the combustion gases in the order of 15 bar. These products of combustion are then supplied to the gas turbine through a gas nozzle. The combustion chamber is also fitted with a non-return valve V_4 which supplies the products of combustion to the nozzle and it closes automatically due to increase and decrease of pressure inside the combustion chamber.

The expansion of these combustion gases take place inside the turbine and thus useful work is developed. The water flows inside the water jackets provided around the combustion chamber and the gas nozzle for cooling purposes. So, the hot water coming out of the water jackets is supplied to the boiler. The temperature of the exhaust gases coming out of the turbine is still quite high so, these exhaust gases are fed to the boiler for steam generation. The generated steam in the boiler is used to run the steam turbine which drives the compressor. The compressor supplies air to the combustion chamber. The main drawback of this turbine is intermittent combustion of fuel which impairs the smooth running of the turbine.

3.20 ANALYSIS OF CONSTANT PRESSURE CLOSED CYCLE (BRAYTON OR JOULE CYCLE) FOR GAS TURBINE

(i) Ideal Case

The constant pressure gas turbine operates on Brayton cycle (or Joule cycle). The ideal Brayton cycle is shown in Figure 3.31 on $p - v$ and $T - s$ diagram. It consists of the following processes: 1-2: Isentropic

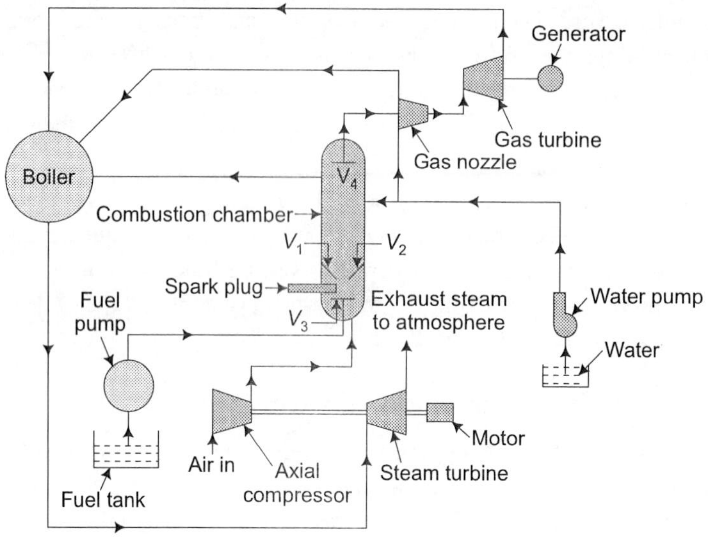

Fig. 3.30 A Constant Volume Gas Turbine Plant (Holzworth Gas Turbine)

compression in the compressor, 2-3: Heat addition at constant pressure in the combustion chamber, 3-4: Isentropic expansion in the turbine, and 4-1: Heat rejection (or cooling) at constant pressure.

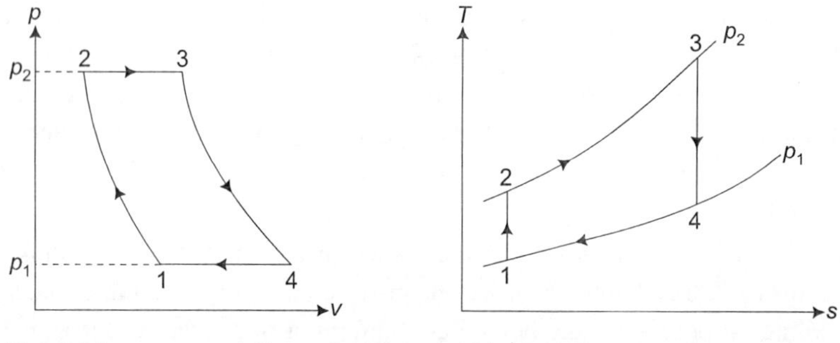

Fig. 3.31 $p - v$ and $T - s$ diagrams for Ideal Closed Cycle Gas Turbine

Expression for Thermal Efficiency

Work consumed by the compressor,

$$w_c = c_p\ (T_2 - T_1)\ \text{kJ/kg}$$

Work done by the turbine,

$$w_t = c_p\ (T_3 - T_4)\ \text{kJ/kg}$$

Net work available,

$$w_{\text{net}} = w_t - w_c = [c_p\ (T_3 - T_4) - c_p\ (T_2 - T_1)]\ \text{kJ/kg}$$

Heat supplied,

$$q_s = c_p\,(T_3 - T_2) \text{ kJ/kg}$$

Thermal efficiency,

$$\eta_{th} = \frac{w_{\text{net}}}{q_s} = \frac{c_p\,(T_3 - T_4) - c_p\,(T_2 - T_1)}{c_p\,(T_3 - T_2)}$$

$$= 1 - \frac{T_4 - T_1}{T_3 - T_2} \qquad \text{(i)}$$

Let pressure ratio,

$$r_p = \frac{p_2}{p_1} = \frac{p_3}{p_4} \qquad (\because p_2 = p_3 \text{ and } p_1 = p_4)$$

For isentropic compression process,

$$\frac{T_2}{T_1} = \left(\frac{p_2}{p_1}\right)^{(\gamma-1)/\gamma} = r_p^{(\gamma-1)/\gamma}$$

or $$T_2 = T_1\, r_p^{(\gamma-1)/\gamma}$$

For isentropic expansion process,

$$\frac{T_3}{T_4} = \left(\frac{p_3}{p_4}\right)^{(\gamma-1)/\gamma} = r_p^{(\gamma-1)/\gamma}$$

or $$T_3 = T_4\, r_p^{(\gamma-1)/\gamma}$$

Substituting the values of T_2 and T_3 in the expression (i), we have,

$$\eta_{th} = 1 - \frac{T_4 - T_1}{T_4\, r_p^{(\gamma-1)/\gamma} - T_1\, r_p^{(\gamma-1)/\gamma}}$$

$$\therefore \qquad \eta_{th} = 1 - \frac{1}{r_p^{(\gamma-1)/\gamma}} = 1 - \left(\frac{1}{r_p}\right)^{(\gamma-1)/\gamma} \qquad (3.39)$$

It is observed from equation (3.39) that the thermal efficiency of an ideal gas turbine cycle is a function of pressure ratio only. The variation of thermal efficiency with pressure ratio is shown in Figure 3.32. It can be seen that the efficiency increases with increase in pressure ratio for a particular gas constant γ and for constant pressure ratio it increases with increase in γ.

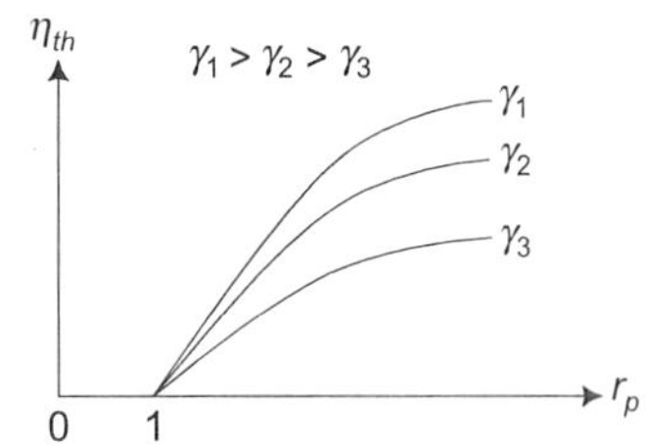

Fig. 3.32 Variation of Efficiency w.r.t. Pressure Ratios

Expression for Work Ratio

Work ratio (r_w) is defined as the ratio of net work (or useful work) obtained from the plant to the turbine work.

$$r_w = \frac{w_t - w_c}{w_t} = \frac{c_p\,(T_3 - T_4) - c_p\,(T_2 - T_1)}{c_p\,(T_3 - T_2)}$$

$$= \frac{(T_3 - T_4) - (T_2 - T_1)}{(T_3 - T_2)} = 1 - \frac{T_2 - T_1}{T_3 - T_4}$$

Substituting the values of T_3 and T_2, we have,

$$r_w = 1 - \frac{T_1\left[r_p^{(\gamma-1)/\gamma} - 1\right] - T_1}{T_4\left[r_p^{(\gamma-1)/\gamma} - 1\right] - T_4} = 1 - \frac{T_1}{T_4} \quad \text{(ii)}$$

Now substituting $T_4 = \dfrac{T_3}{r_p^{(\gamma-1)/\gamma}}$ in the expression (ii), we have,

$$r_w = 1 - \frac{T_1}{T_3}\, r_p^{(\gamma-1)/\gamma} \quad (3.40)$$

It is observed from equation (3.40) that the work ratio of an ideal gas turbine cycle is a function of pressure ratio (r_p), compressor inlet temperature (T_1), and the turbine inlet temperature (T_3). The work ratio increases with the decrease in compressor inlet temperature and pressure ratio, and with the increase in turbine inlet temperature.

Expression for Optimum Pressure Ratio and Maximum Net Work

If
$$r_w = 0$$

$$1 - \frac{T_1}{T_3}\, r_p^{(\gamma-1)/\gamma} = 0$$

Then
$$(r_p)_{max} = \left(\frac{T_3}{T_1}\right)^{\gamma/(\gamma-1)} \quad (3.41)$$

When compressor work and turbine work will be equal the work ratio will be zero which occurs at $r_p = 1$ and at $r_p = (r_p)_{max}$ and thus the plant yields no output. It means work ratio will be maximum at some value of pressure ratio between $r_p = 1$ and $r_p = (r_p)_{max}$ which is called optimum pressure ratio.

For maximum work ratio the net work obtained from the plant should be maximum. The net work is given by,

$$w_{net} = w_t - w_c = [c_p\,(T_3 - T_4) - c_p\,(T_2 - T_1)]$$

$$= c_p\, T_3\left(1 - \frac{T_4}{T_3}\right) - c_p\, T_1\left(\frac{T_2}{T_1} - 1\right)$$

Substituting
$$\frac{T_4}{T_3} = \frac{1}{r_p^{(\gamma-1)/\gamma}}$$

$$\frac{T_2}{T_1} = r_p^{(\gamma-1)/\gamma}, \text{ we get,}$$

$$w_{net} = c_p\, T_3\left(1 - 1/r_p^{(\gamma-1)/\gamma}\right) - c_p\, T_1\left(r_p^{(\gamma-1)/\gamma} - 1\right) \quad \text{(iii)}$$

When T_1 and T_3 are constant then w_{net} depends on r_p only. So, w_{net} will be maximum when

$$\frac{dw_{net}}{dr_p} = 0$$

$$\frac{d}{dr_p}\left[c_p\, T_3\left(1 - 1/r_p^{(\gamma-1)}\right)/\gamma - c_p\, T_1\left(r_p^{(\gamma-1)/\gamma} - 1\right)\right] = 0$$

$$\therefore \quad (r_p)_{optimum} = \left(\frac{T_3}{T_1}\right)^{\gamma/2(\gamma-1)} \tag{3.42}$$

Comparing equations (3.41) and (3.42),

$$(r_p)_{optimum} = \sqrt{(r_p)_{max}} \tag{3.43}$$

Substituting equation (3.43) in the expression (iii), the maximum net work is given by,

$$(w_{net})_{max} = c_p\, T_3\left(1 - \sqrt{\frac{T_1}{T_3}}\right) - c_p\, T_1\left(\sqrt{\frac{T_3}{T_1}} - 1\right)$$

$$(w_{net})_{max} = c_p\left(T_3 - \sqrt{T_3\, T_1}\right) - c_p\left(\sqrt{T_3\, T_1} - T_1\right)$$

$$= c_p\left(T_3 - 2\sqrt{T_3\, T_1} + T_1\right)$$

$$\therefore \quad (w_{net})_{max} = c_p\left(\sqrt{T_3} - \sqrt{T_1}\right)^2 \tag{3.44}$$

From equation (3.44) it can be seen that the net work obtained from the plant increases with the increase in the turbine inlet temperature and with the decrease in compressor inlet temperature.

(ii) Actual Case

In actual Brayton cycle the compression and expansion processes are non-isentropic due to friction and turbulence. Therefore, the temperature at the end of compression and expansion processes are higher than in the case of ideal Brayton cycle. The actual Brayton cycle is shown in Figure 3.33 in which the dotted line shows the isentropic processes.

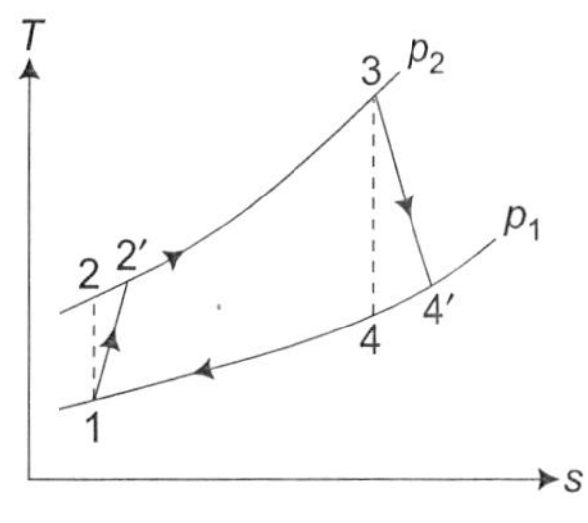

Fig. 3.33 *T – s* Diagram for Actual Closed Cycle Gas Turbine

The isentropic efficiency of compressor,

$$\eta_c = \frac{\text{Ideal work}}{\text{Actual work}}$$

$$= \frac{T_2 - T_1}{T_{2'} - T_1} \tag{3.45}$$

The isentropic efficiency of turbine,

$$\eta_t = \frac{\text{Actual work}}{\text{Ideal work}}$$

$$= \frac{T_3 - T_{4'}}{T_3 - T_4} \tag{3.46}$$

Expression for Optimum Pressure Ratio and Maximum Net Work for Actual Cycle

Net work for actual cycle,

$$(w_{net}) = w_t - w_c = c_p (T_3 - T_{4'}) - c_p (T_{2'} - T_1) \text{ kJ/kg}$$

$$w_{net} = c_p \eta_t (T_3 - T_4) - c_p (T_2 - T_1)/\eta_c$$

Substituting $\frac{T_4}{T_3} = \frac{1}{r_p^{(\gamma - 1)/\gamma}}$

$\frac{T_2}{T_1} = r_p^{(\gamma - 1)/\gamma}$, we get,

$$w_{net} = c_p \eta_t T_3 \left(1 - \frac{1}{r_p^{(\gamma-1)/\gamma}}\right) - \frac{c_p T_1}{\eta_c}\left(r_p^{(\gamma-1)/\gamma} - 1\right) \quad \text{(i)}$$

For maximum work

$$\frac{dw_{net}}{dr_p} = 0$$

$$\frac{d}{dr_p}\left[c_p \eta_t T_3 \left(1 - \frac{1}{r_p^{(\gamma-1)/\gamma}}\right) - \frac{c_p T_1}{\eta_c}\left(r_p^{(\gamma-1)/\gamma} - 1\right)\right] = 0$$

$$\therefore \quad (r_p)_{optimum} = \left(\eta_c \cdot \eta_t \frac{T_3}{T_1}\right)^{\gamma/2(\gamma-1)} \quad (3.47)$$

Equation (3.47) gives the optimum pressure ratio for maximum net work for actual cycle.

If $\eta_c = \eta_t = 1$, then equation (3.47) reduces to equation (3.42) and $T_{2'} = T_2$ & $T_{4'} = T_4$.

The maximum value of net work can be obtained by substituting equation (3.47) in the expression (i) as,

$$(w_{net})_{max} = c_p \eta_t T_3 \left(1 - \frac{1}{\sqrt{\eta_c}\sqrt{\eta_t}}\frac{\sqrt{T_1}}{\sqrt{T_3}}\right) - \frac{c_p T_1}{\eta_c}\left(\sqrt{\eta_c}\sqrt{\eta_t}\frac{\sqrt{T_3}}{\sqrt{T_1}} - 1\right)$$

$$(w_{net})_{max} = c_p \left(\eta_t T_3 - \frac{\sqrt{\eta_t}\sqrt{T_3}\sqrt{T_1}}{\sqrt{\eta_c}} - \frac{\sqrt{\eta_t}\sqrt{T_3}\sqrt{T_1}}{\sqrt{\eta_c}} + \frac{T_1}{\eta_c}\right)$$

$$\therefore \quad (w_{net})_{max} = c_p \left(\sqrt{\eta_t}\sqrt{T_3} - \frac{\sqrt{T_1}}{\sqrt{\eta_c}}\right)^2 \quad (3.48)$$

Expression for Optimum Pressure Ratio for Maximum Actual Cycle Efficiency

For simplification considering the actual heat supplied,

$$q_s = c_p (T_3 - T_{2'}) = c_p (T_3 - T_2) \text{ kJ/kg}$$

Thermal efficiency,

$$\eta_{th} = \frac{w_{\text{net}}}{q_s} = \frac{c_p (T_3 - T_{4'}) - c_p (T_{2'} - T_1)}{c_p (T_3 - T_2)}$$

$$= \frac{\eta_t (T_3 - T_4) - (T_{2'} - T_1)}{(T_3 - T_2)}$$

$$\eta_{th} = \frac{\eta_t T_3 (1 - T_4/T_3) - (T_2 - T_1)/\eta_c}{(T_3 - T_2)}$$

$$= \frac{\eta_t T_3 (1 - T_4/T_3) - (T_1/\eta_c)(T_2/T_1 - 1)}{T_1(T_3/T_1 - T_2/T_1)}$$

$$\eta_{th} = \frac{\eta_t T_3 \left(1 - 1/r_p^{(\gamma-1)/\gamma}\right) - (T_1/\eta_c)\left(r_p^{(\gamma-1)/\gamma} - 1\right)}{T_1\left(T_3/T_1 - r_p^{(\gamma-1)}\right)} \quad \text{(ii)}$$

For the fixed limit of temperatures the efficiency is a function of pressure ratio only.

Therefore, the optimum pressure ratio can be obtained as,

$$\frac{d\eta_{th}}{dr_p} = 0$$

$$\frac{d}{dr_p}\left[\frac{\eta_t T_3 \left(1 - 1/r_p^{(\gamma-1)/\gamma}\right) - (T_1/\eta_c)\left(r_p^{(\gamma-1)/\gamma} - 1\right)}{T_1\left(T_3/T_1 - r_p^{(\gamma-1)}\right)}\right] = 0$$

$$\therefore \quad (r_p)_{\text{optimum}} = \left(\frac{T_3/T_1}{1 + \sqrt{(T_3/T_1 - 1)(1/\eta_c \cdot \eta_t - 1)}}\right)^{\gamma/(\gamma-1)} \quad (3.49)$$

By substituting equation (3.49) in the expression (ii) the maximum thermal efficiency of actual cycle can be obtained.

3.21 ANALYSIS OF CONSTANT PRESSURE OPEN CYCLE (BRAYTON OR JOULE CYCLE) FOR GAS TURBINE

The analysis used for closed cycle gas turbine can also be used for open cycle gas turbine provided properties of exhaust gases are considered similar to the properties of air and the fuel mass compared with air mass is neglected. The actual Brayton cycle for open cycle gas turbine is shown in Figure 3.34 in which the dotted line shows the isentropic processes.

Expression for Thermal Efficiency

Let m_a = Mass of air supplied per second, m_f = Mass of fuel supplied per second, c_{pa} = Specific heat of air, and c_{pg} = Specific heat of gas

Work consumed by the compressor,

$$W_c = [c_{pa} (T_{2'} - T_1)] \text{ kJ}$$

Work done by the turbine,

$$W_t = [(m_a + m_f)\, c_{pg} (T_3 - T_{4'})] \text{ kJ}$$

Net work available,

$$W_{\text{net}} = W_t - W_c = [(m_a + m_f)\, c_{pg}\, (T_3 - T_{4'})] - [c_{pa}\, (T_{2'} - T_1)]$$

Heat supplied,

$$Q_3 = (m_f \times CV) \text{ kJ}$$

where CV is calorific value of fuel in kJ/kg

Thermal efficiency,

$$\eta_{th} = \frac{W_{\text{net}}}{Q_3} = \frac{(m_a + m_f)\, c_{pg}\, (T_3 - T_{4'}) - c_{pa}\, (T_{2'} - T_1)}{m_f \times CV} \quad (3.50)$$

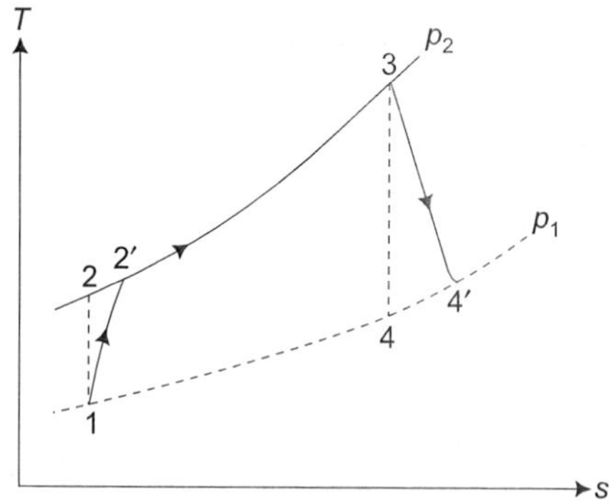

Fig. 3.34 *T – s* Diagram for Actual Open Cycle Gas Turbine

3.22 ADVANTAGES AND DISADVANTAGES OF GAS TURBINES OVER I.C. ENGINES

Advantages

1. **Higher work output** In gas turbines complete expansion occur means expansion is carried up to atmospheric pressure which results in increase in work output.
2. **Higher mechanical efficiency** Due to absence of reciprocating parts the mechanical efficiency of gas turbine is quite high and it will be in order of 95% to 97 %. Whereas I.C. engines have about 85% of mechanical efficiency due to large number of sliding parts.
3. **Less weight per kilowatt developed** The weight of gas turbine per kilowatt developed is less than I.C. engine. It is 0.15 kg/kW for gas turbine whereas for I.C. engine is 2.5 kg/kW.
4. **Perfect balancing** Perfect balancing of rotating parts can be attained in gas turbines.
5. **Uniform torque** The gas turbine is a continuous power developing machine due to which uniform torque is achieved at the crankshaft hence gas turbine does not require any flywheel.
6. **Cheaper fuel can be used** Gas turbine can use cheaper fuels such as kerosene, heavy oil, peat, coal slurry. So, there is no need to use special grade fuels such as petrol and diesel.

7. **No need of cooling arrangement** Simple gas turbines require no cooling. Even modified cycles require very little cooling.
8. **Less maintenance and installation costs** Gas turbine needs less maintenance cost. It has simple rotary mechanism thus lubricating oil consumption is small. In the gas turbines the cost of lubricating oil consumption is about 0.5% to 1% of the total fuel cost, whereas for I.C. engines it may be about 10% to 15% of the total fuel cost.
9. **Light weight components can be used** As maximum pressure used in gas turbine is up to 5 bar which is very low in comparison to I.C. engines which have about 60 bar. Thus, components of gas turbines can be made light in weight.
10. **High speed** Gas turbines can be run at very high speed may be around 40,000 rpm, whereas it is not feasible in the case of I.C. engines.
11. **Capable to operate at high altitude** Gas turbines are quite capable of maintaining their work out put at high altitude, whereas output of I.C. engines is severely affected.
12. **Simplicity** A gas turbine has all the advantages a rotary machine has over a reciprocating engine, like simple mechanism, high speed, small foundation, no balancing problem, easy maintenance, etc.
13. **Flexibility** There is large flexibility in designing, testing, and development of various components used in gas turbines as the process of compression, combustion, and expansion take place in different components. This also allows a great variety in the arrangement depending upon the duty, space, and performance desired.
14. **Lubrication and ignition system is simple** The lubrication and ignition systems used in the case of gas turbines are much simpler than used in I.C. engines.
15. **Fuel consumption does not vary with the period of operation** The fuel consumption in the case of gas turbines does not increase with period of operation as it increases in the case of I.C. engines.
16. **Less pollution** Due to supply of excess air the exhaust from gas turbine is free from smoke and it is less polluting than exhaust of I.C. engine.

Disadvantages

1. **Low thermal efficiency** The thermal efficiency of a simple gas turbine unit is low which varies from 15% to 20%, whereas for I.C. engines it varies from 25% to 30%.
2. **Large exhaust losses and exhaust gases** In gas turbines, the mass flow rate is very high which results in large exhaust losses. It produces about 5 times of exhaust gases as produced by an I.C. engine for the same power output.
3. **Low overall efficiency and increased plant complexity** The overall efficiency of a gas turbine plant is lower than that of a reciprocating plant. To obtain efficiency approximately near as that of I.C. engine it requires some additional devices such as regenerator which increases the cost and complexity of the plant.

4. **Poor part load efficiency** The part load efficiency of a gas turbine is poor. As most of the applications do not operate at full load except for a short duration and the minimum specific fuel consumption for gas turbine occurs near the maximum output thus it is a distinct disadvantage. Whereas minimum specific fuel consumption for I.C. engines occurs at about half the rated output so, its part load efficiency is higher than the gas turbine unit.
5. **Not self-starting** It is not a self-starting machine, so it requires an additional high power motor for starting purpose.
6. **High cost of manufacturing** Manufacturing of gas turbine blades needs high heat resistant materials (such as nickel-chromium alloy) and special manufacturing process which results in high cost of manufacturing. Particularly, low power rating gas turbine unit is very costly in initial as well as in running cost as compared to I.C. engine.
7. **Not permissible to have higher temperature and pressure** In gas turbines due to metallurgical limitations the maximum temperature and pressure are low as compared to I.C. engine. In the case of I.C. engines the highest temperature will remain for a small fraction of the cycle, whereas in gas turbines it will remain constant as gas turbine is a continuous machine.
8. **Low life of the blades and the combustion chamber** As gas turbine is a continuous combustion machine so, the blades and the combustion chamber are subjected continuously to a high temperature of the order of 850°C which results in low life.
9. **Slow response to acceleration** The gas turbine is slow in its response to acceleration.
10. **Requires speed reduction device** Due to high operating speeds of the turbine (around 40,000 rpm) it always requires speed reduction device.
11. **Requires special cooling system** The turbine blades require special cooling system.

3.23 ADVANTAGES OF GAS TURBINES OVER STEAM TURBINES

The main advantages of an open cycle gas turbine plant over steam plant are as follows:

1. **Simple in construction and operation** The gas turbine plant does not have any boiler and condenser thus it is simple in construction and operation. Also it has low weight which makes it more suitable for mobile plants.
2. **Low weight to power ratio** Weight to power ratio for gas turbine plant is low.
3. **Quick in operation** A gas turbine plant can be started up as well as shut down quickly. Thus, it suits well for peak load and standby operations. Under load varying conditions its control is easy and quick.
4. **Cooling water requirement is less** The gas turbine plant does not depend much on water supply so, these can be used in the region where little amount of water is available.
5. **Manpower requirement is less** To operate a gas turbine plant manpower requirement is quite less in comparison to steam power plant.
6. **Smaller in size** Because of its smaller size it requires less space for the same output.

7. **Transmission losses are less** It can be installed at the load centre itself. Thus, transmission losses are very less.
8. **Less costly** The capital cost per kilowatt developed is considerably low as compared to steam power plant. The installation maintenance costs are much less in comparison to steam power plant.
9. **Less warm up time** The gas turbine plant will accelerate from cold start to full load without any warm up time if it has been brought up to the rated speed by the motor and the fuel is ignited.

3.24 ADVANTAGES AND DISADVANTAGES OF CLOSED CYCLE OVER OPEN CYCLE

Advantages

1. **Reduced size** The use of higher pressure throughout the cycle results in reduced size of the plant for the same power output.
2. **Higher thermal efficiency** The thermal efficiency is high for the same temperature working limits.
3. **No contamination problem** As indirect heating is used in closed cycle so, there is no problem of corrosion and erosion of blades due to contamination of working medium.
4. **No loss of working medium** As the working circuit is closed, so there is no loss of working medium.
5. **Use of fuel** The closed cycle employs external heating, so any cheaper fuel like coal or wood can be used.
6. **Reliability** Due to longer useful life the reliability is high.
7. **Low maintenance cost** Maintenance cost is low.
8. **Improved part load efficiency and heat transmission** Part load efficiency and rate of heat transmission of closed cycle is higher.

Disadvantages

1. **Limited use** Due to the utilization of large quantity of water its use is limited to the stationary installation or in marine where plenty of water is available. It cannot be used in aircraft.
2. **Higher installation cost** The requirement of air tight system increases its installation cost.
3. **Not economical** Due to low power to weight ratio it is not economical in moving vehicles.
4. **Bulky and complex system** The use of large air heater and heat exchangers makes it bulky in comparison to open system. The load control of the closed system is complex and thus costly.
5. **Poor response under changing load condition** The response of closed cycle constant pressure gas turbine is relatively poor under changing load conditions.

3.25 METHODS OF IMPROVING THE EFFICIENCY AND WORK OUTPUT OF OPEN CYCLE CONSTANT PRESSURE GAS TURBINES

The performance of open cycle constant pressure gas turbines can be increased (i) by reducing heat addition by regeneration, (ii) by increasing turbine work by reheating, and (iii) by reducing compressor work by intercooling.

3.25.1 Regeneration

The heat of exhaust gases is used in a heat exchanger to preheat the compressed air before entering the combustion chamber and thus the fuel consumption is reduced. The net work in this process remains unchanged but there is substantial reduction in the heat addition in combustion chamber. The gas turbine plant with regenerator is shown in Figure 3.35 (a) and the various processes are shown in $T-s$ diagram in Figure 3.35 (b).

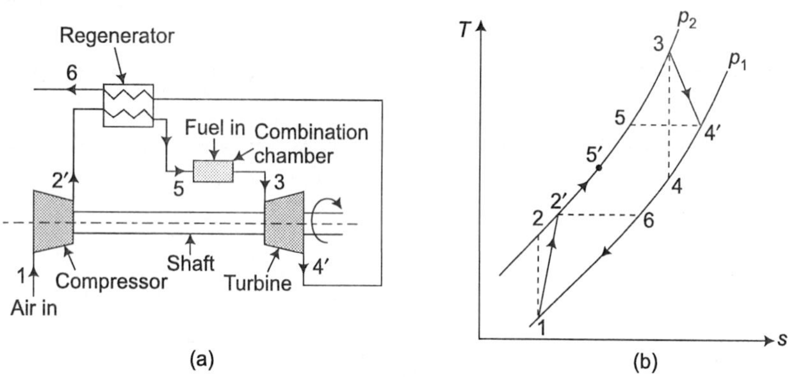

Fig. 3.35 Schematics and $T-s$ Diagram for Brayton Cycle with Regeneration

If the heat exchanger is ideal (i.e., effectiveness, $\varepsilon = 1$) then it is possible to raise the outlet temperature of the compressed air from the compressor from $T_{2'}$ to $T_5 = T_{4'}$.

Thermal efficiency of ideal regenerative cycle is given by,

$$\eta_{th} = \frac{w_{net}}{q_s} = \frac{c_p (T_3 - T_{4'}) - c_p (T_{2'} - T_1)}{c_p (T_3 - T_5)}$$

$$= 1 - \frac{T_{2'} - T_1}{T_3 - T_{4'}} \qquad (\because T_5 = T_{4'})$$

$$\eta_{th} = 1 - \frac{(T_2 - T_1)/\eta_c}{\eta_t (T_3 - T_4)} = 1 - \frac{1}{\eta_c \eta_t} \frac{(T_2 - T_1)}{(T_3 - T_4)}$$

$$= 1 - \frac{1}{\eta_c \eta_t} \frac{T_1 (T_2/T_1 - 1)}{T_3 (1 - T_4/T_3)}$$

Substituting $$\frac{T_4}{T_3} = \frac{1}{r_p^{(\gamma - 1)/\gamma}}$$

$$\frac{T_2}{T_1} = r_p^{(\gamma-1)/\gamma}, \text{ we get,}$$

$$\eta_{ith} = 1 - \frac{1}{\eta_c \eta_t} \frac{T_1}{T_3} \frac{\left(r_p^{(\gamma-1)/\gamma} - 1\right)}{\left(1 - 1/r_p^{(\gamma-1)/\gamma}\right)}$$

$$= 1 - \frac{1}{\eta_c \eta_t} \frac{T_1}{T_3} r_p^{(\gamma-1)/\gamma} \tag{3.51}$$

It can be seen from equation (3.51) that the thermal efficiency of the cycle increases with the increase in T_3 and with decrease in r_p. At high pressure ratios the temperature of compressed air can exceed the exhaust gases temperature thus the regenerative process will not be of any use. Therefore, regeneration is employed only when the pressure ratio is low enough and thus the exhaust gases temperature is higher than the compressor discharge temperature.

In actual practice the effectiveness ε could not be equal to 1. This means it is not possible to raise the outlet temperature of the compressed air from $T_{2'}$ to $T_5 = T_{4'}$. So, the actual temperature of air will be $T_{5'} < T_5$. Effectiveness of heat exchanger can be defined as the ratio of actual temperature gain by air to the maximum possible temperature gain. Mathematically, effectiveness of the regenerator is given by,

$$\varepsilon = \frac{m_a\, c_{pa}\, (T_{5'} - T_{2'})}{m_g\, c_{pg}\, (T_{4'} - T_{2'})}$$

where m_a = Mass of air supplied per second, m_g = Mass of exhaust gases passed to the heat exchanger per second, c_{pa} = Specific heat of air and, c_{pg} = Specific heat of gases.

If $m_a = m_g$ and $c_{pa} = c_{pg}$ then,

$$\varepsilon = \frac{(T_{5'} - T_{2'})}{(T_{4'} - T_{2'})} \tag{3.52}$$

Thermal efficiency of actual regenerative cycle is given by,

$$\eta_{ith} = \frac{w_{\text{net}}}{q_s} = \frac{c_p\,(T_3 - T_{4'}) - c_p\,(T_{2'} - T_1)}{c_p\,(T_3 - T_{5'})} \tag{3.53}$$

3.25.2 Reheating

In reheating the expansion of gases is carried out in two stages with a reheater provided between them. After the expansion of gases in the first turbine these are sent to another combustion chamber to reheat up to the maximum temperature of the cycle and then are allowed to expand further in the second turbine. This produces more work and thus the work ratio improves. In this process thermal efficiency decreases but it can be improved by providing regenerator. The schematic view of simple gas turbine with reheating is shown in Figure 3.36 (a) and the $T - s$ diagram is shown in Figure 3.36 (b).

The work obtained per kg of the working fluid by the two turbines with reheating system is given by,

$$w_t = w_{t1} + w_{t2} = c_{pg}\,(T_3 - T_{4'}) + c_{pg}\,(T_5 - T_{6'})$$

$$= c_{pg}\,\eta_{t1}\,(T_3 - T_4) + c_{pg}\,\eta_{t2}\,(T_5 - T_6)$$

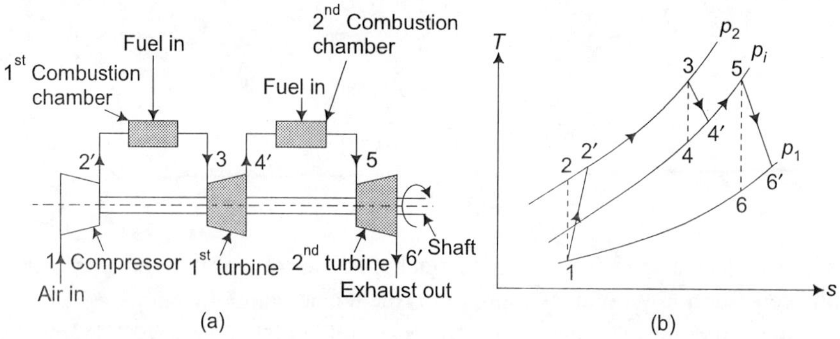

Fig. 3.36 Schematics and $T-s$ Diagram for Brayton Cycle with Reheating

where η_{t1} and η_{t2} are the efficiencies of the first and second turbines respectively.

$$\therefore \quad w_t = c_{pg}\, \eta_{t1}\, T_3\, (1 - T_4/T_3) + c_{pg}\, \eta_{t2}\, T_5\, (1 - T_6/T_5) \qquad \text{(i)}$$

For isentropic expansion processes,

$$\frac{T_3}{T_4} = \left(\frac{p_2}{p_i}\right)^{(\gamma-1)/\gamma}$$

$$\frac{T_5}{T_6} = \left(\frac{p_i}{p_1}\right)^{(\gamma-1)/\gamma}$$

Substituting these values in the expression (i), we have,

$$w_t = c_{pg}\, \eta_{t1}\, T_3 \left[1 - \left(\frac{p_i}{p_2}\right)^{(\gamma-1)/\gamma}\right] + c_{pg}\, \eta_{t2}\, T_5 \left[1 - \left(\frac{p_1}{p_i}\right)^{(\gamma-1)/\gamma}\right]$$

Assuming $k = T_5/T_3$, for maximum work,

$$\frac{dw_t}{dp_i} = 0,$$

$$\frac{d}{dp_i}\left[c_{pg}\, \eta_{t1} \left\{1 - \left(\frac{p_i}{p_2}\right)^{(\gamma-1)/\gamma}\right\} + c_{pg}\, \eta_{t2}\, k \left\{1 - \left(\frac{p_1}{p_i}\right)^{(\gamma-1)/\gamma}\right\}\right] = 0,$$

solving it, we get,

$$p_i = \sqrt{p_1\, p_2 \left(k\, \frac{\eta_{t2}}{\eta_{t1}}\right)^{\gamma/(\gamma-1)}} \qquad (3.54)$$

If $\eta_{t1} = \eta_{t2}$ and $T_5 = T_3$, i.e.,

$$k = \frac{T_5}{T_3} = 1,$$

then

$$p_i = \sqrt{p_1\, p_2} \qquad (3.55)$$

If $c_{pa} = c_{pg}$ and mass of fuel is neglected then thermal efficiency of the reheated gas turbine system is given by,

$$\eta_{th} = \frac{w_{\text{net}}}{q_s} = \frac{(T_3 - T_{4'}) + (T_5 - T_{6'}) - (T_{2'} - T_1)}{(T_3 - T_{2'}) + (T_5 - T_{4'})} \qquad (3.56)$$

3.25.3 Intercooling

The net work of the gas turbine cycle can be increased by saving some compression work by using stages of compression with intercooling of air between stages. The air from the first compressor is cooled in an intercooler ideally to initial temperature before entering the second compressor. Although this consumes more fuel but reduces the specific volume of the air entering the second compressor due to which the work required for overall compression decreases and hence the net work increases. So, intercooling increases the net work and decreases the efficiency for the same compression ratio. But due to intercooling there is possibility of improving the overall efficiency by increasing the maximum pressure ratio due to reduced temperature of compressor discharge. A two-stage compressor with intercooler is shown in Figure 3.37 (a) and the corresponding $T-s$ diagram is shown in Figure 3.37 (b).

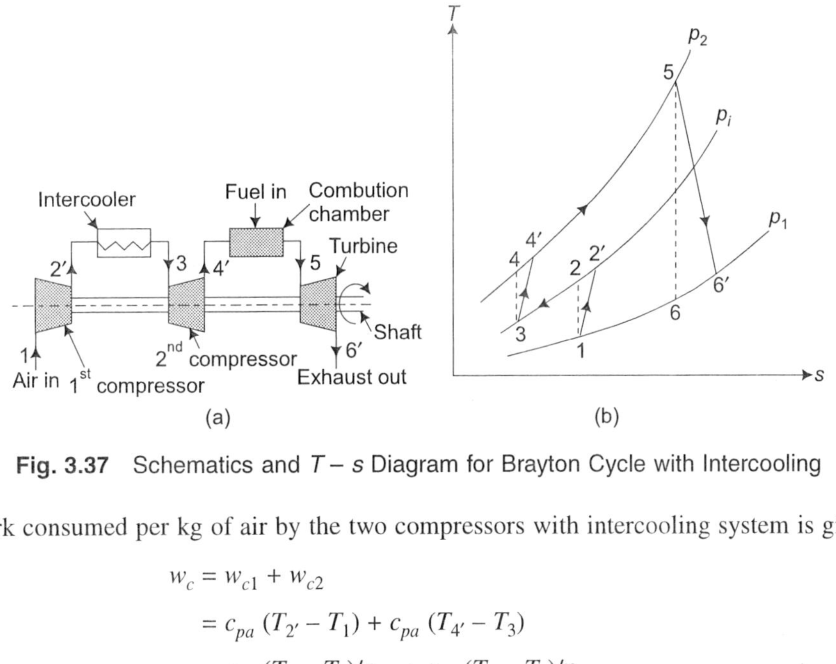

Fig. 3.37 Schematics and $T-s$ Diagram for Brayton Cycle with Intercooling

The work consumed per kg of air by the two compressors with intercooling system is given by,

$$w_c = w_{c1} + w_{c2}$$
$$= c_{pa}\,(T_{2'} - T_1) + c_{pa}\,(T_{4'} - T_3)$$
$$= c_{pa}\,(T_2 - T_1)/\eta_{c1} + c_{pa}\,(T_4 - T_3)/\eta_{c2}$$

where η_{c1} and η_{c2} are the efficiencies of the compressors.

$$\therefore \qquad w_c = \frac{c_{pa}}{\eta_{c1}}\,T_1\,(T_2/T_1 - 1) + \frac{c_{pa}}{\eta_{c2}}\,T_3\,(T_4/T_3 - 1) \qquad \text{(i)}$$

For isentropic compression processes,

$$\frac{T_2}{T_1} = \left(\frac{p_i}{p_1}\right)^{(\gamma-1)/\gamma}$$

$$\frac{T_4}{T_3} = \left(\frac{p_2}{p_i}\right)^{(\gamma-1)/\gamma}$$

Substituting these values in the expression (i), we have,

$$w_c = \frac{c_{pa}}{\eta_{c1}} T_1 \left[\left(\frac{p_i}{p_1}\right)^{(\gamma-1)/\gamma} - 1\right] + \frac{c_{pa}}{\eta_{c2}} T_3 \left[\left(\frac{p_2}{p_i}\right)^{(\gamma-1)/\gamma} - 1\right]$$

Assuming $k = T_3/T_1$, for minimum compressor work,

$$\frac{dw_c}{dp_i} = 0$$

$$\frac{d}{dp_i}\left[\frac{c_{pa}}{\eta_{c1}}\left\{\left(\frac{p_i}{p_1}\right)^{(\gamma-1)/\gamma} - 1\right\} + \frac{c_{pa}}{\eta_{c2}} k \left\{\left(\frac{p_2}{p_i}\right)^{(\gamma-1)/\gamma} - 1\right\}\right] = 0,$$

solving it, we get,

$$\therefore \qquad p_i = \sqrt{p_1 p_2 \left(k \frac{\eta_{c1}}{\eta_{c2}}\right)^{\gamma/(\gamma-1)}} \tag{3.57}$$

If $\eta_{c1} = \eta_{c2}$ and $T_3 = T_1$, i.e., $k = T_3/T_1 = 1$, then

$$p_i = \sqrt{p_1 p_2} \tag{3.58}$$

If mass of fuel is neglected then thermal efficiency of the intercooled turbine plant is given by,

$$\eta_{th} = \frac{w_{net}}{q_s} = \frac{(T_5 - T_{6'}) - [(T_{2'} - T_1) + (T_{4'} - T_3)]}{(T_5 - T_{4'})} \tag{3.59}$$

If $T_3 = T_1$, i.e., perfect intercooling,

$$\eta_{th} = \frac{w_{net}}{q_s} = \frac{(T_5 - T_{6'}) - (T_{2'} - 2T_1 + T_{4'})}{(T_5 - T_{4'})} \tag{3.60}$$

Example 3.32 Air enters the compressor of a closed cycle constant pressure gas turbine at a pressure of 1 bar and 25°C with a pressure ratio of 6. The gases enter the turbine at a temperature of 750°C. Take $c_p = 1.005$ kJ/kgK and $\gamma = 1.4$ for air and for gases. Determine (i) the efficiency of the cycle, (ii) the efficiency of the cycle with regenerator whose effectiveness is 0.65, (iii) the efficiency of the cycle when an intercooler is employed with optimum pressure ratio, and (iv) the efficiency of the cycle when a reheater is employed with optimum pressure ratio.

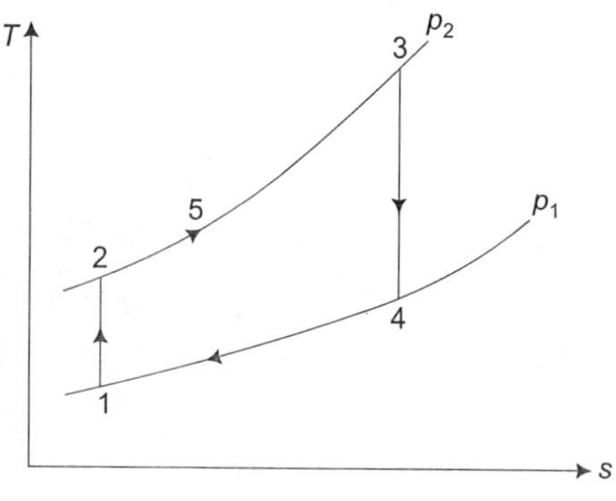

Fig. 3.38

Solution

(i) Refer Figure 3.38.

$$\frac{T_2}{T_1} = \left(\frac{p_2}{p_1}\right)^{(\gamma-1)/\gamma}$$

$$\therefore \qquad T_2 = T_1 \left(\frac{p_2}{p_1}\right)^{(\gamma-1)/\gamma}$$

$$= 298 \times \left(\frac{6}{1}\right)^{(1.4 - 1)/1.4} = 497.22 \text{ K}$$

$$\frac{T_3}{T_4} = \left(\frac{p_2}{p_1}\right)^{(\gamma - 1)/\gamma}$$

$$\therefore \qquad T_4 = T_3 \left(\frac{p_1}{p_2}\right)^{(\gamma - 1)/\gamma}$$

$$= 1023 \times \left(\frac{1}{6}\right)^{(1.4 - 1)/1.4} = 613.12 \text{ K}$$

Work consumed by the compressor,

$$w_c = c_p (T_2 - T_1) = 1.005 \times (497.22 - 298)$$
$$= 200.22 \text{ kJ/kg}$$

Work done by the turbine,

$$w_t = c_p (T_3 - T_4) = 1.005 \times (1023 - 613.12)$$
$$= 411.93 \text{ kJ/kg}$$

Net work available,

$$w_{\text{net}} = w_t - w_c = 411.93 - 200.11$$
$$= 211.71 \text{ kJ/kg}$$

Heat supplied,

$$q_s = c_p (T_3 - T_2) = 1.005 \times (1023 - 697.22)$$
$$= 528.41 \text{ kJ/kg}$$

Thermal efficiency,

$$\eta_{th}' = \frac{w_{\text{net}}}{q_s} = \frac{211.71}{528.41}$$
$$= 0.4006 \text{ or } 40.06\%$$

(ii) Refer Figure 3.38.

Net work available will remain same, i.e.,

$$w_{\text{net}} = w_t - w_c = 411.93 - 200.22$$
$$= 211.71 \text{ kJ/kg}$$

Effectiveness of the regenerator,

$$\varepsilon = \frac{T_5 - T_2}{T_4 - T_2}$$

$$0.65 = \frac{T_5 - 497.22}{613.12 - 497.22}$$

$$\therefore \qquad T_5 = 572.55 \text{ K}$$

Heat supplied,

$$q_s = c_p\,(T_3 - T_5) = 1.005 \times (1023 - 572.55)$$
$$= 450.45 \text{ kJ/kg}$$

Thermal efficiency,

$$\eta_{th} = \frac{w_{net}}{q_s} = \frac{211.71}{450.45}$$
$$= 0.4699 \text{ or } 46.99\%$$

(iii) Refer Figure 3.39.

For optimum pressure ratio,

$$p_i = \sqrt{p_1\,p_2} = \sqrt{1 \times 6}$$
$$= 2.45 \text{ bar}$$
$$\frac{T_2}{T_1} = \left(\frac{p_2}{p_1}\right)^{(\gamma-1)/\gamma}$$

$\therefore$
$$T_2 = T_1\left(\frac{p_2}{p_1}\right)^{(\gamma-1)/\gamma}$$
$$= 298 \times \left(\frac{2.45}{1}\right)^{(1.4-1)1.4} = 384.95 \text{ K}$$

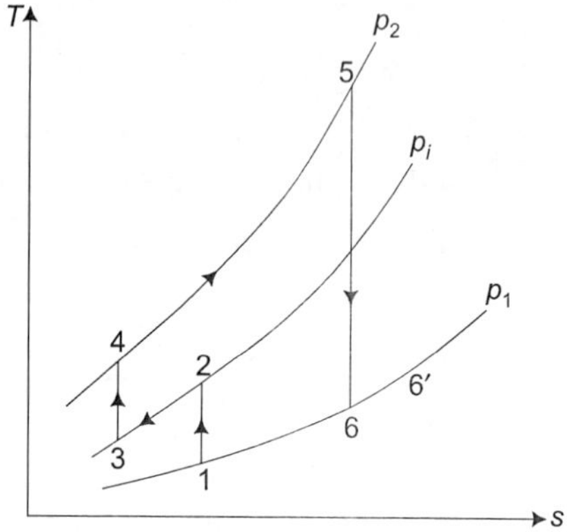

Fig. 3.39

For ideal or perfect intercooling,

$$T_3 = T_1 = 298 \text{ K}$$
$$\frac{T_4}{T_3} = \left(\frac{p_2}{p_1}\right)^{(\gamma-1)/\gamma}$$

$\therefore$
$$T_4 = T_3\left(\frac{p_1}{p_2}\right)^{(\gamma-1)/\gamma}$$
$$= 298 \times \left(\frac{6}{2.45}\right)^{(1.4-1)/1.4} = 384.91 \text{ K}$$

Work done by the turbine,

$$w_t = c_p\,(T_5 - T_6) = 1.005 \times (1023 - 613.12)$$
$$= 411.93 \text{ kJ/kg}$$

Work consumed by the compressor,

$$w_c = c_p\,(T_2 - T_1) + c_p\,(T_4 - T_3) \text{ kJ/kg}$$

$\therefore$
$$w_c = 1.005 \times (384.95 - 298) + 1.005 \times (384.91 - 298)$$
$$= 174.73 \text{ kJ/kg}$$
$$w_{net} = w_t - w_c = 411.93 - 174.73$$
$$= 237.2 \text{ kJ/kg}$$

Heat supplied, $q_s = c_p (T_5 - T_4) = 1.005 \times (1023 - 384.91)$

$= 641.28$ kJ/kg

Thermal efficiency, $\eta_{th} = \dfrac{w_{net}}{q_s} = \dfrac{237.2}{641.28}$

$= 0.3699$ or 36.99%

(iv) Refer Figure 3.40.

For optimum pressure ratio,

$$p_i = \sqrt{p_1 p_2} = \sqrt{1 \times 6}$$

$$= 2.45 \text{ bar}$$

$$\frac{T_3}{T_4} = \left(\frac{p_2}{p_i}\right)^{(\gamma - 1)/\gamma}$$

$$\therefore \quad T_4 = T_3 \left(\frac{p_i}{p_2}\right)^{(\gamma - 1)/\gamma}$$

$$= 1023 \times \left(\frac{2.45}{6}\right)^{(1.4 - 1)1.4}$$

$$= 792.02 \text{ K}$$

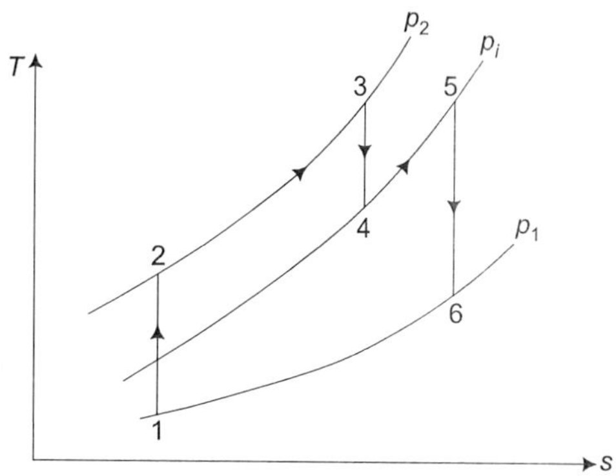

Fig. 3.40

For ideal or perfect reheating,

$$T_3 = T_5 = 1023 \text{ K}$$

$$\frac{T_5}{T_6} = \left(\frac{p_i}{p_1}\right)^{(\gamma - 1)/\gamma}$$

$$\therefore \quad T_6 = T_5 \left(\frac{p_1}{p_i}\right)^{(\gamma - 1)/\gamma}$$

$$= 1023 \times \left(\frac{1}{2.45}\right)^{(1.4 - 1)1.4} = 791.93 \text{ K}$$

Work done by the turbine,

$$w_t = c_p (T_3 - T_4) + c_p (T_5 - T_6)$$

$$\therefore \quad w_t = 1.005 \times (1023 - 792.02) + 1.005 \times (1023 - 791.93)$$

$$= 464.36 \text{ kJ/kg}$$

Work consumed by the compressor,

$$w_c = c_p (T_2 - T_1) = 1.005 \times (497.22 - 298)$$

$$= 200.22 \text{ kJ/kg}$$

$$w_{net} = w_t - w_c = 464.36 - 200.22$$

$$= 264.14 \text{ kJ/kg}$$

Heat supplied, $q_s = c_p\,(T_3 - T_2) + c_p\,(T_5 - T_4)$ kJ/kg

$\therefore$ $q_s = 1.005 \times (1023 - 497.22) + 1.005 \times (1023 - 792.02)$

$= 760.54$ kJ/kg

Thermal efficiency, $\eta_{th} = \dfrac{w_{net}}{q_s} = \dfrac{264.14}{760.54}$

$= 0.3473$ or 34.73%

Example 3.33 Air enters the compressor of a gas turbine plant operating on Brayton cycle at 1.02 bar and 27°C. The pressure of air after the compression is 4.08 bar. Determine the maximum temperature in the cycle and cycle efficiency if the turbine work is 1.5 times of the compressor work. Take $\gamma = 1.4$ for air and gases.

Solution Refer Figure 3.38.

$$\frac{T_2}{T_1} = \left(\frac{p_2}{p_1}\right)^{(\gamma-1)/\gamma}$$

$$\therefore \quad T_2 = T_1\left(\frac{p_2}{p_1}\right)^{(\gamma-1)/\gamma}$$

$$= 300 \times \left(\frac{4.08}{1.02}\right)^{(1.4-1)1.4} = 445.8 \text{ K}$$

$$\frac{T_3}{T_4} = \left(\frac{p_2}{p_1}\right)^{(\gamma-1)/\gamma}$$

$$\therefore \quad T_4 = T_3\left(\frac{p_1}{p_2}\right)^{(\gamma-1)/\gamma}$$

$$= T_3 \times \left(\frac{1.02}{4.08}\right)^{(1.4-1)1.4} = 0.673\,T_3 \text{ K}$$

Now $w_t = 1.5 \times w_c$

$$m \times c_p\,(T_3 - T_4) = 1.5 \times m \times c_p\,(T_2 - T_1)$$

$$(T_3 - 0.673\,T_3) = 1.5 \times (445.8 - 300)$$

$$\therefore \quad T_3 = 668.81 \text{ K}$$

$$\therefore \quad T_4 = 0.673 \times 668.81 = 450.11 \text{ K}$$

Thermal efficiency, $\eta_{th} = \dfrac{w_{net}}{q_s} = \dfrac{(w_t - w_c)}{q_s} = \dfrac{(T_3 - T_4) - (T_2 - T_1)}{(T_3 - T_2)}$

$$\therefore \quad \eta_{th} = \frac{(668.81 - 450.11) - (445.8 - 300)}{(668.81 - 445.8)}$$

$$= \frac{72.85}{223.01} = 0.3267 \text{ or } 32.67\%$$

Example 3.34 Air enters the compressor of an open cycle constant pressure gas turbine at a pressure of 1 bar and 20°C. The pressure of air after the compression is 5 bar. The isentropic efficiencies of compressor and turbine are 0.8 and 0.85 respectively. Maximum cycle temperature is 800°C. Determine the air-fuel ratio used if there is a loss of 10% of calorific value in the combustion chamber. Take $c_p = 1$ kJ/kgK and $\gamma = 1.4$ for air and gases. The calorific value of fuel used is 41750 kJ/kg.

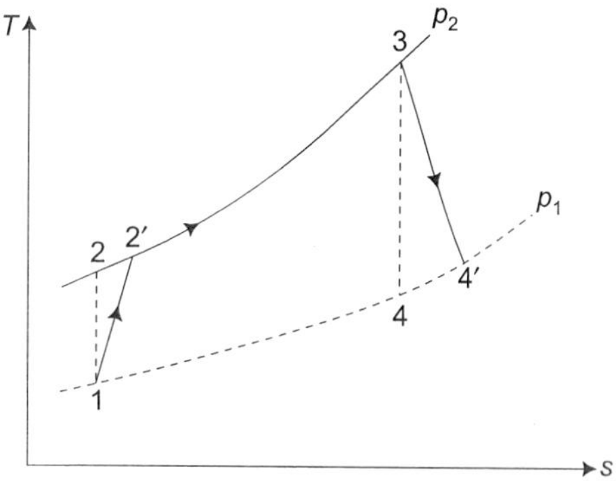

Fig. 3.41

Solution Refer Figure 3.41.

$$\frac{T_2}{T_1} = \left(\frac{p_2}{p_1}\right)^{(\gamma-1)/\gamma}$$

$$\therefore \quad T_2 = T_1\left(\frac{p_2}{p_1}\right)^{(\gamma-1)/\gamma}$$

$$= 293 \times \left(\frac{5}{1}\right)^{(1.4-1)1.4} = 464.06 \text{ K}$$

$$\eta_c = \frac{T_2 - T_1}{T_{2'} - T_1}$$

$$0.8 = \frac{464.06 - 293}{T_{2'} - 293}$$

$$\therefore \quad T_{2'} = 5069.82 \text{ K}$$

Heat supplied by fuel = Heat taken by the burning gas

$$\eta_{\text{combustion}} \times m_f \times CV = (m_a + m_f)\, c_p\, (T_3 - T_{2'})$$

$$\eta_{\text{combustion}} \times CV = \left(1 + \frac{m_a}{m_f}\right) c_p\, (T_3 - T_{2'})$$

$$0.9 \times 41750 = \left(1 + \frac{m_a}{m_f}\right) \times 1 \times (1073 - 506.82)$$

$$\therefore \quad \frac{m_a}{m_f} = 65.36$$

Example 3.35 Air enters the compressor of an open cycle constant pressure gas turbine at a pressure of 1.02 bar and 25°C. The pressure of air after the compression is 4.08 bar. The isentropic efficiencies of compressor and turbine are 0.8 and 0.85 respectively. The air-fuel ratio used is 85 : 1. Determine the power developed and thermal efficiency of the cycle if the flow rate of air is 3.5 kg/s. Take $c_p = 1$ kJ/kgK and $\gamma = 1.4$ for air and gases. The calorific value of fuel used is 42000 kJ/kg

Solution Refer Figure 3.41.

$$\frac{T_2}{T_1} = \left(\frac{p_2}{p_1}\right)^{(\gamma-1)/\gamma}$$

$$\therefore \quad T_2 = T_1 \left(\frac{p_2}{p_1}\right)^{(\gamma - 1)/\gamma}$$

$$= 293 \times \left(\frac{4.08}{1.02}\right)^{(1.4 - 1)1.4} = 442.83 \text{ K}$$

$$\eta_c = \frac{T_2 - T_1}{T_{2'} - T_1}$$

$$0.8 = \frac{442.83 - 298}{T_{2'} - 298}$$

$$\therefore \quad T_{2'} = 479.04 \text{ K}$$

Heat supplied by fuel = Heat taken by the burning gas

$$m_f \times CV = (m_a + m_f)\, c_p\, (T_3 - T_{2'})$$

$$CV = \left(\frac{m_a}{m_f} + 1\right) c_p\, (T_3 - T_{2'})$$

$$42000 = (85 + 1) \times 1 \times (T_3 - 479.04)$$

$$\therefore \quad T_3 = 967.41 \text{ K}$$

$$\frac{T_3}{T_4} = \left(\frac{p_2}{p_1}\right)^{(\gamma - 1)/\gamma},$$

$$\therefore \quad T_4 = T_3 \times \left(\frac{p_1}{p_2}\right)^{(\gamma - 1)/\gamma}$$

$$= 967.41 \times \left(\frac{1.02}{4.08}\right)^{(1.4 - 1)1.4} = 651.02 \text{ K}$$

$$\eta_t = \frac{T_3 - T_{4'}}{T_3 - T_4}$$

$$0.85 = \frac{964.41 - T_{4'}}{967.41 - 651.02}$$

$$\therefore \quad T_{4'} = 698.48 \text{ K}$$

Work consumed by the compressor,

$$w_c = c_{pa}\, (T_{2'} - T_1) = 1 \times (479.04 - 298)$$

$$= 181.04 \text{ kJ/kg}$$

Work done by the turbine,

$$w_t = \left(\frac{m_a + m_f}{m_a}\right) c_{pg}\, (T_3 - T_{4'})$$

$$= \left(\frac{85 + 1}{85}\right) \times 1 \times (967.41 - 698.48)$$

$$= 272.09 \text{ kJ/kg}$$

Net work available, $w_{net} = w_t - w_c = 272.09 - 181.04$

$$= 91.05 \text{ kJ/kg}$$

Power developed, $P = m_a \times w_{net} = (3.5 \text{ kg/s}) \times (91.05 \text{ kJ/kg})$

$$= 318.67 \text{ kW}$$

Heat supplied, $q_s = \dfrac{m_f}{m_a} \times CV = \dfrac{1}{85} \times 42000$

$$= 494.12 \text{ kJ/kg}$$

Thermal efficiency, $\eta_{th} = \dfrac{w_{net}}{q_s} = \dfrac{91.05}{494.12}$

$$= 0.1843 \text{ or } 15.43\%$$

Example 3.36 Air enters the compressor of an open cycle constant pressure gas turbine at a pressure of 1 bar and 20°C. The pressure of air after the compression is 5 bar. The gases enter the turbine at a temperature of 725°C. The isentropic efficiencies of compressor and turbine are 0.8 and 0.83 respectively. There is a pressure drop of 0.1 bar in the combustion chamber and its efficiency is 0.88. Determine (i) the quantity of air circulation if the power developed by the plant is 950 kW, (ii) actual heat supplied per kg of air circulation, and (iii) thermal efficiency of the cycle if the mass of fuel is neglected. Take $c_p = 1$ kJ/kgK and $\gamma = 1.4$ for air and gases.

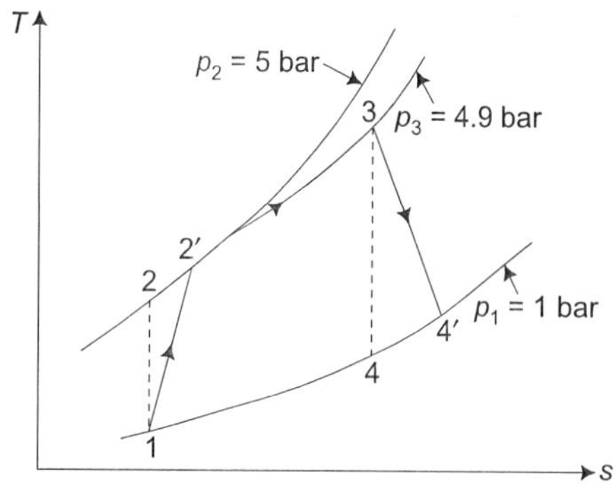

Fig. 3.42

Solution Refer Figure 3.42.

$$\frac{T_2}{T_1} = \left(\frac{p_2}{p_1}\right)^{(\gamma - 1)/\gamma}$$

$$\therefore \quad T_2 = T_1 \left(\frac{p_2}{p_1}\right)^{(\gamma - 1)/\gamma}$$

$$= 293 \times \left(\frac{5}{1}\right)^{(1.4 - 1)1.4} = 464.06 \text{ K}$$

$$\eta_c = \frac{T_2 - T_1}{T_{2'} - T_1}$$

$$0.8 = \frac{464.06 - 298}{T_{2'} - 293}$$

$$\therefore \quad T_{2'} = 506.82 \text{ K}$$

$$\frac{T_3}{T_4} = \left(\frac{p_3}{p_1}\right)^{(\gamma - 1)/\gamma}$$

$$\therefore \qquad T_4 = T_3 \times \left(\frac{p_1}{p_3}\right)^{(\gamma-1)/\gamma}$$

$$= 998 \times \left(\frac{1}{4.9}\right)^{(1.4-1)1.4} = 633.77 \text{ K}$$

$$\eta_t = \frac{T_3 - T_{4'}}{T_3 - T_4}$$

or
$$0.88 = \frac{998 - T_{4'}}{998 - 633.77}$$

$$\therefore \qquad T_{4'} = 677.48 \text{ K}$$

Work consumed by the compressor,

$$w_c = c_{pa}(T_{2'} - T_1) = 1 \times (506.82 - 293)$$

$$= 213.82 \text{ kJ/kg}$$

Work done by the turbine,

$$w_t = c_{pg}(T_3 - T_{4'}) = 1 \times (998 - 677.48)$$

$$= 320.52 \text{ kJ/kg}$$

Net work available, $w_{net} = w_t - w_c = 320.52 - 213.82$

$$= 106.7 \text{ kJ/kg}$$

(i) Power developed, $P = m_a \times w_{net}$

$$\therefore \qquad m_a = \frac{P}{w_{net}} = \frac{(950 \text{ kJ/s})}{(106.7 \text{ kJ/kg})}$$

$$= 8.90 \text{ kg/s}$$

(ii) Actual heat supplied,

$$q_s = \eta_{combustion} \times c_p \times (T_3 - T_{2'})$$

$$= 0.88 \times 1 \times (998 - 506.82)$$

$$= 432.24 \text{ kJ/kg}$$

(iii) Thermal efficiency, $\eta_{th} = \frac{w_{net}}{q_s} = \frac{106.7}{432.24}$

$$= 0.2468 \text{ or } 24.68\%$$

Example 3.37 A gas turbine plant works between the temperature limits of 850°C and 25°C. The isentropic efficiencies of the compressor and turbine are 0.8 and 0.85 respectively. Determine the optimum pressure ratio for maximum work output and for maximum thermal efficiency of the cycle. Assume air is used as working fluid. What is the percentage change in pressure ratios for maximum work output and for maximum thermal efficiency if argon is used as working fluid. Take $\gamma = 1.4$ for air and $c_{pa} = 1$ kJ/kgK for argon.

Solution

(i) Optimum pressure ratio for maximum work output can be given by,

$$[(r_p)_{\text{optimum}}]_{\text{Air}} = \left(\eta_c \cdot \eta_t \frac{T_3}{T_1}\right)^{\gamma/2(\gamma - 1)}$$

$$= \left(0.8 \times 0.85 \times \frac{1123}{298}\right)^{1.4/2(1.4 - 1)}$$

$$= 5.19$$

$$[(r_p)_{\text{optimum}}]_{\text{Argon}} = \left(\eta_c \cdot \eta_t \frac{T_3}{T_1}\right)^{\gamma/2(\gamma - 1)}$$

$$= \left(0.8 \times 0.85 \times \frac{1123}{298}\right)^{1.67/2(1.67 - 1)}$$

$$= 3.23$$

$$\text{Percentage decrease} = \left(\frac{5.19 - 3.23}{5.19}\right) \times 100$$

$$= 37.76\%$$

(ii) Optimum pressure ratio for maximum thermal efficiency can be given by,

$$(r_p)_{\text{optimum}} = \left(\frac{T_3/T_1}{1 + \sqrt{(T_3/T_1 - 1)\,(1/\eta_c \cdot \eta_t - 1)}}\right)^{\gamma/(\gamma - 1)}$$

$$[(r_p)_{\text{optimum}}]_{\text{Air}} = \left(\frac{1123/298}{1 + \sqrt{(1123/298 - 1)\,(1/(0.8 \times 0.85) - 1)}}\right)^{1.4/(1.4 - 1)}$$

$$= 7.23$$

$$[(r_p)_{\text{optimum}}]_{\text{Argon}} = \left(\frac{1123/298}{1 + \sqrt{(1123/298 - 1)\,(1/(0.8 \times 0.85) - 1)}}\right)^{1.67/(1.67 - 1)}$$

$$= 4.09$$

$$\text{Percentage decrease} = \left(\frac{7.23 - 4.09}{7.23}\right) \times 100$$

$$= 43.43\%$$

Example 3.38 A gas turbine plant works between the temperature limits of 800°C and 27°C with a pressure ratio of 6. The air is being drawn at 1 bar in the first compressor. The compression process is completed in two stages with equal pressure ratio with a perfect intercooler in between. The isentropic efficiencies of each compressor stage and of the turbine are 0.80 and 0.85 respectively. If the air circulation is 2.5 kg/s, determine the thermal efficiency of the unit. Take $\gamma = 1.4$ for air and gases and $c_{pa} = 1$ kJ/kgK.

Solution Refer Figure 3.43.

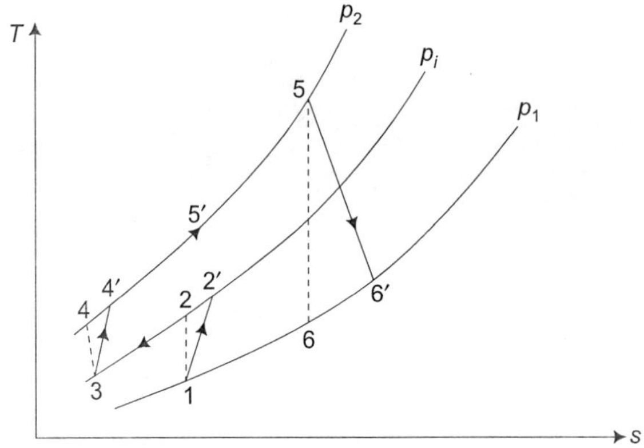

Fig. 3.43

Under the given working conditions for minimum work supplied to the compressor and maximum work developed by the turbine the pressure ratio is given by,

$$p_i = \sqrt{p_1\, p_2} = \sqrt{1 \times 6} = 2.45 \text{ bar}$$

$$\frac{T_2}{T_1} = \left(\frac{p_i}{p_1}\right)^{(\gamma - 1)/\gamma}$$

$$\therefore \quad T_2 = T_1 \left(\frac{p_i}{p_1}\right)^{(\gamma - 1)/\gamma}$$

$$= 300 \times \left(\frac{2.45}{1}\right)^{(1.4 - 1)1.4} = 387.53 \text{ K}$$

$$\eta_c = \frac{T_2 - T_1}{T_{2'} - T_1}$$

$$0.8 = \frac{387.53 - 300}{T_{2'} - 300}$$

$$\therefore \quad T_{2'} = 409.41 \text{ K}$$

As, $\eta_{c1} = \eta_{c2}$

$$\therefore \quad T_{2'} = T_{4'} = 409.41 \text{ K}$$

$$T_1 = T_3 = 300 \text{ K}$$

$$\frac{T_5}{T_6} = \left(\frac{p_2}{p_1}\right)^{(\gamma - 1)/\gamma}$$

$$\therefore \quad T_6 = T_5 \times \left(\frac{p_1}{p_3}\right)^{(\gamma - 1)/\gamma}$$

$$= 1073 \times \left(\frac{1}{6}\right)^{(1.4 - 1)1.4} = 643.09 \text{ K}$$

$$\eta_t = \frac{T_5 - T_{6'}}{T_5 - T_6}$$

$$0.85 = \frac{1073 - T_{6'}}{1073 - 643.09}$$

$$\therefore \quad T_{6'} = 707.58 \text{ K}$$

$$w_{net} = w_t - (w_{c1} + w_{c2})$$

$$= (w_1 - 2w_{c1}) \text{ kJ/s} \qquad [\because w_{c1} = w_{c2}]$$

$$w_{net} = m_a\, c_p\, [(T_5 - T_{6'}) - 2 \times (T_{2'} - T_1)] \text{ kJ/s}$$

$$\therefore \quad w_{net} = 2.5 \times 1 \times [(1073 - 707.58) - 2 \times (409.41 - 300)$$

$$= 366.5 \text{ kJ/kg}$$

Heat supplied,

$$Q_s = m_a\, c_p\, (T_5 - T_{4'}) \text{ kJ/s}$$

$$\therefore \quad Q_s = 2.5 \times 1 \times [(1073 - 409.41)$$

$$= 1658.97 \text{ kJ/s}$$

Thermal efficiency,

$$\eta_{th} = \frac{w_{net}}{Q_s} = \frac{366.5}{1658.97}$$

$$= 0.2209 \text{ or } 22.09\%$$

Example 3.39 A gas turbine plant works between the temperature limits of 850°C and 27°C with a pressure ratio of 6. The isentropic efficiencies of compressor and turbine are 0.82 and 0.88 respectively. The combustion chamber efficiency is 0.9. Determine (i) air-fuel ratio, (ii) power developed if the air circulation is 4.5 kg/s, (iii) the thermal efficiency, and (iv) specific fuel consumption. Take $\gamma = 1.4$ for air and gases, $c_{pa} = 1$ kJ/kgK, $c_{pg} = 1.08$ kJ/kgK and $CV = 42000$ kJ/kg. Neglect the losses in the plant if any.

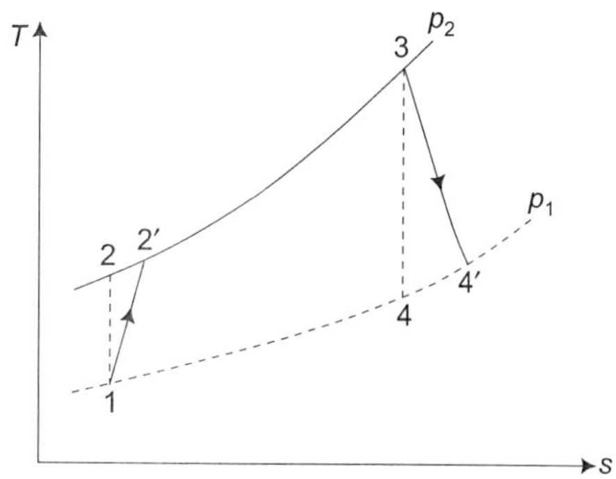

Fig. 3.44

Solution Refer Figure 3.44.

$$\frac{T_2}{T_1} = \left(\frac{p_2}{p_1}\right)^{(\gamma - 1)/\gamma}$$

$$\therefore \quad T_2 = T_1 \left(\frac{p_2}{p_1}\right)^{(\gamma - 1)/\gamma}$$

$$= 300 \times \left(\frac{6}{1}\right)^{(1.4 - 1)1.4} = 500.55 \text{ K}$$

$$\eta_c = \frac{T_2 - T_1}{T_{2'} - T_1}$$

$$0.82 = \frac{500.55 - 300}{T_{2'} - 300}$$

$\therefore$ $$T_{2'} = 544.57 \text{ K}$$

$$\frac{T_3}{T_4} = \left(\frac{p_2}{p_1}\right)^{(\gamma - 1)/\gamma}$$

$\therefore$ $$T_4 = T_3 \times \left(\frac{p_1}{p_3}\right)^{(\gamma - 1)/\gamma}$$

$$= 1123 \times \left(\frac{1}{6}\right)^{(1.4 - 1)1.4} = 673.05 \text{ K}$$

$$\eta_t = \frac{T_3 - T_{4'}}{T_3 - T_4}$$

$$0.88 = \frac{1123 - T_{4'}}{1123 - 673.05}$$

$\therefore$ $$T_{4'} = 727.04 \text{ K}$$

(i) $$m_f \times CV \times \eta_{\text{combustion}} = (m_a + m_f)\, c_{pg}\, (T_3 - T_{2'}) \text{ or}$$

$$CV \times \eta_{\text{combustion}} = \left(1 + \frac{m_a}{m_f}\right) c_{pg}\, (T_3 - T_{2'})$$

$$42000 \times 0.9 = \left(1 + \frac{m_a}{m_f}\right) \times 1.08 \times (1123 - 544.57)$$

$\therefore$ $$\frac{m_a}{m_f} = 59.79$$

(ii) Work consumed by the compressor,

$$w_c = c_{pa}\, (T_{2'} - T_1) = 1 \times (544.57 - 300)$$

$$= 244.57 \text{ kJ/kg}$$

Work done by the turbine,

$$w_t = \left(\frac{m_a + m_f}{m_a}\right) c_{pg}\, (T_3 - T_{2'})$$

$$= \left(\frac{60.79}{59.79}\right) \times 1.08 \times (1123 - 727.04)$$

$$= 434.79 \text{ kJ/kg}$$

Net work available, $$w_{\text{net}} = w_t - w_c = 464.79 - 244.57$$

$$= 190.22 \text{ kJ/kg}$$

Power developed, $P = m_a \times w_{net} = 4.5 \times 190.22$

$= 855.99$ kW

(iii) Heat supplied, $q_s = \dfrac{m_f}{m_a} \times CV \times \eta_{combustion}$

$= \dfrac{1}{59.79} \times 42000 \times 0.9$

$= 632.21$ kJ/kg

Thermal efficiency, $\eta_{th} = \dfrac{w_{net}}{q_s} = \dfrac{190.22}{632.21}$

$= 0.3009$ or 30.09%

(iv) Fuel supplied per hour, M_f = Air supplied/hour × fuel supplied/kg of air

$\therefore$ $M_f = (4.5 \times 3600) \times \dfrac{1}{59.79}$

$= 270.95$ kg/h

Therefore, specific fuel consumption

$$\frac{M_f}{P} = \frac{270.95}{855.99}$$

$= 0.3165$ kg/kWh

Example 3.40 A gas turbine cycle takes in air at 30°C and at 1 bar. The pressure of air after the compression is 4 bar. The isentropic efficiency of compressor is 0.8. The inlet temperature to the turbine is limited to 800°C. Determine the turbine efficiency which would give zero percent overall cycle efficiency. Take $\gamma = 1.4$ for air and gases.

Solution Refer Figure 3.44.

$$\frac{T_2}{T_1} = \left(\frac{p_2}{p_1}\right)^{(\gamma-1)/\gamma}$$

$\therefore$
$$T_2 = T_1\left(\frac{p_2}{p_1}\right)^{(\gamma-1)/\gamma}$$

$$= 303 \times \left(\frac{4}{1}\right)^{(1.4-1)1.4} = 450.26 \text{ K}$$

$$\frac{T_3}{T_4} = \left(\frac{p_2}{p_1}\right)^{(\gamma-1)/\gamma}$$

$\therefore$
$$T_4 = T_3 \times \left(\frac{p_1}{p_3}\right)^{(\gamma-1)/\gamma}$$

$$= 1073 \times \left(\frac{1}{4}\right)^{(1.4-1)1.4} = 722.07 \text{ K}$$

$$\eta_t = \frac{T_3 - T_{4'}}{T_3 - T_4}$$

or $$T_3 - T_{4'} = \eta_t (T_3 - T_4) \qquad \text{(i)}$$

$$\eta_c = \frac{T_2 - T_1}{T_{2'} - T_1}$$

or $$T_{2'} - T_1 = \frac{(T_2 - T_1)}{\eta_c}$$

$$= \frac{450.26 - 303}{0.80} = 184.07 \qquad \text{(ii)}$$

For zero per cent cycle efficiency the net work output is zero, i.e.,

$$w_{net} = w_t - w_c = 0$$

∴ $$w_t = w_c$$

Therefore, $$T_3 - T_{4'} = T_{2'} - T_1 \qquad \text{(iii)}$$

Substituting the values from the expressions (i) and (ii) in (iii), we have,

$$\eta_t (T_3 - T_4) = 184.07$$

$$\eta_t (1073 - 722.07) = 184.07$$

∴ $$\eta_t = \frac{184.07}{350.93}$$

$$= 0.5245 \text{ or } 52.45\%$$

Example 3.41 Air enters the compressor of an open cycle constant pressure gas turbine at a pressure of 1 bar and 27°C. The pressure of air after the compression is 5 bar. The isentropic efficiencies of compressor and turbine are 0.8 and 0.85 respectively. The air-fuel ratio is 80 : 1. Determine power developed and thermal efficiency if the air circulation is 4 kg/s. Take $\gamma = 1.4$ for air and $\gamma = 1.33$ for gases, $c_{pa} = 1$ kJ/kgK, $c_{pg} = 1.075$ kJ/kgK, and $CV = 41800$ kJ/kg. Neglect the losses in the plant if any.

Solution Refer Figure 3.44.

$$\frac{T_2}{T_1} = \left(\frac{p_2}{p_1}\right)^{(\gamma - 1)/\gamma}$$

∴ $$T_2 = T_1 \left(\frac{p_2}{p_1}\right)^{(\gamma - 1)/\gamma}$$

$$= 300 \times \left(\frac{5}{1}\right)^{(1.4 - 1)1.4} = 475.14 \text{ K}$$

$$\eta_c = \frac{T_2 - T_1}{T_{2'} - T_1}$$

$$0.80 = \frac{475.14 - 300}{T_{2'} - 300}$$

∴ $$T_{2'} = 518.92 \text{ K}$$

Heat supplied by fuel = Heat taken by the burning gas

$$m_f \times CV = (m_a + m_f)c_{pg}\,(T_3 - T_{2'})$$

or
$$CV = \left(\frac{m_a}{m_f} + 1\right) c_{pg}\,(T_3 - T_{2'})$$

$$41800 = (80 + 1) \times 1.075 \times (T_3 - 518.92)$$

$\therefore$
$$T_3 = 998.97 \text{ K}$$

$$\frac{T_3}{T_4} = \left(\frac{p_2}{p_1}\right)^{(\gamma - 1)/\gamma}$$

$\therefore$
$$T_4 = T_3 \times \left(\frac{p_1}{p_2}\right)^{(\gamma - 1)/\gamma}$$

$$= 998.97 \times \left(\frac{1}{5}\right)^{(1.4 - 1)1.4} = 630.73 \text{ K}$$

$$\eta_t = \frac{T_3 - T_{4'}}{T_3 - T_4}$$

$$0.85 = \frac{998.97 - T_{4'}}{998.97 - 630.73}$$

$\therefore$
$$T_{4'} = 685.97 \text{ K}$$

Work consumed by the compressor,

$$w_c = c_{pa}\,(T_{2'} - T_1) = 1 \times (518.92 - 300)$$

$$= 218.92 \text{ kJ/kg}$$

Work done by the turbine,

$$w_t = \left(\frac{m_a + m_f}{m_a}\right) c_{pg}\,(T_3 - T_{4'})$$

$$= \left(\frac{81}{80}\right) \times 1.075 \times (998.97 - 685.97)$$

$$= 340.68 \text{ kJ/kg}$$

Net work available, $w_{net} = w_t - w_c = 340.68 - 218.92$

$$= 121.76 \text{ kJ/kg}$$

Power developed, $P = m_a \times w_{net} = 4 \times 121.76$

$$= 487.04 \text{ kW}$$

Heat supplied, $q_s = \dfrac{m_f}{m_a} \times CV = \dfrac{1}{80} \times 41800$

$$= 522.5 \text{ kJ/kg}$$

Thermal efficiency, $\eta_{th} = \dfrac{w_{net}}{q_s} = \dfrac{121.76}{522.5}$

$$= 0.233 \text{ or } 23.3\%$$

Example 3.42 A closed cycle gas turbine consists of a two-stage compressor and a two-stage turbine which uses air as working fluid. The pressure and temperature at the inlet of the first stage compressor are 1.02 bar and 27°C. The maximum cycle temperature and pressure are limited to 800°C and 6 bar. A perfect intercooler is used between the two compressors and an ideal reheater is used between the two turbines. The isentropic efficiencies of compressor and turbines are equal to 0.85. Determine (i) the cycle efficiency without regenerator, (ii) with regenerator having an effectiveness of 0.7, and (iii) the mass of air circulated if the power developed by this system is 325 kW. Take $\gamma = 1.4$ for air and $c_{pa} = 1$ kJ/kgK.

Solution Refer Figure 3.45.

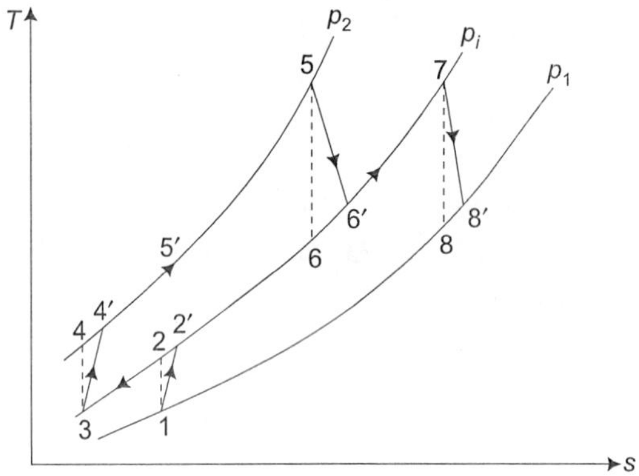

Fig. 3.45

Under the given working conditions for minimum work supplied to the compressor and maximum work developed by the turbine the pressure ratio is given by,

$$p_i = \sqrt{p_1 \, p_2} = \sqrt{1.02 \times 6} = 2.47 \text{ bar}$$

$$\frac{T_2}{T_1} = \left(\frac{p_i}{p_1}\right)^{(\gamma-1)/\gamma}$$

$$\therefore \quad T_2 = T_1 \left(\frac{p_i}{p_1}\right)^{(\gamma-1)/\gamma}$$

$$= 300 \times \left(\frac{2.47}{1.02}\right)^{(1.4-1)1.4} = 386.24 \text{ K}$$

$$\eta_c = \frac{T_2 - T_1}{T_{2'} - T_1}$$

$$0.85 = \frac{386.24 - 300}{T_{2'} - 300}$$

$$\therefore \quad T_{2'} = 401.46 \text{ K}$$

As, $\eta_{c1} = \eta_{c2}$,

$\therefore \quad T_{2'} = T_{4'} = 401.43 \text{ K}$

and $\quad T_1 = T_3 = 300 \text{ K}$

$$\frac{T_5}{T_6} = \left(\frac{p_2}{p_1}\right)^{(\gamma - 1)/\gamma}$$

$$\therefore \quad T_6 = T_5 \times \left(\frac{p_1}{p_2}\right)^{(\gamma - 1)/\gamma}$$

$$= 1073 \times \left(\frac{2.47}{6}\right)^{(1.4 - 1)1.4} = 832.66 \text{ K}$$

$$\eta_{t1} = \frac{T_5 - T_{6'}}{T_5 - T_6}$$

$$0.85 = \frac{1073 - T_{4'}}{1073 - 832.66}$$

$\therefore \quad T_{6'} = 868.71 \text{ K}$

As, $\quad \eta_{c1} = \eta_{c2}$

$\therefore \quad T_{6'} = T_{8'} = 868.71 \text{ K} \quad \text{and} \quad T_5 = T_7 = 1073 \text{ K}$

(i) Net work available,

$$w_{net} = (w_{t1} + w_{t2}) - (w_{c1} + w_{c2})$$

$$= 2(w_{t1} + w_{c1}) \text{ kJ/kg} \qquad [\because w_{t1} = w_{t2} \ \& \ w_{c1} = w_{c2}]$$

$$w_{net} = 2c_p \left[(T_5 - T_{6'}) - (T_{2'} - T_1)\right] \text{ kJ/kg}$$

$$\therefore \quad w_{net} = 2 \times 1 \times [(1073 - 848.71) - (401.46 - 300)]$$

$$= 205.66 \text{ kJ/kg}$$

Heat supplied, $\quad q_s = c_p\,(T_5 - T_{4'}) + c_p\,(T_7 - T_{6'}) \text{ kJ/kg}$

$$\therefore \quad q_s = 1 \times (1073 - 401.46) + 1 \times (1073 - 868.71)$$

$$= 875.83 \text{ kJ/kg}$$

Thermal efficiency without regenerator,

$$\eta_{th} = \frac{w_{net}}{q_s} = \frac{205.66}{875.83}$$

$$= 0.2348 \text{ or } 23.48\%$$

(ii) $$\varepsilon = \frac{(T_{5'} - T_{4'})}{(T_{8'} - T_{4'})},$$

$$0.7 = \frac{T_{5'} - 401.46}{868.71 - 401.46},$$

$\therefore \quad T_{5'} = 728.53 \text{ K}$

Heat supplied, $q_s = c_p (T_5 - T_{5'}) + c_p (T_7 - T_{6'})$ kJ/kg

$\therefore$ $q_s = 1 \times (1073 - 728.53) + 1 \times (1073 - 868.71)$

$= 548.76$ kJ/kg

Thermal efficiency with regenerator,

$$\eta_{th} = \frac{w_{net}}{q_s} = \frac{205.66}{548.76}$$

$$= 0.3748 \text{ or } 37.48\%$$

(iii) Power developed, $P = m_a \times w_{net}$

$$\therefore \quad m_a = \frac{P}{w_{net}} = \frac{325}{205.66}$$

$$= 1.58 \text{ kg/s}$$

Example 3.43 In a gas turbine plant air is taken in the compressor at 25°C and 1.2 bar having a pressure ratio of 6. The hot gases having a maximum temperature of 650°C expand in high pressure (H.P.) turbine which runs the compressors and then these gases are further expanded in low pressure (L.P.) turbine which runs the generator. The gases coming out from the L.P. turbine are exhausted to atmosphere.

The following data are given for this unit: isentropic efficiency of compressor = 0.80, isentropic efficiency of each turbine stage = 0.85, $\gamma = 1.4$ for air, $c_{pa} = 1.005$ kJ/kgK, $\gamma = 1.33$ for gas, $c_{pg} = 1.15$ kJ/kgK. If losses are neglected then determine (i) power developed, (ii) work ratio, and (iii) thermal efficiency.

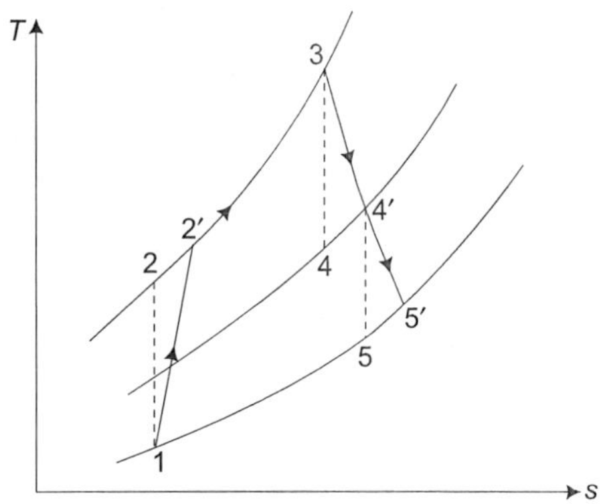

Fig. 3.46

Solution Refer Figure 3.46.

$$\frac{T_2}{T_1} = \left(\frac{p_2}{p_1}\right)^{(\gamma-1)/\gamma}$$

$$\therefore \quad T_2 = T_1 \left(\frac{p_2}{p_1}\right)^{(\gamma-1)/\gamma}$$

$$= 298 \times (6)^{(1.4-1)1.4} = 497.22 \text{ K}$$

$$\eta_{c1} = \frac{T_2 - T_1}{T_{2'} - T_1}$$

$$0.80 = \frac{497.22 - 298}{T_{2'} - 298}$$

$$\therefore \quad T_{2'} = 547.02 \text{ K}$$

Work output of H.P. turbine = Work input to the compressor

$$c_{pg} = (T_3 - T_{4'}) = c_{pa}\,(T_{2'} - T_1)$$

$$1.15 \times (923 - T_{4'}) = 1.005 \times (547.02 - 298)$$

$$\therefore \quad T_{4'} = 705.38 \text{ K}$$

$$\eta_{t1} = \frac{T_3 - T_{4'}}{T_3 - T_4}$$

$$0.85 = 923 - 705.38/923 - T_4$$

$$\therefore \quad T_4 = 666.98 \text{ K}$$

$$p_3 = 1.2 \times 6 = 7.2 \text{ bar}$$

$$\frac{T_4}{T_3} = \left(\frac{p_4}{p_3}\right)^{(\gamma - 1)/\gamma}$$

$$p_4 = p_3 \left(\frac{T_4}{T_3}\right)^{\gamma/(\gamma - 1)}$$

$$= 7.2 \times \left(\frac{666.98}{923}\right)^{1.33/(1.33 - 1)} = 1.94 \text{ K}$$

$$\frac{T_{4'}}{T_5} = \left(\frac{p_4}{p_5}\right)^{(\gamma - 1)/\gamma}$$

$$\frac{705.38}{T_5} = \left(\frac{1.94}{1.2}\right)^{(1.33 - 1)/1.33}$$

$$\therefore \quad T_5 = 626.12 \text{ K}$$

$$\eta_{t2} = \frac{T_{4'} - T_{5'}}{T_{4'} - T_5}$$

$$0.85 = \frac{705.38 - T_{5'}}{705.38 - 626.12}$$

$$\therefore \quad T_{5'} = 638.01 \text{ K}$$

(i) Net work available, $w_{net} = c_{pg}\,(T_{4'} - T_{5'}) = 1.15 \times (705.38 - 638.01)$

$$= 77.47 \text{ kJ/kg}$$

Power developed, $P = m_a \times w_{net} = 1 \times 77.47$

$$= 77.47 \text{ kJ/s} = 77.47 \text{ kW}$$

(ii) Work produced by H.P. turbine,

$$w_{t1} = 1.15 \times (923 - 705.38)$$

$$= 250.26 \text{ kJ/kg}$$

Work ratio, $$r_w = \frac{\text{Net work}}{\text{Total work}}$$

$$= \frac{77.47}{(250.26 + 77.47)} = 0.2364$$

(iii) Heat supplied,

$$q_s = c_{pg}\,(T_3 - T_{2'})$$

$$= 1.15 \times (923 - 547.02) = 432.38 \text{ kJ/kg}$$

Thermal efficiency, $$\eta_{ith} = \frac{w_{\text{net}}}{q_s} = \frac{77.47}{432.38}$$

$$= 0.1792 \text{ or } 7.92\%$$

Example 3.44 In a gas turbine plant air is taken in the low pressure compressor at 25°C and 1.1 bar and after compression it is passed through intercooler where its temperature is reduced to 30°C. The cooled air is again compressed in second compressor and then passed in the combustion chamber where its maximum temperature reaches to 800°C. The hot gases expand in high pressure (H.P) turbine which runs the compressors and these gases are further expanded in low pressure (L.P.) turbine which runs the generator. The gases coming out from the L.P. turbine are used for heating the incoming air from the second compressor and then exhausted to atmosphere.

The following data are given for this unit: pressure ratio of each compressor = 2, isentropic efficiency of each compressor stage = 0.85, isentropic efficiency of each turbine stage = 0.85, effectiveness of heat exchanger = 0.65, airflow = 10 kg/s, CV = 42000 kJ/kg, γ = 1.4 for air and c_{pa} = 1 kJ/kgK, γ = 1.33 for gas and c_{pg} = 1.15 kJ/kgK. If losses are neglected then determine (i) power developed, (ii) thermal efficiency, and (iii) specific fuel consumption.

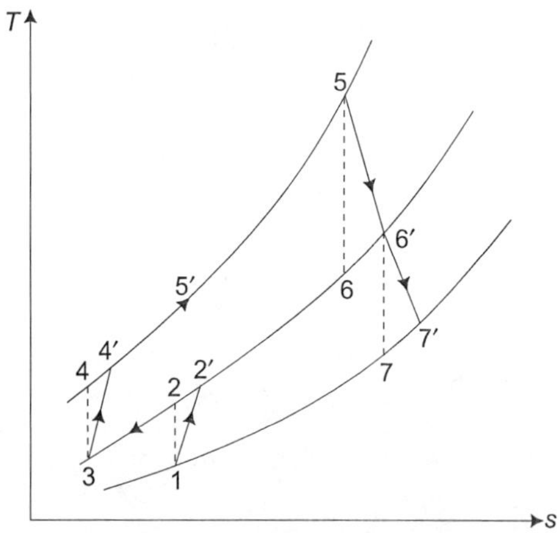

Fig. 3.47

Solution Refer Figure 3.47.

$$\frac{T_2}{T_1} = \left(\frac{p_2}{p_1}\right)^{(\gamma - 1)/\gamma}$$

$$\therefore \quad T_2 = T_1 \left(\frac{p_2}{p_1}\right)^{(\gamma - 1)/\gamma}$$

$$= 298 \times (2)^{(1.4 - 1)1.4} = 363.27 \text{ K}$$

$$\eta_{c1} = \frac{T_2 - T_1}{T_{2'} - T_1}$$

$$0.85 = \frac{363.27 - 298}{T_{2'} - 298}$$

$$\therefore \quad T_{2'} = 374.79 \text{ K}$$

$$\frac{T_4}{T_3} = \left(\frac{p_4}{p_3}\right)^{(\gamma - 1)/\gamma}$$

$$\therefore \quad T_4 = T_3 \left(\frac{p_4}{p_3}\right)^{(\gamma - 1)/\gamma}$$

$$= 303 \times (2)^{(1.4 - 1)/1.4} = 369.36 \text{ K}$$

$$\eta_{c2} = \frac{T_4 - T_3}{T_{4'} - T_3}$$

$$0.85 = \frac{369.36 - 303}{T_{4'} - 303}$$

$$\therefore \quad T_{4'} = 381.07 \text{ K}$$

Work output of H.P. turbine = Work input to the compressor

$$c_{pg} = (T_5 - T_{6'}) = c_{pa} [(T_{2'} - T_1) + (T_{4'} - T_3)]$$

$$115 \times (1073 - T_{6'}) = 1 \times [(374.79 - 298) + (381.07 - 303)]$$

$$\therefore \quad T_{6'} = 938.34 \text{ K}$$

$$\eta_{t1} = \frac{T_5 - T_{6'}}{T_5 - T_6}$$

$$0.85 = \frac{1073 - 938.34}{1073 - T_6}$$

$$\therefore \quad T_6 = 914.58 \text{ K}$$

$$p_5 = 1.1 \times 4 = 4.4 \text{ bar}$$

$$\frac{T_6}{T_5} = \left(\frac{p_6}{p_5}\right)^{(\gamma - 1)/\gamma}$$

$$p_6 = p_5 \left(\frac{T_6}{T_5}\right)^{\gamma/(\gamma - 1)}$$

$$= 4.4 \times \left(\frac{914.58}{1073}\right)^{1.33/(1.33 - 1)}$$

$$= 2.143 \text{ bar}$$

$$\frac{T_{6'}}{T_7} = \left(\frac{p_6}{p_7}\right)^{(\gamma - 1)/\gamma}$$

$$\frac{938.34}{T_7} = \left(\frac{2.143}{1.1}\right)^{(1.33 - 1)/1.33}$$

$$\therefore \quad T_7 = 795.24 \text{ K}$$

$$\eta_{t2} = \frac{T_{6'} - T_{7'}}{T_{6'} - T_7}$$

$$0.85 = \frac{938.34 - T_{7'}}{938.34 - 795.24}$$

$$\therefore \quad T_{7'} = 816.70 \text{ K}$$

(i) Net work available,

$$w_{net} = c_{pg} (T_{6'} - T_{7'}) = 1.15 \times (938.34 - 816.70)$$

$$= 139.89 \text{ kJ/kg}$$

Power developed,

$$P = m_a \times w_{net} = 10 \times 139.89$$

$$= 1398.9 \text{ kJ/s} = 1398.9 \text{ kW}$$

$$\varepsilon = \frac{(T_{5'} - T_{4'})}{(T_{7'} - T_{4'})},$$

$$0.65 = \frac{T_{5'} - 381.07}{816.70 - 381.07},$$

$$\therefore \quad T_{5'} = 664.23 \text{ K}$$

(ii) Heat supplied,

$$q_s = c_{pg} (T_5 - T_{5'}) = 1.15 \times (1073 - 656.08)$$

$$= 479.46 \text{ kJ/kg}$$

Thermal efficiency,

$$\eta_{th} = \frac{w_{net}}{q_s} = \frac{139.89}{479.46}$$

$$= 0.2918 \text{ or } 29.18\%$$

(ii) Let m_f be the mass of fuel per kg of air

Also, $m_f \times 42000 = 479.46$

$$\frac{1}{m_f} = \frac{87.6}{1}$$

Air-fuel ratio, $A : F = 87.6 :1$

So, fuel supplied per hour,

$$M_f = \text{Air supplied/hour} \times \text{fuel supplied/kg of air}$$

$$\therefore \quad M_f = (10 \times 3600) \times \frac{1}{87.6} = 410.96 \text{ kg/h}$$

Therefore, specific fuel consumption

$$= \frac{M_f}{P} = \frac{410.96}{1398.9}$$

$$= 0.294 \text{ kg/kWh}$$

Example 3.45 An open cycle gas turbine plant has a single-stage compressor and a single-stage turbine incorporating a heat exchanger. The air is being sucked at a temperature of 20°C and pressure of 1 bar. For an overall pressure ratio of 4.5 and power output of 4000 kW, the mass flow is 40 kg/s. If the effectiveness of heat exchanger is 0.6 and the isentropic efficiency of compressor is 0.8, find the isentropic efficiency of the gas turbine for a plant thermal efficiency of 0.29. Take γ = 1.4 for air and c_{pa} = 1.005 kJ/kgK, γ = 1.36 for gas and c_{pg} = 1.07 kJ/kgK.

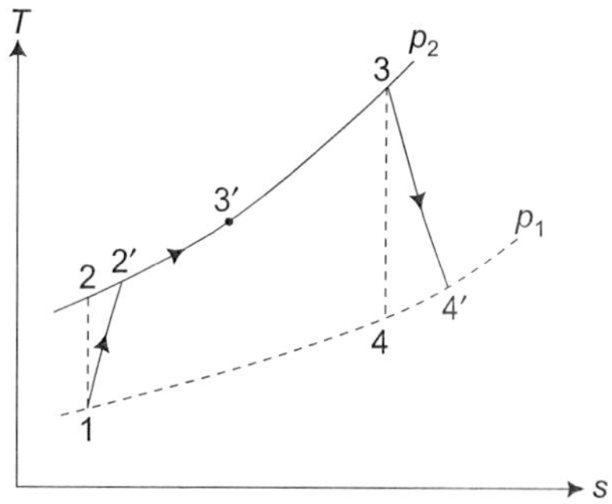

Fig. 3.48

Solution Refer Figure 3.48.

$$\frac{T_2}{T_1} = \left(\frac{p_2}{p_1}\right)^{(\gamma-1)/\gamma}$$

$$\therefore \quad T_2 = T_1 \left(\frac{p_2}{p_1}\right)^{(\gamma-1)/\gamma}$$

$$= 293 \times (4.5)^{(1.4-1)1.4} = 450.62 \text{ K}$$

$$\eta_c = \frac{T_2 - T_1}{T_{2'} - T_1}$$

$$0.8 = \frac{450.62 - 293}{T_{2'} - 293}$$

$$\therefore \quad T_{2'} = 489.62 \text{ K}$$

$$\varepsilon = \frac{c_{pa}(T_{3'} - T_{2'})}{c_{pg}(T_{4'} - T_{2'})}$$

$$= \frac{1.005 \times (T_{3'} - 489.62)}{1.07 \times (T_{4'} - 489.62)} = 0.6$$

$$\frac{T_{3'} - 489.62}{T_{4'} - 489.62} = 0.6388 \qquad \text{(i)}$$

$$w_t = \frac{4000}{40} = 100 \text{ kJ/kg}$$

$$\eta_t = \frac{w_t}{q_s} = \frac{100}{c_{pg}(T_3 - T_{3'})} = 0.29$$

$$T_3 - T_{3'} = \frac{100}{1.07 \times 0.29} = 322.27$$

$$T_{3'} = T_3 - 322.27 \qquad \text{(ii)}$$

Also

$$w_t = c_{pg}(T_3 - T_{4'}) - c_{pa}(T_{2'} - T_1)$$

$$= 100 \text{ kJ/kg}$$

$$1.07 \times (T_3 - T_{4'}) - 1.005 \times (489.62 - 293) = 100$$

or $$T_3 - T_{4'} = 278.13$$

$$T_{4'} = T_3 - 278.13 \qquad \text{(iii)}$$

Substituting the expressions (ii) and (iii) in (i), we have,

$$\frac{(T_3 - 322.27) - 489.62}{(T_3 - 278.13) - 489.61} = 0.6388,$$

$\therefore$ $$T_3 = 889.95 \text{ K}$$

$\therefore$ $$T_{3'} = 889.95 - 322.27 = 567.68 \text{ K}$$

$$T_{4'} = 889.95 - 278.13 = 611.82 \text{ K}$$

$$\frac{T_3}{T_4} = \left(\frac{p_2}{p_1}\right)^{(\gamma - 1)/\gamma}$$

$\therefore$ $$T_4 = T_3 \left(\frac{p_1}{p_2}\right)^{(\gamma - 1)/\gamma}$$

$$= 889.95 \times (1/4.5)^{(1.36 - 1)/1.36} = 597.66 \text{ K}$$

$$\eta_t = \frac{T_3 - T_{4'}}{T_3 - T_4}$$

$$= \frac{889.95 - 611.82}{889.95 - 597.66}$$

$$= 0.9515 \text{ or } 95.15\%$$

EXERCISE

(i) Objective Type Questions

1. Petrol engines are also called:
(a) quality governing (b) quantity governing
(c) hit and miss governing (d) none of the above

2. Diesel engines are also called:
(a) quality governing (b) quantity governing
(c) hit and miss governing (d) none of the above

3. Fuel injector is used in:
(a) diesel engines (b) petrol engines (c) gas engines (d) none of the above

4. Carburettor is used in:
(a) diesel engines (b) gas engines (c) petrol engines (d) none of the above

5. At the starting of petrol engines:
   (a) rich fuel-air ratio is needed
   (b) lean fuel-air ratio is needed
   (c) chemically correct fuel-air ratio is needed
   (d) none of the above
6. Compression ratio for diesel engines lies in the range of:
   (a) 6-10 (b) 10-12 (c) 12-16 (d) 16-20
7. Compression ratio for petrol engines lies in the range of:
   (a) 6-10 (b) 10-12 (c) 12-16 (d) 16-20
8. Thermal efficiency of I.C. engines lies in the range of:
   (a) 10-15% (b) 15-20% (c) 20-25% (d) 30-35%
9. Thermal efficiency of diesel engines in comparison to petrol engines is:
   (a) lower (b) higher (c) equal (d) none of the above
10. Specific fuel consumption of diesel engines in comparison to petrol engines is:
    (a) lower (b) higher (c) equal (d) none of the above
11. Mass of flywheel for 2-stroke petrol engine as compared to 4-stroke petrol engine will be:
    (a) lighter (b) heavier (c) equal (d) none of the above
12. The following is the C.I. engine:
    (a) petrol engine (b) gas engine (c) steam engine (d) diesel engine
13. The following is the S.I. engine:
    (a) petrol engine (b) gas engine (c) steam engine (d) diesel engine
14. In 4-stroke S.I. engines the cam shaft runs at:
    (a) twice the speed of crankshaft (b) half of the speed of crankshaft
    (c) same speed (d) none of the above
15. In 4-stroke C.I. engines during suction stroke:
    (a) only fuel is sucked (b) air fuel mixture is sucked
    (c) only air is sucked (d) none of the above
16. The numbers of revolution of crankshaft in a 2-stroke I.C. engine for completing the operations in a cycle are:
    (a) 1 (b) 2 (c) 3 (d) 4
17. The number of revolutions of crankshaft in a 4-stroke I.C. engine for completing the operations in a cycle are:
    (a) 1 (b) 2 (c) 3 (d) 4
18. Compression ratio of I.C. engines is given by:
    (a) $(v_c + v_s)$ (b) $(v_c + v_s)/v_c$ (c) v_c (d) $v_c/(v_c + v_s)$

19. 2-stroke I.C. engines have:
(a) 1 port (b) 2 ports (c) 3 ports (d) 4 ports

20. In Diesel engines fuel is ignited by:
(a) injector (b) spark plug
(c) ignition coil (d) heat due to compression

21. The ratio of stroke to crank radius is:
(a) 2 (b) 1 (c) 1/2 (d) none

22. Engine indicators are used to measure:
(a) mean temperature (b) mean speed
(c) pressure (d) mean effective pressure

23. Piston head in 2-stroke engines is:
(a) flat (b) swelled (c) crown shaped (d) none

24. Mechanical efficiency is the ratio of:
(a) i.p to b.p (b) i.p to f.p (c) f.p to i.p (d) b.p to i.p

25. Relative efficiency is the ratio of:
(a) η_{ith} to η_{vol} (b) η_{bth} to η_{vol} (c) η_{ith} to η_{air} (d) none of the above

26. Speed of high-speed engines may be:
(a) 200-400 rpm (b) 400-600 rpm (c) 600-800 rpm (d) above 900 rpm

27. Piston speed lies in the range of:
(a) 2-5 m/s (b) 5-7 m/s (c) 7-15m/s (d) none

28. The cycle generally used for gas turbines is:
(a) Otto cycle (b) Diesel cycle (c) Carnot cycle (d) Brayton cycle

29. The ideal efficiency of a simple gas turbine depends on:
(a) cut-off ratio (b) pressure ratio (c) both (a) and (b) (d) none of the above

30. Thermal efficiency of a gas turbine as compared to I.C. engine is:
(a) higher (b) lower (c) equal (d) none of the above

31. Mechanical efficiency of I.C. engines in comparison to gas turbines is:
(a) higher (b) equal (c) higher (d) none

32. Thermal efficiency of a closed cycle gas turbine unit can be increased by:
(a) intercooling (b) reheating (c) regeneration (d) all of the above

33. Thermal efficiency of simple gas turbine with fixed turbine inlet temperature and with increase in pressure ratio:
(a) increases (b) decreases
(c) first increases and then decreases (d) none of the above

34. For maximum work output in two-stage expansion with perfect reheating, the intermediate pressure is:

(a) $p_i = \sqrt{p_1 p_2}$ (b) $p_i = p_1/p_2$ (c) $p_i = 1/p_1 p_2$ (d) none

35. For optimum pressure ratio, maximum net work is:

(a) $w_{net} = c_p \left(\sqrt{T_3} - \sqrt{T_1}\right)^2$ (b) $w_{net} = c_v \left(\sqrt{T_1} - \sqrt{T_3}\right)^2$

(c) $w_{net} = \left(\sqrt{T_3} - \sqrt{T_1}\right)^2$ (d) $w_{net} = c_p \left(\sqrt{T_3} - \sqrt{T_1}\right)$

36. Processes involved in the Brayton cycle are:

(a) isothermal and constant pressure (b) isothermal and constant volume

(c) isentropic and constant volume (d) isentropic and constant temperature

37. In a 2-stage gas turbine, reheating after first stage:

(a) decreases work ratio (b) increases work ratio

(c) does not affect (d) none of the above

38. Pressure ratio for a gas turbine may be in the range of:

(a) 2 to 3 (b) 3 to 5 (c) 5 to 10 (d) 10 to15

39. Thermal efficiency of closed cycle gas turbine is given by:

(a) $1 - 1/(r_p)^{(\gamma-1)/\gamma}$ (b) $(r_p)^{(\gamma-1)/\gamma}$ (c) $1/(r_p)^{(\gamma-1)/\gamma}$ (d) none of the above

40. By intercooling and reheating in a 2-stage gas turbine unit:

(a) thermal efficiency increases but work ratio decreases

(b) both increase

(c) work ratio increases but thermal efficiency decreases

(d) none of the above

ANSWERS

1. (b)	2. (a)	3. (a)	4. (c)	5. (a)	6. (d)	7. (a)	8. (d)	9. (b)	10. (a)
11. (a)	12. (d)	13. (a)	14. (b)	15. (c)	16. (a)	17. (b)	18. (b)	19. (c)	20. (d)
21. (a)	22. (d)	23. (c)	24. (d)	25. (c)	26. (d)	27. (c)	28. (d)	29. (b)	30. (b)
31. (a)	32. (d)	33. (c)	34. (a)	35. (a)	36. (d)	37. (b)	38. (b)	39. (a)	40. (c)

(ii) Review Questions

1. Define an engine and a heat engine.
2. What are the two general classes of combustion engines and how do they basically differ in principle?
3. Discuss the relative merits and demerits of internal combustion and external combustion engines.

4. How is internal combustion engine classified?
5. What are the main part of an engine? Explain with their functions.
6. Define the following:

   (i) bore (ii) dead centers (iii) stroke (iv) swept volume
   (v) clearance volume (vi) cylinder volume (vii) compression ratio.
7. Explain with suitable neat sketches the working of four-stroke S.I. and C.I. engines.
8. Describe with suitable sketches the construction and working of two-stroke S.I. engine.
9. Give a comparison between (i) S.I. engine and C.I. engine and (ii) 2-stroke and 4-stroke engines.
10. Explain the following cycles with $p - v$ and $T - s$ diagrams: (i) Otto cycle, (ii) Diesel cycle, and (iii) dual cycle. Also derive expressions for thermal efficiency.
11. What are the advantages, limitations, and applications of gas turbines?
12. Give the classification of gas turbines.
13. Explain the working of constant volume gas turbine with a neat sketch.
14. Explain the working of open and closed constant pressure gas turbine.
15. Discuss the means of refinement the performance of simple open cycle gas turbine.
16. Derive an expression for (i) thermal efficiency, (ii) work ratio, (iii) optimum pressure ratio, and (iv) maximum work output for ideal constant pressure closed gas turbine cycle.
17. Derive an expression for optimum pressure ratio and maximum network for actual constant pressure closed gas turbine cycle.
18. Derive an expression for optimum pressure ratio for maximum actual constant pressure closed gas turbine cycle.
19. What are the advantages and disadvantages of gas turbines over I.C. engines?
20. What are the advantages of gas turbines over steam turbines?
21. What are the advantages and disadvantages of closed cycle over open cycle gas turbines?
22. Mention the comparisons of Otto, Diesel, and dual cycles by considering different variables.

(iii) Numerical Problems

1. The bore and stroke of a 4-cylinder petrol engine are 6 cm and 7 cm respectively. The compression ratio is 6.8. Determine the suction capacity of the engine and clearance volume.

   [**Ans.** 791.6 cm^3, 34.12 cm^3]
2. The output of an engine is 8 kW and its mechanical efficiency is 0.75. Find indicated power and friction power. [**Ans.** 10.67 kW, 2.67 kW]
3. An engine of capacity 75 kW requires 10 kW to start the engine. Determine its mechanical efficiency. Also determine the mechanical efficiency at 50% and 25% of full load assuming that mechanical losses are same at all the loads. [**Ans.** 88.25%, 78.95%, 65.22%]

4. A single cylinder spark ignition engine has mechanical efficiency 80%, brake thermal efficiency 30% and it uses oil of calorific value 43000 kJ/kg. Determine brake specific fuel consumption and indicated specific fuel consumption. [**Ans.** 0.31 kg/kWh, 0.242 kg/kWh]

5. A 4-stroke S.I. engine develops 425 kW brake power with mechanical efficiency of 88%. The measured fuel and air consumption are 160 kg/hr and 2450 kg/hr respectively. Find (i) indicated power and friction power, (ii) indicated thermal efficiency, and (iii) brake thermal efficiency. Take calorific value of fuel as 42000 kJ/kg. [**Ans.** 482.95 kW, 57.95 kW, 25.87%, 22.08%]

6. A 4-stroke S.I. engine develops 38 kW brake power with mechanical efficiency of 86%. The brake specific fuel consumption is 0.36 kg/kWh and the air fuel ratio is 15.2:1. Find (i) indicated power and friction power, (ii) indicated thermal efficiency, (iii) brake thermal efficiency, and (iv) fuel consumption and air consumption. Take calorific value of fuel as 42000 kJ/kg.
   [**Ans.** 44.19 kW, 6.19 kW, 27.69%, 23.81%, 13.68 kg/h, 207.94 kg/h]

7. A C.I. engine develops 15 kW at 35% indicated thermal efficiency with mechanical efficiency of 86%. Find (i) fuel consumption in kg/hr and liters/hr, and (ii) indicated and brake specific fuel consumption. Take calorific value of fuel as 42000 kJ/kg and specific gravity of fuel as 0.86.
   [**Ans.** 4.32 kg/h, 5.02 litres/h, 0.282 kg/kWh, 0.288 kg/kWh]

8. A two stroke C.I. engine of 6000 kW capacity needs 600 kW to start the engine. The fuel consumption at full load is 2000 kg/hr and air-fuel ratio used is 20:1. Determine (i) indicated power, (ii) mechanical efficiency, (iii) air consumption per hour, (iv) indicated thermal efficiency, and (v) brake thermal efficiency. Take calorific value of fuel as 42000 kJ/kg.
   [**Ans.** 6600 kW, 90.9%, 40000 kg/h, 27.95%, 25.41%]

9. Determine brake thermal efficiency and indicated thermal efficiency of a four-stroke C.I. engine whose power developing capacity is 60 kW. The fuel consumption is 15 litres per hour. Mechanical efficiency of the engine is 86%. Take specific gravity of oil as 0.85 and its calorific value as 42000 kJ/kg. [**Ans.** 27.29%, 31.78%]

10. A four-stroke petrol engine delivers 50 kW at 85% mechanical efficiency. The specific fuel consumption of the engine is 0.3 kg/kWh. The calorific value of fuel is 42500 kJ/kg and air-fuel ratio of the mixture is 15.8:1. Determine (i) indicated power, (ii) friction power, (iii) brake thermal efficiency, and (iv) fuel and air consumption per hour.
    [**Ans.** 58.82 kW, 8.82 kW, 28.23%, 15 kg/h, 237 kg/h]

11. The air-fuel ratio used in S.I. engine of 75 kW capacities is 16:1. Determine the amount of fuel and air consumed by the engine at full load when brake thermal efficiency is 25.7%. Also determine m^3 of fuel and air required per hour if the density of air is 1.21 kg/m^3 and density of fuel is 3.56 kg/m^3. Take calorific value of fuel as 43000 kJ/kg.
    [**Ans.** 24.45 kg/h, 391.2 kg/h, 6.8 m^3/h, 3.23.31 m^3/h]

12. A 4-cylinder 4-stroke petrol engine has cylinder diameter as 8.5 cm and stroke 10.5 cm. The clearance volume per cylinder is 80 cubic centimetres. The fuel consumption is 4.5 gm per second and developed torque is 160 N-m at a speed of 4000 rpm. Find (i) compression ratio, (ii) brake power, (iii) brake mean effective pressure, and (iv) brake thermal efficiency if calorific value of fuel is 43500 kJ/kg. [**Ans.** 8.45, 67.02 kW, 8.44 bar, 34.24%]

13. A four-stroke S.I. engine develops brake power of 450 kW with mechanical efficiency of 76%. The measured fuel and air consumption are 160 kg/hr and 2200 kg/hr respectively. Determine (i) indicated power, (ii) friction power, (iii) indicated thermal efficiency, and (iv) brake thermal efficiency. Take calorific value of fuel as 42500 kJ/kg.

[**Ans.** 592.1 kW, 142.1 kW, 31.35%, 23.83%]

14. A single cylinder four-stroke diesel engine having a swept volume of 800 cm^3 is tested at 300 rpm. When a braking torque 75 Nm is applied, analysis of an indicator diagram gives a mean effective pressure of 12 bar. Calculate (i) b.p. and (ii) η_m of the engine.

[**Ans.** 2.748 kW, 98.14%]

15. A diesel engine develops 6 kW. Its indicated thermal efficiency is 32% and mechanical efficiency is 85%. Determine (i) the fuel consumption of engine in kg/hr and in litres per hour, (ii) the indicated specific fuel consumption, and (iii) the brake specific fuel consumption. Take specific gravity of oil 0.86 and calorific value of oil 43 MJ/kg.

[**Ans.** 1.85 kg/h, 0.26 kg/kWh, 0.31 kg/kWh]

16. Determine bore and stroke of a 4-cylinder, 4-stroke petrol engine for the following data: b.p = 400 kW, compression ratio = 8, speed = 300 rpm, stroke = 1.25 times the bore, $\eta_m = 0.85$, i.m.e.p. = 12 bar, $\eta_{rel} = 0.60$, and CV = 42500 kJ/kg. Also determine the fuel consumption in kg/s and in kg/kWh. Take $\gamma = 1.4$. [**Ans.** 34.14 cm, 40.97 cm, 0.032 kg/s, 0.288 kg/kWh]

17. Following observations were taken during the testing of single cylinder four-stroke oil engine: brake wheel diameter = 70 cm, rope diameter = 2.5 cm, speed = 400 rpm, load on rope = 300 N, spring balance reading = 80 N, area and length of indicator diagram = 4.5 cm^2 and 6.5 cm respectively, spring stiffness = 11.5 bar/cm, bore = 10.5 cm, stroke = 15.5 cm, specific fuel consumption = 0.32 kg/kWh, and calorific value of fuel = 42000 kJ/kg. Determine b.p., i.p., η_{ith} and η_{bth}. [**Ans.** 3.38 kW, 3.72 kW, 29.12%, 26.46%]

18. The air-fuel ratio used in S.I. engine of 80 kW capacity is 14:1. Determine the amount of air consumed by the engine at full load when brake thermal efficiency is 25%. Also determine m^3 of air and m^3 of fuel required per hour if density of air is 1.2 kg/m^3 and density of fuel is 3.45 kg/m^3. If the calorific value of fuel is 42500 kJ/kg, how many m^3 per hour of the mixture is required? [**Ans.** 379.46 kg/h, 316.22 m^3/h, 7.66 m^3/h, 323.88 m^3/h]

19. A two-stroke C.I. engine delivers 5000 kW while using 1000 kW to overcome friction losses. It consumes 2000 kg of fuel per hour at an air fuel ratio of 16:1. The calorific value of fuel is 42000 kJ/kg. Determine i.p, η_m, η_{ith}, and η_{bth}. [**Ans.** 6000 kW, 83.33%, 25.7%, 21.4%]

20. A 4-stroke, 4-cylinder diesel engine running at 2250 rpm develops 75 kW. Its brake thermal efficiency is 26%, mechanical efficiency is 85%, and calorific value of fuel is 42500 kJ/kg. Engine has a bore of 15 cm and stroke of 12 cm. Take density of air as 1.2 kg/m^3 and air-fuel ratio = 17:1. Determine (i) the fuel consumption of engine in kg/min, (ii) air consumption in m^3/s, (iii) the indicated thermal efficiency, (iv) volumetric efficiency, (v) brake mean effective pressure, and (vi) mean piston speed.

[**Ans.** 6.78×10^{-3} kg/s, 0.096 m^3/s, 30.62%, 60.37%, 4.71 bar, 9 m/s]

21. Compare the changes in efficiency of an Otto engine when the compression ratio changes from 5 to 6. Take $\gamma = 1.4$ for air. [**Ans.** 8.15]

22. An engine works on the Otto cycle which is supplied with air at 100 kN/m^2, 25°C. The compression ratio is 6. Heat supplied is 1750 kJ/kg. Calculate (i) cycle efficiency, (ii) the maximum pressure and temperature of the cycle, and (iii) mean effective pressure. Take $\gamma = 1.4$, and $R = 0.287$ kJ/kgK for air. [**Ans.** 51.16%, 3047.52 K, 6135.99 kN/m^2, 1256.56 kN/m^2]

23. In an Otto cycle air at 20°C and 1 bar is compressed adiabatically until the pressure is 12 bar. Heat is added at constant volume until the pressure rises to 30 bar. Calculate (i) compression ratio, (ii) air standard efficiency, and (iii) mean effective pressure. Take $c_v = 0.718$ kJ/kgK, and $R = 8.314$ kJ/kmol K. [**Ans.** 5.89, 50.8%, 4.69 bar]

24. In a S.I. engine working on ideal Otto cycle compression ratio is 10. The pressure and temperature at the beginning of compression are 1 bar and 30°C respectively. The peak pressure is 40 bar. Calculate temperature and pressure at the salient points, the air standard efficiency, and mean effective pressure. Take $\gamma = 1.4$ for air.

[**Ans.** 761.83 K, 1211.9 K, 482.46 K, 60.18%, 2.48 bar]

25. In a constant volume cycle the temperature at the beginning and end of the compression are 30°C and 350°C respectively. Calculate the air standard efficiency and compression ratio. Take $\gamma = 1.4$ for air. A petrol engine with the above compression ratio develops 20 kW indicated power and consumes 6 liters of fuel per hour. The specific gravity of fuel is 0.75 and its calorific value is 43 MJ/kg. Determine indicated thermal efficiency and the relative efficiency.

[**Ans.** 37.21%, 72.4%]

26. A gas engine works on Otto cycle has a cylinder of diameter 15 cm and the stroke of 20 cm. The clearance volume is 1200 cm^3. Determine the air standard efficiency. Take $\gamma = 1.4$ for air.

[**Ans.** 42.2%]

27. A petrol engine working on Otto cycle has initial pressure 1 bar and pressure after compression 10 bar. Calculate compression ratio, percentage clearance, and thermal efficiency. If the initial temperature is 30°C, find the temperature at the end of compression, and mean effective pressure if the maximum pressure in the cycle is 16 bar.

[**Ans.** 5.17, 19.3.4, 48.16%, 584.57 K, 1.83 bar]

28. A diesel engine has a compression ratio of 20 and cut-off takes place at 6% of the stroke. Determine the air standard efficiency. [**Ans.** 64.05%]

29. A diesel engine operating on a standard Diesel cycle has 250 mm bore and 350 mm stroke. The clearance volume is 450 cm^3. The fuel is injected at constant pressure for 8% of the stroke. Determine the air standard efficiency, if the cut-off is delayed from 8% to 10%. What will be the percentage loss in efficiency? Compression ratio is same in both the cases. [**Ans.** 2.6%]

30. In an air standard diesel cycle, the compression ratio is 18, and at the beginning of compression the temperature is 25°C and the pressure is 0.1 MPa. Heat is added until the temperature at the end of the constant pressure process is 1450°C. Calculate (i) cut-off ratio, (ii) heat supplied per kg of air, (iii) cycle efficiency, and (iv) mean effective pressure.

[**Ans.** 1.82; 779.94 kJ/kg; 64%, 6.18 bar]

31. A Diesel engine has a compression ratio of 18. The fuel is cut-off at 0.06 stroke. The relative efficiency is 58%. If the calorific value of fuel is 42000 kJ/kg calculate the mass of fuel required per kW/h. [**Ans.** 0.234 kg/kWh]

32. An oil engine works on dual cycle has cylinder bore of 20 cm and stroke 40 cm. The compression ratio is 16 and explosion ratio is 1.54. Cut-off occurs at 6% of the stroke. Calculate the air standard efficiency. [**Ans.** 62.89%]

33. An air standard dual cycle has a compression ratio of 15. The pressure and temperature at the beginning of compression are 1 bar, 27°C. The maximum pressure reached is 50 bar and the maximum temperature is 1427°C. Calculate the temperature at the end of constant volume heat addition, work done per kg of air, and the cycle efficiency. Take $\gamma = 1.4$ for air.
[**Ans.** 1001.46 K, 489.24 kJ/kg, 62.41%]

34. A dual combustion cycle has an adiabatic compression volume ratio of 17: 1. The condition at the commencement of the compression are 1 bar, 27°C, and 0.12 m^3. The maximum pressure of the cycle is 60 bar and the maximum temperature is 1600°C. If $c_p = 0.717$ kJ/kgK and $\gamma = 1.4$, determine the pressure, volume, and temperature at each point of the cycle and the thermal efficiency.
[**Ans.** 7.059×10^{-3}, 52.79 bar, 931.75 K, 1059.01 K, 3.26 bar, 59.33%]

35. Evaluate the efficiency of Brayton cycle if the temperature and pressure before compression are 30°C and 1 bar and the temperature at the end of combustion is 1200°C. The pressure ratio is 4 and $c_p = 1.005$ kJ/kgK. [**Ans.** 32.71%]

36. In a gas turbine plant the air enters the compressor at 1 bar and 25°C. The pressure ratio in the cycle is 4. The temperature at the turbine inlet is 750°C. If the mass flow rate is 4.5 kg/s, determine the thermal efficiency. Take $c_p = 1.005$ kJ/kgK. [**Ans.** 40.03%]

37. Air enters the compressor of a gas turbine plant operating on Brayton cycle at 1 bar and 27°C. The pressure ratio in the cycle is 6. Determine the maximum temperature in the cycle and the cycle efficiency when turbine work is 2.5 times of compressor work. Take $\gamma = 1.4$.
[**Ans.** 1251.25 K, 40.07%]

38. A closed cycle gas turbine uses helium as working fluid. The gas enters the compressor at 4 bar and 47°C and discharges at 16 bar and 317°C. It then enters a regenerator having an effectiveness of 0.7. The turbine inlet condition is 15.5 bar and 1127°C. The turbine exhausts at 4.2 bar and 587°C and enters the regenerator. Determine (i) the compressor efficiency, (ii) the turbine efficiency, and (iii) the thermal efficiency of cycle. [**Ans.** 88.2%, 94.6%, 43.5%]

39. An open gas turbine works between the temperature limits of 27°C and 1427°C and pressure limits of 1 bar and 23 bar. The efficiencies of turbine and compressors are 0.86 and 0.85 respectively. If the fuel consumption is 10 kg/s and the calorific value of fuel is 42000 kJ/kg, determine the actual thermal efficiency. [**Ans.** 42.81%]

40. A closed cycle gas turbine has a two-stage compression with perfect intercooling. Each stage has a pressure ratio equal to 3 and isentropic efficiency equal to 0.85. The turbine is also a two-stage with equal pressure ratio with inter-stage reheat to original temperature. The isentropic efficiency of each turbine is 0.90. If the turbine inlet temperature is 827°C and the compressor inlet is 27°C then determine the work ratio, and the overall cycle efficiency. Take the properties of working fluid as c_p = 0.5207 kJ/kgK and γ = 1.4. [**Ans.** 0.446, 32.8%]

41. In a gas turbine plant, the following data have been collected: inlet pressure to compressor = 1 bar, turbine inlet temperature = 1227°C, pressure ratio = 64, compressor inlet temperature = 27°C, turbine efficiency = 0.88, compressor efficiency = 0.87, and regenerator effectiveness = 0.65. Determine the thermal efficiency and work ratio. [**Ans.** 46.62%, 0.515]

42. A gas turbine receives air at 1 bar and 27°C and compresses it isentropically to 6.2 bar. The compressor and turbine efficiencies are 0.88 and 0.90 respectively. If the fuel air (F/A) ratio used is 0.017 and the calorific value of fuel is 44186 kJ/kg, determine the thermal efficiency of the cycle. Take c_p = 1.005 kJ/kgK & γ = 1.4 for air, and c_p = 1.147 kJ/kgK & γ = 1.333 for gases. [**Ans.** 25.24%]

43. A gas turbine plant works between the temperature limits of 750°C and 27°C. The isentropic efficiencies of compressor and turbine are 0.82 and 0.86 respectively. Determine the optimum pressure ratio for maximum work output and for maximum thermal efficiency of the cycle. Assume air is used as working fluid. What is the percentage change in pressure ratios for maximum work output and for maximum thermal efficiency if argon is used as working fluid. Take γ = 1.4 for air and γ = 1.67 for argon. [**Ans.** 35.77%, 41.16%]

44. An open cycle gas turbine plant has a single stage-compressor and a single-stage turbine incorporating a heat exchanger. The air is being sucked at a temperature of 20°C and pressure of 1 bar. For an overall pressure ratio of 4.5 and power output of 4000 kW, the mass flow is 40 kg/s. If the effectiveness of heat exchanger is 0.6 and the isentropic efficiency of compressor is 0.8, find the isentropic efficiency of the gas turbine for a plant thermal efficiency of 0.29. Take γ = 1.4 for air & c_{pa} = 1.005 kJ/kgK, γ = 1.36 for gas & c_{pg} = 1.07 kJ/kgK. [**Ans.** 72.99%]

45. In a gas turbine plant air is taken in compressor at 20°C and 1 bar having a pressure ratio of 5. The hot gases having a maximum temperature of 750°C expand in high pressure (H.P.) turbine which runs the compressors and then these gases are further expanded in low pressure (L.P.) turbine which runs the generator. The gases coming out from the L.P. turbine are exhausted to atmosphere. The following data are given for this unit: isentropic efficiency of compressor = 0.80, isentropic efficiency of each turbine stage = 0.85, γ = 1.4 for air, c_{pa} = 1.005 kJ/kgK, γ = 1.33 for gas, c_{pg} = 1.075 kJ/kgK. If losses are neglected then determine (i) power developed, (ii) work ratio, and (iii) thermal efficiency. [**Ans.** 96.73 kW, 0.3104, 17.43%]

46. A gas turbine plant works between the temperature limits of 850°C and 25°C with a pressure ratio of 6. The air is being drawn at 1 bar in the first compressor. The compression process is completed in two stages with equal pressure ratio with a perfect intercooler in between. The isentropic efficiencies of each compressor stage and of the turbine are 0.75 and 0.80 respectively. If the air circulation is 4 kg/s, determine the thermal efficiency of the unit. Take γ = 1.4 for air and gases and c_{pa} = 1 kJ/kgK. [**Ans.** 18.06%]

47. In a gas turbine plant air is taken in low pressure compressor at 27°C and 1.02 bar and after compression it is passed through intercooler where its temperature is reduced to 32°C. The cooled air is again compressed in second compressor and then passed in the combustion chamber where its maximum temperature reaches to 850°C. The hot gases expand in high pressure (H.P.) turbine which runs the compressors and these gases are further expanded in low pressure (L.P.) turbine which runs the generator. The gases coming out from the L.P. turbine are used for heating the incoming air from the second compressor and then exhausted to atmosphere. The following data are given for this unit: pressure ratio of each compressor = 3, isentropic efficiency of each compressor stage = 0.8, isentropic efficiency of each turbine stage = 0.85, effectiveness of heat exchanger = 0.6, air flow = 5 kg/s, $CV = 41800$ kJ/kg, $\gamma = 1.4$ for air and $c_{pa} = 1$ kJ/kgK, $\gamma = 1.33$ for gas and $c_{pg} = 1.15$ kJ/kgK. Determine (i) power developed, (ii) thermal efficiency, and (iii) specific fuel consumption. **[Ans.** 605.82 kW, 21.96%, 0.392 kg/kWh]

4

Water Turbines, Pumps, and Hydraulic Devices

4.1 INTRODUCTION

Hydraulic energy means the energy possessed by water in the form of potential energy, kinetic energy, and intermolecular energy. The power transmitted by a rotating shaft is usually known as mechanical energy. The hydraulic machines are energy conversion devices in which either hydraulic energy converts into mechanical energy or mechanical energy converts into hydraulic energy. The hydraulic machines which convert the hydraulic energy into mechanical energy are called turbines, whereas the hydraulic machines which convert the mechanical energy into hydraulic energy are called pumps.

In general, a hydraulic turbine consists of a wheel called runner (or rotor) having a number of evenly spaced vanes (blades or buckets) on its periphery. When the water falls through a certain height its potential energy is converted into kinetic energy which is further converted into mechanical energy by allowing the water to flow through the turbine runner. The mechanical energy so developed is utilized to run an electric generator which is coupled to the turbine shaft and thus electric energy is generated. The electric power which has been obtained from hydraulic energy is known as hydroelectric power.

4.2 GENERAL LAYOUT OF A HYDROELECTRIC POWER PLANT

Figure 4.1 shows a general layout of a hydroelectric power plant.

A hydroelectric power plant consists of the following components:

1. **Dam** A dam is built across a river for two reasons: (i) To impound the river water from storage, and (ii) To create the head of water.

   The height through which water is stored behind the dam above the turbine level is called head of water. The water surface of the stored water is called head race level or simply head race.

2. **Penstocks** Penstocks are the pipes of large diameter which carry water under pressure from the storage reservoir to the turbine. Penstocks are usually made of steel, wood, or reinforced concrete.

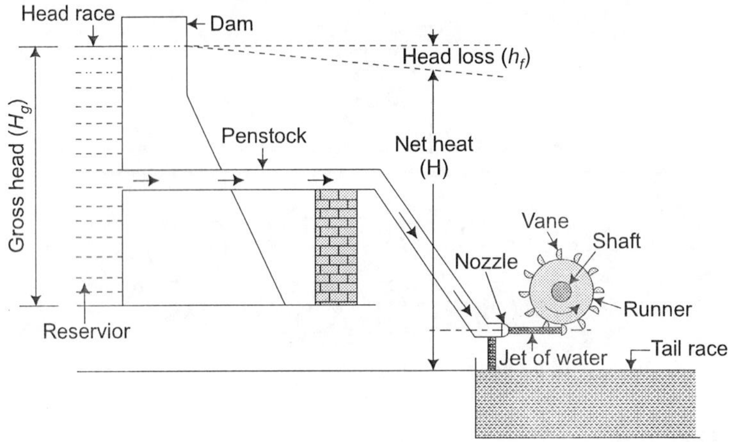

Fig. 4.1 General Layout of a Hydroelectric Power Plant

3. **Turbines** Turbines are the machines which converts the hydraulic energy into mechanical energy.
4. **Tail race** Tail race is a passage for discharging water after it has passed through the turbines into the river.

4.3 CLASSIFICATION OF HYDRAULIC TURBINES

The hydraulic turbines may be classified as follows:

1. According to the type of energy available at the inlet
   (i) **Impulse Turbine** If at the inlet of the turbine the water has only kinetic energy then the turbine is known as impulse turbine, for example, Pelton turbine, Banki turbine, and Jonval turbine.
   (ii) **Reaction Turbine** If at the inlet of the turbine the water has both kinetic energy and pressure energy the turbine is known as reaction turbine, for example, Francis turbine, Propeller turbine, and Kaplan turbines
2. According to the direction of flow through the runner
   (i) **Tangential Flow Turbine** If the water flows along the tangent to the path of rotation of the runner then the turbine is known as tangential flow turbine, for example, Pelton wheel.
   (ii) **Radial Flow Turbine** If the water flows in the radial direction through the runner the turbine is known as radial flow turbine. The radial flow turbine either is inward radial flow type or outward radial flow type.

   If the water enters at the outer circumference and flows inwards radially towards the centre of the runner the turbine is called inward radial flow turbine, for example, old Francis turbine, and Thomson turbine.

If the water enters at the centre and flows radially outwards towards the outer periphery of the runner the turbine is called outward radial flow turbine, for example, Fourneyron turbine.

(iii) **Axial Flow Turbine** If the water flows through the runner along the direction parallel to the axis of rotation of the runner the turbine is known as axial flow turbine, for example, propeller turbine, Kaplan turbine, and Jonval turbine.

(iv) **Mixed Flow Turbine** If the water enters the runner at the outer periphery in the radial direction and leaves it at the centre in the direction parallel to the axis of rotation of the runner is called mixed flow turbine, for example, modern Francis turbine.

3. According to the head available at the inlet of turbine

(i) **High Head Turbine** High head turbines are capable of working under very high heads usually more than 250 m, for example, Pelton turbine. These turbines require relatively less quantity of water.

(ii) **Medium Head Turbine** The turbines which are capable of working under medium heads ranging from 60 m to 250 m are called medium head turbines, for example, modern Francis turbine. These turbines require relatively large quantity of water.

(iii) **Low Head Turbine** The turbines which are capable of working under the heads less than 60 m are known as low head turbines, for example, Kaplan and propeller turbines. These turbines require a large quantity of water.

4. According to the specific speed of the turbine

(i) **Low Specific Speed Turbine** Specific speed of turbines varies from 8.5 to 50. If the specific speed of turbines varies from 8.5 to 30 then it will be Pelton wheel with single jet and if it varies from 30 to 50 then it will be Pelton wheel with double jet.

(ii) **Medium Specific Speed Turbine** Specific speed varies from 50 to 250, for example, Francis turbine.

(iii) **High Specific Speed Turbine** Specific speed varies from 255 to 860, for example, Kaplan and propeller turbines.

5. According to the name of the originator

(i) **Pelton Turbine** It is named after Lester A. Pelton, an American engineer. It is the only impulse type of turbine now in common use which is used for high head and low discharge.

(ii) **Francis Turbine** It is named in the honour of James B. Francis, an American engineer, who developed an inward radial flow turbine but later on it was modified. The modern Francis turbine is a mixed flow reaction turbine which is used for medium head and medium discharge.

(iii) **Kaplan Turbine** It was named after Austrian engineer V. Kaplan. It is an axial flow reaction turbine and is used for low heads. It requires large quantity of water to produce large amount of power.

6. According to the disposition of the turbine shaft

The turbines may be disposed with either vertical or horizontal shafts. Commonly, vertical disposition of shafts are used for turbines. Pelton wheel is an example of horizontal shaft, whereas Kaplan turbine is of vertical shaft.

4.4 COMPARISON BETWEEN IMPULSE AND REACTION TURBINES

Impulse Turbine	*Reaction Turbine*
1. All the available hydraulic energy is converted into kinetic energy by a nozzle.	1. Only a part of the available hydraulic energy is converted into kinetic energy before the water enters the runner of the turbine.
2. Impulse turbines operate under atmospheric conditions thus the pressure remains constant throughout the action of water on the runner.	2. Reaction turbines operate under varying pressures different from the atmospheric pressure and the pressure of water drops partly in the rotor blades and partly in the stator blades.
3. It is not essential to have a casing as it does not perform any hydraulic function.	3. Air-and water-tight casing is quintessential which maintains the pressure in the turbine passage.
4. Water may be allowed to enter a part or whole of the wheel circumference.	4. Water is admitted over the whole circumference of the wheel.
5. Air has free access to the runner as it does not run full.	5. Air has no access to the runner as all the passages are completely filled by the water.
6. It is always installed above the tail race and it does not require any draft tube.	6. It is connected to the tail race through a draft tube and it may be installed above or below the tail race.
7. Flow regulation is attained by means of a needle valve (spear) fitted into the nozzle which is possible without any loss.	7. Flow regulation is attained by means of guide vane assembly which is always accompanied by loss.
8. Some of the impulse turbines are Pelton wheel, Banki turbine, Jonval turbine, Girard turbine, and Turgo- impulse wheel.	8. Some of the reaction turbines are propeller turbine, Kaplan turbine, Francis turbine, Fourneyron turbine, and Thomson turbine.
9. These turbines are suitable for high head, low discharge, and low specific speed conditions.	9. These turbines are suitable for low to medium head and specific speed, and high discharge.
10. Relative velocity of water either remains constant or reduces slightly due to the presence of blade friction.	10. The relative velocity of water increases due to the continuous drop in pressure during flow through the blades.

4.5 HEADS AND EFFICIENCIES OF A HYDRAULIC TURBINE

1. **Gross Head** Gross head is the difference between the head race level and the tail race level when no water is flowing (Figure 4.1). It is denoted by H_g.
2. **Net Head** It is the head available at the inlet of the turbine. It is also known as effective head. When water flows from the head race to the turbine inlet certain head loss occurs mainly due to friction between water and the penstocks. If h_f is the head loss due to friction then the net head is given by,

$$H = H_g - h_f \tag{4.1}$$

where
$$h_f = \frac{4 \times f \times L \times V^2}{D \times 2g} \tag{4.2}$$

where, f = Coefficient of friction or friction factor, L = Length of penstock, V = Velocity of flow in penstock, D = Diameter of penstock, and g = Gravitational acceleration.

3. **Efficiencies of a Hydraulic Turbine** The important efficiencies of a hydraulic turbine are given below:

   (i) **Hydraulic efficiency (η_h)** It is defined as the ratio of power developed by the turbine runner to the power supplied by the water jet at the inlet of the turbine. The hydraulic losses are taken into account by the hydraulic efficiency of a turbine and it is given as,

$$\eta_h = \frac{\text{Power developed by the runner}}{\text{Power supplied at the turbine inlet}}$$

$$= \frac{\text{Runner Power}}{\text{Water Power}} = \frac{RP}{WP} \tag{4.3}$$

   (ii) **Mechanical efficiency (η_m)** It is defined as the ratio of the power available at the turbine shaft to the power developed by the runner. The mechanical losses are taken into account by the mechanical efficiency. Mathematically, it is given as,

$$\eta_m = \frac{\text{Power available at the turbine shaft}}{\text{Power developed by the runner}}$$

$$= \frac{\text{Shaft Power}}{\text{Runner Power}} = \frac{SP}{RP} \tag{4.4}$$

   (iii) **Volumetric efficiency (η_v)** It is defined as the ratio of the volume of the water actually striking the runner to the volume of water supplied to the turbine by the jet. The leakage losses are taken into account by volumetric efficiency. Mathematically, it is given by,

$$\eta_v = \frac{\text{Volume of water actually striking the runner}}{\text{Volume of water supplied to the jet}} \tag{4.5}$$

   (iv) **Overall efficiency (η_o)** It is defined as the ratio of power available at the turbine shaft to the power available from the water jet at the turbine inlet. The judgment of the performance of a hydraulic turbine is made by its overall efficiency. Mathematically, it is given as,

$$\eta_o = \frac{\text{Power available at the turbine shaft}}{\text{Power available at the turbine inlet from the water jet}}$$

$$= \frac{SP}{WP} \tag{4.6}$$

$$\eta_o = \frac{SP}{WP} \times \frac{RP}{RP} \qquad \text{(Multiplying and dividing by runner power, } RP\text{)}$$

$$\therefore \qquad \eta_o = \frac{SP}{RP} \times \frac{RP}{WP}$$

$$= \eta_m \times \eta_h \qquad \left(\because \eta_m = \frac{SP}{RP} \;\&\; \eta_h = \frac{RP}{WP}\right) \tag{4.7}$$

When η_v is also considered,

$$\eta_o = \eta_m \times \eta_h \times \eta_v \tag{4.8}$$

4.6 PELTON TURBINE (OR PELTON WHEEL)

The Pelton turbine (or Pelton wheel) is a tangential flow impulse turbine. It is the only impulse type of turbine now in common use and is named after L.A. Pelton, an American engineer. This turbine is used for high heads and is generally used for heads in excess of 250 m. So far, Pelton turbine has been used under a head of 1770 m.

The main parts of a Pelton turbine are given below:

1. **Penstock** Penstock is a large sized pipe through which water flows from the reservoir to the turbine. On the basis of water head available, a penstock may consist of wood, concrete, or steel. It is equipped with screens called trash racks at the inlet which prevent the debris from entering into it. For regulating the flow from the reservoir to the turbine penstocks are fitted with control valves.

2. **Spear and Nozzle Arrangement** The nozzle fitted at the end of the penstock is provided with a spear which controls the quantity of water striking the buckets of runner as shown in Figure 4.2. The spear is a conical needle having streamlined head. The axial movement of spear in the case of smaller units is controlled by a wheel while in the case of larger units it is controlled automatically by a governor. When the spear moves in the forward direction, the flow area decreases thus the amount of water striking the runner is reduced. On the other hand, if the spear moves in backward direction, the flow area increases and the amount of water striking the runner increases.

3. **Runner with Buckets** The runner consists of a large circular disc on which a number of buckets (always 15) are evenly spaced around its periphery as shown in Figure 4.2. The buckets have a shape of double hemispherical cup or bowl as shown in Figure 4.3. Each bucket is divided into two symmetrical parts by a sharp edged ridge called a splitter. The jet of water strikes on the splitter, which divides the jet into two equal parts, each of which after flowing around the inner smooth bucket surface leaves at its outer edge as shown in Figure 4.3.

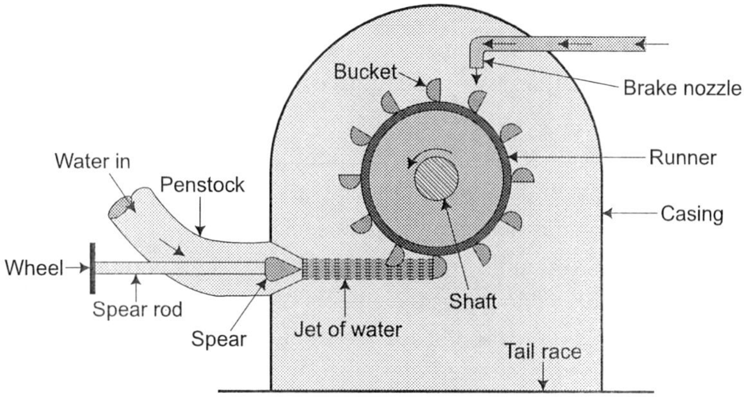

Fig. 4.2 Pelton Turbine (or Pelton Wheel)

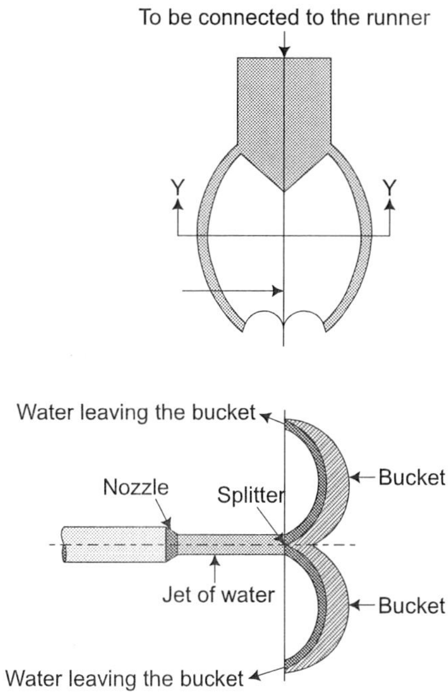

Fig. 4.3 Pelton Turbine (Wheel) Bucket

The advantage of double shaped buckets is that the axial thrust produced by the water in each half neutralize each other. Thus, the bearing supporting the wheel shaft are not subjected to any axial or end thrust. The buckets are so shaped that the jet of water gets deflected through 160° to 170°. For obtaining maximum change in momentum of the fluid the deflection of water should be 180° but in practice the deflection is limited to about 165° so that water leaving the bucket does not hit the back of the following bucket. Depending upon the head at the inlet of turbine buckets are made of different materials. Generally, for low heads the buckets are made of cast steel, bronze, or stainless steel. When the Pelton turbine shaft is horizontal then not more than two jets are used, but for vertical shaft turbine it is limited to six jets.

4. **Casing** The casing does not perform any hydraulic function. It is provided only to prevent splashing of water and to guide it to the tail race. It also acts as a safeguard against accidents. It is made of cast iron or fabricated steel plates.
5. **Braking Jet** In order to shut down the turbine the nozzle is completely closed by moving the spear in the forward direction, thus the amount of water striking the runner becomes zero. But the runner still goes on revolving for a long time due to inertia. To stop the runner in a short time, a small braking nozzle is provided which directs the jet of water on the back of the buckets. This jet of water is called braking jet which is shown in Figure 4.2.
6. **Governing Mechanism** As turbines are directly coupled to the electric generators hence, the speed of the turbine runner has to be maintained constant so that the generator always runs at constant speed under all working conditions. It is achieved by regulating the quantity of water flowing through the runner according to the changing load conditions on the turbine. Such an operation by which the speed of the turbine is kept constant under all working conditions is known as governing of a turbine and is usually done automatically by a governor as shown in Figure 4.4.

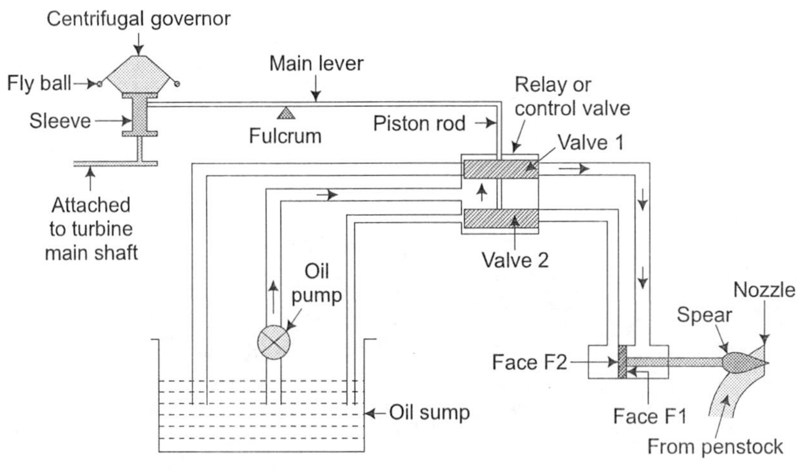

Fig. 4.4 Governing Mechanism of a Pelton Turbine

The speed of the turbine runner decreases when the load on turbine increases and the fly balls of the governor move downwards. Due to which sleeve moves downward and the main lever turns about the fulcrum, pulls the piston rod of relay valve in the upward direction. This closes the valve 2 and opens the valve 1. The oil pump supplies oil under pressure to the relay cylinder through valve 1. The oil exerts a force on the face F1 of the relay cylinder piston which results in movement of the spear towards left. This increases the area of flow of water at the nozzle outlet which allows a larger quantity of water to strike on the runner. Consequently, the turbine speed increases till the turbine is restored to its normal position.

Conversely, the speed of turbine runner increases when the load decreases and the fly balls of the governor move upward. Due to which sleeve moves upward and the main lever turns about the fulcrum, pushes the piston rod of relay valve in downward direction. This closes the valve 1 and opens the valve 2. The oil exerts force on the face F2 of the relay cylinder piston which results in movement of the piston towards right. This decreases the area of flow of water at the nozzle outlet which allows smaller quantity of water to strike on the runner. Consequently, the turbine speed decreases till the speed of the turbine becomes normal.

Modern Pelton turbines are provided with double regulation that is the combined spear and deflector control. Deflectors are used along with the spear to prevent the water hammer problem due to sudden reduction in the flow rate. Deflectors are simply plates which deflect the jet of water so that the entire flow does not reach the bucket. Then the spear is moved to its new position which reduces the rate of flow in pipeline gradually.

4.6.1 Velocity Triangles, Work Done, and Efficiency of the Pelton Turbine

The jet of water from the nozzle strikes the bucket at the splitter, which splits up the jet into two parts. The parts of the jet glide over the inner surfaces and come out at the outer edge. The splitter is the inlet tip and the outer edge of the bucket is the outer tip of the bucket. Figure 4.5 shows the velocity triangles at the tips of the bucket of a Pelton turbine.

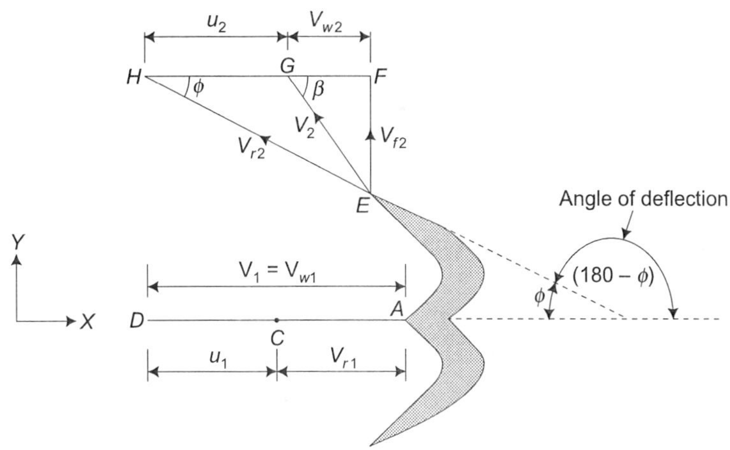

Fig. 4.5 Velocity Triangles

Let V_1 and V_2 = Absolute velocity of the jet at the inlet and outlet respectively,

u_1 and u_2 = Velocity of bucket at the inlet and outlet respectively,

V_{r1} and V_{r2} = Relative velocity of the jet and bucket at the inlet and outlet respectively,

V_{w1} and V_{w2} = Velocity of whirl at the inlet and outlet respectively,

V_{f1} and V_{f2} = Velocity of flow at the inlet and outlet respectively,

α = Guide blade angle or angle between the direction of jet and direction of motion of the bucket,

θ = Vane angle at inlet or angle made by relative velocity V_{r1} with the direction of motion at inlet,

β = Angle between absolute velocity V_2 and the direction of motion of the bucket,

ϕ = Vane angle at the outlet or angle between relative velocity V_{r2} and the direction of motion of the bucket at the outlet. It is also called clearance angle.

Since the velocities V_1 and u_1 are collinear, the velocity triangle at the inlet tip of the bucket is a straight line where

$$V_{r1} = V_1 - u_1 = V_1 - u \quad (\because \text{ vane is smooth, } \therefore\ u_1 = u_2 = u)$$

$$V_{w1} = V_1, \quad \alpha = 0, \quad \theta = 0$$

From the velocity triangle at outlet ($\beta < 90°$), we have,

$$V_{r1} = V_{r2}$$

(Bucket surfaces are perfectly smooth and energy losses due to impact of jet at the splitter are neglected).

$$V_{w2} = V_{r2} \cos \phi - u_2$$

The force exerted by the jet of water in the direction of motion of the bucket is given by,

$$F_x = \frac{\text{Mass}}{\text{Time}} \times [\text{Initial velocity in the jet direction} - \text{Final velocity in the jet direction}]$$

$$\text{Rate of mass flow} = \text{density} \times \text{area} \times \text{velocity of jet} = \rho \times A \times V_1$$

Area of jet, $A = (\pi/4)D^2$

$$F_x = \rho A V_1 \times [(V_{w1} - u_1) - \{-(V_{w2} + u_2)\}]$$

$$= \rho A V_1 \times [V_{w1} - u_1 + V_{w2} + u_2]$$

$$\therefore \quad F_x = \rho A V_1 \times [V_{w1} + V_{w2}] \quad (\because u = u_1 = u_2 \text{ and } \beta < 90°) \tag{4.9}$$

Depending upon the magnitude of the peripheral speed u the runner may be classified as: (i) Slow runner ($\beta < 90°$ and V_{w2} is negative), (ii) Medium runner ($\beta = 90°$ and $V_{w2} = 0$), and (iii) Fast runner ($\beta > 90°$ and V_{w2} is positive).

Now the work done by the jet on the runner per second,

$$W = F_x \times u$$

$$= \rho AV_1 \times [V_{w1} + V_{w2}] \times u \text{ N/ms} \quad (4.10)$$

Power given to the runner by the jet,

$$P = \frac{W}{1000}$$

$$= \frac{\rho AV_1 \times [V_{w1} + V_{w2}] \times u}{1000} \text{ kW} \quad (4.11)$$

Work done per second per unit weight of water striking per sec,

$$W = \frac{\rho AV_1 \times [V_{w1} + V_{w2}] \times u}{\text{Weight of water striking per sec}}$$

$$= \frac{\rho AV_1 \times [V_{w1} + V_{w2}] \times u}{\rho AV_1 \times g}$$

$$\therefore \quad W = \frac{1}{g} \times [V_{w1} + V_{w2}] \times u \quad (4.12)$$

Hydraulic efficiency,

$$\eta_h = \frac{\text{Work done/sec by jet on the runner}}{\text{Initial kinetic energy}}$$

$$= \frac{W}{(1/2)\, mV_1^2}$$

$$\therefore \quad \eta_h = \frac{\rho AV_1 \times [V_{w1} + V_{w2}] \times u}{(1/2) \times (\rho AV_1) \times V_1^2} \quad (4.13)$$

Now $\quad V_{w1} = V_1,$

$$V_{r1} = V_1 - u,$$

and $\quad V_{r2} = V_{r1}$

$\therefore \quad V_{r2} = V_1 - u$

$$V_{w2} = V_{r2} \cos \phi - u_2 = V_{r2} \cos \phi - u$$

$$= (V_1 - u) \cos \phi - u$$

Substituting the values of V_{w1} and V_{w2} in equation (4.13), we have,

$$\eta_h = \frac{2 \times [V_1 + \{(V_1 - u) \cos \phi - u\}] \times u}{V_1^2}$$

$$= \frac{2\, [(V_1 - u) + (V_1 - u) \cos \phi] \times u}{V_1^2}$$

$$\therefore \quad \eta_h = \frac{2(V_1 - u)\,[1 + \cos\phi] \times u}{V_1^2} \tag{4.14}$$

When the effect of friction is considered (i.e., blade is not smooth) then friction factor $K = V_{r2}/V_{r1}$ will be less than unity and the expression for efficiency of turbine becomes,

$$\eta_h = \frac{2(V_1 - u)\,[1 + K\cos\phi] \times u}{V_1^2} \tag{4.14a}$$

For maximum efficiency,

$$\frac{d(\eta_h)}{du} = 0$$

$$\frac{d}{du}\left[\frac{2(V_1 - u)\,[1 + \cos\phi] \times u}{V_1^2}\right] = 0$$

$$\frac{1 + \cos\phi}{V_1^2} \cdot \frac{d}{du}(2uV_1 - 2u_1^2) = 0$$

$$1 + \cos\phi/V_1^2 \,.\, (2V_1 - 4u) = 0$$

But $\quad 1 + \cos\phi/V_1^2 \neq 0$

$$\therefore \quad 2V_1 - 4u = 0$$

$$4u = 2V_1$$

$$\therefore \quad u = \frac{V_1}{2}$$

Thus, the hydraulic efficiency is maximum when the bucket speed is half of the velocity of jet. The variation of hydraulic efficiency with speed ratio is illustrated in Figure 4.6. Theoretically, the maximum efficiency occurs at $u = V_1/2$, i.e., $\varphi = 0.5$, whereas in actual practice the maximum value of efficiency occurs when speed ratio is slightly less than 0.5 and is about 0.46.

Substituting the value of $u = (V_1/2)$ in equation (4.14), we have,

$$(\eta_h)_{max} = \frac{2\left(V_1 - \frac{V_1}{2}\right)[1 + \cos\phi] \times \frac{V_1}{2}}{V_1^2}$$

$$= \frac{2 \times \frac{V_1}{2} \times \frac{V_1}{2} \times [1 + \cos\phi]}{V_1^2}$$

$$\therefore \quad (\eta_h)_{max} = \frac{[1 + \cos\phi]}{2} \tag{4.15}$$

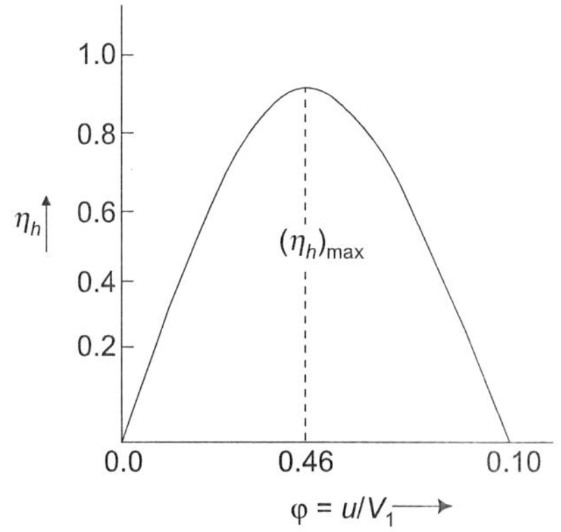

Fig. 4.6 Hydraulic Efficiency vs. Speed Ratio

When the friction factor is considered the expression for maximum efficiency becomes,

$$\therefore \qquad (\eta_h)_{max} = \frac{[1 + K \cos \phi]}{2} \tag{4.15a}$$

4.6.2 Design Aspects of Pelton Turbine

The following points should be considered while designing a Pelton turbine:

1. **Number of Jets** It is obtained by dividing the total rate of flow through the turbine by the rate of flow through a single jet. Ordinarily, a Pelton turbine has one jet. However, a number of jets may be employed for providing more power with the same turbine. Theoretically, six jets can be used with one Pelton turbine.
2. **Jet Ratio** It is defined as the ratio of pitch diameter D of the Pelton wheel to the jet diameter d. It is denoted by m and is given as,

$$m = \frac{D}{d} \tag{4.16}$$

   For maximum efficiency the jet ratio varies from 11 to 14 but in practice for most of the cases it is taken as 12.
3. **Number of Buckets** The number of buckets for a Pelton turbine should be such that no water escapes without striking the buckets. The number of buckets is usually more than 15. For determining the number of buckets Taygun empirical formula is widely used which is given by,

$$Z = 15 + \frac{D}{2d}$$
$$= 15 + 0.5\, m \tag{4.17}$$

   where, Z = Number of buckets, D = Pitch diameter of the Pelton wheel, d = Jet diameter, and m = Jet ratio = D/d.

4. **Depth and Width of Buckets** Depth and width of a bucket are expressed in terms of jet diameter as:

   Depth of bucket varies from 0.8*d* to 1.2*d* but in general it is taken as 1.2*d*.

   Width of bucket varies from 4*d* to 5*d* but in general it is taken as 5*d*.

5. **Speed Ratio** It is defined as the ratio of peripheral (linear) velocity of buckets to the theoretical (spouting) velocity of the jet. The value of speed ratio varies from 0.43 to 0.47. It is denoted by K_u or φ. Mathematically, speed ratio is given by,

$$\varphi = \frac{\text{Peripheral velocity}}{\text{Theoretical velocity}}$$

$$= \frac{u}{\sqrt{2gH}} \quad (4.18)$$

   where, H = Net head, and $u = \dfrac{\pi DN}{60}$

6. **Velocity of Jet** Due to friction loss the velocity of jet at the inlet will be slightly less than the theoretical velocity in the nozzle.

   Velocity of jet at inlet is given as,

$$V_1 = C_v \sqrt{2gH} \quad (4.19)$$

   where, C_v = Coefficient of velocity which varies between 0.97 and 0.99, and H = net head on the turbine.

Example 4.1 A Pelton wheel is having a mean bucket diameter of 1 m and is running at 1000 rpm. The net head on the Pelton wheel is 700 m. If the side clearance angle is 15° and discharge is 0.1 m^3/s, find (i) power available at the nozzle, and (ii) hydraulic efficiency of the turbine. Take coefficient of velocity = 1.

Solution

Tangential velocity of wheel,

$$u = \frac{\pi DN}{60} = \frac{\pi \times 1 \times 1000}{60}$$

$$= 52.36 \text{ m/s}$$

Velocity of jet at inlet,

$$V_1 = C_v \sqrt{2gH} = 1 \times \sqrt{2 \times 9.81 \times 700}$$

$$= 117.19 \text{ m/s}$$

(i) Power available at the nozzle,

$$WP = \frac{\rho \times g \times Q \times H}{1000}$$

$$= \frac{1000 \times 9.81 \times 0.1 \times 700}{1000}$$

$$= 686.7 \text{ kW}$$

(ii) Hydraulic efficiency,

$$\eta_h = \frac{2\,(V_1 - u)\,[1 + \cos\phi] \times u}{V_1^2}$$

$$\therefore \qquad \eta_h = \frac{2 \times (117.19 - 52.36) \times (1 + \cos 15^\circ) \times 52.36}{(1170.19)^2}$$

$$= 0.9718 \text{ or } 97.18\%$$

Example 4.2 The mean bucket speed of a Pelton turbine is 14 m/s. The rate of flow of water supplied by the jet under a head of 45 m is 800 litres per second. If the jet is deflected by the buckets at an angle of 160°, find the power and efficiency of the turbine. Assume coefficient of velocity as 0.985.

Solution Refer Figure 4.7.

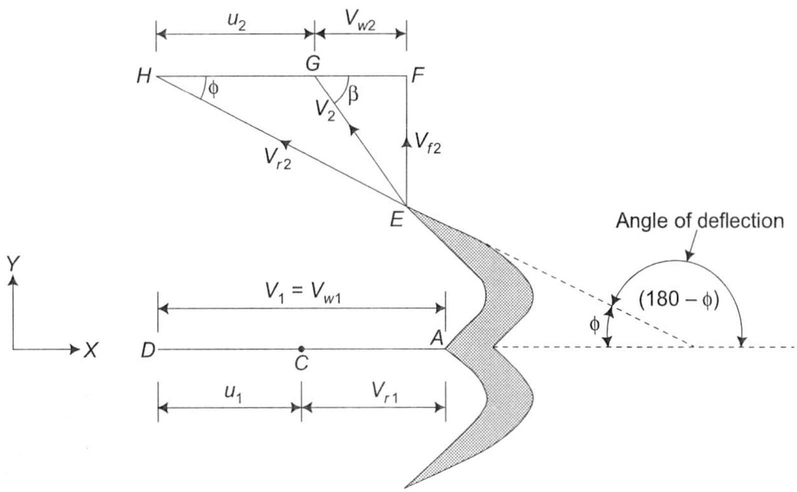

Fig. 4.7

Discharge, $\quad Q = 800 \text{ litres/s} = 800 \times 0.001$

$$= 0.8 \text{ m}^3\text{/s } (1 \text{ litre} = 0.001 \text{ m}^3)$$

Angle of deflection,

$$(180 - \phi) = 160^\circ$$

$$\therefore \qquad \phi = 180 - 160 = 20^\circ$$

The velocity of jet,

$$V_1 = C_v \sqrt{2gH} = 0.985 \times \sqrt{2 \times 9.81 \times 45}$$

$$= 29.27 \text{ m/s}$$

From inlet velocity triangle,

$$V_{r1} = V_1 - u = 29.27 - 14 = 15.27 \text{ m/s}$$

$$V_{w1} = V_1 = 29.27 \text{ m/s}$$

From outlet velocity triangle,

$$V_{r2} = V_{r1} = 15.27 \text{ m/s, and}$$

$$V_{w2} = V_{r2} \cos \phi - u_2$$

$$= 15.27 \times \cos 20° - 14 = 0.35 \text{ m/s}$$

Power given to the turbine,

$$P = \frac{\rho\, AV_1 \times [V_{w1} + V_{w2}] \times u}{1000}$$

$$= \frac{\rho \times Q_1 \times [V_{w1} + V_{w2}] \times u}{1000} \text{ kW} \quad (\because AV_1 = Q)$$

$$\therefore \quad P = \frac{1000 \times 0.8 \times [29.27 + 0.35] \times 14}{1000}$$

$$= 331.74 \text{ kW}$$

Hydraulic efficiency of the turbine,

$$\eta_h = \frac{2 \times [V_{w1} + V_{w2}] \times u}{V_1^2}$$

$$\therefore \quad \eta_h = \frac{2 \times [29.27 + 0.35] \times 14}{(29.27)^2}$$

$$= 0.968 \text{ or } 96.8\%$$

Example 4.3 The penstock supplies water from a reservoir to the Pelton wheel with a gross head of 500 m. One-third of the gross head is lost in the penstock. The rate of flow through the nozzle fitted at the end of the penstock is 2 m^3/s. The angle of deflection of the jet is 165°. Determine the power given by the water to the runner and the hydraulic efficiency of the Pelton wheel. Take speed ratio as 0.45 and coefficient of velocity as 1.

Solution Refer Figure 4.7.

Head lost in the friction,

$$h_f = \frac{H_g}{3} = \frac{500}{3}$$

$$= 166.7 \text{ m}$$

Net head, $H = H_g - h_f$

$$= 500 - 166.7 = 333.3 \text{ m}$$

Angle of deflection,

$$(180 - \phi) = 165°$$

$$\therefore \quad \phi = 180 - 165 = 15°$$

The velocity of jet,

$$V_1 = C_v \sqrt{2gH} = 1 \times \sqrt{2 \times 9.81 \times 333.3}$$

$$= 80.86 \text{ m/s}$$

$$u = \varphi \sqrt{2gH} = 0.45 \times \sqrt{2 \times 9.81 \times 333.3}$$

$$= 36.39 \text{ m/s}$$

From inlet velocity triangle,

$$V_{r1} = V_1 - u = 80.86 - 36.39$$

$$= 44.47 \text{ m/s}$$

$$V_{w1} = V_1 = 80.86 \text{ m/s}$$

From outlet velocity triangle,

$$V_{r2} = V_{r1} = 44.47 \text{ m/s, and}$$

$$V_{w2} = V_{r2} \cos \phi - u_2$$

$$= 44.47 \times \cos 15° - 36.39 = 6.57 \text{ m/s}$$

Power given to the turbine,

$$P = \frac{\rho\, AV_1 \times [V_{w1} + V_{w2}] \times u}{1000}$$

$$= \frac{\rho \times Q_1 \times [V_{w1} + V_{w2}] \times u}{1000} \text{ kW} \quad (\because AV_1 = Q)$$

$$\therefore \quad P = \frac{1000 \times 2 \times [80.86 + 6.57] \times 36.39}{1000}$$

$$= 6362.63 \text{ kW}$$

Hydraulic efficiency of the turbine,

$$\eta_h = \frac{2 \times [V_{w1} + V_{w2}] \times u}{V_1^2}$$

$$\therefore \quad \eta_h = \frac{2 \times [80.86 + 6.57] \times 36.39}{(80.86)^2}$$

$$= 0.9732 \text{ or } 97.32\%$$

Example 4.4 A Pelton wheel is to be designed for the following specifications: shaft power = 11,772 kW, head = 380 m, speed = 750 rpm, overall efficiency = 86%. Jet diameter is not to exceed one-sixth of the wheel. Determine: (i) the wheel diameter, (ii) number of jets required, and (iii) diameter of the jet. Take $C_v = 0.985$ and $\varphi = 0.45$.

Solution The velocity of jet,

$$V_1 = C_v \sqrt{2gH} = 0.3985 \times \sqrt{2 \times 9.81 \times 380}$$

$$= 85.05 \text{ m/s}$$

The velocity of wheel,

$$u = u_1 = u_2$$

$$u = \varphi \sqrt{2gH} = 0.45 \times \sqrt{2 \times 9.81 \times 380}$$

$$= 38.85 \text{ m/s}$$

$$u = \frac{\pi DN}{60}$$

$$38.85 = \frac{\pi \times D \times 750}{60}$$

$$\therefore \quad D = \frac{38.85 \times 60}{\pi \times 750} = 0.989 \text{ m}$$

$$\therefore \quad d = \frac{D}{6} = \frac{0.989}{6}$$

$$= 0.165 \text{ m}$$

Discharge of one jet,

$$Q_1 = \left(\frac{\pi}{4} d^2\right) \times V_1 = \frac{\pi}{4} \times (0.165)^2 \times 85.05$$

$$= 1.818 \text{ m}^3/\text{s}$$

Overall efficiency,

$$\eta_o = \frac{SP}{WP} = \frac{SP}{\dfrac{\rho \times g \times Q \times H}{1000}}$$

where Q = Total discharge

$$0.86 = \frac{11772 \times 1000}{1000 \times 9.81 \times Q \times 380}$$

$$\therefore \quad Q = \frac{11772 \times 1000}{1000 \times 9.81 \times 0.86 \times 380}$$

$$= 3.672 \text{ m}^3/\text{s}$$

$$\therefore \text{ Number of jets} = \frac{Q}{Q_1} = \frac{3.672}{1.818}$$

$$= 2.02 = 2 \text{ jets}$$

Example 4.5 Two jets strike the bucket of a Pelton wheel which is having shaft power as 15450 kW. The diameter of each jet is given as 200 mm. If the net head on the turbine is 400 m, find the overall efficiency of the turbine. Assume coefficient of velocity as 1.

Solution The velocity of jet,

$$V_1 = C_v \sqrt{2gH} = 1 \times \sqrt{2 \times 9.81 \times 400}$$

$$= 88.58 \text{ m/s}$$

Discharge of each jet,

$$Q_1 = \left(\frac{\pi}{4} d^2\right) \times V_1 = \frac{\pi}{4} \times (0.2)^2 \times 88.58$$

$$= 2.78 \text{ m}^3/\text{s}$$

Therefore, total discharge,

$$Q = 2 \times 2.78 = 5.56 \text{ m}^3/\text{s}$$

Power at the inlet of turbine,

$$WP = \frac{\rho \times g \times Q \times H}{1000}$$

$$= \frac{1000 \times 9.81 \times 5.56 \times 400}{1000}$$

$$= 21817.44 \text{ kW}$$

Overall efficiency,

$$\eta_o = \frac{SP}{WP} = \frac{15450}{21817.44}$$

$$= 0.708 \text{ or } 70.8\%$$

Example 4.6 A double jet Pelton wheel operates under 50 m head and develops 750 kW brake power when running at 475 rpm. Evaluate the flow rate and the diameter of the nozzle jet if the overall efficiency and coefficient of velocity are 86% and 0.98 respectively.

Solution

Overall efficiency, $\eta_o = \dfrac{SP}{WP} = \dfrac{SP}{\dfrac{\rho \times g \times Q \times H}{1000}}$

$$0.86 = \frac{750 \times 1000}{1000 \times 9.81 \times Q \times 50}$$

$$\therefore \quad Q = \frac{750 \times 1000}{1000 \times 9.81 \times 0.86 \times 50}$$

$$= 1.78 \text{ m}^3/\text{s}$$

The velocity of jet, $V_1 = C_v \sqrt{2gH} = 0.98 \times \sqrt{2 \times 9.81 \times 100}$

$$= 30.69 \text{ m/s}$$

Therefore, total discharge,

$$Q = \text{Number of jets} \times \text{Area of jet} \times \text{Velocity of jet}$$

$$0.178 = 2 \times \pi/4 \times d^2 \times 30.69$$

$$d = \sqrt{\frac{1.78 \times 4}{2 \times \pi \times 30.69}}$$

$\therefore$ Diameter of nozzle, $d = 0.19$ m or 19 cm

Example 4.7 The following data are obtained from a test on a Pelton wheel: head at the base of the nozzle = 40 m, discharge of the nozzle = 0.2 m^3/s, area of the jet = 7550 mm^2, power available at the shaft = 60 kW, mechanical efficiency = 94%. Calculate the power lost (i) in the nozzle, (ii) in the runner, and (iii) in the mechanical friction.

Solution Power at the base of nozzle,

$$WP = \frac{\rho \times g \times Q \times H}{1000}$$

$$= \frac{1000 \times 9.81 \times 0.2 \times 40}{1000}$$

$$= 78.48 \text{ kW}$$

Velocity of the nozzle,

$$V_1 = \frac{\text{Discharge of jet}}{\text{Area of jet}} = \frac{0.2}{7550 \times 10^{-6}}$$

$$= 26.49 \text{ m/s}$$

Power at the nozzle exit = Kinetic energy of the jet

$$= \frac{1}{2} mV_1^2 = \frac{1}{2} (\rho\, AV_1)V_1^2$$

Therefore, power at the exit of nozzle,

$$= \frac{(1/2)\,\rho AV_1^3}{1000}$$

$$= \frac{1000 \times 0.00755 \times (26.49)^3}{1000}$$

$$= 70.17 \text{ kW}$$

$\therefore$ Power lost in the nozzle = 78.48 – 70.17

= 8.31 kW

Power supplied to the runner = Kinetic energy of the jet = 70.17 kW

Mechanical efficiency,

$$\eta_m = \frac{SP}{RP}$$

$$\therefore \quad RP = \frac{SP}{\eta_m} = \frac{60}{0.94}$$

$$= 63.83 \text{ kW}$$

$\therefore$ Power lost in the runner = 70.17 – 63.83 = 6.34 kW

$\therefore$ Power lost in mechanical friction = 63.83 – 60 = 3.83 KW

Example 4.8 A Pelton wheel having semicircular buckets functions under a head of 150 m and consumes 50 liters of water per second. If 60 cm diameter wheel turns 600 revolutions per minute, calculate (i) power available at the nozzle, and (ii) hydraulic efficiency of the wheel. Take coefficient of velocity as 0.99.

Solution Due to semicircular buckets, angle at outlet,

$$\phi = 0°$$

Discharge, $Q = 50 \text{ litres}/s = 0.05 \text{ m}^3/\text{s}$

Velocity of wheel, $u = \dfrac{\pi DN}{60} = \dfrac{\pi \times 0.6 \times 600}{60}$

$$= 18.85 \text{ m/s}$$

(i) Power available at the nozzle

$$= WP = \frac{\rho \times g \times Q \times H}{1000}$$

$$\therefore \qquad WP = \frac{1000 \times 9.81 \times 0.05 \times 150}{1000}$$

$$= 73.57 \text{ kW}$$

(ii) The velocity of jet,

$$V_1 = C_v \sqrt{2gH} = 0.99 \times \sqrt{2 \times 9.81 \times 150}$$

$$= 53.71 \text{ m/s}$$

Hydraulic efficiency of the wheel,

$$\eta_h = \frac{2\,(V_1 - u)\,[1 + \cos\phi] \times u}{V_1^{\,2}}$$

$$\therefore \qquad \eta_h = \frac{2 \times (53.71 - 18.85) \times (1 + \cos 0°) \times 18.85}{(53.71)^2}$$

$$= 0.9111 \text{ or } 91.11\%$$

4.7 FRANCIS TURBINE

The Francis turbine is named in the honour of J.B. Francis an American engineer, who was the first to develop an inward radial flow type of reaction turbine as shown in Figure 4.8. Later on, it was modified in which the water enters the runner radially at its outer periphery and leaves axially at its centre. Thus, the modern Francis turbine is a mixed flow type turbine. It is a reaction turbine and hence only a part of the available head is converted into the velocity head before the water enters the runner. The pressure head goes on decreasing as the water flows over the runner blades. The static pressure at the runner exit may be less than the atmospheric pressure and as such, water completely fills all the passages of the runner. Francis turbines may have either a horizontal shaft or a vertical shaft arrangement. The vertical shaft arrangement is most common particularly on large sizes which help in economy in building space. A Francis turbine is most suitable for medium heads (60 m to 250 m) and requires a relatively large quantity of water. Francis turbine is a medium specific speed turbine which varies from 50 to 255.

Main parts of a Francis turbine are given below.

1. Penstock Penstock is a large sized conduit which conveys water from the reservoir to the runner. It is provided with the control valves and trash racks to control the quantity of water and to prevent the entry of debris and other foreign matter. In Francis turbine, large sized penstocks are needed due to large volume of water flow.

2. Scroll Casing The water from the penstock enters the scroll casing which is of spiral shape, so it is also called spiral casing. It completely surrounds the runner of the turbine and maintains the even distribution of water around the circumference of the runner. The area of scroll casing goes on decreasing gradually which keeps the velocity of water constant throughout its path around

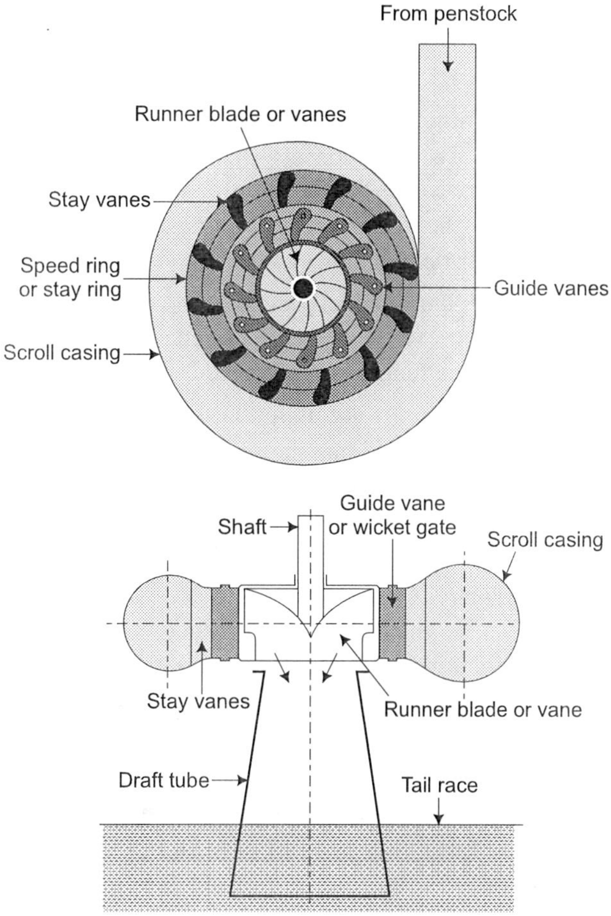

Fig. 4.8 Francis Turbine

the runner. In Francis turbine, the casing and runner are always full of water. The casing is made of cast steel, concrete, or plate steel.

3. Guide Mechanism From the casing water flows through a speed ring (or stay ring) having fixed stay vanes which direct the water to the guide vanes. Generally, the number of stay vanes is taken as half of the number of guide vanes. The guide vanes are adjustable and are also called wicket gates. The guide vanes are airfoil shaped and are fixed on a stationary circular wheel all around the periphery of the turbine runner. The guide vanes direct the water to strike the vanes fixed on the runner without shock at inlet. The guide vanes also regulate the quantity of water supplied to the runner by changing the width between two adjacent vanes either by means of a wheel or automatically by a governor.

4. Runner Runner is also called rotor. It is a circular wheel on which a series of radial curved vanes are evenly fixed. Generally, the number of vanes varies in between 16 and 24. The radial curved vanes are so shaped that water enters the runner radially at the outer periphery and leaves it axially at the inner periphery without shock. To minimize the frictional resistance the surface of these vanes is made very smooth. The direction of water flow in the runner changes from radial to axial. Due to this change of direction of flow a circumferential force is produced on the runner which makes the runner to rotate. The runner is keyed to the shaft which is further coupled to the generator shaft. The runners are made of stainless steel, cast iron, or cast steel.

5. Draft Tube A draft tube is a pipe of gradually increasing cross-sectional area. One end of the draft tube is connected to the runner exit while the other end is submerged deep into the tail race. Thus, water after passing through the runner is discharged to the tail race through the draft tube.

The draft tube has to perform the following important functions:

1. The incorporation of a draft tube permits the turbine runner to be installed above the tail race without any loss of available head by maintaining a negative or suction head at the outlet of the runner. This causes increase in the net effective head and thereby the output of the turbine.
2. Draft tube reduces the velocity of the discharged water by which the loss of kinetic energy at the runner outlet minimizes. In other words, a large proportion of kinetic energy rejected at the runner exit is converted into useful pressure energy by which efficiency of the turbine increases.
3. The draft tube prevents the splashing of water at the runner outlet and guides it to the tail race.

The important type of draft tubes which are commonly used in reaction turbines are: (i) Conical draft tube or straight divergent tube (Figure 4.9a), (ii) Moody spreading tube (Figure 4.9b), (iii) Simple elbow tubes (Figure 4.9c), and (iv) Elbow tube having circular inlet and rectangular outlet (Figure 4.9d).

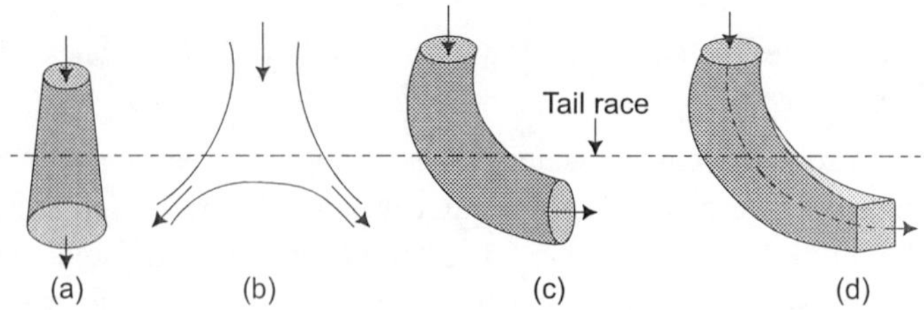

Fig. 4.9 Draft Tubes

The conical draft tube and moody spreading draft tubes are the most efficient, but the simple elbow and draft tubes with circular inlet and rectangular outlet require relatively lesser excavation for their installation.

6. Governing Mechanism The quantity of water in a Francis turbine is controlled by varying the area of flow between adjacent guide vanes by rotating them. These guide vanes are pivoted

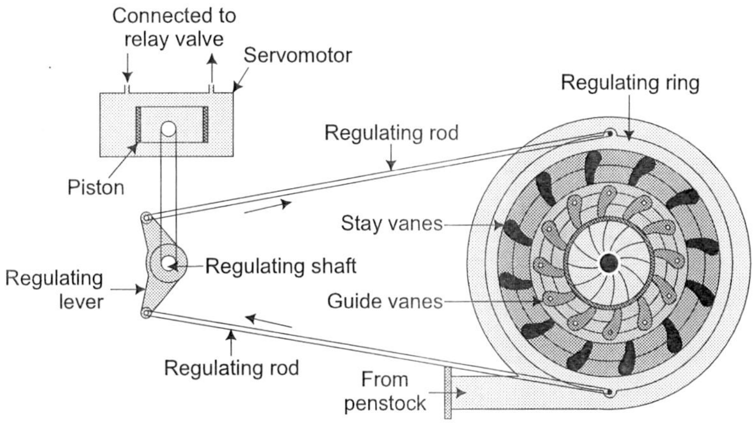

Fig. 4.10 Governing Mechanism of Francis Turbine

and connected by levers and links to the regulating ring. The regulating ring is connected to the regulating shaft through regulating lever and two regulating rods as shown in Figure 4.10.

The regulating shaft is connected to the servomotor. The servomotor, control valve, oil sump, and piping system, etc., are similar to that in Pelton wheel. However, these components are relatively stronger as greater energy is required to move the guide vanes.

With the variation in load, speed of the turbine changes which causes the movement of servomotor piston either to the left or to the right. This rotates the regulating lever clockwise or anticlockwise. This rotation is further transmitted to the regulating ring through regulating rods. The regulating ring rotates in the same direction as the regulating lever and thus closes or opens the passage between the two adjacent guide vanes as per the requirement. The penstock is provided with a pressure regulator or relief valve which has to perform same function as that of deflector in Pelton turbine. When there is a sudden decrease in load on the turbine, the relief valve opens and diverts the water to the tail race and thus it prevents sudden closure of guide vanes. The rapid closure of guide vanes is not desirable because a sudden reduction of the flow rate in the penstock may result in serious water hammer problems.

4.7.1 Velocity Triangles, Work Done, and Efficiency of the Francis Turbine

The velocity triangles at inlet and outlet are shown in Figure 4.11.

Let R_1 and R_2 = Radius of wheel at inlet and outlet of the vane respectively.

ω = Angular speed of the wheel.

Then $u_1 = \omega R_1$ and $u_2 = \omega R_2$

All other notations are usual as given for Pelton turbine.

The mass of the water striking per second for series of vanes

$$= \rho \times A \times V_1$$

Momentum of water at inlet per second

$$= \rho \times A \times V_1 \times V_{w1}$$

$$[V_{w1} = V_1 \cos \alpha = \text{component of } V_1 \text{ in tangential direction}]$$

Similarly, momentum of water at outlet per second

$$= \rho \times A \times V_1 \times (-V_2 \cos \beta)$$

$$= -\rho \times A \times V_1 \times V_{w2} \qquad [\because V_{w2} = V_2 \cos \beta]$$

Negative sign is taken because V_2 at outlet is in opposite direction.

Now angular momentum per second at inlet

$$= \rho \times A \times V_1 \times V_{w1} \times R_1$$

Similarly, angular momentum per second at outlet

$$= -\rho \times A \times V_1 \times V_{w2} \times R_2$$

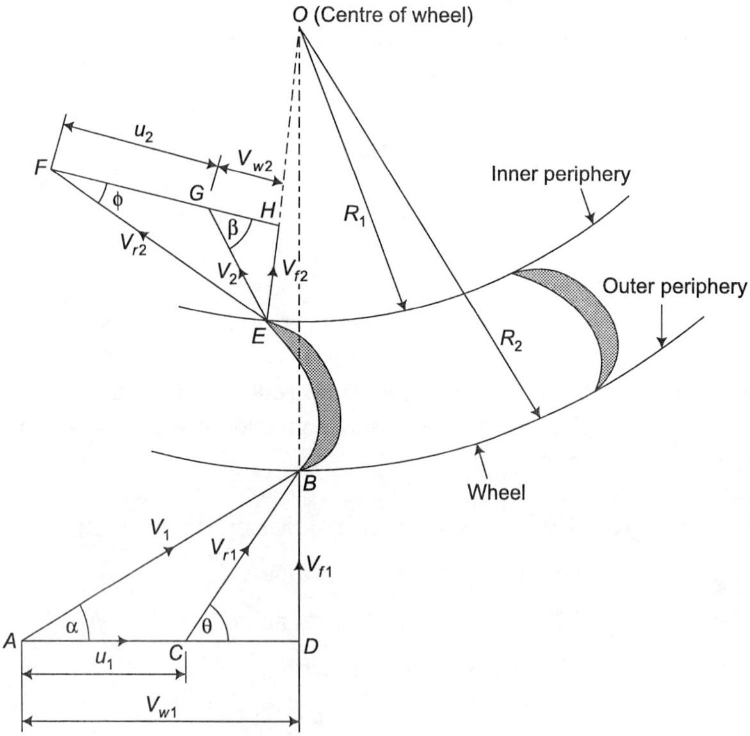

Fig. 4.11 Velocity Triangles

Torque exerted by the water on the wheel,

$$T = \text{Rate of change of angular momentum}$$

$$T = \text{Initial angular momentum per second} - \text{Final angular momentum per second}$$

$$\therefore \quad T = \rho\, AV_1 \times V_{w1} \times R_1 - (-\rho\, AV_1 \times V_{w2} \times R_2)$$

$$= \rho\, AV_1\, [V_{w1} \times R_1 + V_{w2} \times R_2]$$

Work done/sec on the wheel,

$$W = T \times \omega$$

$$W = \rho\, AV_1\, [V_{w1} \times R_1 + V_{w2} \times R_2] \times \omega$$

$$= \rho\, AV_1\, [V_{w1} \times R_1 \times \omega + V_{w2} \times R_2 \times \omega]$$

$$\therefore \quad W = \rho\, AV_1\, [V_{w1}\, u_1 + V_{w2}\, u_2]$$

If the angle $\beta > 90°$, the work done will be given by,

$$W = \rho\, AV_1\, [V_{w1}\, u_1 - V_{w2}\, u_2]$$

Therefore, the general expression for work done/sec,

$$W = \rho\, AV_1\, [V_{w1}\, u_1 \pm V_{w2}\, u_2] \tag{4.20}$$

$$W = \rho\, Q\, [V_{w1}\, u_1 \pm V_{w2}\, u_2] \qquad [\because AV_1 = Q] \tag{4.20 a}$$

Work done/sec per unit weight of water striking/sec,

$$= \frac{\text{Work done/sec}}{\text{Weight of water striking/sec}}$$

$$= \frac{\rho\, Q\, [V_{w1}\, u_1 \pm V_{w2}\, u_2]}{\rho\, Q \times g}$$

$\therefore$ Work done/sec per unit weight

$$= \frac{1}{g} \times [V_{w1}\, u_1 \pm V_{w2}\, u_2] \tag{4.21}$$

Equation (4.21) is known as Euler momentum equation.

If the discharge is radial at outlet then $\beta = 90°$ and output will be maximum.

$$\therefore \quad \text{Work done} = \frac{1}{g} \times [V_{w1}\, u_1] \tag{4.22}$$

$$\text{Hydraulic efficiency,} \quad \eta_h = \frac{\rho\, Q\, [V_{w1}\, u_1 \pm V_{w2}\, u_2]}{\rho \times Q \times g \times H}$$

$$\therefore \quad \eta_h = \frac{[V_{w1}\, u_1 \pm V_{w2}\, u_2]}{g \times H} \tag{4.23}$$

For radial flow at outlet,

$$V_{w2} = 0$$

$$\therefore \qquad \eta_h = \frac{V_{w1}\, u_1}{g \times H} \qquad (4.24)$$

The hydraulic efficiency of the Francis turbine varies from 85% to 90%.

4.7.2 Definitions and Working Proportions of a Francis Turbine

1. **Speed Ratio** The speed ratio varies from 0.6 to 0.9. It is defined as,

   Speed ratio, $$K_u = \frac{u_1}{\sqrt{2gH}} \qquad (4.25)$$

   where u_1 = Tangential velocity

2. **Flow Ratio** The flow ratio varies from 0.15 to 0.30. It is given by,

   Flow ratio, $$K_f = \frac{V_{f1}}{\sqrt{2gH}} \qquad (4.26)$$

   where V_{f1} = Velocity of flow, and H = Head on the turbine

3. **Discharge of the Turbine** The discharge through a Francis turbine is given by,

   $$Q = \pi \times D_1 \times B_1 \times V_{f1} \qquad (4.27)$$

   where D_1 = Diameter of runner at inlet, B_1 = Width of runner at inlet, and V_{f1} = Velocity of flow at inlet

   The discharge through a radial flow reaction turbine is given by,

   $$Q = \pi \times D_1 \times B_1 \times V_{f1}$$
   $$= \pi \times D_2 \times B_2 \times V_{f2} \qquad (4.28)$$

   where D_2 = Diameter of runner at outlet, B_2 = Width of runner at outlet, and V_{f2} = Velocity of flow at outlet.

4. **Radial Discharge** Radial discharge at outlet means $\beta = 90°$ and $V_{w2} = 0$, while radial discharge at inlet means $\alpha = 90°$ and $V_{w1} = 0$.

5. **Ratio of Width to Diameter** The ratio of width of the wheel (B_1) to its diameter (D_1) is given as $n = B_1/D_1$. Its values vary from 0.10 to 0.45.

Example 4.9 A Francis turbine works at 500 rpm under a head of 150 m. Its diameter at inlet is 1 m and the flow area is 0.3 m^2. The guide blade angle and vane angle at inlet are 15° and 60° respectively. Determine (i) the volume flow rate, (ii) the power developed, and (iii) hydraulic efficiency.

Solution Refer Figure 4.12.

(i) Tangential velocity of wheel,

$$u_1 = \frac{\pi D_1 N}{60} = \frac{\pi \times 1 \times 500}{60} = 26.18 \text{ m/s}$$

From inlet velocity triangle,

$$\tan \alpha = \frac{V_{f1}}{V_{w1}}$$

$$V_{f1} = V_{w1} \tan \alpha$$

$$= V_{w1} \tan (15)$$

$$= 0.2679\ V_{w1}$$

$$\tan \theta = \frac{V_{f1}}{V_{w1} - u_1} = \frac{0.2679\ V_{w1}}{V_{w1} - u_1}$$

$$\tan (60) = \frac{0.2679\ V_{w1}}{V_{w1} - u_1}$$

$$1.732 \times (V_{w1} - 26.18) = 0.2679\ V_{w1}$$

$$(1.732 - 0.2679) \times V_{w1} = 1.732 \times 26.18$$

$$\therefore \quad V_{w1} = 30.97 \text{ m/s}$$

$$\therefore \quad V_{f1} = 0.2679 \times 30.97$$

$$= 8.29 \text{ m/s}$$

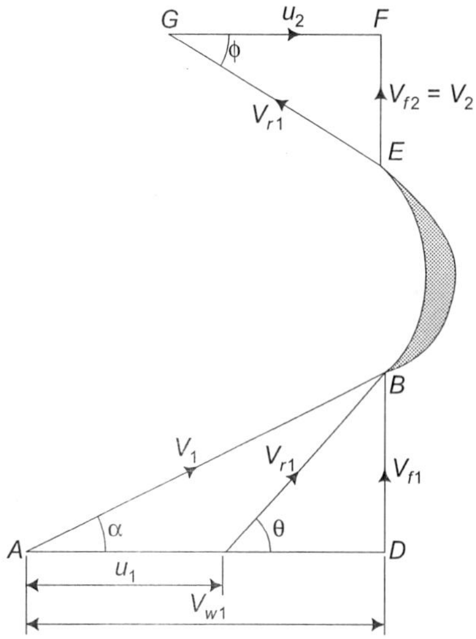

Fig. 4.12

Volume flow rate is given by,

$$Q = \pi \times D_1 \times B_1 \times V_{f1}$$

But $\quad \pi \times D_1 \times B_1 = 0.3 \text{ m}^2$

$$\therefore \quad Q = 0.3 \times 8.29 = 2.487 \text{ m}^3/\text{s}$$

(ii) The power developed,

$$P = \frac{\rho\ Q[V_{w1}\ u_1]}{1000}$$

$$\therefore \quad P = \frac{1000 \times 2.487 \times [30.97 \times 26.18]}{1000}$$

$$= 2016.45 \text{ kW}$$

(iii) The hydraulic efficiency,

$$\eta_h = \frac{V_{w1}\ u_1}{g \times H}$$

$$= \frac{30.97 \times 26.18}{9.81 \times 150}$$

$$= 0.5509 \text{ or } 55.09\%$$

Example 4.10 A Francis turbine with an overall efficiency of 75% is required to produce 150 kW power. It is working under a head of 10 m. The peripheral velocity = $0.25\sqrt{2gH}$ and the radial velocity of flow at inlet is = $0.95\sqrt{2gH}$. The wheel runs at 200 rpm and the hydraulic losses in the turbine are 20% of the available energy. Assuming radial discharge, determine (i) the guide blade angle, (ii) the wheel vane angle at inlet, (iii) diameter of the wheel at inlet, and (iii) width of the wheel at inlet.

Solution Refer Figure 4.12.

Peripheral velocity, $u_1 = 0.25\sqrt{2gH} = 0.25 \times \sqrt{2 \times 9.81 \times 10}$

$= 3.5$ m/s

Velocity of flow at inlet,

$$V_{f1} = 0.95\sqrt{2gH} = 0.95 \times \sqrt{2 \times 9.81 \times 10}$$

$= 13.31$ m/s

The hydraulic efficiency,

$$\eta_h = \frac{\text{Total head at inlet} - \text{Hydraulic loss}}{\text{Total head at inlet}}$$

$$= \frac{H - 0.2\,H}{H} = 0.8$$

But $\eta_h = \dfrac{V_{w1}\,u_1}{g \times H}$

$\therefore$ $\dfrac{V_{w1}\,u_1}{g \times H} = 0.8$

$\therefore$ $V_{w1} = \dfrac{0.8 \times g \times H}{u_1} = \dfrac{0.8 \times 9.81 \times 10}{3.5}$

$= 22.42$ m/s

(i) From inlet velocity triangle,

$$\tan\alpha = \frac{V_{f1}}{V_{w1}} = \frac{13.31}{22.42}$$

$\therefore$ $\alpha = \tan^{-1}\dfrac{13.31}{22.42} = 30.69°$

(ii) From inlet velocity triangle,

$$\tan\theta = \frac{V_{f1}}{V_{w1} - u_1} = \frac{13.31}{22.42 - 3.5}$$

$\therefore$ $\theta = \tan^{-1}\dfrac{13.31}{19.42} = 35.12°$

(iii) $u_1 = \dfrac{\pi D_1 N}{60}$

$\therefore$ $D_1 = \dfrac{u_1 \times 60}{\pi \times N} = \dfrac{3.5 \times 60}{\pi \times 200}$

$= 0.3342$ m

(iv) Overall efficiency, $\eta_o = \dfrac{SP}{WP} = \dfrac{SP \times 1000}{\rho \times g \times Q \times H}$

$$Q = \frac{SP \times 1000}{\rho \times g \times \eta_o \times H}$$

$$\therefore \quad Q = \frac{150 \times 1000}{1000 \times 9.81 \times 0.75 \times 10} = 2.039 \text{ m}^3/\text{s}$$

But $Q = \pi \times D_1 \times B_1 \times V_{f1}$

$$B_1 = \frac{Q}{\pi \times D_1 \times V_{f1}}$$

$$\therefore \quad B_1 = \frac{2.039}{\pi \times 0.3342 \times 13.31}$$

$$= 0.1459 \text{ m}$$

Example 4.11 The following data is given for a Francis turbine: net head = 65 m, speed = 500 rpm, shaft power = 300 kW, overall efficiency = 80%, hydraulic efficiency = 90%, flow ratio = 0.2, breadth ratio = 0.1, outer diameter of runner = 2 times of inner diameter of runner, the thickness of vanes occupy 6% of circumferential area of runner, velocity of low is constant at inlet and outlet, and discharge is radial at outlet. Determine (i) diameter of runner at outlet and inlet, (ii) width of wheel at inlet, (iii) guide blade angle, and (iv) runner vane angle at inlet and outlet.

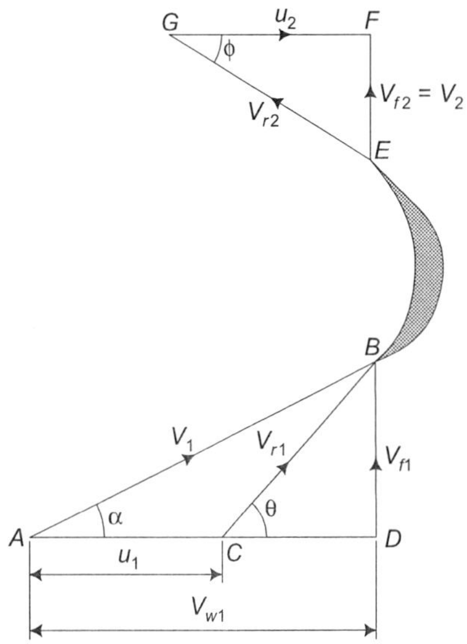

Fig. 4.13

Solution Refer Figure 4.13.

(i) Flow ratio $= 0.2 = \dfrac{V_{f1}}{\sqrt{2gH}}$

$$V_{f1} = 0.2 \times \sqrt{2gH}$$

$$\therefore \quad V_{f1} = 0.2 \times \sqrt{2 \times 9.81 \times 65H}$$

$$= 7.138 \text{ m/s}$$

Thickness of vanes = 6% of circumferential area of runner

$$\therefore \quad \text{Actual area of flow} = \left(1 - \frac{6}{100}\right) \times \pi \times D_1 \times B_1 \times V_{f1}$$

$$= 0.94 \times \pi \times D_1 \times B_1 \times V_{f1}$$

Overall efficiency, $\eta_o = \dfrac{SP}{WP} = \dfrac{SP \times 1000}{\rho \times g \times Q \times H}$

$$Q = \frac{SP \times 1000}{\rho \times g \times \eta_o \times H}$$

$$Q = \frac{300 \times 1000}{1000 \times 9.81 \times 0.8 \times 65}$$

$$= 0.588 \text{ m}^3/\text{s}$$

$\therefore$ $$0.588 = 0.94 \times \pi \times D_1 \times B_1 \times V_{f1}$$

$$0.94 \times \pi \times D_1 \times (0.1\ D_1) \times 7.138 = 0.588 \qquad (\because B_1/D_1 = 0.1)$$

$\therefore$ $$D_1 = \sqrt{\frac{0.588}{0.94 \times \pi \times 0.1 \times 7.138}}$$

$$= 0.528 \text{ m}$$

$\therefore$ $$D_1 = 2 \times D_2$$

$\therefore$ $$D_2 = (D_1/2) = (0.528/2)$$

$$= 0.264 \text{ m}$$

(ii) $$\frac{B_1}{D_1} = 0.1$$

$\therefore$ $$B_1 = 0.1 \times D_1 = 0.1 \times 0.528$$

$$= 0.0528 \text{ m or } 52.8 \text{ mm}$$

(iii) $$u_1 = \frac{\pi D_1 N}{60}$$

$$= \frac{\pi \times 0.528 \times 500}{60}$$

$$= 13.82 \text{ m/s}$$

$$\eta_h = \frac{V_{w1}\ u_1}{g \times H}$$

$$V_{w1} = \frac{\eta_h \times g \times H}{u_1}$$

$$= \frac{0.9 \times 9.81 \times 65}{13.82}$$

$$= 41.525 \text{ m/s}$$

$$\tan \alpha = \frac{V_{f1}}{V_{w1}} = \frac{7.138}{41.525},$$

$\therefore$ $$\alpha = \tan^{-1} \frac{7.138}{41.525}$$

$$= 9.76^\circ$$

(iv) From inlet velocity triangle,

$$\tan \theta = \frac{V_{f1}}{V_{w1} - u_1} = \frac{7.138}{41.525 - 13.82}$$

$$\therefore \qquad \theta = \tan^{-1} \frac{7.138}{27.705} = 14.45^\circ$$

$$u_2 = \frac{\pi D_2 N}{60} = \frac{\pi \times 0.264 \times 500}{60}$$

$$= 6.91 \text{ m/s}$$

From outlet velocity triangle,

$$\tan \phi = \frac{V_{f2}}{u_2} = \frac{V_{f1}}{u_2} = \frac{7.138}{6.91}$$

$$\therefore \qquad \phi = \tan^{-1} \frac{7.138}{6.91}$$

$$= 45.93^\circ$$

Example 4.12 An inward flow reaction turbine has an external and internal diameter as 1 m and 0.5 m respectively. The turbine is running at 180 rpm and width of turbine at inlet is 250 mm. The velocity of flow through the runner is constant and is equal to 2 m/s. The guide blade angle makes an angle of 10° to the tangent of the wheel and the discharge at the outlet is radial. If the net head on the turbine is 15 m, determine (i) the absolute velocity of water at inlet of runner, (ii) the velocity of whirl at inlet, (iii) the relative velocity at inlet, (iv) runner blade angles, (v) width of the runner at outlet, (vi) mass of water flowing through the runner/second, (vii) power developed, and (viii) hydraulic efficiency.

Solution Refer Figure 4.13.

(i)

$$u_1 = \frac{\pi D_1 N}{60} = \frac{\pi \times 1 \times 180}{60}$$

$$= 9.42 \text{ m/s}$$

$$\therefore \qquad u_2 = \frac{\pi D_2 N}{60} = \frac{\pi \times 0.5 \times 180}{60}$$

$$= 4.71 \text{ m/s}$$

From inlet velocity triangle,

$$V_1 \sin \alpha = V_{f1}$$

$$V_1 = \frac{V_{f1}}{\sin \alpha} = \frac{2}{\sin (10)}$$

$$= 11.52 \text{ m/s}$$

(ii) From inlet velocity triangle,

$$V_{w1} = V_1 \sin \alpha$$
$$= 11.52 \times \cos (10) = 11.34 \text{ m/s}$$

(iii) From inlet velocity triangle,

$$V_{r1} = \sqrt{V_{f1}^2 + (V_{w1} - u_1)^2}$$
$$= \sqrt{(2)^2 + (11.34 - 9.42)^2}$$
$$= 2.77 \text{ m/s}$$

(iv) From inlet velocity,

$$\tan \theta = \frac{V_{f1}}{V_{w1} - u_1} = \frac{2}{11.34 - 9.42}$$

$$\therefore \qquad \theta = \tan^{-1} \frac{2}{1.92}$$
$$= 46.17°$$

From outlet velocity,

$$\tan \theta = \frac{V_{f2}}{u_2} = \frac{2}{4.17}$$

$$\therefore \qquad \phi = \tan^{-1} \frac{2}{4.71}$$
$$= 23.01°$$

(v)
$$\pi \times D_1 \times B_1 \times V_{f1} = \pi \times D_2 \times B_2 \times V_{f2}$$
$$D_1 \times B_1 = D_2 \times B_2 \qquad (\because V_{f1} = V_{f2})$$
$$B_2 = \frac{D_1 \times B_1}{D_2} = \frac{1 \times 250}{0.5}$$
$$= 500 \text{ mm}$$

(vi)
$$Q = \pi \times D_1 \times B_1 \times V_{f1}$$
$$= \pi \times 1 \times 0.25 \times 2$$
$$= 1.57 \text{ m}^3\text{/s}$$

Therefore, mass per second,

$$m = \rho \times Q = 1000 \times 1.57$$
$$= 1570 \text{ kg/s}$$

(vii) The power developed, $P = \dfrac{\rho\, Q\, [V_{w1}\, u_1]}{1000}$

$$\therefore \qquad P = \frac{1000 \times 1.57 \times [11.34 \times 9.42]}{1000}$$

$$= 167.71 \text{ kW}$$

(viii) The hydraulic efficiency,

$$\eta_h = \frac{V_{w1}\, u_1}{g \times H} = \frac{11.34 \times 9.42}{9.81 \times 15}$$

$$= 0.7259 \text{ or } 72.59\%$$

4.8 PROPELLER AND KAPLAN TURBINES

The propeller and Kaplan turbines are important types of axial flow reaction turbines. Due to the highest specific speed these turbines are suitable for low heads and large flow of water. In these turbines, water flows parallel to the axis of the rotation of the shaft. The shaft of axial flow turbine is vertical. The lower end of the shaft is made larger which is known as 'hub' or 'boss'. The hub of these turbines has either fixed blades or variable pitch blades and it acts as a runner for axial flow turbine. When the blades are fixed to the hub and they are non-adjustable, the turbine is known as propeller turbine. But if the blades are of variable pitch and they are adjustable on the hub then the turbine is known as the Kaplan turbine after the name of Dr. V. Kaplan, an Austrian engineer. Due to the adjustable pitch of the blades, the Kaplan turbine can deal with a wide range of operating heads. Practically all Kaplan turbines have a single runner. The runner of a Kaplan turbine and propeller turbine is shown in Figure 4.14.

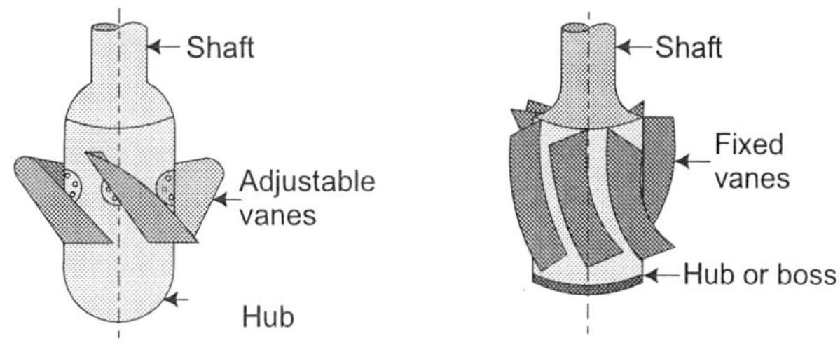

(a) Kaplan turbine runner (b) Propeller turbine runner

Fig. 4.14

The propeller turbines are easy to construct but their efficiency falls sharply at reduced loads, therefore, this should be kept fully loaded for maximum efficiency. A propeller turbine is most suitable when load on the turbine remains constant. At part load its efficiency is very low. The blades are fixed, so water enters with a shock and eddies are formed which reduces the efficiency under part load condition. In propeller turbines, as in the Francis turbine, the runner blades are fixed.

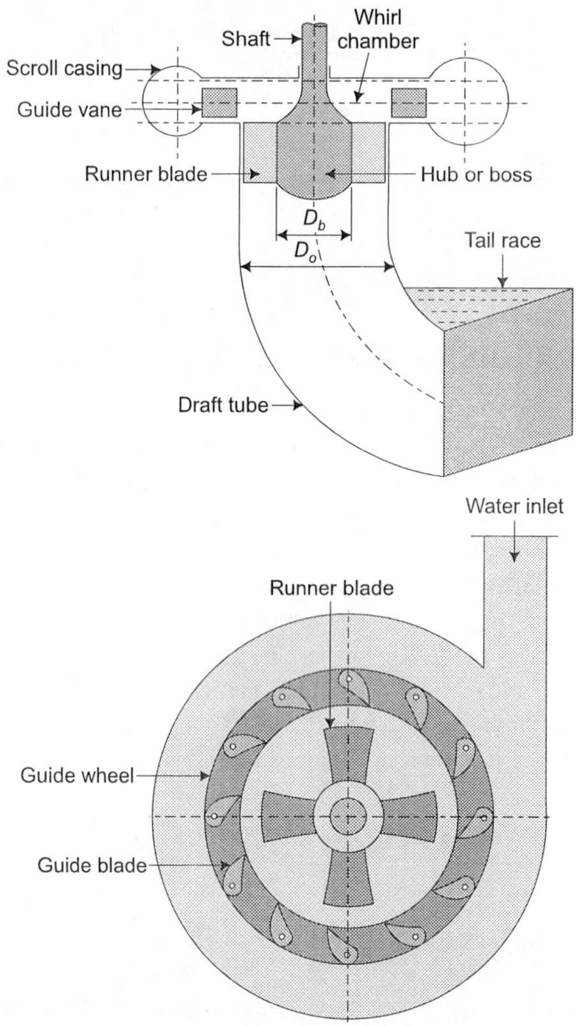

Fig. 4.15 Kaplan Turbine

Figure 4.15 shows the outline of a Kaplan turbine. The main parts of a Kaplan turbine are:

1. Scroll casing,
2. Stay ring,
3. Guide vanes mechanism,
4. Hub with vanes or runner of the turbine, and
5. Draft tube.

The casing, stay ring, guide mechanism, and draft tube are similar to that of a Francis turbine. The Kaplan turbine also runs full means it is a reaction type of turbine which operates in a closed conduit from the inlet to the tail race. A space has been provided between the ends of guide vanes and the leading edges of the runner. This space is called whirl chamber in which flow direction changes from radial to axial. Thus, the runner of a Kaplan turbine varies from the runner of a Francis turbine on these points:

(i) In Kaplan turbine water enters the runner axially, while in Francis turbine it enters radially.

(ii) The number of blades in Kaplan turbine varies from 4 to 8, whereas in Francis turbine it varies from 16 to 24.

(iii) The blades of the Kaplan turbine are hollow and attached to the hub in such a way that they are able to move on their axis.

The water from the penstock enters the scroll casing and then moves to the guide vanes. From the guide vanes the water turns through 90° and flows axially through the runner as shown in Figure 4.15. The discharge through the runner is given by,

$$Q = \frac{\pi}{4}(D_o^2 - D_b^2) \times V_{f1} \tag{4.29}$$

where D_o = Outer diameter of the runner, D_b = Diameter of the hub, and V_{f1} = Velocity of flow at inlet.

4.8.1 Governing of Kaplan Turbine

In the case of Kaplan turbine in addition to the guide vanes the runner vanes are also adjustable. Therefore, the governor is required to operate simultaneously both sets of vanes. The runner vanes are also operated by a separate servomotor mechanism. The servomotor mechanisms both for runner and guide vanes are interconnected to ensure that for a given guide vane opening there shall be a definite runner vane inclination.

4.8.2 Some Important Points and Working Properties of Kaplan Turbine

1. The discharge Q flowing through the runner is given by,

$$Q = \frac{\pi}{4}(D_o^2 - D_b^2) \times V_{f1} \tag{4.30}$$

2. The ratio of the hub diameter to runner diameter usually varies from 0.35 to 0.6. It is given by,

$$n = (D_b/D_o) \tag{4.31}$$

3. The peripheral velocity at inlet and outlet are equal, i.e.,

$$u_1 = u_2 = \frac{\pi D_o N}{60} \tag{4.32}$$

4. Velocity of flow remains constant, i.e.,

$$V_{f1} = V_{f2}$$

5. Area of flow is constant, i.e.,

Area of flow at inlet = Area of flow at outlet

$$= \frac{\pi}{4}(D_o^2 - D_b^2) \tag{4.33}$$

4.9 COMPARISON BETWEEN FRANCIS AND KAPLAN TURBINES

Francis Turbine	Kaplan Turbine
1. The Francis turbine is either radially inward or mixed flow turbine.	1. The Kaplan turbine is purely axial flow turbine.
2. Large number of blades in runner which varies from 16 to 24.	2. Small number of blades in runner which varies from 4 to 8.
3. The runner blades are not adjustable.	3. The runner blades are adjustable.
4. The disposition of shaft may be horizontal or vertical.	4. The disposition of shaft is always vertical.
5. It is a medium head turbine which varies between 60 m and 250 m.	5. It is a low head turbine and is capable to work under the head less than 60 m.
6. It works under medium discharge.	6. It works under high discharge.
7. The specific speed varies from 50 to 250.	7. Specific speed varies from 255 to 860.
8. Resistance to be overcome is large due to large number of vanes and greater area of contact with water.	8. Resistance to be overcome is less due to small number of vanes and less contact area.
9. Ordinary governor is used.	9. Heavy duty governor is used.

4.10. SPECIFIC SPEED

It is defined as the speed of a turbine geometrically similar to the actual turbine but of such a size that under corresponding conditions it will develop unit power when working under unit head. It is denoted by N_s.

Specific speed may be derived as follows:

Overall efficiency of any turbine,

$$\eta_o = \frac{SP}{WP} = \frac{P \times 1000}{\rho \times g \times Q \times H} \tag{i}$$

$$P = \frac{\eta_o \times \rho \times g \times Q \times H}{1000}$$

or

$$P \propto Q \times H \qquad (\because \rho, g, \eta_o = \text{Constant}) \tag{ii}$$

The absolute velocity, tangential velocity, and head on the turbine are related as,

$$u \propto V \qquad \text{and} \qquad V \propto \sqrt{H}$$

or $$u \propto \sqrt{H} \quad \text{(iii)}$$

But $$u = (\pi DN/60)$$

or $$u \propto DN \quad \text{(iv)}$$

From equation (iii) and (iv), we have,

$$\sqrt{H} \propto DN$$

or $$D \propto \frac{\sqrt{H}}{N} \quad \text{(v)}$$

The discharge through the turbine is given by,

$$Q = \text{Area} \times \text{Velocity}$$

But $$\text{Area} \propto B \times D$$

∴ $$\text{Area} \propto D^2 \qquad (\because B = \text{width} \ \& \ B \propto D)$$

$$\text{Velocity} \propto \sqrt{H}$$

∴ $$Q \propto D^2 \times \sqrt{H}$$

Using equation (v), we have,

$$Q \propto \left(\frac{\sqrt{H}}{N}\right)^2 \times \sqrt{H}$$

or $$Q \propto \frac{H^{3/2}}{N^2} \quad \text{(vi)}$$

Substituting value of Q in equation (ii), we have,

$$P \propto \frac{H^{3/2}}{N^2} \times H$$

or $$P \propto \frac{H^{5/2}}{N^2}$$

∴ $$P = k\,\frac{H^{5/2}}{N^2} \qquad (k = \text{Constant}) \quad \text{(vii)}$$

If $P = 1$ kW, $H = 1$ m, then $N = N_s$, substituting these values in (vii), we have,

$$1 = k\,\frac{(1)^{5/2}}{N_s^{\,2}}$$

or $$N_s^{\,2} = k$$

∴ $$P = N_s^{\,2}\,\frac{H^{5/2}}{N^2}$$

or $$N_s^2 = \frac{N^2 P}{H^{5/2}}$$

$\therefore$ $$N_s = \sqrt{\frac{N^2 P}{H^{5/2}}} = \frac{N\sqrt{P}}{H^{5/4}} \qquad (4.34)$$

4.10.1 Significance of Specific Speed

Specific speed plays an important role in the selection of the type of turbine. The performance of a turbine can also be predicted by knowing the specific speed of the turbine. The higher specific speed of a turbine results in the reduction of the runner diameter as well as the overall size of the runner, due to which the weight and the cost of the runner reduces. The types of turbine for different specific speed are given below:

Pelton wheel with single jet: 8.5 to 30,

Pelton wheel with two or more jets: 30 to 50,

Francis turbine: 51 to 250, and

Propeller and Kaplan turbine: 255 to 860.

Example 4.13 A Kaplan turbine develops 50,000 kW under a net head of 30 m with an overall efficiency of 85%. Taking the value of speed ratio = 2, flow ratio = 0.6, and diameter of the hub = 0.35 times of the diameter of the runner, calculate (i) diameter of the runner, (ii) speed of the turbine, and (iii) specific speed of the turbine.

Solution

(i) $$\text{Speed ratio} = 2 = \frac{u_1}{\sqrt{2gH}}$$

$$u_1 = 2 \times \sqrt{2gH}$$

$$= 2 \times \sqrt{2 \times 9.81 \times 30}$$

$$= 48.52 \text{ m/s}$$

$$\text{Flow ratio} = 0.6 = \frac{V_{f1}}{\sqrt{2gH}}$$

$$V_{f1} = 0.6 \times \sqrt{2 \times 9.81 \times 30}$$

$$= 14.56 \text{ m/s}$$

Overall efficiency, $$\eta_o = \frac{SP}{WP} = \frac{SP \times 1000}{\rho \times g \times Q \times H}$$

$$Q = \frac{SP \times 1000}{\rho \times g \times \eta_o \times H}$$

$\therefore$ $$Q = \frac{50{,}000 \times 1000}{1000 \times 9.81 \times 0.85 \times 30}$$

$$= 199.876 \text{ m}^3\text{/s}$$

Also $$Q = \frac{\pi}{4}[D_o^2 - D_b^2] \times V_{f1}$$

$$= \frac{\pi}{4}[D_o^2 - (0.35\,D_o)^2] \times 14.56 \qquad (\because D_b = 0.35\,D_o)$$

$$\therefore \quad \frac{\pi}{4}[D_o^2 - (0.35\,D_o)^2] \times 14.56 = 199.876$$

$$10.03\,D_o^2 = 199.876$$

$$\therefore \quad D_o = \sqrt{\frac{199.876}{10.03}} = 4.464 \text{ m}$$

(ii) $$u_1 = \frac{\pi\,D_o\,N}{60}$$

$$\therefore \quad N = \frac{u_1 \times 60}{\pi \times D_o} = \frac{48.52 \times 60}{\pi \times 4.464}$$

$$= 207.58 \text{ rpm}$$

(iii) $$N_s = \frac{N\sqrt{P}}{H^{5/4}} = \frac{207.58 \times \sqrt{50000}}{(30)^{5/4}}$$

$$= 661.10$$

Example 4.14 A Kaplan turbine working under a head of 25 m develops 11770 kW shaft power. The outlet diameter of the runner is 3 m and the hub diameter is 1.5 m. The guide blade angle at the extreme edge of the runner is 36°. The hydraulic and overall efficiency of the turbine are 85% and 80% respectively. If the velocity of the whirl is zero at outlet, determine runner vane angles at the extreme edge of the runner, and speed of the turbine.

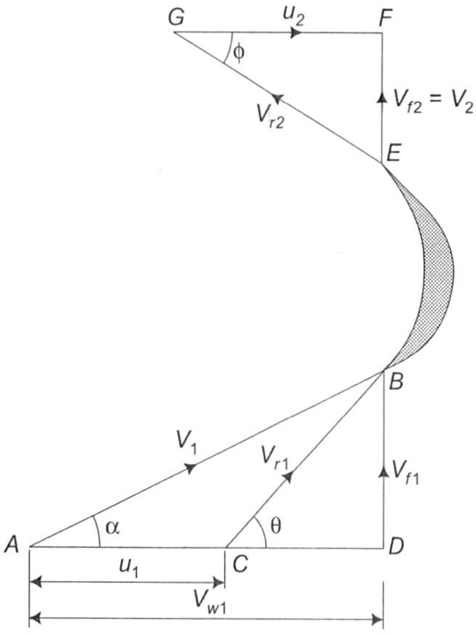

Fig. 4.16

Solution Refer Figure 4.16.

Overall efficiency, $$\eta_o = \frac{SP}{WP} = \frac{P \times 1000}{\rho \times g \times Q \times H}$$

$$Q = \frac{P \times 1000}{\rho \times g \times \eta_o \times H}$$

$$\therefore \quad Q = \frac{11{,}770 \times 1000}{1000 \times 9.81 \times 0.5 \times 25}$$

$$= 59.989 \text{ m}^3/\text{s}$$

Also $$Q = \frac{\pi}{4}[D_o^2 - D_b^2] \times V_{f1}$$

$$= \frac{\pi}{4}[(3)^2 - (1.5)^2] \times V_{f1}$$

$$= 56.989$$

$$\therefore \quad V_{f1} = 11.32 \text{ m/s}$$

From the inlet velocity triangle,

$$\tan \alpha = \frac{V_{f1}}{V_{w1}}$$

$$\therefore \quad V_{w1} = \frac{V_{f1}}{\tan \alpha} = \frac{11.32}{\tan (36)}$$

$$= 15.58 \text{ m/s}$$

$$\eta_h = \frac{V_{w1}\, u_1}{g \times H}$$

$$u_1 = \frac{\eta_h \times g \times H}{V_{w1}}$$

$$= \frac{0.85 \times 9.81 \times 25}{15.58} = 13.38 \text{ m/s}$$

$$\tan \theta = \frac{V_{f1}}{V_{w1} - u_1} = \frac{11.32}{15.58 - 13.38}$$

$$\therefore \quad \theta = \tan^{-1} \frac{11.32}{2.2} = 79^\circ$$

For Kaplan turbine, $u_1 = u_2 = 13.38$ m/s

and $V_{f1} = V_{f2} = 11.32$ m/s

From outlet velocity,

$$\tan \phi = \frac{V_{f2}}{u_2} = \frac{11.32}{13.38},$$

$$\therefore \quad \phi = \tan^{-1} \frac{11.32}{13.38} = 40.23^\circ$$

$$u_1 = u_2 = \frac{\pi D_o N}{60}$$

$$N = \frac{u_1 \times 60}{\pi \times D_o} = \frac{13.38 \times 60}{\pi \times 3}$$

$$= 85.18 \text{ rpm}$$

Example 4.15 The hub diameter of a Kaplan turbine working under a head of 15 m is 0.3 times the diameter of the runner. The turbine is running at 90 rpm. The velocity of wheel at outlet is zero. If the vane angle of the extreme edge of the runner at outlet is 15° and flow ratio is 0.6, determine (i) diameter of runner, (ii) diameter of boss, and (iii) discharge through the runner.

Solution Refer Figure 4.16.

(i) $$\text{Flow ratio} = 0.6 = \frac{V_{f1}}{\sqrt{2gH}}$$

$$V_{f1} = 0.6 \times \sqrt{2 \times 9.81 \times 15} = 10.29 \text{ m/s}$$

For Kaplan turbine, $V_{f1} = V_{f2} = 10.29$ m/s

From outlet velocity,

$$\tan \phi = \frac{V_{f2}}{u_2}$$

$$u_2 = \frac{V_{f2}}{\tan \phi} = \frac{10.29}{\tan (15)}$$

$$= 38.40 \text{ m/s}$$

$$u_1 = u_2 = 13.38 \text{ m/s}$$

$$u_1 = \frac{\pi D_o N}{60}$$

$$\therefore \quad D_o = \frac{u_1 \times 60}{\pi \times N} = \frac{13.38 \times 60}{\pi \times 90}$$

$$= 8.148 \text{ m}$$

(ii) Diameter of boss, $D_b = 0.3 \times 3.8148$

$$= 2.444 \text{ m}$$

(iii) Discharge through the turbine is given by,

$$Q = \frac{\pi}{4} [D_o^2 - D_b^2] \times V_{f1}$$

$$= \frac{\pi}{4} [(8.148)^2 - (2.444)^2] \times 10.29$$

$$= 488.27 \text{ m}^3\text{/s}$$

4.11 UNIT QUANTITIES

Turbines are often required to work under varying conditions of head, speed, output, and gate opening. So, in order to predict the behaviour of a turbine working under the varying conditions and to facilitate comparison between the performances of the turbines of the same type but of different sizes, it is often convenient to express the results in terms of certain quantities which may be obtained when the head on the turbine is reduced to unity. It is assumed that the efficiency of the turbine remains unchanged when the turbine is working under unit head. The important unit quantities are: (i) Unit speed, (ii) Unit power, and (iii) Unit discharge.

4.11.1 Unit Speed

It is defined as the speed of the turbine working under unit head. It is denoted by N_u.

Expression for Unit Speed

Let H = Head under which turbine is working,

N = Speed of a turbine under a head H m, and

u = Tangential velocity

The tangential velocity, absolute velocity of water, and head on the turbine are related as,

$$u \propto V \qquad \text{and} \qquad V \propto \sqrt{H}$$

$$\therefore \quad u \propto \sqrt{H} \qquad \text{(i)}$$

Also $u = (\pi DN/60)$

$$\text{or} \quad u \propto N \qquad \text{(ii)}$$

From equations (i) and (ii),

$$N \propto \sqrt{H} \qquad \text{(iii)}$$

$$\therefore \quad N = k_1 \sqrt{H} \qquad (k_1 = \text{constant}) \qquad \text{(iv)}$$

If head on the turbine becomes unity, i.e., $H = 1$ m then $N = N_u$

$$\therefore \quad N_u = k_1 \sqrt{1} = k_1 \qquad \text{(v)}$$

From (iv) and (v), we have,

$$N = N_u \sqrt{H}$$

$$\therefore \quad N_u = \frac{N}{\sqrt{H}} \qquad (4.35)$$

4.11.2 Unit Power

It is defined as the power developed by a turbine working under a unit head. It is denoted by P_u.

Expression for unit power

Let H = Head under which turbine is working,

P = Power developed by the turbine under a head H m, and

Q = Discharge through the turbine under a head H m.

Overall efficiency of any turbine,

$$\eta_o = \frac{SP}{WP} = \frac{P \times 1000}{\rho \times g \times Q \times H}$$

$$P = \frac{\eta_o \times \rho \times g \times Q \times H}{1000}$$

$$\text{or} \quad P \propto Q \times H \qquad (\because \rho, g, \eta_o = \text{Constant}) \qquad \text{(i)}$$

But $\quad Q \propto \sqrt{H}$

so, $\quad P \propto \sqrt{H} \times H$

$$P \propto H^{3/2}$$

or $\quad P = k_2\, H^{3/2} \qquad (k_2 = \text{Constant}) \qquad \text{(ii)}$

If head on the turbine becomes unity, i.e., $H = 1$ m then $P = P_u$

$\therefore \quad P_u = k_2\,(1)^{3/2} = k_2 \qquad \text{(iii)}$

From (ii) and (iii), we have,

$$P = P_u\, H^{3/2}$$

$\therefore$
$$P_u = \frac{P}{H^{3/2}} \qquad (4.36)$$

4.11.3 Unit Discharge

It is defined as the discharge passing through a turbine which is working under a unit head. It is denoted by Q_u.

Expression for unit discharge

Let $\quad H$ = Head under which turbine is working and

Q = Discharge through the turbine under a head H m.

The discharge through the turbine is given by,

$$Q = \text{Area} \times \text{Velocity}$$

But $\quad$ Area = Constant

Velocity $\propto \sqrt{H}$

or $\quad Q \propto \sqrt{H}$

or $\quad Q = k_3 \sqrt{H} \qquad (k_3 = \text{Constant}) \qquad \text{(i)}$

If head on the turbine becomes unity, i.e., $H = 1$ m then $Q = Q_u$

$$Q_u = k_3 \sqrt{1} = k_3 \qquad \text{(ii)}$$

From equations (i) and (ii), we have,

$$Q = Q_u \sqrt{H}$$

$\therefore$
$$Q_u = \frac{Q}{\sqrt{H}} \qquad (4.37)$$

4.11.4 Use of Unit Quantities (N_u, Q_u, P_u)

If a turbine is working under different heads the behavior of the turbine can be easily known from the value of the unit quantities as follows:

Let H_1 and H_2 = Heads under which turbine works,
Q_1 and Q_2 = Corresponding discharges,
N_1 and N_2 = Corresponding speeds, and
P_1 and P_2 = Corresponding powers developed.

Then using equations (4.35), (4.36), and (4.37), respectively, we have,

$$N_u = \frac{N_1}{\sqrt{H_1}} = \frac{N_2}{\sqrt{H_2}} \tag{4.38}$$

$$P_u = \frac{P_1}{H_1^{3/2}} = \frac{P_2}{H_2^{3/2}} \tag{4.39}$$

$$Q_u = \frac{Q_1}{\sqrt{H_1}} = \frac{Q_2}{\sqrt{H_2}} \tag{4.40}$$

4.12 MODEL RELATIONSHIP

The model is the small-scale replica of the actual machine. The actual machine is called prototype. For complete similarity to exist between the model and the prototype turbines, the following conditions may be required to be satisfied:

(i) Head coefficient The tangential velocity of the runner is given by,

$$u = K_u \sqrt{2gH} = \frac{\pi DN}{60}$$

or

$$N = \frac{60 \times K_u \sqrt{2gH}}{\pi D}$$

or

$$N \propto \frac{\sqrt{H}}{D}$$

$$ND \propto \sqrt{H} \tag{i}$$

$\therefore$

$$\frac{H}{N^2 D^2} = \text{Constant} \tag{4.41}$$

The parameter $\frac{H}{N^2 D^2}$ is called head coefficient which is denoted by C_H.

(ii) Capacity or flow coefficient The discharge through the turbine is given by,

$$Q = \text{Area} \times \text{Velocity}$$

But Area = D^2 and Velocity $\propto \sqrt{H}$

$\therefore$

$$Q \propto D^2 \sqrt{H} \tag{ii}$$

Substituting $ND \propto \sqrt{H}$ in equation (ii), we have,

$$Q \propto D^2 ND$$

or

$$Q \propto ND^3 \tag{iii}$$

$$\therefore \qquad \frac{Q}{ND^3} = \text{Constant} \tag{4.42}$$

The parameter $\frac{Q}{ND^3}$ is called the capacity or flow coefficient which is denoted by C_Q.

(iii) Power coefficient

Overall efficiency of any turbine,

$$\eta_o = \frac{SP}{WP} = \frac{P \times 1000}{\rho \times g \times Q \times H}$$

$$P = \frac{\eta_o \times \rho \times g \times Q \times H}{1000}$$

or $\qquad P \propto Q \times H \qquad (\because \rho, g, \eta_o = \text{Constant}) \qquad$ (iv)

But $\qquad \sqrt{H} \propto ND$

or $\qquad H \propto N^2D^2$

and $\qquad Q \propto ND^3$

Substituting the values of H and Q in the expression (iv), we have,

$$P \propto ND^3 \times N^2D^2$$

or $\qquad P \propto N^3D^5$

$$\therefore \qquad \frac{P}{N^3D^5} = \text{Constant} \tag{4.43}$$

The parameter P/N^3D^5 is called the power coefficient which is denoted by C_P.

With the use of above relations it is possible to present the behavior of a prototype from the test runs made on geometrically similar model. The model is presumed to have the same values of speed ratio K_u, flow ratio K_f, and specific speed N_s. A group of geometrically similar machines are said to belong to a homologous series. All machines of such a series have the same values of C_H, C_Q, and C_P or their combinations.

Example 4.16 A turbine is to operate under a head of 25 m at 200 rpm. The discharge is 9 m^3/s. If the efficiency is 90%, determine (i) power generated, (ii) specific speed of the turbine, and (iii) type of turbine.

Solution

(i) Overall efficiency, $\qquad \eta_o = \frac{SP}{WP} = \frac{P \times 1000}{\rho \times g \times Q \times H}$

$$\therefore \qquad P = \frac{\eta_o \times \rho \times g \times Q \times H}{1000}$$

$$= \frac{0.9 \times 1000 \times 9.81 \times 9 \times 25}{1000}$$

$$= 1986.5 \text{ kW}$$

(ii) $$N_s = \frac{N\sqrt{P}}{H^{5/4}} = \frac{200 \times \sqrt{1986.5}}{(25)^{5/4}}$$

$$= 159.46$$

(iii) As the specific speed lies in the range of 51 to 250, thus the turbine is a Francis turbine.

Example 4.17 In a hydroelectric station, water is available at the rate of 175 m^3/s under a head of 18 m. The turbine runs at a speed of 150 rpm with an overall efficiency of 82%. Find the number of turbines required if they have maximum specific speed of 460.

Solution

$$N_s = \frac{N\sqrt{P}}{H^{5/4}}$$

$$P = \left(\frac{N_s \times H^{5/4}}{N}\right)^2$$

$\therefore$ $$P = \left(\frac{460 \times (18)^{5/4}}{150}\right)^2$$

$$= 12927.5 \text{ kW}$$

Overall efficiency, $$\eta_o = \frac{SP}{WP} = \frac{P \times 1000}{\rho \times g \times Q \times H}$$

Power available from these turbines,

$$P = \frac{\eta_o \times \rho \times g \times Q \times H}{1000}$$

$$= \frac{0.82 \times 1000 \times 9.81 \times 175 \times 18}{1000}$$

$$= 25339.23 \text{ kW}$$

$\therefore$ Number of turbines required

$$= \frac{25339.23}{12927.5} = 1.96 = 2$$

Example 4.18 A turbine is to operate under a head of 25 m at 200 rpm. The discharge is 9 m^3/s. If the efficiency is 90%, determine the performance of turbine under a head of 20 m.

Solution

Overall efficiency, $$\eta_o = \frac{SP}{WP} = \frac{P \times 1000}{\rho \times g \times Q \times H}$$

$$P = \frac{\eta_o \times \rho \times g \times Q \times H}{1000}$$

$$= \frac{0.90 \times 1000 \times 9.81 \times 9 \times 25}{1000} = 1986.53 \text{ kW}$$

Using equation (4.38),

$$\frac{N_1}{\sqrt{H_1}} = \frac{N_2}{\sqrt{H_2}}$$

$$\therefore \quad N_2 = \frac{N_1 \sqrt{H_2}}{\sqrt{H_1}}$$

$$= \frac{200 \times \sqrt{20}}{\sqrt{25}} = 178.89 \text{ rpm}$$

Using equation (4.39),

$$\frac{P_1}{H_1^{3/2}} = \frac{P_2}{H_2^{3/2}}$$

$$\therefore \quad P_2 = \frac{P_1 H_2^{3/2}}{H_1^{3/2}}$$

$$= \frac{1986.53 \times (20)^{3/2}}{(25)^{3/2}} = 1421.45 \text{ kW}$$

Using equation (4.40),

$$\frac{Q_1}{\sqrt{H_1}} = \frac{Q_2}{\sqrt{H_2}}$$

$$\therefore \quad Q_2 = \frac{Q_1 \sqrt{H_2}}{\sqrt{H_1}}$$

$$= \frac{9 \times \sqrt{20}}{\sqrt{25}} = 8.05 \text{ rpm}$$

Example 4.19 A turbine is working under a head of 10 m. If speed ratio based on outer diameter = 2, flow ratio = 0.6, boss diameter = 0.3 times of external diameter, and overall efficiency = 80%, determine (i) the diameter, (ii) the speed, and (iii) specific speed of a propeller turbine to develop 6000 kW.

Solution

(i) Overall efficiency, $$\eta_o = \frac{SP}{WP} = \frac{P \times 1000}{\rho \times g \times Q \times H}$$

$$Q = \frac{P \times 1000}{\rho \times g \times \eta_o \times H}$$

$$\therefore \quad Q = \frac{6000 \times 1000}{1000 \times 9.81 \times 0.8 \times 10} = 76.45 \text{ m}^3/\text{s}$$

$$V_{f1} = K_f \sqrt{2gH}$$

$$= 0.6 \times \sqrt{2 \times 9.81 \times 10} = 8.4 \text{ m/s}$$

Discharge through the turbine is given by,

$$Q = \frac{\pi}{4} [D_o^2 - D_b^2] \times V_{f1}$$

$$= \frac{\pi}{4} [(D_o^2 - (0.3\, D_o)^2] \times 8.4$$

$$= 76.45 \text{ m}^3/\text{s}$$

$$\therefore \quad D_o = \sqrt{\frac{76.45 \times 4}{\pi \times 8.4 \times 0.91}} = 3.57 \text{ m}$$

(ii) $$u = K_u \sqrt{2gH}$$

$$= 2 \times \sqrt{2 \times 9.81 \times 10} = 28.01 \text{ m/s}$$

Also $$u = \frac{\pi DN}{60}$$

$$\therefore \quad N = \frac{u \times 60}{\pi \times D} = \frac{28.01 \times 60}{\pi \times 3.57}$$

$$= 149.85 \text{ rpm}$$

(iii) $$N_s = \frac{N\sqrt{P}}{H^{5/4}} = \frac{149.85 \times \sqrt{6000}}{(10)^{5/4}}$$

$$= 652.73$$

Example 4.20 A hydraulic turbine is to develop 1000 kW when running at 150 rpm under a head of 15 m. Determine the specific speed and maximum flow rate for the turbine if the overall efficiency is 90%. In order to predict its performance a 1:10 scale model is tested under a head of 8 m. What would be the speed, power output, and water consumption of the model if it runs under the condition similar to the prototype?

Solution

Overall efficiency, $$\eta_o = \frac{SP}{WP} = \frac{P \times 1000}{\rho \times g \times Q \times H}$$

$$Q = \frac{P \times 1000}{\rho \times g \times \eta_o \times H}$$

$$\therefore \quad Q = \frac{1000 \times 1000}{1000 \times 9.81 \times 0.9 \times 15}$$

$$= 7.55 \text{ m}^3/\text{s}$$

$$N_s = \frac{N\sqrt{P}}{H^{5/4}} = \frac{150 \times \sqrt{1000}}{(15)^{5/4}}$$
$$= 160.69$$

The two runners will be similar if

$$\left(\frac{H}{N^2 D^2}\right)_m = \left(\frac{H}{N^2 D^2}\right)_p$$
$$\left(\frac{Q}{ND^3}\right)_m = \left(\frac{Q}{ND^3}\right)_p$$
$$\left(\frac{P}{N^3 D^5}\right)_m = \left(\frac{P}{N^3 D^5}\right)_p$$

Model speed,
$$N_m = N_p \left(\frac{D_p}{D_m}\right)\left(\frac{H_m}{H_p}\right)^{1/2}$$

$$\therefore \quad N_m = 150 \times \left(\frac{10}{1}\right)\left(\frac{6}{10}\right)^{1/2}$$
$$= 1161.9 \text{ rpm}$$

Discharge in the model,

$$Q_m = Q_p \left(\frac{N_m}{N_p}\right)\left(\frac{D_m}{D_p}\right)^3$$

$$\therefore \quad Q_m = 7.55 \times \left(\frac{1161.9}{150}\right) \times \left(\frac{1}{10}\right)^3$$
$$= 0.0585 \text{ m}^3/\text{s}$$

Power developed by the model,

$$P_m = P_p \left(\frac{N_m}{N_p}\right)^3 \left(\frac{D_m}{D_p}\right)^5$$

$$\therefore \quad P_m = 1000 \times \left(\frac{1161.9}{150}\right)^3 \times \left(\frac{1}{10}\right)^5$$
$$= 4.65 \text{ kW}$$

4.13 SELECTION OF TURBINES

The following factors should be considered while selecting right type of hydraulic turbine.

1. **Specific Speed** As a general rule, it may be stated that as far as possible a turbine with highest permissible specific speed should be chosen. It will not only be the cheapest in itself but its relatively small size and high rotational speed will reduce the size of the generator

as well as powerhouse. But higher specific speed is responsible for cavitation which should be avoided. The specific speed should be high when the head is low and output is large. The types of turbine for different specific speed are: (i) Pelton wheel: 8.5 to 50, (ii) Francis turbine: 51 to 250, and (iii) Propeller and Kaplan turbine: 255 to 860.

2. **Head** The range of head for which each type of turbine is most suitable is given as: (i) Pelton wheel with Single or Multiple jet: 300 m or more, (ii) Pelton or Francis turbine: 150 m to 300 m, (iii) Francis turbine: 60 m to 150 m, and (iv) Kaplan or Propeller turbine: Less than 60 m.
3. **Part Load Operation** The turbines are required to work under variable load conditions. As the load varies from the normal working load, the efficiency would also vary. At part load, the performance of Kaplan and Pelton turbines is better in comparison to that of Francis and propeller turbines. For higher range of heads (i.e., 150 m to 300 m) Pelton turbine is preferable for part load operation in comparison to Francis turbine, and for heads below 30 m Kaplan turbine is preferable in comparison to propeller turbine.
4. **Disposition of Turbine Shaft** Generally, vertical shaft arrangement is preferable for large sized reaction turbine which is almost universally adopted. And in case of large sized impulse turbine, horizontal shaft arrangement is employed.
5. **Overall Cost** The overall cost which includes the initial cost and the running cost should also be considered while selecting a turbine.
6. **Cavitation** The alternate formation and collapse of vapor bubbles in a flowing fluid due to local fall in fluid pressure is called cavitation. Cavitation may cause severe damage to the surface. The surface becomes badly scored and pitted which ultimately fails by fatigue. In Pelton turbines there is possibility of cavitation occurring at the nozzle. In reaction turbines the cavitation commences at the top portion of the draft tubes. The installation of hydraulic reaction turbines over the tail race is affected by cavitation. According to Prof. Thoma, cavitation can be avoided if the turbine is installed in such a way that the cavitation factor σ is always greater than $\sigma_{critical}$. The value of $\sigma_{critical}$ for different turbines may be calculated by different empirical relations. $\sigma_{critical}$ is found to be a function of the specific speed of the turbine.

4.14 HYDRAULIC PUMPS

The hydraulic machines which convert the mechanical energy into hydraulic energy are called pumps. The hydraulic energy is in the form of pressure energy. Pump lifts liquids from a lower level to a higher level by absorbing power. So, a pump is a power absorbing device used to increase the pressure energy of a liquid. The pressure energy is increased by creating a region of low pressure (less than the atmospheric pressure) near the inlet of the pump and a higher pressure at the outlet of the pump. Due to this low pressure at inlet the liquid reaches to the pump from a low level reservoir. The high pressure created inside the pump forces the liquid out of the pump to be delivered to the reservoir. The liquid is more often water in the domestic and agriculture spheres. In industries, chemicals, petroleum products, and slurries have to be moved by the pumps.

Pumps Applications Pump finds important applications in the following fields: agriculture and irrigation works, water supply plants, sewage and drainage system, steam power plants, oil refineries, chemical plants and steel mills, food processing industries, mines, etc.

4.15 CLASSIFICATION OF PUMPS

Pumps can be broadly classified into two categories:

1. Rotodynamic pumps (or dynamic pressure pumps), and
2. Positive displacement pumps.

1. Rotodynamic Pumps (or Dynamic Pressure Pumps) The rotodynamic pumps have a rotating element called impeller. Angular momentum changes when the liquid passes through impeller due to which the pressure energy of the liquid increases. According to the flow direction rotodynamic pumps can be of three types, namely, radial flow pumps, axial flow pumps, and mixed flow pumps (Figure 4.17).

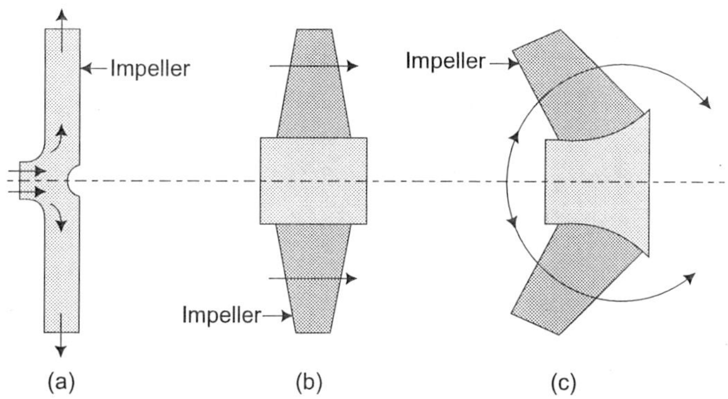

Fig. 4.17

(i) **Radial Flow Pumps** In radial flow pumps liquid flows through the impeller in radial direction only (Figure 4.17a). The radial flow (or centrifugal) pumps handle lower volumes at higher pressures. The centrifugal pumps are the most common examples of rotodynamic pumps.

(ii) **Axial Flow Pumps** In axial pumps liquid flows in axial direction only (Figure 4.17b). Axial flow pumps can handle very large volume but at limited pressures.

(iii) **Mixed Flow Pumps** In mixed flow pumps liquid flows through the impeller axially as well as radially (Figure 4.17c). Mixed flow pumps handle comparatively larger volumes at medium range of pressures.

2. Positive Displacement Pumps In the positive displacement pumps the liquid is sucked and actually pushed or bodily displaced due to the thrust exerted on it by a moving member which lifts the liquid to the required height. The liquid inside the positive displacement pumps may be subjected either to a reciprocating motion or to a rotary (or circular) motion. So, based on the motion of the liquid these pumps are of two types.

(i) Reciprocating Pumps The most common example of the positive displacement pumps is reciprocating pump. Reciprocating pumps first trap the liquid in a cylinder by suction and then push the liquid against pressure. These pumps are limited by the low speed of operation and small volume it handles.

(ii) Rotary Positive Displacement Pumps Rotary positive displacement pumps also trap the liquid in a volume and push the same out against pressure. These pumps are limited by lower pressures of operation and small volumes it handles. Its main types are gear, vane, and robe pumps.

4.16 CENTRIFUGAL PUMPS

Centrifugal pumps are so-called because energy is imparted to the fluid by centrifugal action of moving blades from the inner radius to the outer radius. In general, all the rotodynamic pumps resemble the reaction type of hydraulic turbines and they may be called reversed reaction turbines. Thus, the action of the centrifugal pump is just the reverse of an inward flow reaction turbine. Thus, the flow in centrifugal pumps is in the radial outward direction.

4.16.1 Main Parts of a Centrifugal Pump

Figure 4.18 shows the main parts of a centrifugal pump which are described below:

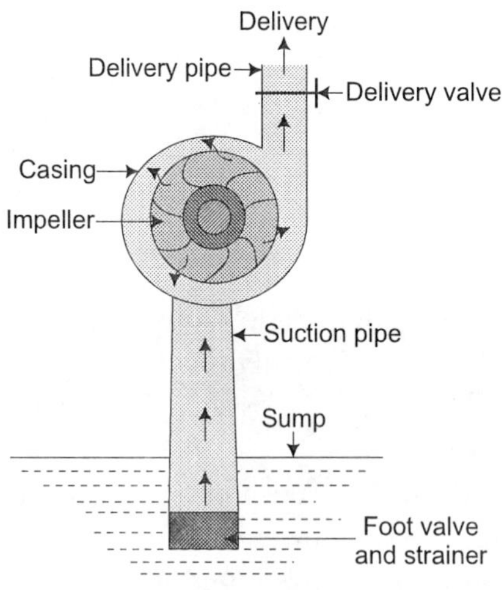

Fig. 4.18

I. Impeller It is a wheel or rotor having a series of backward curved blades or vanes. So, impeller is a rotating part of a centrifugal pump. It is mounted on a shaft which is coupled to the shaft of an electric motor. The impellers are of three types (Figure 4.19), namely, (i) Shrouded or closed impeller, (ii) Semi-open impeller, and (iii) Open impeller.

(i) Shrouded or Closed Impeller A shrouded or closed impeller is that whose vanes are covered on both sides with metal plates (Figure 4.19a). These metal plates or shrouds are known as crown plate and base plate. The closed impeller provides better guidance for the liquid and is more efficient. This type of impeller is most suited when the liquid to be pumped is pure and free from debris.

(ii) Semi-open Impeller If the vanes have only the base plate and no crown plate then the impeller is known as semi-open type impeller (Figure 4.19b). Such an impeller can be used even if the liquid has some debris.

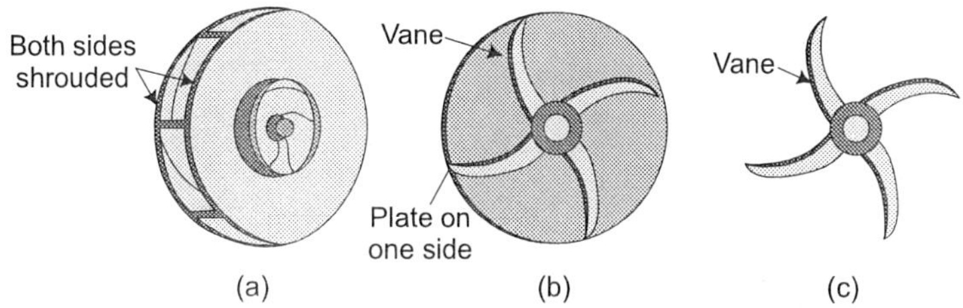

Fig. 4.19 Types of Impeller

(iii) Open Impeller An open impeller is that whose vanes have neither the crown plate nor the base plate, i.e., the vanes are open on both sides (Figure 4.19c). Such impellers are used for pumping the liquids containing suspended solid matter, such as paper pulp, sewage, and water containing sand or grit.

2. Casing The casing is an airtight chamber which surrounds the impeller and is designed in such a way that the kinetic energy of the water discharged at the outlet of the impeller is converted into pressure energy before the water leaves the casing and enters the delivery pipe. It is similar to the casing of a reaction turbine. The following three types of casing (Figure 4.20) are commonly used and the pump is named after the casing it uses: (i) Volute casing, (ii) Vortex casing, and (iii) Casing with guide blades.

(i) Volute Casing The volute casing is shown in Figure 4.20 (a). It is of spiral shape in which area of flow increases gradually from the impeller outlet to the delivery pipe. The increase in area of flow decreases the velocity of flow with corresponding increase in the pressure of water flowing through the casing. Single-stage pumps are mostly having volute casing. The volute casing has greater eddy losses which results in lower overall efficiency. The pumps having volute casing are known as volute pumps.

(ii) Vortex Casing If a circular chamber is provided between the impeller and the casing as shown in Figure 4.20 (b), such a casing is known as vortex casing. The circular chamber is also known as vortex or whirlpool chamber. The vortex chamber reduces the eddy formation to a considerable extent and thus the efficiency of the pump will be more than that of volute casing pump. The pumps using vortex casing are known as volute pump with vortex chamber.

(iii) Casing with Guide Blades The impeller is surrounded by a series of fixed guide blades which are mounted on a ring known as diffuser as shown in Fig. 4.20 (c). The guide vanes are designed in such a way that the water from the impeller enters it without shock. The liquid leaving the impeller passes through the passage between the guide vanes whose area increases. This reduces the velocity of flow through guide vanes and consequently increases the pressure of water. The water from the guide vanes then passes through the casing which is in most of the cases will be concentric with the rotor. The pumps having guide vanes are called diffuser pumps or turbine pumps. This type of casing is employed when the pump impellers are to be connected in series. The pumps with diffuser blades have maximum efficiency but are less satisfactory when a wide range of operating conditions is required.

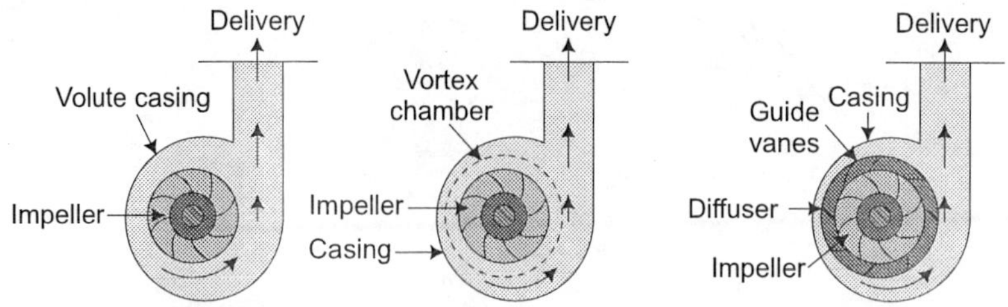

(a) Volute casing (b) Vortex casing (c) Casing with guide blades

Fig. 4.20 Types of Casing

3. Suction Pipe with Foot Valve and a Strainer It is a pipe whose upper end is connected to the inlet of the pump (or to the centre of the impeller which is known as eye) and the lower end dips into the liquid in the sump from which the liquid is to be pumped. The lower end of the suction pipe is fitted with a foot valve and a strainer. The foot valve is a non-return or one-way valve which opens in the upward direction only. It does not allow the liquid to flow in downwards direction back to the sump. The liquid first enters the strainer which removes the debris. The suction pipe should always be fitted in such a way that water always flows in the upward direction and all the pipe fittings must be airtight so that no air pockets are formed and the pumps works smoothly.

4. Delivery Pipe and Delivery Valve It is a pipe whose lower end is connected to the outlet of the pump and its upper end delivers liquid to the required height. Just near the outlet of the pump on delivery pipe a regulating valve is fitted which controls the flow of liquid from the pump into the delivery pipe.

4.16.2 Working of a Centrifugal Pump

The working of a centrifugal pump is explained stepwise below:

(i) The first step in the working operation of a centrifugal pump is priming. During the priming operation delivery valve is kept closed. Priming is the operation in which the suction pipe, casing of the pump, and the portion of the delivery pipe up to the delivery valve are completely filled with the liquid which is to be pumped so that no air pocket is left. Even the presence of a small air pocket in any part of the pump may cause no delivery of the liquid from the pump. So, it is essential to properly prime a pump before it can be started.

(ii) The delivery valve is still kept closed to reduce the starting torque and the electric motor is started to rotate the impeller.

(iii) The rotation of the impeller in the casing full of liquid produces a forced vortex which provides a centrifugal head to the liquid and thus results in an increase of pressure throughout the liquid. The rise in pressure head at any point of the rotating liquid is proportional to the square of the tangential velocity of the liquid at that point. Thus, if the speed of the impeller of the pump is high enough the pressure of the liquid surrounding the impeller increases considerably. When the delivery valve is opened the liquid flows in an outward radial direction with a high velocity and pressure.

(iv) The rotation of the impeller due to centrifugal action causes a partial vacuum at the eye of the impeller which causes the suction of the liquid from the sump through the suction pipe. The sucked liquid replaces the liquid which is being discharged from the circumference of the impeller.

(v) The high pressure of the liquid leaving the impeller is utilized in lifting the liquid to the required height.

4.16.3 Velocity Triangle and Work Done by the Centrifugal Pump

The expression for the work done by the impeller on the water is obtained by drawing velocity triangles at inlet and outlet of the impeller in the same way as for a turbine. For drawing the velocity triangles, the same notations are used as that of turbines. The velocity triangles are shown in Figure 4.21.

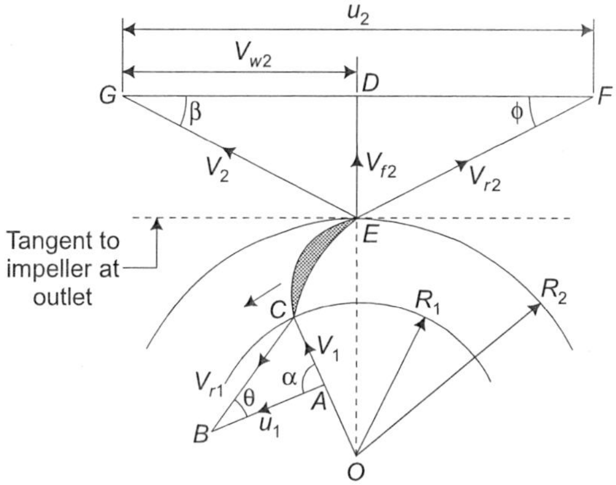

Fig. 4.21 Velocity Triangles at Inlet and Outlet

The water enters the impeller radially at inlet for best efficiency of the pump, hence $\alpha_1 = 0$. A centrifugal pump is the reverse of a radially inward flow reaction turbine. In the case of radially inward flow reaction turbine, the work done by water on the runner per sec per unit weight is given by Euler equation as,

$$W = \frac{1}{g}\,[V_{w1}\, u_1 - V_{w2}\, u_2]$$

∴ Work done by impeller on water per sec per unit weight of water striking per sec,

$$W = -\text{ [Work done by the turbine]}$$

$$= -\left[\frac{1}{g}\,(V_{w1}\, u_1 - V_{w2}\, u_2)\right]$$

Because water enters radially at inlet,

$$\therefore \qquad V_{w1} = 0$$

∴ Work done by impeller on water per sec per unit weight of water striking per sec,

$$W = \frac{1}{g}[V_{w2}\, u_2] \tag{4.44}$$

And work done by impeller on water per sec,

$$W = \frac{w}{g}[V_{w2}\, u_2] \tag{4.45}$$

where w = Weight of water

$$= \rho \times g \times Q$$

$$Q = \pi\, D_1\, B_1 \times V_{f1} = \pi\, D_2\, B_2 \times V_{f2}$$

where B_1 and B_2 are widths of impeller at inlet and outlet, and

V_{f1} and V_{f2} are the velocities of flow at inlet and outlet of the impeller.

4.16.4 Efficiencies of a Centrifugal Pump

(i) Manometric Efficiency It is defined as the ratio of manometric head developed by the pump to the head imparted by the impeller to the liquid. It is denoted by η_{man} and is given by,

$$\eta_{man} = \frac{\text{Manometric head developed}}{\text{Head imparted by impeller to water}}$$

$$= \frac{H_m}{(V_{w2}\, u_2/g)}$$

$$\therefore \quad \eta_{man} = \frac{g\, H_m}{V_{w2}\, u_2} \tag{4.46}$$

Manometric head is defined as the head against which a centrifugal pump has to work.

(ii) Mechanical Efficiency It is defined as the ratio of the power available at the impeller to the power at the shaft of the centrifugal pump. It is denoted by η_m and is given by,

$$\eta_m = \frac{\text{Power at the impeller}}{\text{Power at the shaft}}$$

$$= \frac{(V_{w2}\, u_2/g)}{P} \tag{4.47}$$

(iii) Overall Efficiency It is defined as the ratio of power output of the pump to the power input to the pump. It is denoted by η_o.

Also

$$\eta_o = \eta_{man} \times \eta_m \tag{4.48}$$

4.16.5 Classification of Centrifugal Pump

Centrifugal pumps may be classified in several ways. On the basis of their characteristic features, utility, design, and constructional features centrifugal pumps may be classified as follows:

1. On the basis of type of casing
   (a) Volute pump,
   (b) Volute pump with vortex chamber, and
   (c) Diffuser pump or turbine pumps.
2. On the basis of working head
   (a) Low Head They work against heads up to 15 m and below.
   (b) Medium Head Work between the ranges 15 m to 40 m.
   (c) High Head These pumps are used to build up heads more than 40 m.
3. On the basis of relative direction of flow through impeller
   Radial Flow Liquid flows through the impeller in radial direction only.
   Axial Flow Pump Liquid flows through the impeller in axial direction only.
   Mixed Flow Liquid flows through the impeller axially as well as radially.
4. On the basis of number of stages
   (a) Single-stage pump, and
   (b) Multistage pump.

   If a centrifugal pump consists of two or more impellers, the pump is called multistage centrifugal pump. The impellers may be mounted on the same shaft or on different shafts. Multistage pumps are required to perform two functions: (1) Produce high head, (2) Discharge large quantity of water.

   To develop a high head the impellers are connected in series or mounted on the same shaft, whereas for discharging large quantity of water the impellers are connected in parallel.
5. On the basis of liquid handled
   (a) Closed impeller pump,
   (b) Semi-open impeller pump, and
   (c) Open impeller pump.
6. On the basis of number of entrances to the impeller (Figure 4.22)

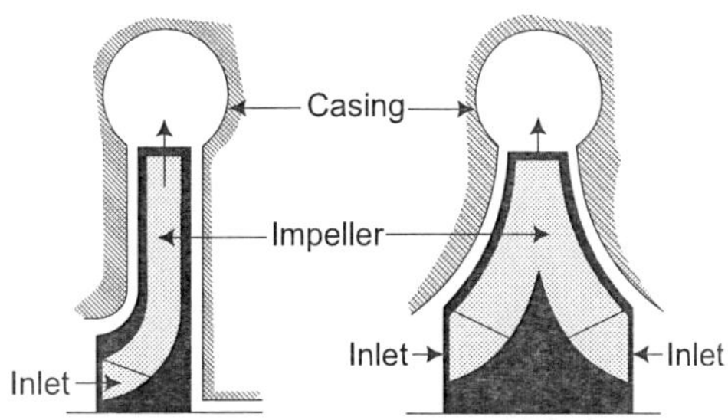

(a) Single entry (b) Double entry

Fig. 4.22

(a) **Single Entry or Single Suction Pump** Liquid is admitted from a suction pipe from one side of impeller (Figure 4.22a).

(b) **Double Entry or Double Suction Pump** Liquid enters from the both sides of impeller (Figure 4.22b).

7. On the basis of specific speed
   (a) Low speed radial flow pump: Specific speed varies from 10 to 30
   (b) Medium speed radial flow pump: Specific speed varies from 30 to 50
   (c) High speed radial flow pump: Specific speed varies from 50 to 80
   (d) Mixed flow pump: Specific speed varies from 80 to 160
   (e) Axial flow pump: Specific speed varies from 100 to 450
8. On the basis of disposition of shaft
   (a) **Horizontal Shaft** Generally, the pumps have horizontal shafts.
   (b) **Vertical shafts** For deep wells and mines pumps have vertical shafts.

4.16.6 Specific Speed of a Centrifugal Pump

The specific speed (N_s) of a centrifugal pump is defined as the speed of a geometrically similar pump which would deliver 1 m^3 of liquid per second against a head of 1 m. Specific speed is used for the classification of pumps on the basis of their performance and dimensions regardless of their actual size or speed at which they operate.

Specific speed may be derived as follows:

$$Q = \pi \times D \times B \times V_f$$

or $$Q \propto D \times B \times V_f$$

or $$Q \propto D^2 \times V_f \qquad (\because B \propto D) \qquad \text{(i)}$$

$$u \propto DN$$

also $$u \propto V_f \propto \sqrt{H_m}$$

$\therefore$ $$\sqrt{H_m} \propto DN$$

or $$D \propto \frac{\sqrt{H_m}}{N}$$

Substituting values of D and V_f in the expression (i), we have,

$$Q \propto \left(\frac{\sqrt{H_m}}{N}\right)^2 \times \sqrt{H_m}$$

or $$Q \propto \frac{H_m^{3/2}}{N^2}$$

or $$Q = k\,\frac{H_m^{3/2}}{N^2} \qquad (k = \text{Constant}) \qquad \text{(ii)}$$

If $H_m = 1$ m, $Q = 1$ m^3/s, then $N = N_s$, substituting these values in (ii), we have,

$$1 = k\,\frac{(1)^{3/2}}{N_s^{\,2}}$$

or
$$N_s^{\,2} = k.$$

Substituting the value of k in the expression (ii), we have,

$$Q = N_s^{\,2}\,\frac{H_m^{3/2}}{N^2}$$

or
$$N_s^{\,2} = \frac{N^2 Q}{H_m^{3/2}}$$

$\therefore$
$$N_s = \frac{N\sqrt{Q}}{H_m^{3/4}} \tag{4.49}$$

where N = Pump speed in revolution per minute, Q = Discharge in m^3/s, and H_m = Manometric head in m.

Specific speeds of different type of pump impellers are given below:

(i) Radial flow pump: Specific speed varies from 10 to 80 (10 to 30 slow speed, 30 to 35 medium speed, and 50 to 80 high speed).

(ii) Mixed flow pump: Specific speed varies from 80 to 160.

(iii) Axial flow pump: Specific speed varies from 100 to 450.

Example 4.21 The following data are given for a centrifugal pump: internal diameter = 250 mm, external diameter = 500 mm, speed = 1250 rpm, vane angle at inlet = 15°, vane angle at outlet = 25°, net head = 60 m, the whirl velocity at inlet is zero, and the flow velocity is constant. Determine the work done by the impeller per unit weight of water and the manometric efficiency.

Solution Refer Figure 4.23. Tangential velocity of impeller at inlet and outlet,

$$u_1 = \frac{\pi D_1 N}{60} = \frac{\pi \times 0.25 \times 1250}{60}$$

$$= 16.36 \text{ m/s}$$

$$u_2 = \frac{\pi D_2 N}{60} = \frac{\pi \times 0.5 \times 1250}{60}$$

$$= 32.72 \text{ m/s}$$

From inlet velocity triangle,

$$\tan\theta = \frac{V_{f1}}{u_1}$$

$\therefore$
$$V_{f1} = u_1 \times \tan\theta = 16.36 \times \tan(15) = 4.38 \text{ m/s}$$

$$V_{f1} = V_{f2} = 4.38 \text{ m/s}$$

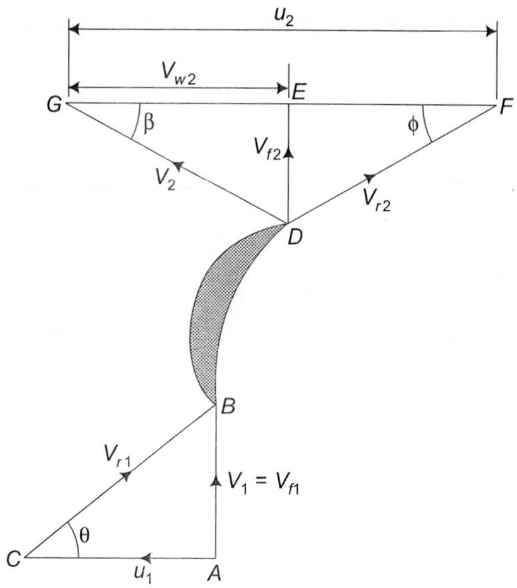

Fig. 4.23

From outlet velocity,

$$\tan \phi = \frac{V_{f2}}{u_2 - V_{w2}}$$

$$V_{w2} = u_2 - \frac{V_{f2}}{\tan \phi} = \frac{32.72 - 4.38}{\tan (25)}$$

$$= 23.33 \text{ m/s}$$

Work done by impeller on water per sec per unit weight is given by,

$$= \frac{1}{g} [V_{w2} \, u_2] = \frac{1}{9.81} \times [23.33 \times 32.72]$$

$$= 77.81 \text{ Nm/N}$$

Manometric efficiency is given by,

$$\eta_{\text{man}} = \frac{g \, H_m}{V_{w2} \, u_2}$$

$$\therefore \qquad \eta_{\text{man}} = \frac{9.81 \times 60}{23.33 \times 32.72}$$

$$= 0.7711 \text{ or } 77.11\%$$

Example 4.22 A centrifugal pump is to discharge 0.12 m^3/s at a speed of 1400 rpm against a head of 20 m. The impeller diameter is 200 rpm, its width at outlet is 40 mm and manometric efficiency is 80%. Determine the vane angle at the outer periphery of the impeller.

Solution Refer Figure 4.23.

Tangential velocity of impeller outlet,

$$u_2 = \frac{\pi D_2 N}{60} = \frac{\pi \times 0.2 \times 1400}{60}$$

$$= 14.66 \text{ m/s}$$

Discharge is given by,

$$Q = \pi \times D_2 \times B_2 \times V_{f2}$$

$$V_{f2} = \frac{Q}{\pi \times D_2 \times B_2}$$

$$\therefore \quad V_{f2} = \frac{0.12}{\pi \times 0.2 \times 0.04} = 4.77 \text{ m/s}$$

Manometric efficiency is given by,

$$\eta_{\text{man}} = \frac{g H_m}{V_{w2}\, u_2}$$

$$V_{w2} = \frac{g H_m}{\eta_{\text{man}}\, u_2}$$

$$\therefore \quad V_{w2} = \frac{9.81 \times 20}{0.8 \times 14.66} = 12.55 \text{ m/s}$$

From outlet velocity,

$$\tan \phi = \frac{V_{f2}}{u_2 - V_{w2}}$$

$$\phi = \tan^{-1}\left(\frac{V_{f2}}{u_2 - V_{w2}}\right)$$

$$\therefore \quad \phi = \tan^{-1}\left(\frac{4.77}{14.66 - 12.55}\right)$$

$$= 66.13°$$

4.17 RECIPROCATING PUMPS

The reciprocating pumps are positive displacement pumps in which a certain volume of liquid is taken in an enclosed volume and then it is forced out against pressure to the required application. The mechanical energy is converted into pressure energy by sucking the liquid into a cylinder in which a piston is reciprocating which exerts the thrust on the liquid and increases its pressure energy. Reciprocating pumps are now obsolete due to their high capital and maintenance cost relative to the centrifugal pumps. Reciprocating pumps are being replaced by centrifugal pumps where large quantities of liquid are to be handled. Reciprocating pumps are still used for oil drilling industries, and where there is no electricity. These are also used for pneumatic pressure systems. The hand pumps are used to lift water from deep wells for drinking purposes.

4.17.1 Classification of Reciprocating Pumps

The reciprocating pumps may be classified as:

1. According to the water being in contact with piston
   - (i) **Single Acting Pump** If the water is in contact with one side of piston the pump is known as single acting pump.
   - (ii) **Double Acting Pump** If the water is in contact with both sides of the piston the pump is known as double acting pump.
2. According to the number of cylinders provided
   - (a) Single cylinder pump,
   - (b) Double cylinder pump, and
   - (c) Multi cylinder pump

4.17.2 Main Parts of a Reciprocating Pump

The main parts of the reciprocating pump are: (i) Cylinder, (ii) Piston (or plunger), (iii) Piston rod, (iv) Crank, (v) Connecting rod, (vi) Suction pipe and suction valve, and (vii) Delivery pipe and delivery valve which are shown in Figure 4.24.

4.17.3 Working of Reciprocating Pump

Figure 4.24 shows a single acting reciprocating pump. It consists of a piston (or plunger) which moves to and fro in close fitting cylinder. The cylinder is connected to the suction and delivery pipes, each of which is provided with a one way valve called suction valve and delivery valve respectively. The suction valve allows water from the suction pipe to the cylinder while delivery valve allows water from the cylinder to delivery pipe only. The piston is connected to a crank by means of a connecting rod. The action of reciprocating pump is similar to that of reciprocating engines. When the crank is rotated by a driving engine or motor, the piston moves to and fro in the cylinder. When the crank rotates from $\theta = 0°$ to $\theta = 180°$ the piston which is at the extreme left position '*A*' moves to extreme right position '*B*'. This outward movement of the piston from position '*A*' to position '*B*' is called suction stroke. Due to outward movement of the piston partial vacuum is created inside the cylinder. But atmospheric pressure is acting on the surface of liquid in the sump which is more than the pressure inside the cylinder. Thus, liquid is forced in the suction pipe which opens the suction valve and fills the cylinder.

When the crank is further rotated from $\theta = 180°$ to $\theta = 360°$ the piston moves inwardly from position '*B*' to position '*A*'. This inward movement of the piston raises the pressure of the liquid inside the cylinder above the atmospheric pressure due to which the suction valve closes and the delivery valve opens. Thus, the liquid is then forced into delivery pipe and raised to the required

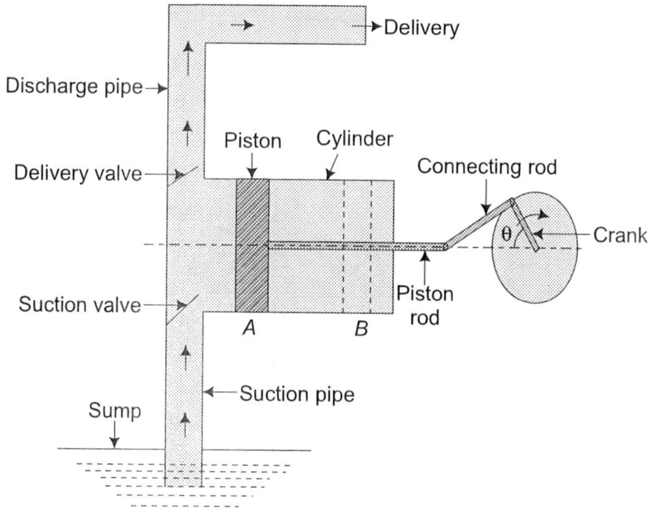

Fig. 4.24 Main parts of a Reciprocating Pump

height. This inward movement of the piston which causes the delivery of the liquid to the required height is known as delivery stroke.

Let, r = Radius of crank, N = Crank speed in rpm, $L = 2r$ = Length of the stroke or cylinder, D = Diameter of the piston, $A = (\pi D^2)/4$ = Area of the piston or cylinder, h_s = Height of the axis of the cylinder from water surface in sump, h_d = Delivery head or height of delivery outlet above the cylinder axis.

Discharge (or volume) of water swept in one revolution = $A \times L$

Number of revolutions per second = $N/60$

So, discharge of the pump per second,

$$Q = \frac{(A \times L \times N)}{60} \tag{4.50}$$

Weight of water discharged per second,

$$w = \rho\, gQ = \frac{\rho\, g\, (A \times L \times N)}{60} \tag{4.51}$$

Work done per second,

$$W = \frac{\rho\, g\, (A \times L \times N)}{60} \times (h_s + h_d) \tag{4.52}$$

Power required to drive the pump,

$$P = \frac{W}{1000} = \frac{\rho\, g\, (A \times L \times N)}{60 \times 1000} \times (h_s + h_d) \text{ kW} \tag{4.53}$$

For double acting pump discharge becomes two times that of single acting and so the work done and power required. The discharge, work done, and power required to drive the double acting reciprocating pump are given below:

$$Q = \frac{2 \times (A \times L \times N)}{60} \tag{4.54}$$

$$W = \frac{2 \times \rho\, g\, (A \times L \times N)}{60} \times (h_s + h_d) \tag{4.55}$$

$$P = \frac{2 \times W}{1000} = \frac{2 \times \rho\, g\, (A \times L \times N)}{60 \times 1000} \times (h_s + h_d) \text{ kW} \tag{4.56}$$

4.17.4 Slip of Reciprocating Pump

The difference between the theoretical discharge (Q_{th}) and actual discharge (Q_{act}) is called slip. Due to leakage the actual discharge of a pump is less than the theoretical discharge. Mathematically, slip is given by,

$$\text{Slip} = Q_{th} - Q_{act} \tag{4.57}$$

Generally, slip is expressed as percentage slip which is given by,

$$\%\text{Slip} = \frac{Q_{th} - Q_{act}}{Q_{th}} \times 100$$

$$= \left(1 - \frac{Q_{act}}{Q_{th}}\right) \times 100$$

$$= (1 - C_d) \times 100 \tag{4.58}$$

where $C_d = (Q_{act}/Q_{th})$ is the coefficient of discharge.

If the delivery pipe is small in length more water may be delivered through the delivery valve because of late closing of the suction valve. In such cases the slip will become negative slip.

Example 4.23 A single acting reciprocating pump has a piston diameter of 0.4 m and a stroke of 0.3 m. The suction head is 10 m and the delivery head is 15 m. The speed of the pump is 45 rpm. Actual discharge of the pump is 0.026 m^3/s. Determine the slip and percentage of slip. Also calculate the theoretical power required to drive the pump.

Solution Theoretical discharge is given by,

$$Q_{th} = \frac{(A \times L \times N)}{60}$$

$$= \frac{\pi}{4} \times D^2 \times L \times \frac{N}{60}$$

$$= \frac{\pi}{4} \times (0.4^2) \times 0.3 \times \frac{45}{60} = 0.0283 \text{ m}^3/\text{s}$$

$$\text{Slip} = Q_{th} - Q_{\text{act}} = 0.0283 - 0.026$$
$$= 0.0023 \text{ m}^3/\text{s}$$
$$\%\text{Slip} = \frac{Q_{th} - Q_{\text{act}}}{Q_{th}} \times 100 = \frac{0.0023}{0.0283} \times 100$$
$$= 8.13\%$$

Theoretical power required to drive the pump,

$$P = \frac{\rho\, g\,(A \times L \times N)}{60 \times 1000} \times (h_s + h_d)$$
$$= \frac{1000 \times 9.81 \times 0.283}{1000} \times (10 + 15)$$
$$= 6.94 \text{ kW}$$

4.18 COMPARISON BETWEEN RECIPROCATING AND CENTRIFUGAL PUMPS

Centrifugal Pumps	*Reciprocating Pumps*
1. The centrifugal pumps can run at higher speeds and hence the discharge capacity is high.	1. The speed is less and hence the discharge capacity is low.
2. Suitable for large volumes of discharge at moderate pressures in a single stage.	2. Suitable for fairly low volumes of flow at high pressures.
3. As there are no valves, so these pumps can handle highly viscous liquids, grit, slurry, etc.	3. These pumps can handle only pure water and less viscous liquids free from impurities.
4. The wear and tear is less because of less moving parts.	4. The wear and tear is more because of more moving parts.
5. Simple in construction and its initial cost is low.	5. More complex due to several moving parts and its initial cost is high.
6. Flow is smooth and continuous.	6. Flow is fluctuating.
7. Requires less maintenance cost and occupies less space.	7. Requires high maintenance cost and occupies more space.
8. The efficiency is high and torque is uniform.	8. The efficiency is low and torque is not uniform.
9. It requires priming.	9. It does not require priming.
10. The delivery valve should be closed before switching on the pump.	10. The delivery valve should not be closed before switching on the pump.
11. It can be operated at very high speeds without any danger of separation and cavitation.	11. The maximum speed is limited from the considerations of separation and cavitation.

4.19 HYDRAULIC PRESS

The hydraulic press is a device used for lifting heavy loads by the application of much smaller force. The first hydraulic press was built in 1795 by Joseph Bramah and it is still in use. It is based on Pascal's law which states that intensity of pressure is transmitted equally in all directions through a mass of fluid at rest.

Construction The hydraulic press consists of two cylinders say 'C_1' and 'C_2' of different diameters. The larger diameter cylinder 'C_1' has ram while the smaller diameter cylinder 'C_2' has a plunger as shown in Figure 4.25. These two cylinders are connected by a chamber having fluid in it through which pressure is transmitted.

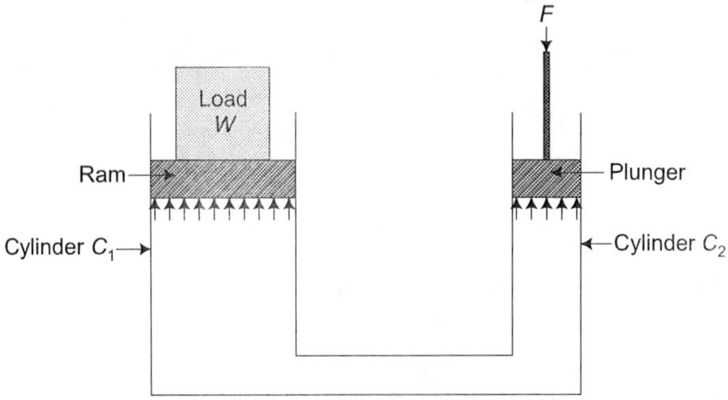

Fig. 4.25 Hydraulic Press

Working Principle When a small force 'F' is applied on the plunger in the downward direction a pressure is produced on the liquid in contact with the plunger. This pressure is transmitted equally in all directions and acts on the ram in the upward direction. Thus, the heavier load placed on the ram is lifted up.

Let W = Weight to be lifted, F = Force applied on the plunger, A = Area of ram, a = Area of plunger, and pressure intensity produced by force, $p = F/a$.

According to Pascal's law, the pressure intensity p will be equally transmitted in all directions.

∴ The pressure intensity on ram,

$$p = F/a \quad \text{(i)}$$

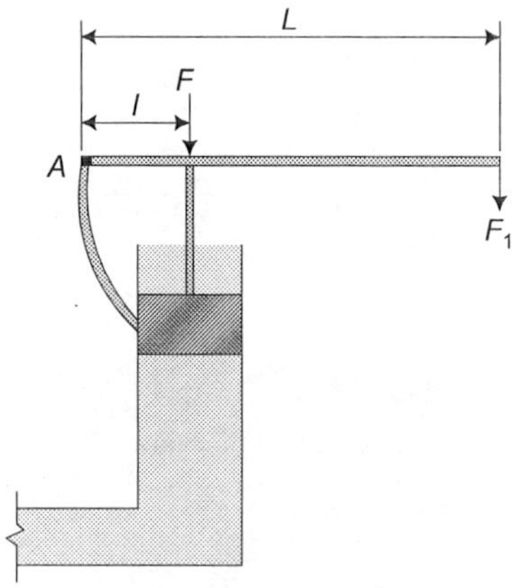

Fig. 4.26

But pressure intensity on ram is also equal to W/A (ii)

From equations (i) and (ii),

$$\frac{F}{a} = \frac{W}{A}$$

$$\therefore \quad W = \frac{F}{a} \times A \quad (4.59)$$

$$\text{Mechanical advantage (MA)} = \frac{\text{Load lifted}}{\text{Force applied}}$$

$$= \frac{W}{F} = \frac{A}{a} \tag{4.60}$$

If a force is applied by the lever on the plunger then a force F_1 smaller than F can lift the load W (Figure 4.26).

Taking moments about A,

$$F_1 \times L = F \times l$$

$$F = F_1 \times \frac{L}{l} \tag{4.61}$$

Substituting the value of F in equation (4.59), we have,

$$W = F_1 \times \frac{L}{l} \times \frac{A}{a} \tag{4.62}$$

or

$$\frac{W}{F_1} = \frac{L}{l} \times \frac{A}{a}$$

$\therefore$

$$MA = \frac{L}{l} \times \frac{A}{a} \tag{4.63}$$

where $(L/l)$is called the leverage of the hydraulic press.

Commonly, the hydraulic presses are used in sheet metal press work, die sinking, cotton press, forging, etc.

Example 4.24 A hydraulic press has a ram of 250 mm diameter and a plunger of 25 mm diameter. (i) If the applied force on the plunger is 40 N, determine the weight lifted. (ii) If a lever with the leverage is 1/10 is used for applying force on the plunger, determine the force needed at the end of the lever.

Solution

(i) Area of ram,

$$A = (\pi/4)\, D^2 = (\pi/4)\, (0.25)^2$$

$$= 0.04909 \text{ m}^2$$

Area of plunger,

$$a = (\pi/4)\, d^2 = (\pi/4)\, (0.025)^2$$

$$= 0.000491 \text{ m}^2$$

The weight is given by,

$$W = \frac{F}{a} \times A = \frac{40}{0.000491} \times 0.04909$$

$$= 3999.18 \text{ N}$$

(ii) Let F_1 be the force applied at the end of the lever which can be evaluated by,

$$W = F_1 \times \frac{L}{l} \times \frac{A}{a}$$

or $$F_1 = W \times \frac{l}{L} \times \frac{a}{A}$$

∴ $$F_1 = 3999.18 \times \frac{1}{10} \times \frac{0.000491}{0.04909}$$

$$= 3.999 \text{ N}$$

4.20 HYDRAULIC JACK

The hydraulic jack is a useful hydraulic device which is extensively used for lifting heavy loads by applying a much small effort. Its working principle is similar to that of a hydraulic press. Its main parts are: (i) Fixed ram, (ii) Sliding cylinder having a screw at its top, (iii) Cistern having oil and a reciprocating pump, and (iv) A cover with seat for lifting load.

The sliding cylinder can slide on a fixed ram as shown in Figure 4.27.

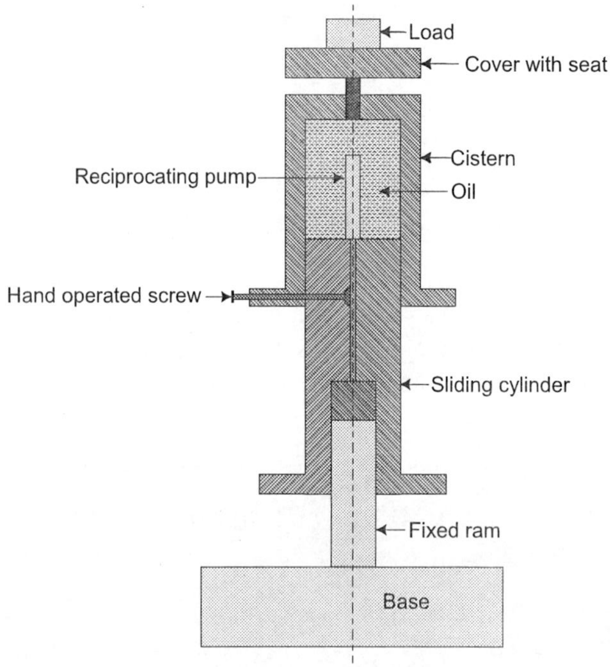

Fig. 4.27 Hydraulic Jack

For lifting the load the hydraulic oil is pumped in the space provided between the ram and the cylinder with the help of a reciprocating pump. The screw at the top of the cylinder acts like a delivery valve. For lowering the load the delivery valve is opened which makes the oil to flow back in the cistern.

4.21 HYDRAULIC LIFT

The hydraulic lift is a device which is used for carrying the persons or goods from one floor to another in multistoreyed buildings. The hydraulic lifts are of two types, namely, direct acting hydraulic lift, and suspended hydraulic lift.

(i) **Direct Acting Hydraulic Lift** It consists of a ram sliding in a fixed cylinder as shown in Figure 4.28.

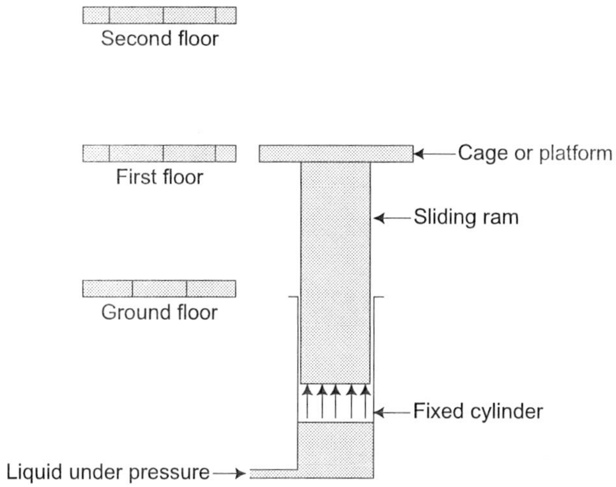

Fig. 4.28 Direct Acting Hydraulic Lift

A cage or platform is fitted at the end of the ram on which passengers may stand or goods may be placed. When the fluid under pressure is admitted into the cylinder the ram moves vertically up and the cage is lifted to the required height. The cage moves in the downward direction by removing the liquid from the fixed cylinder.

(ii) **Suspended Hydraulic Lift** The suspended hydraulic lift is a modified form of direct acting hydraulic lift. Modern lifts are generally of suspended type and have a high velocity ratio. The lifting speeds vary from 100 m/min to 200 m/min. It consists of a cage which is suspended by wire ropes, a jigger consists of a fixed cylinder, a sliding ram, and a set of two pulleys, namely, fixed pulley block, and movable pulley block as shown in Figure 4.29.

When liquid under pressure is admitted into the fixed cylinder of the jigger the sliding ram is forced to move towards left. The movable pulley block which is connected to the sliding ram also moves towards left. This increases the gap between the two pulley blocks. As a result, the wire rope connected to the cage is pulled and the cage is lifted up.

When the liquid from the fixed cylinder is removed the cage is lowered. The removal of liquid causes the movement of the sliding ram towards right. This decreases the gap between the two blocks and thus the cage is lowered due to increase in length of the rope.

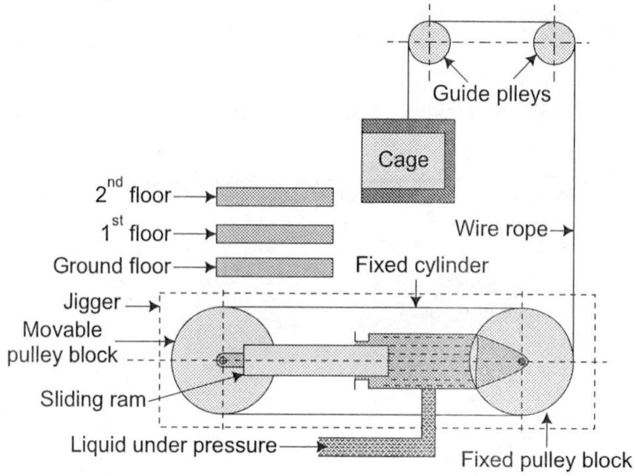

Fig. 4.29 Suspended Hydraulic Lift

EXERCISE

(i) Objective Type Questions

1. An impulse turbine:
  (a) always makes use of a draft tube
  (b) always operates submerged
  (c) operates by initial complete conversion to kinetic energy
  (d) suits to low head installations

2. Francis turbine is:
  (a) a reaction turbine (b) an impulse turbine
  (c) a tangential flow turbine (d) an axial flow turbine

3. Kaplan turbine is:
  (a) a high head axial flow turbine (b) a low head axial flow turbine
  (c) an impulse turbine (d) a mixed flow turbine

4. Cavitation is caused by:
  (a) high pressure (b) low barometric pressure
  (c) high velocity (d) low pressure

5. Specific speed of a turbine:
  (a) $N\sqrt{P}/H^{5/4}$ (b) $N\sqrt{P}/H^{3/4}$ (c) $N/H^{5/4}$ (d) none

6. Unit power of a turbine:
  (a) $P/H^{5/4}$ (b) $P/H^{3/4}$ (c) $P/H^{1/4}$ (d) $P/H^{3/2}$

7. Pure reaction turbine is:
(a) Kaplan turbine (b) Lawn sprinkler (c) Francis turbine (d) None of the above

8. Pelton turbine ideally suits:
(a) low head and low discharge (b) high head and high discharge
(c) high head and low discharge (d) medium head and medium discharge

9. Draft tube in a reaction turbine helps to:
(a) convert the kinetic energy to pressure head
(b) prevent debris
(c) increase flow rate
(d) prevent air to enter the turbine

10. Modern Francis turbine is:
(a) axial flow (b) radial flow (c) mixed flow (d) none

11. For pumping mustard oil the suitable pump is:
(a) gear pump (b) vane pump
(c) reciprocating pump (d) centrifugal pump

12. For flood control and irrigation purposes suitable pump is:
(a) gear pump (b) vane pump
(c) reciprocating pump (d) centrifugal pump

13. Reciprocating pump is started with the delivery valve:
(a) partially open (b) open (c) closed (d) none

14. Slip of the reciprocating pump is:
(a) $(Q_{\text{act}} - Q_{th})/Q_{\text{act}}$ (b) $(Q_{th} - Q_{\text{act}})/Q_{\text{act}}$
(c) $(Q_{th} - Q_{\text{act}})/Q_{th}$ (d) none of the above

15. The hydraulic machine which converts the mechanical energy into pressure energy is called
(a) turbine (b) steam engine (c) gas turbine (d) pump

16. When the specific speed of a turbine is more than 300, the type of turbine is:
(a) Pelton with multi-jets (b) Francis turbine
(c) Kaplan turbine (d) Pelton turbine

17. Cavitation may occur in:
(a) Francis turbine (b) centrifugal pump
(c) Pelton turbine, reciprocating pump (d) both (a) and (b)

18. Mechanical efficiency of a centrifugal pump is the ratio of:
(a) shaft power to power at impeller (b) power at impeller to shaft power
(c) water power to impeller power (d) water power to shaft power

19. Manometric efficiency of centrifugal pump is given by:

(a) $(gH_m/V_{w2}\,u_2)$ (b) $(H_m/V_{w2}\,u_2)$

(c) $(g/V_{w2}\,u_2)$ (d) $(V_{w2}\,u_2/\,gH_m)$

20. Manometric head of a centrifugal pump is given by:

(a) total head at inlet – total head at outlet

(b) pressure head at inlet – pressure head at outlet

(c) total head at outlet – total head at inlet

(d) none of the above

21. For discharging a large quantity of water by multistage centrifugal pumps, the impellers are connected in:

(a) in parallel (b) in series

(c) in parallel and in series (d) none

22. For producing a high head of water by multistage centrifugal pumps, the impellers are connected in:

(a) in parallel (b) in series

(c) in parallel and in series (d) none

23. Speed ratio of Pelton turbine varies from:

(a) 0.8 to 0.9 (b) 0.6 to 0.7 (c) 0.3 to 0.4 (d) 0.45 to 0.50

24. Mechanical efficiency of a turbine is the ratio of:

(a) shaft power to water power (b) water power to shaft power

(c) shaft power to runner power (d) none

25. Hydraulic efficiency of a turbine is the ratio of:

(a) shaft power to water power (b) water power to shaft power

(c) runner power to water power (d) none

26. Overall efficiency of a turbine is:

(a) η_m/η_h (b) $\eta_m/(\eta_h \times \eta_{vol})$ (c) $(\eta_h \times \eta_{vol})/\eta_m$ (d) $\eta_h \times \eta_{vol} \times \eta_m$

27. Jet ratio is:

(a) ratio of jet diameter to wheel diameter

(b) ratio of wheel diameter to jet diameter

(c) product of jet diameter to wheel diameter

(d) none

28. Jet ratio varies from:

(a) 20 to 18 (b) 17 to 15 (c) 14 to 11 (d) 10 to 7

29. Number of blades in a Francis turbine runner varies from:

(a) 30 to 26 (b) 26 to 24 (c) 24 to 20 (d) 24 to 16

30. Specific speed of a Francis turbine varies from:
(a) 8.5 to 30 (b) 30 to 50 (c) 51 to 250 (d) 255 to 860

31. Specific speed of axial flow pump varies from:
(a) 30 to 50 (b) 50 to 80 (c) 80 to 160 (d) 100 to 450

32. Hydraulic efficiency of Kaplan turbine in comparison to Francis turbine is:
(a) equal (b) less (c) more (d) none of the above

33. Slip in reciprocating pump may be:
(a) zero (b) positive (c) negative (d) both (b) & (c)

34. For maximum efficiency of a Pelton turbine, blade velocity is:
(a) half of jet velocity (b) double the jet velocity
(c) equal to jet velocity (d) none of the above

35. Unit speed of a turbine is:
(a) $N/H^{1/2}$ (b) $N/H^{3/4}$ (c) $N/H^{1/4}$ (d) $N/H^{3/2}$

36. Unit discharge of a turbine is:
(a) $Q/H^{1/2}$ (b) $Q/H^{3/4}$ (c) $Q/H^{1/4}$ (d) $Q/H^{3/2}$

37. Spouting velocity of jet means:
(a) ideal velocity of jet (b) actual velocity of jet
(c) half of actual velocity (d) half of ideal velocity

38. Axial flow turbine is:
(a) old Francis turbine (b) Thomson turbine
(c) Fourneyron turbine (d) Jonval turbine

39. Following are impulse turbines except:
(a) Jonval turbine (b) Banki turbine
(c) Girard turbine (d) Thomson turbine

40. Following are reaction turbines except:
(a) Thomson turbine (b) Fourneyron turbine
(c) Banki turbine (d) Kaplan turbine

ANSWERS

1. (c)	2. (a)	3. (b)	4. (d)	5. (a)	6. (d)	7. (b)	8. (c)	9. (a)	10. (c)
11. (a)	12. (d)	13. (b)	14. (c)	15. (d)	16. (c)	17. (a)	18. (b)	19. (a)	20. (c)
21. (a)	22. (b)	23. (d)	24. (c)	25. (c)	26. (d)	27. (b)	28. (c)	29. (d)	30. (c)
31. (d)	32. (c)	33. (d)	34. (a)	35. (a)	36. (a)	37. (a)	38. (d)	39. (d)	40. (c)

(ii) Review Questions

1. Define a hydraulic turbine, and a pump.
2. Give the classification of turbines and the pumps.
3. Give a general layout of a hydroelectric power plant.
4. Give comparison between impulse turbine and reaction turbine.
5. Define the gross head, and net head.
6. Define the hydraulic efficiency, mechanical efficiency, volumetric efficiency, and overall efficiency of a turbine.
7. Draw a schematic diagram of a Pelton turbine and explain briefly its construction, and working.
8. Define governing of turbine. Explain the governing mechanisms of impulse and reaction hydraulic turbines.
9. Give the design aspects of a Pelton turbine.
10. Give the constructional and working details of a Francis turbine with a neat sketch.
11. Give the functions of a draft tube.
12. Define the speed ratio, and flow ratio.
13. Draw a schematic diagram of a propeller and Kaplan turbine and compare these two.
14. Give the differences between Francis and Kaplan turbines.
15. Define specific speed of a turbine and derive an expression for the same. Also give its physical significance.
16. Define the unit speed, unit power, unit discharge and also mention their use.
17. Define head coefficient, flow coefficient, and power coefficient.
18. Give the selection criteria for hydraulic turbines.
19. Draw the main parts of a centrifugal pump also explain how it works?
20. Differentiate between volute and vortex casing of a pump.
21. Define the following in context with a centrifugal pump:
    (i) manometric efficiency (ii) mechanical efficiency (iii) overall efficiency.
22. How are centrifugal pumps classified?
23. Describe the working principle of a reciprocating pump with a neat diagram.
24. Make comparison between reciprocating and centrifugal pumps.
25. Describe the following hydraulic devices with neat diagrams:
    (i) hydraulic press (ii) hydraulic jack (iii) hydraulic lift.

(iii) Numerical Problems

1. A Pelton wheel is to be designed for the following specifications: shaft power = 6000 kW, head = 300 m, speed = 550 rpm, overall efficiency = 85%. Jet diameter is not to exceed

one-tenth of the wheel. Determine: number of jets, diameter of the jet, the wheel diameter, and quantity of water required. Take $C_v = 0.98$ & $\varphi = 0.46$.

[**Ans.** 3, 0.1225 m, 1.225 m, 2.398 m^3/s]

2. The following data were obtained from a test on a Pelton wheel: head at the base of the nozzle = 85 m, discharge of the nozzle = 0.25 m^3/s, diameter of the jet = 95 mm, power available at the shaft = 140 kW, power absorbed in mechanical resistance = 2.5 kW. Calculate (i) the power lost in nozzle and (ii) power lost due to hydraulic resistance in the runner.

   [**Ans.** 15.82 kW, 13.35 kW]

3. The nozzle diameter of a Pelton wheel is 200 mm and it works under a head of 220 m. The wheel operates at 250 rpm and develops 3.8 MW. Blade outlet angle is 15° and the speed ratio is 0.45. If the coefficient of velocity is 0.99, calculate overall efficiency of the turbine. [**Ans.** 86.31%]

4. A Pelton wheel is having a mean bucket diameter of 1.2 m and is running at 1200 rpm. The net head on the Pelton wheel is 850 m. If the side clearance angle is 15° and discharge is 0.15 m^3/s, find power available at the nozzle, and hydraulic efficiency of the turbine. Take coefficient of velocity = 1. [**Ans.** 1250.77 kW, 95.54%]

5. A Pelton wheel is to be designed for the following specifications: shaft power = 10,000 kW, head = 350 m, speed = 700 rpm, overall efficiency = 88%. Jet diameter is not to exceed one-sixth of the wheel. Determine: the wheel diameter, number of jets required, and diameter of the jet. Take $C_v = 0.985$ & $\varphi = 0.45$. [**Ans.** 1.02 m, 2, 0.169 m]

6. Two jets strike the bucket of a Pelton wheel which is having shaft power as 6500 kW. The diameter of each jet is given as 100 mm. If the net head on the turbine is 500 m, find the overall efficiency of the turbine. Assume coefficient of velocity as unity. [**Ans.** 86.89%]

7. A double jet Pelton wheel operates under 75 m head and develops 1000 kW brake power when running at 525 rpm. Evaluate the flow rate and the diameter of the nozzle jet if the overall efficiency and coefficient of velocity are 85% and 0.98 respectively.

   [**Ans.** 1.599 m^3/s, 0.16 m]

8. A Francis turbine works at 450 rpm under a head of 125 m. Its diameter at inlet is 1 m and the flow area is 0.25 m^2. The guide blade angle and vane angle at inlet are 15° and 60° respectively. Determine the volume flow rate, the power developed, and hydraulic efficiency. [**Ans.** 1.867, 1225.9, 53.55%]

9. A Francis turbine with an overall efficiency of 80% is required to produce 160 kW power. It is working under a head of 20 m. The peripheral velocity = $0.25\sqrt{2gH}$, and the radial velocity of flow at inlet is = $0.95\sqrt{2gH}$. The wheel runs at 210 rpm and the hydraulic losses in the turbine are 10% of the available energy. Assuming radial discharge, determine the guide blade angle, the wheel vane angle at inlet, diameter of the wheel at inlet, and width of the wheel at inlet. [**Ans.** 27.82°, 31.49°, 0.4502 m, 3.83 cm]

10. An inward flow reaction turbine has an external and internal diameter as 1.2 m and 0.6 m respectively. The turbine is running at 130 rpm and width of turbine at inlet is 260 mm. The velocity of flow through the runner is constant and is equal to 2.5 m/s. The guide blade angle

makes an angle of 15° to the tangent of the wheel and the discharge at the outlet is radial. If the net head on the turbine is 10 m, determine the absolute velocity of water at inlet of runner, the velocity of whirl at inlet, the relative velocity at inlet, runner blade angles, width of the runner at outlet, mass of water flowing through the runner/second, power developed, and hydraulic efficiency.

[**Ans.** 9.66, 9.33, 2.76, 65.11°, 31.49°, 520 mm, 2450.44 kg/s, 186.75 kW, 77.7%]

11. A Kaplan turbine develops 60 MW under a net head of 25 m with an overall efficiency of 86%. Taking the value of speed ratio = 2, flow ratio = 0.6, diameter of the hub = 0.3 times of the diameter of the runner, calculate diameter of the runner, speed of the turbine, and specific speed of the turbine. [**Ans.** 5.47 m, 154.64 rpm, 677.59 rpm]

12. The hub diameter of a Kaplan turbine working under a head of 20 m is 0.35 times the diameter of the runner. The turbine is running at 95 rpm. The velocity of wheel at outlet is zero. If the vane angle of the extreme edge of the runner at outlet is 15° and flow ratio is 0.6, determine diameter of runner, diameter of boss, and discharge through the runner.

[**Ans.** 8.914 m, 3.119 m, 650.63 rpm]

13. A turbine is to operate under a head of 20 m at 250 rpm. The discharge is 10 m^3/s. If the efficiency is 85%, determine power generated, specific speed of the turbine, and type of turbine. [**Ans.** 1667.7 kW, 241.38 rpm, Francis turbine]

14. A turbine is to operate under a head of 30 m at 200 rpm. The discharge is 10 m^3/s. If the efficiency is 88%, determine the performance of turbine under a head of 25 m.

[**Ans.** 164.32 rpm, 1970.16 kW, 9.13 m^3/s]

15. A turbine is working under a head of 15 m. If speed ratio based on outer diameter = 2, flow ratio = 0.6, boss diameter = 0.3 times of external diameter, and overall efficiency = 90%, determine the diameter, the speed, and specific speed of a propeller turbine to develop 7500 kW. [**Ans.** 7.7 m, 85.1 rpm, 249.66 rpm]

16. A centrifugal pump is to discharge 0.1118 m^3/s at a speed of 1450 rpm against a head of 25 m. The impeller diameter is 250 rpm, its width at outlet is 50 mm, and manometric efficiency is 75%. Determine the vane angle at the outer periphery of the impeller.

[**Ans.** 59.74°]

17. A centrifugal pump delivers water against a head of 15 m at a speed of 975 m. The impeller diameter is 350 mm and outlet width is 50 mm. If the vane angle at outlet is 30° and manometric efficiency is 85%, determine the discharge of the pump. [**Ans.** 0.277 m^3/s]

18. The following data are given for a centrifugal pump: internal diameter = 200 mm, external diameter = 400 mm, speed = 1200 rpm, vane angle at inlet = 20°, vane angle at outlet = 30°, the whirl velocity at inlet is zero, and the flow velocity is constant. Determine the work done by the impeller per unit weight of water. [**Ans.** 44.1 Nm/N]

5

Simple Lifting Machines

5.1 INTRODUCTION

A machine may be defined as a device which receives energy in some available form and uses it to do some particular useful work. For example, internal combustion engine which converts the chemical energy of petrol or diesel into the rotary motion of the crankshaft. The machines may be simple machines or complex machines. The machines in which there is only one point for the application of load and one point for the effort application are called simple machines. For example, lever, pulley, inclined plane, screw, wedge, and wheel and axle. The machines in which there is more than one point for the application of effort and load are called compound machines. In other words, compound machines comprise several simple machines and are capable of doing complex tasks, for example, milling machine, lathe machine, shaper, and planer.

A lifting machine may be defined as a device capable of lifting heavy loads using an effort much smaller in magnitude. This objective is achieved by applying a small effort moving through a larger distance and the heavy load moving through the smaller distance. Some important lifting machines are simple wheel and axle, pulleys, differential pulley blocks, worm and worm wheel, winch crabs, and screw jack.

5.2 BASIC CONCEPTS AND DEFINITIONS

1. **Load** A load is defined as the force or weight which is to be lifted by the machine. It is denoted by W.
2. **Effort** Effort is defined as the force required by a machine to lift a load. It is denoted by P.
3. **Mechanical Advantage** Mechanical advantage is the ratio of the load lifted to the effort applied. It is denoted by MA and is given by,

$$MA = \frac{\text{Load Lifted}}{\text{Effort Applied}} = \frac{W}{P} \tag{5.1}$$

The effort applied is generally smaller than the load lifted thus the mechanical advantage will always be greater than one. Mechanical advantage of a machine is not a constant but it varies within a narrow range.

4. **Velocity Ratio** Velocity ratio is the ratio of the distance moved by an effort to the distance moved by the load. It is denoted by *VR* and is given by,

$$VR = \frac{\text{Distance moved by effort}}{\text{Distance moved by load}} = \frac{y}{x} \tag{5.2}$$

5. **Input** It is the work spent on the machine and is given by the product of the effort and the distance it travels.

$$\therefore \quad \text{Input} = \text{effort} \times \text{distance moved by effort}$$

$$= P \times y \tag{5.3}$$

6. **Output** It is the useful work obtained from the machine and is given by the product of load lifted and the distance through which it moves.

$$\therefore \quad \text{Output} = \text{load lifted} \times \text{distance moved by load}$$

$$= W \times x \tag{5.4}$$

7. **Efficiency of a Machine** This is the ratio of output to input. It is denoted by η and is given by,

$$\eta = \frac{\text{Output}}{\text{Input}} = \frac{W\,x}{P\,y}$$

$$= \left(\frac{W}{P}\right) \times \frac{1}{(y/x)} = \frac{MA}{VR} \tag{5.5}$$

Machine efficiency is a fraction less than one and is generally expressed as percentage. No machine exists on this earth whose efficiency is 100%. It means work spent on machine is always larger than the useful work obtained from it.

8. **Ideal Machine** A machine is said to be ideal if its efficiency is 100%. It is not practically possible to get 100% efficiency thus ideal machine is a hypothetical machine.

In the case of ideal machines, friction losses are neglected, so for an ideal machine,

$$\text{Output} = \text{Input}$$

i.e.,

$$P\,y = W\,x$$

$$\frac{W}{P} = \frac{y}{x}$$

$$MA = VR$$

Thus, for an ideal machine its efficiency (η) is 100% or $MA = VR$.

9. **Ideal Effort** An ideal effort is defined as the effort required for lifting a load by an ideal machine whose efficiency is 100%. It is denoted by P_{ideal} or P_i.

For an ideal machine, $\eta = 1$

$$\therefore \quad \frac{W\,x}{P_i\,y} = 1$$

$$P_i = \frac{W}{(y/x)} = \frac{W}{VR} \tag{5.6}$$

For actual machine,

$$\eta = \frac{MA}{VR} = \frac{W}{P} \times \frac{1}{VR}$$

$$P_{\text{actual}} = \frac{W}{\eta} \times \frac{1}{VR} \tag{5.7}$$

Thus, the extra effort has to be applied to lift the load due to the presence of friction. So, effort wasted in friction can be given as,

$$P_{\text{friction}} = P_{\text{actual}} - P_{\text{ideal}}$$

$$= \frac{W}{\eta} \times \frac{1}{VR} - \frac{W}{VR}$$

$$= \frac{W}{VR} \times \left(\frac{1}{\eta} - 1\right) \tag{5.8}$$

Equation (5.8) represents loss of effort due to friction in terms of effort for a particular load.

10. **Ideal Load** An ideal load is defined as the load that can be lifted by an ideal machine by applying an effort. It is denoted by W_{ideal} or W_i.

Since, for an ideal machine

$$\eta = 1$$

$$\therefore \quad \frac{W_i\,x}{Py} = 1$$

$$W_i = P \times (y/x) = P \times VR \tag{5.9}$$

For actual machine,

$$\eta = \frac{MA}{VR} = \frac{W}{P} \times \frac{1}{VR}$$

$$\therefore \quad W_{\text{actual}} = \eta \times P \times VR \tag{5.10}$$

Thus, smaller load will be lifted due to the presence of friction. So, the decrease in load lifted due to the presence of friction can be given as,

$$W_{\text{friction}} = W_{\text{ideal}} - W_{\text{actual}}$$

$$= P \times VR - \eta \times P \times VR$$

$$= P \times VR \times (1 - \eta) \tag{5.11}$$

Equation (5.11) represents the loss of load due to friction in terms of load for a particular effort.

5.3 REVERSIBLE AND IRREVERSIBLE MACHINES

On removal of the effort completely if the load falls down and the machine works in reverse direction, such a machine is known as reversible machine, for example, rope and pulley arrangement. On the other hand, if the load does not fall down on removal of the effort and the machine does not work in reverse direction such a machine is known as irreversible machine, for example, a screws jack. The irreversible machine is also known as self-locking machine.

Condition for Irreversibility of Machine

Let an effort P is applied through a distance y to lift a load W through a distance x, then

$$\text{Input} = P \times y$$

and $$\text{Output} = W \times x$$

Work lost in friction or frictional work

$$= \text{Input} - \text{Output} = Py - Wx$$

In the case of irreversible machine, the load does not fall back on removal of the effort, in other words, frictional work is greater than the output.

$$Py - Wx > Wx$$

or $$Py > 2Wx$$

or $$\frac{1}{2} > \frac{Wx}{Py}$$

or $$\frac{Wx}{Py} < \frac{1}{2}$$

or $$\frac{W}{P} \cdot \frac{1}{(y/x)} < \frac{1}{2}$$

$$\frac{MA}{VR} < \frac{1}{2}$$

$\therefore$ $$\eta < \frac{1}{2} \text{ or } 50\% \quad (5.12)$$

Thus, the condition for self-locking or irreversibility of a machine is that the efficiency of machine will always be less than 50%.

Condition for Reversibility of Machine

From the definition of a reversible machine it is clear that the machine works in reverse direction only when the output of the machine is greater than the frictional work.

Thus, $$\text{output} > \text{work lost in friction}$$

i.e., $$Wx = Py - Wx$$

or $$2Wx > Py$$

$$\frac{Wx}{Py} > \frac{1}{2}$$

or $$\frac{W}{P} \cdot \frac{1}{(y/x)} > \frac{1}{2}$$

$$\frac{MA}{VR} > \frac{1}{2}$$

$\therefore$ $$\eta > 1/2 \text{ or } 50\% \tag{5.13}$$

Hence, a machine is said to be reversible if its efficiency is greater than 50%.

Example 5.1 A load of 500 N moved through a distance of 0.08 m by an effort of 225 N which moved through a distance of 7 m. Determine the mechanical advantage, velocity ratio, and efficiency of the machine. Also comment on the reversibility of this machine.

Solution

$$MA = \frac{\text{Load Lifted}}{\text{Effort Applied}} = \frac{5000}{225} = 22.22$$

$$VR = \frac{y}{x} = \frac{7}{0.08} = 87.5$$

$$\eta = \frac{MA}{VR} = \frac{22.22}{87.5}$$

$$= 0.2539 \text{ or } 25.39\%$$

Since efficiency is less than 50% hence, machine is irreversible.

Example 5.2 In a lifting machine a load of 500 N is moved through a distance of 0.05 m by an effort moved through a distance of 0.045 m. If the efficiency of this machine is 60%, determine the value of the effort.

Solution

$$VR = \frac{y}{x} = \frac{0.045}{0.05} = 9$$

$$\eta = \frac{MA}{VR}$$

$$MA = \eta \times VR$$

$$\frac{W}{P} = \eta \times VR$$

$$P = \frac{W}{\eta \times VR} = \frac{500}{0.60 \times 9}$$

$$= 92.59 \text{ N}$$

Example 5.3 The velocity ratio of a machine is 12 and its efficiency is 56 %. Determine the load which can be raised on application of an effort of 60 N.

Solution

$$\eta = \frac{MA}{VR}$$

or $$MA = \eta \times VR$$

$$= 0.56 \times 12 = 6.72$$

But $$MA = \frac{W}{P}$$

∴ $$W = MA \times P = 6.72 \times 60$$

$$= 403.2 \text{ N}$$

Example 5.4 An effort of 60 N is applied to a lifting machine to lift a load of 1000 N. If the distance moved by the effort is 7.5 m and by the load is 0.375 m. Determine (i) MA, (ii) VR, (iii) η, (iv) friction-wise load, and (v) frictional effort lost.

Solution

(i) $$MA = \frac{W}{P} = \frac{1000}{60} = 16.67$$

(ii) $$VR = \frac{y}{x} = \frac{7.5}{0.375} = 20$$

(iii) $$\eta = \frac{MA}{VR} = \frac{16.67}{20} = 83.35\%$$

(iv) $$\text{Friction-wise load} = P \times VR \times (1 - \eta)$$

$$= 60 \times 20 \times (1 - 0.8335) = 199.8 \text{ N}$$

(v) $$\text{Frictional effort lost} = \frac{W}{VR} \times \left(\frac{1}{\eta} - 1\right)$$

$$= \frac{1000}{20} \times \left(\frac{1}{0.8335} - 1\right)$$

$$= 9.98 \text{ N}$$

Example 5.5 In a lifting machine 20% of the effort applied to raise a load is lost in friction. What load can be lifted through a height of 0.75 m if an effort of 180 N is moved through 8.5 m. Also determine the mechanical advantage and efficiency of this machine.

Solution

$$VR = \frac{y}{x} = \frac{8.5}{0.75} = 11.33$$

$$\text{Effort lost in friction} = 0.20 \times 180 = 36 \text{ N}$$

The extra effort required or the effort wasted is given by,

$$P_{friction} = P_{actual} - P_{ideal}$$

$$= P_{actual} - \frac{W}{VR}$$

$$36 = (180 - 36) - \frac{W}{11.33}$$

$$\therefore \quad W = 1223.64 \text{ N}$$

$$MA = \frac{W}{P} = \frac{1223.64}{180} = 6.798$$

$$\eta = \frac{MA}{VR} = \frac{6.798}{11.33}$$

$$= 0.60 \text{ or } 60\%$$

Example 5.6 A machine can lift a load of 4500 N with an effort of 300 N. The velocity ratio of the machine is 20. Determine the machine friction in terms of effort and load. Also comment on the reversibility of the machine.

Solution

$$MA = \frac{W}{P} = \frac{4500}{300} = 15$$

$$\eta = \frac{MA}{VR} = \frac{15}{20} = 0.75 \text{ or } 75\%$$

$$\therefore \quad \eta > 50\%$$

$\therefore$ machine is reversible

The friction in terms of effort is given by,

$$P_{friction} = P_{actual} - P_{ideal}$$

$$= \frac{W}{\eta} \times \frac{1}{VR} - \frac{W}{VR}$$

$$= \frac{W}{VR} \times \left(\frac{1}{\eta} - 1\right)$$

$$\therefore \quad P_{friction} = \frac{4500}{20} \times \left(\frac{1}{0.75} - 1\right) = 75 \text{ N}$$

The friction in terms of load is given by,

$$W_{friction} = W_{ideal} - W_{actual}$$

$$= P \times VR - \eta \times P \times VR$$

$$= P \times VR \times (1 - \eta)$$

$$\therefore \quad W_{friction} = 300 \times 20 \times (1 - 0.75) = 1500 \text{ N}$$

Example 5.7 In a lifting machine, an effort of 300 N moved through a distance of 300 cm to lift a load of 6000 N through a distance of 10 cm. Is the machine reversible? If so, what effort should be applied so that the machine is at the point of reversing?

Solution

$$MA = \frac{W}{P} = \frac{6000}{300} = 20$$

$$VR = \frac{y}{x} = \frac{300}{10} = 30$$

$$\eta = \frac{MA}{VR} = \frac{20}{30} = 0.6667 \text{ or } 66.67\%$$

$\therefore$ $\eta > 50\%$, therefore, machine is reversible.

The machine will be at the point of reversing when its efficiency is 50%.

Let P_1 be the effort required to raise the load of 6000 N at the time of reversing.

$$MA = \frac{W}{P_1} = \frac{6000}{P_1}$$

$$\eta = \frac{MA}{VR} = \frac{W}{P_1} \times \frac{1}{VR}$$

$$0.50 = \frac{6000}{P_1} \times \frac{1}{30}$$

$$\therefore \quad P_1 = \frac{6000}{0.50} \times \frac{1}{30}$$

$$= 400 \text{ N}$$

5.4 LAW OF MACHINE

The law of machine gives the relationship between the effort applied and the load lifted. It is obtained by plotting a graph between various efforts required to raise the corresponding loads.

Since P varies linearly with W, thus the relationship can be presented mathematically by an equation of a straight line.

$$\therefore \quad P = mW + C \quad (5.14)$$

Equation (5.14) is known as law of machine.

where P = Effort applied on machine, W = Load lifted by machine, C = Intercept of the line on P-axis, and m = Slope of straight line.

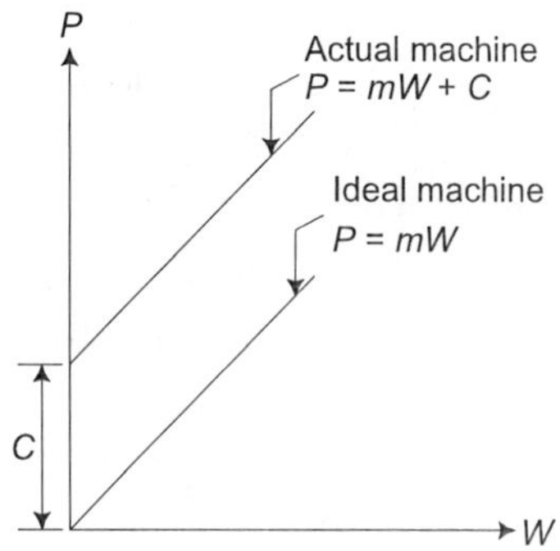

Fig. 5.1 Variation of Effort against Load

For an actual machine, the graph does not pass through the origin but makes an intercept on the P-axis which shows the effort lost in overcoming the friction (Figure 5.1). But for an ideal machine, the line passes through the origin O, means $C = 0$, as there is no frictional force.

Maximum Mechanical Advantage $(MA)_{max}$

Mechanical advantage

$$MA = \frac{W}{P} = \frac{W}{mW + C} = \frac{1}{m + (C/W)} \quad [\because P = mW + C]$$

With the increase in load W the value of (C/W) decreases and it becomes negligible for larger values of W. Thus, for maximum advantage, (C/W) tends to zero.

$$\therefore \quad (MA)_{max} = \frac{1}{m} \tag{5.15}$$

The variation of mechanical advantage with load is illustrated in Figure 5.2.

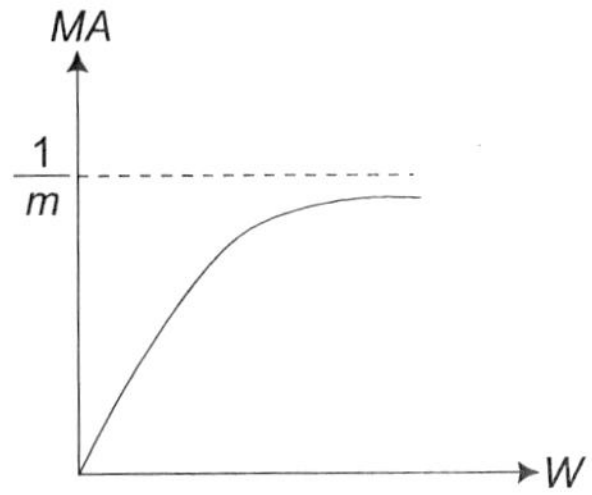

Fig. 5.2 MA vs. Load

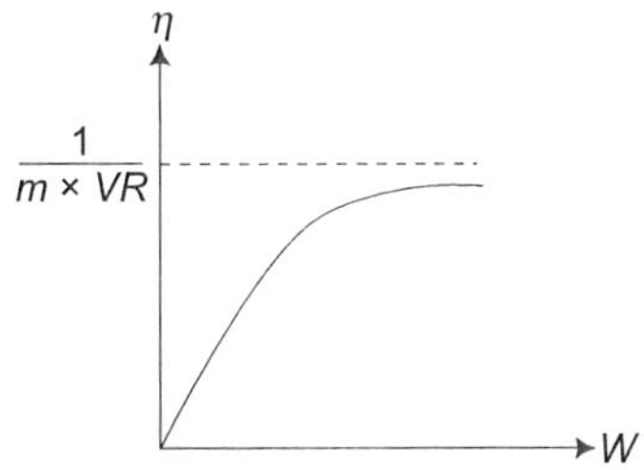

Fig. 5.3 Efficiency vs. Load

Maximum Efficiency (η_{max}) The efficiency of a machine is given by,

$$\eta = \frac{MA}{VR} = \frac{W}{P} \times \frac{1}{VR}$$

$$= \frac{W}{(mW + C)} \times \frac{1}{VR}$$

$$= \frac{1}{(m + C/W)} \times \frac{1}{VR}$$

By increasing W, the value of (C/W) decreases and in turn efficiency increases.

Thus, maximum efficiency will be achieved when (C/W) tends to zero.

$$\therefore \quad \eta_{max} = \frac{1}{m \times VR} \tag{5.16}$$

The variation of efficiency with load is illustrated in Figure 5.3.

Example 5.8 In a lifting machine an effort of 16 N is required for lifting a load of 800 N at an efficiency of 60%. The same machine requires an effort of 25 N to lift a load of 1500 N. Determine the law of machine and calculate the maximum MA and maximum efficiency.

Solution

$$MA = \frac{W}{P} = \frac{800}{16} = 50$$

$$\eta = \frac{MA}{VR}$$

$$\therefore \qquad VR = \frac{MA}{\eta} = \frac{50}{0.60} = 83.33$$

The law of machine becomes

$$P = mW + C$$

Substituting the values of P and W, we have,

$$16 = m \times 800 + C \qquad \text{(i)}$$

$$25 = m \times 1500 + C \qquad \text{(ii)}$$

Solving equations (i) & (ii), we get,

$$m = (9/700) = 0.01286$$

and $$C = 5.71$$

∴ The law of machine becomes,

$$P = 0.01286\ W + 5.71$$

$$(MA)_{max} = \frac{1}{m} = \frac{1}{0.01286} = 77.76$$

$$\eta_{max} = \frac{1}{m \times VR} = \frac{1}{0.01286 \times 83.33}$$
$$= 93.32\%$$

Example 5.9 A machine requires an effort of 18 kN for lifting a load of 90 kN. The same machine requires an effort of 15 kN for lifting a load of 72 kN. If the velocity ratio of the machine is 20, determine (a) law of machine, (b) the limiting mechanical advantage, (c) the limiting efficiency, and (d) the effort required to lift a load of 10 kN, also determine the MA & η of the machine for this load.

Solution

(a) The law of machine is

$$P = mW + C$$

Substituting the values of P and W, we have,

$$18 = m \times 90 + C \qquad \text{(i)}$$

$$15 = m \times 72 + C \qquad \text{(ii)}$$

Solving equations (i) & (ii), we get, $m = 0.167$ and $C = 2.97$

∴ The law of machine becomes,

$$P = 0.167\ W + 2.97$$

(b) Limiting values of mechanical advantage is given by,

$$(MA)_{max} = \frac{1}{m} = \frac{1}{0.167} = 5.99$$

(c) Limiting values of efficiency is given by,

$$\eta_{max} = \frac{1}{m \times VR} = \frac{1}{0.167 \times 20}$$

$$= 0.2994 \text{ or } 29.94\%$$

(d) Effort required to lift a load of 10 kN:

$$P = 0.167 \times 10 + 2.97 = 4.64 \text{ kN}$$

$$MA = \frac{W}{P} = \frac{10}{4.64} = 2.16$$

$$\therefore \quad \eta = \frac{MA}{VR} = \frac{2.16}{20}$$

$$= 0.108 \text{ or } 10.8\%$$

Example 5.10 An effort of 90 N is applied to raise a load by a lifting machine. Determine the load if velocity ratio of machine is 20 and efficiency is 65%. Also determine the law of machine if it is observed that an effort of 100 N is required to lift a load of 2000 N, and calculate the effort required to run the machine at a load of 1000 N.

Solution

$$\eta = \frac{MA}{VR} = \frac{(W/P)}{VR}$$

Substituting the values,

$$0.65 = \frac{(W/90)}{20}$$

$$0.65 \times 20 = \frac{W}{90}$$

$$\therefore \quad W = 1170 \text{ N}$$

The law of machine is

$$P = mW + C$$

Substituting the values of P and W, we have,

$$90 = m \times 1170 + C \quad \text{(i)}$$

$$100 = m \times 2000 + C \quad \text{(ii)}$$

Solving equations (i) and (ii), we get, $m = 0.012$ and $C = 75.96$

∴ The law of machine becomes,

$$P = 0.012\ W + 75.96$$

Effort required to lift a load of 1000 N:

$$P = 0.012 \times 1000 + 75.96 = 87.96 \text{ N}$$

Example 5.11 The velocity ratio of a lifting machine is 20 and efficiency is 65%. Determine the effort required to lift a load of 125 N. What is the law of machine if the frictional resistance of the machine is constant. Calculate the effort required to run the machine at no load and at a load of 75 N.

Solution

$$\eta = \frac{MA}{VR} = \frac{(W/P)}{VR}$$

Substituting the values,

$$0.75 = \frac{(125/P)}{20}$$

$$0.65 \times 20 = \frac{125}{P}$$

$$\therefore \quad P = 9.62 \text{ N}$$

The law of machine is

$$P = mW + C$$

The intercept C denotes the effort required to overcome friction and equals the effort wasted in friction during lifting a load.

$$\therefore \quad C = P_{\text{friction}} = P_{\text{actual}} - P_{\text{ideal}}$$

$$= 9.62 - (125/20) = 3.37$$

For lifting a load of 125 N the effort required is 9.62 N

Thus, law of machine becomes,

$$9.62 = m \times 125 + 3.37$$

$$\therefore \quad m = 0.05$$

∴ Law of machine becomes,

$$P = 0.05\ W + 3.37$$

Effort required at no load:

$$P = 0.05 \times (0) + 3.37 = 3.37 \text{ N}$$

Also effort required to lift a load of 75 N:

$$P = 0.05 \times 75 + 3.37 = 7.12 \text{ N}$$

5.5 SIMPLE WHEEL AND AXLE

A simple wheel and axle unit consists of a larger diameter cylinder '*A*' known as wheel and a smaller diameter cylinder '*B*' known as axle. Both the wheel and axle are mounted on the same shaft '*C*' and are free to rotate. The shaft is mounted on ball bearings to minimize the frictional resistance. A string is wound around the axle '*B*' which carries the load. A second string is wound around the wheel '*A*' and the effort is applied to it. The two strings are wound in opposite directions so that when effort is applied in downward direction the load moves up as shown in Figure 5.4.

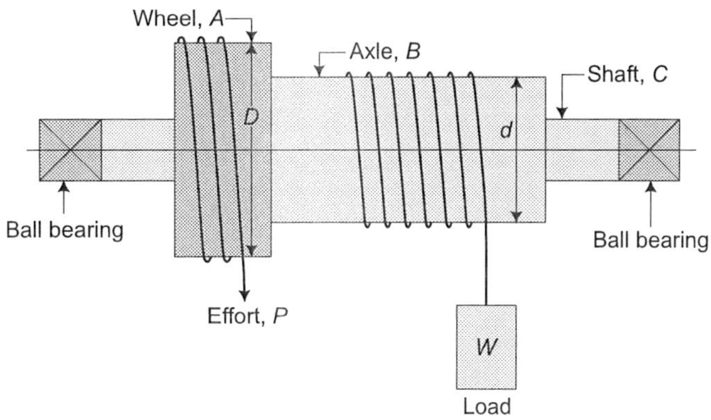

Fig. 5.4 Simple Wheel and Axle

Let D = Diameter of wheel '*A*', d = Diameter of wheel '*B*', W = Load lifted, and P = Effort applied.

As both wheel and axle are keyed to the same shaft, so when the wheel completes one revolution the axle will also turn one revolution.

Let us assume one revolution is completed by wheel and axle then,

Distance moved by load = πd, and distance moved by effort = πD

$$VR = \frac{\pi D}{\pi d} = \frac{D}{d} \tag{5.17}$$

If the diameter of rope is also considered then

$$VR = \frac{D + t_1}{d + t_2} \tag{5.17a}$$

where t_1 = Diameter of rope on wheel, and t_2 = Diameter of rope on axle.

$$MA = \frac{\text{Load lifted}}{\text{Effort applied}} = \frac{W}{P}$$

and

$$\eta = \frac{MA}{VR}$$

Example 5.12 In a simple wheel and axle machine the diameter of wheel is 20 cm. The machine is capable to lift a load of 60 N on applying an effort of 15 N at an efficiency of 80%, determine the diameter of the axle.

Solution

$$MA = \frac{W}{P} = \frac{60}{15} = 4$$

$$\eta = \frac{MA}{VR}$$

$$VR = \frac{MA}{\eta} = \frac{4}{0.80} = 5$$

Also $$VR = \frac{\text{Diameter of wheel}}{\text{Diameter of axle}} = \frac{D}{d}$$

$$5 = \frac{20}{d}$$

$\therefore$ $$d = 4 \text{ cm}$$

Example 5.13 In a simple wheel and axle machine a load of 120 N can be lifted on applying 25 N effort at an efficiency of 60%. Determine the ratio of the axle to the wheel diameters.

Solution

$$MA = \frac{W}{P} = \frac{120}{25} = 4.8$$

$$\eta = \frac{MA}{VR}$$

$$VR = \frac{MA}{\eta} = \frac{4.8}{0.6} = 8$$

$$VR = \frac{\text{Wheel diameter}}{\text{Axle diameter}} = \frac{D}{d}$$

$$\frac{d}{D} = \frac{1}{VR} = \frac{1}{8}$$

$$\frac{d}{D} = 0.125$$

Example 5.14 In a wheel and axle machine, the axle diameter is 160 mm and velocity ratio is 3. If the thickness of cord on axle is 6 mm and that on wheel is 10 mm determine the diameter of wheel. This machine is lifting a load of 1170 N through a distance of 15 cm with an efficiency of 65%. Calculate the energy input to the machine. Also determine the effort lost in friction.

Solution

$$VR = \frac{D + t_1}{d + t_2}$$

$$3 = \frac{D + 10}{160 + 6}$$

$$\therefore \quad D = 488 \text{ mm}$$

$$\eta = \frac{MA}{VR} = \frac{W/P}{VR}$$

$$P = \frac{W}{\eta \times VR} = \frac{1170}{0.65 \times 3} = 600 \text{ N}$$

Distance moved by effort

$$= \text{velocity ratio} \times x = 3 \times 15$$

$$= 45 \text{ cm} = 0.45 \text{ m}$$

Energy input to the machine = effort applied × distance moved by load

Therefore, energy input = 600 × 0.45 = 270 Nm

Effort lost in friction = $P_{\text{friction}} = P_{\text{actual}} - P_{\text{ideal}}$

$$\therefore \quad P_{\text{friction}} = P - \frac{W}{VR} = 600 - \frac{1170}{3} = 210 \text{ N}$$

5.6 DIFFERENTIAL WHEEL AND AXLE

In a differential wheel and axle unit, the axle consists of two cylinders '*A*' and '*B*' of different diameters and a wheel '*C*' which is of a larger diameter, as shown in Figure 5.5.

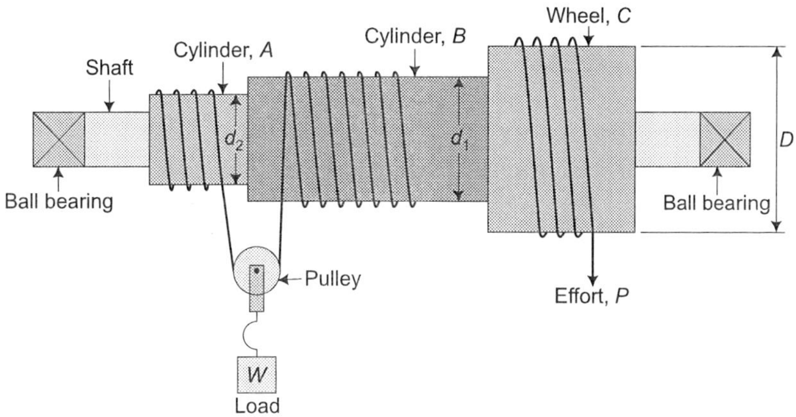

Fig. 5.5 Differential Wheel and Axle

All these three members are mounted on the same shaft and this shaft is mounted on ball bearings to minimize the frictional resistance. A string is wound around the wheel '*C*' to which effort is applied. Another string is wound around the axle '*B*' which after passing around the pulley goes around the axle 'A'. It is to be noted that the string on wheel 'C' and the smaller diameter axle 'A' are wound in the same direction but on axle 'B' it is in opposite direction. As a result when effort is applied in the downward direction there is unwinding of string on wheel 'C' and the axle 'A' but the string winds on the axle 'B' due to which load is lifted in upward direction. The load is carried by the pulley.

Let D = Diameter of wheel 'C', d_1 = Diameter of bigger axle 'B', d_2 = Diameter of smaller axle 'A', P = Effort applied, and W = Load lifted.

For one revolution of the shaft,

Distance moved by effort = πD

Length of string that winds on axle 'B' = πd_1

Length of string that unwinds on axle 'A' = πd_2

Thus, net length of string which will wound the axle 'B' = $\pi d_1 - \pi d_2$

Therefore, distance moved by load = $\dfrac{\pi d_1 - \pi d_2}{2} = \dfrac{\pi (d_1 - d_2)}{2}$

$$VR = \frac{\text{Distance moved by effort}}{\text{Distance moved by load}}$$

$$= \frac{\pi D}{\pi (d_1 - d_2)/2} = \frac{2D}{d_1 - d_2} \tag{5.18}$$

If diameter of string is also considered,

$$VR = \frac{2(D + t_1)}{(d_1 + t_2) - (d_2 + t_2)}$$

$$= \frac{2(D + t_1)}{d_1 - d_2} \tag{5.18a}$$

where t_1 = Thickness of string wound around the cylinder 'C' and

t_2 = Thickness of string wound around the axle 'A' and 'B'.

$$MA = \frac{\text{Load lifted}}{\text{Effort applied}} = \frac{W}{P}$$

and $$\eta = \frac{MA}{VR}$$

Example 5.15 In a differential wheel and axle, the wheel diameter is 25 cm. The larger and smaller diameters of the differential axle are 10 cm and 8 cm respectively. An effort of 60 N is applied to lift a load of 1200 N. Determine (i) velocity ratio, (ii) mechanical advantage, and (iii) efficiency.

Solution

(i) $$VR = \frac{2D}{d_1 - d_2} = \frac{2 \times 25}{10 - 8} = 25$$

(ii) $$MA = \frac{W}{P} = \frac{1200}{60} = 20$$

(iii) $$\eta = \frac{MA}{VR} = \frac{20}{25} = 0.8 \text{ or } 80\%$$

Example 5.16 In a differential wheel and axle machine the diameters of wheel and axle are 12 cm and 9 cm respectively. If a load of 2000 N is lifted by an effort of 100 N at an efficiency of 75%. Determine the diameter of the wheel.

Solution

$$MA = \frac{W}{P} = \frac{2000}{100} = 20$$

$$VR = \frac{2D}{d_1 - d_2} = \frac{2D}{12 - 9} = \frac{2D}{3}$$

$$\eta = \frac{MA}{VR}$$

$$0.75 = \frac{20}{(2/3)\,D}$$

$$D = \frac{3 \times 20}{2 \times 0.75}$$

$\therefore$ $$D = 40 \text{ cm}$$

Example 5.17 In a differential wheel and axle unit, the diameters of the wheel is 30 cm and that of axles are 12 cm and 10 cm. Find the effort required to lift a load of 720 N if the machine efficiency is 60%. Also determine the number of revolutions made by the machine in lifting the load through a distance of 20 cm.

Solution

$$VR = \frac{2D}{d_1 - d_2} = \frac{2 \times 30}{12 - 10} = 30$$

$$\eta = \frac{MA}{VR} = \frac{W}{P} \times \frac{1}{VR}$$

or $$P = \frac{W}{\eta} \times \frac{1}{VR} = \frac{720}{0.6} \times \frac{1}{30} = 40 \text{ N}$$

Distance moved by effort = $VR \times$ distance moved by load

$$= 30 \times 20 = 600 \text{ cm}$$

Distance moved by effort wheel in one revolution

$$= \pi D = \pi \times 30 = 30\pi \text{ cm}$$

Therefore, number of revolutions made by effort wheel

$$= 600/30\pi = 6.4 \text{ revolution}$$

Example 5.18 In a differential wheel and axle arrangement, the diameter of the wheel is 25 cm and that of the axles are 15 cm and 10 cm. The effort required to lift a load of 2250 N is 500 N while an effort of 750 N can lift a load of 3500 N. Find (i) the law of machine, (ii) effort lost in friction at the load of 3500 N, and (iii) maximum efficiency.

Solution

(i) The law of machine is

$$P = mW + C$$

Substituting the values of P and W, we have,

$$500 = m \times 2250 + C \qquad \text{(i)}$$

$$750 = m \times 3500 + C \qquad \text{(ii)}$$

Solving equation (i) & (ii), $m = 0.2$ and $C = 50$

Therefore, the law of machine becomes,

$$P = 0.2W + 50$$

(ii)
$$VR = \frac{2D}{d_1 - d_2} = \frac{2 \times 25}{15 - 10} = 10$$

Effort lost in friction is given by,

$$P_{\text{friction}} = P_{\text{actual}} - P_{\text{ideal}} = P - \frac{W}{VR}$$

$\therefore$
$$P_{\text{friction}} = \frac{750 - 3500}{10} = 400 \text{ N}$$

(iii)
$$\eta_{\max} = \frac{1}{m \times VR} = \frac{1}{0.2 \times 10}$$
$$= 0.5 \text{ or } 50\%$$

5.7 DIFFERENTIAL PULLEY BLOCK

Differential pulley block is also known as Weston pulley block. It consists of two blocks, namely, upper block 'A' and lower block 'B'. The upper block 'A' has two concentric pulleys say 1 and 2 which have a common axis fixed to a support. The pulley 1 has slightly larger diameter than the pulley 2. The lower block 'B' carries a movable pulley say 3 to which the load W is attached. An endless chain first passes over the pulley 1 then around the movable pulley 3 and finally over the pulley 2. The remaining part of the chain hangs slack and it is joined to first portion of the chain as shown in Figure 5.6.

The effort P is applied to the chain passing over the pulley 1. To prevent any slip of the chain teeth are provided in the grooves of the pulleys.

Let D = Diameter of pulley 1, d = Diameter of pulley 2, W = Weight lifted, and P = Effort applied.

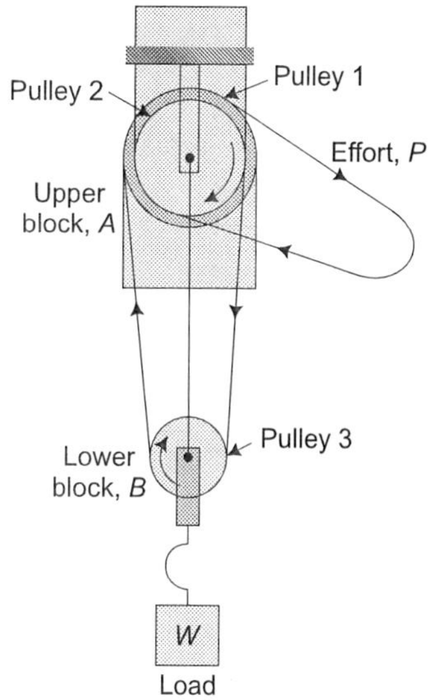

Fig. 5.6 Differential Pulley Block

For one revolution of the upper pulley block the length of chain passes over pulley 1 towards right will be equal to the distance moved by the effort.

Thus, distance moved by effort = πD

Since the pulley 2 also turns with pulley 1 hence a length passes over pulley 2 towards left will be equal to πd

The increase of chain length on effort side and decrease of chain length on load side lifts the load in the upward direction.

Hence, net shortening of the chain

$$= \pi D - \pi d = \pi (D - d)$$

This shortening of the chain is equally divided in two sides of chain wound around movable pulley 3.

Therefore, distance moved by load in upward direction

$$= \frac{\pi (D - d)}{2}$$

$$VR = \frac{\text{Distance moved by effort}}{\text{Distance moved by load}}$$

$$= \frac{\pi D}{\pi (D - d)/2} = \frac{2D}{D - d} \tag{5.19}$$

When the pulleys are provided with teeth,

$$VR = \frac{2T_1}{T_1 - T_2} \tag{5.19a}$$

where T_1 = Number of teeth on pulley 1, and T_2 = Number of teeth on pulley 2.

$$MA = \frac{\text{Load lifted}}{\text{Effort applied}} = \frac{W}{P}$$

and $$\eta = \frac{MA}{VR}$$

Example 5.19 The upper block of the Westorn differential pulley block has two pulleys of diameters 300 mm and 250 mm. If the efficiency of this system is 65%, calculate the effort required to lift a load of 200 N.

Solution

$$VR = \frac{2D}{D - d} = \frac{2 \times 300}{300 - 250} = 12$$

$$\eta = \frac{MA}{VR} = \frac{W}{P} \times \frac{1}{VR}$$

∴ Effort required, $$P = \frac{W}{\eta} \times \frac{1}{VR} = \frac{200}{0.65} \times \frac{1}{12} = 25.64 \text{ N}$$

Example 5.20 A differential pulley block consists of an upper pulley block and a lower pulley block. The upper pulley block has two pulleys one of which has a radius of 165 mm and the other a radius of 150 mm. If the efficiency of the machine is 60% calculate the effort required to lift a load of 1980 N

Solution

$$VR = \frac{2D}{D-d} = \frac{2 \times 165}{165-150}$$

$$= 22$$

$$\eta = \frac{MA}{VR} = \frac{W}{P} \times \frac{1}{VR}$$

$$\therefore \quad P = \frac{W}{\eta} \times \frac{1}{VR} = \frac{1980}{0.60} \times \frac{1}{22}$$

$$= 150 \text{ N}$$

Example 5.21 In a Weston differential pulley block the pulleys are of diameters 300 mm and 250 mm. Calculate the effort required to lift a load of 1.5 kN if the efficiency is 50%. Find the length of chain required to lift the load to a height of 40 mm.

Solution

$$VR = \frac{2D}{D-d} = \frac{2 \times 300}{300-250}$$

$$= 12$$

$$\eta = \frac{MA}{VR} = \frac{W}{P} \times \frac{1}{VR}$$

$$\therefore \quad P = \frac{W}{\eta} \times \frac{1}{VR} = \frac{1500}{0.50} \times \frac{1}{12}$$

$$= 250 \text{ N}$$

Let L = Length of chain required

Distance moved by effort = $L = VR \times$ distance moved by load $\quad (\because VR = y/x)$

$$\therefore \quad L = 12 \times 40 = 480 \text{ mm}$$

Example 5.22 In a differential pulley block a load of 2000 N is raised by an effort of 125 N. The numbers of teeth on larger and smaller pulleys are 10 and 9 respectively. Find the velocity ratio, mechanical advantage, and efficiency of the machine.

Solution

$$VR = \frac{2D}{D-d} = \frac{2 \times 10}{10-9}$$

$$= 20$$

$$MA = \frac{W}{P} = \frac{2000}{125} = 16$$

$$\eta = \frac{MA}{VR} = \frac{16}{20}$$
$$= 0.8 \text{ or } 80\%$$

5.8 WORM AND WORM WHEEL

A worm and worm wheel unit is shown in Figure 5.7 which consists of a square threaded screw called worm and a toothed wheel called worm wheel. The worm and worm wheel are meshed together. A wheel of larger radius R is mounted on the same shaft as the worm, called effort wheel. The effort is applied to the rope wound round the effort wheel. A load drum of smaller radius r concentric to the worm wheel is securely attached to it. The load W is carried by the rope wound round the load axle.

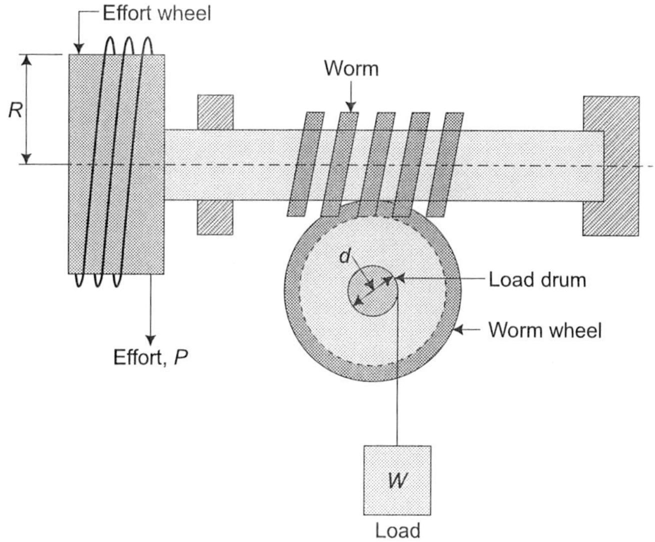

Fig. 5.7 Worm and Worm Wheel

Let R = Radius of the effort wheel, $r = d/2$ = Radius of the load drum, W = Load lifted, P = Effort applied, and T = Number of teeth on the worm wheel.

If effort P is applied to the effort wheel and it completes one revolution,

Distance covered by effort = $2\pi R$

Further, if the worm is single threaded then for one revolution of the effort wheel the worm pushes the worm wheel through one tooth.

It means the load drum will move through $1/T$ revolution.

∴ Distance moved by load

$$= 2\pi\, r \times \frac{1}{T} = \frac{2\pi\, r}{T}$$

$$VR = \frac{\text{Distance moved by effort}}{\text{Distance moved by load}}$$

$$= \frac{2\pi R}{2\pi r/T}$$

$$= \frac{RT}{r} \qquad (5.20)$$

If the worm is double threaded then for the one revolution of the effort wheel, worm will push the worm wheel through two teeth.

It means the load drum will move through $2/T$ revolution.

∴ Distance moved by load

$$= 2\pi\, r \times \frac{2}{T} = \frac{4\pi\, r}{T}$$

$$VR = \frac{2\pi R}{4\pi r/T}$$

$$= \frac{RT}{2r} \qquad (5.21)$$

In general, if the worm is n threaded,

$$VR = \frac{RT}{nr} \qquad (5.22)$$

$$MA = \frac{\text{Load lifted}}{\text{Effort applied}} = \frac{W}{P}$$

and

$$\eta = \frac{MA}{VR}$$

Example 5.23 A double threaded worm wheel has 60 teeth. The diameters of the effort wheel and load drum are 60 cm and 20 cm respectively. The efficiency of the worm and worm wheel unit is 25%. Determine the load lifted when 30 N effort is applied. Also calculate the velocity ratio.

Solution

$$VR = \frac{RT}{2r} = \frac{30 \times 60}{2 \times 10} = 90$$

$$\eta = \frac{MA}{VR} = \frac{W}{P} \times \frac{1}{VR}$$

$$0.25 = \frac{W}{30} \times \frac{1}{90}$$

∴

$$W = 675 \text{ N}$$

Example 5.24 The number of teeth on the worm wheel of a single threaded worm and worm wheel unit is 40 and this machine requires an effort of 400 N to lift a load of 10 kN. If the diameters of effort wheel and load drum are 200 mm and 80 mm respectively, calculate the velocity ratio, mechanical advantage, and efficiency of the machine.

Solution

$$VR = \frac{RT}{2r} = \frac{100 \times 40}{80} = 50$$

$$MA = \frac{W}{P} = \frac{10000}{400} = 25$$

$$\therefore \quad \eta = \frac{MA}{VR} = \frac{25}{50}$$

$$= 0.5 \text{ or } 50\%$$

5.9 WINCH CRAB

A portable winch is known as crab winch and it is used to lift heavy loads by the application of small efforts. Due to high mechanical advantage the crab winches are widely used. To increase the velocity ratio these machines use a gear system. If one set of gear is used, it is called single purchase crab winch but if two gears are used then it is called double purchase crab winch.

5.9.1 Single Purchase Crab Winch

The single purchase crab winch is shown in Figure 5.8 which comprises two axles, namely, effort axle and the load axle. Both these axles are mounted in bearings in a rigid frame. The effort axle carries a smaller toothed wheel called the pinion and the load axle carries a larger gear wheel called the spur gear or spur wheel. Both the pinion and the spur gear are meshed together. The spur wheel is rigidly keyed to the drum of the load axle. A rope is wound around the load drum which carries the load. The effort is applied at the end of the lever which rotates the pinion and the spur gear and thus the load is lifted up.

Let P = Effort applied, W = Load lifted, T_1 = Number of teeth on the pinion, T_2 = Number of teeth on spur gear, l = Length of the lever arm, and d = Diameter of load drum.

In one revolution of the lever, the distance moved by the effort is $2\pi l$.

The pinion also makes one revolution so, the number of revolutions made by the spur gear and the load axle is T_1/T_2

$\therefore$ Distance moved by load = $\pi d \times (T_1/T_2)$

$$VR = \frac{\text{Distance moved by effort}}{\text{Distance moved by load}}$$

$$= \frac{2\pi l}{\pi d \times (T_1/T_2)} = \frac{2l}{d} \times \frac{T_2}{T_1} \qquad (5.23)$$

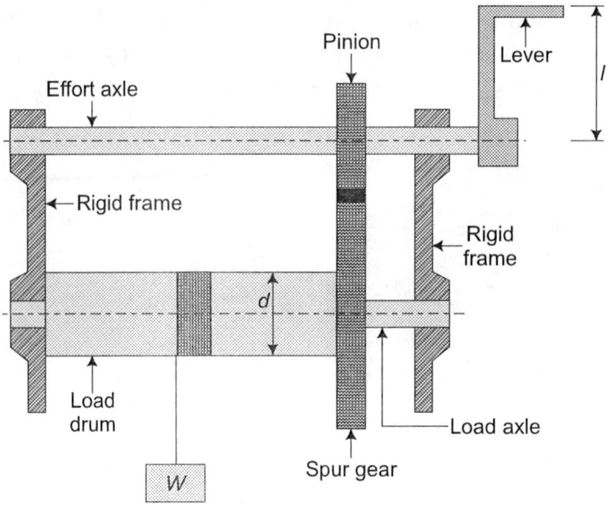

Fig. 5.8 Single Purchase Winch Crab

$$MA = \frac{\text{Load lifted}}{\text{Effort applied}} = \frac{W}{P}$$

and
$$\eta = \frac{MA}{VR}$$

5.9.2 Double Purchase Crab Winch

In a double purchase crab winch two sets of gears are used which gives a greater velocity ratio in comparison to single purchase crab winch. This unit has three axles, namely, an effort axle, the load axle, and an intermediate axle. The effort axle carries a pinion which meshes with the spur gear (or spur wheel) of the intermediate axle. The pinion of the intermediate axle meshes with the spur gear (or spur wheel) of the load axle. So, in this unit two pinion and two spur gears are used. The spur wheel is rigidly keyed to the drum of the load axle. A rope is wound around the load drum which carries the load. The effort is applied at the end of the lever as shown in Figure 5.9.

Let P = Effort applied, W = Load lifted, l = Length of the lever arm, d = Diameter of the load drum, $T_1 \equiv$ Number of teeth on the pinion of the effort axle, T_2 = Number of teeth on spur gear of intermediate axle, T_3 = Number of teeth on the pinion of intermediate axle, and T_4 = Number of teeth on the spur gear of the load axle.

Now for one revolution of the lever,

The distance moved by the effort is $2\pi l$

Number of revolutions made by the pinion of effort axle = 1

Number of revolutions made by spur wheel and pinion of intermediate axle = $1 \times T_1/T_2$

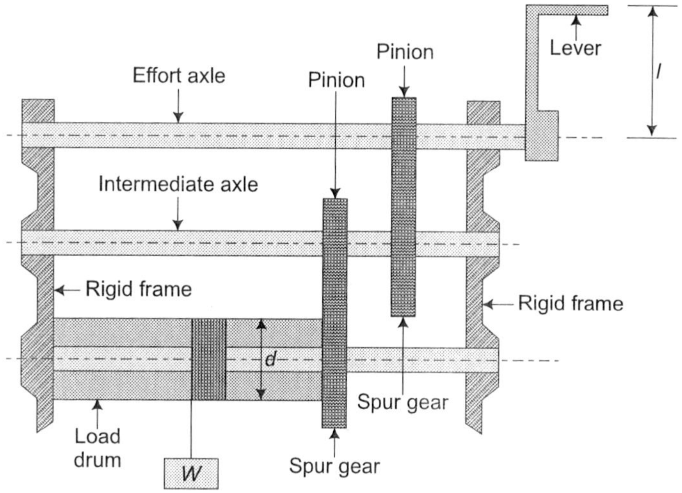

Fig. 5.9 Double Purchase Winch Crab

Number of revolutions made by spur wheel and load drum mounted on the load axle

$$= 1 \times \frac{T_1}{T_2} \times \frac{T_3}{T_4}$$

$$\therefore \text{ Distance moved by load} = \pi d \times \left(1 \times \frac{T_1}{T_2} \times \frac{T_3}{T_4}\right)$$

$$= \pi d \times \frac{T_1}{T_2} \times \frac{T_3}{T_4}$$

$$VR = \frac{\text{Distance moved by effort}}{\text{Distance moved by load}}$$

$$= \frac{2\pi l}{\pi d \times (T_1/T_2) \times (T_3/T_4)}$$

$$= \frac{2l}{d} \times \frac{T_2}{T_1} \times \frac{T_4}{T_3} \qquad (5.24)$$

$$MA = \frac{\text{Load lifted}}{\text{Effort applied}} = \frac{W}{P}$$

and $$\eta = \frac{MA}{VR}$$

Example 5.25 In a single purchase crab winch the number of teeth in the pinion and spur wheels are 20 and 140 respectively. The diameter of the load axle is 25 cm and the lever arm is 50 cm. A load of 4 kN is lifted by applying an effort of 200 N. Find the velocity ratio and the efficiency of the machine.

Solution

Velocity ratio, $$VR = \frac{2l}{d} \times \frac{T_2}{T_1} = \frac{2 \times 50}{25} \times \frac{140}{20} = 28$$

$$MA = \frac{\text{Load lifted}}{\text{Effort applied}} = \frac{W}{P}$$

$$= \frac{4000}{200} = 20$$

Efficiency, $$\eta = \frac{MA}{VR} = \frac{20}{28}$$

$$= 0.7143 \text{ or } 71.43\%$$

Example 5.26 In a single purchase crab winch the number of teeth on pinion and spur wheels are 25 and 100 respectively. The diameters of load axle and effort wheel are 125 mm and 450 mm respectively. At no load condition friction in the machine is 30 N. Limiting efficiency is 75%, find the effort needed to lift a load of 250 N. Also determine the load at which the machine is at the point of reversing.

Solution Radius of effort wheel,

$$l = (450/2) = 225 \text{ mm}$$

$$VR = \frac{2l}{d} \times \frac{T_2}{T_1} = \frac{2 \times 225}{125} \times \frac{100}{25} = 14.4$$

$$\eta = \frac{MA}{VR} = \frac{W}{P} \times \frac{1}{VR}$$

$$P = \frac{W}{\eta} \times \frac{1}{VR} = \frac{250}{0.75} \times \frac{1}{14.4} = 23.15 \text{ N}$$

Also, no load friction = 30 N

∴ Effort needed to lift a load of 250 N

$$= 23.15 + 30 = 53.15 \text{ N}$$

The machine will be at the point of reversing when its efficiency is 50 %.

$$\eta = \frac{MA}{VR} = \frac{W}{P} \times \frac{1}{VR}$$

$$0.5 = \frac{W}{53.15} \times \frac{1}{14.4}$$

∴ $$W = 382.68 \text{ N}$$

Example 5.27 In a double purchase crab winch the length of the lever arm and the diameter of the drum are 450 mm and 120 mm respectively. The number of teeth on the pinion of effort axle and intermediate axle are 16 and 20 respectively. And the number of teeth on spur wheel of intermediate axle and load axle are 60 and 80 respectively. This unit can lift loads of 300 N and 1000 N on applying effort of 25 N and 45 N respectively. Find (i) the law of machine, (ii) effort required to lift a load of 750 N and efficiency at this load, and (iii) maximum efficiency of this machine.

Solution

(i) The law of machine is

$$P = mW + C.$$

Substituting the values of P & W

$$25 = m \times 300 + C \qquad \text{(i)}$$

$$45 = m \times 1000 + C \qquad \text{(ii)}$$

Solving the equations (i) and (ii), we get, $m = 0.0286$, and $C = 16.42$

∴ The law of machine becomes

$$P = 0.0286W + 16.42$$

(ii) Effort required to lift 750 N:

$$P = 0.0286 \times 750 + 16.42 = 37.87 \text{ N}$$

$$VR = \frac{2l}{d} \times \frac{T_2}{T_1} \times \frac{T_4}{T_3}$$

$$= \frac{2 \times 450}{120} \times \frac{60}{16} \times \frac{80}{20} = 112.5$$

$$MA = \frac{W}{P} = \frac{750}{37.87} = 19.8,$$

$$\eta = \frac{MA}{VR} = \frac{19.8}{112.5} \times 100 = 17.6\%$$

(iii) Maximum efficiency can be given as,

$$\eta_{\max} = \frac{1}{m \times VR} = \frac{1}{0.0286 \times 112.5}$$

$$= 0.3108 \text{ or } 31.08\%$$

5.10 SCREW JACK

A screw jack is a simple lifting machine used for lifting heavy loads with the help of a small effort applied at its handle. It is generally used to lift the vehicle for any maintenance work such as changing of the wheel or tyre. Generally, the loads to be lifted are centrally placed on it. The working principle of a screw jack is similar to that of raising and lowering a load on an inclined plane by the action of a horizontal force as shown in Figures 5.10a and 5.10b respectively.

Screw jacks are of two types: (i) Simple screw jack, and (ii) Differential screw jack.

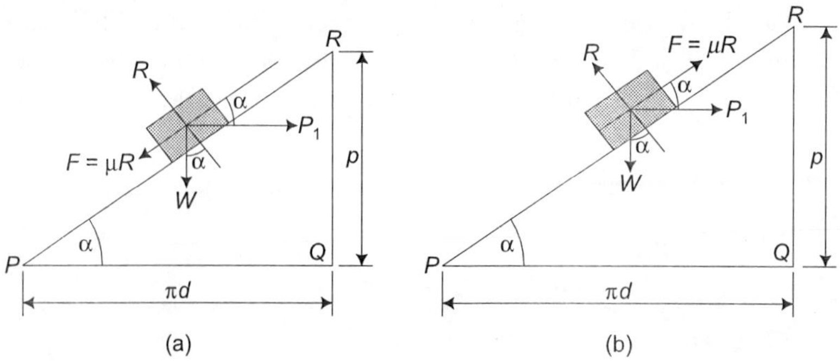

Fig. 5.10

5.10.1 Simple Screw Jack

A simple screw jack consists of a square threaded screw which moves inside a heavy nut which forms the base of the unit. A handle is fitted to the screw head. The load to be lifted is placed on the head of screw which is lifted or lowered by the application of an effort at the end of the handle as shown in Figure 5.11.

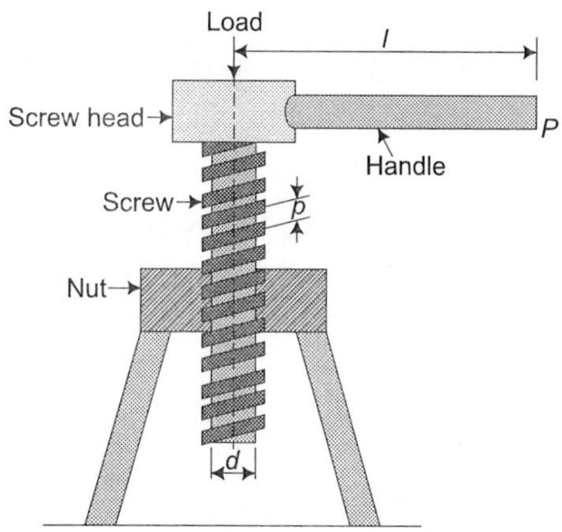

Fig. 5.11 Simple Screw Jack

Let d = Mean diameter of the screw = Average diameter of the root diameter and outer diameter of the screw, p = Pitch of the screw = Distance between two successive threads, α = Helix angle or screw angle, ϕ = Angle of friction, l = Length of the handle, W = Load lifted, P = Effort applied at the end of the handle, and

$$\mu = \tan \phi = \text{Coefficient of friction}$$

In one revolution of the screw the load is moved by a distance equal to the pitch of the screw.

$\therefore$ Distance moved by load = p

Distance moved by effort = $2\pi l$

Velocity ratio, $$VR = \frac{\text{Distance moved by effort}}{\text{Distance moved by load}}$$

$$= \frac{2\pi l}{p} \quad (5.25)$$

If the screw is double threaded then distance moved by load = $2p$

Therefore, $$VR = \frac{2\pi l}{2p} = \frac{\pi l}{p} \quad (5.26)$$

$$MA = \frac{\text{Load lifted}}{\text{Effort applied}} = \frac{W}{P}$$

and $$\eta = \frac{MA}{VR}$$

Expression for Effort Required to Lift the Load

When the handle completes one revolution, the screw is also rotated through one turn and load is lifted by a height equal to pitch of the screw. The development of one complete revolution of a screw thread is shown in Figure 5.10 (a). The distance PQ will be equal to the circumference (πd) and the distance QR will be equal to the pitch (p) of the screw. From the geometry of the Figure (5.10 a), we have,

$$\tan \alpha = \frac{QR}{PQ} = \frac{p}{\pi d} \tag{5.27}$$

Let P_1 is the effort applied horizontally at the mean radius of the screw jack to lift the load W, R is the normal reaction, and F is the force of friction = μR

When the load W is lifted upwards, the force of friction F acts downwards.

Resolving the forces along the inclined plane,

$$F + W \sin \alpha = P_1 \cos \alpha$$

or $$\mu R + W \sin \alpha = P_1 \cos \alpha \quad (\because F = \mu R) \tag{i}$$

Resolving the forces normal to the inclined plane,

$$R = W \cos \alpha + P_1 \sin \alpha \tag{ii}$$

Substituting equation (ii) in (i),

$$\mu (W \cos \alpha + P_1 \sin \alpha) + W \sin \alpha = P_1 \cos \alpha$$

But $$\mu = \tan \phi = \frac{\sin \phi}{\cos \phi}$$

$\therefore$ $$\frac{\sin \phi}{\cos \phi} (W \cos \alpha + P_1 \sin \alpha) + W \sin \alpha = P_1 \cos \alpha$$

or $$W \frac{\sin \phi \cos \alpha}{\cos \phi} + P_1 \frac{\sin \phi \sin \alpha}{\cos \phi} + W \sin \alpha = P_1 \cos \alpha$$

Multiplying by $\cos \phi$,

$$W \sin \phi \cos \alpha + P_1 \sin \phi \sin \alpha + W \sin \alpha \cos \phi = P_1 \cos \alpha \cos \phi$$

$$W [\sin \phi \cos \alpha + \sin \alpha \cos \phi] = P_1 [\cos \alpha \cos \phi - \sin \phi \sin \alpha]$$

or $$W \sin (\alpha + \phi) = P_1 \cos (\alpha + \phi)$$

$\therefore$ $$P_1 = W \frac{\sin (\alpha + \phi)}{\cos (\alpha + \phi)}$$

$$= W \tan (\alpha + \phi) \tag{5.28}$$

P_1 is the effort applied at the mean radius of the screw jack but actually the effort is applied at the end of the handle as shown in Figure 5.11.

Moment of P_1 about the axis of the screw = $P_1 \times (d/2)$

Moment of P about the axis of the screw = $P \times l$

Equating the two moments, we have,

$$P_1 \times \frac{d}{2} = P \times l$$

or

$$P_1 = P \times \frac{2l}{d}$$

Substituting the value of P_1 in Eq. (5.28), we have,

$$P \times \frac{2l}{d} = W \tan (\alpha + \phi)$$

$\therefore$

$$P = \frac{d}{2l} \times W \tan (\alpha + \phi) \quad (5.29)$$

Equation (5.29) gives the relation between the effort required and the load lifted.

The value of P in terms of pitch p and coefficient of friction μ:

We know that

$$P = \frac{d}{2l} \times W \tan (\alpha + \phi)$$

$$= \frac{Wd}{2l} \times \frac{\tan \alpha + \tan \phi}{1 - \tan \alpha \tan \phi}$$

or

$$P = \frac{Wd}{2l} \times \frac{(p/\pi d) + \mu}{1 - (p/\pi d) \cdot \mu}$$

$\therefore$

$$P = \frac{Wd}{2l} \left(\frac{p + \mu \pi d}{\pi d - p\mu} \right) \quad (5.30)$$

Equation (5.30) gives the value of P in terms of μ and p.

Expression for Effort Required to Lower the Load

When the load is lowered by the screw jack, the force of friction F will act upwards as shown in Figure 5.10 (b).

Resolving the forces along the inclined plane,

$$F + P_1 \cos \alpha = W \sin \alpha$$

or

$$\mu R + P_1 \cos \alpha = W \sin \alpha \quad \text{(i)}$$

Resolving the forces normal to the inclined plane,

$$R = W \cos \alpha + P_1 \sin \alpha$$

Substituting the values of R in equation (i),

$$\mu (W \cos \alpha + P_1 \sin \alpha) + P_1 \cos \alpha = W \sin \alpha$$

$$\mu W \cos \alpha + \mu P_1 \sin \alpha + P_1 \cos \alpha = W \sin \alpha$$

or

$$\mu P_1 \sin \alpha + P_1 \cos \alpha = W \sin \alpha - \mu W \cos \alpha$$

$$P_1 (\mu \sin \alpha + \cos \alpha) = W (\sin \alpha - \mu \cos \alpha)$$

But

$$\mu = \tan \phi = \frac{\sin \phi}{\cos \phi}$$

$\therefore$

$$P_1 \left(\frac{\sin \phi}{\cos \phi} \sin \alpha + \cos \alpha \right) = W \left(\sin \alpha - \frac{\sin \phi}{\cos \phi} \cos \alpha \right)$$

Multiplying by $\cos \phi$, we get,

$$P_1 (\sin \phi \sin \alpha + \cos \alpha \cos \phi) = W (\sin \alpha \cos \phi - \sin \phi \cos \alpha)$$

or

$$P_1 \cos (\phi - \alpha) = W \sin (\phi - \alpha)$$

$\therefore$

$$P_1 = W \frac{\sin (\phi - \alpha)}{\cos (\phi - \alpha)}$$

$$= W \tan (\phi - \alpha) \tag{5.31}$$

If $\alpha > \phi$ then

$$P_1 = W \tan (\alpha - \phi) \tag{5.32}$$

P_1 is the effort applied at the mean radius of the screw jack but in actual practice the effort P is applied at the end of the handle of the jack. When the effort P is applied at the end of the handle as shown in Figure 5.11, then equating the moments of P and P_1 about the axis of the jack, we get,

$$P \times l = P_1 \times \frac{d}{2}$$

or

$$P = \frac{d}{2l} \times P_1$$

Now substituting the value of P_1 from equation (5.31), we have,

$$P = \frac{d}{2l} \times W \tan (\phi - \alpha) \tag{5.33}$$

Equation (5.33) gives the relation between effort required at the end of the handle to lower the load.

The value of P in terms of pitch p and coefficient of friction μ:

We know that

$$P = \frac{d}{2l} \times W \tan (\phi - \alpha)$$

$$= \frac{Wd}{2l} \times \frac{\tan \phi - \tan \alpha}{1 + \tan \phi \tan \alpha}$$

$$= \frac{Wd}{2l} \times \frac{\mu - (p/\pi d)}{1 + \mu (p/\pi d)}$$

$$\therefore \quad P = \frac{Wd}{2l}\left(\frac{\mu \pi d - p}{\pi d + \mu p}\right) \quad (5.34)$$

Expression for Efficiency of a Screw Jack

The actual effort P required to lift the load W is given by equation (5.29),

$$P = \frac{d}{2l} \times W \tan (\alpha + \phi) \quad \text{(i)}$$

In the absence of friction ϕ will become zero. In that case, effort P will become ideal effort P_i,

$$\therefore \quad P_i = \frac{d}{2l} \times W \tan \alpha \quad \text{(ii)}$$

$$\eta = \frac{\text{Ideal effort}}{\text{Actual effort}}$$

$$= \frac{P_i}{P_a} = \frac{(d/2l) \times W \tan \alpha}{(d/2l) \times W \tan (\alpha + \phi)}$$

$$\therefore \quad \eta = \frac{\tan \alpha}{\tan (\alpha + \phi)} \quad (5.35)$$

Expression for Maximum Efficiency

The condition for maximum efficiency can be obtained by,

$$\frac{d\eta}{d\alpha} = 0$$

or

$$\frac{d}{d\alpha}\left[\frac{\tan \alpha}{\tan (\alpha + \phi)}\right] = 0$$

$$\frac{\tan (\alpha + \phi) \sec^2\alpha - \tan \alpha \sec^2(\alpha + \phi)}{\tan^2(\alpha + \phi)} = 0$$

or

$$\tan (\alpha + \phi) \sec^2\alpha - \tan \alpha \sec^2(\alpha + \phi) = 0$$

$$\frac{\sin (\alpha + \phi)}{\cos (\alpha + \phi)} \times \frac{1}{\cos^2\alpha} = \frac{\sin \alpha}{\cos \alpha} \times \frac{1}{\cos^2(\alpha + \phi)}$$

or

$$\sin (\alpha + \phi) \cos (\alpha + \phi) = \sin \alpha \cos \alpha$$

$$\sin 2(\alpha + \phi) = \sin 2\alpha = \sin (\pi - 2\alpha)$$

or

$$2(\alpha + \phi) = (\pi - 2\alpha)$$

or

$$4\alpha = \pi - 2\phi$$

$$\alpha = \frac{\pi}{4} - \frac{\phi}{2} = 45° - \frac{\phi}{2} \tag{5.36}$$

Hence, maximum efficiency can be obtained by substituting the value of α in equation (5.35), we get,

$$\eta_{max} = \frac{\tan(45° - \phi/2)}{\tan(45° + \phi/2)} = \frac{\dfrac{1 - \tan\phi/2}{1 + \tan\phi/2}}{\dfrac{1 + \tan\phi/2}{1 - \tan\phi/2}}$$

$$= \frac{(1 - \tan\phi/2)^2}{(1 + \tan\phi/2)^2} = \frac{\left(\dfrac{\cos\phi/2 - \sin\phi/2}{\cos\phi/2}\right)^2}{\left(\dfrac{\cos\phi/2 + \sin\phi/2}{\cos\phi/2}\right)^2}$$

$$= \frac{(\cos\phi/2 - \sin\phi/2)^2}{(\cos\phi/2 + \sin\phi/2)^2}$$

$$= \frac{\cos^2\frac{\phi}{2} + \sin^2\frac{\phi}{2} - 2\cos\frac{\phi}{2}\sin\frac{\phi}{2}}{\cos^2\frac{\phi}{2} + \sin^2\frac{\phi}{2} + 2\cos\frac{\phi}{2}\sin\frac{\phi}{2}}$$

$$\therefore \quad \eta_{max} = \frac{1 - \sin\phi}{1 + \sin\phi} \quad \left(\because \cos^2\frac{\phi}{2} + \sin^2\frac{\phi}{2} = 1,\ 2\cos\frac{\phi}{2}\sin\frac{\phi}{2} = \sin\phi\right) \tag{5.37}$$

5.10.2 DIFFERENTIAL SCREW JACK

A differential screw jack is shown Figure 5.12.

It consists of two threaded parts say '*A*' and '*B*'. Part '*A*' has only external threads while part '*B*' has both internal as well as external threads. The threads of screw '*A*' engage inside the internal threads of screw '*B*', while the external threads of screw '*B*' engage inside the threads of the nut '*C*' which forms the main body of this unit. So, part '*B*' acts as a nut for screw *A* and acts as a screw for nut '*C*'.

Let W = Load lifted, P = Effort applied, l = Length of the handle, p_a = Pitch of screw A, and p_b = Pitch of screw B

Let the handle complete one revolution,

Distance moved by effort = $2\pi l$

Distance moved by screw B in upward direction = p_b

Distance moved by screw A in downward direction = p_a

∴ The net distance through which load is lifted = $p_b - p_a$

$$VR = \frac{\text{Distance moved by effort}}{\text{Distance moved by load}}$$

$$= \frac{2\pi l}{p_b - p_a} \quad (5.38)$$

$$MA = \frac{\text{Load lifted}}{\text{Effort applied}} = \frac{W}{P}$$

and $$\eta = \frac{MA}{VR}$$

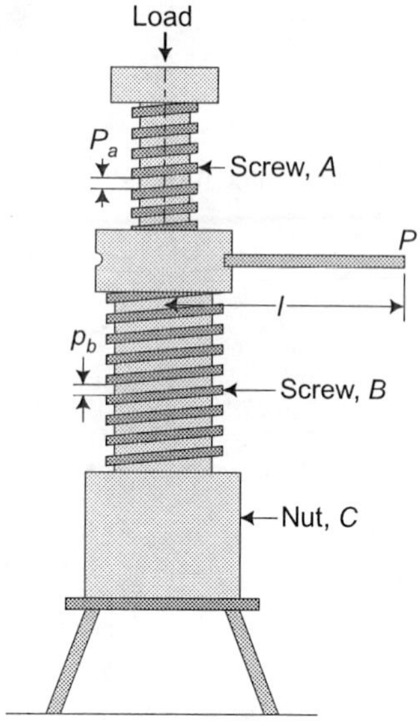

Fig. 5.12 Differential Screw Jack

Example 5.28 To lift a load of 3000 N an effort of 75 N is applied at the end of a handle of length 150 mm. Determine the efficiency of the single screw jack if the pitch of the screw is 10 mm.

Solution

$$MA = \frac{W}{P} = \frac{3000}{7.5} = 40$$

Distance moved by effort,

$$y = 2\pi l = 2\pi \times 50$$

Distance moved by load,

$$x = p = 10$$

$$VR = (y/x)$$

$$= (2\pi \times 150)/10 = 94.25$$

$$\eta = \frac{MA}{VR} = \frac{40}{94.25}$$

$$= 0.4244 \text{ or } 42.44\%$$

Example 5.29 A differential screw jack has the following specifications: pitch of the smaller screw (upper part A) = 10 mm, pitch of the larger screw (lower part B) = 15 mm, length of the handle from the centre = 250 mm. If an effort of 100 N is applied at the end of the handle to lift a load of 3500 N, determine its velocity ratio and efficiency.

Solution

$$VR = \frac{2\pi l}{p_b - p_a}$$

$$= \frac{2\pi \times 250}{15 - 10} = 314.16$$

$$MA = \frac{W}{P} = \frac{3500}{100} = 35$$

$$\therefore \quad \eta = \frac{35}{314.16} = 0.1114 \text{ or } 11.14\%$$

Example 5.30 The efficiency of a screw jack is 54%, when a load of 1000 N is lifted by an effort applied at the end of the handle of length 45 cm. Find out the effort applied if the pitch of the screw thread is 10 mm.

Solution For one complete turn of the handle,

Distance moved by load, x = pitch of the screw,

$$p = \frac{10}{100} \text{ m} = 0.01 \text{ m}$$

Distance moved by effort,

$$y = 2\pi l = 2\pi \times 0.45 = 2.83 \text{ m}$$

$$\therefore \quad VR = \frac{y}{x} = \frac{2.83}{0.01} = 282.7$$

$$MA = \frac{W}{P} = \frac{1600}{P}$$

$$\eta = \frac{MA}{VR} = \frac{1600}{P} \times \frac{1}{282.7}$$

$$0.54 = \frac{1600}{P} \times \frac{1}{282.7}$$

$$\therefore \quad P = \frac{1600}{0.54 \times 282.7} = 10.48 \text{ N}$$

Example 5.31 The mean diameter of the screw jack having pitch 1 cm is 6 cm. The coefficient of friction at the screw thread is 0.05. (i) Find the tangential force to be applied at the end of the handle of length 35 cm to lift a load of 500 N, (ii) State whether the jack is self-locking, if so, find the torque necessary to lower the load. If the jack is not self-locking, find the torque, which must be applied to keep the load from descending.

Solution

(i) Effort required to lift the load,

$$P = \frac{Wd}{2l}\left(\frac{p + \mu\pi d}{\pi d - p\mu}\right)$$

$$= \frac{500 \times 0.06}{2 \times 0.35} \times \left(\frac{0.01 + 0.05 \times \pi \times 0.06}{\pi \times 0.06 - 0.01 \times 0.05}\right)$$

$$\therefore \quad P = 42.86\left(\frac{0.01942}{0.18799}\right) = 4.43 \text{ N}$$

(ii)
$$\tan \alpha = \frac{p}{\pi d} = \frac{0.01}{\pi \times 0.06} = 0.05305$$

and
$$\tan \phi = \mu = 0.05$$

$$\tan (\alpha + \phi) = \frac{\tan \alpha + \tan \phi}{1 - \tan \alpha \tan \phi}$$

$$= \frac{0.05305 + 0.05}{1 - 0.05305 \times 0.05} = 0.10332$$

$$\eta = \frac{\tan \alpha}{\tan (\alpha + \phi)} = \frac{0.05305}{0.10332}$$

$$= 0.5135 \text{ or } 51.35\%$$

The efficiency is more than 50%, therefore, the jack is not self-locking.

(iii) Because the screw jack is not self-locking, the weight ($W = 500$ N) will start coming down. The effort P_1 applied at the mean radius of the screw jack to balance the load which is descending is given by,

$$P_1 = W \tan (\alpha - \phi)$$

$$= W \frac{\tan \alpha - \tan \phi}{1 + \tan \alpha \tan \phi}$$

$$= 500 \times \left(\frac{0.05305 - 0.05}{1 + 0.05305 \times 0.05} \right)$$

$$= 1.52 \text{ N}$$

$$\therefore \quad \text{Torque required} = P_1 \times (d/2) = 1.52 \times (0.06/2)$$

$$= 0.0456 \text{ Nm}$$

Example 5.32 A screw jack has a square thread of mean diameter 7.5 cm and pitch 1.5 cm. The co-efficient of friction at the screw thread is 0.075. A load of 2500 N is to be lifted through 10 cm. Determine (i) the torque required and the work done in lifting the load through 10 cm, and (ii) efficiency of the screw jack.

Solution

(i)
$$\tan \alpha = \frac{p}{\pi d} = \frac{0.015}{\pi \times 0.075}$$

$$= 0.06366$$

$$\tan \phi = \mu = 0.075$$

The effort applied at the mean radius of the screw jack for lifting the given load is given by,

$$P_1 = W \tan (\alpha + \phi)$$

$$= W \left(\frac{\tan \alpha + \tan \phi}{1 - \tan \alpha \tan \phi} \right)$$

$$= 25000 \times \left(\frac{0.06366 + 0.075}{1 - 0.06366 \times 0.075}\right)$$

$$= 348.31 \text{ N}$$

$$\text{Therefore, torque required} = P_1 \times \frac{d}{2} = 348.31 \times \frac{0.075}{2}$$

$$= 13.06 \text{ Nm}$$

(ii) $$\tan(\alpha + \phi) = \frac{\tan\alpha + \tan\phi}{1 - \tan\alpha\tan\phi}$$

$$= \frac{0.06366 + 0.075}{1 - 0.06366 \times 0.075}$$

$$= 0.139933$$

$\therefore$ $$\eta = \frac{\tan\alpha}{\tan(\alpha + \phi)} = \frac{0.06366}{0.13933}$$

$$= 0.4569 \text{ or } 45.69\%$$

Example 5.33 A square threaded screw jack has a pitch of 12 mm and mean diameter of screw is 50 mm. Coefficient of friction between the screw and nut is 0.12. Determine (i) the magnitude of the torque applied on the screw to raise a load of 25 kN, assuming that the load rotates with screw, (ii) the ratio of torque applied to raise the load to the torque applied to lower the load, and (iii) the efficiency of the screw jack.

Solution

(i) $$\tan\alpha = \frac{p}{\pi d}$$

$\therefore$ $$\alpha = \tan^{-1}\left(\frac{p}{\pi d}\right)$$

$$= \tan^{-1}\left(\frac{12}{\pi \times 50}\right) = 4.37^\circ$$

$$\tan\phi = 0.12$$

$$\phi = \tan^{-1}(0.12) = 6.84^\circ$$

Force applied on the screw to raise the load is given by,

$$P_1 = W\tan(\alpha + \phi)$$

$\therefore$ $$P_1 = 25\tan(4.37 + 6.84) = 4.95 \text{ kN}$$

Torque applied on the screw to raise the load,

$$T_1 = P_1 \times \frac{d}{2} = 4.95 \times \frac{0.05}{2}$$

$$= 0.12375 \text{ kNm} = 123.75 \text{ Nm}$$

(ii) Force applied to lower the load is given by,

$$P_1 = W \tan(\alpha + \phi)$$

$$\therefore \quad P_1 = 25 \tan(6.84 - 4.37) = 1.0784 \text{ kN}$$

Torque applied on the screw to lower the load,

$$T_2 = P_1 \times \frac{d}{2} = 1.0784 \times \frac{0.05}{2}$$

$$= 0.02696 \text{ kNm} = 26.96 \text{ Nm}$$

$$\therefore \quad \text{Ratio of torque} = (T_1/T_2) = (123.75/26.96) = 4.59$$

(iii) Efficiency of the screw jack is given by,

$$\eta = \frac{\tan \alpha}{\tan(\alpha + \phi)}$$

$$= \frac{\tan(4.37)}{\tan(4.37 + 6.84)}$$

$$= 0.3856 \text{ or } 38.56\%$$

5.11 PULLEYS

A pulley is a metallic or wooden wheel which is able to rotate about its axis. It has a groove along its periphery to rest a rope or chain. The pulley is mounted on an axle which is supported in a framework of metal or wood known as block. Pulley is used to lift weights. Pulleys are of two types, namely, fixed pulley, and movable pulley.

If the block is fixed then the pulley is known as fixed pulley (Figure 5.13 a) and if the block is movable then the pulley is known as movable pulley (Figure 5.13 b). The assumptions for the analysis of pulley systems are: (i) The weight of pulley is small as compared to the weight to be

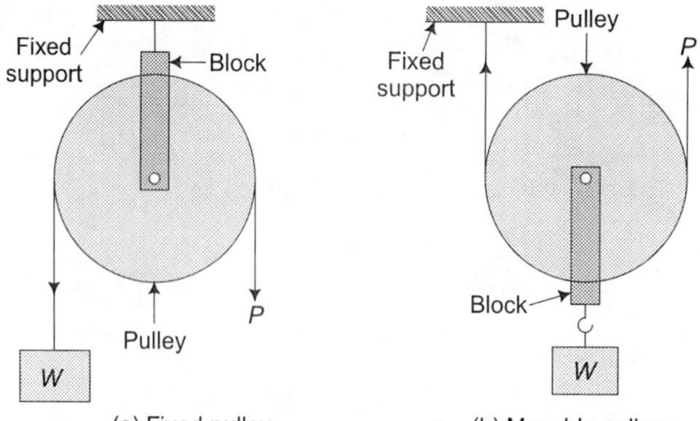

(a) Fixed pulley (b) Movable pulley

Fig. 5.13 Types of Pulleys

lifted and hence the former can be neglected in calculations, and (ii) The friction between the pulley surface and the rope is negligible and thus the tensions in the rope on either side of a pulley are same.

5.11.1 Single Fixed Pulley

In a single fixed pulley, the weight is attached at one end and effort is applied at the other end of the rope.

Let W = Weight lifted, T = Tension in the rope and, P = Effort applied.

As per the assumptions, $T = W$ and because the tension in the rope is same,

$$\therefore \quad T = P = W$$

$$\therefore \quad MA = \frac{W}{P} = 1 \tag{5.39}$$

The distance moved by the load is the same as that moved by the effort,

$$\therefore \quad VR = 1 \tag{5.40}$$

5.11.2 A Movable and a Fixed Pulley

In practical applications to apply force easily it becomes important to use a combination of a fixed pulley and movable pulley as shown in Figure 5.14.

The weight W is shared equally by two portions of the rope as shown in Figure 5.14.

$$\therefore \quad P = \frac{W}{2}$$

Mechanical advantage,

$$MA = \frac{W}{P} = \frac{W}{W/2} = 2 \tag{5.41}$$

It is clear from Figure 5.14 that to lift a load through a distance, the effort has to move double the distance.

$$\therefore \quad VR = 2 \tag{5.42}$$

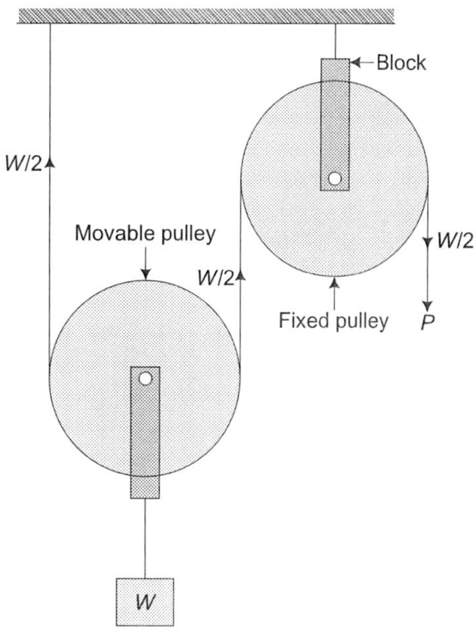

Fig. 5.14 A Movable and a Fixed Pulley

5.12 PULLEY SYSTEMS

When two or more pulleys are combined together to obtain higher mechanical advantage such a system is known as pulley systems or systems of pulleys. Three pulleys systems are commonly used: (i) First system of pulleys, (ii) Second system of pulleys, and (iii) Third system of pulleys.

5.12.1 First System of Pulleys

In this system, one end of each rope is tied to a fixed support while the other end passing around the periphery of the bottom pulley is tied to the block of the next higher pulley. So, there are as many fixed ropes as the number of movable pulleys. The rope of the last movable pulley after passing over its periphery wound around over the fixed pulley. The effort is applied at the end of the rope passing over the fixed pulley and the load is attached to the lowermost pulley as shown in Figure 5.15. Movable pulleys are used to increase the mechanical advantage. Fixed pulley provides convenience in the application of effort.

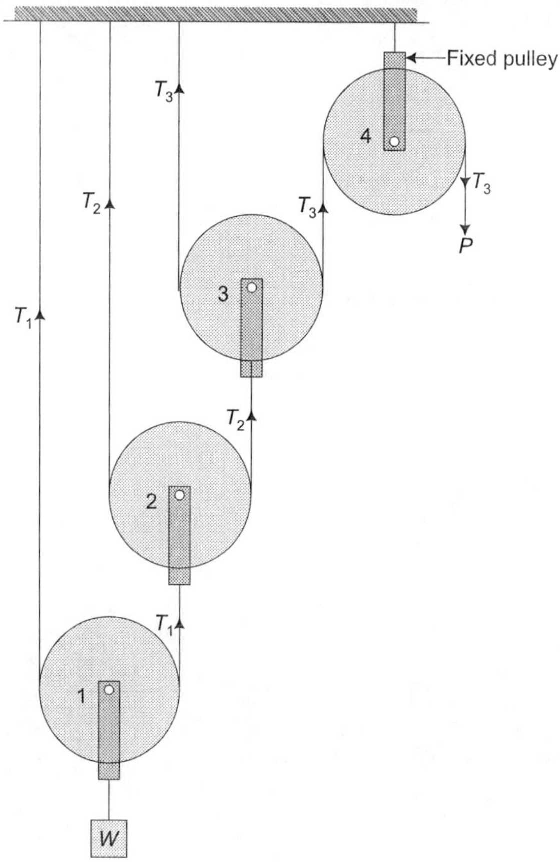

Fig. 5.15 First System of Pulleys

Let W = Load applied, P = Effort applied, and y = Distance moved by effort

Pulleys 1, 2, 3 are movable while pulley 4 is a fixed pulley.

For pulley 1: $2T_1 = W$

$\therefore$ $T_1 = W/2$

For pulley 2: $2T_2 = T_1 = W/2$

$\therefore \quad T_2 = W/2^2$

For pulley 3: $2T_3 = T_2 = W/2^2$

$\therefore \quad T_3 = P = W/2^3$

Similarly, for n number of pulleys,

$$T_n = P = W/2^n$$

$$MA = \frac{\text{Load}}{\text{Effort}} = \frac{W}{W/2^n} = 2^n \tag{5.43}$$

When effort moves a distance y, Pulley 3 moves = $y/2$

$$\text{Pulley 2 moves} = \frac{1}{2} \times \frac{y}{2} = \frac{y}{2^2}$$

Therefore, with n number of pulleys, distance moved = $\dfrac{y}{2^n}$

$$VR = \frac{\text{Distance moved by effort}}{\text{Distance moved by load}}$$

$$= \frac{y}{(y/2^n)} = 2^n \tag{5.44}$$

$$\therefore \quad MA = VR = 2^n \tag{5.45}$$

Efficiency, $$\eta = \frac{MA}{VR} = \frac{2^n}{2^n} = 1 \tag{5.46}$$

5.12.2 Second System of Pulleys

This system consists of an upper fixed pulley block and a lower movable pulley block as shown in the Figure 5.16.

Both pulley blocks may have equal number of pulleys as shown in the Figure 5.16 (a) or the upper block may have one pulley more than the lower block as shown in Figure 5.16 (b). The same rope is passed around all the pulleys. In the first case when blocks have equal number of pulleys one end of the rope is tied to the fixed block, whereas in the second case when blocks do not have equal number of pulleys then one end of the rope is tied to the movable block. The other end of the rope passes over the topmost pulley and will remain free. The effort is applied to this free end.

Let P = Effort applied, W = Load lifted, T = Tension in the rope, and n = Number of segments supporting the lower block = Total number of pulleys in the complete system

As pulleys are smooth, the tension in the rope throughout will remain same,

$\therefore \quad T = P.$

Under equilibrium condition,

$$W = nT = nP$$

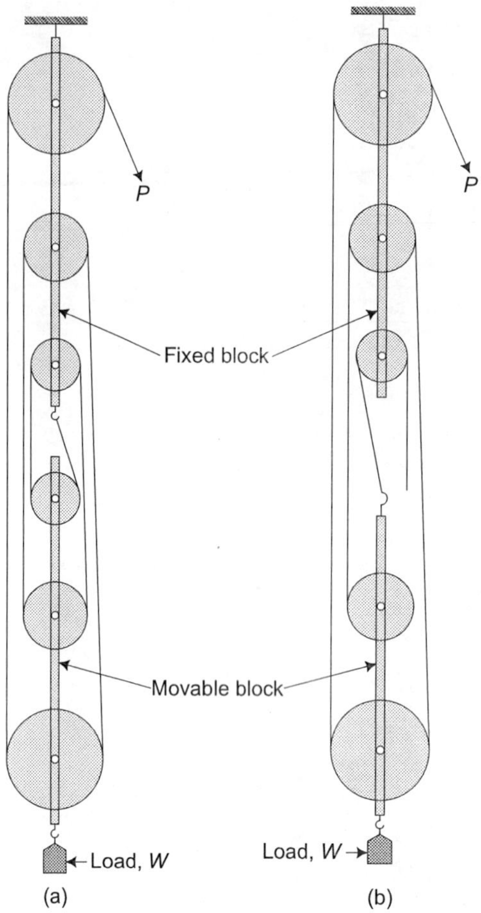

Fig. 5.16 Second System of Pulleys

$$\therefore \qquad MA = \frac{W}{P} = \frac{nP}{P} = n \qquad (5.47)$$

Under ideal conditions,

$$MA = VR$$

$$\therefore \qquad VR = 1 \qquad (5.48)$$

5.12.3 Third System of Pulleys

In this system, the topmost pulley is fixed while all other pulleys are movable. Each pulley has a segment of rope passing over it. One end of each rope segment is attached to a common block carrying the load while the other end is tied to the next movable pulley, except of lowermost pulley. The effort is applied to the free end of the rope passing over the last pulley as shown in Figure 5.17.

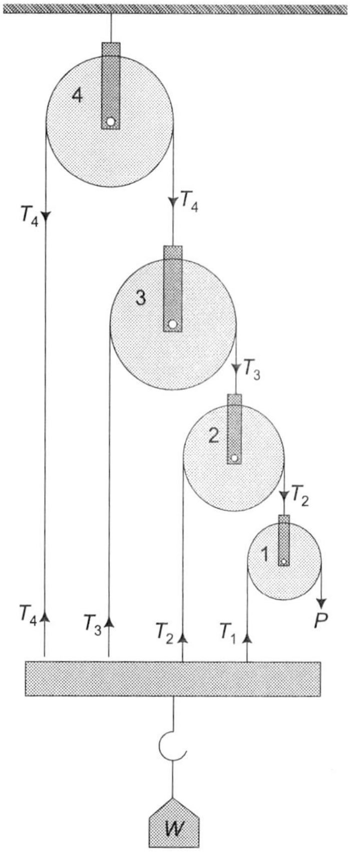

Fig. 5.17 Third System of Pulleys

Let W = Load lifted, P = Effort applied, and T_1, T_2, T_3, and T_4 are the tensions in the ropes (Figure 5.17).

Neglecting friction and weight of pulleys, for equilibrium of pulleys,

$$T_1 = P,\ T_2 = 2T_1 = 2P,$$

$$T_3 = 2T_2 = 2(2P) = 2^2P,$$

and

$$T_4 = 2T_3 = 2(2^2P) = 2^3P$$

Also,

$$W = T_1 + T_2 + T_3 + T_4$$

$$= P + 2P + 2^2P + 2^3P$$

$$= (1 + 2 + 2^2 + 2^3)P$$

or

$$\frac{W}{P} = (1 + 2 + 2^2 + 2^3)$$

or

$$MA = (1 + 2 + 2^2 + 2^3)$$

If there are n number of pulleys,

$$MA = (1 + 2 + 2^2 + 2^3 + \ldots + 2^{n-1})$$

Summing up, the geometrical progression gives,

$$MA = 2^n - 1 \qquad (5.49)$$

Under ideal conditions,

$$MA = VR$$

$\therefore$

$$VR = 2^n - 1 \qquad (5.50)$$

Example 5.34 In the first system of pulleys there are four movable pulleys. If an effort of 120 N is applied to lift a load of 1500 N, determine (i) the efficiency of the machine, (ii) load wasted in friction, and (iii) effort wasted in friction.

Solution

(i)

$$MA = \frac{W}{P} = \frac{1500}{120} = 12.5$$

$$VR = 2^n = 2^4 = 16$$

$\therefore$

$$\eta = \frac{MA}{VR} = \frac{12.5}{16}$$

$$= 0.7812 \text{ or } 78.12\%$$

(ii) For an ideal machine,

$$MA = VR$$

$$\frac{W_i}{P} = VR$$

$$\therefore \quad W_i = P \times VR = 120 \times 16$$

$$= 1920 \text{ N}$$

$\therefore$ Load wasted in friction = ideal load – actual load

$$= 1920 - 1500 = 420 \text{ N}$$

(iii) For an ideal machine,

$$MA = VR$$

$$\frac{W}{P_i} = VR$$

$$\therefore \quad P_i = \frac{W}{VR} = \frac{1500}{16} = 93.75 \text{ N}$$

$\therefore$ Effort wasted in friction

$$= \text{actual effort} - \text{ideal effort}$$

$$= 120 - 93.75 = 26.25 \text{ N}$$

Example 5.35 In a second system of pulleys there are three pulleys in upper block and two pulleys in the lower block. If the efficiency of the pulley system is 75%, determine the effort required to lift a load of 1.75 kN.

Solution

$$MA = \frac{W}{P} = \frac{1750}{P}$$

$$VR = n = 3 + 2 = 5$$

$$\eta = \frac{MA}{VR} = \frac{(W/P)}{VR}$$

$$0.75 = \frac{(1750/P)}{5}$$

$$\therefore \quad P = \frac{1750}{0.75 \times 5} = 466.67 \text{ N}$$

Example 5.36 In the second system of pulleys there are three pulleys in the upper block and two pulleys in the lower block. At zero loading, friction effort required is 25 N and friction wasted increases at the rate of 2% of the load lifted during pulling operation. Determine, (i) the effort required to raise a load of 2500 N, and (ii) efficiency of the system at the given load.

Solution

(i) Total friction at 2500 N

$$= 25 + \frac{2 \times 2500}{100} = 75 \text{ N}$$

Under ideal conditions,

$$W = nP$$

$$2500 = 5 \times P$$

$$\therefore \quad P = \frac{2500}{5} = 500 \text{ N}$$

$\therefore$ Total effort required = 500 + 75 = 575 N

(ii)
$$\eta = \frac{\text{Effort without friction}}{\text{Effort with friction}}$$

$$= \frac{500}{575} = 0.8696 \text{ or } 86.96\%$$

Example 5.37 In the third system of pulleys there are four pulleys. The efficiency of the system is 75%. Find the weight lifted by an effort of 100 N.

Solution

$$VR = 2^n - 1 = 2^4 - 1 = 15$$

$$MA = \frac{W}{P} = \frac{W}{100}$$

$$\eta = \frac{MA}{VR} = \frac{(W/P)}{VR}$$

$$0.75 = \frac{(W/100)}{15}$$

$$\therefore \quad W = 0.75 \times 15 \times 100 = 1125 \text{ N}$$

EXERCISE

(i) Objective Type Questions

1. In ideal machines, mechanical advantage is:

(a) equal to velocity ratio (b) greater than velocity ratio
(c) less than velocity ratio (d) unity

2. In actual machines, mechanical advantage is:

(a) equal to velocity ratio (b) greater than velocity ratio
(c) less than velocity ratio (d) unity

3. If ϕ is the angle of friction, the coefficient of friction is:

(a) ϕ radians (b) tan ϕ (c) sin ϕ (d) cos ϕ

4. The maximum frictional force that acts when a body is about to slide over a surface is called:
   (a) rolling friction (b) sliding friction (c) kinetic friction (d) limiting friction
5. The angle of friction is the:
   (a) angle between the normal reaction and the resultant of the normal reaction and limiting friction
   (b) frictional force when the body is in motion
   (c) angle between the friction force and the tangent
   (d) none of the above
6. In a lifting machine:
   (a) the effort is always greater than the load
   (b) $VR > 1$ enables large loads to be lifted by a smaller effort
   (c) VR is always less than unity
   (d) none
7. Velocity ratio of an inclined plane inclined at an angle α to the horizontal and the weight being pulled up the inclined plane by vertical effort is:
   (a) $\cos \alpha$ (b) $\sin \alpha$ (c) $1/\cos \alpha$ (d) $1/\sin \alpha$
8. A machine is said to be reversible, if its efficiency is:
   (a) equal to 50% (b) more than 50% (c) less than 50% (d) none of the above
9. Velocity ratio is the ratio of:
   (a) distance travelled by load to distance travelled by effort
   (b) distance travelled by effort to distance travelled by load
   (c) none
10. Which one of the following is the simple machine:
    (a) stapler (b) screw jack (c) scissors (d) bicycle
11. If a machine is lubricated then its:
    (a) efficiency decreases (b) velocity decreases
    (c) velocity ratio increases (d) mechanical advantage increases
12. Condition of irreversible machine is that efficiency will be:
    (a) equal to 50% (b) more than 50% (c) less than 50% (d) none of the above
13. If P is the effort and W is the load, the ideal load is:
    (a) $P \times VR$ (b) VR/P (c) P/VR (d) W/P
14. If W is the load, the part of effort P lost due to friction is:
    (a) $P \times VR \times (1 - \eta)$ (b) $W/VR \times [(1/\eta) - 1]$
    (c) P/VR (d) W/P

15. The law of machine relates:
(a) W & η (b) P & η (c) MA & VR (d) P & W

16. A self-locking machine is one for which:
(a) $MA = VR$ (b) $\eta = 100\%$ (c) $\eta < 100\%$ (d) $\eta < 50\%$

17. Velocity ratio of the first system of pulleys is:
(a) 2^n (b) $2^n - 1$ (c) n (d) $n - 1$

18. Velocity ratio of the second system of pulleys is:
(a) 2^n (b) $2^n - 1$ (c) n (d) $n - 1$

19. Velocity ratio of the third system of pulleys is:
(a) 2^n (b) $2^n - 1$ (c) n (d) $n - 1$

20. Law of machine is given by:
(a) $W = mP + C$ (b) $W = mP$ (c) $P = mW + C$ (d) none

21. Maximum efficiency of a simple screw jack is:
(a) $\dfrac{1 + \sin \phi}{1 - \sin \phi}$ (b) $\dfrac{1 + \cos \phi}{1 - \cos \phi}$ (c) $\dfrac{1 - \sin \phi}{1 + \sin \phi}$ (d) $\dfrac{1 - \cos \phi}{1 + \cos \phi}$

22. A simple screw jack in comparison to compound screw jack:
(a) gives a higher VR (b) gives a lower VR
(c) is more efficient (d) requires less effort to lift the load

23. Velocity ratio of a simple screw jack is:
(a) $2\pi p/l$ (b) $\pi p/l$ (c) $l/2\pi p$ (d) $2\pi l/p$

24. Velocity ratio of a compound screw jack is:
(a) $2\pi p/l$ (b) $\pi p/l$ (c) $l/2\pi p$ (d) $2\pi l/(p_b - p_a)$

25. Velocity ratio of a worm and worm wheel is:
(a) RT/nr (b) r/RT (c) l/nR (d) $2\pi l/r$

26. Velocity ratio of a differential pulley block is:
(a) $2D/(D - d)$ (b) $D/(D - d)$ (c) $(D - d)/D$ (d) none

27. Velocity ratio of a simple wheel and axle is:
(a) $D \times d$ (b) D/d (c) d/D (d) none

28. Velocity ratio of adifferential wheel and axle is:
(a) $2D/(d_1 - d_2)$ (b) $(d_1 - d_2)/2D$ (c) $(D - d)/D$ (d) none

29. Velocity ratio of a single purchase winch crab is:
(a) $2D/(d_1 - d_2)$ (b) $(2l \times T_2)/(d \times T_1)$
(c) $(d \times T_1)/(2l \times T_2)$ (d) none

30. Velocity ratio of a double purchase winch crab is:
(a) $2D/(d_1 - d_2)$ (b) $(d \times T_1 \times T_3)/(2l \times T_2 \times T_4)$
(c) $(2l \times T_2 \times T_4)/(d \times T_1 \times T_3)$ (d) none

ANSWERS

1. (a)	2. (c)	3. (b)	4. (d)	5. (a)	6. (b)	7. (d)	8. (b)	9. (b)	10. (b)
11. (d)	12. (c)	13. (a)	14. (b)	15. (d)	16. (d)	17. (a)	18. (c)	19. (b)	20. (c)
21. (c)	22. (b)	23 (d)	24. (d)	25. (a)	26. (a)	27. (b)	28. (a)	29. (b)	30. (c).

(ii) Review Questions

1. Define a machine and give the difference between an ideal machine and an actual machine.
2. Define mechanical advantage, velocity ratio, and efficiency of a machine.
3. Clearly explain the condition for the reversibility and self-locking of a machine.
4. Derive an expression for the law of machine.
5. Derive an expression for the velocity ratio of the following lifting machines: (i) wheel and axle, (ii) single and double purchase winch crabs, (iii) single start worm and worm wheel.
6. Explain with sketches the operation of simple wheel and axle and differential wheel and axle.
7. Sketch a single purchase winch crab and derive an expression for mechanical advantage, velocity ratio, and efficiency.
8. Explain with sketches the constructional and working details of a simple screw jack and a differential screw jack.
9. Explain how the Western's differential pulley block works? Also derive an expression for its velocity ratio.
10. Explain the constructional and working details of a double purchase winch crab with a neat sketch. Also derive an expression for mechanical advantage, velocity ratio, and efficiency for the same.

(iii) Numerical Problems

1. In a lifting machine, 20% of the effort applied to raise a load is lost in friction. Find out what load can be lifted through a distance of 100 cm if an effort of 250 N is applied through 15 m. Also find out the mechanical advantage and efficiency.

   [**Ans.** 2250 N, 11.25, 75%]
2. An effort of 65 N is applied to lift a load of 2000 N. The distance moved by effort is 4 m and by the load is 0.10 m. Determine (i) mechanical advantage, (ii) velocity ratio, (iii) efficiency, and (iv) ideal effort required. [**Ans.** 30.77 N, 40, 76.92%, 49.99 N]
3. A machine needs an effort of 16 N for lifting a load of 800 N at an efficiency of 60%. The same machine requires an effort of 30 N for lifting a load of 1500 N. Evaluate the law of machine, the maximum MA, and maximum η. [**Ans.** $P = 0.02\ W + 0$, 50, 60%]

4. In a simple wheel and axle machine the diameter of the axle is 4 cm. The machine is capable to lift a load of 60 N on applying an effort of 10 N, at the efficiency of 75%, determine the diameter of the wheel. [**Ans.** 32 cm]

5. A load of 500 N is to be raised by means of a simple wheel and axle unit. The axle diameter is 5 cm and wheel diameter is 50 cm. If a force of 100 N is to applied to the wheel, evaluate (i) MA, (ii) VR, and (iii) η [**Ans.** 5, 10, 50%]

6. In case of a differential wheel and axle unit, an effort of 50 N can lift a load of 500 N. If the efficiency of the machine is 65% at this load, determine velocity ratio of the machine. Also determine the diameters of the axles if the diameter of the effort wheel is 500 mm and the sum of the diameters of axle is 300 mm. [**Ans.** 7.69, 215 mm, 85 mm]

7. In a differential wheel and axle unit the diameters of the wheel is 20 cm and that of the axle are 10 cm and 5 cm. If the efficiency of the machine is 60%, determine the effort required to lift a load of 250 N. Also determine the number of revolutions the machine will complete in lifting the load through distance of 20 cm. [**Ans.** 52.08 N, 2.55 revolutions]

8. A Weston differential pulley has two pulleys of diameters 25 cm and 20 cm. Find the efficiency of the machine if an effort of 20 N is required to lift a load of 125 N. [**Ans.** 62.5%]

9. In a differential pulley block the upper block has two pulleys of diameters 25 cm and 24 cm. Determine the effort required to lift a load of 20,000 N if the efficiency of the machine is 40%. Also determine the length of the rope required to pull the load to lift through a height of 4 cm. [**Ans.** 1000 N, 200 cm]

10. In a single purchase winch crab the diameters of effort wheel and load drum are 200 cm and 25 cm respectively. The number of teeth on the pinion and spur gear are 10 and 100, respectively. Determine the velocity ratio of the machine, if efforts of 90 N and 150 N are required to lift loads of 2500 N and 8000 N respectively. Find the law of machine and the efficiencies at the above loads. [**Ans.** $P = 0.01\ W + 6.5$, 34.72%, 66.67%]

11. In a single purchase crab winch the number of teeth on spur wheel and pinion are 125 and 40 respectively. The diameter of load axle is 100 mm and the length of handle is 20 cm. At no load friction in machine is 20 N. Limiting efficiency is 80%. Find the effort required to lift a load of 150 N and the load at which the machine becomes reversible. [**Ans.** 35 N, 218.75 N]

12. A double purchase crab winch has the following particulars: number of teeth on pinion = 20 and 25, number of teeth on spur wheel = 55 and 45, length of handle = 450 mm, effective diameter of load drum = 150 mm. What load will be lifted by an effort of 250 N applied at the end of effort handle, if the efficiency of the machine is 45%? [**Ans.** 3341.25 N]

13. In a double purchase crab winch the diameter of the drum and length of lever arm are 15 cm and 35 cm respectively. Number of teeth on pinion are 20 and 30 and number of teeth on spur wheel are 75 and 90. This machine requires 80 N and 125 N efforts to lift the loads of 1750 N and 3200 N respectively. Find (i) the law of machine, (ii) effort required

to lift a load of 4000 N, (iii) efficiencies at the given load, and (iv) maximum efficiency of the machine. [**Ans.** $P = 0.03\ W$, 27.5, 147.5 N, 51.65%, 63.49%].

14. In a double threaded worm and worm wheel the number of teeth on worm wheel is 50. The diameter of the effort wheel and the load drum are 14 cm and 9 cm respectively. Find out the VR and η of the machine if it needs an effort of 30 N to lift a load of 400 N. [**Ans.** 38.89, 34.28%]

15. A screw jack lifts a load of 15 kN. The pitch of the screw is 10 mm. Find the effort required at the end of 25 cm arm, if the efficiency of the screw jack is 45%. Determine the MA of the screw jack. [**Ans.** 212.19 N, 70.69]

16. In a differential screw jack the pitches of the outer and inner screws are 10 mm and 6 mm respectively. Determine the load lifted by an effort of 90 N, if the efficiency of the machine is 75%. Length of handle is given 400 m. [**Ans.** 42.41 kN]

17. The following data apply for a screw jack: pitch of screw = 12.5 mm, mean diameter of screw = 50 mm, and coefficient of friction between screw and nut = 0.13. Determine (i) the magnitude of the torque applied on the screw to raise a load of 20 kN, assuming that the load rotates with the screw, (ii) the ratio of torque applied to raise the load to the torque applied to lower the load, and (iii) the efficiency of the screw jack. [**Ans.** 105 Nm, 25 Nm, 4.2, 37.6%]

18. A screw jack has a square thread of mean diameter 6 cm and pitch 9 mm. The coefficient of friction at the screw thread is 0.09. A load of 2000 N is to be lifted through 0.1 m. Determine the torque required and work done in lifting the load through 0.1 m. Also find the efficiency of the jack. [**Ans.** 8.21, 5792 kN, 34.53%]

6

Power Transmission Methods and Devices

6.1 INTRODUCTION

Power drives are used to transmit power from one shaft to another shaft. These can be divided into two categories, namely, flexible drive and direct drive. Flexible drive includes belt drive, rope drive, and chain drive, whereas direct drive includes gear drive. Belt, rope, and chain drives are used when the distance between the two shafts is large, whereas gear drive is preferred for small distance. When the power to be transmitted exceeds the frictional force the belt or rope will slip over the pulley. Due to straining action and slipping behavior these drives are also called non-positive drives means their velocity ratio is not constant. The chains are made up of small rigid links having relative motion between the two links. The chain and gear drives are positive or no slip drives, means these drives have constant velocity ratio.

A clutch is a device used to transmit power between two coaxial shafts when desired. Different types of clutches are friction clutch, centrifugal clutch, hydraulic clutch, electromagnetic clutch, and vacuum clutch. Friction clutches include single plate clutch, multiplate clutch, and cone clutch. A brake is a mechanical device used to apply frictional resistance to a moving body either to stop or retard it by absorbing its kinetic energy. Dynamometer is a brake which permits measurements of the frictional force or torque by which the power of the engine can be known. Dynamometers are of two types, namely, absorption dynamometers, and transmission dynamometers.

6.2 BELT DRIVE

A belt is used to transmit power by means of friction between two parallel shafts which are at a considerable distance apart. Pulleys are mounted on the shafts and are connected by a continuous belt or rope passing over them. The pulley which rotates first is called the driver pulley and the pulley that is driven by the driver is called follower or driven pulley. In Figure 6.1 (a), the pulley '*A*' which is mounted over the rotating shaft rotates first is called the driver pulley and the pulley '*B*' which is driven by pulley '*A*' is called follower or driven pulley.

The driver pulley pulls the belt from one side (lower side) and delivers it to the upper side. The tension in the lower side of the belt is more than that in the upper side of the belt. Thus, the lower

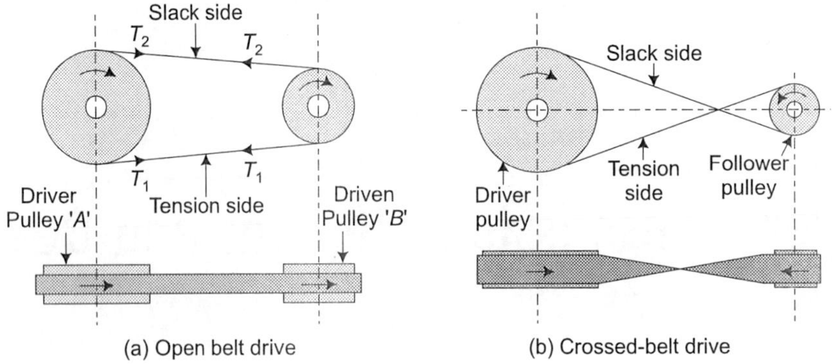

Fig. 6.1

side of the belt is known as tension side (or tight side) and the upper side of the belt is known as slack side.

6.2.1 Types of Belt Drives

The different types of belt drives are given below:

(i) **Open Belt Drive** In this arrangement, both the driver and the follower pulleys rotate in the same direction as shown in Figure 6.1 (a). Open belt drive is used for a maximum centre distance of 15 m.

(ii) **Crossed Belt Drive** In a crossed belt drive, the driver and the follower pulleys rotate in the opposite direction as shown in Figure 6.1 (b). A crossed belt drive can transmit more power than the open belt drive. The point where the belt crosses rubs against each other due to which excessive wear and tear takes place at that point. In order to avoid this phenomenon, the belt speed should be less than 15 m/s and the shafts should be placed at a maximum distance of 20 times of the width of the belt.

(iii) **Quarter Turn Belt Drive** This drive is also known as right angle belt drive as shown in Figure 6.2. This drive is used when the shafts are arranged at right angles.

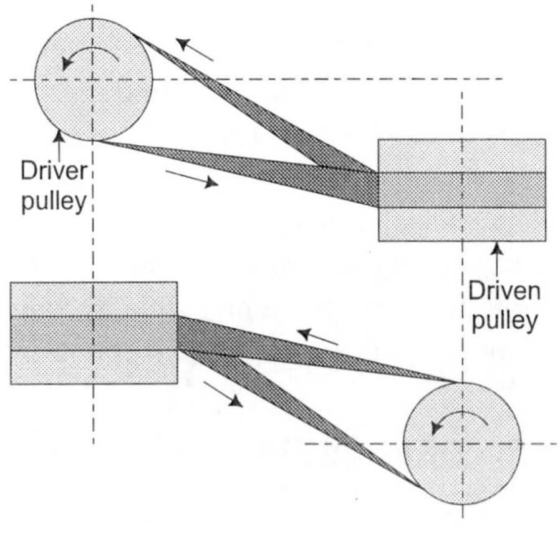

Fig. 6.2 Quarter Turn Belt Drive

(iv) **Compound Drive** When the power is transmitted from one shaft to another shaft through a number of pulleys then the drive is called a compound belt drive (Figure 6.3).

6.2.2 Types of Belts

The commonly used belts are: (i) Flat belt, (ii) V-belt, and (iii) Round belt.

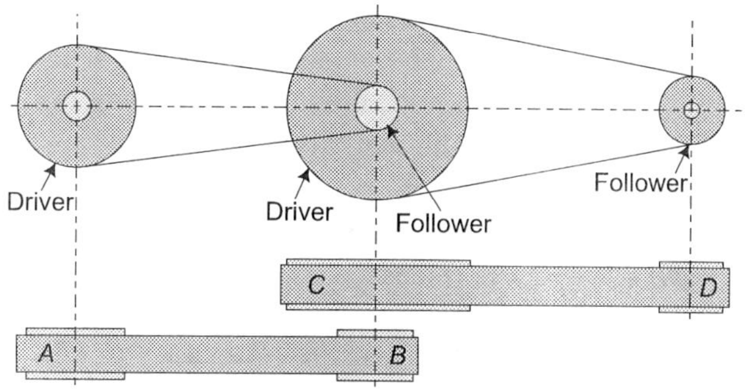

Fig. 6.3 Compound Belt Drive

(i) **Flat Belt** The flat belts have a narrow rectangular cross-section. These are used where moderate amount of power is to be transmitted from one shaft to another. They can be used for long centre distances, even up to 15 m and are efficient for high speed.

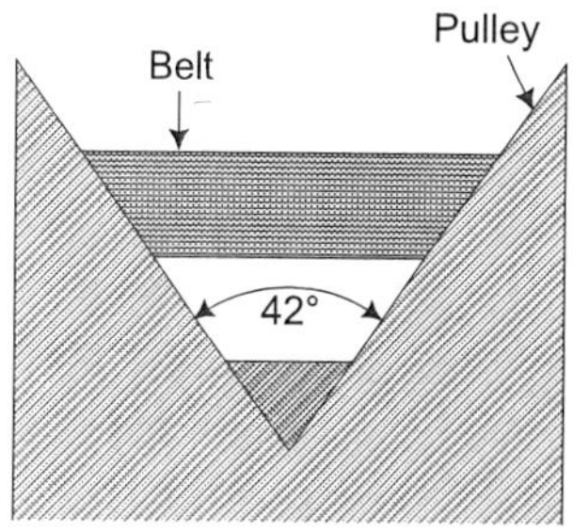

Fig. 6.4 V- Belt

(ii) **V-Belt** V-belts are used when the two pulleys are very close to each other. They can transmit more power without slip as compared to flat belts. V-belts have a trapezoidal cross section and it runs on pulley with groove cut to match the V-belt. The usual angle between the sides of groove is 42° as shown in Figure 6.4.

(iii) **Round Belt** The round belts are used to transmit low power such as in instruments, table type machine tools, clothing machines, and household appliances. These belts are made up of leather, canvas, and rubber. The diameter of the round belts varies from 4 mm to 8 mm.

6.2.3 Materials for Belts

The materials used for manufacturing the belts are: (i) Leather, (ii) Cotton or fabric, (iii) Rubber, and (iv) Balata.

(i) **Leather Belts** Either oak-tanned or chrome-tanned leather is used for making flat belts. Chrome-tanned leather belts are most suitable for exposure to water, steam, or oil. For heavy duty belts, number of leather strips are cemented together to increase the thickness of the belt and are specified according to the number of plies used (e.g., single, double, or triple ply belts) and according to the resulting thickness of the belt (e.g., light, medium, or heavy belts).

(ii) **Cotton or Fabric Belts** Cotton belts are made from canvas or cotton duck. These have three or more number of plies stitched together. In order to make the belts waterproof, these are treated with linseed oil which also prevents the fibres from any injury. These belts require little maintenance and are quite cheaper. These belts are suitable under hot or dry conditions and are used for conveyor belts.

(iii) Rubber Rubber belts are made of layers of canvas duck or fabrics impregnated with rubber compound and have a thin layer of rubber on the faces. These belts are suitable where they are exposed to moisture, e.g., paper mill. It is less durable and is cheaper than the leather belts. These are affected by heat, light, and oil.

(iv) Balata Belt These belts are made up from plies of stitched canvas impregnated with balata gum which is acidic in nature and waterproof. However, it is severely affected by temperature in excess of 40°C and mineral oils.

6.2.4 Advantages and Disadvantages of Flat Belts

Advantages

(i) Flat belts are relatively cheap and easy to maintain.

(ii) These do not require precise alignment of shafts and pulleys.

(iii) These are easy to replace when worn out.

(iv) These are quite smooth and silent in operation.

(v) These are flexible and are capable to absorb shocks and vibrations due to slipping action. This gives protection to the machine under sudden and temporary overload conditions.

(vi) These can be used for long centre distances, even up to 15 m, where other types of drives cannot be used.

(vii) These belts are capable to transmit power between the shafts arranged in a variety of ways in spaces.

(viii) The design of flat belt drives is quite simple and inexpensive.

(ix) A belt drive can be used as a clutch by shifting the belt from the loose pulley to a tight pulley and vice versa.

(x) Different velocity ratios can be obtained by using a stepped pulley where the belt is shifted from one step to another having different diameters.

(xi) These belts can be used in dusty and abrasive atmosphere and require no closed casing.

Disadvantages

(i) Flat belt is not a positive drive, thus slip and creep always occur.

(ii) Velocity ratio is not constant, so the belt drive cannot be used in applications where constant velocity ratio is required.

(iii) The belt drive has large dimensions and occupies more space.

(iv) The belt drive requires a tensioning device and has a relatively short life for high-speed applications.

(v) Belts are pre-tensioned to avoid slip which results in increase in bearing reactions as well as stress in the belt cross section.

(vi) As flat belts are hinged, so these belts make noise at high operating speeds.

6.2.5 Advantages and Disadvantages of V-belts over Flat Belts

Advantages

(i) The wedge action permits a smaller arc of contact which increases the pulling capacity of the belt and consequently, power transmission capacity of belt increases.

(ii) As V-belts are used for short-centre distances thus it is quite compact in construction.

(iii) It can be easily shielded with guards.

(iv) V-belt permits high-speed reduction even up to seven to one is possible.

(v) V-belt drive can be used in any position, even when the belt is vertical.

(vi) V-belts are endless due to which these are smooth and quite in operation.

(vii) Due to wedge action slip is negligible.

Disadvantages

(i) The construction of V-grooved pulley is complicated.

(ii) These are costlier as compared to flat belt drive.

(iii) Creep is high.

(iv) The efficiency of V-belt is lower than that of flat belt.

(v) Due to its constructional features bending stress increases which adversely affects its durability.

6.3 ROPE DRIVE

The rope drives are used where a large amount of power is to be transmitted over long centre distance between the shafts. However, long distances are avoided due to which use of ropes has been limited. Rope drives have two main advantages: (i) Ropes are cheaper than the belts, and (ii) A number of separate drives may be taken from one driving pulley. Grooved pulleys are used for transmitting power by ropes as shown in Figure 6.5.

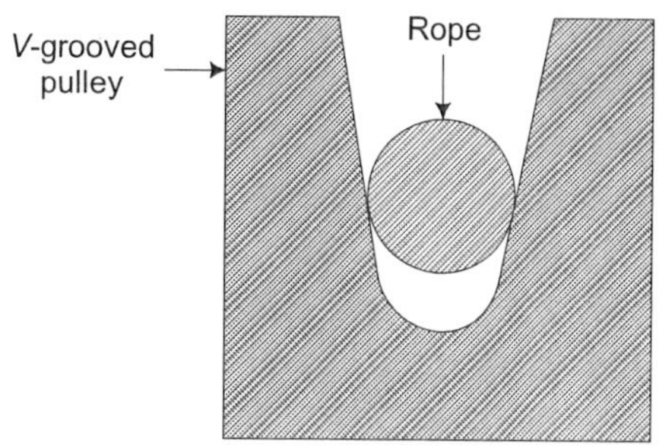

Fig. 6.5 Rope Drive

The groove angle varies from 40° to 60° but generally it is kept 45°. The ropes are made of cotton or wires of iron and steel which are usually of circular shape.

Rope drives are of two types: (i) Fibre ropes, and (ii) Wire ropes.

(i) **Fibre Ropes** These ropes are made of fibrous materials such as cotton, hemp, and manila. Hemp and manila fibres are not very flexible and these have poor mechanical properties. The hemp ropes have less strength, so these are used only for hand operated hoisting machines and as tie ropes for hooks, etc. Manila ropes are stronger than cotton ropes. The cotton ropes are very soft and smooth and are costlier than manila ropes.

(ii) Wire Ropes These ropes consist of strands of iron and steel and are used to transmit large amount of power over long distances from one pulley to another. These ropes are mostly used in conveyors, cranes, elevators, and suspension bridges.

The major advantages of wire ropes over cotton ropes are: (i) Wire ropes are more reliable and can withstand shock loads, (ii) Wire ropes are more durable and do not fail suddenly, and (iii) Their efficiency is quite high.

6.4 VELOCITY RATIO OF BELT DRIVES

Velocity ratio is the ratio of speed of driven pulley to that of the driving pulley.

Let N_1 and N_2 = Rotational speeds of the driving and driven pulleys respectively, d_1 and d_2 = Diameters of the driving and driven pulleys respectively, t = Thickness of the belt.

Assuming there is no slip between the belt and the pulley, and the belt is inelastic. Then,

peripheral velocity of the belt on driver = Peripheral velocity of the belt on follower,

$$\frac{\pi d_1 N_1}{60} = \frac{\pi d_2 N_2}{60}$$

or

$$\frac{N_2}{N_1} = \frac{d_1}{d_2}$$

But

$$\frac{N_2}{N_1} = \text{Velocity ratio}$$

∴

$$VR = \frac{N_2}{N_1} = \frac{d_1}{d_2} \tag{6.1}$$

If thickness of the belt is considered,

$$VR = \frac{N_2}{N_1} = \frac{d_1 + t}{d_2 + t} \tag{6.1a}$$

It is observed from equation (6.1) that speed of a pulley is inversely proportional to its diameter.

If a number of pulleys are used to transmit power when

d_1, d_3 and d_5 = Diameters of the drivers and

d_2, d_4 and d_6 = Diameters of the driven pulleys.

Then,

$$\frac{N_{\text{last follower}}}{N_{\text{first driver}}} = \frac{d_1}{d_2} \times \frac{d_3}{d_4} \times \frac{d_5}{d_6} \tag{6.2}$$

6.5 SLIP AND ITS EFFECTS ON VELOCITY RATIO

The relative motion between the pulley and belt is called slip. Belt drive is not a positive drive. Due to insufficient frictional grip between the belt and pulley there occurs some forward motion of the driver without carrying the belt with it. So, there is a difference between the linear speeds of the pulley rim and the belt. This difference in speeds is a measure of slip and is generally expressed as a percentage.

Let S_1 = Percentage slip between the pulley and the belt, and S_2 = Percentage slip between the belt and the driven belt.

Peripheral velocity of driver pulley = $\pi\, d_1\, N_1$ (Neglecting belt thickness)

Due to the slip between the driver pulley and the belt, the belt will run at a linear speed

$$= \left(\frac{100 - S_1}{100}\right) \text{ times } \pi\, d_1\, N_1$$

$$\therefore \qquad \text{Belt speed} = \left(\frac{100 - S_1}{100}\right) \times \pi\, d_1\, N_1$$

Due to slip the driven pulley will run at a peripheral speed of $\left(\frac{100 - S_2}{100}\right)$ times the belt speed.

∴ Peripheral speed of the driven pulley

$$= \left(\frac{100 - S_2}{100}\right) \times \left(\frac{100 - S_1}{100}\right) \times \pi\, d_1\, N_1$$

Again the peripheral speed of the driven pulley

$$= \pi\, d_2\, N_2$$

$$\therefore \qquad \pi\, d_2\, N_2 = \left(\frac{100 - S_2}{100}\right)\left(\frac{100 - S_1}{100}\right) \pi\, d_1\, N_1$$

or

$$\frac{N_2}{N_1} = \left(1 - \frac{S_2}{100}\right)\left(1 - \frac{S_1}{100}\right) \times \frac{d_1}{d_2}$$

$$\frac{N_2}{N_1} = \left[1 - \frac{S_1}{100} - \frac{S_2}{100} + \frac{S_1\, S_2}{10000}\right] \times \frac{d_1}{d_2}$$

$$= \left[1 - \frac{S_1}{100} - \frac{S_2}{100} + \frac{0.01\, S_1\, S_2}{100}\right] \times \frac{d_1}{d_2}$$

$$\frac{N_2}{N_1} = \left[1 - \frac{S_1 + S_2 - S_1\, S_2}{100}\right] \times \frac{d_1}{d_2}$$

If 'S' is the total percentage slip, i.e.,

$$S = S_1 + S_2 - 0.01\, S_1.S_2 \qquad (6.3)$$

Then,

$$\frac{N_2}{N_1} = \left(1 - \frac{S}{100}\right) \times \frac{d_1}{d_2}$$

$$\therefore \qquad VR = \frac{N_2}{N_1} = \left(1 - \frac{S}{100}\right) \times \frac{d_1}{d_2} \qquad (6.4)$$

If the thickness of the belt is considered,

$$\therefore \quad VR = \frac{N_2}{N_1} = \left(1 - \frac{S}{100}\right) \times \frac{d_1 + t}{d_2 + t} \tag{6.5}$$

If a number of pulleys are used to transmit the power,

$$\frac{N_{\text{last follower}}}{N_{\text{first driver}}} = \frac{d_1}{d_2} \times \frac{d_3}{d_4} \times \frac{d_5}{d_6} \times \left[1 - \frac{S}{100}\right] \tag{6.6}$$

where d_1, d_3 and d_5 = Diameters of the drivers, and d_2, d_4 and d_6 = Diameters of the driven pulleys.

6.6 CREEP

Material of the belt is elastic. When the belt passes from the slack side to the tight side, a certain portion of the belt extends and it contracts when the belt passes from the tight side to slack side. These uneven changes of the length cause a relative motion of the belt on the pulley. This relative motion between the belt and pulley is known as the creep in the belt. The net effect of creep in belts is to reduce the speed of the driven pulley and consequently, reduces the power output.

When creep is considered,

$$VR = \frac{N_2}{N_1} = \frac{d_1}{d_2} \times \frac{E + \sqrt{\sigma_2}}{E + \sqrt{\sigma_1}} \tag{6.7}$$

where, σ_1 = Stress in belt on the tight side, σ_2 = Stress in belt on the slack side, and E = Young's modulus of elasticity for the belt material.

Example 6.1 A horizontal shaft running at 240 rpm is to drive a parallel shaft at 300 rpm. The diameter of the pulley on the driving shaft is 600 mm. If the belt is 6 mm thick, evaluate the diameter of the follower under the following condition (i) neglecting belt thickness, (ii) taking belt thickness into account, and (iii) the belt thickness is 6 mm and a slip of 2%.

Solution

(i)
$$\frac{N_2}{N_1} = \frac{d_1}{d_2}$$

$$\therefore \quad d_2 = d_1 \times \frac{N_1}{N_2} = 600 \times \frac{240}{300}$$

$$= 480 \text{ mm}$$

(ii)
$$\frac{N_2}{N_1} = \frac{d_1 + t}{d_2 + t}$$

$$\therefore \quad d_2 + t = (d_1 + t) \times \frac{N_1}{N_2}$$

$$= (600 + 6) \times \frac{240}{300} = 484.8 \text{ mm}$$

$\therefore \qquad d_2 = 484.8 - t = 484.8 - 6$

$$= 478.8 \text{ mm}$$

(iii) $$\frac{N_2}{N_1} = \left(1 - \frac{S}{100}\right) \times \frac{d_1 + t}{d_2 + t}$$

where $$S = S_1 + S_2 - 0.01\, S_1 S_2$$

$$= 2 + 2 - 0.01 \times 2 \times 2 = 3.96$$

or $$d_2 + t = \left(1 - \frac{S}{100}\right) \times (d_1 + t) \times \frac{N_1}{N_2}$$

$$= \left(1 - \frac{3.96}{100}\right) \times (600 + 6) \times \frac{240}{300} = 465.6$$

$\therefore \qquad d_2 = 465.6 - t = 465.6 - 6$

$$= 459.4 \text{ mm}$$

Example 6.2 An engine shaft is running at 300 rpm having a pulley of diameter 750 mm and transmitting power to another pulley of diameter 1800 mm by means of a belt. Determine the loss in speed of the follower due to creep, if the stress on the tight and slack side of the belt is 1.2×10^6 N/m^2 and 0.4×10^6 N/m^2 respectively. Take the Young's modulus of elasticity for the belt material as 80×10^6 N/m^2.

Solution When there is no creep, we have,

$$\frac{N_2}{N_1} = \frac{d_1}{d_2}$$

$\therefore$ $$N_2 = \frac{d_1}{d_2} \times N_1 = \frac{750}{1800} \times 300$$

$$= 125 \text{ rpm}$$

When creep is considered,

$$\frac{N_2}{N_1} = \frac{d_1}{d_2} \times \frac{E + \sqrt{\sigma_2}}{E + \sqrt{\sigma_1}}$$

$\therefore$ $$N_2 = \frac{d_1}{d_2} \times N_1 \times \frac{E + \sqrt{\sigma_2}}{E + \sqrt{\sigma_1}}$$

$$= \frac{750}{1800} \times 300 \times \frac{80 \times 10^6 + \sqrt{0.4 \times 10^6}}{80 \times 10^6 + \sqrt{1.2 \times 10^6}}$$

$$= 124.99 \text{ rpm}$$

$\therefore$ Loss in follower speed due to creep

$$= 125 - 124.99 = 0.01 \text{ rpm}$$

6.7 LENGTH OF BELT

(i) Length of Open Belt Drive In an open belt drive both the pulleys rotate in the same direction as shown in Figure 6.6.

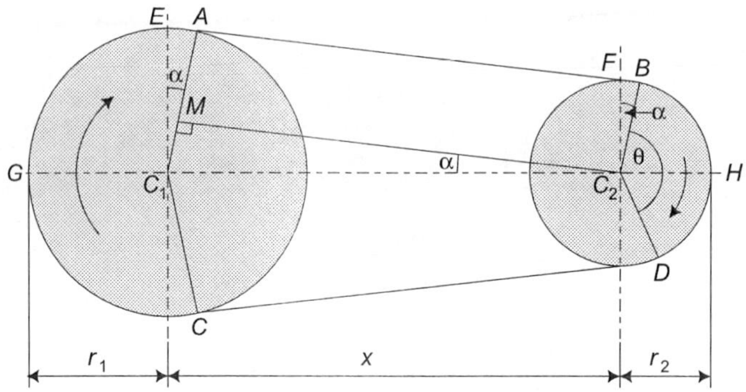

Fig. 6.6 Length of an Open Belt System

Let r_1 and r_2 = Radii of large and smaller pulleys respectively, x = Distance between the centres of the pulleys, L = Total length of the belt, and θ = Angle of lap or angle of contact = $(180 - 2\alpha)$.

The belt leaves the larger pulley at '*A*' and '*C*' and the smaller pulley at '*B*' and '*D*'.

Line C_2M is drawn parallel to AB which will be perpendicular to C_1A.

$$\angle EC_1A = \angle BC_2F = \angle MC_2C_1 = \alpha$$

$$C_1M = r_1 - r_2$$

$$\sin\alpha = \frac{C_1M}{C_1C_2} = \frac{C_1A - AM}{C_1C_2}$$

$$= \frac{r_1 - r_2}{x} \qquad [\because AM = BC_2 = r_2]$$

Since α is very small,

$$\therefore \qquad \sin\alpha = \alpha = \frac{r_1 - r_2}{x}$$

The length of the belt,

$$L = \text{Arc } CGA + AB + \text{Arc } BHD + DC$$

$$= 2(\text{Arc } GA + AB + \text{Arc } BH)$$

$$\text{Arc } GA = r_1 \times \left(\frac{\pi}{2} + \alpha\right),$$

and

$$\text{Arc } BH = r_2 \times \left(\frac{\pi}{2} - \alpha\right)$$

$$AB = MC_2 = \sqrt{(C_1C_2)^2 - (C_1M)^2}$$

$$= \sqrt{x^2 - (r_1 - r_2)^2}$$

$$= x \times \sqrt{1 - \left(\frac{(r_1 - r_2)}{x}\right)^2}$$

Expanding this by using binomial expression, we have,

$$AB = x \times \left[1 - \frac{1}{2}\left[\frac{(r_1 - r_2)}{x}\right]^2 + \ldots\right]$$

$$= x - \frac{(r_1 - r_2)^2}{2x} \qquad \text{[Neglecting higher powers]}$$

$$\therefore \qquad L = 2\left[r_1\left(\frac{\pi}{2} + \alpha\right) + x - \frac{(r_1 - r_2)^2}{2x} + r_2\left(\frac{\pi}{2} - \alpha\right)\right]$$

$$= 2\left[r_1 \times \frac{\pi}{2} + r_1\,\alpha + x - \frac{(r_1 - r_2)^2}{2x} + r_2 \times \frac{\pi}{2} - r_2\alpha\right]$$

$$= 2\left[\frac{\pi}{2}(r_1 + r_2) + \alpha\,(r_1 - r_2) + x - \frac{(r_1 - r_2)^2}{2x}\right]$$

$$= \pi\,(r_1 + r_2) + 2\alpha\,(r_1 - r_2) + 2x - \frac{(r_1 - r_2)^2}{x}$$

Now substituting the value of

$$\alpha = \frac{r_1 - r_2}{x}$$

$$L = \pi\,(r_1 + r_2) + 2\left(\frac{r_1 - r_2}{x}\right)(r_1 - r_2) + 2x - \frac{(r_1 - r_2)^2}{x}$$

$$= \pi\,(r_1 + r_2) + 2\,\frac{(r_1 - r_2)^2}{x} + 2x - \frac{(r_1 - r_2)^2}{x}$$

$$\therefore \qquad L = \pi\,(r_1 + r_2) + 2x + \frac{(r_1 - r_2)^2}{x} \qquad (6.8)$$

If d_1 and d_2 are the diameters of the pulleys,

$$L = \frac{\pi}{2}(d_1 + d_2) + 2x + \frac{(d_1 - d_2)^2}{4x} \qquad (6.9)$$

As the belt begins to slip on the smaller pulley, so the angle of lap θ must be taken into consideration on the smaller pulley.

$$\therefore \qquad \theta = (180 - 2\alpha)$$

(ii) Length of a Crossed Belt Drive In a crossed belt drive both the pulleys rotate in opposite direction as shown in Figure 6.7.

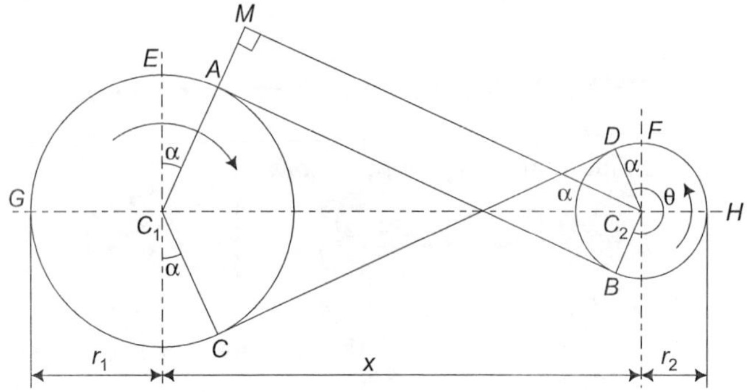

Fig. 6.7 Length of a Crossed Belt System

$$\angle EC_1A = \angle MC_2C_1 = \angle FC_2D = \alpha$$

$$\sin\alpha = \frac{C_1M}{C_1C_2} = \frac{C_1A + AM}{C_1C_2}$$

$$= \frac{r_1 + r_2}{x} \qquad [\because AM = BC_2 = r_2]$$

The length of the belt,

$$L = \text{Arc } CGA + AB + \text{Arc } BHD + DC$$

$$= 2(\text{Arc } GA + AB + \text{Arc } BH)$$

$$\text{Arc } GA = r_1 \times \left(\frac{\pi}{2} + \alpha\right)$$

and

$$\text{Arc } BH = r_2 \times \left(\frac{\pi}{2} + \alpha\right)$$

$$AB = MC_2 = \sqrt{(C_1C_2)^2 - (C_1M)^2}$$

$$= \sqrt{x^2 - (r_1 + r_2)^2}$$

$$AB = x \times \sqrt{1 - \left(\frac{r_1 + r_2}{x}\right)^2}$$

Expanding this by using binomial expression,

$$AB = x \times \left[1 - \frac{1}{2}\left(\frac{r_1 + r_2}{x}\right)^2 + \ldots\right]$$

$$= x - \frac{(r_1 + r_2)^2}{2x} \qquad \text{[Neglecting higher powers]}$$

$$\therefore \quad L = 2\left[r_1\left(\frac{\pi}{2}+\alpha\right) + x - \frac{(r_1+r_2)^2}{2x} + r_2\left(\frac{\pi}{2}+\alpha\right)\right]$$

$$= 2\left[r_1 \times \frac{\pi}{2} + r_1\,\alpha + x - \frac{(r_1+r_2)^2}{2x} + r_2 \times \frac{\pi}{2} + r_2\alpha\right]$$

$$= 2\left[\frac{\pi}{2}(r_1+r_2) + \alpha\,(r_1+r_2) + x - \frac{(r_1+r_2)^2}{2x}\right]$$

$$= \pi\,(r_1+r_2) + 2\alpha\,(r_1+r_2) + 2x - \frac{(r_1+r_2)^2}{x}$$

Now substituting the value of

$$\alpha = \frac{(r_1+r_2)}{x}$$

$$L = \pi\,(r_1+r_2) + 2\left(\frac{r_1+r_2}{x}\right)(r_1+r_2) + 2x - \left(\frac{(r_1+r_2)^2}{x}\right)$$

$$= \pi\,(r_1+r_2) + 2\,\frac{(r_1+r_2)^2}{x} + 2x - \frac{(r_1+r_2)^2}{x}$$

$$\therefore \quad L = \pi\,(r_1+r_2) + 2x + \frac{(r_1+r_2)^2}{x} \tag{6.10}$$

If d_1 and d_2 are the diameters of the pulleys,

$$L = \frac{\pi}{2}(d_1+d_2) + 2x + \frac{(d_1+d_2)^2}{4x} \tag{6.11}$$

It is noted from equations (6.8), (6.9), (6.10), and (6.11) that the length of crossed belt depends only on the sum of the pulley radii (or diameters) and the central distance, whereas in the case of open belts, the length depends on the sum as well as difference of the pulley radii (or diameters) apart from the centre distance. Further, if other conditions remain same then the length of a crossed belt is more than that of an open belt.

Angle of lap, $\theta = (180 - 2\alpha)$,

where $\sin\alpha = \dfrac{r_1+r_2}{x}$

Example 6.3 Two parallel shafts are connected by a crossed belt running over pulleys of diameters 60 cm and 40 cm respectively. If the centre distance between the pulleys is 6 m, determine the length of belt for (i) open belt system, and (ii) crossed belt system.

Solution

(i) For open belt system,

$$L = \frac{\pi}{2}(d_1+d_2) + 2x + \frac{(d_1-d_2)^2}{4x}$$

$$\therefore \quad L = \frac{\pi}{2} \times (60 + 40) + 2 \times 600 + \frac{(60 - 40)^2}{4 \times 600}$$

$$= 1357.25 \text{ cm}$$

(ii) For crossed belt system,

$$L = \frac{\pi}{2}(d_1 + d_2) + 2x + \frac{(d_1 + d_2)^2}{4x}$$

$$\therefore \quad L = \frac{\pi}{2} \times (60 + 40) + 2 \times 600 + \frac{(60 + 40)^2}{4 \times 600}$$

$$= 1364.25 \text{ cm}$$

It is observed from the results that under the same conditions, length of the crossed belt drive is more than that of the open belt drive.

6.8 POWER TRANSMITTED BY BELTS

Let T_1 and T_2 = Tensions on the tight and slack sides of the belt respectively in N

V = Linear velocity of the belt in m/s

Effective tension = $(T_1 - T_2)$ N

Work done by belt = $(T_1 - T_2) \times V$ m/s

$$\therefore \quad \text{Power transmitted} = \frac{(T_1 - T_2) \times V}{1000} \text{ kW} \qquad (6.12)$$

6.9 RATIO OF DRIVING TENSIONS

(i) Flat Belt Drive Consider a driven pulley, rotating in a clockwise direction as shown in Figure 6.8.

Let T_1 and T_2 = Tensions on the tight and slack sides of the belt respectively,

θ = Angle of lap or angle of contact of the belt over the pulley in radian, and

μ = Coefficient of friction between the belt and the pulley.

Consider a small portion of the belt AB, subtending an angle $\delta\theta$ at the centre of the pulley as shown in Figure 6.8.

The belt part AB is in equilibrium under the action of the following forces: (1) Tension T in the belt at A, i.e., tension on the slack side, (2) Tension $T + \delta T$ in the belt at B, i.e., tension on the tight side, (3) Normal reaction R between the element length of the belt and the pulley, and (4) Frictional force $F = \mu R$, where 'μ' is the coefficient of friction between the belt and pulley.

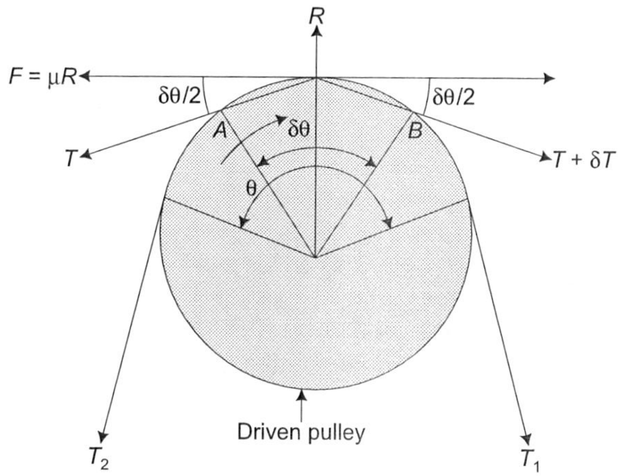

Fig. 6.8 Ratio of Driving Tensions for a Flat Belt

Resolving the forces horizontally,

$$R = T \sin(\delta\theta/2) + (T + \delta T) \sin(\delta\theta/2)$$

$$= T \times (\delta\theta/2) + (T + \delta T)(\delta\theta/2)$$

$$[\because \delta\theta \text{ is very small, } \therefore \sin(\delta\theta/2) = \delta\theta/2)]$$

$$= T \times (\delta\theta/2) + T \times (\delta\theta/2) + \delta T \times (\delta\theta/2)$$

$$= 2T \times (\delta\theta/2) \qquad [\text{Neglecting, } \delta T \times (\delta\theta/2)]$$

$$\therefore \qquad R = T \times \delta\theta \qquad (6.13)$$

Now resolving the forces vertically,

$$\mu R + T\cos(\delta\theta/2) = (T + \delta T) \times \cos(\delta\theta/2)$$

$$\mu R = (T + \delta T) \times \cos\frac{\delta\theta}{2} - T\cos\frac{\delta\theta}{2}$$

$$= T \times \cos\frac{\delta\theta}{2} + \delta T \times \cos\frac{\delta\theta}{2} - T \times \cos\frac{\delta\theta}{2}$$

$$= \delta T \,.\, \cos(\delta\theta/2)$$

$$\mu R = \delta T \qquad [\delta\theta \text{ is very small, } \cos(\delta\theta/2) = 1]$$

$$R = \frac{\delta T}{\mu} \qquad (6.14)$$

From equations (6.13) and (6.14), we have,

$$T\,\delta\theta = \frac{\delta T}{\mu}$$

or

$$\frac{\delta T}{T} = \mu \,.\, \delta\theta$$

Integrating both sides between the limits T_2 and T_1 and from 0 to θ respectively,

$$\int_{T_2}^{T_1} \frac{\delta T}{T} = \mu \int_0^{\theta} \delta\theta$$

or
$$\log_e \frac{T_1}{T_2} = \mu \,.\, \theta$$

∴
$$\frac{T_1}{T_2} = e^{\mu \,.\, \theta} \tag{6.15}$$

It is noted that the equation (6.15) is valid only when the belt is on the point of slipping on the pulleys. It means when larger and smaller pulleys are used then slip will occur first on the smaller pulley, thus angle of contact 'θ' is to be taken at the smaller pulley.

(ii) For V-Belt or Rope Drive A V-belt or a rope runs into a grooved pulley as shown in Figure 6.9

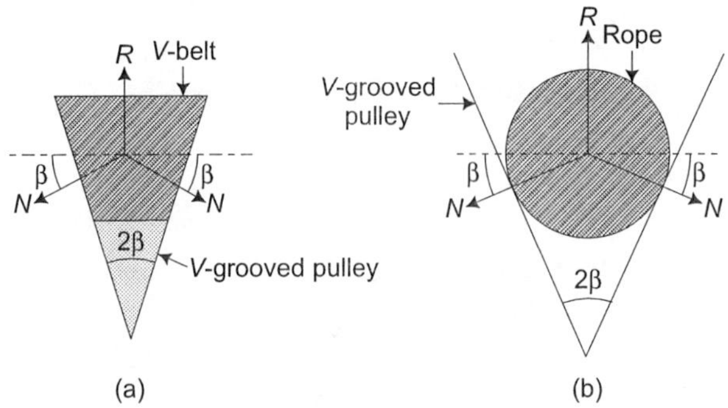

Fig. 6.9 V-Belt and Rope Drive

Let 2β = Angle of groove, N = Normal reaction between the belt and sides of V-grooved pulley, R = Total reaction in the plane of groove, and μ = Coefficient of friction between belt and groove.

Resolving forces vertically,

$$R = N \sin \beta + N \sin \beta = 2N \sin \beta$$

∴
$$N = \frac{R}{2 \sin \beta} = \frac{R}{2} \text{cosec } \beta$$

Frictional resistance,
$$F = \mu N + \mu N = 2\mu N$$

∴
$$F = 2\mu \times \frac{R}{2} \text{cosec } \beta$$

$$= \mu R \text{ cosec } \beta$$

In the previous article for flat belt derive we have observed that $F = \mu R$, whereas in the case of V- belt it is $F = \mu R$ cosec β.

Thus, the ratio of tensions T_1/T_2 in the case of V- belt can be obtained by replacing μ by μ cosec β in equation (6.15).

$$\therefore \quad \frac{T_1}{T_2} = e^{(\mu \text{ cosec } \beta)\, \theta}$$

$$= e^{\mu\theta \text{ cosec } \beta} \tag{6.16}$$

The ratio of tensions in rope drive will also be same as for the V-belt.

Example 6.4 Two pulleys, one 35 cm diameter and the other 80 cm diameter have been mounted on parallel shafts 2 m apart and are connected by a belt 160 mm wide. If the larger pulley rotates at 550 m/min and coefficient of friction between the belt and pulley is 0.20, determine the power transmitted in case of (i) open belt drive, and (ii) cross belt drive. The maximum permissible tension in the belt is 15 N/mm width of belt.

Solution

(i) Open belt drive

$$\sin \alpha = \frac{r_1 - r_2}{x} = \frac{0.40 - 0.175}{2}$$

$$\therefore \quad \alpha = \sin^{-1}\left(\frac{0.40 - 0.175}{2}\right) = 6.46°$$

Angle of lap,

$$\theta = 180 - 2\alpha = 180 - 2 \times 6.46$$

$$= 167.08°$$

$$\text{Maximum tension} = 15 \times 160 = 2400 \text{ N}$$

$$\frac{T_1}{T_2} = e^{\mu\theta}$$

$$\therefore \quad T_2 = \frac{T_1}{e^{\mu\theta}} = \frac{2400}{e^{0.2 \times \left(167.08 \times \frac{\pi}{180}\right)}}$$

$$= 673.78 \text{ N}$$

Power transmitted,

$$P = \frac{(T_1 - T_2) \times V}{1000}$$

$$= \frac{(2400 - 673.78) \times 550}{1000 \times 60}$$

$$= 15.82 \text{ kW}$$

(ii) Crossed belt drive

$$\sin \alpha = \frac{r_1 + r_2}{x} = \frac{0.40 + 0.175}{2}$$

$$\therefore \quad \alpha = \sin^{-1}\left(\frac{0.40 + 0.175}{2}\right) = 16.71°$$

Angle of lap, $\theta = 180 + 2\alpha = 180 + 2 \times 16.71$

$= 213.42°$

$$\frac{T_1}{T_2} = e^{\mu\theta}$$

$$\therefore \quad T_2 = \frac{T_1}{e^{\mu\theta}} = \frac{2400}{e^{0.2 \times \left(213.42 \times \frac{\pi}{180}\right)}}$$

$= 1139.39$ N

Power transmitted, $$P = \frac{(T_1 - T_2) \times V}{1000}$$

$$= \frac{(2400 - 1139.39) \times 550}{1000 \times 60}$$

$= 11.55$ kW

Example 6.5 A crossed flat leather belt is running at a speed of 250 rpm over a pulley having diameter of 1.2 m. The belt transmits 10 kW power when the angle of lap is 165° and coefficient of friction between the belt and pulley is 0.25. If maximum safe tension of the belt 25 N/mm width of belt, determine the required width of the belt.

Solution

$$\frac{T_1}{T_2} = e^{\mu\theta}$$

$$\therefore \quad T_1 = T_2 \times e^{\mu\theta}$$

$$= e^{0.25 \times \left(165 \times \frac{\pi}{180}\right)}$$

$= 2.05\ T_2$

$$V = \frac{\pi DN}{60} = \frac{\pi \times 1.2 \times 250}{60}$$

$= 15.71$ m/s

Power transmitted,

$$P = \frac{(T_1 - T_2) \times V}{1000}$$

$$10 = \frac{(2.05\ T_2 - T_2) \times 15.71}{1000}$$

$\therefore \quad T_2 = 606.23$ N

and $\quad T_1 = 2.05 \times 606.23 = 1242.77$ N

Therefore, required width of belt

$= (1242.77/25) = 49.71$ mm

6.10 CENTRIFUGAL TENSION

When a belt is running over a pulley centrifugal force acts on the belt due to its own weight which causes tension in both sides of the belt. The tension caused in the running belt by the centrifugal force is known as centrifugal tension. Due to symmetry centrifugal force induces equal tension on tight and slack sides of the belt.

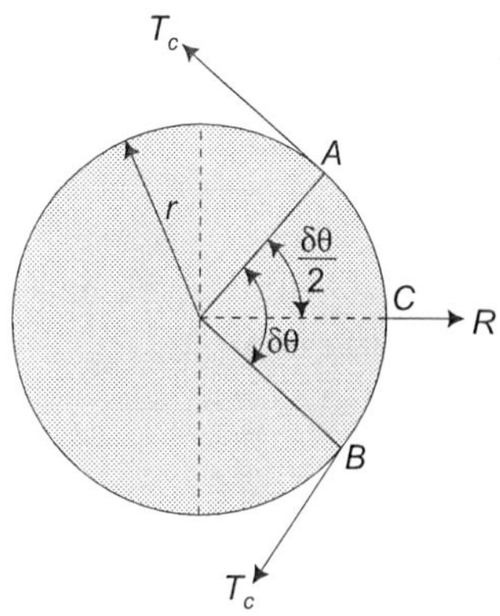

Fig. 6.10

Consider a small element AB of the belt subtending an angle $\delta\theta$ at the centre of the pulley as shown in Figure 6.10.

Let m = Mass of the belt per metre length, M = Mass of the small element AB, V = Velocity of belt in m/s, r = Radius of pulley, R = Centrifugal force acting radially outward, and T_c = Centrifugal tension acting at A and B tangentially.

The centrifugal force R acting radially outwards is balanced by the component of T_c acting radially inwards.

Length of small element

$$AB = r \times \delta\theta$$

Mass of small element AB,

$$M = \text{mass}/m \text{ length} \times \text{length } AB$$

$$= m \times r \times \delta\theta \tag{6.17}$$

Centrifugal force, $R = M \times (V^2/r)$ (6.18)

Substituting equation (6.17) in equation (6.18),

$$R = m \times r \times \delta\theta \times (V^2/r)$$

Now resolving the forces horizontally and equating the same, we have,

$$T_c \sin(\delta\theta/2) + T_c \sin(\delta\theta/2) = R$$

$$2T_c \sin(\delta\theta/2) = m \times r \times \delta\theta \times (V^2/r)$$

Since the angle $\delta\theta$ is very small, hence

$$\sin(\delta\theta/2) = (\delta\theta/2)$$

Then the above expression becomes,

$$2T_c \times (\delta\theta/2) = m \times r \times \delta\theta \times (V^2/r)$$

$$\therefore \quad T_c = mV^2 \tag{6.19}$$

From equation (6.19), it is seen that the centrifugal tension depends on the velocity of the belt and is independent of T_1 and T_2. When the centrifugal tension is taken into consideration then maximum tension on the tight side and slack side of the belt is given by,

$$T_{max} = T_1 + T_c \quad \text{(For tight side)} \tag{6.20}$$

$$T_{max} = T_2 + T_c \quad \text{(For slack side)} \tag{6.21}$$

For lower belt speed (i.e., belt speed less than 10 m/s), the centrifugal tension is very small and may be neglected. Centrifugal tension reduces the driving power capacity of the belt.

Maximum tension can also be given as,

$$T_{max} = \text{Maximum safe stress} \times \text{Cross-sectional area of the belt}$$

If b = Width of the belt in m, t = Thickness of the belt in m and σ = Maximum safe stress in the belt in N/m^2

Then
$$T_{max} = \sigma \times (b \times t) \tag{6.22}$$

At high speed (i.e., belt speed greater than 10 m/s) the centrifugal tension is to be considered, then maximum tension becomes,

$$T_{max} = T_1 + T_c \tag{6.23}$$

$$T_{max} = T_1 + T_c$$

$$= \sigma \times b \times t \tag{6.23 a}$$

6.11 MAXIMUM POWER TRANSMITTED BY A BELT

Let T_1 = Tension on tight side, T_2 = Tension on slack side, and V = Linear velocity of belt

Then the power transmitted is given by,

$$P = (T_1 - T_2) \times V \tag{i}$$

$$\frac{T_1}{T_2} = e^{\mu\theta}$$

∴
$$T_2 = \frac{T_1}{e^{\mu\theta}} \tag{ii}$$

Now substituting the expression (ii) in the expression (i), we get,

$$P = \left(T_1 - \frac{T_1}{e^{\mu\theta}}\right) \times V$$

$$= T_1\left(1 - \frac{1}{e^{\mu\theta}}\right) \times V \tag{6.24}$$

$$P = T_1 \times K \times V \tag{iii}$$

Assuming
$$1 - \frac{1}{e^{\mu\theta}} = K$$

$$T_{max} = T_1 + T_c$$

or
$$T_1 = T_{max} - T_c$$

Substituting the value of T_1 in equation (iii),

$$P = K \times (T_{max} - T_c) \times V$$

$$P = K \times (T_{max} - mV^2) \times V$$
$$= K \times (T_{max} \times V - mV^3) \qquad (\because T_c = mV^2) \qquad \text{(iv)}$$

For maximum power,

$$\frac{dP}{dV} = 0$$

or
$$\frac{d}{dV}\left[K \times (T_{max} \times V - mV^3)\right] = 0$$
$$T_{max} - 3mV^2 = 0$$

or
$$3mV^2 = T_{max}$$

$\therefore$
$$V = \sqrt{\frac{T_{max}}{3m}} \qquad (6.25)$$

Equation (6.25) gives the velocity of the belt at which maximum power is transmitted.

$$3mV^2 = T_{max}$$

or
$$T_{max} = 3T_c \qquad [\because T_c = mV^2]$$
$$T_c = \frac{T_{max}}{3} \qquad (6.26)$$

Thus, when transmitted power is maximum, centrifugal tension would be one-third of the maximum tension.

$$T_{max} = T_1 + T_c = T_1 + \frac{T_{max}}{3}$$
$$T_1 = T_{max} - \frac{T_{max}}{3}$$
$$= \frac{2}{3} T_{max} \qquad (6.27)$$

$\therefore$ Conditions for the transmission of maximum power are,

(i) $T_c = \dfrac{T_{max}}{3}$ (ii) $T_1 = \dfrac{2}{3} T_{max}$

Maximum power transmitted can be obtained by substituting equation (6.25) and (6.27) in equation (6.24)

$\therefore$ Maximum power, $$P_{max} = \frac{2}{3} T_{max}\left(1 - \frac{1}{e^{\mu\theta}}\right) \times \sqrt{\frac{T_{max}}{3m}} \qquad (6.28)$$

6.12 INITIAL TENSION

When a belt is fitted on the surface of the driver and the follower an initial tension T_o is given to the belt when the system is stationary which is called initial tension. When the belt is running and transmitting power, the tension on the tight side increases to T_1 and that on slack side decreases to T_2.

Now for belt calculation, the following two assumptions are made:

(i) Belt motion is perfectly elastic, and

(ii) Total length of the belt remains unchanged,

whether the belt is stationary or transmitting power.

On the tight side the tension increases from T_o to T_1 and on the slack side it decreases from T_o to T_2. If this change in the tension is δT,

Tension on tight side,

$$T_1 = T_o + \delta T \quad \text{(i)}$$

Tension on slack side,

$$T_2 = T_o - \delta T \quad \text{(ii)}$$

Solving the expressions (i) and (ii), we have,

$$T_o = \frac{T_1 + T_2}{2} \quad (6.29)$$

Thus, equation (6.29) shows that under ideal conditions, initial tension should be equal to the mean of the limiting values of tensions on tight and slack sides.

If centrifugal tension is considered,

$$T_o = \frac{T_1 + T_2 + 2T_c}{2} \quad (6.30)$$

Example 6.6 A belt 8 mm × 200 mm runs at a speed of 28 m/s and angle of contact is 165°. Its density is 1000 kg/m^3. If the maximum permissible stress in the belt is not to exceed 2.5 MN/m^2 and coefficient of friction is 0.25, determine centrifugal tension, and power transmitted.

Solution Mass of belt per metre of length,

$$m = \rho \times b \times t \times L$$

$$= 1000 \times 0.2 \times 0.008 \times 1 = 1.6 \text{ kg}$$

$$T_c = mV^2 = 1.6 \times (28)^2 = 1254.4 \text{ N}$$

Maximum tension, $T_{max} = \sigma \times b \times t$

$$= 2.5 \times 10^6 \times 0.2 \times 0.008 = 4000 \text{ N}$$

$$T_{max} = T_1 + T_c$$

$$\therefore \quad T_1 = T_{max} - T_c = 4000 - 1254.4$$

$$= 2745.6 \text{ N}$$

$$\frac{T_1}{T_2} = e^{\mu\theta}$$

$$T_2 = \frac{T_1}{e^{\mu\theta}} = \frac{2745.6}{e^{0.25 \times \left(\frac{\pi}{180} \times 165\right)}}$$

$$= 1336.49 \text{ N}$$

Power transmitted,

$$P = \frac{(T_1 - T_2) \times V}{1000}$$

$$= \frac{(2745.6 - 1336.49) \times 28}{1000}$$

$$= 39.45 \text{ kW}$$

Example 6.7 Determine the maximum power that can be transmitted by using a belt of dimensions: thickness = 10 mm and width = 100 mm with an angle of lap 160°. The density of belt is 1.1 Mg/m^3 and coefficient of friction is 0.25. The tension in the belt should not exceed 1.5 MN/m^2.

Solution Maximum tension,

$$T_{max} = \sigma \times b \times t = 1.5 \times 10^6 \times 0.1 \times 0.01$$

$$= 1500 \text{ N}$$

$$\frac{T_1}{T_2} = e^{\mu\theta} = e^{0.25 \times \left(\frac{\pi}{180} \times 160\right)} = 2$$

$$T_2 = \frac{T_1}{2}$$

For maximum power transmission,

$$T_c = \frac{T_{max}}{3} = \frac{1500}{3}$$

$$= 500 \text{ N}$$

$$T_{max} = T_1 + T_c$$

$$\therefore \quad T_1 = T_{max} - T_c = 1500 - 500$$

$$= 1000 \text{ N}$$

$$\therefore \quad T_2 = (T_1/2) = (1000/2)$$

$$= 500 \text{ N}$$

Mass of belt per metre of length,

$$m = \rho \times b \times t \times L$$

$$= 1.1 \times \frac{10^6}{10^3} \times 0.1 \times 0.01 \times 1$$

$$= 1.1 \text{ kg}$$

For maximum power transmitted,

$$V = \sqrt{\frac{T_{max}}{3m}} = \sqrt{\frac{1500}{3 \times 1.1}}$$

$$= 21.32 \text{ m/s}$$

Power transmitted, $$P = \frac{(T_1 - T_2) \times V}{1000}$$

$$= \frac{(1000 - 500) \times 21.32}{1000}$$

$$= 10.66 \text{ kW}$$

Example 6.8 A flat belt is required to transmit 25 kW from a pulley of 2 m effective diameter running at 300 rpm. The angle of contact spreads over 11/25 of the circumference and the coefficient of friction between belt and pulley surface is 0.25. Determine the width of the belt while considering centrifugal tension. It is given that the belt thickness is 10 mm, density of its material is 1000 kg/m^3, and related permissible working stress is 2.5 MN/m^2.

Solution Velocity of the belt,

$$V = \frac{(\pi \times D \times N)}{60}$$

$$= \frac{(\pi \times 2 \times 300)}{60} = 31.42 \text{ m/s}$$

Power transmitted, $$P = \frac{(T_1 - T_2) \times V}{1000}$$

$$(T_1 - T_2) = P \times 1000/V$$

$\therefore$ $$(T_1 - T_2) = (25 \times 1000)/31.42 = 795.67 \quad \text{(i)}$$

$$(T_1/T_2) = e^{\mu\theta} = e^{0.25 \times \left(\frac{11}{25} \times \frac{\pi}{180} \times 160\right)}$$

$$= 1.99 \quad \text{(ii)}$$

Solving equations (i) and (ii), we get,

$$T_1 = 1599.36 \text{ N}$$

$$T_2 = 803.7 \text{ N}$$

$$m = \rho \times b \times t \times L$$

$$= 1000 \times (b/1000) \times 0.01 \times 0.1$$

$$= 0.01\, b \text{ kg}$$

$$T_c = mV^2 = 0.01\, b \times (31.42)^2$$

$$= 9.872\, b \text{ N}$$

$$T_{max} = T_1 + T_c = \sigma \times b \times t$$

$$T_1 + T_c = \sigma \times b \times t$$

$$1599.36 + 9.872\ b = 2.5 \times 10^6 \times 0.001\ b \times 0.01$$

$$1599.36 + 9.872\ b = 25\ b$$

∴ Width, $b = 105.72$ mm

Example 6.9 A belt 10 mm × 150 mm is used to transmit power at a speed of 25 m/s. Determine the power transmitted if the net driving tension is 2.3 times the tension on the slack side. It is given that density of belt measured is 1000 kg/m^3 and permissible working stress is 2.56 N/mm^2. Also determine the maximum power and the corresponding belt speed.

Solution

$$T_{max} = \sigma \times b \times t$$

$$= 2.56 \times 150 \times 10 = 3840 \text{ N}$$

Mass of belt per metre of length,

$$m = \rho \times b \times t \times L$$

$$= 1000 \times 0.15 \times 0.01 \times 1 = 1.5 \text{ kg}$$

Centrifugal tension induced on each side of the belt,

$$T_c = mV^2 = 1.5 \times (25)^2$$

$$= 937.5 \text{ N}$$

$$T_{max} = T_1 + T_c$$

$$\therefore \quad T_1 = T_{max} - T_c$$

$$= 3840 - 937.5 = 2902.5 \text{ N}$$

Now, net driving tension = 2.3 × Tension on slack side

$$\therefore \quad T_1 - T_2 = 2.3\ T_2$$

$$T_2 = (T_1/3.3) = (2902.5/3.3)$$

$$= 879.54 \text{ N}$$

Power transmitted, $$P = \frac{(T_1 - T_2) \times V}{1000}$$

$$= \frac{(2902.5 - 879.54) \times 25}{1000} = 50.57 \text{ KW}$$

For maximum power transmitted,

$$V = \sqrt{\frac{T_{max}}{3m}} = \sqrt{\frac{3840}{3 \times 1.5}}$$

$$= 29.21 \text{ m/s}$$

$$T_c = (T_{max}/3) = (3840/3)$$
$$= 1280 \text{ N}$$

$\therefore$ $$T_1 = T_{max} - T_c$$
$$= 3840 - 1280 = 2560 \text{ N}$$
$$T_2 = \frac{T_1}{3.3} = \frac{2560}{3.3}$$
$$= 775.76 \text{ N}$$

Power transmitted, $$P = \frac{(T_1 - T_2) \times V}{1000}$$
$$= \frac{(2560 - 775.76) \times 29.21}{1000}$$
$$= 52.12 \text{ kW}$$

Example 6.10 A rope drive transmits 100 kW through a 45° grooved pulley of 1.75 m diameter running at 250 rpm. Coefficient of friction between the rope and the pulley groove is 0.25 and the angle of lap is 150°. The mass of each rope is 0.65 kg/m and it can withstand a tension of 800 N safely. Taking centrifugal tension into account, determine the number of ropes required for the drive and initial rope tension.

Solution

Velocity of the belt, $$V = \frac{(\pi \times D \times N)}{60}$$
$$= \frac{(\pi \times 1.75 \times 250)}{60} = 2291 \text{ m/s}$$

Centrifugal tension, $$T_c = mV^2 = 0.65 \times (22.91)^2$$
$$= 341.16 \text{ N}$$

$\therefore$ $$T_1 = T_{max} - T_c = 800 - 341.16$$
$$= 458.84 \text{ N}$$
$$\frac{T_1}{T_2} = e^{\mu\theta \text{ cosec } \beta}$$
$$T_2 = \frac{T_1}{e^{\mu\theta \text{ cosec } \beta}} = \frac{458.84}{e^{0.25 \times \frac{\pi}{180} \times 150 \times \text{cosec } \frac{45}{2}}}$$
$$= 82.97 \text{ N} \qquad \left[\beta = \frac{45}{2}\right]$$

Power transmitted, $$P = \frac{(T_1 - T_2) \times V}{1000}$$

$$= \frac{(458.84 - 82.97) \times 22.91}{1000}$$

$$= 8.61 \text{ kW}$$

$$\text{Number of ropes} = \frac{\text{Total power transmitted}}{\text{Power transmitted per rope}}$$

$$= \frac{100}{8.61} = 11.6 = 12$$

Initial rope tension, $$T_o = \frac{T_1 + T_2 + 2T_c}{2}$$

$$= \frac{458.84 + 82.97 + 2 \times 341.16}{2}$$

$$= 612.06 \text{ N}$$

Example 6.11 An open belt drive transmits 3 kW power. The linear velocity of belt is 5 m/s. The angle of lap of the smaller pulley is 160°. The coefficient of friction is 0.3. Determine the effect on power transmission if the (i) initial tension in the belt is increased by 10%, (ii) coefficient of friction is increased by 10% by suitable dressing to the friction surface of the belt, and (iii) angle of lap is increased by 10% by the use of an idle pulley, for the same speed and tension on the tight side.

Solution

Power transmitted, $$P = \frac{(T_1 - T_2) \times V}{1000}$$

$$(T_1 - T_2) = \frac{P \times 1000}{V} = \frac{3 \times 1000}{5}$$

$$\therefore \quad T_1 - T_2 = 600 \qquad \text{(i)}$$

$$(T_1/T_2) = e^{\mu\theta} = e^{\left(\frac{\pi}{180} \times 160\right)} = 2.31 \qquad \text{(ii)}$$

Solving (i) and (ii), we have,

$$T_1 = 1058.02 \text{ N}$$

$$T_2 = 458.02 \text{ N}$$

Initial tension, $$T_o = \frac{(T_1 + T_2)}{2}$$

$$= \frac{(1058.02 + 458.02)}{2} = 758.02 \text{ N}$$

(i) When initial tension is increased by 10%

$$T_o = 1.10 \times 758.02 = 833.82 \text{ N}$$

$$\frac{T_1 + T_2}{2} = 833.82 \qquad \text{(iii)}$$

As μ and θ remains same, $e^{\mu\theta}$ or (T_1/T_2) is same.

Therefore, solving (ii) and (iii), we get,

$$T_1 = 1163.82 \text{ N}$$

$$T_2 = 503.82 \text{ N}$$

Power transmitted, $$P = \frac{(T_1 - T_2) \times V}{1000}$$

$$= \frac{(1163.82 - 503.82) \times 5}{1000} = 3.3 \text{ kW}$$

$$\text{Increase in power} = \left(\frac{3.3 - 3}{3}\right) \times 100 = 10\%$$

(ii) When coefficient of friction is increased by 10%

T_1 remains same but θ increases by 10%

$$\therefore \quad \frac{1058.02}{T_2} = e^{\mu\theta} = e^{0.3 \times 1.10 \times \left(\frac{\pi}{180} \times 160\right)}$$

$$= 2.51$$

$$\therefore \quad T_2 = 421.52 \text{ N}$$

Power transmitted, $$P = \frac{(T_1 - T_2) \times V}{1000}$$

$$= \frac{(1058.02 - 432.52) \times 5}{1000} = 3.18 \text{ kW}$$

$$\text{Increase in power} = \left(\frac{3.18 - 3}{3}\right) \times 100 = 6\%$$

(iii) When angle of lap is increased by 10%

$$\therefore \quad \frac{1052.02}{T_2} = e^{\mu\theta} = e^{0.3 \times \left(\frac{\pi}{180} \times 160 \times 1.10\right)}$$

$$= 2.51$$

$$T_1 = 2.51\, T_2$$

Initial tension, $$T_o = \frac{(T_1 + T_2)}{2}$$

$$T_1 + T_2 = 2 \times T_o = 2 \times 758.02$$

$$= 1516.04$$

Solving, $$T_1 + T_2 = 1516.04$$

and $T_1 = 2.51\, T_2$, we have,

$$T_1 = 1084.12 \text{ N}$$

$$T_2 = 431.92 \text{ N}$$

Power transmitted, $P = \dfrac{(T_1 - T_2) \times V}{1000}$

$$= \frac{(1084.12 - 431.92) \times 5}{1000} = 3.26 \text{ kW}$$

$$\text{Increase in power} = \left(\frac{3.26 - 3}{3}\right) \times 100$$

$$= 8.67\%$$

6.13 CHAIN DRIVES

A chain drive consists of an endless chain wrapped around two sprockets, namely, driven sprocket, and driving sprocket as shown in Figure 6.11.

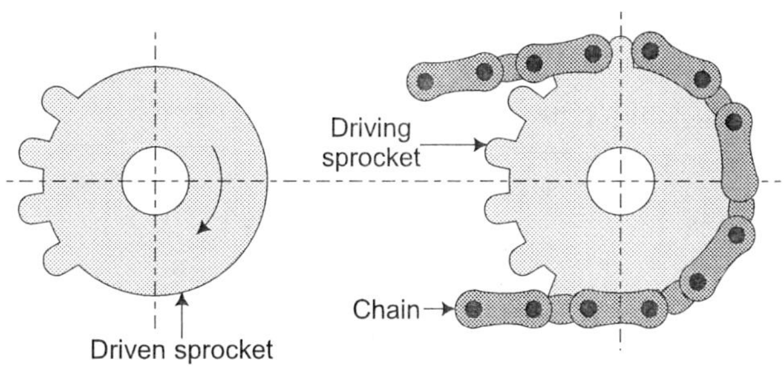

Fig. 6.11 Chain Drive

The chain consists of a number of rigid links connected by pin joints, while the sprockets are toothed wheels. The chain drive transmits motion between parallel shafts only. The chain drive is intermediate between belt and gear drives, means it has some features of belt drive and some of gear drives. When the power is transmitted over a short distance (0.5 m to 10 m) then chain drives are used. It is popular for transmission of power in bicycles, motorcycles, agricultural machinery, rod rollers, and automobile vehicles. They are used in metal and wood working machinery for transmission of power. They are also used in textile, building construction, and material handling machinery. In general, the chain drives are recommended to transmit power up to 100 kW. Chain drives are used for velocity ratio less than 10:1 and chain velocities up to 25 m/s.

In the case of belt and rope drives slipping may occur. Steel chains are used to avoid slipping. Thus, chain drives are used where positive action is required and these are becoming popular to obtain constant velocity ratio.

6.14 ADVANTAGES AND DISADVANTAGES OF CHAIN DRIVE

Following are the advantages and disadvantages of chain drive over belt or rope drive:

Advantages

1. Chain drive is a positive drive having no slip, hence perfect velocity ratio is obtained.
2. It gives high transmission efficiency (up to 98%).
3. It is small in size and thus takes up less space than a belt drive or rope drive.
4. Initial tension is not required in chain drive, whereas in belt drive it is essential.
5. In case of chain drive, wear and tear is less and thus maintenance cost is less in comparison to belt drives.
6. It gives fewer loads on the shafts and is able to carry heavy loads.
7. It gives effective transmission of power over considerable distances.
8. A single chain can transmit motion to several shafts.
9. Chain drive can operate under adverse temperature and atmospheric conditions.

Disadvantages

1. Its design is more complicated thus the production cost is high.
2. The operation is noisy and velocity fluctuations are present.
3. It needs accurate monitoring and careful maintenance.
4. It requires precise alignment of shafts, accurate monitoring, and careful maintenance.

6.15 CLASSIFICATION OF CHAINS

The chains may be primarily classified on the basis of use as: (i) Hoisting and hauling chains, (ii) Conveyor chains, and (iii) Power transmitting chains.

1. Hoisting and Hauling Chains These chains include oval link and square link as shown in Figure 6.12.

(i) **Chain with Oval Links** An oval link chain is a common form of hoisting chain. It consists of oval links and is also known as coil chain as shown in Figure 6.12(a). Such types of chains are used for low speeds only as in chain hoists and in anchors for machine works. This type of chain does not kink easily.

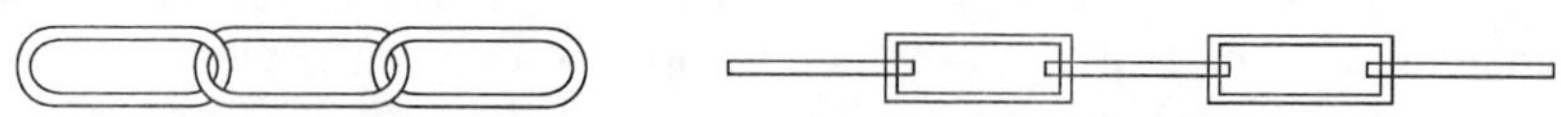

(a) Chain with oval link (b) Chain with square link

Fig. 6.12 Hoisting and Hauling Chains

(ii) **Chain with Square Links** The links of this type of chain are of square shape as shown in Figure 6.12(b). These are employed for hoists, cranes, and dredgers. The manufacturing cost of this type of chain is less than that of chain with oval links but these are easily subjected to kinks on overloading.

2. Conveyor Chains The conveyor chains are used for elevating and conveying the materials continuously. It runs at slow speed of about 50 m/min to 200 m/min. These are made of malleable cast iron and do not have smooth running qualities. The conveyor chains are of two types: (i) Detachable or hook joint type as shown in Figure 6.13(a), and (ii) Closed joint type as shown in Figure 6.13(b).

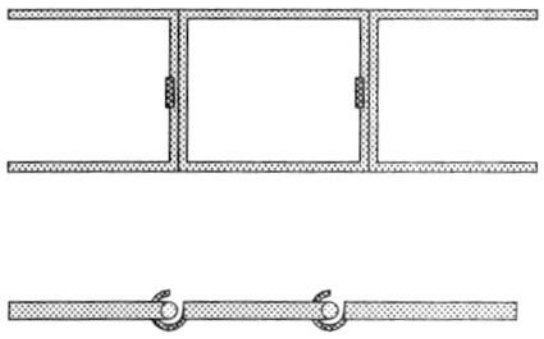

(a) Detachable hook joint chain

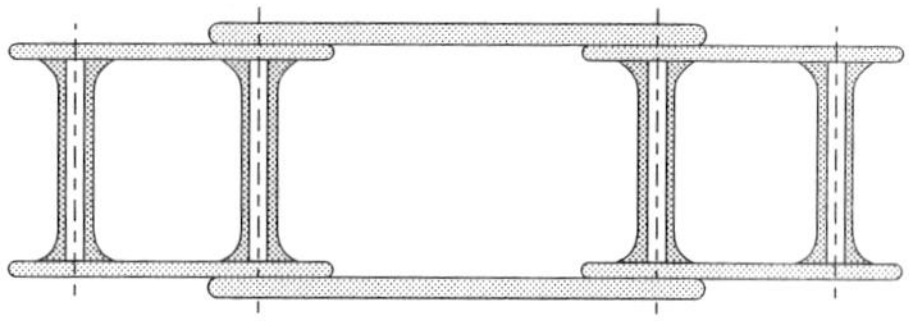

(b) Closed joint chain

Fig. 6.13 Conveyor Chains

3. Power Transmitting Chains These chains are used for high-speed performance. It has provision for efficient lubrication. These chains are of three types: (i) Block chains, (ii) Roller chains, and (iii) Silent chains.

(i) Block Chains These chains belong to the earlier stage of development in the power transmission. A block chain is shown in Figure 6.14. These are also known as bush chains. Due to rubbing action between the links and the teeth its operation is noisy. At low speed these chains are used to some extent as conveyor chains.

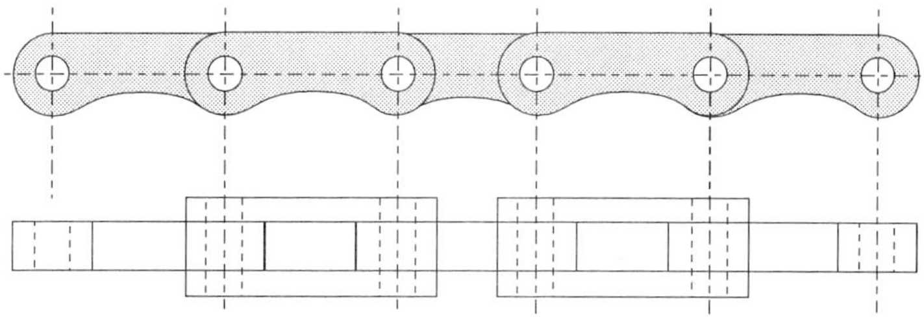

Fig. 6.14 Block Chain

(ii) Roller Chains These chains were primarily used in bicycles but nowadays these are used for so many other purpose. The roller chain as the name implies has a roller around the bush and also is held between roller links plates as shown in Figure 6.15.

There is a central pin which passes through the bush. A pin link plate is provided on both sides to join the two central pins and keep them in position. A good roller chain is quieter and wears less as compared to a block chain.

(iii) Silent Chains These chains are also called inverted tooth chains as shown in Figure 6.16 (a). The credit for its advancement goes to H. Renold. When it is aimed to achieve an almost silent performance coupled with transmission of high power, these chains

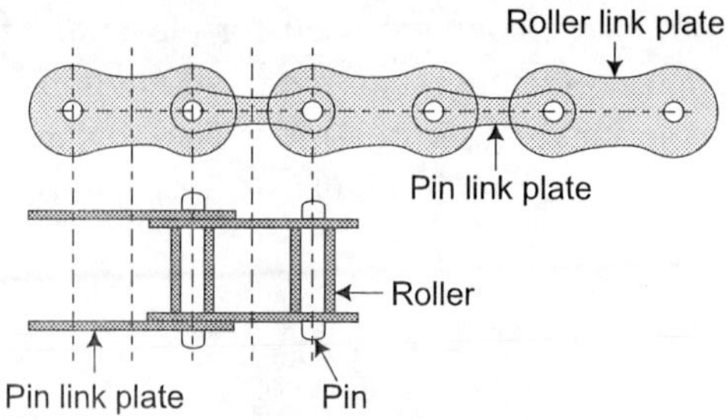

Fig. 6.15 Roller Chain

are preferred. Silent chains do not have any rollers but have a series of flat laminations or links of peculiar V-shape with included angle $\theta = 60°$ or $75°$ as shown in Figure 6.16 (b). The links are shaped to engage directly with the sprocket teeth. The edges E_1, E_2 and the corresponding faces of the teeth are straight and there occurs no relative motion between the links and sprocket teeth. Silent chains can be used for high speed applications. But applications are limited due to (i) heavy in weight, (ii) difficult to manufacture, and (iii) cost.

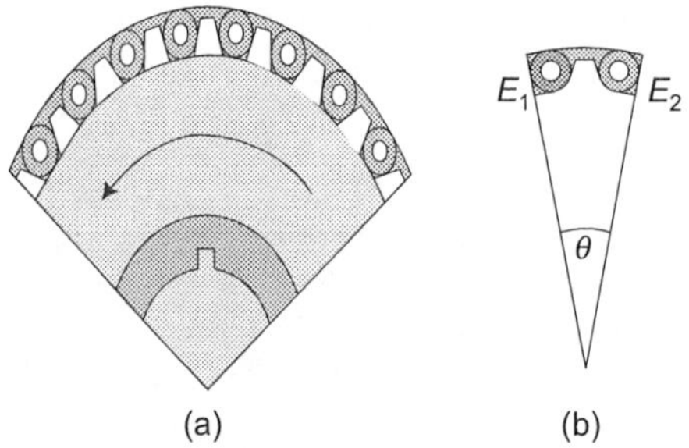

Fig. 6.16 Silent Chain

6.16 PULLEY

Pulleys are used to transmit power from one shaft to another by means of belts or ropes running over them. These are fitted on the shaft by means of a key. The pulley may be made up of cast iron, cast steel, pressed steel, or wood. The main parts of pulley are: (i) Rim, (ii) Hub or bush, and (iii) Rib as shown in Figure 6.17.

(i) Rim It is the outer portion of the pulley on which rope or belt is mounted. The slight convexity on the rim is called crowning which avoids the slip off the belt on the pulley.

(ii) Hub It is the central portion of the pulley which has a bore of shaft diameter on which it mounts.

(iii) Ribs Ribs are provided to add strength and stiffness to pulleys and it connects the hub and rim of the pulleys.

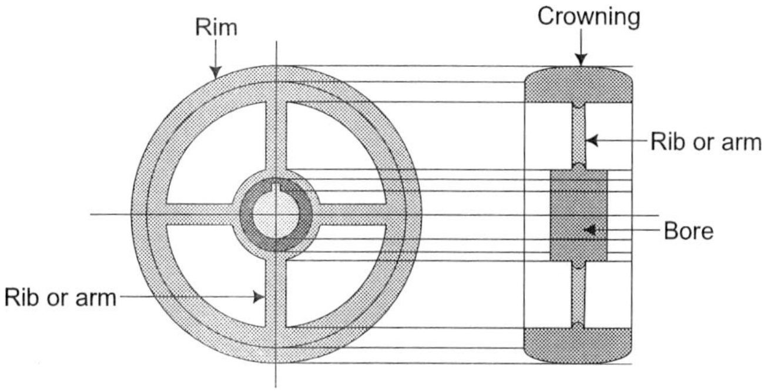

Fig. 6.17 Solid Cast Iron Pulley

The pulley may be either solid one piece or split pulley in two halves which helps in mounting the pulley on a shaft which is already carrying pulleys, etc.

6.16.1 Types of Pulleys

(i) **Stepped or Cone Pulleys** A stepped pulley consists of a number of pulleys of different diameters cast together as one piece as shown in Figure 6.18.

These pulleys are used to obtain variable speeds of a driven machine. Stepped pulleys changes the speed of a machine while the shaft runs at constant speed, e.g., in a lathe machine.

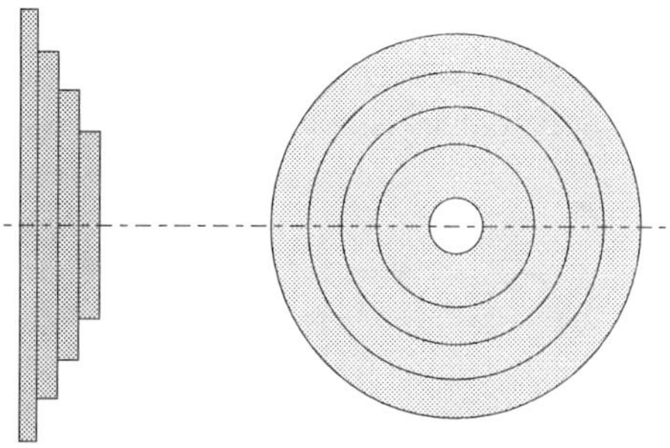

Fig. 6.18 Stepped Pulley

(ii) **Idler Pulley** Idler pulleys are used for increasing the tension, taking up stretch in the belt and for increasing angle of lap of the two pulleys. The idler pulley reduces the slippage of belt and thus increases its life. The idler pulley is free to rotate on its axis. Idler pulleys are shown in Figure 6.19.

(iii) **Split Pulleys** Sometimes it becomes necessary to mount a pulley on a shaft which already have pulleys, etc. or have its ends swelled then it is very easy to use a split pulley. Split pulleys are made in two halves. There is a clearance in between the faces and the two halves are tightened on the shaft by the bolts.

(iv) **Fast and Loose Pulleys** It is often necessary to drive several machines from a single main shaft. The main shaft keeps running continuously but the other driven machines are working intermittently. This is accomplished by a set of loose and fast pulleys. A fast pulley is keyed to the machine shaft and thus it runs at the shaft speed but loose pulley is not keyed to the machine shaft and thus it can rotate freely on the shaft and is incapable of transmitting any power. The loose pulley is generally mounted on a cast iron or gun metal sleeve with a collar at one end to prevent axial motion. The sleeve is secured to the shaft with the help

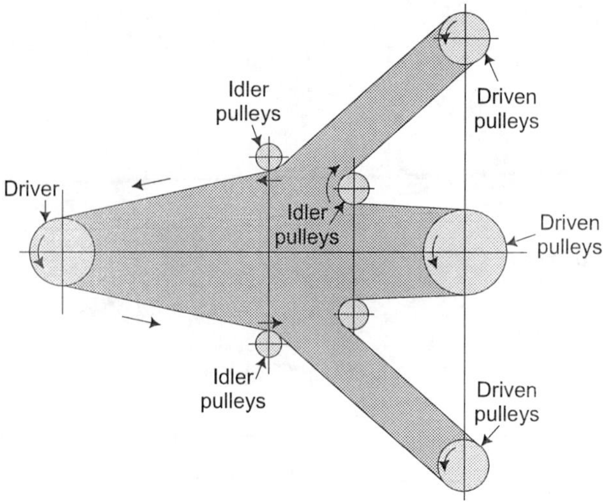

Fig. 6.19 Idler Pulleys

of a set screw as shown in the Figure 6.20. When it is desired to stop a particular driven machine its belt is pushed on its loose pulley by means of a sliding bar.

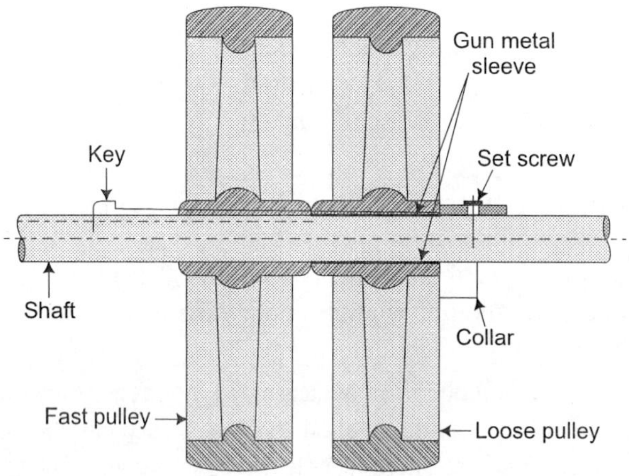

Fig. 6.20 Fast and Loose Pulleys

(v) **Grooved Pulleys** These pulleys are used to accommodate V-belts or rope. The V-groove is provided in the pulley rim.

6.17 GEAR DRIVE

A gear is a toothed wheel which is used for transmitting power and rotary motion from one shaft to another shaft or between a shaft and a slider by means of successive engagement of teeth.

Gears transmit power and motion from parallel, intersecting, or skew shafts under the following conditions:

(i) When the centre distance is short,
(ii) Speed of shaft is low and belt drive is not recommended,
(iii) Velocity ratio is constant,
(iv) High torque is to be transmitted,
(v) When the speed is to be stepped up or down, and
(vi) Exact and precise timing is required.

The gears are always used in pairs, the larger one is called gear and smaller one is called pinion and combination of these two is termed gear drive. Gear drives find wide applications in automobiles, tractors, metal cutting machine tools, rolling mills, etc.

6.17.1 Advantages and Disadvantages of Gear Drive

The following are the advantages and disadvantages of the gear drive as compared to belt, rope, and chain drives:

Advantages

1. It is a positive and smooth drive which transmits exact velocity ratio.
2. It is compact in construction due to small centre distance between the shafts.
3. It can transmit very large power which is beyond the reach of belt and chain drive.
4. It gives very high transmission efficiency even up to 99% in the case of spur gears.
5. It can transmit motion at very low velocity which is not possible with belt drives.
6. It can be used where precise timing is required.
7. A provision can be made in the gearbox for gear shifting, thus a wide range of velocity ratio can be attained.
8. It has longest service life as compared to other drives, thus it is highly reliable.
9. It can be used for transmitting wide range of loads and power.

Disadvantages

1. Gear drives are costly and their maintenance cost is also higher.
2. The manufacturing processes for gear are complicated and highly specialized.
3. Gear drives require careful attention for lubrication and cleanliness.
4. They also require precise alignment of the shafts.
5. They are noisy in operation at considerable speeds.

6.17.2 Types of Gears

Main types of gears are discussed below:

1. Spur Gears In spur gears the teeth are cut parallel to the axis of the shaft. The spur gears are used to connect two parallel and coplanar shafts only. The profile of the gear tooth is in the shape of an involute curve and it remains identical along the entire width of the gear wheel. As the teeth are parallel to the axis of the shaft, so these are not subjected to axial thrust due to tooth load rather they impose radial loads on the shafts. They are used as sliding gears for speed change mechanism in gearboxes of a lathe machine. These gears are able to give 96% to 98% power transmission efficiency. The spur gears may have three important arrangements: (i) External spur gear, (ii) Internal spur gear, and (iii) Rack and pinion.

In Figure 6.21 (a) the gears have external teeth on the outer surface and the two shafts rotate in opposite directions.

In an internal spur gear, the teeth are cut on the inner surface of an annular ring. A smaller gear called pinion having external teeth meshes with the internal gear and the two shafts rotate in the same direction as shown in Figure 6.21 (b).

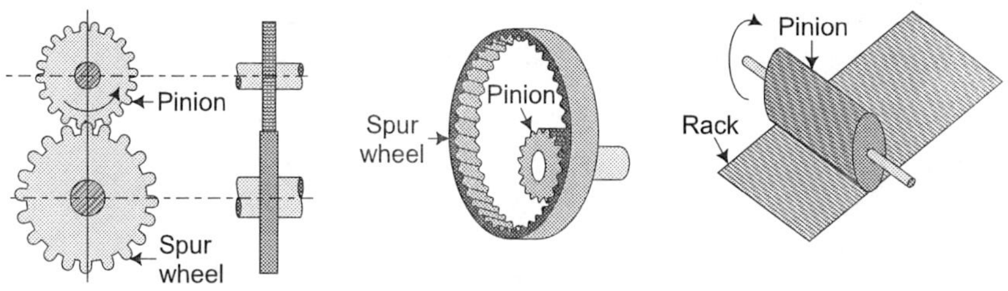

(a) External spur gear (b) Internal spur gear (c) Spur rack and pinion

Fig. 6.21

Spur rack is a spur gear of infinite radius. The pitch line of a spur rack moves in a straight path. The spur rack and pinion combination converts rotary motion into translatory motion or vice versa as shown in Figure 6.21 (c). It is used mostly in a lathe machine where rack is used to transmit motion to the saddle. In spur gears the contact occurs across a line thus spur gears have line contact.

2. Helical Gears In helical gears teeth are cut at an angle with the axis of the shaft means teeth are curved and each teeth is helical in shape as shown in Figure 6.22 (a and b). Two helical gears are used in the same way as spur gear and are used to connect two parallel shafts.

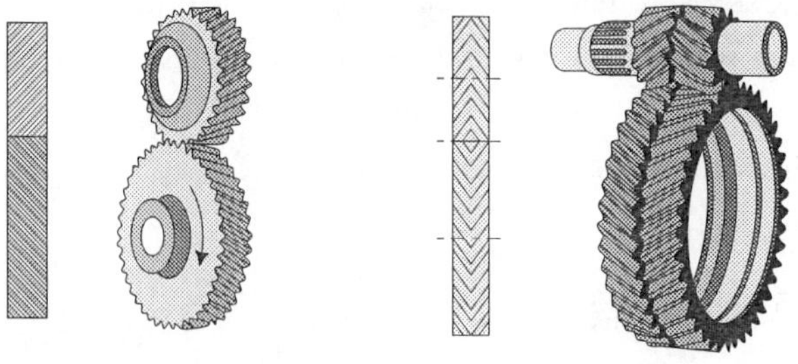

(a) Single helical gear (b) Double helical gear

Fig. 6.22

Two mating gears have the same helix angle, but have teeth of opposite hands. A right hand pinion meshes with a left hand gear and vice versa. Helical gears have an involute profile similar to that of spur gears. At the beginning of engagement, the contact occurs at a point of curved teeth and afterwards it extends along a diagonal line across the teeth. Thus, the load application is gradual which causes reduction in noise and low impact stresses. Hence, the helical gears may be used at higher velocities than spur gears and have greater load carrying capacity. Helical gears impose radial and thrust loads on the shafts. These gears are used for speed reduction in automobile gearboxes and in steam and gas turbine units.

Axial thrust which occurs in case of single helical gears is eliminated in double helical gears. This is because the axial thrusts of the two rows of teeth cancel each other. As the name indicates these gears are the equivalent of two helical gears secured together one having a right-hand helix and other a left-hand helix, but they are manufactured as one piece as shown in Figure 6.22 (b). The teeth may be continuous or separated by a groove. The groove is required for hobbing and grinding operations. If there is no groove and the left and right inclinations of double-helical gear meet at a common apex the gear is known as herring bone gear.

3. Bevel Gears Bevel gears have the shape of a truncated cone and are used to connect two intersecting shafts as shown in Figure 6.23 (a). Bevel gears are normally used for shafts which are at right angles to each other but the angle can be slightly more or less than 90°. If the shafts are at right angles and gears are equal in size, the bevel gears are called mitre gears. If the teeth on the gears are straight radial to the point of intersection of the shaft axis then gears are known as straight bevel (Figure 6.23 b). But when teeth are inclined then gears are known as helical bevel gears or spiral bevel gears (Figure 6.23 c). Spiral bevel gears are used in the differential of an automobile.

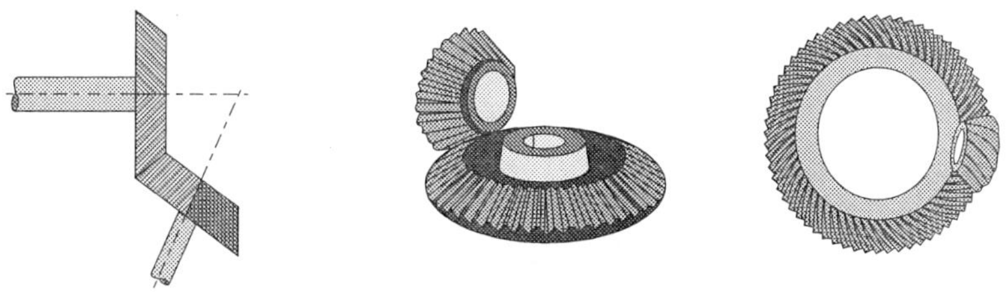

(a) (b) (c)

Fig. 6.23

4. Worm Gears The worm gears consist of a worm and a worm wheel as shown in Figure 6.24. The worm is in the form of a threaded screw which meshes with the teeth on a gear wheel called the worm wheel. The threads on the worm can be single- or multi-start and usually have a small lead. The worm gear is used for connecting two non-parallel, non-intersecting shafts which are usually at right angles. Worm gears are the best choice for high-speed reduction. These are quite smooth in operation and thus are used in material handling equipment.

5. Spiral Gears Spiral gears are used to connect two non-parallel and non-intersecting shafts. These gears are identical to helical gears except there is a point contact between two gears rather than line contact. These gears are suitable for transmitting low power. Helical gears mounted on parallel shafts will be of opposite hand but spiral gears may be of same hand or of opposite hand.

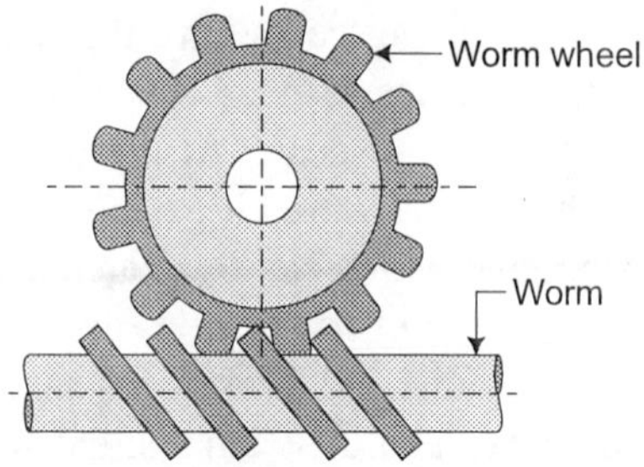

Fig. 6.24 Worm Gear

6.17.3 Classification of Gears

Gears can be classified as follows:

1. According to the position of the axes of the shafts: The axis of the two shafts between which the motion is to be transmitted may be (i) parallel, (ii) intersecting, and (iii) non-parallel and non-intersecting.
   (i) **Gears with parallel axes** Spur gears, helical gears, herring bone gears, rack and pinion, and internal gears.
   (ii) **Gears with intersecting axes** Bevel gears, straight, and spiral gears.
   (iii) **Gears with non-parallel and non-intersecting axes** Worm and spiral gears.
2. According to the peripheral velocity of the gears.
   (i) **Low velocity gears** Velocity < 3 m/s,
   (ii) **Medium velocity gears** Velocity varies from 3 m/s to 15 m/s, and
   (iii) **High velocity gears** Velocity > 15 m/s.
3. According to the type of gearing: (i) External gearing, (ii) Internal gearing, and (iii) Rack and pinion.
4. According to the type of teeth: (i) Straight teeth, (ii) Inclined teeth, and (iii) Curved teeth.

6.17.4 Terminology Used in Gears

Figure 6.25 shows the profile of a gear having important terms used in the study of gears.

1. **Pitch Circle and Pitch Circle Diameter** It is an imaginary circle which by pure rolling action produces the same motion as the toothed gear wheel. The diameter of pitch circle is called pitch circle diameter. The size of the gear is usually specified by the pitch circle diameter. It is also known as pitch diameter.
2. **Pitch Point** It is the point of contact of two pitch circles of mating gears.
3. **Pitch Surface** It is the surfaces of the rolling cylinders which the meshing gears replace are called pitch surfaces.
4. **Circular Pitch** It is the distance measured along the circumference of the pitch circle from a point on one tooth to a corresponding point on the adjacent tooth. It is denoted by p_c. It is equal to the pitch circle circumference divided by the number of teeth on the wheel.

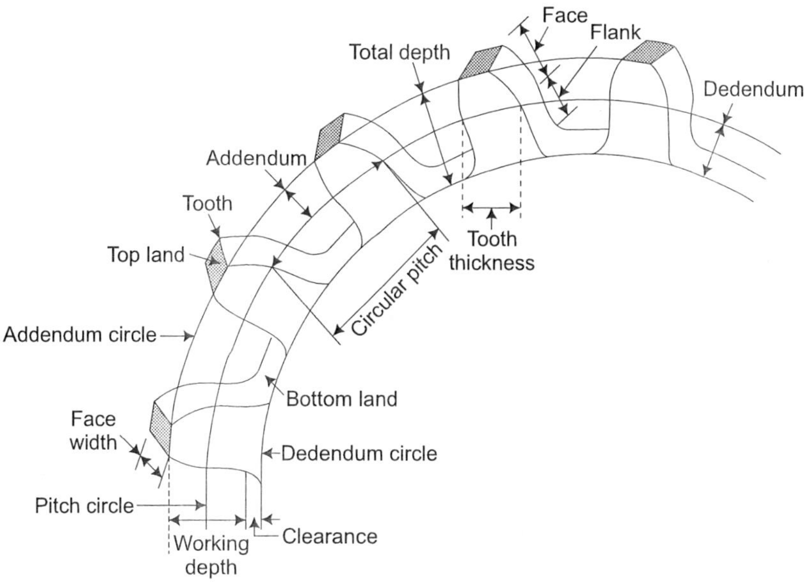

Fig. 6.25 Gear Terminology

$$\text{Circular pitch} = \frac{\text{Circumference of pitch}}{\text{Number of teeth}}$$

$$\therefore \quad p_c = \frac{\pi D}{T} \tag{6.31}$$

where, D = Pitch diameter, and T = Number of teeth.

5. **Diametral Pitch** It is equal to the number of teeth per unit length of pitch circle diameter. It is denoted by p_d.

$$\therefore \quad p_d = \frac{T}{D} \tag{6.32}$$

From equations (6.31) and (6.32), we have,

$$p_c \times p_d = \pi \tag{6.33}$$

Hence, if circular pitches of two gears are equal then their diametral pitches will also be equal.

6. **Module** It is the ratio of pitch circle diameter to the number of teeth. Thus, module is the reciprocal of diametral pitch. It is denoted by m. Module is generally expressed in millimetre.

$$\therefore \quad m = \frac{D}{T} \tag{6.34}$$

7. **Addendum Circle** It is the circle that passes through the top of teeth.

Diameter of addendum circle = D + 2 m (6.35)

8. **Dedendum or Root Circle** It is the circle that passes through the bottom of teeth.

   Diameter of dedendum circle = $D - 2 \times 1.157$ m (6.36)

9. **Addendum** It is the radial distance from the pitch circle to the addendum circle. Its value is generally 1 module.
10. **Dedendum** It is the radial distance from the pitch circle to the dedendum root circle. Its value is 1.157 m
11. **Clearance** It is the radial difference between the addendum and dedendum of a tooth.

$$\therefore \quad \text{Clearance} = 1.157 \text{ m} - 1 \text{ m}$$
$$= 0.157 \text{ m}$$

12. **Total Depth of Teeth** It is the total radial depth of the tooth space.

   Total depth = addendum + dedendum

13. **Working Depth of Teeth** The maximum depth to which a tooth penetrates into the tooth space of the mating gear is called working depth of teeth.

   Working depth = Sum of addendums of the two gears.

14. **Tooth Thickness** The width of the tooth measured along the pitch circle is called the tooth thickness.
15. **Tooth Space** The width of space between two adjacent teeth measured along the pitch circle is called the tooth space.
16. **Face** It is that part of the tooth which is above the pitch circle.
17. **Flank** It is that part of the tooth surface which is below the pitch circle.
18. **Gear Ratio** It is the ratio of the number of teeth on gear to that on the pinion.
19. **Velocity Ratio** It is the ratio of the angular velocity of the follower to the angular velocity of the driving gear.
20. **Pressure Angle** It is the angle between the common normal at the point of contact and the common tangent at the pitch point. The pressure angle will be either 14.5° or 20°.

6.18 GEAR TRAINS

Gear train is a combination of two or more gears by means of which power and motion is transmitted from driving shaft to driven shaft. Thus, the function of a gear train is similar to that of belt, rope, and chain drives. Gear train enables the speed of the driven shaft to be stepped up or down. The main types of gear trains are: (1) Simple gear train, (2) Compound gear train, (3) Reverted gear train, and (4) Epicyclic gear train.

1. Simple Gear Train If the axes of all the shafts on which gears are mounted remain fixed relative to each other and each shaft carries one gear only, the gear train is known as simple gear train. Figure 6.26 shows a simple gear train.

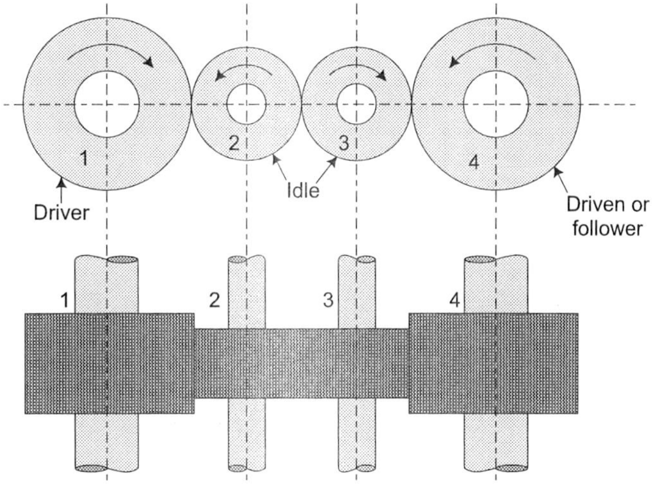

Fig. 6.26 Simple Gear Train

Let N_1, N_2, N_3, and N_4 be the speeds of gears 1, 2, 3, and 4 respectively in rpm.

T_1, T_2, T_3, and T_4 be the number of teeth on gears 1, 2, 3, and 4 respectively.

D_1, D_2, D_3, and D_4 be the diameters of gears 1, 2, 3, and 4 respectively.

Velocity (or speed) ratio between the gears 1 and 2: Since gears 1 and 2 are in mesh together,

$$\therefore \quad \pi D_1 N_1 = \pi D_2 N_2$$

or
$$\frac{N_1}{N_2} = \frac{D_2}{D_1} \qquad \text{(i)}$$

Diametral pitch of gear 1 = Diametral pitch of gear 2

$$\frac{T_1}{D_1} = \frac{T_2}{D_2}$$

or
$$\frac{T_2}{T_1} = \frac{D_2}{D_1} \qquad \text{(ii)}$$

From equations (i) and (ii),

$$\frac{N_1}{N_2} = \frac{T_2}{T_1} \qquad \text{(iii)}$$

Similarly, velocity ratio between gears 2 and 3,

$$\frac{N_2}{N_3} = \frac{T_3}{T_2} \qquad \text{(iv)}$$

Velocity ratio between gears 3 and 4,

$$\frac{N_3}{N_4} = \frac{T_4}{T_3} \qquad \text{(v)}$$

Multiplying (iii), (iv) and (v),

$$\frac{N_1}{N_2} \times \frac{N_2}{N_3} \times \frac{N_3}{N_4} = \frac{T_2}{T_1} \times \frac{T_3}{T_2} \times \frac{T_4}{T_3}$$

∴ Velocity ratio, $$\frac{N_1}{N_4} = \frac{T_4}{T_1} \qquad (6.37)$$

Train value of gear train is the reciprocal of the velocity ratio.

In a simple gear train, the following points have been observed:

(i) A pair of meshing gears moves in opposite directions.

(ii) All odd number gears rotate in one direction (i.e., gears 1 and 3) and even number gears in opposite direction (i.e., gears 2 and 4).

(iii) The speed ratio or train value of the gear train is not affected by intermediate gears. Thus, the intermediate gears are called idlers. Here gear 2 and 3 are idlers. Idlers are used to bridge the distance between the shafts with large centre distance and to control the direction of gears.

(iv) All gears can be in a straight line or in a zigzag fashion. A simple gear train may also have bevel gears.

2. Compound Gear Train In a compound gear train, each intermediate shaft carries two gears which are rigidly fixed so that both gears have the same speed. Figure 6.27 shows a compound gear train in which gear 1 drives the gear 2 mounted on the intermediate shaft. On the intermediate shaft another gear 3 is also mounted which meshes with the gear 4 mounted on the driven shaft. As gears 2 and 3 are mounted on the same shaft they rotate at same speed.

Let N_1, N_2, N_3, and N_4 be the speeds of gears 1, 2, 3, and 4 respectively in rpm.

T_1, T_2, T_3, and T_4 be the number of teeth on gears 1, 2, 3, and 4 respectively.

Since gear 1 is in mesh with gear 2, therefore, its speed ratio is,

$$\frac{N_1}{N_2} = \frac{T_2}{T_1}$$

or $$N_2 = N_1 \times \frac{T_1}{T_2}$$

But $$N_2 = N_3$$

∴ $$N_2 = N_3 = N_1 \times \frac{T_1}{T_2} \qquad \text{(i)}$$

Similarly, for gear 3 and 4,

$$\frac{N_3}{N_4} = \frac{T_4}{T_3}$$

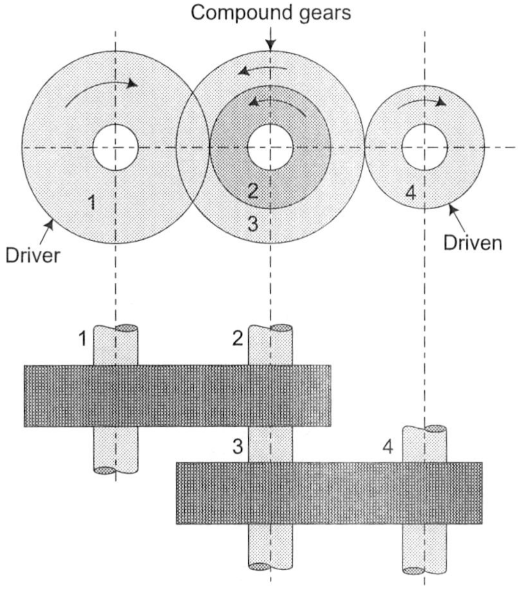

Fig. 6.27 Compound Gear Train

or $$N_3 = N_4 \times \frac{T_4}{T_3} \quad \text{(ii)}$$

From equations (i) and (ii),

$$N_1 \times \frac{T_1}{T_2} = N_4 \times \frac{T_4}{T_3}$$

∴ Speed ratio, $$\frac{N_1}{N_4} = \frac{T_2}{T_1} \times \frac{T_4}{T_3} \quad (6.38)$$

Comparėd to simple gear trains, compound train gear trains provide a larger velocity ratio in limited space thus it is a much compact system. Compound gear train is used in back gear of lathe headstock.

3. Reverted Gear Train Reverted gear train is a special type of compound gear train in which the axes of the first gear and last gear are coaxial as shown in Figure 6.28. Here gears 1 and 4 are mounted on the driver and driven shafts respectively whose axes are collinear. The gears 2 and 3 are mounted on the same intermediate shaft, thus they form a compound gear and will rotate in the same direction. We can observe that gear 1 drives the gear 2 and gear 3 in the opposite direction. The gear 3 drives the gear 4 in opposite direction. Thus, the motion of gear 4 and gear 1 is in same direction as shown in the Figure 6.28.

Let T_1, T_2, T_3, and T_4 be the number of teeth on gears 1, 2, 3, and 4 respectively.

r_1, r_2, r_3, and r_4 be the radii of gears 1, 2, 3, and 4 respectively.

Since the distance between the centres of the shaft of gears 1 and 2 as well as 3 and 4 is same,

$$\therefore \qquad r_1 + r_2 = r_3 + r_4$$

It is assumed that circular pitch or module of all the gears is same, so the number of teeth on each gear is proportional to its radius.

$$\therefore \qquad T_1 + T_2 = T_3 + T_4$$

Speed ratio,
$$\frac{N_1}{N_2} = \frac{T_2}{T_1} \times \frac{T_4}{T_3} \qquad (6.39)$$

The reverted gear trains are used in clocks, industrial speed reducers, lathe back gears, and in automotive transmissions.

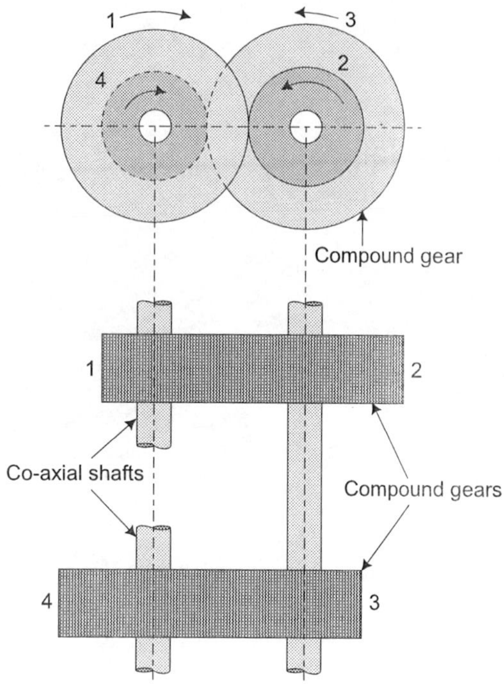

Fig. 6.28 Reverted Gear Train

4. Epicyclic (or Planetary) Gear Train In the simple and compound gear trains, the axes about which gears rotate are fixed. But, if there is a relative motion of axis in a gear train then it is called an epicyclic gear train. Hence, in an epicyclic train, at least one of the gear axes is in motion relative to fixed axis or frame. Figure 6.29 shows two gears G_1 and G_2, the axes of which are connected by an arm A. If the arm is fixed the gears form the simple train. However, if gear G_2 is fixed so that the arm can rotate about the axis of G_2, the gear G_1 would also move around G_2. Thus, it is an epicyclic gear train.

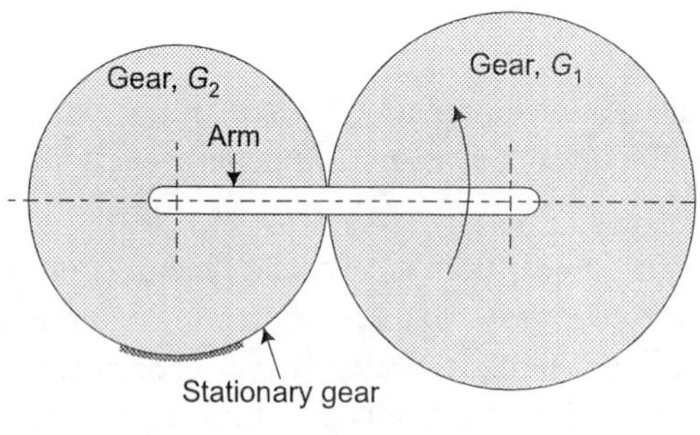

Fig. 6.29 Epicyclic Gear Train

The gear G_1 rolls outside the gear G_2 and traces an epicyclic path hence it is called an epicyclic gear train. Generally, the wheel G_1 is called epicyclic wheel.

Let N_1 = Speed of gear G_1 in rpm, N_2 = Speed of gear G_2 in rpm, N_A = Speed of the arm in rpm, T_1 = Number of teeth on gear G_1, and T_2 = Number of teeth on gear G_2.

If the speed of arm A is known then speed of the gear G_1 can be calculated by using the relation,

$$\frac{N_1}{N_A} = 1 + \frac{T_2}{T_1} \tag{6.40}$$

$$\therefore \quad N_1 = N_A \left[1 + \frac{T_2}{T_1}\right] \tag{6.40 a}$$

With epicyclic gear trains large speed reductions are possible and if fixed wheel is annular then more compact unit can be obtained. Important applications of epicyclic gears are in differential gear of automobile, wrist watches, hoists, pulley blocks, etc. Unlike in simple and compound gear train, it is not possible to determine the direction of rotation from the look of gear train in an epicyclic gear train.

Example 6.12 A toothed gear has 48 teeth and circular pitch of 16, find the following (i) pitch diameter, (ii) diametral pitch, and (iii) module of the gear.

Solution

(i)
$$p_c = \frac{\pi D}{T}$$

$$\therefore \quad D = \frac{p_c \times T}{\pi} = \frac{16 \times 48}{\pi}$$

$$= 244.46 \text{ mm}$$

(ii)
$$p_c \times p_d = \pi$$

$$\therefore \quad p_d = \frac{\pi}{p_c} = \frac{\pi}{16} = 0.196 \text{ teeth/mm}$$

(iii)
$$m = \frac{D}{T} = \frac{244.46}{48}$$

$$= 5.09 \text{ mm per tooth}$$

Example 6.13 Two parallel shafts are connected with the help of two gears one gear on each shaft. The number of teeth on one gear is 40 and speed of the shaft is 400 rpm. If the speed ratio is equal to 4 and circular pitch of the gear is 24 mm, find (i) number of teeth and speed of the other shaft, and (ii) centre distance between the two shafts.

Solution

(i) Speed ratio,
$$\frac{N_1}{N_2} = \frac{T_2}{T_1} = 4$$

$$\therefore \quad N_2 = \frac{N_1}{4} = \frac{400}{4} = 100 \text{ rpm}$$

$$T_2 = 4 \times T_1 = 4 \times 40 = 160$$

(ii)
$$p_c = \frac{\pi D_1}{T_1}$$

$$D_1 = \frac{p_c \times T_1}{\pi} = \frac{24 \times 40}{\pi}$$

$$= 305.58 \text{ mm}$$

Similarly,
$$D_2 = \frac{p_c \times T_2}{\pi} = \frac{24 \times 160}{\pi}$$

$$= 1222.31 \text{ mm}$$

Therefore, centre distance $= \dfrac{D_1 + D_2}{2} = \dfrac{305.58 + 1222.31}{2}$

$$= 763.95 \text{ mm}$$

6.19 CLUTCHES AND ITS TYPES

A clutch is a mechanical device used in automobiles to engage or disengage the driving and driven shafts instantaneously as and when required by the driver. The clutch is fitted between the engine crankshaft and the gearbox. So, it transmits the power from engine shaft to gearbox shaft. The axes of these two shafts are coincident. A clutch permits the engine to run without driving the vehicle which is desirable when the engine is to be started or stopped, or when the gears are to be changed for obtaining different speeds.

The driving and driven elements of a clutch are flywheel and pressure plate which are mounted on crankshaft and transmission shaft respectively. The clutches are of the following types:

1. Friction clutches:
   (i) Single plate clutch, (ii) Multiplate plate clutch,
   (iii) Conical clutch
2. Jaw clutch or Dog clutch:
   (i) Expanding type clutch, (ii) Non-Expanding type clutch
3. Hydraulically operated clutch,
4. Centrifugal clutch,
5. Semi-centrifugal clutch,
6. Electromagnetic clutch, and
7. Vacuum clutch.

6.19.1 Single Plate Clutch

Single plate clutch is commonly used in most of the cars and small commercial vehicles. It consists of a clutch plate mounted on a hub which is splined from inside and thus it is free to slide over the gearbox shaft. The clutch plate is held between the flywheel and the pressure plate. The clutch plate is made of steel and has a ring of friction lining on each side. The flywheel is mounted on

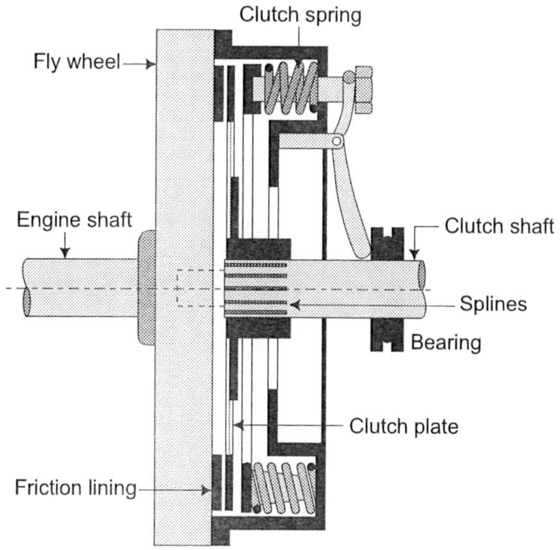

Fig. 6.30 Single Plate Clutch

the engine crankshaft and rotates with it. In order to keep the clutch in engaged position, springs are arranged circumferentially which provide axial force as shown in Figure 6.30.

Thus, clutch plate is held firmly against the flywheel due to the spring loaded pressure plate. A pedal is provided to pull the pressure plate against the spring force whenever it is required to be disengaged. Generally, it remains in engaged position. Because of the friction between the flywheel, clutch plate, and pressure plate, the clutch plate revolves with the flywheel. As the clutch plate revolves, the clutch shaft also revolves. The clutch shaft is connected to the transmission. Thus, the engine power is transmitted from the crankshaft to the clutch shaft.

When the clutch pedal is pressed, the pressure plate is moved to the right against the force of the springs. This is achieved by means of a suitable arrangement of linkage and a thrust bearing. Due to this movement of the pressure plate, the clutch plate becomes free and the clutch is disengaged. The flywheel rotates without driving the clutch plate and thus the driven shaft.

When the force is taken off the clutch pedal, the pressure on the thrust bearing is released. As a result, the springs become free which force the pressure plate to bring it in contact with the clutch plate. The clutch plate slides on the splined hub and is tightly gripped between the pressure plate and the flywheel. The friction between the linings on the clutch plate and the flywheel on one side and the pressure plate on the other, cause the clutch plate and hence the driven shaft to rotate.

6.19.2 Multiplate Clutches

The multiplate clutch is an extension to the single plate clutch in which the number of friction linings and the metal plates are increased. The increased number of friction surfaces increases

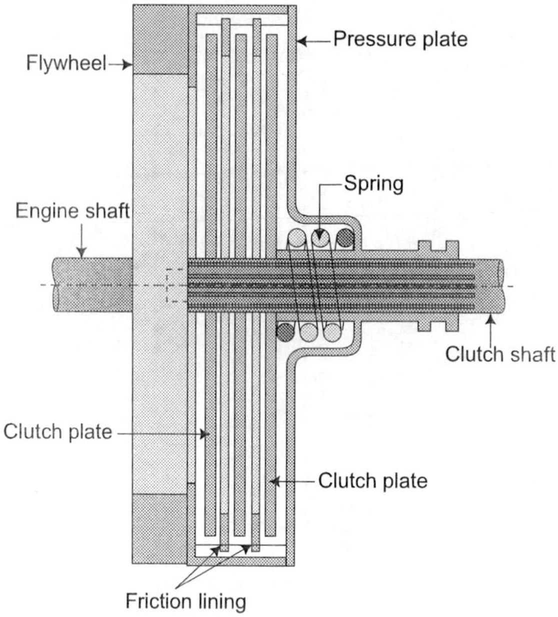

Fig. 6.31 Multiplate Clutch

the capacity of the clutch to transmit torque, though the size of the clutch remains almost same. These clutches are used in heavy transport vehicles and racing cars where high torque is to be transmitted. It is also used in case of motorcycle and scooters where space availability is limited. Figure 6.31 shows simplified schematics of a multiplate clutch.

The construction is similar to that of single plate type except each alternate plate slides in grooves on the flywheel and the other slides on splines on the pressure plate hub. It means one set of the alternate plate slides in grooves on the flywheel and the other one slide on the pressure plate hub. The spring presses the plates into contact with the friction linings and thus the torque is transmitted between the engine shaft and the driven shaft. It may be dry or wet. When the clutch plate is operated in an oil bath, it is called a wet clutch. When the clutch is operated dry, it is called a dry clutch.

6.20 BRAKES AND ITS TYPES

A brake is a mechanical device used to apply frictional resistance to a moving body either to stop or retard it by absorbing its kinetic energy. The kinetic energy absorbed gets converted into the heat energy which is dissipated into the atmosphere to avoid the overheating of brake lining.

The factors on which capacity of brakes depends are: (i) Coefficient of friction, (ii) The unit pressure between the breaking surfaces, (iii) Peripheral velocity of the brake drum, (iv) Projected area of the friction surfaces, and (v) The heat dissipation rate.

Brakes and clutches have a similar construction and work on the same principle. However, there is a functional difference between the two. A clutch connects two moving members, whereas a brake connects a moving member to a stationary member.

Some of the important characteristics for brake lining materials are: (i) High coefficient of friction, (ii) Low wear rate, (iii) High heat resistance, (iv) High heat dissipation ability, (v) High mechanical strength, and (vi) Should not be affected by moisture, oil, and temperature.

The commonly used materials for lining of brakes are cast iron on cast iron, bronze on cast iron, steel on cast iron, wood on cast iron, fibre on metal, leather on metal, wire asbestos on metal, asbestos on metal, and metal on cast iron.

Based on the means used for transforming the energy by the braking elements, brakes are classified as: (i) Mechanical brakes, (ii) Hydraulic brakes, (iii) Electric brakes, and (iv) Magnetic brakes.

The important types of mechanical brakes are:

(i) Block or shoe brake:
  (a) Single block or shoe brake, (b) Double block or shoe brake.
(ii) Band brake,
(iii) Band and block brake, and
(iv) Internal expanding shoe brake.

6.20.1 Single Block or Shoe Brake

A block or shoe brake consists of a block or shoe which is pressed against a rotating drum. The block is rigidly fixed to the lever which has one end pivoted to a fixed fulcrum and the other end is free as shown in Figure 6.32.

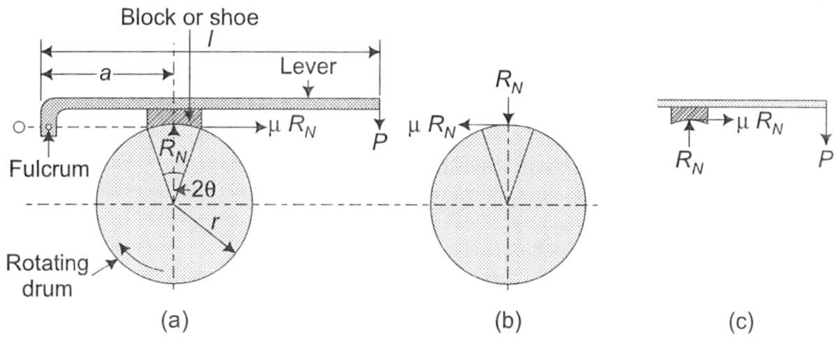

Fig. 6.32 Single Block or Shoe Brake

The blocks or shoes are made of softer material than that of the drum so that the block can be replaced easily and cheaply after wear. For light and slow speed vehicles, blocks are made of rubber or hard wood, whereas for heavy duty and fast running vehicles these are made of cast iron. As the force is applied to the lever the block is pressed against the rotating drum. The block exerts a radial force on the drum which passes through the centre of the drum. The drum will exert a normal reaction on the block. This causes a braking torque due to frictional force between the drum and the block which will retard or stop the motion of the drum.

In Figure 6.32, the drum is rotating clockwise hence the frictional force on the drum will act in the opposite direction. Let the line of action of the frictional force pass through the fulcrum 'O' of the lever.

Let P = Force or effort applied at the free end of the lever, r = Radius of the drum, μ = Coefficient of friction, R_N = Normal reaction on the block, $F_f = \mu R_N$ = Frictional force acting on the block, 2θ = Angle made by contact surface of the block at the centre of the drum (Assuming angle of contact, $2\theta \leq 40°$), T_b = Braking torque.

Braking torque on the drum = Frictional force × Radius

$$\therefore \quad T_b = F_f \times r = \mu R_N \times r \tag{6.41}$$

The braking torque can be calculated if the normal reaction R_N is known. We consider equilibrium of the block for determining the value of R_N. The forces acting on the block are: (i) Frictional force, $F_f = \mu R_N$, (ii) Normal reaction, R_N, and (iii) Applied force, P.

Taking moments about the fulcrum 'O',

$$R_N \times a = P \times l$$

$$\therefore \quad R_N = \frac{P \times l}{a}$$

Substituting the value of R_N in equation (6.41),

$$T_b = \mu \times \frac{P \times l}{a} \times r \tag{6.42}$$

The braking torque will remain same even if the drum rotates anticlockwise.

It is not necessary that the line of action of the frictional force should pass through the fulcrum of the lever. The line of action of the frictional force may be at a distance b below or above the fulcrum 'O'.

Case I When the line of action of the frictional force is at a distance b below the fulcrum 'O' and brake drum rotates clockwise as shown in Figure 6.33 (a).

The various forces acting on the block are:

(i) R_N, (ii) μR_N, (iii) P

Taking moments about the fulcrum 'O',

$$R_N \times a + \mu R_N \times b = P \times l$$

or

$$R_N (a + \mu \times b) = P \times l$$

$$\therefore \quad R_N = \frac{P \times l}{(a + \mu b)}$$

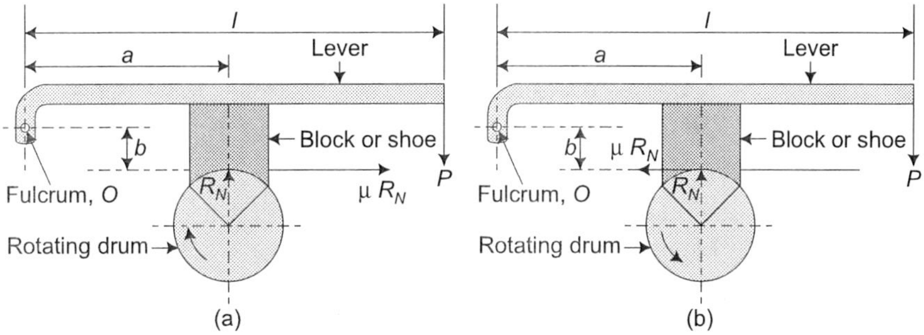

Fig. 6.33

Substituting the value of R_N in equation (6.41),

$$T_b = \mu \times \frac{P \times l}{(a + \mu b)} \times r$$

$$\therefore \qquad T_b = \mu \times \frac{P \times l \times r}{a + \mu b} \qquad (6.43)$$

If the drum is rotating in anticlockwise direction then the frictional force will also act in anticlockwise direction as shown in Figure 6.33 (b).

Now taking the moment of all forces acting on the block about the fulcrum, we have,

$$R_N \times a = P \times l + \mu R_N \times b$$

or

$$R_N \times a - \mu R_N \times b = P \times l$$

$$R_N (a - \mu \times b) = P \times l$$

$$\therefore \qquad R_N = \frac{P \times l}{a - \mu \times b} \qquad (6.44)$$

Therefore, braking torque becomes,

$$T_b = \frac{\mu \times P \times l \times r}{(a - \mu \times b)} \qquad (6.45)$$

Self-locking brake From equation (6.44), the expression for the effort P required to apply brakes is given by,

$$P = \frac{R_N (a - \mu b)}{l}$$

If $a \leq \mu b$, the effort P becomes negative or zero. Thus no effort is needed to apply brakes and hence the brake is self locking. Thus condition for self locking is,

$$a \leq \mu b \qquad (6.46)$$

Self-energised brakes or self activating brakes Again by equation (6.44)

$$P = \frac{R_N (a - \mu b)}{l} = \frac{(R_N a - \mu R_N b)}{l} \qquad (6.47)$$

It may be noted from equation (6.47) that $R_N \cdot a$ is the moment of R_N about fulcrum 'O' and $\mu R_N \cdot b$ is the moment of frictional force about fulcrum 'O' which is negative. Thus effort required for applying brake decreases due to frictional force. Thus frictional force helps to enhance the brake. Such types of brakes are known as self-energised brakes. If $P = 0$, the brake is self-locking brake but if $P > 0$, the brake is self energised brake.

Case 2 When the line of action of the frictional force F_f is at a distance b above the fulcrum 'O' and the brake drum rotates clockwise as shown in Figure 6.34 (a).

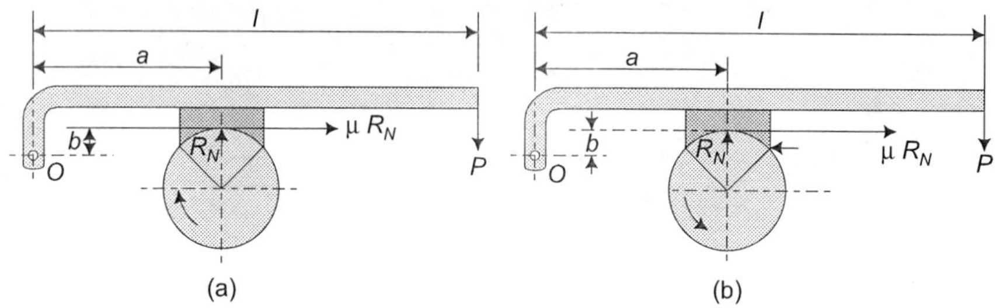

Fig. 6.34

The various forces acting on the block are

(i) R_N, (ii) μR_N, (iii) P.

Taking moments about the fulcrum 'O',

$$R_N \times a = P \times l + \mu R_N \times b$$

$$R_N (a - \mu \times b) = P \times l$$

$$\therefore \quad R_N = \frac{P \times l}{(a - \mu b)} \tag{6.48}$$

Thus, substituting the value of R_N in equation (6.41), we get,

$$T_b = \mu \times \frac{P \times l}{(a - \mu b)} \times r$$

$$\therefore \quad T_b = \frac{\mu \times P \times l \times r}{(a - \mu b)} \tag{6.49}$$

In this case also, the brake may be self-locking or self-energised.

If $P = 0$, the brake is self-locking and if $P > 0$ the brake is self-energised.

If the drum is rotating in anticlockwise direction then the frictional force will also act in anticlockwise direction as shown in Figure 6.34 (b).

Taking moments about the fulcrum 'O', we have,

$$R_N \times a + \mu R_N \times b = P \times l$$

or

$$R_N (a + \mu \times b) = P \times l$$

$$\therefore \quad R_N = \frac{P \times l}{(a + \mu b)}$$

Substituting the value of R_N in equation (6.41), we have,

$$T_b = \mu \times \frac{P \times l}{(a + \mu b)} \times r$$

$$\therefore \quad T_b = \frac{\mu \times P \times l \times r}{(a + \mu b)} \tag{6.50}$$

If the angle of contact is more than 40° then μ (actual coefficient of friction) has to be replaced by an equivalent coefficient of friction μ′ as given by,

$$\mu' = \frac{4\mu \sin \theta}{2\theta + \sin 2\theta} \tag{6.51}$$

6.20.2 Double Block or Shoe Brake

If a single block or shoe is pressed against a rotating drum then due to normal reaction a side thrust will act on the bearing of the shaft supporting the drum. This causes bending of the shaft. This can be avoided by using two blocks on the opposite sides of the drum as shown in Figure 6.35.

In such arrangement of brakes braking torque becomes two times and it is given by,

$$T_b = \mu R_{N1} \times r + \mu R_{N2} \times r$$
$$= (\mu R_{N1} + \mu R_{N2}) \times r \tag{6.52}$$

The value of R_{N1} is obtained by taking moments of the forces R_{N1}, μ R_{N1}, and P about fulcrum O_1. Similarly, the value of R_{N2} is obtained by taking moments of the forces R_{N2}, μ R_{N2}, and P about fulcrum O_2. Such brakes are commonly used in railway trains and road vehicles. These are applied by fluid pressure.

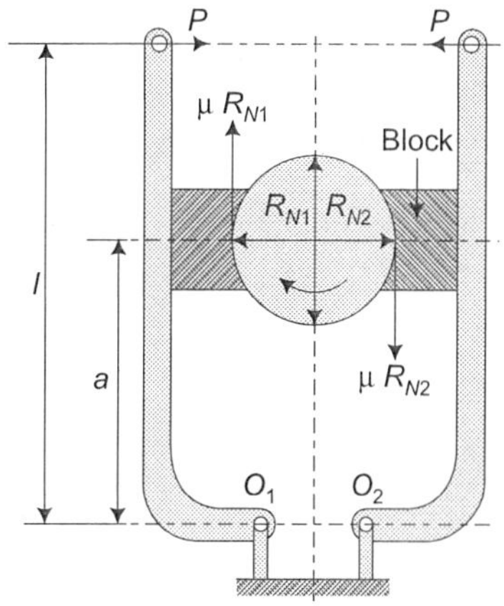

Fig. 6.35 Double Block or Shoe Brake

6.20.3 Band Brake

Band brakes are of two types: (i) Simple band brake, and (ii) Differential band brakes.

(i) Simple Band Brake It consists of flexible belt or band of leather or one or two ropes of flexible steel band lined with frictional material which embraces a part of the circumference of the rotating drum. Figure 6.36 shows a simple band brake in which one end of the band is attached with the fixed fulcrum '*O*' of the lever, whereas the other end is attached to the lever at a point 'A' on the lever at a distance *a* from the fulcrum.

The force *P* applied at the free end of the lever tighten the band on the drum and thus brakes are applied. Friction force between the band and the drum provides the braking force.

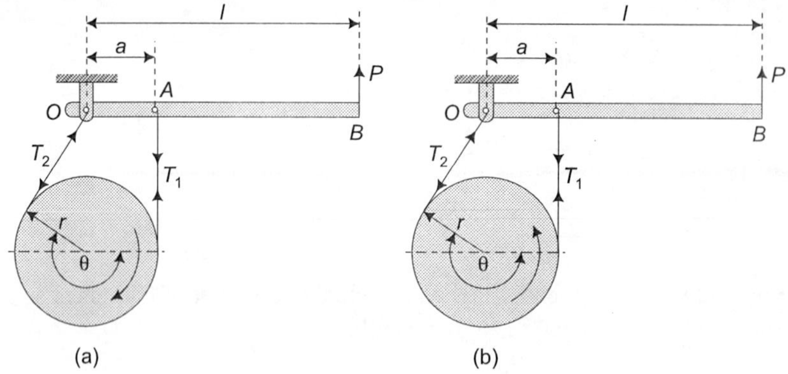

Fig. 6.36 Simple Band Brake

Let θ = Angle of lap of the band or the drum, T_1 = Tension on the tight side of the band, T_2 = Tension on the slack side of the band, r = Radius of the drum, t = Thickness of the band, r_e = Effective radius of the drum = $r + t/2$, μ = Coefficient of friction between the drum and the band, and P = Force or effort applied at the end of the lever.

Limiting ratio of the tensions is given by the relation,

$$\frac{T_1}{T_2} = e^{\mu\theta}$$

or

$$\mu\theta = \ln\left(\frac{T_1}{T_2}\right)$$

Braking torque on the drum is given by,

$$T_b = (T_1 - T_2) \times r \quad \text{when thickness of the belt is neglected} \qquad (6.53a)$$

$$T_b = (T_1 - T_2) \times r_e \quad \text{when thickness of the belt is considered} \qquad (6.53\text{ b})$$

For clockwise rotation of the drum as shown in Figure 6.36 (a) the value of P is obtained as:

Taking the moments about the fulcrum 'O',

$$P \times l = T_1 \times a$$

$$\therefore \quad P = \frac{a}{l} \times T_1 \qquad (6.54)$$

For anticlockwise rotation of the drum as shown in Figure 6.36(b), the value of P is obtained as:

Taking the moments about the fulcrum 'O',

$$P \times l = T_2 \times a$$

$$\therefore \quad P = \frac{a}{l} \times T_2 \qquad (6.55)$$

(ii) Differential Band Brake Figure 6.37 shows a differential band brake in which the band is attached to the two points '*A*' and '*B*' of the lever at a distance *a* and *b* which are on different sides of the fulcrum '*O*'. Thus in differential band brakes no end of the band is connected to the fulcrum.

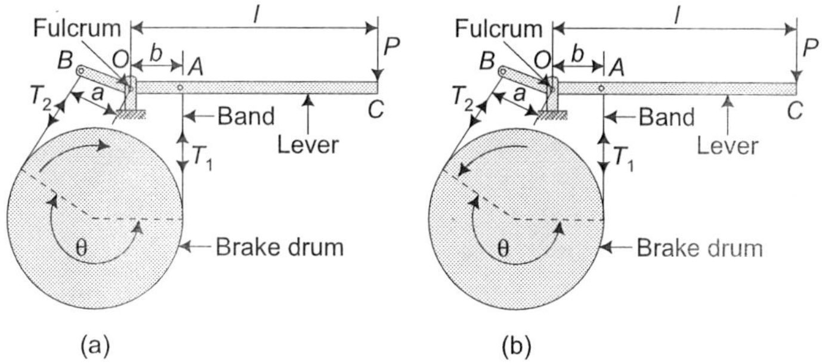

Fig. 6.37 Differential Band Brake

Let P = Force or effort applied at the end of the lever, a = Perpendicular distance from fulcrum '*O*' on the line of action of tension T_2, b = perpendicular distance from fulcrum '*O*' on the line of action of tension T_1, and l = length of lever from '*O*'.

Taking the moments about the fulcrum '*O*',

$$P \times l + T_1 \times b = T_2 \times a$$

$$P \times l = T_2 \times a - T_1 \times b$$

$$\therefore \qquad P = \frac{T_2 \times a - T_1 \times b}{l} \qquad (6.56)$$

In equation (6.56), the force P will be zero or negative if

$$T_2 \times a \leq T_1 \times b$$

or

$$\frac{T_2}{T_1} \leq \frac{b}{a}$$

If the force P is zero or negative, then the brake will become self-locking.

Thus, for self-locking of the brake when the drum rotates clockwise (6.37a) the condition becomes,

$$\frac{T_2}{T_1} \leq \frac{b}{a} \qquad (6.57)$$

When the drum rotates anticlockwise as shown in Figure 6.37(b)

Taking the moments about the fulcrum '*O*',

$$P \times l + T_2 \times b = T_1 \times a$$

$$P \times l = T_1 \times a - T_2 \times b$$

$$\therefore \qquad P = \frac{T_1 \times a - T_2 \times b}{l} \tag{6.58}$$

For self-locking of the brake, the force P should be zero or negative. Force P will be zero or negative when

$$T_1 \times a \le T_2 \times b$$

or
$$\frac{T_1}{T_2} \le \frac{b}{a} \tag{6.59}$$

6.20.4 Band and Block Brake

Figure 6.38 (a) shows the band and block arrangement which is a modification of the band brakes. It consists of a number of wooden blocks fixed inside a flexible steel band. The friction between blocks and the drum produces the brake torque. When the brake is applied, the wooden blocks are pressed against the drum and due to higher coefficient of friction of the wood it provides good braking action. Each block subtends a small angle at the centre of the drum.

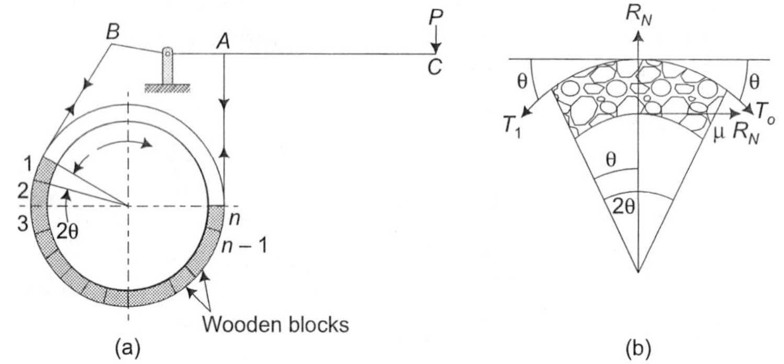

Fig. 6.38

Let n = Number of blocks, T_n = Tension after n^{th} block on tight side of the whole brake, T_o = Tension on slack side, T_1 = Tension on the tight side after one block, T_2, T_3, T_4, ... etc. = Tension after 2nd, 3rd, 4th … block etc., μ = Coefficient of friction, 2θ = Angle subtended by one block at the centre, and R_N = Normal reaction of drum on block

Consider the equilibrium of one block as shown in Figure 6.38 (b).

Resolving the tensions T_o and T_1 tangentially and radially, we have,

$$(T_1 - T_o)\cos\theta = \mu R_N \tag{i}$$

and
$$(T_1 + T_o)\sin\theta = R_N \tag{ii}$$

Dividing equation (i) by equation (ii), we have,

$$\frac{(T_1 - T_o)}{(T_1 + T_o)}\cot\theta = \mu$$

or $$\frac{(T_1 - T_o)}{(T_1 + T_o)} = \mu \tan \theta$$

$$\frac{(T_1 + T_o) + (T_1 - T_o)}{(T_1 + T_o) - (T_1 - T_o)} = \frac{1 + \mu \tan \theta}{1 - \mu \tan \theta}$$

or $$\frac{2T_1}{2T_o} = \frac{1 + \mu \tan \theta}{1 - \mu \tan \theta}$$

$\therefore$ $$\frac{T_1}{T_o} = \frac{1 + \mu \tan \theta}{1 - \mu \tan \theta}$$

Similarly, $$\frac{T_2}{T_1} = \frac{1 + \mu \tan \theta}{1 - \mu \tan \theta}$$

and $$\frac{T_n}{T_{n-1}} = \frac{1 + \mu \tan \theta}{1 - \mu \tan \theta}$$

$\therefore$ $$\frac{T_n}{T_o} = \frac{T_n}{T_{n-1}} \cdot \frac{T_{n-1}}{T_{n-2}} \cdots \frac{T_1}{T_o}$$

$$= \frac{1 + \mu \tan \theta}{1 - \mu \tan \theta} \tag{6.60}$$

where θ = Half of the angle subtended by each block on the centre of the drum.

Now braking torque on the drum is given by,

$$T_b = (T_n - T_o) \times r$$

where r = effective radius of band.

6.20.5 Internal Expanding Shoe Brake

Figure 6.39 shows an internal expanding shoe brake.

This consists of two semicircular shoes S_1 and S_2 which are lined with a friction material such as ferodo. This increases the coefficient of friction and prevents the wearing away of the metal drum. Each shoe is pivoted at one end about a fixed fulcrum O_1 and O_2 and the other end made contact with a cam. When the cam rotates, the shoes are pressed outwards against the rim of the drum. The friction between shoes and drum produces the braking torque and thus stops or reduces the speed of the drum. The shoes are normally held in off position by a spring. The drum encloses all the parts to keep the brake system free from dust and moisture.

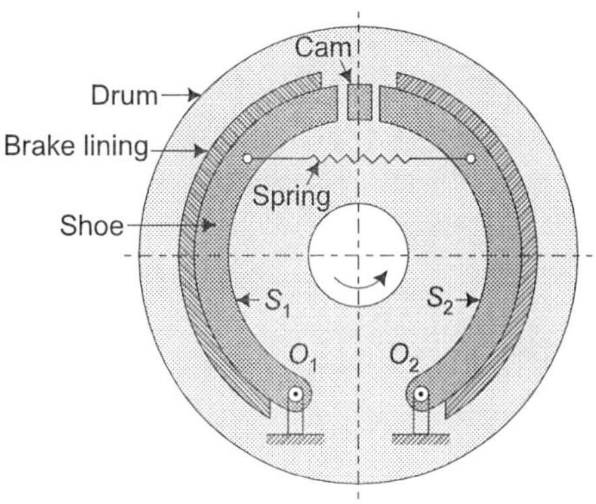

Fig. 6.39 Internally Expanding Shoe Brake

Earlier the shoes were operated mechanically by cam but nowadays the brakes are actuated by means of hydraulic pistons moving in a wheel cylinder which gets oil under pressure from the master cylinder connected to the brake pedal. The internal expanding shoe brakes are commonly used in motor cars and light trucks. These provide good friction and great braking torque without excessive use of pedal pressure, so internally expanding shoe brakes have replaced band brakes.

6.20.6 Hydraulic Brakes

Mechanical linkage is completely eliminated in hydraulic brakes and the pedal force is transmitted by the fluid to the brake shoe. Generally, the fluid used is a mixture of glycerine and alcohol, or caster oil and denatured alcohol. The hydraulic brake system consists of a master cylinder in which the piston is forced by the brake pedal and thus the pressure is developed in the complete system. This pressure is transmitted to the wheel cylinder on each of the four wheel brakes instantaneously as shown in Figure 6.40.

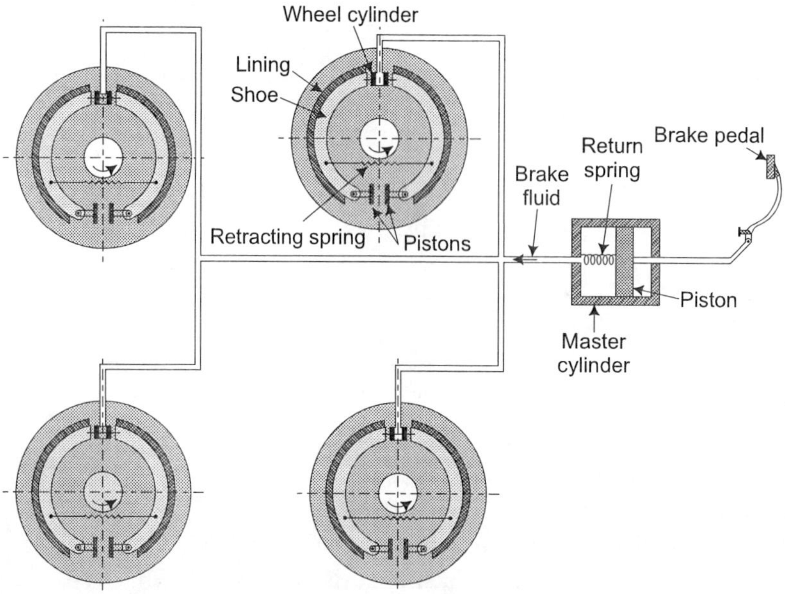

Fig. 6.40 Hydraulic Brake System

The brake lining and shoes are pressed by the pistons attached to the shoes. When the pedal is released, the pressure drops and the return spring in the master cylinder pushes the piston and pedal in the original position. And the retracting springs attached to the shoes also bring the shoes to the original position. Each wheel brake consists of a cylinder and brake drum which is mounted on the inner side of the wheel and revolves with it.

6.21 DYNAMOMETERS AND ITS TYPES

A dynamometer is an apparatus used to measure frictional force or frictional torque due to braking action. Thus, dynamometer is a brake which permits measurement of the frictional force or torque by which the power of the engine can be obtained.

The two types of dynamometer used for measuring the brake power of the engine are:

1. **Absorption Dynamometers** In absorption dynamometers the work done is converted into heat by friction during the measurement process. These are used for the measurement of moderate powers only. These are of the following types:
   (a) Mechanical absorption dynamometers: (i) Prony brake dynamometer, (ii) Rope brake dynamometer, and (iii) Band brake dynamometer
   (b) Hydraulic or fluid absorption dynamometers
2. **Transmission Dynamometers** In transmission dynamometers the work done is transmitted for use after it has been measured. So, work is not absorbed in the measurement process but is used for doing work. These are used for measuring large powers and for heat run tests. These are of the following types:
   (i) Epicyclic train dynamometer,
   (ii) Belt transmission dynamometer, and
   (iii) Torsion dynamometer.

6.21.1 Prony Brake Dynamometer

It consists of two wooden blocks placed around a pulley fixed to the shaft of an engine whose power is to be measured. The blocks are clamped by means of two bolts and nuts. A helical spring is provided between the nut and the upper block to adjust the pressure on the pulley to control its speed. The revolving pulley is called brake drum. The upper block has a long lever attached to it which carries adjustable weight W at its outer end. A counterweight is placed at the other end of the lever which balances the brake when unloaded. Two stops S_1 and S_2 are provided to limit the motion of the lever as shown in Figure 6.41.

When the brakes are to be applied, the lever is loaded with suitable weight W and the nuts are tightened until the engine shaft runs at a constant speed and the lever is in horizontal position. Under these conditions, the moment due to weight W will balance the moment of the frictional resistance between the blocks and the pulley.

Let W = Weight at the end of the lever, L = Horizontal distance of weight W from the centre of the pulley, N = Speed of the shaft in rpm, F = Frictional resistance between the blocks and the pulley in N, r = Radius of pulley in m, and T = Torque on the shaft.

Torque on the shaft = Moment of the frictional resistance

$$\therefore \quad T = W \times l = F \times r \text{ Nm}$$

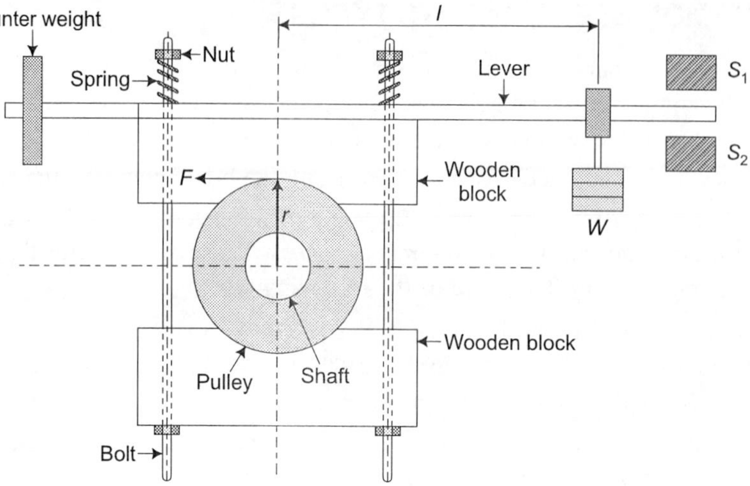

Fig. 6.41 Prony Brake Dynamometer

Power of the engine = Torque × Angular speed

$$\therefore \quad P = T \times \omega = T \times \frac{2\pi N}{60} \quad \left[\because \ \omega = \frac{2\pi N}{60}\right]$$

$$\text{or} \quad P = W \times l \times \frac{2\pi N}{60} \text{ watts} \tag{6.61}$$

From the above expression it is observed that the power of the engine is independent of: (i) Radius of the pulley r, (ii) Coefficient of friction μ between the pulley and blocks, and (iii) Pressure exerted due to tightening of the nuts.

6.21.2 Rope Brake Dynamometer

Figure 6.42 (a) shows a rope brake dynamometer. It is extensively used for testing the power of the engines. It consists of one, two, or more turns of rope wound around the flywheel or rim of a pulley fixed rigidly to the shaft of the engine whose power is to be measured. The upper end of the rope is attached to a spring balance while the lower end carries the dead weight W. Four wooden blocks are placed around the circumference of the flywheel to avoid slipping of the rope and to keep them spaced evenly across the width of the rim. A cooling arrangement becomes necessary if excessive heat is being generated due to friction between the rope and the pulley. For this, the channel of the flywheel usually has flanges turned inside in which cold water from a pipe is supplied. An outlet pipe with a flattened end takes the water out as shown in Figure 6.42 (b).

For measuring the power of an engine, it is to be run at constant speed. Under this condition, the frictional torque due to rope must be equal to the torque transmitted by the engine.

Let W = Dead weight in N, N = Speed of the engine shaft in rpm, S = Spring balance reading in N, D = Diameter of flywheel in m, and T = Torque.

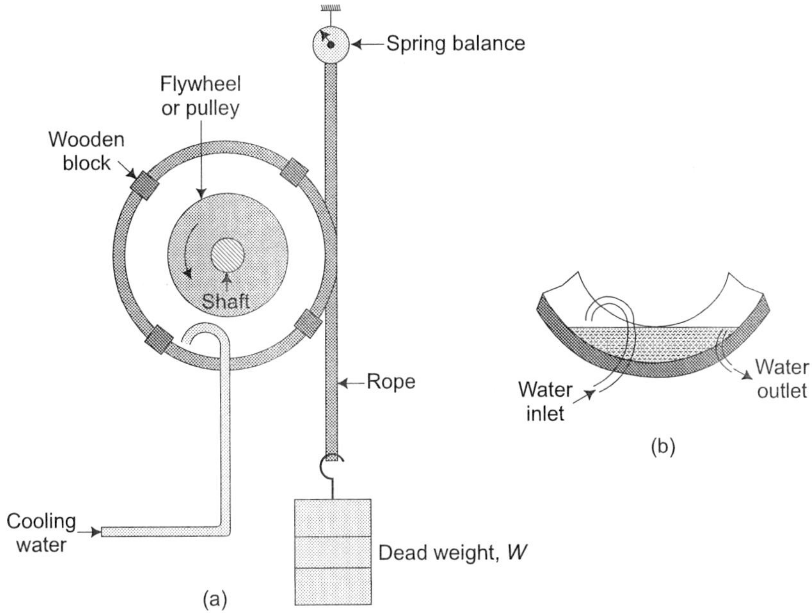

Fig. 6.42 Rope Brake Dynamometer and Cooling Arrangement

Then the net load on brake = $(W - S)$

∴ Frictional torque due to rope

= Net load on brake × Distance of load line from the centre of shaft

$= (W - S) \times (D/2)$

$= (W - S) \times R$

where R is the radius of the flywheel.

But torque transmitted by engine at constant speed = Friction torque due to rope

$$\therefore \quad T = (W - S) \times R$$

Brake power of the engine = Torque × Angular speed

$= T \times \omega$

$$\therefore \quad P = (W - S) \times R \times \frac{2\pi N}{60} \text{ watts} \qquad \left[\because \omega = \frac{2\pi N}{60}\right] \tag{6.62}$$

If diameter of rope d is also considered then equation (6.62) becomes,

$$P = (W - S) \times \left(\frac{D + d}{2}\right) \times \frac{2\pi N}{60} \tag{6.63}$$

6.21.3 Band Brake Dynamometer

The band brake dynamometers may be available in many varieties depending upon their arrangement of loading and fixing the spring balance. One arrangement is shown in Figure 6.43.

Let W = Weight on the right side, w = Weight on the left side, S = Spring balance reading, N = Speed in rpm, and r = Effective radius of brake wheel.

∴ Torque is given by,

$$T = [(W + S) - w] \times r$$

Thus, power is given by,

$$P = T \times \omega$$

$$\therefore \quad P = [(W + S) - w] \times r \times \frac{2\pi N}{60} \text{ watts} \tag{6.64}$$

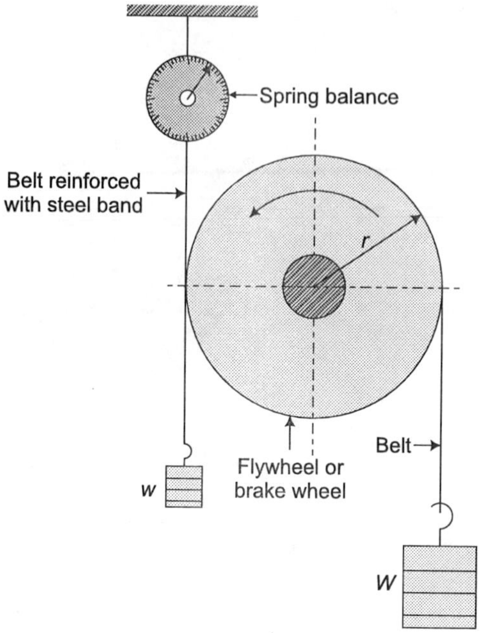

Fig. 6.43 Band Brake Dynamometer

6.21.4 Hydraulic or Fluid Dynamometer

Hydraulic or fluid dynamometers can be divided into two categories: (i) Friction type, e.g., Tesla fluid friction dynamometer and (ii) Agitator type, e.g., Froude water vortex dynamometer.

In friction type fluid dynamometers the coupling force arises from the viscous shearing of fluid between the rotor and stator, while in the agitator type of dynamometer the coupling force arises from the change in momentum of fluid as it is transported from the rotor vanes to the stator vanes and back again.

(i) Tesla Fluid Friction Dynamometer It consists of a disc keyed to the revolving shaft and enclosed in a stationary casing as shown in Figure 6.44.

The space between the casing and the disc is filled with water or any other viscous liquid. A lever is mounted on the casing which carries weight at its free end. Relative motion between disc and casing causes viscous forces in the liquid which are used for braking the disc and the shaft.

$$\text{Torque} = \text{Viscous force} \times \text{Radius of disc}$$

$$\therefore \quad T = F \times r \tag{6.65}$$

$$\text{Power} = \text{Torque} \times \text{Angular speed}$$

$$\therefore \quad P = T \times \omega \tag{6.66}$$

This type of dynamometer is used for measuring power when the speed is high and the viscous force is small.

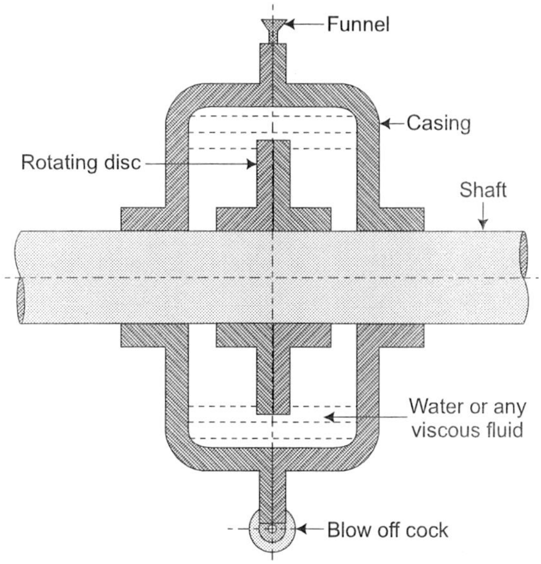

Fig. 6.44 Tesla Fluid Friction Dynamometer

(ii) Froud Water Vortex Dynamometer It consists of a rotating disc having many semi-elliptical blades or vanes which form brake chambers. The disc is rigidly mounted on the driving shaft of the engine whose power is to be measured. It revolves inside a stationary casing having counter blades or vanes projecting inside to form ring chambers as shown in Figure 6.45.

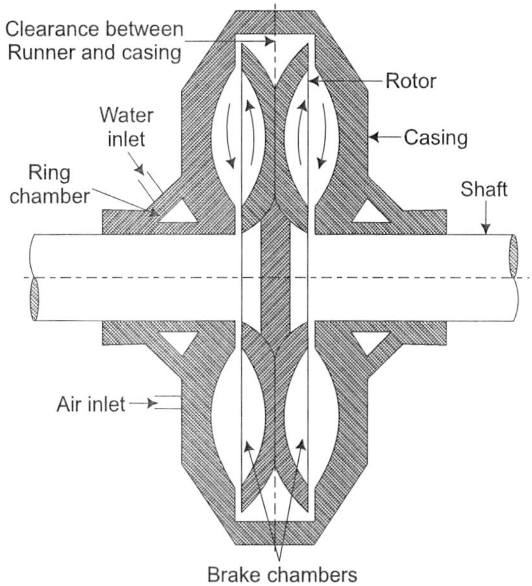

Fig. 6.45 Forced Water Vortex Dynamometer

Water is supplied continuously through a flexible metal hose pipe to the ring chamber of the casing and allowed to flow into the brake chamber of disc. The vanes on the rotor direct the water outward towards the stator vanes which redirect it into the rotor. This turbulent process is repeated again and again. The change in momentum experienced by the water due to change in direction results in a reaction force on the casing. Air which is supplied to the casing through another hose is also passed on to the runner. When the disc rotates, vortices and eddy currents are set up in water enabling the mechanical power to be dissipated as heat.

The engine power is absorbed by the water which circulates through the dynamometer. This energy absorption results in a rise in the temperature of the water. Heated water comes out at the clearance between the runner and the casing as shown in Figure 6.45. So, enough cooling capacity must be available to dissipate the rated power. The casing carries a lever balanced at the end against known weights. The braking action can be controlled by varying either the amount of water and its pressure or the distance between the rotor and the casing. The maximum absorption of water is obtained when the casing is full of water. The power absorbed by hydraulic dynamometer varies approximately as the cube of rotational speed and as the fifth power of rotor diameter.

Let W = Dead weight in N, t = Effective lever arm in m, N = rpm of the brake drum or disc

Torque, $$T = W \times l$$

$$P = T \times \omega = \frac{W \times l}{1000} \times \frac{2\pi N}{60}$$

$$= \frac{WN}{K} \text{ kW} \tag{6.67}$$

where $$K = \frac{60}{2\pi \times l \times 1000} = \text{Constant}$$

These dynamometers are used to measure the power of automotive, aircraft, and marine engines having capacity from 67 kW to about 13000 kW which operate from very low speed of about 50 rpm to more than 20,000 rpm.

6.21.5 Epicyclic Train Dynamometer

Figure 6.46 shows an epicyclic train dynamometer. It consists of a simple epicyclic train gear, i.e., a spur gear '*A*', an annular gear '*B*' having internal teeth, and a pinion '*C*'. The gear wheel '*A*' is keyed to the driving shaft or the output shaft of the engine and rotates in anticlockwise direction. The annular gear wheel '*B*' is keyed to the driven shaft and rotates in clockwise direction. The pinion or the intermediate gear '*C*' meshes with both the driver gear wheel and the driven annular gear wheel. Thus the power is transmitted from gear wheel '*A*' to gear wheel '*B*' through the pinion '*C*'. The pinion is free to rotate on a pin fixed to the lever arm. The lever is pivoted at point '*O*' through which the common axis of wheel '*A*' and '*B*' passes. When the dynamometer is at rest, the small jockey weight w balances the lever arm. The movement of lever is limited by two stops S_1 and S_2.

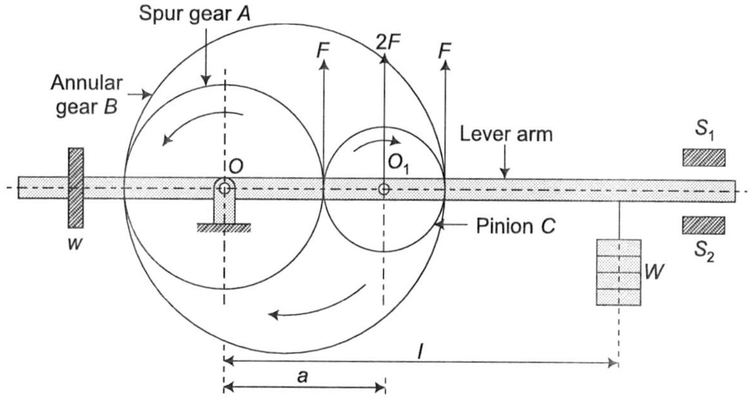

Fig. 6.46 Epicyclic Train Dynamometer

When the dynamometer is operating the tangential force F is exerted by the gear wheel 'A' on the pinion 'C', and the tangential reaction of the annular gear 'B' on the pinion 'C' in the upward direction as shown in Figure 6.46. If friction of the pin on which pinion 'C' revolves is neglected, then these two forces will be equal. Thus a force of magnitude $2F$ acts on the lever arm through the axis 'O_1' of the wheel 'C'. This force tends to rotate the lever arm about its fulcrum 'O' and a torque $2F \times a$ is exerted on the lever in clockwise direction. This torque will be balanced by the torque due to dead weight W placed at the end of the lever.

Consider the equilibrium of the lever arm,

Torque due to force $2F$ = Torque due to dead weight W

$$2F \times a = W \times l$$

$$\therefore \quad F = \frac{W \times l}{2a} \tag{6.68}$$

Let r = Radius of spur gear 'A' and

N = Speed of the rotation of driving shaft or speed of engine.

Torque transmitted by engine,

$$T = F \times r = \frac{W \times l}{2a} \times r$$

Power transmitted,

$$P = T \times \omega$$

$$\therefore \quad P = \frac{W \times l}{2a} \times r \times \frac{2\pi N}{60} \times \frac{1}{1000}$$

$$= W\,N\,K \text{ kW} \tag{6.69}$$

where $$K = \frac{2\pi l r}{2a \times 60 \times 1000} = \text{Constant}$$

6.21.6 Belt Transmission Dynamometer

When a belt transmits power from one pulley to another there will be a difference in tensions between the tight and slack sides. A belt transmission dynamometer measures this difference directly between the tensions on the tight and slack sides of a belt when it is running.

A belt transmission dynamometer (Tatham dynamometer) is shown in Figure 6.47. It consists of a driving pulley 'A', driven pulley 'B', and intermediate pulleys 'C' and 'D'. The driving pulley 'A' is rigidly fixed to the shaft of an engine whose power is to be measured. The driven pulley 'B' is mounted on another shaft to which the power from pulley 'A' is to be transmitted. The pulleys 'A' and 'B' are connected by means of a continuous belt passing around the two intermediate loose pulleys 'C' and 'D'. The intermediate pulleys 'C' and 'D' rotate on pins fixed to the lever. The lever is pivoted at 'E' which is the midpoint of two intermediate pulley centres and its movement is controlled by two stops S_1 and S_2.

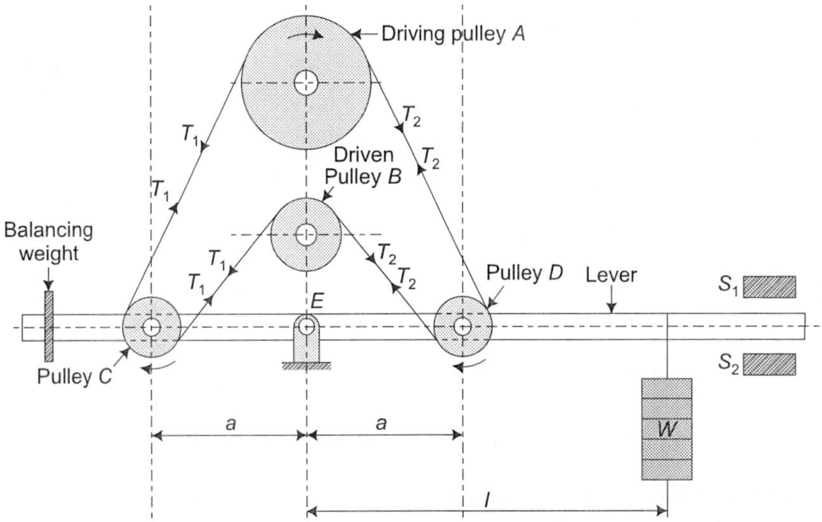

Fig. 6.47 Belt Transmission Dynamometer (Tatham Dynamometer)

Let the driving pulley 'A' rotate in clockwise direction. The total upward force acting on pulley 'C' is $2T_1$, whereas the total downward force on pulley 'D' is $2T_2$. As $2T_1$ is greater than $2T_2$, hence the lever starts rotating about 'E' in clockwise direction. In order to balance it, a dead weight W is attached at a distance l from 'E' on the lever end.

Under equilibrium conditions the balance of moments about fulcrum 'E' gives,

$$2T_1 \times a = 2T_2 \times a + W \times l$$

or $$2T_1 \times a - 2T_2 \times a = W \times l$$

or $$2a\,(T_1 - T_2) = W \times l$$

$\therefore$ $$T_1 - T_2 = \frac{W \times l}{2a} \qquad (6.70)$$

Let $V = (\pi \times d \times N)/60$ = Belt speed in m/s, d = Diameter of driving pulley 'A' in m, N = Speed of engine shaft in rpm.

We know that when difference of tensions is multiplied by the speed of belt it gives power transmitted,

$$P = \frac{(T_1 - T_2) \times V}{1000}$$

$$= \frac{W \times l}{2a} \times \frac{V}{1000} = W\,N\,K \text{ kW} \tag{6.71}$$

where $$K = \frac{\pi \times d \times l}{2a \times 60 \times 1000} = \text{Constant}$$

6.21.7 Torsion Dynamometer

A torsion dynamometer is used for measuring large powers when the power is transmitted along the shaft. It works on the fact that the torque transmitted is directly proportional to the angle of twist of the given shaft. This twist is measured accurately and then the corresponding transmitted torque can be calculated by the following equation,

$$\frac{T}{J} = \frac{C \times \theta}{l} \tag{6.72}$$

where, T = Torque transmitted, C = Modulus of rigidity of the shaft material, θ = Angle of twist in radian, l = Length of shaft and, J = Polar moment of inertia of shaft

$$J = \frac{\pi}{32} d^4 \qquad \text{(For solid shafts)}$$

and $$J = \frac{\pi}{32} (d_o^4 - d_i^4) \qquad \text{(For hollow shafts)}$$

where d_o = External diameter, and d_i = Internal diameter of the hollow shaft.

$$\therefore \quad T = \frac{CJ}{l} \times \theta = K\,\theta \tag{6.73}$$

where, $K = \frac{CJ}{l}$ = Constant

The power transmitted is given by,

$$P = \frac{T \times 2\pi N}{60} \text{ watts}$$

where, N = Speed in rpm

A number of dynamometers are used to measure the angle of twist, one of which, the Bevis-Gibson flash light torsion dynamometer, is discussed here. In practice, the angle of twist is measured for a small length of the shaft, thus some magnifying device must be used in the dynamometer for accurate measurement of angle of twist.

Bevis-Gibson Flash Light Torsion Dynamometer Figure 6.48 (a) shows a Bevis- Gibson torsion dynamometer which consists of two discs '*A*' and '*B*' fixed on a shaft at a measured convenient distance. Each disc has a small radial slot and these two slots are in line when no torque is transmitted along the shaft. A bright electric lamp '*L*', behind the disc '*A*' is fixed on the bearing of the shaft. The lamp is masked having a slot directly opposite the slot of disc '*A*'. At every revolution of the shaft, a flash light is projected through the slot in the disc '*A*' towards the disc '*B*' in the direction parallel to the shaft. Behind the disc '*B*', an eyepiece '*E*' is fitted on the shaft bearing. This eyepiece is capable of slight circumferential adjustment.

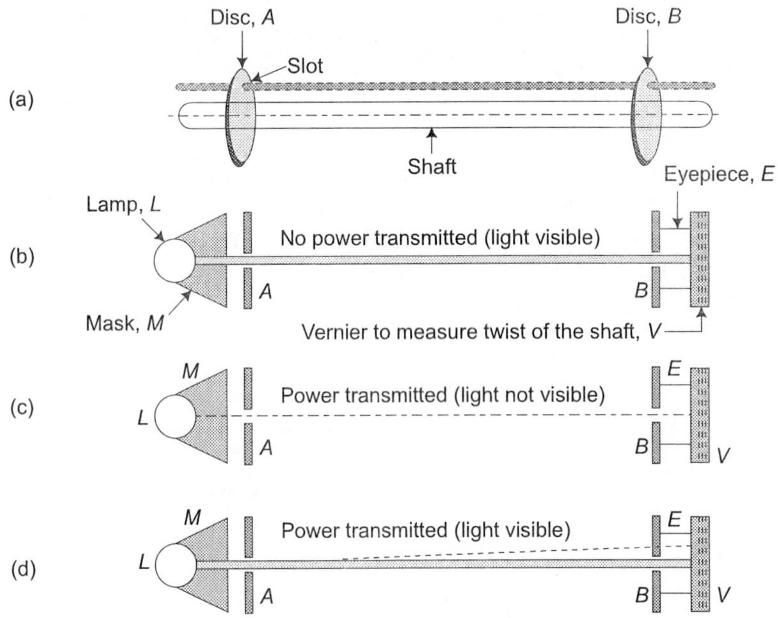

Fig. 6.48 Bevis-Gibson Flash Light Torsion Dynamometer

When the shaft is at rest it means it does not transmit any torque, the eyepiece is adjusted so as to receive a flash of light which passes from the lamp through the slots in the two discs. When the shaft rotates but it does not transmit any torque, a flash of light will be received in the eyepiece once per revolution as shown in Figure 6.48 (b). When shaft is rotating and torque is transmitted, the shaft twists and the slot position of '*B*' with respect to '*A*' changes and no light is visible by the eye-piece as shown in Figure 6.48 (c). If the eyepiece is now moved around by an amount equal to the lag of disc '*B*', then the flash of light will be visible in the eyepiece as shown in Figure 6.48(d). The eyepiece is moved by operating a micrometer spindle and by means of vernier. The angle of twist can be measured up to $1/100^{th}$ of a degree.

Example 6.14 Calculate the brake power of an engine which is running at a constant speed of 400 rpm. A rope brake dynamometer is used to absorb the power output of the engine. The dead weight on the engine and spring balance readings are 500 N and 75 N respectively. The diameters of flywheel and rope are 150 cm and 1.5 cm respectively.

Solution

The brake power is given by,

$$P = (W - S) \times \left(\frac{D + d}{2}\right) \times \frac{2\pi N}{60}$$

$$\therefore \quad P = (500 - 75) \times \left(\frac{1.5 + 0.015}{2}\right) \times \frac{2 \times \pi \times 400}{60} \times \frac{1}{1000}$$

$$= 13.48 \text{ kW}$$

Example 6.15 A torsion dynamometer is used to measure the power of a marine engine. It is observed that the shaft twist 1.5° in a length of 15 m at 100 rpm. If the shaft is hollow with 300 mm external diameter and 200 mm internal diameter, calculate the power of the engine. If the modulus of rigidity for the shaft material is 8×10^7 kN/m^2.

Solution

$$J = \frac{\pi}{32}(d_o^4 - d_i^4)$$

$$= \frac{\pi}{32} \times [(0.3)^4 - (0.2)^4]$$

$$= 0.0006 \text{ m}^4$$

Torque, $$T = \frac{C\,J}{l} \times \theta$$

$$= \frac{8 \times 10^{10} \times 0.0006}{15} \times \frac{1.5 \times \pi}{180}$$

$$= 83775.8 \text{ Nm}$$

Power is given by, $$P = \frac{T \times 2\pi N}{60 \times 1000} = \frac{83775.8 \times 2 \times \pi \times 100}{60 \times 1000}$$

$$= 877.3 \text{ kW}$$

EXERCISE

(i) Objective Type Questions

1. A belt drive is most effective if:
   (a) the friction is maximum (b) belt is thin
   (c) the friction is least (d) none of the above
2. If θ is the angle of contact of a belt with a pulley, the ratio of tensions is:
   (a) ln (μθ) (b) μ ln (μθ) (c) $e^{\mu\theta}$ (d) $e^{\mu \text{ cosec } \theta}$

3. In belt drives, the effect of centrifugal force is:
   (a) to reduce the driving power
   (b) to allow the belt to leave the pulley and to increase the power
   (c) to allow the belt to stay on the pulley and to increase the power
   (d) none of the above
4. A rack is a gear of infinite:
   (a) module (b) pitch (c) number of teeth (d) diameter
5. In a belt drive the condition for maximum power is:
   (a) $T_{max} = T_c/3$ (b) $T_{max} = 2T_c$ (c) $T_{max} = T_c/2$ (d) $T_{max} = 3T_c$
6. The ratio of tensions in a V-belt having angle of groove β is:
   (a) $\ln (T_1/T_2) = e^{\mu\theta}$ (b) $\ln (T_1/T_2) = e^{\mu \text{ cosec } \beta}$
   (c) $\ln (T_1/T_2) = e^{\mu\theta \text{ cosec } \beta}$ (d) none
7. The ratio of tensions in the tight and slack side of a flat belt drive increases:
   (a) directly proportional to increase in angle of lap
   (b) directly proportional to increase in width of belt
   (c) directly proportional to increase in coefficient of friction
   (d) exponentially as the angle of lap increases
8. In a belt system, slip has the effect of:
   (a) reducing belt speed (b) increasing friction
   (c) increasing belt speed (d) none of the above
9. Centrifugal tension in a belt drive is proportional to:
   (a) ratio of belt tensions (b) square of belt velocity
   (c) gravitational acceleration (d) angle of contact
10. Which of the following is correct:
    (a) grip is maximum in V-belt system
    (b) belt length is maximum in open belt system
    (c) angle of contact is highest in open belt system
    (d) belt length is least in open belt system
11. Two shafts whose axes are coplanar but intersecting are connected by:
    (a) spur gear (b) spiral gear (c) helical gear (d) bevel gear
12. For constant velocity ratio positive drive with large distance between shafts the suitable drive is:
    (a) gear drive (b) rope drive (c) chain drive (d) none
13. Standard angle of a V-belt is:
    (a) 30° (b) 20° (c) 10° (d) 42°

14. If T_1 and T_2 are the normal belt tensions, the initial tension in belt in a belt drive system is:

(a) $(T_1 - T_2)/2$ (b) $(T_1 + T_2)/2$ (c) $(T_1 \times T_2)/2$ (d) none

15. Dynamometers are used to measure:

(a) pump power (b) crankshaft power
(c) engine brake power (d) none

16. In belt drive system which one correct:

(a) $N_2/N_1 = D_1/D_2$ (b) $N_2 \times N_1 = D_1 \times D_2$
(c) $N_1/N_2 = D_1/D_2$ (d) $N_1/N_2 = D_2/D_1$

17. The vertical feed mechanism in a drilling machine has:

(a) spur gears (b) helical gears (c) bevel gears (d) rack and pinion

18. In crossed belt drive the power is transmitted between pulleys rotating in:

(a) same direction (b) opposite direction
(c) both (a) & (b) (d) none

19. The power transmitted by a belt is given by:

(a) $(T_1 - T_2) \times V$ (b) $(T_1 \times T_2) \times V$ (c) $(T_1/T_2) \times V$ (d) none

20. The following belts are acid- and waterproof:

(a) balata (b) leather (c) rubber (d) fibre

21. The following drives have constant velocity ratio:

(a) rope drives (b) belt drives
(c) gear and chain drives (d) none

22. Idler pulleys are used to:

(a) slow down the belt speed
(b) to increase the belt speed
(c) to not change the belt speed
(d) to change the direction of motion of the belt

23. Which one of the following belts has the best pulling capacity:

(a) balata (b) fibre (c) leather (d) rubber

24. Train value of a simple gear train is:

(a) reciprocal of its velocity ratio (b) equal to its velocity ratio
(c) equal to one (d) none

25. Reverse of diametral pitch is termed:

(a) circular pitch (b) clearance (c) addendum (d) module

26. Length of an open belt is:

(a) $\pi (r_1 + r_2) + 2x + (r_1 - r_2)^2/x$ (b) $\pi (r_1 + r_2) + (r_1 - r_2)^2/x$
(c) $\pi (r_1 + r_2) + 2x + (r_1 + r_2)^2/x$ (d) none

27. Length of a crossed belt is:

(a) $\pi (r_1 + r_2) + 2x + (r_1 - r_2)^2/x$ (b) $\pi (r_1 + r_2) + (r_1 - r_2)^2/x$

(c) $\pi (r_1 + r_2) + 2x + (r_1 + r_2)^2/x$ (d) none

28. Creep in belts is due to:

(a) expansion and contraction of belts (b) misalignment of pulleys

(c) material of belt (d) none

29. The following are the conditions for maximum power transmission by a belt:

(a) $T_c = T_{max}/3$ (b) $T_1 = 2T_{max}/3$

(c) $T_1 = 2T_{max}$ (d) (a) & (b) both (e) none

30. Low velocity gears have velocity:

(a) less than 3 m/s (b) between 3 m/s to 15 m/s

(c) greater than 15 m/s (d) none of the above

31. Gear train used in clocks:

(a) reverted (b) simple (c) compound (d) none

32. The following clutch plate is used in racing cars:

(a) single plate clutch (b) multi-plate clutch

(c) conical clutch (d) none of the above

33. The clutch is fitted between the engine and:

(a) gearbox (b) rear axle (c) differential (d) none

34. For connecting two non-parallel, non-intersecting shafts the following gears are used:

(a) bevel (b) spur (c) worm (d) spiral

35. Froud water vortex dynamometer is used to measure power in the range of:

(a) 6 to 10 kW (b) 10 to 50 kW (c) 67 to 13000 kW (d) none

36. Which one of the following is correct:

(a) gear drive: positive drive

(b) high velocity ratio: worm gear

(c) clocks: reverted gear train

(d) non-parallel and non-intersecting shafts: spiral gears

(e) all of the above

37. The following material is generally used for clutch facing:

(a) rubber (b) leather (c) plastic (d) asbestos

38. For heavy duty and high-speed vehicles the shoe brakes are made of:

(a) wood (b) hard rubber (c) cast iron (d) none of the above

39. Condition for self-locking brakes is:

(a) $a \geq \mu b$ (b) $a \leq \mu b$ (c) $a \geq \mu b/l$ (d) none of the above

40. The driving and driven shafts are coaxial in the following gear train:

(a) simple gear train (b) compound gear train

(c) reverted gear train (d) none

41. The following gears are used in the differential of an automobile:

(a) spur gears (b) helical gears

(c) spiral bevel gears (d) worm gears

42. The following gears are used for transmitting power between two parallel shafts:

(a) spur gears (b) helical gears

(c) spiral bevel gears (d) worm gears

ANSWERS

1. (a)	2. (c)	3. (a)	4. (d)	5. (d)	6. (c)	7. (d)	8. (a)	9. (b)	10. (a)
11. (d)	12. (c)	13. (d)	14. (b)	15. (c)	16. (d)	17. (d)	18. (b)	19. (a)	20. (a)
21. (c)	22. (d)	23. (c)	24. (a)	25. (d)	26. (a)	27. (c)	28. (a)	29. (d)	30. (a)
31. (a)	32. (b)	33. (a)	34. (d)	35. (c)	36. (e)	37. (d)	38. (c)	39. (b)	40. (c)
41. (c)	42. (a)								

(ii) Review Questions

1. Give the name of the different means by which power and motion can be transmitted from one shaft to the other.
2. Explain the various types of belt drives with neat sketches.
3. Briefly explain the various types of belts used for power transmission. Also give the material used for belts.
4. Give the relative advantages and disadvantages of flat belts over V-belts.
5. Briefly explain the rope drives used for power transmission.
6. What do you mean by velocity ratio of belt drives? Give the effects of slip and creep on velocity ratio.
7. Derive an expression for the length of belt in the case of (i) open belt drive, and (ii) crossed belt drive.
8. Derive the expression for the flat belt drive $T_1/T_2 = e^{\mu\theta}$, where T_1 & T_2 are the tensions on the tight and slack sides of the belt, θ is the angle of contact, and μ is the coefficient of friction between the belt and the pulley.
9. What is centrifugal tension? Derive an expression for the same.
10. Derive an expression for maximum power transmitted by a belt.
11. What do you mean by initial tension of belt?
12. Explain chain drives with their advantages and disadvantages.

13. Give the classification of chains and their applications.
14. Explain the various systems of pulleys with the help of sketches.
15. What is a gear? How gears can be classified? Also give the advantages and disadvantages of gear drives?
16. Explain the various types of gear drives with their applications.
17. Define the following terms:
    (i) Circular pitch, (ii) Diametral pitch,
    (iii) Module, (iv) Pitch circle,
    (v) Addendum, (vi) Dedendum,
    (vii) Working depth.
18. Describe pulley with its parts. What are the various types of pulleys?
19. Give the classifications of gear trains. Also give their applications.
20. What is the function of a clutch? What are its types? Give their working principle.
21. Explain: (i) Single plate clutch, and (ii) Multiplate clutch with neat sketches.
22. What is the function of a brake? What are its types?
23. Explain the following with the help of neat sketches:
    (i) Internal expanding shoe brake, (ii) Hydraulic brakes.
24. What is the function of a dynamometer? Write its types.
25. Explain the following with neat sketches:
    (i) Prony brake dynamometer, (ii) Rope brake dynamometer.
26. Write short notes on the following:
    (i) Tesla fluid friction dynamometer, (ii) Froud water vortex dynamometer,
    (iii) Band brake dynamometer.
27. Explain the following:
    (i) Epicyclic train dynamometer,
    (ii) Bevis-Gibson flash light tension dynamometer.

(iii) Numerical Problems

1. Two parallel shafts are connected by crossed belt, running over pulleys of diameters 80 cm and 60 cm respectively. If the centre distance between the pulleys is 6 m, determine the length of belt for (i) open belt system, and (ii) crossed belt system.
   [**Ans.** 1420.08 cm, 1428.08 cm]
2. A crossed flat leather belt is running at a speed of 300 rpm over a pulley having diameter of 1 m. The belt transmits 8 kW power when the angle of lap is 165° and coefficient of friction between the belt and pulley is 0.25. If maximum safe tension of the belt 20 N/mm width of belt, determine the required width of the belt. [**Ans.** 39.77 mm]

3. Determine the maximum power that can be transmitted by using a belt of dimensions: thickness = 8 mm and width = 80 mm with an angle of lap 165°. The density of belt is 1.1×10^6 g/m^3 and coefficient of friction is 0.25. The tension in the belt should not exceed 2.5 MN/m^2. [**Ans**. 10.5 kW]
4. A rope drive transmits 80 kW through a 1.5 m diameter 46° grooved pulley running at 200 rpm. Coefficient of friction between the rope and the pulley groove is 0.25 and the angle of lap is 160°. The mass of each rope is 0.6 kg/m and it can withstand a tension of 750 N safely. Taking centrifugal tension into account, determine the number of ropes required for the drive and initial rope tension. [**Ans**. 11, 499.45 N]
5. A toothed gear has 40 teeth and circular pitch of 20, find the following (i) pitch diameter, (ii) diametral pitch, and (iii) module of the gear. [**Ans**. 254.65 mm, 0.157, 12.73]
6. A torsion dynamometer is used to measure the power of a marine engine. It is observed that the shaft twist 1° in a length of 10 m at 75 rpm. If the shaft is hollow with 200 mm external diameter and 100 mm internal diameter and the modulus of rigidity for the shaft material is 8×10^{10} N/m^2, calculate the power of the engine. [**Ans**. 28.51 kW]
7. A flat belt is required to transmit 20 kW from a pulley of 1.5 m effective diameter running at 250 rpm. The angle of contact spreads over 11/25 of the circumference and the coefficient of friction between belt and pulley surface is 0.25. Determine the width of the belt while considering centrifugal tension. It is given that the belt thickness is 8 mm, density of its material is 1000 kg/m^3, and related permissible working stress is 2 MN/m^2. [**Ans**. 158.07 mm]

7

Stresses and Strains

7.1 INTRODUCTION

When an external force or load is applied on a body, it tends to undergo some deformation. The property by virtue of which the body resists the deformation is called strength of material. The materials may be classified as elastic, plastic, or rigid materials. An elastic material undergoes a deformation when an external load is applied to it, and regains its original dimensions on the removal of the load. A plastic material undergoes continuous deformation so long as the external load acts on it and does not regain its original dimensions even after removal of the load. A rigid material does not undergo any deformation under the action of an external load applied to it. Generally, a material body is handled within elastic stage in which the resisting force equals the applied load. All the structural members are designed so as to remain in the elastic stage under the action of the working loads, otherwise the deformation will continue until failure takes place.

This unit deals with the concepts of simple stress, strain, and the effect of various forces on the deformation of a body. Bars of varying cross sections, composite bars subjected to direct axial stresses due to change in temperature and due to impact load are discussed. The stresses acting on an inclined plane are analyzed. The determination of principal stresses, principal strain, and elastic constants are also explained.

7.2 STRESS

When an external force or load acts on a body, internal resisting forces are set up then the body is said to be in a state of stress. The force of resistance offered by the body to deformation is known as stress. The load is applied on the body while the stress is induced in the material of the body. Stress is usually measured by the intensity of force per unit area.

Mathematically, stress is given by,

$$\sigma = \frac{R}{A} = \frac{P}{A}$$

where σ = Intensity of stress or simply stress, R = Resistance to deformation, A = Cross sectional area, and P = External force or load. Resistance R is equal and opposite to the load P.

In the S.I. units, the force is expressed in Newton (N) and area is expressed in m^2. Therefore, the unit of stress becomes N/m^2. Stress is also expressed in the forms such as N/mm^2, kN/m^2, and kN/mm^2.

The large and small quantities of various magnitudes are: *kilo* = 10^3 denoted by *k*, *mega* = 10^6 denoted by *M*, *giga* = 10^9 denoted by *G*, *tera* = 10^{12} denoted by *T*, *peta* = 10^{15} denoted by *P*, *exa* = 10^{18} denoted by *E*, *milli* = 10^{-3} denoted by *m*, *micro* = 10^{-6} denoted by μ, *nano* = 10^{-9} denoted by η, *pico* = 10^{-12} denoted by *p*, *femto* = 10^{-15} denoted by *f*, *atto* = 10^{-18} denoted by *a*.

7.3 STRAIN

Strain is the measure of the deformation caused in the material body by external loading. It is given by the ratio of change of dimension (deformation) to the original length. It is dimensionless, means it has no unit. It is denoted by *e* or ε.

7.4 TYPES OF STRESSES

On the basis of force or load acting on a material body, stresses are of three types: (i) Tensile stress, (ii) Compressive stress, and (iii) Shear stress.

(i) Tensile Stress The stress induced in a material body, when it is subjected to two equal and opposite pulls (Figure 7.1) which cause increase in length of the body is called tensile stress.

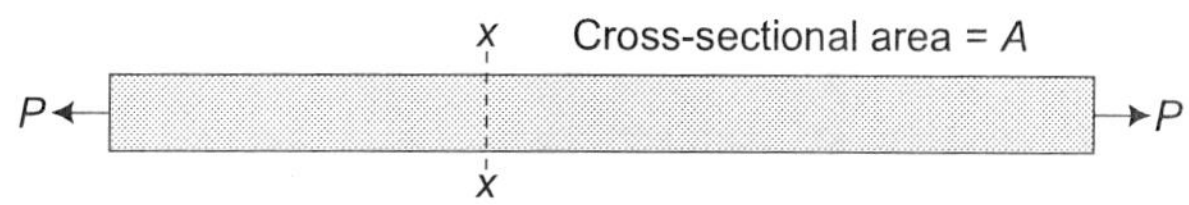

Fig. 7.1

Let P = Pull or force applied, A = Cross sectional area, σ = Stress induced, l = Original length, dl = Increase in length, and e = Strain or tensile strain.

The tensile stress at any cross section $x - x$ is given by,

$$\sigma = \frac{P}{A} \tag{7.1}$$

The strain corresponding to tensile stress is called tensile strain and is given by,

$$e = \frac{\text{Increase in length}}{\text{Original length}}$$

$$= \frac{dl}{l} \tag{7.2}$$

(ii) Compressive Stress The stress induced in a material body, when it is subjected to two equal and opposite pushes (Figure 7.2) which causes decrease in length of the body, is called compressive stress.

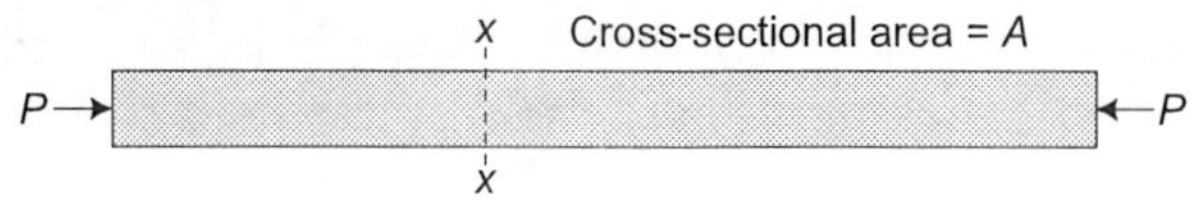

Fig. 7.2

Let P = Push or force applied, A = Cross sectional area, σ = Stress induced, l = Original length, dl = Decrease in length, and e = Strain or compressive strain.

The compressive stress at any cross section $x - x$ is given by,

$$\sigma = \frac{P}{A}$$

The strain corresponding to compressive stress is called compressive strain and is given by,

$$\varepsilon = \frac{\text{Decrease in length}}{\text{Original length}}$$

$$= \frac{dl}{l}$$

(iii) Shear Stress The stress produced by a force tangential to the surface of a body is known as shear stress. It is denoted by τ. The strain corresponding to shear stress is called shear strain which is denoted by ϕ.

A riveted joint shown in Figure 7.3 is a typical example, where the rivet resists shear failure across its cross sectional area.

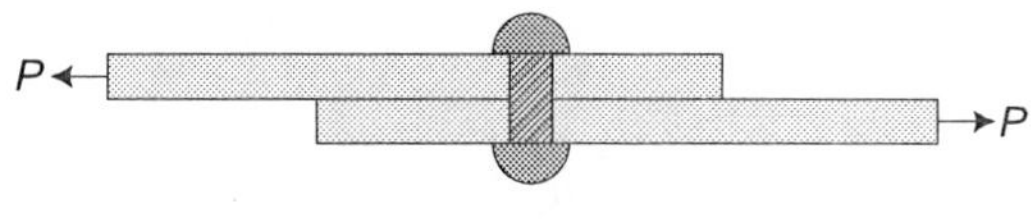

Fig. 7.3

Figure 7.4 shows a rectangular block of length l, width b, and height h fixed at the bottom plane $ABCD$ and a tangential force P is acting on the top plane $EFGH$. Under the action of this force the block is distorted to a new shape $ABCDE'F'G'H'$.

Shear stress, $$\tau = \frac{\text{Shear force } P}{\text{Area of plane } EFGH}$$

$$= \frac{P}{l \times b}$$

A shear stress is always accompanied by a balancing shear stress across the planes at right angle and this balancing stress is called complementary stress.

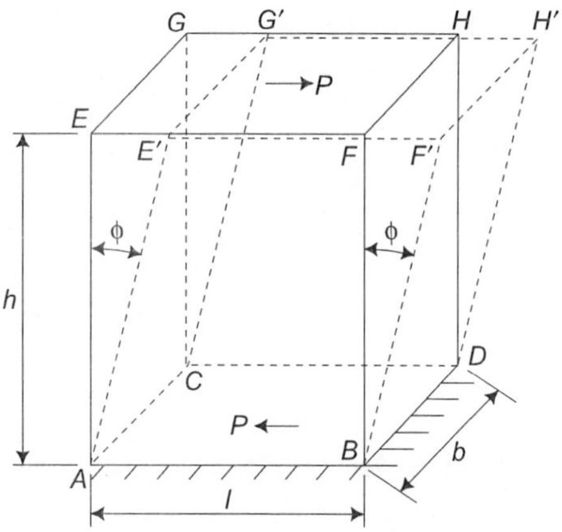

Fig. 7.4

Reaction at the bottom plane will be equal and opposite to the applied force P as shown in Figure 7.4.

The shear force τ on any plane parallel to plane $EFGH$ will remain same.

Shear strain,
$$\phi = \frac{\text{Tranverse displacement}}{\text{Distance from the lower face}}$$
$$= \frac{\text{Displacement } EE'}{EA} = \tan\phi$$

Within the elastic limit of the material the angle ϕ is very small.

$$\therefore \qquad \text{Shear strain} = \tan\phi \approx \sin\phi \approx \phi \qquad (7.3)$$

The angle ϕ is measured in radian.

7.5 ELASTIC LIMIT

When a body is subjected to an external load, it undergoes some deformation. If the load is removed and it regains its original shape and size, the body is known as elastic body. This property by virtue of which the body regains its original dimensions is known as elasticity.

As we know that the material body subjected to an external force or load offers resistance to deformation called stress. When the load is removed, the stress disappears and deformation also disappears. But this is true as long as the deformation caused by external load is within a certain limit. The value of stress corresponding to this limiting load or force is known as the elastic limit of the body.

If the external load acting on the body is so large that the resulting stress exceeds the elastic limit, the material of the body loses to some extent its property of elasticity. It means the body does not regain its original shape and size after removal of the exceeding load and thus a residual strain or permanent set remains in the body.

7.5.1 Hooke's Law

Hooke's law states that when a material is loaded within the elastic limit, the stress is directly proportional to the strain produced by the stress.

Thus,
$$\text{Stress } \alpha \text{ Strain}$$

or
$$\frac{\text{Stress}}{\text{Strain}} = \text{Constant} \qquad (7.4)$$

$$\therefore \qquad \frac{\text{Tensile stress}}{\text{Tensile strain}} = \text{Constant}$$

and
$$\frac{\text{Compressive stress}}{\text{Compressive strain}} = \text{Constant}$$

Tensile stress and compressive stress are called longitudinal stress, and tensile strain and compressive strain are called longitudinal (or linear) strain.

Therefore, within elastic limit,

$$\frac{\text{Longitudinal stress}}{\text{Longitudinal strain}} = \text{Constant.}$$

This constant is known as Young's modulus of elasticity and is usually denoted by E. For some common engineering materials, the values of modulus of elasticity are given in Table 7.1.

Table 7.1 Values of modulus of elasticity for some common materials

Material	Aluminium	Cast Iron & Brass	Bronze	Mild Steel	Alloys
E, Gpa	68-70	100-110	110-120	200-210	200-600

7.5.2 Modulus of Rigidity or Shear Modulus

The ratio of shear stress to the corresponding shear strain within elastic limit is called Modulus of Rigidity or Shear Modulus. This constant is usually denoted by G (or C or N).

Thus, $$G = \frac{\tau}{\phi} \tag{7.5}$$

7.6 STRESS-STRAIN DIAGRAM

A stress-strain diagram obtained for a mild steel specimen subjected to a tensile test on a universal testing machine is shown in Figure 7.5(a) and for cast iron it is shown in Figure 7.5(b).

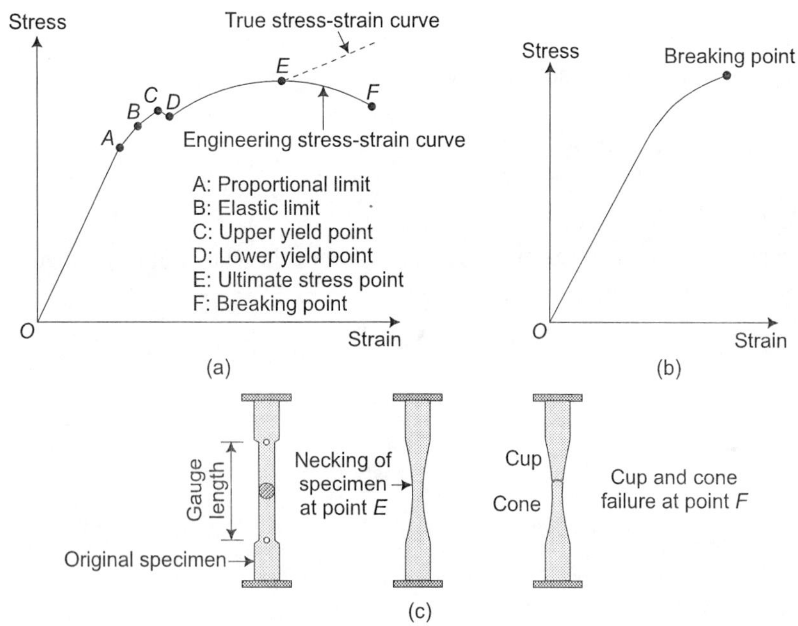

Fig. 7.5

Mild steel is commonly used in machine members and in structural applications. The cylindrical shaped specimen is clamped in the fixtures of a testing machine as shown in Figure 7.5(c). Collars are provided at both ends to firmly grip the specimen in the fixtures of the machine. The central portion where the section is uniform is called gauge length. The specimen is gradually extended by applying gradual load due to which the internal resistance offered by the specimen increases gradually. During the initial stage up to which Hooke's law is valid extensometer is used to measure the change in length. After this stage, the vernier scale on the machine is used to measure the extension. The applied load and the extension are simultaneously recorded till the specimen breaks into two pieces.

The salient features in the stress-strain diagram are given below:

1. **Proportional Limit or Limit of Proportionality** The straight line from O to A represents the range of extension in which the stress is proportional to strain and Hooke's law is valid for this range. The stress corresponding to the point A is called the limit of proportionality or proportional limit.
2. **Elastic Limit** Beyond point A (i.e., proportionality limit), stress and strain does not follow straight-line relationship, but the material still remains elastic up to state point B. The stress at point B is called the elastic limit.
3. **Yield Point** If the specimen is extended beyond the elastic limit, plastic deformation occurs. Beyond point B, i.e., at state point C the material shows considerable extension without further increase of stress, and the strain is not fully recovered when the load is withdrawn from the material sample. The onset of plastic deformation is called yielding and it pertains to extension region C to D. The point C is called upper yield point and the stress at this point is called upper yield strength. At state point D, again the internal resistance of the material increases up to state point E. The point D is known as lower yield point which is also called the point of plastic instability.
4. **Ultimate Tensile Stress** After yielding, the material becomes strain hardened and an increase in load is required to reach the material to its maximum stress point E. The stress corresponding to state point E is called the ultimate stress (or ultimate tensile stress) of the material. This stress is about 370 N/mm^2 to 400 N/mm^2. At point E necking develops in the specimen.
5. **Breaking Point** Beyond state point E, the extension increases, the load decreases and the specimen breaks at the state point F making a cup and cone type fracture (Figure 7.5c) which is a typical fracture for a ductile material. The stress at the breaking point is called breaking stress or stress at failure.

$$\text{Nominal breaking strength} = \frac{\text{Load at fracture}}{\text{Original cross sectional area}} \tag{7.6}$$

$$\text{Actual breaking strength} = \frac{\text{Load at fracture}}{\text{Area at neck}} \tag{7.7}$$

$$\%\ \text{Reduction in area} = \left(\frac{A-a}{A}\right) \times 100 \tag{7.8}$$

$$\% \text{ elongation} = \frac{\delta l}{l} \times 100 \tag{7.9}$$

Percentage elongation and percentage reduction in area give the estimation about the ductility of the material.

The material which suffers a considerable increase in length till it breaks is called ductile material, for example, mild steel. On the contrary, a material which fails after a very small elongation (Figure 7.5b) is called brittle material, for example, cast iron. An arbitrary percentage elongation of 5% is generally considered the dividing line between ductile and brittle materials. In some ductile materials which do not show a definite yield point, a proof stress (at 0.2% strain) is determined to find the onset of yielding.

Generally, short specimens are used for the compression test as there are chances of buckling of long specimen. In the case of ductile materials, the stress strain diagram up to the yield point follows exactly the same path as in the tensile test but it diverges for the larger values. Further, there will not be necking formation during the compression test. In the brittle materials the presence of flaws and cracks weaken it in tension but do not affect its strength in compression. Therefore, the ultimate compressive stress in compression will be more than in tension.

7.7 FACTOR OF SAFETY

Before defining factor of safety let us know about working stress. Working stress is the maximum stress allowed to be set up in a material in actual practice. It is much less than ultimate stress.

Factor of Safety It is the ratio of ultimate stress to the working (or 'permissible' or 'safe' or 'allowable') stress.

Mathematically, it is given by,

$$\text{Factor of safety} = \frac{\text{Ultimate stress}}{\text{Working stress}} \tag{7.10}$$

The factor of safety must be greater than unity. It depends on nature of the material, type of loading, extent of damage caused by failure parts, amount of deformation due to corrosion and fatigue of the material, etc. For general engineering applications, the factor of safety is taken from 3 to 12.

7.8 LONGITUDINAL STRAIN, LATERAL STRAIN, AND POISSON'S RATIO

7.8.1 Longitudinal and Lateral Strain

When a cylindrical rod is subjected to an axial tensile load P, the length of the rod increases but the diameter of the rod decreases as shown in Figure 7.6. It means a longitudinal stress not only produces a strain in its own direction but also produces a lateral strain.

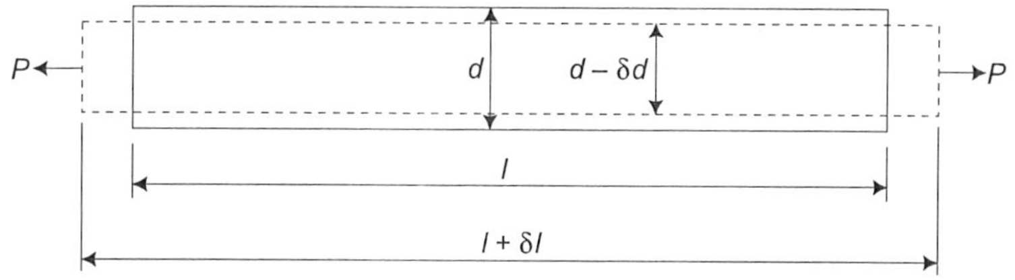

Fig. 7.6

Let P = Tensile force or load acting on the body, l = Original length, and δl = Increase in length.

Then, $$\text{longitudinal strain} = \frac{\delta l}{l}$$

Similarly, let a rectangular bar of length l, breadth b, and depth d be subjected to an axial tensile load P as shown in Figure 7.7. The length of the bar increases while the breadth and depth decrease.

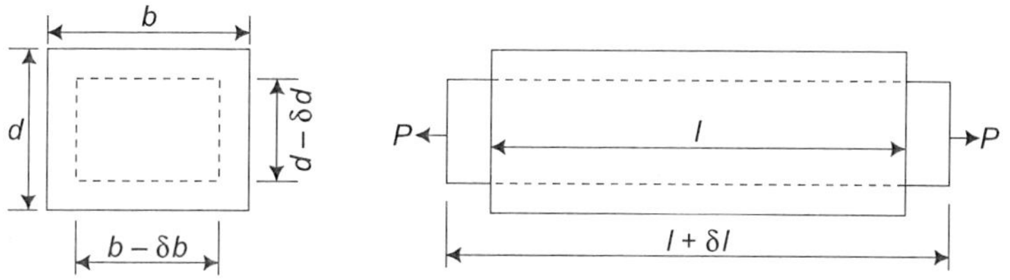

Fig. 7.7

Let δl = Increase in length, δb = Decrease in breadth, and δd = Decrease in diameter.

Then $$\text{longitudinal strain} = \frac{\delta l}{l}$$

and $$\text{lateral strain} = \frac{\delta b}{b} \text{ or } \frac{\delta d}{d} \tag{7.11}$$

7.8.2 Poisson's Ratio

When the deformation of a material body is within elastic limit, the ratio of the lateral strain to the longitudinal strain is a constant for a given material. This constant is called Poisson's ratio and is usually denoted by ν. The value of Poisson's ratio for most of the metals varies from 0.25 to 0.33, for rubber 0.45 to 0.50, for steel 0.3, and for concrete 0.15.

Mathematically, Poisson's ratio is give by,

$$\nu = \frac{\text{Lateral strain}}{\text{Longitudinal strain}} \tag{7.12}$$

7.9 VOLUMETRIC STRAIN AND BULK MODULUS

When a body is subjected to some external force or load, it undergoes change in its dimensions and thus its volume will also change. The ratio of change in volume of the body to the original volume is called its volumetric strain. It is usually denoted by e_v and mathematically, it is given by,

$$e_v = \frac{\text{Change in volume}}{\text{Original volume}} \tag{7.13}$$

The ratio of average normal stress to volumetric strain is called the bulk modulus of elasticity which is constant for a given material. It is denoted by K.

Thus,
$$K = \frac{\text{Average normal stress}}{\text{Volumetric strain}}$$

$$= \frac{\sigma_{av}}{e_v} \tag{7.14}$$

7.10 CHANGE IN LENGTH DUE TO AXIAL LOAD

Let l = Original length and δl = Change in length,

Then strain,
$$e = \frac{\delta l}{l} \tag{i}$$

Let σ = Stress induced due to applied load, and E = Young's modulus of the material,

Then stress,
$$E = \frac{\sigma}{e}$$

or
$$e = \frac{\sigma}{E} \tag{ii}$$

From equations (i) and (ii), we have,

$$\frac{\delta l}{l} = \frac{\sigma}{E}$$

or
$$\delta l = \frac{\sigma l}{E} \tag{iii}$$

But
$$\sigma = \frac{P}{A}$$

where P = Force or load applied, and A = Cross sectional area

Therefore, equation (iii) becomes,

$$\delta l = \frac{Pl}{AE} \tag{7.15}$$

Example 7.1 Determine the elongation of a mild steel rod 25 mm in diameter and 2 m long when subjected to an axial pull of 100 kN. Take Young's modulus of the material as 200 kN/mm^2.

Solution

Cross sectional area,

$$A = \frac{\pi}{4} \times (25^2) = 490.87 \text{ mm}^2$$

The stress,
$$\sigma = \frac{P}{A} = \frac{100 \times 10^3}{490.874} = 203.72 \text{ N/mm}^2$$

The strain ,
$$e = \frac{P}{E} = \frac{203.72}{200 \times 10^3} = 1.0186 \times 10^{-3}$$

Elongation,
$$dl = e \times l = 1.0186 \times 10^{-3} \times 2 \times 10^3$$
$$= 2.0372 \text{ mm}$$

Example 7.2 A load of 10 kN is to be raised with the help of a steel wire. Determine the minimum diameter of the wire if the stress is not to exceed 150 MN/m^2.

Solution Let d be the required diameter of the wire in meter.

$$\sigma = \frac{P}{A}$$

$$150 \times 10^6 = \frac{10 \times 10^3}{(\pi/4) \times (d^2)}$$

$$d^2 = \frac{10 \times 10^3 \times 4}{\pi \times 150 \times 10^6}$$

$$\therefore \quad d = 0.009213 \text{ m}$$

Example 7.3 A mild steel rod is subjected to a pull of 200 kN. If the ultimate tensile stress of mild steel is 450 N/mm^2, find the necessary diameter of the rod if factor of safety is 3.

Solution

$$\text{Factor of safety} = \frac{\text{Ultimate stress}}{\text{Working stress}}$$

Therefore, working stress,

$$\sigma = \frac{450 \times 10^{-3}}{3} = 0.15 \text{ kN/mm}^2$$

Let d be the required diameter of the rod in meter.

$$\sigma = \frac{P}{A}$$

$$0.15 = \frac{200}{(\pi/4) \times (d^2)}$$

or $$d^2 = \frac{200 \times 4}{\pi \times 0.15}$$

$\therefore$ $$d = 41.2 \text{ m}$$

Example 7.4 Find the maximum and minimum stresses produced in the stepped bar shown in Figure 7.8 due to an axially applied compressive load of 10 kN.

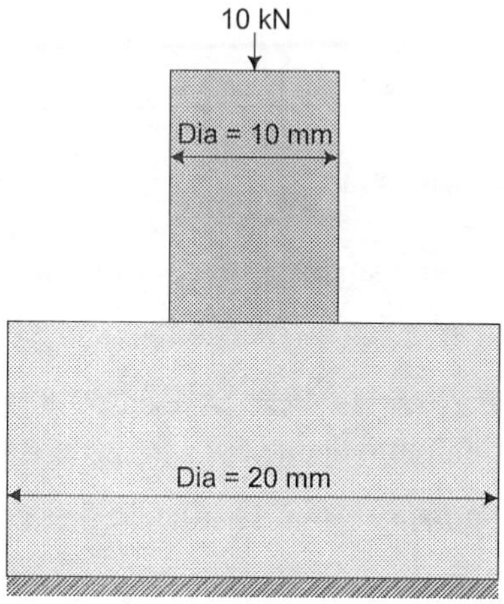

Fig. 7.8

Solution Refer Figure 7.8.

The intensity of stress will be more in the part having smaller area and vice versa.

Thus, maximum stress,

$$\sigma_{max} = \frac{P}{A} = \frac{10 \times 10^3}{(\pi/4) \times (10^2)}$$
$$= 127.32 \text{ N/mm}^2$$

Minimum stress,

$$\sigma_{min} = \frac{P}{A} = \frac{20 \times 10^3}{(\pi/4) \times (20^2)}$$
$$= 31.83 \text{ N/mm}^2$$

Example 7.5 A load of 5 kN has to be raised at the end of a steel wire. If the unit stress in the wire must not exceed 100 N/mm^2, determine the minimum diameter required. Also determine the extension of 2.5 metre length of wire. Take the Young's modulus for elasticity as 0.2 MN/mm^2.

Solution Let d be the minimum diameter of the rod in metre.

$$\sigma = \frac{P}{A}$$

$$100 = \frac{5 \times 10^3}{(\pi/4) \times (d^2)}$$

$$d^2 = \frac{5 \times 10^3 \times 4}{\pi \times 100}$$

$\therefore$ $$d = 7.98 \text{ mm}$$

The strain, $$e = \frac{P}{E} = \frac{100}{0.2 \times 10^6}$$
$$= 5 \times 10^{-4}$$

Elongation, $$dl = e \times l = 5 \times 10^{-4} \times 2.5 \times 10^3$$
$$= 1.25 \text{ mm}$$

Example 7.6 A steel wire 3 m long and 5 mm in diameter is extended by 0.6 mm due to weight. If the same weight is suspended from the brass wire 3.5 m long and 3 mm in diameter, it is elongated by 4.5 mm. Determine the Young's modulus of elasticity of brass if that of steel is 0.2 MN/mm^2.

Solution

$$\delta l = \sigma = \frac{Pl}{AE}$$

So, for the steel wire, we have,

$$0.6 = \frac{P \times 3000}{(\pi/4) \times (5^2) \times (0.2 \times 10^6)}$$

$$\therefore \quad P = \frac{0.6 \times \pi \times (5^2) \times (0.2 \times 10^6)}{4 \times 3000}$$

$$= 785.39 \text{ N}$$

For the brass rod, we have,

$$4.5 = \frac{785.39 \times 3500}{(\pi/4) \times (3^2) \times E_{brass}}$$

$$\therefore \quad E_{brass} = \frac{4 \times 785.39 \times 3500}{4.5 \times \pi \times (3^2)}$$

$$= 86418.85 \text{ N/mm}^2$$

Example 7.7 The wire working on a railway signal is 7.5 mm in diameter and 265 m long. If the movement at the signal end is to be 17.5 cm, make calculations for the movement which should be given to the end of the wire at the signal box. A pull of 1650 N on the wire is applied. Take modulus of elasticity of wire material as 0.2 MN/mm^2.

Solution The induced stress in the wire due to pull is given by,

$$\sigma = \frac{P}{A} = \frac{1650}{(\pi/4) \times (7.5^2)} = 37.35 \text{ N/mm}^2$$

Then strain, $$e = \frac{\sigma}{E} = \frac{37.35}{0.2 \times 10^6} = 1.8675 \times 10^{-4}$$

Elongation, $$dl = e \times l = 1.8675 \times 10^{-4} \times 265 \times 10^3$$

$$= 49.49 \text{ mm} \approx 4.95 \text{ cm}$$

Therefore, total movement required = 17.5 + 4.95 = 22.45 cm

Example 7.8 A compression load of 500 kN acts axially on a hollow tube of external diameter 125 mm. If the yield stress of the material is 300 N/mm^2, determine the thickness of the tube.

Solution The yield stress is given by,

$$\sigma = \frac{P}{A}$$

$$A = \frac{P}{\sigma} = \frac{500 \times 10^3}{300} = 2000 \text{ mm}^2$$

Let D and d be the external and internal diameters of the tube.

Then $$A = 2000 = \frac{\pi}{4}(D^2 - d^2)$$

$$2000 = \frac{\pi}{4}(125^2 - d^2)$$

$$2000 = \frac{\pi}{4}(125^2 - d^2)$$

$$d^2 = 13078.52$$

$\therefore$ $$d = 114.36$$

$\therefore$ Thickness, $$t = \frac{D - d}{2} = \frac{125 - 114.36}{2}$$

$$= 5.32 \text{ mm}$$

Example 7.9 The ultimate stress for a hollow steel column which carries an axial load of 1.5 MN is 450 MN/m^2. If the external diameter of the column is 18 cm, determine the internal diameter and its thickness. Consider the factor of safety as 3.

Solution Factor of safety = $\frac{\text{Ultimate stress}}{\text{Working stress}}$

Therefore, working stress,

$$\sigma = (450/3) = 150 \text{ N/mm}^2$$

The working stress is given by,

$$\sigma = \frac{P}{A}$$

$$A = \frac{P}{\sigma} = \frac{1.5 \times 10^6}{150} = 10000 \text{ mm}^2$$

Let D and d be the external and internal diameters of the tube.

Then $$A = 10000 = \frac{\pi}{4}(D^2 - d^2)$$

$$10000 = \frac{\pi}{4}(180^2 - d^2)$$

$$d^2 = 19667.605$$

$\therefore$ $$d = 140.24 \text{ mm}$$

$\therefore$ Thickness, $t = \dfrac{D-d}{2} = \dfrac{180 - 140.24}{2}$

$= 19.88$ mm

Example 7.10 A hollow cast iron right circular cylinder has dimensions as outer diameter = 80 mm, inside diameter = 65 mm, and height = 650 mm. It is subjected to axial compressive load of 60 kN. Find the normal stress and shortening in length of the cylinder if the possibility of buckling is neglected. Take modulus of elasticity for cast iron material as 110 GPa.

Solution The stress $\sigma = \dfrac{P}{A} = \dfrac{60 \times 10^3}{(\pi/4) \times (80^2 - 65^2)}$

$= 35.12$ N/mm^2

Shortening in length is given by,

$$dl = e \times l = \frac{\sigma}{E} \times l = \frac{35.12}{110 \times 10^3} \times 650$$

$= 0.2075$ mm

Example 7.11 A bar 3 m long and 100 mm × 200 mm in section is subjected to a pull of 65 kN. If Young's modulus of the material of the bar is 200 kN/mm^2, determine (i) stress set up in the bar material, (ii) the strain produced, (iii) elongation of the bar, and (iv) the work done.

Solution

(i) The stress set up in the bar,

$$\sigma = \frac{65 \times 10^3}{(100 \times 200)} = 3.25 \times 10^{-3} \text{ kN/mm}^2$$

(ii) Elongation of the bar,

$$\delta l = \frac{Pl}{AE} = \frac{65 \times 3000}{2 \times 10^4 \times 200} = 0.049$$

(iii) Work done = Average force × δl

$$= \left(\frac{0 + 65}{2}\right) \times 0.049$$

$= 1.592$ kN/mm

Example 7.12 A mild steel bar 40 mm in diameter, 35 cm long was tested for tensile test. The following data were recorded during the test: extension at a load of 125 kN = 0.20 mm, load at elastic limit = 240 kN, maximum load = 3650 kN, total extension = 55 mm, and diameter of the rod at failure = 25 mm. Determine (i) the modulus of elasticity, (ii) the stress at elastic limit, (iii) the percentage elongation, and (iv) the percentage decrease in area.

Solution Cross sectional area,

$$A = (\pi/4) \times (40^2) = 1256.64 \text{ mm}^2$$

$$\sigma = \frac{P}{A} = \frac{125 \times 10^3}{1256.64} = 99.47 \text{ N/mm}^2$$

The strain, $$\varepsilon = \frac{\delta l}{l} = \frac{0.20}{350} = 5.71 \times 10^{-4}$$

(i) Modulus of elasticity,

$$E = \frac{\sigma}{e} = \frac{99.47}{5.71 \times 10^{-4}} = 1.74 \times 10^5 \text{ N/mm}^2$$

(ii) The stress at elastic limit,

$$\sigma_e = \frac{240 \times 10^3}{1256.64} = 190.98 \text{ N/mm}^2$$

(iii) The percentage elongation $= \frac{55}{350} \times 100 = 15.71\%$

(iv) The percentage decrease in area is given by,

$$= \frac{(\pi/4)\,(d_1^{\,2} - d_2^{\,2})}{(\pi/4)\,(d_1^{\,2})} \times 100$$

$$= \frac{(d_1^{\,2} - d_2^{\,2})}{(d_1^{\,2})} \times 100$$

$$= \frac{(40^2 - 25^2)}{(40^2)} \times 100$$

$$= 60.94\%$$

Example 7.13 The following data are related to a tensile test conducted on a tensile test specimen: original diameter and gauge length = 15 mm and 60 mm respectively, minimum diameter at fracture = 10 mm, distance between gauge points at fracture = 75 mm, and load at yield point and fracture point = 50 kN and 55 kN respectively. Maximum load that can be applied on specimen = 75 kN. Determine (i) the yield stress or strength, (ii) ultimate tensile stress, (iii) breaking stress, (iv) the percentage elongation and the percentage decrease in area after fracture, and (v) nominal and true stress at fracture.

Solution

Original cross sectional area = $(\pi/4) \times (15^2) = 176.71 \text{ mm}^2$

Final cross sectional area at fracture = $(\pi/4) \times (10^2) = 78.54 \text{ mm}^2$

(i) Yield stress or strength $= \frac{50 \times 10^3}{176.71} = 282.95 \text{ N/mm}^2$

(ii) Ultimate tensile stress $= \frac{75 \times 10^3}{176.71} = 424.42 \text{ N/mm}^2$

(iii) Breaking stress $= \dfrac{55 \times 10^3}{176.71} = 311.24 \text{ N/mm}^2$

(iv) The percentage elongation $= \dfrac{75 - 60}{75} \times 100 = 20\%$

(v) The percentage decrease in area $= \dfrac{176.71 - 78.54}{176.71} \times 100 = 55.55\%$

(vi) The nominal stress $= \dfrac{\text{Load at fracture}}{\text{Original area}} = \dfrac{55 \times 10^3}{176.71} = 311.24 \text{ N/mm}^2$

It should be noted that breaking and nominal stresses are same.

$$\text{True stress} = \frac{\text{Load at fracture}}{\text{Final area}} = \frac{55 \times 10^3}{78.54} = 700.28 \text{ N/mm}^2$$

7.11 VARYING CROSS SECTIONS AND PRINCIPLE OF SUPERPOSITION

7.11.1 Stepped Bar

Let a bar be made of different lengths and of different cross sections as shown in Figure 7.9. This bar is subjected to an axial load P. The Young's modulus for the material of the bar is E.

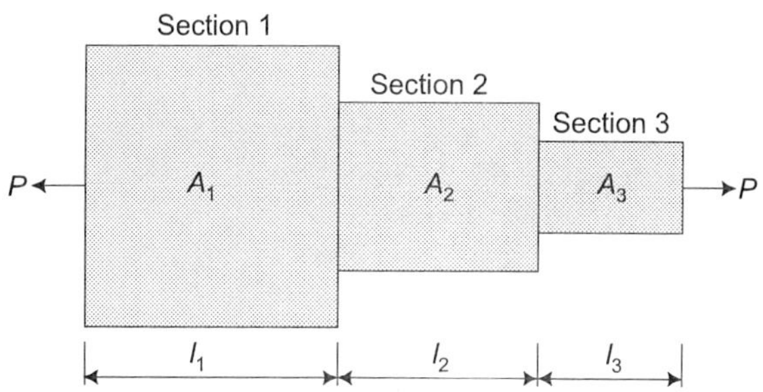

Fig. 7.9

l_1, l_2, and l_3 are the lengths of sections 1, 2, and 3 respectively,

d_1, d_2, and d_3 are the diameters of sections 1, 2, and 3 respectively,

$A_1 = (\pi/4)d_1^2$, $A_2 = (\pi/4)d_2^2$, and $A_3 = (\pi/4)d_3^2$ are the areas of sections 1, 2, and 3 respectively, and

δl_1, δl_2, and δl_3 are the elongations of sections 1, 2, and 3 respectively which can be given by,

$$\delta l_1 = \frac{Pl_1}{A_1E}, \qquad \delta l_2 = \frac{Pl_2}{A_2E}, \quad \text{and} \qquad \delta l_3 = \frac{Pl_3}{A_3E}$$

σ_1, σ_2, and σ_3 are the stresses in the sections 1, 2, and 3 respectively which can be given by,

$$\sigma_1 = (P/A_1), \quad \sigma_2 = (P/A_2), \quad \text{and} \quad \sigma_3 = (P/A_3)$$

The total change in length of the bar will be obtained by adding the changes in the length of individual sections, i.e.,

$$\delta l = \delta l_1 + \delta l_2 + \delta l_3$$

$$\therefore \qquad \delta l = \frac{Pl_1}{A_1E} + \frac{Pl_2}{A_2E} + \frac{Pl_3}{A_3E}$$

or

$$\delta l = \frac{P}{E}\left[\frac{l_1}{A_1} + \frac{l_2}{A_2} + \frac{l_3}{A_3}\right] \qquad (7.16)$$

Principle of Superposition The principle of superposition is applicable when the proportionality between stresses and strains in a material follows linear relationship. As per the principle of superposition, when a number of loads act on a body, the resulting strain will be the algebraic sum of strains caused by individual loads.

While applying this principle for an elastic body which is subjected to a number of tensile or compressive forces at different sections along its length, the forces are split up and their effects are considered on individual sections by drawing free body diagram. The deformation of each section is calculated. The total or resulting deformation of the body is then given by the algebraic sum of the deformation of the individual sections.

The bar AB in Figure 7.10(a) is acted upon by various forces. The free body diagrams for AB, BC, and CD are shown in Figure 7.10(b), 7.10(c), and 7.10(d) respectively.

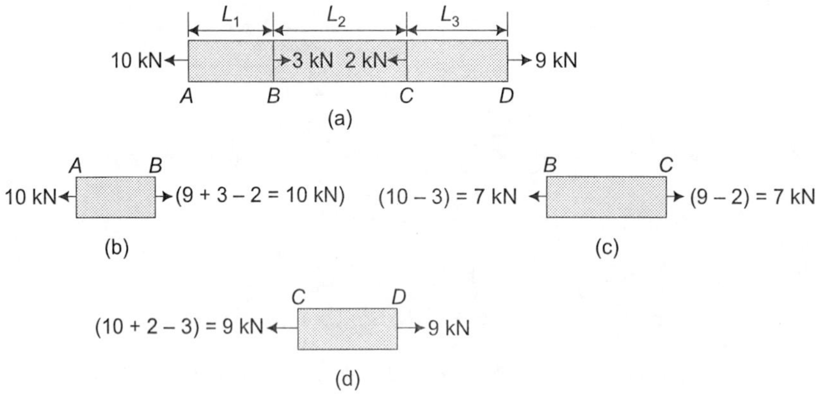

Fig. 7.10

To draw the free body diagram for AB, consider all the forces to the left of the normal section at A. The net force is 10 kN tensile. Considering all the forces to the right of the normal section at B, the net force is $(9 + 3 - 2) = 10$ kN. Thus the section AB of length L_1 is subjected to a tensile force of 10 kN. Similarly, for the section BC of length L_2 considering the algebraic sum of the forces to the left of the normal section B, sum of the forces to the left of the normal section C, the net force is (10 – 3) = 7 kN. Thus, the section BC of length L_2 is subjected to a tensile force of 7 kN. Now consider the algebraic sum of the forces to the left of the normal section C, the net force is (10 – 3 + 2) = 9 kN. Thus, the section CD of length L_3 is subjected to a tensile force of 9 kN.

7.11.2 Tapered Bar

Figure 7.11 shows a bar of length l, uniformly tapering from a diameter d_1 at A to d_2 at B. Let the bar is subjected to an axial load P, and E is the Young's modulus of elasticity of the bar material.

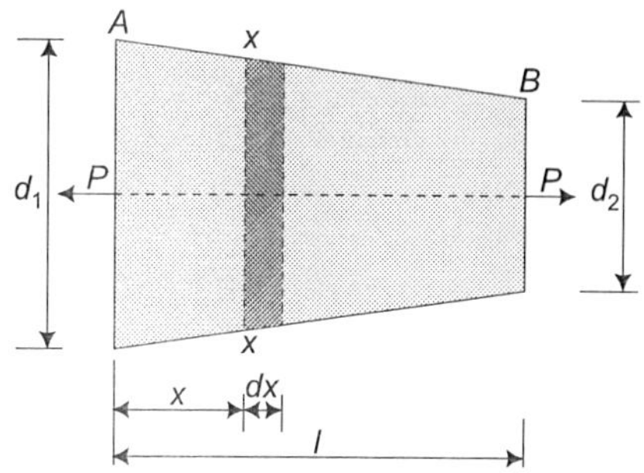

Fig. 7.11

Consider a section $x - x$ and an elemental length dx at a distance x from A. Let the diameter of the bar be d' at the section.

$$\therefore \qquad d' = d_1 - \left(\frac{d_1 - d_2}{l}\right) . x$$

Let
$$\frac{d_1 - d_2}{l} = k$$

So,
$$d' = d_1 - kx$$

Stress on the section,

$$\sigma_x = \frac{P}{A'} = \frac{4P}{\pi(d_1 - kx)^2}$$

and strain on the section,

$$e' = \frac{\sigma_x}{E} = \frac{4P}{\pi E\,(d_1 - kx)^2}$$

Extension of the elemental length,

$$dx = e'\,dx = \frac{4P}{\pi E\,(d_1 - kx)^2}\,dx$$

Total extension of the bar,

$$\delta l = \frac{4P}{\pi E}\int_0^l \frac{dx}{(d_1 - kx)^2} = \frac{4P}{\pi E}\left[\frac{(d_1 - kx)^{-1}}{-(k)}\right]_0^l$$

$$\delta l = -\frac{4P}{\pi E k}\left[\frac{1}{d_1 - kl} - \frac{1}{d_1}\right]$$

Substituting the value of k, we have,

$$= \frac{4P}{\pi E\left(\frac{d_1 - d_2}{l}\right)}\left(\frac{1}{d_1 - \left(\frac{d_1 - d_2}{l}\right) l} - \frac{1}{d_1}\right)$$

$$= \frac{4Pl}{\pi E\,(d_1 - d_2)}\left[\frac{1}{d_1 - d_1 + d_2} - \frac{1}{d_1}\right]$$

$$= \frac{4Pl}{\pi E\,(d_1 - d_2)}\left[\frac{1}{d_2} - \frac{1}{d_1}\right]$$

$$\therefore \qquad \delta l = \frac{4Pl}{\pi E\,(d_1 - d_2)}\left[\frac{d_1 - d_2}{d_1\,d_2}\right]$$

$$= \frac{4Pl}{\pi E\,d_1\,d_2} \qquad (7.17)$$

7.11.3 Tapered Flat Bar

Figure 7.12 shows a bar of length l having a flat thickness of t is uniformly flat tapered from breadth b_1 to b_2. Let the bar be subjected to an axial load P, and E is the Young's modulus of elasticity of the bar material.

Consider a section $x - x$ and an elemental length dx at a distance x from end A. Let the breadth of the bar be b at the section.

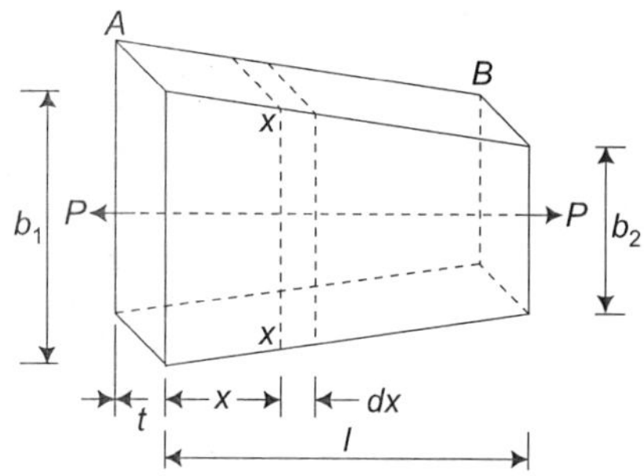

Fig. 7.12

$$\therefore \qquad b' = b_1 - \left(\frac{b_1 - b_2}{l}\right).\,x$$

Let $$\frac{b_1 - b_2}{l} = k$$

So, $$b' = b_1 - kx$$

Area of cross section at the section $x - x$,

$$A_x = b' \times t = (b_1 - kx) \times t$$

Stress on the section,

$$\sigma_x = \frac{P}{A'} = \frac{P}{(b_1 - kx) \times t}$$

Strain on the section,

$$e' = \frac{\sigma_x}{E} = \frac{P}{Et(b_1 - kx)}$$

Extension of the elemental length,

$$dx = e'\,dx = \frac{P}{Et(b_1 - kx)}\,dx$$

Total extension of the bar,

$$\delta l = \int_0^l \frac{P}{Et(b_1 - kx)}\,dx = \frac{P}{Et} \times \frac{1}{(-k)}\,[\ln\,(b_1 - kx)]_0^l$$

$$\delta l = -\frac{P}{Etk}[\ln(b_1 - kl) - \ln b_1]$$

$$= \frac{P}{Etk}[\ln b_1 - \ln(b_1 - kl)]$$

$$= \frac{P}{Etk}\left[\ln\left(\frac{b_1}{b_1 - kl}\right)\right]$$

Substituting the value of k,

$$\delta l = \frac{P \times l}{Et\,(b_1 - b_2)}\left[\ln\left(\frac{b_1}{b_1 - \left\{\frac{b_1 - b_2}{l}\right\} \times l}\right)\right]$$

$$\therefore \qquad \delta l = \frac{P \times l}{Et\,(b_1 - b_2)} \ln\left(\frac{b_1}{b_2}\right) \qquad (7.18)$$

Example 7.14 An axial pull of 25 kN is acting on a bar consisting of three lengths as shown in Figure 7.13. If the Young's modulus for the material of the bar is 200 kN/mm^2, find (i) stresses in each section, and (ii) total extension of the bar.

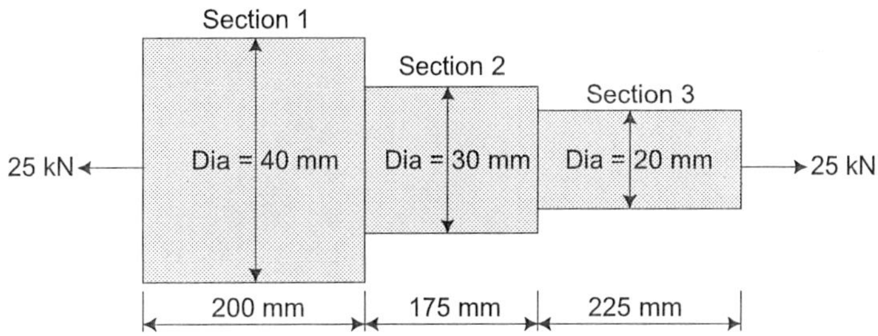

Fig. 7.13

Solution The stresses in the sections 1, 2, and 3 respectively can be given by,

$$\sigma_1 = \frac{P}{A_1} = \frac{25 \times 10^3}{(\pi/4) \times (40^2)} = 19.89 \text{ kN/mm}^2$$

$$\sigma_2 = \frac{P}{A_2} = \frac{25 \times 10^3}{(\pi/4) \times (30^2)} = 35.37 \text{ kN/mm}^2$$

$$\sigma_3 = \frac{P}{A_3} = \frac{25 \times 10^3}{(\pi/4) \times (20^2)} = 79.58 \text{ kN/mm}^2$$

The total change in length of the bar will be given by,

$$\delta l = \frac{P}{E}\left[\frac{l_1}{A_1} + \frac{l_2}{A_2} + \frac{l_3}{A_3}\right]$$

$$\therefore \qquad \delta l = \frac{25}{200} \times \left[\frac{200}{(\pi/4) \times 40^2} + \frac{175}{(\pi/4) \times 30^2} + \frac{225}{(\pi/4) \times 20^2}\right]$$

$$= 0.1404 \text{ mm}$$

Example 7.15 A straight bar 400 mm long is 15 mm in diameter for the first 150 mm of its length and 25 mm diameter for the remaining length. If the bar is subjected to an axial push of 15 kN and the Young's modulus for the material of the bar is 200 kN/mm^2 then find the decrease in length of the bar.

Solution Refer Figure 7.14.

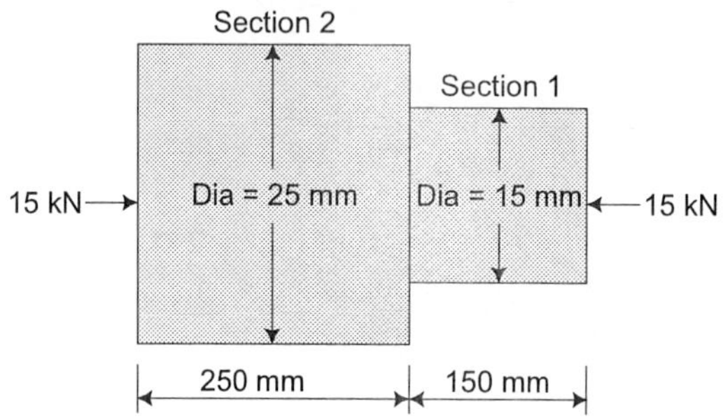

Fig. 7.14

$$A_1 = (\pi/4) \times (15^2) = 176.71 \text{ mm}^2$$

$$A_2 = (\pi/4) \times (25^2) = 490.87 \text{ mm}^2$$

Let δl_1 and δl_2 be the contractions of sections 1 and 2 respectively which can be given by,

$$\delta l_1 = \frac{Pl_1}{A_1 E}$$

$$\delta l_2 = \frac{Pl_2}{A_2 E}$$

The total decrease in length of the bar will be obtained by adding the changes in length of individual section, i.e.,

$$\delta l = \delta l_1 + \delta l_2$$

$$\therefore \qquad \delta l = \frac{Pl_1}{A_1 E} + \frac{Pl_2}{A_2 E}$$

$$\delta l = \frac{P}{E}\left[\frac{l_1}{A_1} + \frac{l_2}{A_2}\right]$$

$$\therefore \qquad \delta l = \frac{15 \times 10^3}{2 \times 10^5}\left[\frac{150}{176.71} + \frac{250}{490.87}\right]$$

$$= 0.075 \times [0.084885 + 0.5093] = 0.1019 \text{ mm}$$

Example 7.16 The bar shown in Figure 7.15 is subjected to an axial pull of 100 kN. If the stress in the middle portion is limited to 125 kN/mm^2, determine the diameter of the middle portion. Determine the length of the middle portion if total elongation of the bar is 0.20 mm. The Young's modulus for the material of the bar is 200 kN/mm^2.

Solution Refer Figure 7.15.

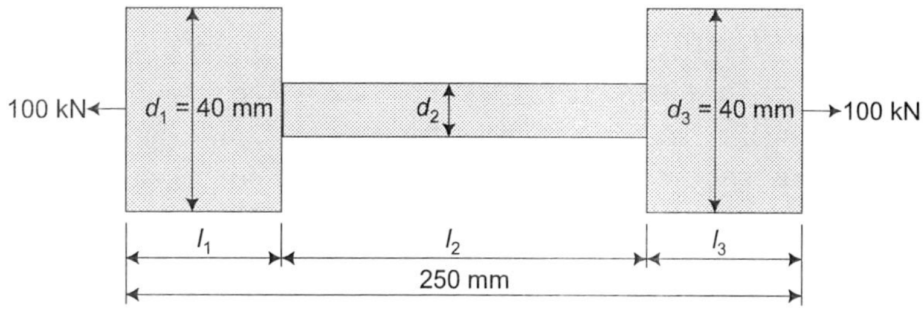

Fig. 7.15

Each section is subjected to axial pull of 100 kN. Stress in the middle portion is given by,

$$\sigma_2 = \frac{P}{A_2}$$

$$125 = \frac{100 \times 10^3}{(\pi/4) \times (d_2^{\,2})}$$

$$d_2^{\,2} = \frac{100 \times 10^3 \times 4}{\pi \times 125} = 1018.59$$

$$\therefore \quad d_2 = \sqrt{1018.59} = 31.9 \text{ mm}$$

Stress in the end portions is given by,

$$\sigma_2 = \sigma_3 = \frac{P}{A_3} = \frac{100 \times 10^3}{(\pi/4) \times (40^2)}$$

$$= 79.58 \text{ N/mm}^2$$

The total change in length of the bar will be obtained by adding the changes in length of individual section, i.e.,

$$\delta l = \delta l_1 + \delta l_2 + \delta l_3$$

$$\therefore \quad \delta l = \frac{Pl_1}{A_1 E} + \frac{Pl_2}{A_2 E} + \frac{Pl_3}{A_3 E}$$

$$= \frac{\sigma_1 l_1}{E} + \frac{\sigma_2 l_2}{E} + \frac{\sigma_3 l_3}{E}$$

Applying the given values,

$$\sigma_1 = \sigma_3 = 79.58 \text{ N/mm}^2$$

and $\quad l_1 + l_3 = (250 - l_2)$

$$0.20 = \frac{79.58}{2 \times 10^5} \times (250 - l_2) + \frac{125 \times l_2}{2 \times 10^5}$$

$$0.20 \times 2 \times 10^5 = 79.58 \times (250 - l_2) + 125 \times l_2$$

$$40000 = 19895 - 79.58\, l_2 + 125\, l_2$$

$$40000 - 19895 = 45.42\, l_2$$

$$\therefore \quad l_2 = \frac{20105}{45.42} = 442.65 \text{ mm}$$

Example 7.17 A steel tie rod of 40 mm in diameter and 2 m long is subjected to an axial pull of 80 kN. Determine the percentage change in elongation produced under the same pull by boring the rod centrally for 0.8 m length of 20 mm diameter. The Young's modulus for the material of the bar is 200 kN/mm^2.

Solution Refer Figure 7.16.

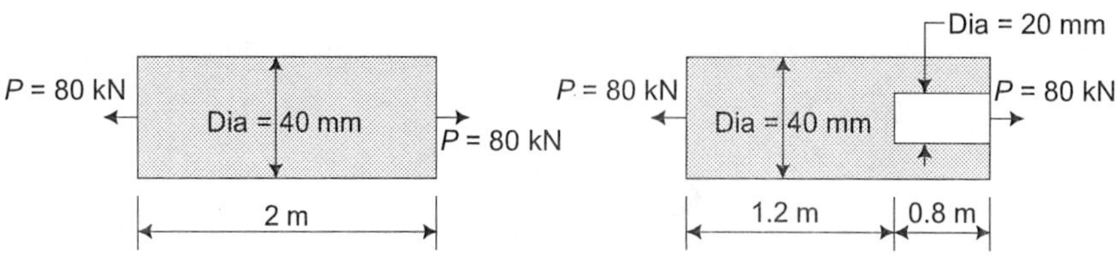

Fig. 7.16

When the rod is solid:

$$A = \frac{\pi}{4} \times (40^2) = 1256.64 \text{ mm}^2$$

$$\text{Stress} = \frac{80 \times 10^3}{1256.64} = 63.66 \text{ N/mm}^2$$

The elongation is given by,

$$\delta l = \frac{Pl}{AE} = \frac{\sigma l}{E}$$

$$= \frac{63.66 \times 2000}{200 \times 10^3} = 0.6366 \text{ mm}$$

When the rod is bored:

$$A = \frac{\pi}{4} \times (40^2 - 20^2) = 942.48 \text{ mm}^2$$

$$\text{Stress} = \frac{80 \times 10^3}{942.48} = 84.88 \text{ N/mm}^2$$

Total elongation = elongation of solid part + elongation of bored part

$$\delta l = \left(\frac{Pl}{AE}\right)_{\text{solid}} + \left(\frac{Pl}{AE}\right)_{\text{bored}}$$

$$= \left(\frac{\sigma l}{E}\right)_{\text{solid}} + \left(\frac{\sigma l}{E}\right)_{\text{bored}}$$

$$\delta l = \left(\frac{63.66 \times 1200}{200 \times 10^3}\right)_{\text{solid}} + \left(\frac{84.88 \times 800}{200 \times 10^3}\right)_{\text{bored}}$$

$$= 0.3819 + 0.3395 = 0.8214 \text{ mm}$$

$$\therefore \quad \%\text{increase in elongation} = \left(\frac{0.7214 - 0.6366}{0.6366}\right) \times 100$$

$$= 13.32\%$$

Example 7.18 Two bars of different sizes and materials are fastened together to support a downward vertical load of 60 kN. The upper bar is made of steel having mass density 7750 kg/m^3, length 7.5 m, and cross sectional area 55 cm^2. The lower bar is made of brass having mass density 9000 kg/m^3, length 5 m, and cross sectional area 35 cm^2. Determine maximum stress in each material. The Young's modulus for the material of the steel bar is 200 GN/m^2 and for brass bar is 100 GN/m^2.

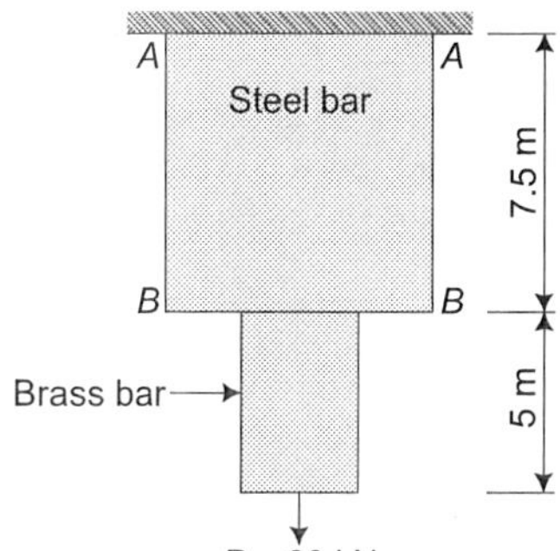

Fig. 7.17

Solution Refer Figure 7.17. The maximum stress in the brass bar occurs at junction B-B due to the weight of brass bar and the load of 60 kN.

Weight of brass bar, $W_b = \rho_b \times g \times A_b \times l_b$

$$= 9000 \times 9.81 \times 35 \times 10^{-4} \times 5$$

$$= 1565.07 \text{ N}$$

The stress at junction $B - B$ is given by,

$$\sigma_{B-B} = \sigma_B = \frac{P + W_b}{A_b} = \frac{60 \times 10^3 + 1545.07}{35 \times 10^{-4}} \times \frac{1}{10^6}$$

$$= 1758 \text{ MN/m}^2$$

In steel bar, the maximum stress occurs at the junction $A - A$ due to the weight of brass bar, steel bar, and the load of 60 kN.

Weight of steel bar, $W_s = \rho_s \times g \times A_s \times l_s$

$$= 7750 \times 9.81 \times 55 \times 10^{-4} \times 7.5$$

$$= 3136.13 \text{ N}$$

The stress at junction $A - A$ is given by,

$$\sigma_{A-A} = \sigma_A = \frac{P + W_b + W_s}{A_s}$$

$$= \left(\frac{60 \times 10^3 + 1545.07 + 3136.13}{55 \times 10^{-4}}\right) \times \frac{1}{10^6}$$

$$= 11.76 \text{ MN/m}^2$$

Example 7.19 A bar 3 m long is divided into three sections *AB*, *BC*, and *CD* as shown in Figure 7.18. The cross sectional area of the bar is 1200 mm^2 and is subjected to axial forces. Determine the total elongation of the bar if the Young's modulus for the material of the bar is 200 kN/mm^2.

Solution Refer Figure 7.18(a). The force of 80 kN acting at *B* is split into three forces of 50 kN, 20 kN, and 10 kN. The part *AB* of the bar is subjected to a tensile load of 50 kN, part *BC* is subjected to a compressive load of 20 kN, and part *CD* is subjected to a compressive load of 10 kN as shown by free body diagram in Figure 7.18(b).

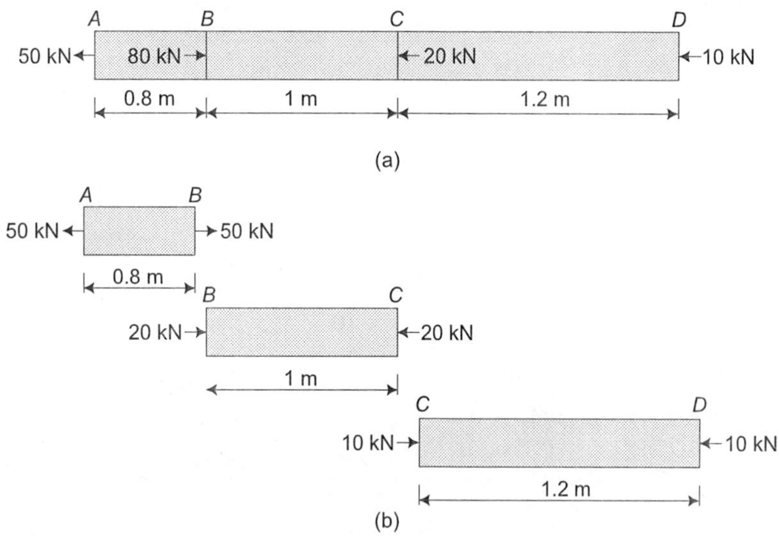

Fig. 7.18

Part AB The section of the bar is subjected to a tensile load of 50 kN.

The (increase) elongation of

$$AB = \frac{P_1 l_1}{AE} = \frac{50 \times 10^3 \times 800}{1200 \times 200 \times 10^3}$$

$$= 0.1667 \text{ mm}$$

Part BC The section of the bar is subjected to a compressive load of 20 kN.

The (decrease) contraction of

$$BC = \frac{P_2 l_2}{AE} = \frac{20 \times 10^3 \times 1000}{1200 \times 200 \times 10^3}$$
$$= 0.0833 \text{ mm}$$

Part CD The section of the bar is subjected to a compressive load of 10 kN.

The (decrease) contraction of

$$CD = \frac{P_3 l_3}{AE} = \frac{10 \times 10^3 \times 1000}{1200 \times 200 \times 10^3}$$
$$= 0.0417 \text{ mm}$$

Therefore, change in length = 0.1667 – 0.0833 – 0.0417 = 0.0417 mm

Example 7.20 A member *ABCD* is subjected to point loads P_1, P_2, P_3, and P_4 as shown in Figure 7.19. Calculate the force P_3 necessary for equilibrium if P_1 = 125 kN, P_2 = 225 kN, and P_4 = 165 kN. Determine the total elongation of the member if the Young's modulus for the material of the bar is 2×10^5 N/mm^2.

Solution Refer Figure 7.19. Resolving the forces on the member along its axis, we have,

$$P_1 + P_3 = P_2 + P_4$$
$$125 + P_3 = 225 + 165$$
$$\therefore \quad P_3 = 265 \text{ kN}$$

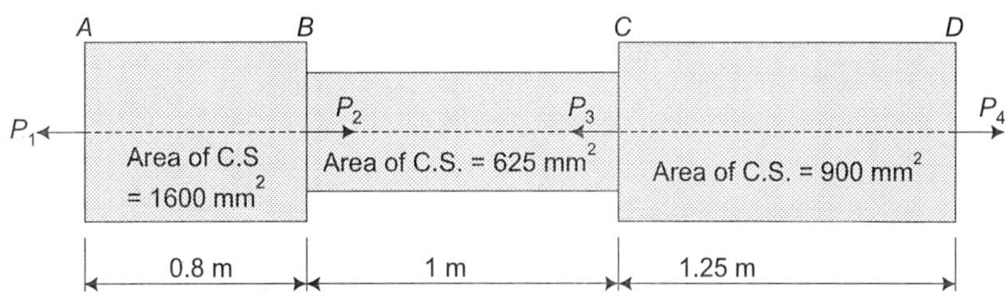

Fig. 7.19

Part AB Section *AB* is subjected to a tensile force of 125 kN.

The (increase) elongation of

$$AB = \frac{Pl}{AE} = \frac{125 \times 10^3 \times 800}{1600 \times 2 \times 10^5}$$
$$= 0.3125 \text{ mm}$$

Part BC Section *BC* is subjected to a force = (125 – 225) = – 100 kN (compressive)

The (decrease) contraction of

$$BC = -\frac{100 \times 10^3 \times 1000}{625 \times 2 \times 10^5}$$
$$= -0.80 \text{ mm}$$

Part CD Section CD is subjected to a tensile force of 165 kN.

The (increase) elongation of

$$CD = \frac{165 \times 10^3 \times 1.25}{900 \times 2 \times 10^5}$$
$$= 1.1458 \text{ mm}$$

Therefore, change in length = 0.3125 – 0.80 + 1.1458 = 0.6583 mm

Example 7.21 A cylindrical bar of 100 mm in diameter is rigidly supported vertically between two rigid plates and loaded by an axial pull equal to 400 kN as shown in Figure 7.20. Find the reactions at the end A and C and the elongation of the portion AB. Take the Young's modulus for the material of the bar as 2×10^5 N/mm^2.

Solution Refer Figures 7.20(a) and 7.20(b).

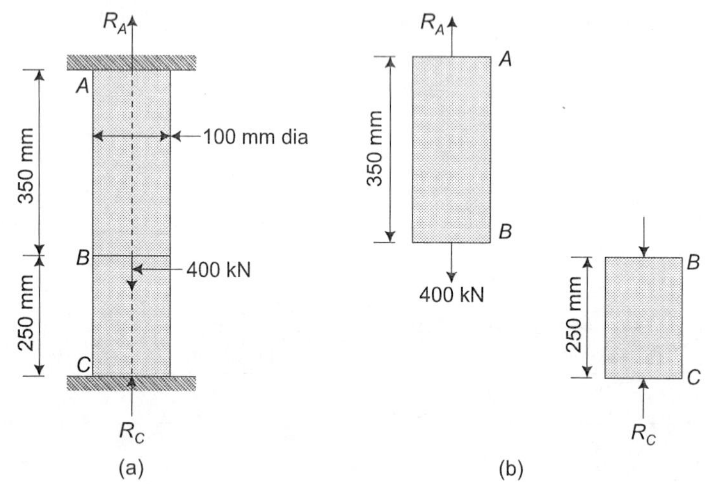

Fig. 7.20

Cross sectional area of the bar = 30 × 30 = 900 mm^2.

Part AB will be subjected to tensile force and part BC will be subjected to compressive force.

Part AB Section AB is subjected to a tensile force of 400 kN.

Cross sectional area = $(\pi/4) \times (100^2) = 7853.98$ mm^2

The (increase) elongation of

$$AB = \frac{Pl}{AE} = \frac{R_A \times 350}{7853.98 \times 2 \times 10^5} \text{ mm} \qquad \text{(i)}$$

Part BC Section *BC* is subjected to a compressive force.

The (decrease) contraction of

$$BC = \frac{R_C \times 250}{7853.98 \times 2 \times 10^5} \text{ mm} \qquad \text{(ii)}$$

Since the bar is held between rigid supports. Therefore, equating (i) and (ii), we have,

$$\frac{R_A \times 350}{7853.98 \times 2 \times 10^5} = \frac{R_C \times 250}{7853.98 \times 2 \times 10^5}$$

$$\therefore \quad R_C = 1.4\ R_A$$

But while considering equilibrium of bar *ABC* as a whole, we have,

$$P = R_A + R_C$$

$$\therefore \quad 400 \times 10^3 = R_A + 1.4\ R_A$$

$$\therefore \quad R_A = \left(\frac{400 \times 10^3}{2.4}\right) \times 10^{-3} = 166.67 \text{ kN}$$

$$R_C = 1.4 \times 166.67 = 233.34 \text{ kN}$$

$$\text{Elongation of part } AB = \frac{166.67 \times 10^3 \times 350}{7853.98 \times 2 \times 10^5}$$

$$= 0.037 \text{ mm}$$

Example 7.22 Find the reaction of the vertical wall on the bar shown in Figure 7.21. Also find the total extension of the bar if the Young's modulus for the material of the bar is 2×10^5 N/mm^2.

Solution Refer Figures 7.21(a) and 7.21(b).

The required reaction,

$$R = 150 + 250 = 400 \text{ kN.}$$

Part *CD* will be subjected to tensile force of 250 kN and part *CA* will be subjected to tensile force of 400 kN.

Part CD Section *CD* is subjected to a tensile load of 250 kN.

The (increase) elongation of

$$AB = \frac{Pl}{AE} = \frac{250 \times 10^3 \times 4000}{(\pi/4) \times 150^2 \times 2 \times 10^5}$$

$$= 0.2829 \text{ mm}$$

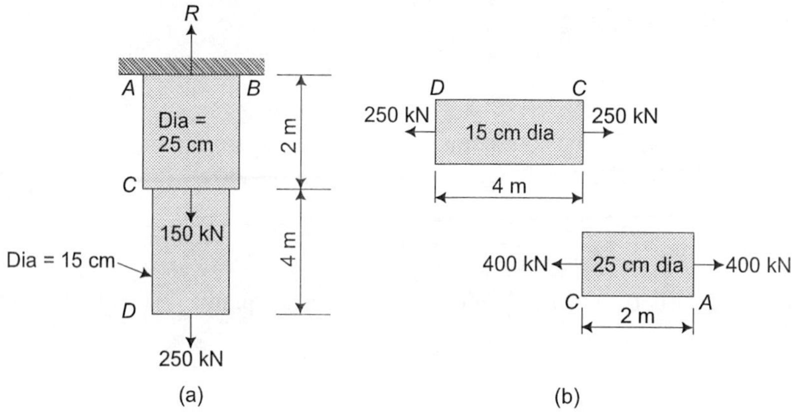

Fig. 7.21

Part AC Section AC is subjected to a tensile load of 400 kN.

The (increase) elongation of

$$AB = \frac{Pl}{AE} = \frac{400 \times 10^3 \times 2000}{(\pi/4) \times 250^2 \times 2 \times 10^5}$$
$$= 0.0815 \text{ mm}$$

Therefore, change in length of bar

$$ACD = 0.2829 + 0.0815 = 0.3644 \text{ mm}$$

Example 7.23 A steel tie rod of 40 mm in diameter and 4 m long is subjected to an axial pull of 40 kN. To what length the bar should be bored centrally so that the total extension will increase by 30% under the same pull, the size of the bore is 20 mm in diameter. The Young's modulus for the material of the bar is 200 kN/mm^2.

Solution Refer Figure 7.22.

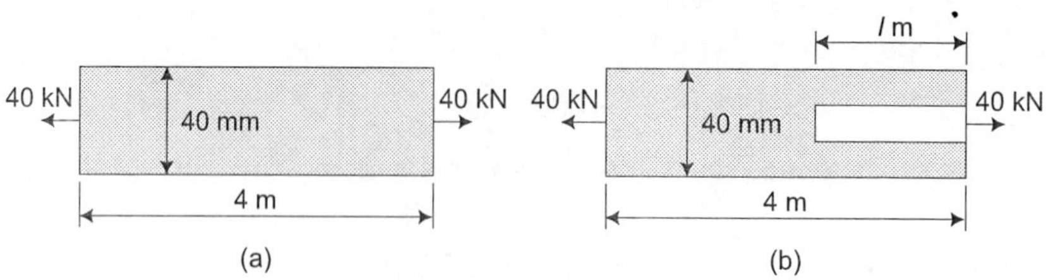

Fig. 7.22

When the rod is solid:

$$A = \frac{\pi}{4} \times (40^2) = 400\,\pi \text{ mm}^2$$

$$\text{Stress} = \frac{40 \times 10^3}{400\,\pi} = \frac{100}{\pi}\ \text{N/mm}^2$$

The elongation is given by,

$$\delta l = \frac{\sigma l}{E} = \frac{100 \times 4000}{\pi \times 200 \times 10^3} = \frac{2}{\pi}\ \text{mm}$$

The extension after the bore is made

$$= 1.3 \times \frac{2}{\pi} = \frac{2.6}{\pi}\ \text{mm} \qquad \text{(i)}$$

When the rod is bored to a length of l m:

$$A = \frac{\pi}{4} \times (40^2 - 20^2) = 300\,\pi\ \text{mm}$$

$$\text{Stress} = \frac{40 \times 10^3}{300\,\pi} = \frac{400}{3\pi}\ \text{N/mm}^2$$

$$\therefore \qquad \text{Extension of rod} = \frac{100 \times (4 - l) \times 10^3}{\pi \times 200 \times 10^3} + \frac{400 \times l \times 10^3}{3\pi \times 200 \times 10^3}$$

$$= \frac{(4 - l)}{2\pi} + \frac{4l}{6\pi}\ \text{mm} \qquad \text{(ii)}$$

From equations (i) and (ii), we have,

$$\frac{(4 - l)}{2\pi} + \frac{4l}{6\pi} = \frac{2.6}{\pi}$$

$$12 - 3l + 4l = 15.6$$

$$\therefore \qquad l = 15.6 - 12 = 3.6\ \text{m}$$

Example 7.24 A bar *ABCD* rigidly fixed at the ends *A* and *B* is subjected to two equal and opposite forces $P = 40$ kN at *C* and *D* as shown in Figure 7.23. Determine axial stresses in each section of the bar.

Solution Refer Figure 7.23. The free body diagrams are shown in Figure 7.23(b).

Part AC Section *AC* is subjected to a tensile force P_1.

The (increase) elongation of *AB*

$$= \frac{Pl}{AE} = \frac{P_1 \times 500}{500 \times E} = \frac{P_1}{E}\ \text{mm} \qquad \text{(i)}$$

Part CD Section *CD* is subjected to a compressive force $(P - P_1)$.

The (decrease) contraction of *CD*

$$= \frac{(P - P_1) \times 800}{800 \times E} = \frac{(P - P_1)}{E}\ \text{mm} \qquad \text{(ii)}$$

Part DB Section *DB* is subjected to a tensile force P_1.

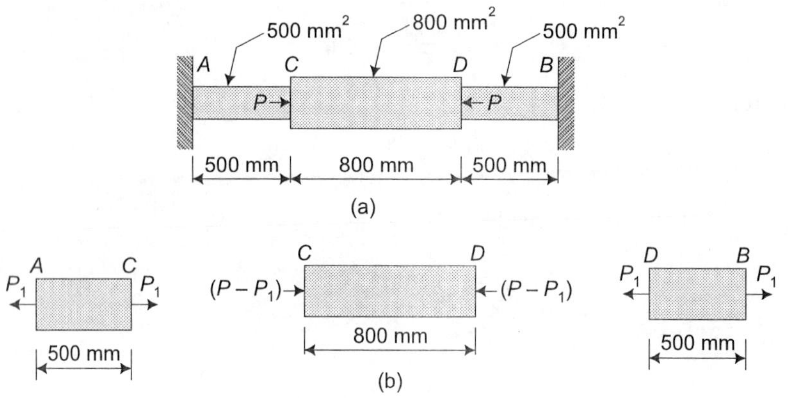

Fig. 7.23

The (increase) elongation of *AB*

$$= \frac{Pl}{AE} = \frac{P_1 \times 500}{500 \times E} = \frac{P_1}{E} \text{ mm} \qquad \text{(iii)}$$

Since the bar is held between rigid supports. Therefore, equating (i), (ii) and (ii), we have,

$$\frac{P_1}{E} - \frac{(P - P_1)}{E} + \frac{P_1}{E} = 0$$

$$P_1 = \frac{P}{3}$$

$$\therefore \qquad P_1 = \frac{30}{3} = 10 \text{ kN}$$

$$P - P_1 = 30 - 10 = 20 \text{ kN}$$

Therefore, tensile stress in section *AC* and *DB*

$$= \frac{30 \times 10^3}{500} = 60 \text{ N/mm}^2$$

Compressive stress in section *CD*

$$= \frac{20 \times 10^3}{800} = 25 \text{ N/mm}^2$$

Example 7.25 A bar *ABCD* rigidly fixed at the ends *A* and *D* is subjected to loads as shown in Figure 7.24. Determine axial forces in each section of the bar and also determine the displacements of the ends *B* and *C* if Young's modulus for the material of the bar is 200 kN/mm^2.

Solution Refer Figure 7.24. Free body diagrams are shown in Figure 7.24(b). R_A and R_D are the reactions at *A* and *D* respectively.

$$R_A + R_D = 300 + 400 = 700 \qquad \text{(i)}$$

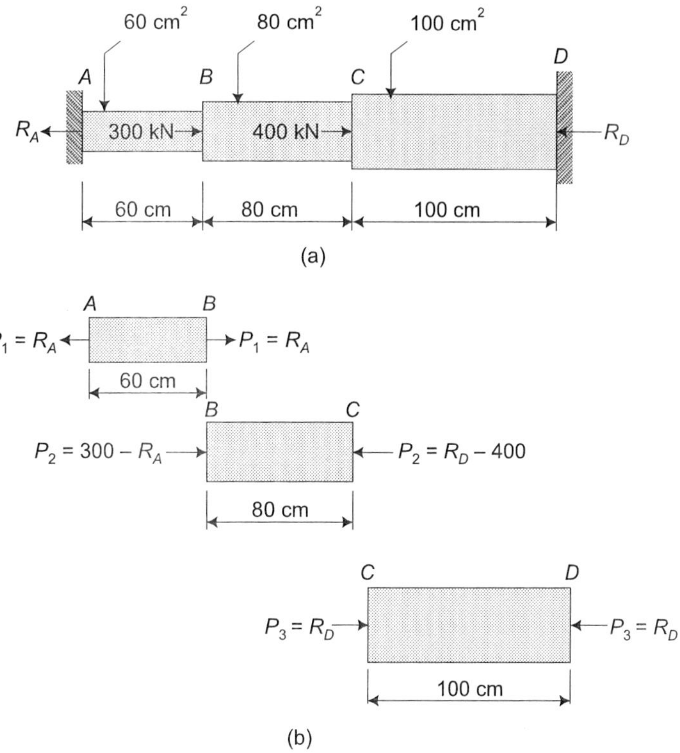

Fig. 7.24

Extension of section AB = Compression of BC + Compression of CD

$$\therefore \qquad \frac{P_1 l_1}{A_1 E} = \frac{P_2 l_2}{A_2 E} + \frac{P_3 l_3}{A_3 E}$$

Substituting the values,

$$\frac{R_A \times 600}{60 \times 10^4 \times E} = \frac{(300 - R_A) \times 800}{80 \times 10^4 \times E} + \frac{R_D \times 1000}{100 \times 10^4 \times E}$$

$$R_A = (300 - R_A) + R_D$$

$$2R_A - R_D = 300 \qquad \text{(ii)}$$

From equations (i) and (ii), we have,

$$R_A = (1000/3) = 333.33 \text{ kN}$$

$$R_D = 700 - 333.33 = 366.67 \text{ kN}$$

$$P_1 = R_A = 333.33 \text{ kN (Tensile)}$$

$$P_2 = 300 - R_A = 300 - 333.33$$

$$= -\ 33.33 \text{ kN (Compressive)},$$

$$P_3 = R_D = 366.67 \text{ kN (Compressive)}$$

Displacement of $B =$ $$\delta l_1 = \frac{P_1\, l_1}{A_1 E} = \frac{333.33 \times 10^3 \times 600}{60 \times 10^4 \times 200 \times 10^3}$$

$$= 1.667 \times 10^{-3} \text{ mm}$$

$$\delta l_2 = \frac{P_2\, l_2}{A_2 E} = \frac{33.33 \times 10^3 \times 800}{80 \times 10^4 \times 200 \times 10^3}$$

$$= 1.667 \times 10^{-4} \text{ mm}$$

Displacement of C,

$$\delta l_1 + \delta l_2 = 1.667 \times 10^{-3} + 1.667 \times 10^{-4}$$

$$= 1.834 \times 10^{-3} \text{ mm}$$

Example 7.26 Determine the reaction produced by the lower support on the bar as shown in Figure 7.25. Take Young's modulus for the material of the bar is 2×10^5 N/mm^2. Also determine the stresses in the bar.

Solution Refer Figure 7.25. Free body diagrams are shown in Figure 7.25(b). R_1 and R_2 are the reactions at upper and lower fixed plates respectively.

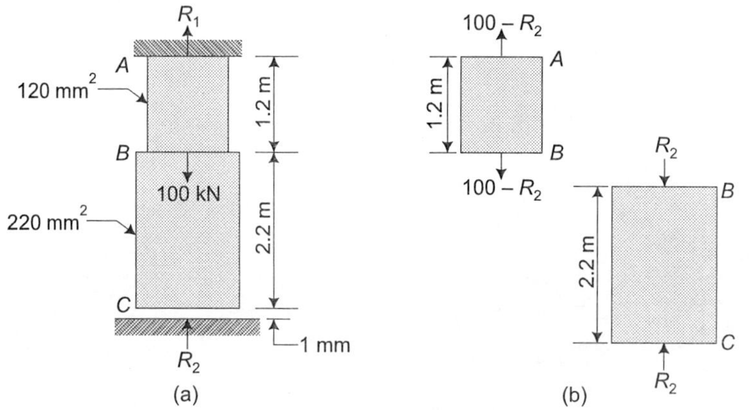

Fig. 7.25

Extension of AB,

$$\delta l_1 = \frac{Pl}{AE} = \frac{(100000 - R_2) \times 1200}{120 \times 2 \times 10^5} \text{ mm}$$

Contraction of BC,

$$\delta l_2 = \frac{Pl}{AE} = \frac{R_2 \times 2200}{220 \times 2 \times 10^5} \text{ mm}$$

According to the given condition,

$$\delta l_1 + \delta l_2 = 1$$

$$\therefore \quad \frac{(100000 - R_2) \times 1200}{120 \times 2 \times 10^5} - \frac{R_2 \times 2200}{220 \times 2 \times 10^5} = 1$$

[–ve sign, $\because$ δl_2 decreases in length]

$$\therefore \quad \frac{(100000 - R_2)}{2 \times 10^4} - \frac{R_2}{2 \times 10^4} = 1$$

$$100000 - R_2 - R_2 = 2 \times 10^4$$

$$\therefore \quad R_2 = (80000/2) = 40000 \text{ N} = 40 \text{ kN}$$

$$R_1 = (100000 - 40000) = 60000 \text{ N} = 60 \text{ kN}$$

$$\text{Stress in } AB = (60000/120) = 500 \text{ N/mm}^2$$

$$\text{Stress in } BC = 40000/220 = 181.82 \text{ N/mm}^2$$

Example 7.27 A rod tapers uniformly from 40 mm to 20 mm diameter in a length of 400 mm. If the rod be subjected to an axial load of 80 kN, determine the extension of the rod. Take Young's modulus for the material of the bar is 2×10^5 N/mm^2.

Solution The extension of the bar is given by equation (7.17) as,

$$\delta l = \frac{4Pl}{\pi E d_1 \, d_2}$$

Substituting the values of given parameters, we have,

$$\delta l = \frac{4 \times 80 \times 10^3 \times 400}{\pi \times 2 \times 10^5 \times 40 \times 20}$$

$$= 0.2546 \text{ mm}$$

Example 7.28 A straight bar of steel rectangular in section is 5000 mm long and is of uniform thickness 20 mm. The width of the rod uniformly varies from 120 mm to 50 mm. If the rod is put to an axial load of 40 kN, determine the extension of the rod. Take Young's modulus for the material of the bar is 2×10^5 N/mm^2.

Solution The extension of the bar is given by equation (7.18) as,

$$\delta l = \frac{P \times l}{Et\,(b_1 - b_2)} \ln\left(\frac{b_1}{b_2}\right)$$

Substituting the values of given parameters, we have,

$$\delta l = \frac{40 \times 10^3 \times 5000}{2 \times 10^5 \times 20 \times (120 - 50)} \ln\left(\frac{120}{50}\right)$$

$$= 0.625 \text{ mm}$$

7.12 EXTENSION UNDER OWN WEIGHT OF BAR

A bar of weight W, cross sectional area A, and length l is suspended vertically with its upper end rigidly fixed as shown in Figure 7.26. The weight density and the Young's modulus of the bar material is w and E respectively.

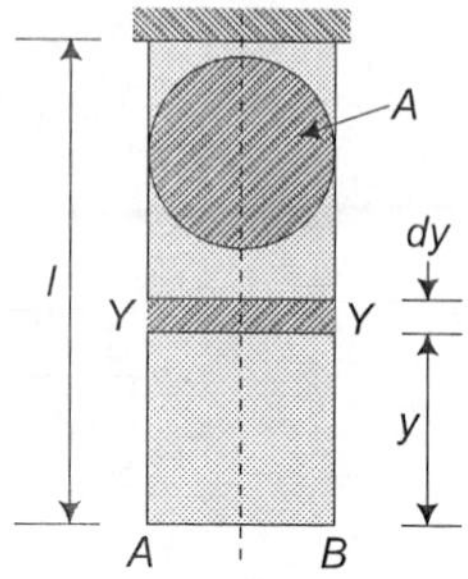

Fig. 7.26

Consider a section $Y - Y$ at a distance y from the lower end.

Weight of the lower portion $AB\ YY = A \times y \times w$

Stress at the section YY,

$$\sigma = \frac{Ayw}{A} = yw$$

Strain at the section YY,

$$\varepsilon = \frac{yw}{E}$$

Change in length of elementary strip of length dy,

$$\delta\, dy = \frac{wy\, dy}{E}$$

Total change in length,

$$\delta l = \int_0^l \frac{wy\, dy}{E} = \frac{wl^2}{2E}$$

or

$$\delta l = \frac{wl^2}{2E} \times \frac{A}{A}$$

$$= \frac{(wlA)l}{2AE} = \frac{Wl}{2AE} \quad (7.19)$$

7.13 BAR OF UNIFORM STRENGTH

Consider a bar of varying cross section such that the same stress σ is developed when an axial load P is applied at its lower end as shown in Figure 7.27. Let W be the weight, w is the weight density, and H is the height of the bar. A_1 is the cross sectional area at the bottom edge and A_2 is the cross sectional area at the top edge.

Consider an elementary strip of length dy at a distance of y from the bottom edge and σ is the stress developed in the bar. $A + dA$ is the area of upper edge of elementary strip and A is of lower edge.

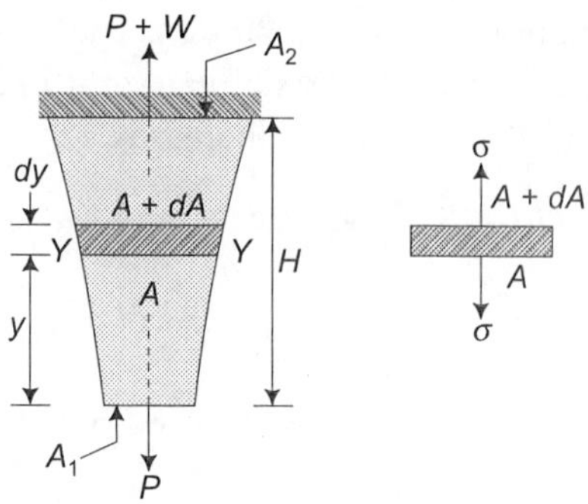

Fig. 7.27

Weight of elementary strip = $A \times dy \times w$

For equilibrium,

$$(\sigma A + A \times dy \times w) = \sigma \times (A + dA)$$

or

$$A \times dy \times w = \sigma \times dA$$

$\therefore$

$$\frac{dA}{A} = \frac{w}{\sigma} \cdot dy \qquad \text{(i)}$$

Integrating equation (i) on both sides, we have,

$$\int_{A_1}^{A_2} \frac{dA}{A} = \int_0^H \frac{w}{\sigma} \cdot dy$$

$$\ln \frac{A_2}{A_1} = \frac{w}{\sigma} \times H$$

or

$$A_2 = A_1\, e^{wH/\sigma}$$

So, at any distance y from the lower edge,

$$A = A_1\, e^{wH/\sigma} \qquad (7.20)$$

Example 7.29 A bar of 40 mm diameter, 2 m long has hole of 20 mm diameter up to a length of 1 m as shown in Figure 7.28. If the weight density of the bar is 0.075×10^{-3} N/mm^3 and Young's modulus for the material of the bar is 2×10^5 N/mm^2, determine the extension of the bar due to its own weight.

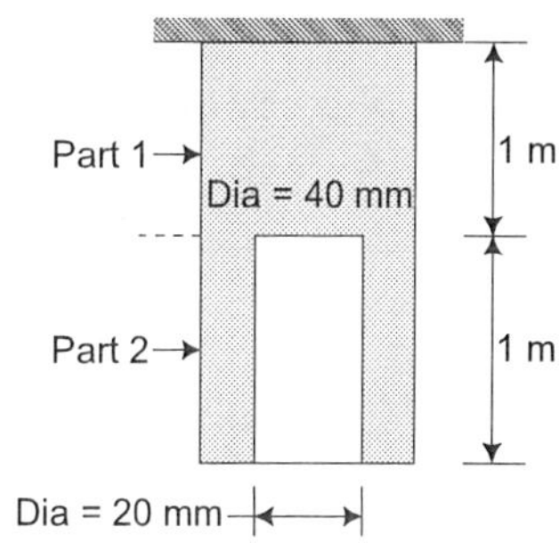

Fig. 7.28

Solution Refer Figure 7.28.

Change in length of the bar due to lower portion is given by,

$$\delta l_2 = \frac{wl^2}{2E} = \frac{0.075 \times 10^{-3} \times (1000^2)}{2 \times 2 \times 10^5}$$

$$= 0.1875 \times 10^{-3} \text{ mm} \qquad \text{(i)}$$

Weight of the lower part

$$= (\pi/4)\,(40^2 - 20^2) \times 1000 \times 0.075 \times 10^{-3}$$

$$= 70.686 \text{ N}$$

Area of cross section of upper part

$$= (\pi/4)\,(40^2) = 1256.64 \text{ mm}^2$$

Change in length of the upper portion due to weight of the lower part is given by,

$$\delta l'_1 = \frac{Wl}{2AE} = \frac{70.686 \times 1000}{2 \times 1256.64 \times 2 \times 10^5}$$

$$= 0.1406 \times 10^{-3} \text{ mm} \qquad \text{(ii)}$$

Change in length of the upper part of the bar due to its own weight is given by,

$$\delta l_1 = \frac{wl^2}{2E} = \frac{0.075 \times 10^{-3} \times (1000^2)}{2 \times 2 \times 10^5}$$

$$= 0.1875 \times 10^{-3} \text{ mm} \qquad \text{(i)}$$

Therefore, total change in the length of the bar,

$$\delta l = \delta l_2 + \delta l'_1 + \delta l_1$$

$$= (0.1875 + 0.1406 + 0.1875) \times 10^{-3}$$

$$= 0.5156 \times 10^{-3} \text{ mm}$$

Example 7.30 A vertical circular bar 4 m long is subjected to a uniform stress of 0.5 N/mm^2 throughout its length. The diameter of the bottom edge of the bar is 120 mm. The weight density of the bar is 0.075×10^{-3} N/mm^3. Determine its diameters at the middle section and at the top section if the bar is rigidly fixed in the ceiling.

Solution Area of cross section at bottom

$$= (\pi/4)\,(120^2) = 11309.73 \text{ mm}^2$$

Area of cross section at the top,

$$A_2 = A_1\, e^{\frac{wH}{\sigma}} = 11309.79 \times e^{\frac{0.075 \times 10^{-3} \times 4000}{0.5}}$$

$$= 20607.67 \text{ mm}^2$$

Therefore, diameter at the top,

$$d_2 = \sqrt{\frac{4 \times 20607.67}{\pi}} = 161.98 \text{ mm}$$

Area of cross section at the middle,

$$A = A_1\, e^{\frac{wH}{\sigma}} = 11309.79 \times e^{\frac{0.075 \times 10^{-3} \times 2000}{0.5}}$$

$$= 15266.54 \text{ mm}^2$$

Therefore, diameter at the middle,

$$d_{mid} = \sqrt{\frac{4 \times 15266.54}{\pi}}$$

$$= 139.42 \text{ mm}$$

7.14 COMPOSITE SECTIONS WITH INITIAL EQUAL LENGTH

A bar made up of two or more bars of different materials rigidly fixed with each other and all the bars strain together under the applied loads is called composite bar. For example, (i) A column or a slab consists of concrete reinforced with steel rods, and (ii) Bimetallic strips used for temperature control in thermal devices.

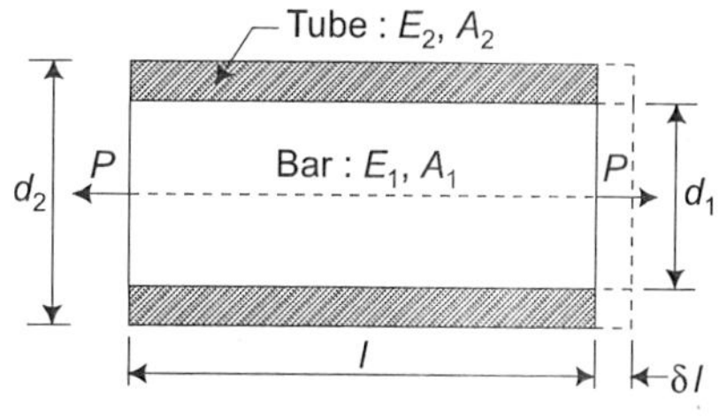

Fig. 7.29

Let us take the case of a solid bar enclosed in the hollow tube and are perfectly fixed at the interface as shown in Figure 7.29. The composite bar is of length l and is subjected to an axial tensile force P. Let suffix 1 be for the bar and 2 for the tube. A_1 and A_2 are the area of the bar and tube respectively. E_1 and E_2 are the Young's modulus of the bar and the tube respectively. P_1 and P_2 are the loads shared by inner bar and outer tube respectively. σ_1 and σ_2 are the unit stresses on the inner bar and outer tube respectively.

The total load on the composite bar is equal to the sum of the load carried by the inner bar and outer tube.

$$\therefore \quad P = P_1 + P_2$$

Stress in bar, $$\sigma_1 = \frac{P_1}{A_1} = \frac{4P_1}{\pi d_1^2}$$

and stress in tube, $$\sigma_2 = \frac{P_2}{A_2} = \frac{4P_2}{\pi(d_2^2 - d_1^2)}$$

Strain in bar, $$e_1 = \frac{\sigma_1}{E_1}$$

Strain in tube, $$e_2 = \frac{\sigma_2}{E_2}$$

Change in length of bar,

$$\delta l_1 = e_1 l = \sigma_1 l/E_1$$

Change in length of tube,

$$\delta l_2 = e_2 l = \sigma_2 l/E_2$$

But, $$\delta l_1 = \delta l_2 = \delta l$$

$$\therefore \quad \frac{\sigma_1 l}{E_1} = \frac{\sigma_2 l}{E_2}$$

$$\frac{\sigma_1}{E_1} = \frac{\sigma_2}{E_2}$$

$$\sigma_1 = \frac{E_1}{E_2}\sigma_2 = m\sigma_2 \quad (7.19)$$

where m is the modular ratio and is given by, $\frac{E_1}{E_2}$.

Force in bar $$P_1 = \sigma_1 A_1$$

and force in tube $\quad P_2 = \sigma_2 A_2$

Therefore, total force,

$$P = P_1 + P_2 = \sigma_1 A_1 + \sigma_2 A_2 \tag{7.21}$$

7.15 COMPOSITE SECTIONS WITH UNEQUAL LENGTHS

A bar of material 1 having diameter d and length l is placed coaxially in a tube of material 2 of outside diameter d_2, inside diameter d_1, and of length l_1 as shown in Figure 7.30.

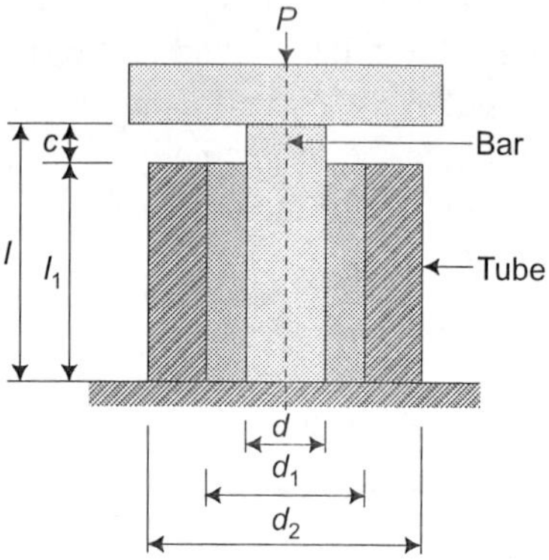

Fig. 7.30

The length l_1 is slightly less than l, i.e., $l = l_1 + c$, where c is the clearance between the bar and the tube. E_1 and E_2 are the Young's modulus of elasticity of the bar and the tube respectively. P_1 and P_2 are the loads shared by the inner bar and the outer tubes respectively. σ_1 and σ_2 are the stresses developed in the bar and the tube respectively.

Change in length of bar,

$$\delta l = e_1 l = \frac{\sigma_1 l}{E_1}$$

Change in length of tube,

$$\delta l_1 = e_2 l_1 = \frac{\sigma_2 l_1}{E_2}$$

But, $\quad \delta l = \delta l_1 + c$

$$\therefore \quad \frac{\sigma_1 l}{E_1} = \frac{\sigma_2 l_1}{E_2} + c$$

$$\sigma_1 = \sigma_2 \times \frac{E_1}{E_2} \times \frac{l_1}{l} + c \times \frac{E_1}{l} \tag{i}$$

Total load, $\quad P = P_1 + P_2 = \sigma_1 A_1 + \sigma_2 A_2$

$$= \left(\sigma_2 \times \frac{E_1}{E_2} \times \frac{l_1}{l} \times A_1 + c \times \frac{E_1}{l} \times A_1\right) + \sigma_2 A_2$$

$$\therefore \quad P = \left(\sigma_2 \times \frac{E_1}{E_2} \times \frac{l_1}{l} \times A_1 + c \times \frac{E_1}{l} \times A_1\right) + \sigma_2 A_2 \tag{ii}$$

From equation (ii) stress σ_2 in tube can be calculated and then from equation (i) stress σ_1 in the bar can be calculated.

Example 7.31 A steel bar of 60 mm diameter is completely enclosed in a brass tube of 90 mm outside diameter. The length of the composite bar is 0.5 m. Determine the magnitude of the compressive force if the composite bar is subjected to a compressive force such that the stress developed in brass tube is 60 N/mm^2. Also determine the change in the length of the composite bar. Take modulus of elasticity of steel, E_s = 208 kN/mm^2, and for brass, E_b = 110 kN/mm^2.

Solution Area of cross section of steel bar = $(\pi/4) \times (30^2) = 706.86$ mm^2

Area of cross section of brass tube = $(\pi/4) \times (45^2 - 30^2) = 883.57$ mm^2

Let the stresses in steel bar and brass tube be σ_s and σ_b respectively.

Strain in steel bar = Strain in brass tube

$$\frac{\sigma_s}{E_s} = \frac{\sigma_b}{E_b}$$

$$\sigma_s = \frac{E_s}{E_b} \times \sigma_b$$

$$\therefore \quad \sigma_s = \frac{208}{110} \times 60 = 113.45 \text{ N/mm}^2$$

Total compressive force,

$$P = 60 \times 883.47 + 113.45 \times 706.86$$

$$= 133207.47 \text{ N} = 133.21 \text{ kN}$$

Change in the length of composite bar,

$$\delta l = \delta l_s = \delta l_b = \frac{\sigma_b}{E_b} \times l$$

$$= \frac{60}{110 \times 10^3} \times 500$$

$$= 0.273 \text{ mm}$$

Example 7.32 A steel bar of 2 cm diameter and 1.5 m long is placed coaxially in an aluminium tube of outer diameter 6 cm and inner diameter 3 cm, and 1500.5 mm long. A compressive force of 100 kN is applied as shown in the Figure 7.31. Find the stresses developed in steel bar and aluminium tube. Take modulus of elasticity of steel, E_s = 210 kN/mm^2, and for aluminium, E_a = 70 kN/mm^2.

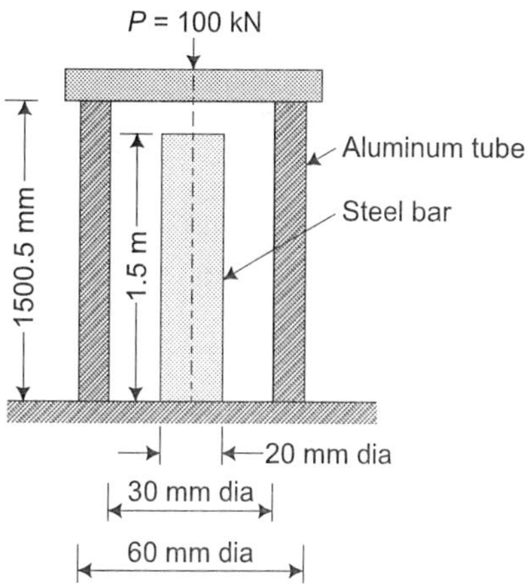

Fig. 7.31

Solution Refer Figure 7.31.

Area of cross section of steel bar,

$$A_s = (\pi/4) \times (20^2)$$

$$= 314.16 \text{ mm}^2$$

Area of cross section of aluminium tube,

$$A_c = (\pi/4) \times (60^2 - 30^2)$$

$$= 2120.58 \text{ mm}^2$$

Let δl_a and δl_s be the change in lengths of aluminium tube and steel bar respectively. Each steel bar and aluminium tube deform by the same amount,

$$\delta l_a = \delta l_s + 0.5$$

$$\therefore \quad \frac{\sigma_a}{E_a} \times l_a = \frac{\sigma_s}{E_s} \times l_s + 0.5$$

$$\sigma_a = \sigma_s \times \frac{E_a}{E_s} \times \frac{l_s}{l_a} + 0.5 \times \frac{E_a}{l_a}$$

Substituting the given values,

$$\sigma_a = \sigma_s \times \frac{70 \times 10^3}{210 \times 10^3} \times \frac{1500}{1500.5} + \frac{0.5 \times 70 \times 10^3}{1500.5}$$

$$\therefore \quad \sigma_a = 0.3332 \times \sigma_s + 12.325$$

The load shared by aluminium tube and steel bar,

$$P = P_a + P_s = \sigma_a \times A_a + \sigma_s \times A_s$$

$$\therefore \quad 100 \times 10^3 = \sigma_a \times 2120.58 + \sigma_s \times 314.16$$

$$100 \times 10^3 = (0.3332 \times \sigma_s + 23.325) \times 2120.58 + \sigma_s \times 314.16$$

$$1020.74 \times \sigma_s = 50537.47$$

$$\therefore \quad \sigma_s = \frac{50537.47}{1020.74} = 49.51 \text{ N/mm}^2$$

$$\sigma_a = 0.3332 \times 49.51 + 23.325 = 39.822 \text{ N/mm}^2$$

Example 7.33 Two copper rods and one steel rod lie in a vertical plane and together support a load of 100 kN as shown in Figure 7.32. Each rod is 50 mm in diameter. Length of steel rod is 6 m and length of each copper rod is 4 m. Find the stress induced in each rod and the load carried by them. Take modulus of elasticity of steel, E_s = 210 kN/mm² and for copper, E_c = 105 kN/mm².

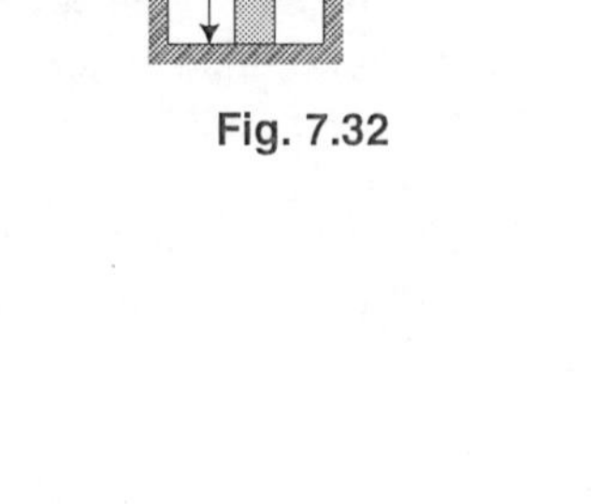

Fig. 7.32

Solution Refer Figure 7.32.

Area of cross section of steel rod,

$$A_s = (\pi/4) \times (50^2) = 1963.49 \text{ mm}^2$$

Area of cross section of copper rod,

$$A_c = (\pi/4) \times (50^2) = 1963.49 \text{ mm}^2$$

Strain in steel = Strain in copper,

$$\frac{\sigma_s}{E_s} \times l_s = \frac{\sigma_c}{E_c} \times l_c$$

$$\sigma_s = \sigma_c \times \frac{E_s}{E_c} \times \frac{l_c}{l_s}$$

$$\therefore \quad \sigma_s = \sigma_c \times \frac{210 \times 10^3}{105 \times 10^3} \times \frac{4}{6}$$

$$= 1.33 \times \sigma_c$$

The load shared by steel bar and copper bars,

$$P = 2 \times P_c + P_s = 2 \times \sigma_c \times A_c + \sigma_s \times A_s$$

$$\therefore \quad 100 \times 10^3 = 2 \times \sigma_c \times 1963.49 + (1.33 \times \sigma_c) \times 1963.49$$

$$\therefore \quad \sigma_c = \frac{100 \times 10^3}{6538.42} = 15.29 \text{ N/mm}^2$$

$$\sigma_s = 1.33 \times 15.29 = 20.33 \text{ N/mm}^2$$

The load carried by steel bar,

$$P_s = \sigma_s \times A_s = 20.33 \times 1963.49$$

$$= 39917.75 \text{ N/mm}^2 = 39.92 \text{ kN}$$

The load carried by copper bars,

$$P_c = 100 - 39.92 = 60.08 \text{ kN}$$

Example 7.34 A steel bar of 6 cm diameter is enclosed centrally in a hollow copper tube of external diameter 10 cm and internal diameter of 8 cm. The composite bar is then subjected to an axial pull of 90 kN. If the length of each bar is equal to 30 cm, determine the stresses in the bar and tube. Take modulus of elasticity of steel, $E_s = 210$ kN/mm^2, and for copper $E_c = 110$ kN/mm^2.

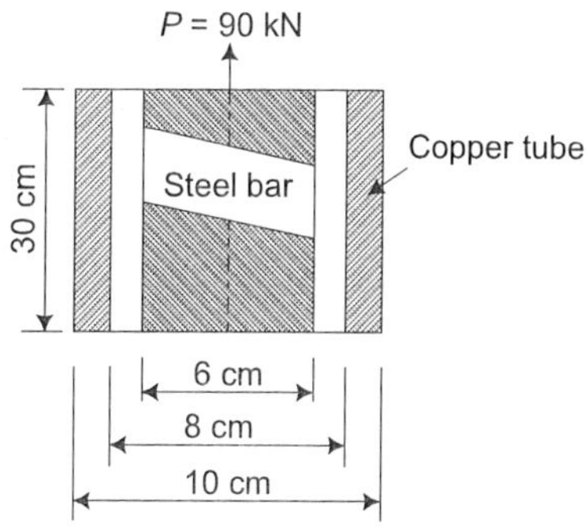

Fig. 7.33

Solution Refer Figure 7.33.

Area of cross section of steel rod,

$$A_s = (\pi/4) \times (60^2)$$

$$= 2827.43 \text{ mm}^2$$

Area of cross section of copper rod,

$$A_c = (\pi/4) \times (100^2 - 80^2)$$

$$= 2827.43 \text{ mm}^2$$

Each rod will deform by the same amount,

$$\therefore \quad \frac{\sigma_s}{E_s} \times l_s = \frac{\sigma_c}{E_c} \times l_c$$

$$\sigma_s = \sigma_c \times \frac{E_s}{E_c}$$

$$\therefore \quad \sigma_s = \sigma_c \times \frac{210 \times 10^3}{110 \times 10^3} = 1.91 \times \sigma_c$$

The load shared by steel bar and copper tube,

$$P = P_c + P_s = \sigma_c \times A_c + \sigma_s \times A_s$$

$$\therefore \quad 90 \times 10^3 = \sigma_c \times 2827.43 + (1.91 \times \sigma_c) \times 2827.43$$

$$\therefore \quad \sigma_c = \frac{90 \times 10^3}{8227.82} = 10.94 \text{ N/mm}^2$$

$$\sigma_s = 1.91 \times 10.94 = 20.89 \text{ N/mm}^2$$

Example 7.35 A reinforced concrete column is 250 mm × 250 mm in section. The column is provided with 8 bars of 15 mm diameter. The column carries a load of 300 kN. Determine the stresses in concrete and the steel bars if modulus of elasticity of steel, $E_s = 210$ kN/mm^2, and for concrete, $E_c = 14$ kN/mm^2.

Solution Area of cross section of steel bars,

$$A_s = 8 \times (\pi/4) \times (15^2) = 1413.72 \text{ mm}^2$$

Actual area of concrete,

$$A_c = (250 \times 250) - 1413.72 = 61086.28 \text{ mm}^2$$

Strain in steel = Strain in concrete

$$\therefore \quad \frac{\sigma_s}{E_s} \times l_s = \frac{\sigma_c}{E_c} \times l_c$$

$$\sigma_s = \sigma_c \times \frac{E_s}{E_c}$$

$$\therefore \quad \sigma_s = \sigma_c \times \frac{210 \times 10^3}{14 \times 10^3} = 15 \times \sigma_c$$

The load shared by steel bar and concrete,

$$P = \sigma_c \times A_c + \sigma_s \times A_s$$

$$\therefore \quad 300 \times 10^3 = 15 \times \sigma_c \times 1413.72 + \sigma_c \times 61086.28$$

$$\therefore \quad \sigma_c = \frac{300 \times 10^3}{82292.08} = 3.64 \text{ N/mm}^2$$

$$\sigma_s = 15 \times 3.64 = 54.6 \text{ N/mm}^2$$

Example 7.36 A load of 300 kN is applied on a short concrete column 250 mm × 250 mm in section. The column is reinforced with 8 bars of 16 mm diameter. Determine the stresses in concrete, and the steel bars if modulus of elasticity of steel is 15 times that of concrete. If the stress in concrete does not exceed 5 N/mm^2, determine the area of steel required so that the column could support a load of 500 kN.

Solution

(i) When the column carries a load of 300 kN

Area of cross section of steel bars,

$$A_s = 8 \times (\pi/4) \times (16^2) = 1608.49 \text{ mm}^2$$

Actual area of concrete,

$$A_c = (250 \times 250) - 1608.49 = 60891.51 \text{ mm}^2$$

$$\text{Strain in steel} = \text{Strain in concrete}$$

$$\therefore \quad \frac{\sigma_s}{E_s} \times l_s = \frac{\sigma_c}{E_c} \times l_c$$

$$\sigma_s = \sigma_c \times \frac{E_s}{E_c}$$

$$\therefore \quad \sigma_s = \sigma_c \times 15$$

The load shared by steel bar and concrete,

$$P = \sigma_c \times A_c + \sigma_s \times A_s$$

$$\therefore \quad 300 \times 10^3 = 15 \times \sigma_c \times 160.49 + \sigma_c \times 60891.51$$

$$\therefore \quad \sigma_c = \frac{300 \times 10^3}{85018.86} = 3.53 \text{ N/mm}^2$$

$$\sigma_s = 15 \times 3.53 = 52.95 \text{ N/mm}^2$$

(ii) When the column carries a load of 500 kN

Let A_s mm^2 be the area of steel bars

So, area of concrete $= (250 \times 250 - A_s) = (62500 - A_s)$ mm^2

$$\text{Stress in concrete} = 5 \text{ N/mm}^2$$

$$\text{Stress in steel} = 15 \times 5 = 75 \text{ N/mm}^2$$

The load shared by steel bar and concrete,

$$P = \sigma_c \times A_c + \sigma_s \times A_s$$

$$\therefore \quad 500 \times 10^3 = 75 \times A_s + 5 \times (62500 - A_s)$$

$$\therefore \quad A_s = \frac{500000 - 312500}{70}$$

$$= 2678.57 \text{ mm}^2$$

Example 7.37 A rigid beam weighing 400 N is held in horizontal position by three wires of the same length. The outer wires are of brass of 1 mm diameter and attached to each one of the beam. The central wire is of steel of 0.5 mm diameter and attached to the middle of the horizontal beam. Determine the stresses induced in each of the wire if E for steel = 2×10^5 N/mm^2 and for brass = 0.8×10^5 N/mm^2.

Solution Refer Figure 7.34.

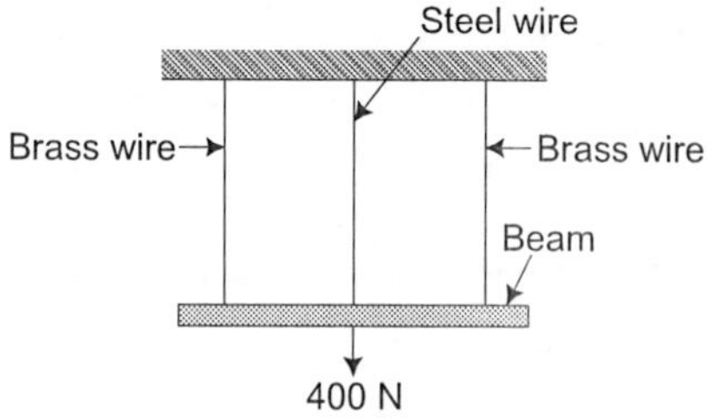

Fig. 7.34

Area of cross section of steel wire,

$$A_s = (\pi/4) \times (0.5^2)$$
$$= 0.196 \text{ mm}^2$$

Area of cross section of brass wire,

$$A_b = (\pi/4) \times (1^2)$$
$$= 0.785 \text{ mm}^2$$

As beam remains horizontal, so, strain in steel wire = strain in brass wire

$$\therefore \quad \frac{\sigma_s}{E_s} = \frac{\sigma_b}{E_b}$$

$$\sigma_s = \sigma_b \times \frac{E_s}{E_b}$$

$$\therefore \quad \sigma_s = \sigma_b \times \frac{2 \times 10^5}{0.8 \times 10^5} = 2.5 \times \sigma_b$$

The load shared by steel and brass wires,

$$P = 2 \times \sigma_b \times A_b + \sigma_s \times A_s$$

$$\therefore \quad 4000 = 2 \times \sigma_b \times 0.785 + 2.5 \times \sigma_b \times 0.196$$

$$\therefore \quad \sigma_b = (400/2.06) = 194.17 \text{ N/mm}^2$$

$$\sigma_s = 2.5 \times 194.17 = 485.42 \text{ N/mm}^2$$

Example 7.38 Two vertical wires of copper and steel of initial length 4 m are suspended at a distance of 0.6 m apart as shown in Figure 7.35. The upper ends of wires are firmly secured and the lower ends carry a horizontal beam on which load P is acting. The copper wire diameter is 1.5 mm and that of steel is 0.8 mm. Take E for steel = 2 $\times 10^5$ N/mm^2 and for copper = 1.2 $\times 10^5$ N/mm^2. Determine the position of the load on the beam if it remains horizontal after being loaded. Also determine the slope of rigid beam if a load of 250 N is hung at the centre of the beam. Neglect the weight of the beam.

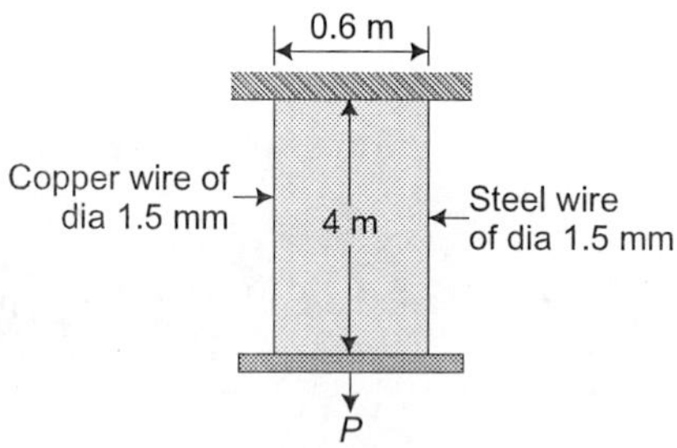

Fig. 7.35

Solution Refer Figure 7.35.

Area of cross section of steel wire,

$$A_s = (\pi/4) \times (0.8^2) = 0.503 \text{ mm}^2$$

Area of cross section of copper wire,

$$A_c = (\pi/4) \times (1.5^2) = 1.767 \text{ mm}^2$$

As beam remains horizontal, so, strain in steel wire = strain in copper wire

$$\therefore \quad \frac{\sigma_s}{E_s} = \frac{\sigma_c}{E_c}$$

$$\sigma_s = \sigma_c \times \frac{E_s}{E_c}$$

$$\therefore \quad \sigma_s = \sigma_c \times \frac{2 \times 10^5}{1.2 \times 10^5} = 1.67 \times \sigma_c$$

The load shared by steel and copper wires,

$$P = \sigma_c \times A_c + \sigma_s \times A_s$$

$$\therefore \quad P = \sigma_c \times 1.767 + 1.67 \times \sigma_c \times 0.503$$

$$\sigma_c = 0.384 \times P$$

Tension in copper wire,

$$T_c = \sigma_c \times A_c = 0.384 \times P \times 1.767$$

$$= 0.678\ P$$

Tension in steel wire,

$$T_s = P - 0.678\ P = 0.322\ P$$

Let the distance of P from the copper wire be x mm

Taking moments about copper wire,

$$0.322\ P \times 600 = P \times x$$

$$\therefore \quad x = 193.2 \text{ mm}$$

When a load of 250 N is hung at the centre of the beam, load in each wire is 125 N.

Extension in copper wire,

$$\delta l_c = \frac{125 \times 4000}{1.2 \times 10^5 \times 1.767}$$

$$= 2.358 \text{ mm}$$

Extension in steel wire,

$$\delta l_s = \frac{125 \times 4000}{0.8 \times 10^5 \times 0.503}$$

$$= 4.970 \text{ mm}$$

Let θ be the inclination of the beam with the horizontal, then

$$\tan \theta = \frac{4.970 - 2.358}{600} = 0.004353$$

$$\therefore \quad \theta = 0^\circ\ 25'$$

Example 7.39 Three rods made of copper, zinc, and aluminium are of equal length and have cross section 600 mm^2, 800 mm^2, and 1000 mm^2 respectively. These bars are rigidly connected to their ends and this unit is subjected to a longitudinal pull of 500 kN. If E for copper = 1.3 ×10^5 N/mm^2, for zinc = 1.0 ×10^5 N/mm^2, and for aluminium = 0.8 ×10^5 N/mm^2 then determine the stress induced and the load carried by each rod.

Solution Strain in copper rod = Strain in zinc rod = Strain in aluminium rod, i.e.,

$$\frac{\sigma_c}{E_c} = \frac{\sigma_z}{E_z} = \frac{\sigma_a}{E_a}$$

$$\therefore \quad \sigma_c = \frac{E_c}{E_a} \times \sigma_a = \frac{1.3 \times 10^5}{0.8 \times 10^5} \times \sigma_a$$

$$= 1.625 \times \sigma_a$$

$$\sigma_z = \frac{E_z}{E_a} \times \sigma_a = \frac{1.0 \times 10^5}{0.8 \times 10^5} \times \sigma_a$$

$$= 1.25 \times \sigma_a$$

The load shared by copper, zinc, and aluminium rods,

$$P = \sigma_c \times A_c + \sigma_z \times A_z + \sigma_a \times A_a$$

$$\therefore \quad 500 \times 10^3 = 1.625\, \sigma_a \times 600 + 1.25\, \sigma_a \times 800 + \sigma_a \times 1000$$

$$2975\, \sigma_a = 500 \times 10^3$$

$$\therefore \quad \sigma_a = \frac{500 \times 10^3}{2975} = 168.07 \text{ N/mm}^2$$

$$\sigma_z = 1.25 \times \sigma_a = 1.25 \times 168.07 = 210.09 \text{ N/mm}^2$$

$$\sigma_c = 1.625 \times \sigma_a = 1.625 \times 168.07 = 273.11 \text{ N/mm}^2$$

Load carried by aluminium rod,

$$P_a = \sigma_a \times A_a = 168.07 \times 1000 = 168070 \text{ N}$$

Load carried by zinc rod,

$$P_z = \sigma_z \times A_z = 210.09 \times 800 = 168072 \text{ N}$$

Load carried by copper rod,

$$P_c = \sigma_c \times A_c = 273.11 \times 600 = 163866 \text{ N}$$

Example 7.40 Two copper rods and one steel rod lie in a vertical plane and together support a load as shown in Figure 7.36. If the stresses in copper and steel are not to exceed 60 N/mm^2 and 120 N/mm^2, find the safe load that can be supported. Each rod is 50 mm in diameter. Young's modulus for steel is twice that of copper. The cross sectional area of steel rod is 1600 mm^2 and of each copper rod is 900 mm^2.

Solution Refer Figure 7.36.

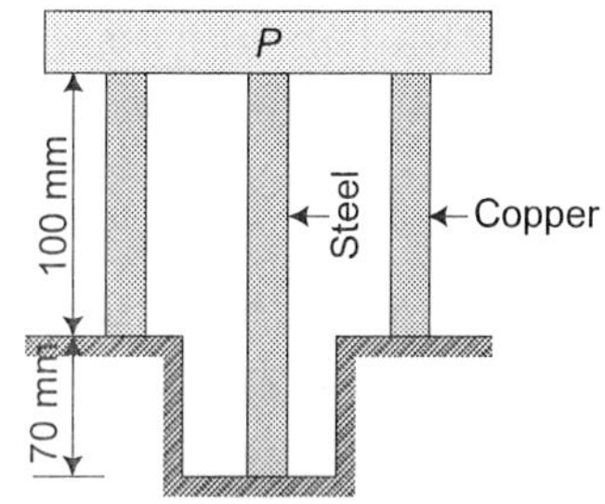

Fig. 7.36

Strain in steel = Strain in copper

$$\therefore \quad \frac{\sigma_s}{E_s} \times l_s = \frac{\sigma_c}{E_c} \times l_c$$

$$\sigma_s = \sigma_c \times \frac{E_s}{E_c} \times \frac{l_c}{l_s}$$

$$\therefore \quad \sigma_s = \sigma_c \times \frac{2}{1} \times \frac{100}{170} = 1.176 \times \sigma_c$$

Let the steel rod be allowed to reach its safe stress of 120 N/mm^2

The corresponding stress in copper rod,

$$\sigma_c = \frac{120}{1.176} = 102.04 \text{ N/mm}^2$$

This value 102.04 N/mm^2 exceeds the safe stress of 60 N/mm^2 for copper.

Therefore, let the copper rod be allowed to reach its safe stress of 60 N/mm^2

The corresponding stress in steel rod,

$$\sigma_s = 60 \times 1.176 = 70.56 \text{ N/mm}^2.$$

The load shared by steel bar and copper bars,

$$P = 2 \times P_c + P_s = 2 \times \sigma_c \times A_c + \sigma_s \times A_s$$

$$\therefore \quad P = 2 \times 60 \times 900 + 70.56 \times 1600$$

$$= 109670.56 \text{ N} = 109.67 \text{ kN}$$

Example 7.41 A steel rod 20 mm in diameter passes centrally through a steel tube 30 mm in internal diameter and 35 mm in external diameter. The tube is 800 mm long and is closed by rigid washers of negligible thickness which are fastened by nuts threaded on the rod. The nuts are tightened until the compressive load on the tube is 10 kN. Determine the stresses in the tube and the rod. Also determine the increase in these stresses when one nut is tightened by one quarter of a turn relative to the other. There are 4 threads per 10 mm. Young's modulus for steel is 200 kN/mm^2.

Solution Refer Figure 7.37.

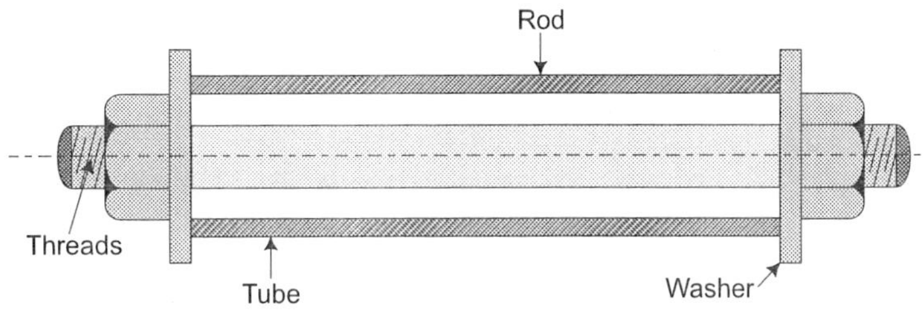

Fig. 7.37

Area of cross section of steel rod,

$$A_r = (\pi/4) \times (20^2) = 100\ \pi\ \text{mm}^2$$

Area of cross section of steel tube,

$$A_t = (\pi/4) \times (35^2 - 30^2) = (325\ \pi/4)\ \text{mm}^2$$

When the nuts are tightened the tube will be compressed and the rod will be elongated. As no external force is acting, so, the compressive load on the tube will be equal to the tensile load.

Let σ_t and σ_r be the stresses in the tube and the rod respectively.

$$\therefore \qquad \sigma_t \times A_t = \sigma_r \times A_r$$

$$\therefore \qquad \sigma_r = \sigma_t \times \frac{A_t}{A_r} = \frac{325\ \pi}{4} \times \frac{1}{100\ \pi} \times \sigma_t$$

$$= \frac{13}{16}\sigma_t$$

(i) When the compressive load on the tube is 10 kN

Stress in the tube, $$\sigma_t = \frac{P}{A_t} = \frac{10 \times 10^3 \times 4}{325\ \pi}$$

$$= 39.18\ \text{N/mm}^2 \quad \text{(Compressive)}$$

Stress in the rod, $$\sigma_r = \frac{13}{16} \times \sigma_t = \frac{13}{16} \times 39.18$$

$$= 31.83\ \text{N/mm}^2 \quad \text{(Tensile)}$$

(ii) When one nut is tightened by one quarter of a turn

Decrease in the length of tube

$$= \frac{\sigma_t}{E} \times l = \frac{\sigma_t}{200 \times 10^3} \times 800$$

$$= 0.008\ \sigma_t\ \text{mm}$$

Increase in the length of rod

$$= \frac{\sigma_r}{E} \times l = \frac{13}{16} \cdot \frac{\sigma_t}{200 \times 10^3} \times 800$$

$$= 0.0065\ \sigma_t\ \text{mm}$$

Decrease in the length of tube + Increase in the length of rod = Axial advance of nut

$$0.008\ \sigma_t + 0.0065\ \sigma_t = \frac{1}{4} \times \frac{10}{4} = 0.625$$

$$\therefore \qquad \sigma_t = (0.625/0.0145) = 43.10\ \text{N/mm}^2$$

$$\sigma_r = \frac{13}{16} \times 43.10 = 35.02\ \text{N/mm}^2$$

7.16 RELATIONS BETWEEN ELASTIC CONSTANTS

(i) Relation between E, K, and ν Consider a cube *ABCDEFGH* of side l subjected to three mutually perpendicular tensile stresses of equal intensity σ along x, y, and z directions as shown in Figure 7.38.

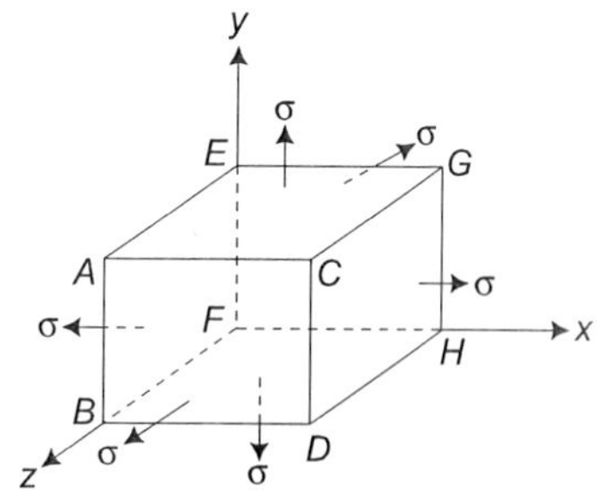

Fig. 7.38

Let us consider the strain of AC one of the sides of the cube under the action of the three mutually perpendicular tensile stresses. This side AC will suffer the following three strains:

1. Strain of AC due to stresses on the faces *AEFB* and *CGHD*. This strain is tensile which is given by, σ/E.
2. Strain of AC due to stresses on the faces *AEGC* and *BFHD*. This strain is compressive lateral strain which is given by $-\,\nu(\sigma/E)$.
3. Strain of AC due to stresses on the faces *ACDB* and *EGHF*. This strain is also compressive lateral strain which is given by, $-\,\nu\,(\sigma/E)$.

Therefore, total strain of AC,

$$\frac{dl}{l} = \frac{\sigma}{E} - \nu\frac{\sigma}{E} - \nu\frac{\sigma}{E} = \frac{\sigma}{E}(1-2\nu)$$

(i)

If δl is the change in length, then δv is the change in volume.

The volume of the cube,

$$v = l^3 \qquad \text{(ii)}$$

Differentiating equation (ii) w.r.t. l,

$$dv = 3l^2 \,.\, dl \qquad \text{(iii)}$$

Dividing equation (iii) by equation (ii),

$$\frac{dv}{v} = \frac{3l^2 \,.\, dl}{l^3} = \frac{3 \,.\, dl}{l} \qquad \text{(iv)}$$

From equations (i) and (iv),

$$\frac{dv}{v} = \frac{3\sigma}{E}(1-2\nu)$$

Bulk modulus, $$K = \frac{\sigma}{(dv/v)} = \frac{\sigma}{(3\sigma/E)(1-2\nu)}$$

∴ $$K = \frac{E}{3(1-2\nu)}$$

or $$E = 3K(1-2\nu) \qquad (7.22)$$

From equation (7.22), the Poisson's ratio becomes,

$$\nu = \frac{3K-E}{6K} \qquad (7.23)$$

(ii) Relation between E, G, and ν Consider a cube *ABCD* fixed at the bottom and subjected to shear stress τ at the top. The cube is deformed as shown in Figure 7.39. A complementary shear stress τ is induced at an angle of 90° to the applied shear stress. The angle of shear ϕ, within the elastic limit is very small and it is taken as 45°. Due to shear stress the diagonal DB is increased in length to DB' and the diagonal AC is decreased to $A'C$. $DB = DE$ and BE is perpendicular to the diagonal DB'.

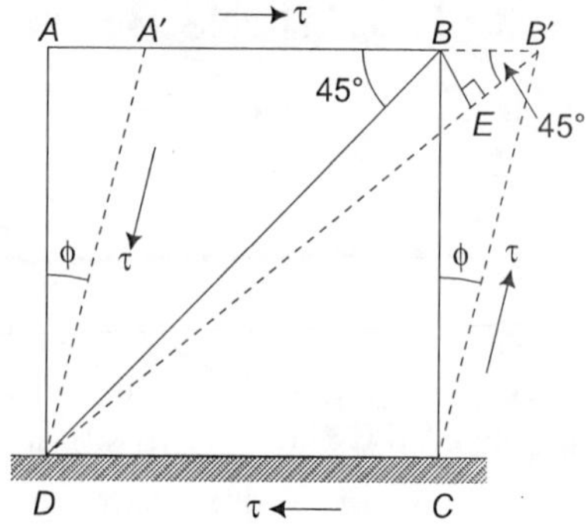

Fig. 7.39

$$\text{Change in length} = EB' = BB' \times \cos 45^\circ$$
$$= \frac{BB'}{\sqrt{2}}$$

Tensile strain along DB,

$$\frac{\text{Change in length}}{\text{Original length}} = \frac{BB'}{\sqrt{2}\ DB} = \frac{BB'}{\sqrt{2}\ \sqrt{2}\ BC}$$
$$= \frac{BB'}{2BC} \quad (DB = \sqrt{2}\ BC)$$
$$= \frac{1}{2} \times \text{Shear strain} = \frac{1}{2} \times \frac{\text{Shear stress}}{G}$$
$$= \frac{1}{2} \times \frac{\tau}{G} \quad \left(\because \frac{\text{Shear stress}}{\text{Shear strain}} = G\right) \tag{7.24}$$

Tensile strain in diagonal DB due to tensile stress τ along

$$DB = \frac{\tau}{E} \tag{i}$$

Tensile strain in diagonal DB due to compressive stress τ along

$$AC = \frac{\nu \times \tau}{E} \tag{ii}$$

So, from (i) and (ii), tensile strain along diagonal

$$DB = \frac{\tau}{E} + \frac{\nu \times \tau}{E} = \frac{\tau}{E}(1 + \nu) \tag{7.25}$$

From equations (7.24) and (7.25),

$$\frac{\tau}{E}(1 + \nu) = \frac{1}{2} \times \frac{\tau}{G}$$

$$\therefore \quad E = 2G(1 + \nu) \tag{7.26}$$

or

$$G = \frac{E}{2(1 + \nu)} \tag{7.27}$$

(iii) Relation between E, G, and K

From equation (7.26),

$$\nu = \frac{E}{2G} - 1 \qquad \text{(i)}$$

Substituting the value ν from equation (i) in equation (7.22),

$$E = 3K\left[1 - 2 \times \left(\frac{E}{2G} - 1\right)\right]$$

$$= 3K\left[1 - \left(\frac{E}{G} - 2\right)\right] = 3K\left[3 - \frac{E}{G}\right]$$

$$= 9K - \frac{3KE}{G}$$

or $$E + \frac{3KE}{G} = 9K$$

or $$E\left(1 + \frac{3K}{G}\right) = 9K$$

or $$E\left(\frac{G + 3K}{G}\right) = 9K$$

∴ $$E = \frac{9KG}{G + 3K} \qquad \text{(ii)}$$

From equations (7.22) and (7.26),

$$E = 2G\,(1 + \nu) = 3K\,(1 - 2\nu) \qquad \text{(iii)}$$

Equating (ii) and (iii), we have,

$$E = 2G\,(1 + \nu) = 3K\,(1 - 2\nu)$$

$$= \frac{9KG}{G + 3K} \qquad (7.28)$$

Example 7.42 When a bar of 30 mm in diameter is subjected to an axial pull 70 kN, the extension on gauge length of 300 mm is observed as 0.2 mm and change in diameter is found 0.005 mm. Determine the Young's modulus, Poisson's ratio, and bulk modulus for the given bar.

Solution Tensile stress,

$$\sigma = \frac{P}{A} = \frac{70 \times 10^3 \times 4}{\pi \times (30^2)}$$

$$= 55.70 \text{ N/mm}^2$$

Longitudinal strain, $$e = \frac{\delta l}{l} = \frac{0.2}{300}$$

$$= 6.67 \times 10^{-4}$$

Young's modulus, $E = \frac{\sigma}{e} = \frac{55.70}{6.67 \times 10^{-4}}$

$= 83508.24 \text{ N/mm}^2$

Lateral strain, $= \frac{\delta d}{d} = \frac{0.005}{40} = 1.25 \times 10^{-4}$

Poisson's ratio, $\nu = \frac{\text{Lateral strain}}{\text{Longitudinal strain}}$

$= \frac{1.25 \times 10^{-4}}{6.67 \times 10^{-4}} = 0.187$

Bulk modulus, $K = \frac{E}{3(1-2\nu)} = \frac{83508.24}{3 \times (1 - 2 \times 0.187)}$

$= 44466.58 \text{ N/mm}^2$

Example 7.43 The diameter of a metallic rod is observed to reduce by 0.003 mm under an axial pull of 25 kN. The original diameter of the rod is measured as 20 mm. If the rigidity modulus for the metal rod be 5×10^4 N/mm^2, determine the Young's modulus, Poisson's ratio, and bulk modulus.

Solution Tensile stress,

$$\sigma = \frac{P}{A} = \frac{25 \times 10^3 \times 4}{\pi \times (20^2)}$$

$$= 0.07958 \text{ kN/mm}^2$$

Longitudinal strain, $e = \frac{\sigma}{E} = \frac{0.07958}{E}$

Lateral strain $= \frac{\delta d}{d} = \frac{0.003}{20} = 1.5 \times 10^{-4}$

Poisson's ratio, $\nu = \frac{\text{Lateral strain}}{\text{Longitudinal strain}}$

$= \frac{1.5 \times 10^{-4}}{(0.07958/E)}$

$= 1.88 \times 10^{-3} E$

Now, $E = 2G(1 + \nu)$

$$E = 2 \times 50 \times (1 + 1.88 \times 10^{-3} \times E)$$

$$E = 100 + 0.188 E$$

$\therefore$ $E = 123.15 \text{ kN/mm}^2$,

$\therefore$ $\nu = 1.88 \times 10^{-3} \times 123.15 = 0.23$

Bulk modulus, $$K = \frac{E}{3(1-2\nu)} = \frac{123.15}{3 \times (1 - 2 \times 0.23)}$$

$$= 76.02 \text{ kN/mm}^2$$

Example 7.44 The measured value of E and G for a particular metal are 2×10^5 N/mm^2 and 0.8×10^5 N/mm^2 respectively. Evaluate the Poisson's ratio and bulk modulus. If both the moduli are liable to an error of ± 2%, determine the maximum percentage error in the evaluated value of Poisson's ratio.

Solution

$$E = 2G(1 + \nu)$$

$$\nu = \frac{E}{2G} - 1$$

$$\therefore \quad \nu = \frac{2 \times 10^5}{2 \times 0.8 \times 10^5} - 1 = 0.25$$

Now, bulk modulus, $$K = \frac{E}{3(1-2\nu)} = \frac{2 \times 10^5}{3 \times (1 - 2 \times 0.25)}$$

$$= 133333.33 \text{ N/mm}^2$$

Error in the determination of E and G is ± 2%

So, the maximum value of

$$\nu = \frac{2 \times 10^5 \times 1.02}{2 \times 0.8 \times 10^5 \times 0.98} - 1$$

$$= 0.30$$

The minimum value of

$$\nu = \frac{2 \times 10^5 \times 0.98}{2 \times 0.8 \times 10^5 \times 1.02} - 1$$

$$= 0.20$$

So, error in the calculation of

$$\nu = (0.30 - 0.25 = +0.05) \text{ to } (0.20 - 0.25 = -0.05)$$

$$\therefore \quad \% \text{ error} = \frac{0.05}{0.25} \times 100$$

$$= \pm 20\%$$

Example 7.45 If the longitudinal strain in the bar during a tensile test is observed four times of the lateral strain, determine the modulus of rigidity and bulk modulus of a cylindrical bar of diameter 25 mm and of length 1000 mm. Also determine the change in volume if the bar is subjected to a hydrostatic pressure of 0.1 kN/mm^2 and Young's modulus is 100 kN/mm^2.

Solution Poisson's ratio,

$$\nu = \frac{\text{Lateral strain}}{\text{Longitudinal strain}} = \frac{1}{4}$$

$$= 0.25$$

$$E = 2G\,(1 + \nu)$$

$$G = \frac{E}{2 \times (1 + \nu)}$$

$$\therefore \quad G = \frac{100 \times 10^3}{2 \times (1 + 0.25)} = 40000 \text{ N/mm}^2$$

$$E = 2G\,(1 + \nu)$$

$$K = \frac{E}{3 \times (1 - 2\nu)}$$

$$\therefore \quad K = \frac{E}{3 \times (1 - 2\nu)} = \frac{100 \times 10^3}{3 \times (1 - 2 \times 0.25)}$$

$$= 66666.67 \text{ N/mm}^2$$

Also,

$$K = \frac{P}{\text{Volumetric strain}}$$

$$= \frac{p}{(dv/v)}$$

$$66666.67 = \frac{100}{(dv/v)}$$

$$\left(\frac{dv}{v}\right) = \frac{100}{66666.67} = 1.5 \times 10^{-3}$$

Volume,

$$v = (\pi/4) \times (d^2) \times l = (\pi/4) \times (25^2) \times 1000$$

$$= 490873.85 \text{ mm}^3$$

$$\therefore \quad dv = 1.5 \times 10^{-3} \times v = 1.5 \times 10^{-3} \times 490873.85$$

$$= 736.31 \text{ mm}^3$$

Example 7.46 For a particular material the Young's modulus and rigidity modulus are 2×10^5 N/mm^2 and 0.85×10^5 N/mm^2 respectively. Find its bulk modulus value. If the length of the rod is 5000 mm, its original diameter is 40 mm, and stretch in the longitudinal direction is 2 mm, evaluate the lateral strain in the given rod.

Solution

$$E = \frac{9KG}{G + 3K}$$

$$2 \times 10^5 = \frac{9 \times K \times 0.85 \times 10^5}{0.85 \times 10^5 + 3K}$$

$$2 \times 10^5 \times [0.85 \times 10^5 + 3K] = 9 \times K \times 0.85 \times 10^5$$

$$3.825\,K - 3\,K = 0.85 \times 10^5$$

$$\therefore \quad K = 103030.30 \text{ N/mm}^2$$

Now $$E = 2G\,(1 + \nu)$$

$$\nu = \frac{E}{2G} - 1$$

$$\therefore \quad \nu = \frac{2 \times 10^5}{2 \times 0.85 \times 10^5} - 1 = 0.18$$

Longitudinal strain, $$= \frac{2}{5000} = 4 \times 10^{-4}$$

$$\nu = \frac{\text{Lateral strain}}{\text{Longitudinal strain}}$$

$$\therefore \quad \text{Lateral strain} = 0.18 \times \text{Longitudinal strain}$$

$$= 0.18 \times 4 \times 10^{-4} = 7.2 \times 10^{-5}$$

$$\text{Lateral strain} = \frac{\delta d}{d}$$

$$\therefore \quad \delta d = d \times \text{Lateral strain}$$

$$= 40 \times 7.2 \times 10^{-5} = 0.0029 \text{ mm}$$

7.17 TEMPERATURE OR THERMAL STRESSES

Temperature or thermal stresses induce in a body when the temperature of the body is raised or lowered and it is not allowed to expand or contract freely. But, when the body is allowed to expand or contract freely there will be no temperature stress in it.

(i) Bar of Uniform Section Let us consider a rod of length l rigidly fixed at the ends A and B as shown in Figure 7.40.

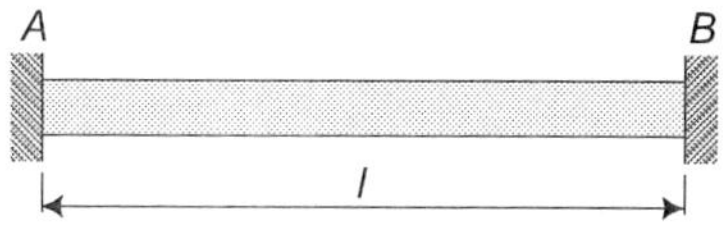

Fig. 7.40

Suppose the temperature of the rod increases through T degree.

The length of the rod increases by

$$\delta l = \alpha T l$$

where α is the coefficient of linear thermal expansion.

Due to external conditions, the rod is not free to expand and, therefore, from Hooke's law

$$\delta l = \frac{Pl}{E}$$

where P is the thermal stress which is compressive in nature and E is the Young's modulus of the rod.

$$\therefore \qquad P = \frac{\delta l E}{l} = \alpha T l \times \frac{E}{l} = \alpha T E \quad (\because \delta l = \alpha T l) \tag{7.29}$$

$$\text{Thermal strain} = \frac{\alpha T l}{l} = \alpha T \tag{7.30}$$

Similarly, if the temperature decreases, the thermal stress is αTE which is tensile in nature.

If the end supports yield by an amount δ, then the net expansion

$$= \delta l - \delta = \alpha T l - \delta$$

$$\therefore \qquad \text{Actual strain} = \frac{(\alpha T l - \delta)}{l}$$

$$\text{Actual stress} = \frac{(\alpha T l - \delta)}{l} \times E \tag{7.31}$$

(ii) Bar of Composite Section Let a brass and a steel bar be permanently fixed together so as to form a composite bar as shown in Figure 7.41. The length of composite bar is l. The area of cross section of brass bar is A_b and that of steel is A_s. The coefficient of linear expansion of brass bar is α_b and that of steel is α_s. The Young's modulus coefficient of brass bar is E_b and that of steel is E_s. Let the temperature of composite bar be increased by T degree due to which its length increases by δl.

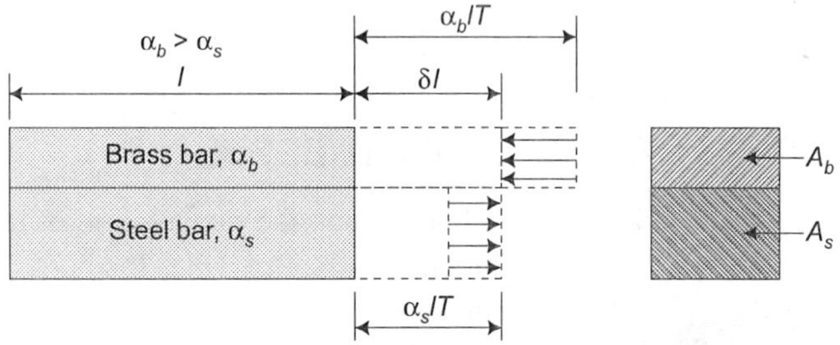

Fig. 7.41

If the bars are free to expand independently then, expansion of brass bar = $\alpha_b\ lT$ and expansion of steel bar = $\alpha_s\ lT$. But in the composite bar, both the bars expand by the same amount δl, such that

$$\alpha_b\ lT > \delta l > \alpha_s\ lT.$$

So, contraction in the free expended length of brass bar = $\alpha_b\ lT - \delta l$

$$\text{Compressive strain in brass bar, } e_b = \frac{\alpha_b\ lT - \delta l}{l} = \alpha_b\ T - e \tag{i}$$

Extension in the free expended length of steel bar = $\delta l - \alpha_s\ lT$

Tensile strain in steel bar,

$$e_s = \frac{\delta l - \alpha_s\ lT}{l} = e - \alpha_s\ T \tag{ii}$$

where $e = (\delta l/l)$ is actual strain of composite bar.

Adding (i) and (ii),

$$e_b + e_s = \alpha_b T - e + e - \alpha_s T$$

$$\therefore \quad e_b + e_s = (\alpha_b - \alpha_s) \times T \qquad (7.32)$$

Let the temperature stresses in brass and steel bars be σ_{bT} and σ_{sT} respectively, then

$$\frac{\sigma_{bT}}{E_b} + \frac{\sigma_{sT}}{E_s} = (\alpha_b - \alpha_s) \times T \qquad (7.33)$$

For equilibrium, compressive force in brass bar = Tensile force in steel bar

$$\sigma_{bT} \times A_b = \sigma_{sT} \times A_s \qquad (7.34)$$

From equations (7.33) and (7.34) stresses σ_{bT} and σ_{sT} can be determined.

Example 7.47 A gun metal rod of diameter 32 mm is screwed at the ends passes through a steel tube of external and internal diameters of 40 mm and 35 mm respectively. The assembly is heated to 150°C and the nuts are then screwed tightly at the ends of the tube. Determine the intensity of stress in rod and tube when the common temperature falls to 20°C. For steel take $\alpha_s = 12 \times 10^{-6}$ per °C, $E_s = 2.1 \times 10^5$ N/mm^2 and for gun metal take $\alpha_g = 20 \times 10^{-6}$ per °C, $E_g = 0.92 \times 10^5$ N/mm^2.

Solution Rise of temperature,

$$T = 150 - 20 = 130°C$$

Area of cross section of steel tube,

$$A_s = (\pi/4) \times (40^2 - 35^2) = 294.52 \text{ mm}^2$$

Area of cross section of gun metal rod,

$$A_g = (\pi/4) \times (32^2) = 804.25 \text{ mm}^2$$

Under equilibrium, compressive force in steel tube = tensile force in gun metal rod

$$\sigma_{sT} \times A_s = \sigma_{gT} \times A_g$$

$$\sigma_{gT} = \sigma_{sT} \times \frac{A_s}{A_g}$$

$$\therefore \quad \sigma_{gT} = \sigma_{sT} \times \frac{294.52}{804.25} = 0.366\, \sigma_{sT}$$

Now

$$\frac{\sigma_{gT}}{E_g} + \frac{\sigma_{sT}}{E_s} = (\sigma_g - \sigma_s) \times T$$

$$\frac{0.366\, \sigma_{sT}}{0.92 \times 10^5} + \frac{\sigma_{sT}}{2.1 \times 10^5} = (20 \times 10^{-6} - 12 \times 10^{-6}) \times 130$$

$$\sigma_{sT} \times [0.476 + 0.398] = 104$$

$$\therefore \qquad \sigma_{sT} = \frac{104}{0.874} = 118.99 \text{ N/mm}^2$$

$$\sigma_{gT} = 0.366 \times 118.99 = 43.55 \text{ N/mm}^2$$

Hence, intensity of stress in gun metal rod,

$$\sigma_{gT} = 43.55 \text{ N/mm}^2 \quad \text{(tensile)}$$

Intensity of stress in steel tube,

$$\sigma_{sT} = 118.99 \text{ N/mm}^2 \quad \text{(compressive)}$$

Example 7.48 A steel bar is placed between two copper bars each having the same area and length as the steel bar at 25°C. These bars are permanently fixed so as to behave as a composite bar. When the temperature is raised to 325°C, the length of the composite bar increases by 2 mm. Find the original length and the final stresses in the bars. For steel take $\alpha_s = 12 \times 10^{-6}$ per °C, $E_s = 2.1 \times 10^5$ N/mm^2 and for copper $\alpha_c = 18 \times 10^{-6}$ per °C, $E_c = 1.0 \times 10^5$ N/mm^2.

Solution Rise of temperature,

$$T = 325 - 25 = 300°\text{C}$$

Let area of cross section of steel tube = A_s mm^2

So, area of cross section of copper tubes = $2A_s$ mm^2

If the bars are free to expand independently then, expansion of copper bar $\alpha_c\, lT$, and expansion of steel bar = $\alpha_s\, lT$. But, in the composite bar, both the bars expand by the same amount δl, such that

$$\alpha_c\, lT > \delta l > \alpha_s\, lT.$$

Under equilibrium,

tensile force in steel bar = compressive force in copper bars

$$\sigma_{sT} \times A_s = \sigma_{cT} \times (2A_s)$$

$$\therefore \qquad \sigma_{sT} = 2\sigma_{cT}$$

Now
$$\frac{\sigma_{cT}}{E_c} + \frac{\sigma_{sT}}{E_s} = (\alpha_c - \alpha_s) \times T$$

$$\frac{\sigma_{cT}}{1.0 \times 10^5} + \frac{2\sigma_{cT}}{2.1 \times 10^5} = (18 \times 10^{-6} - 12 \times 10^{-6}) \times 300$$

$$4.1\, \sigma_{cT} = 378$$

$$\therefore \qquad \sigma_{cT} = \frac{378}{4.1} = 92.19 \text{ N/mm}^2$$

$$\sigma_{sT} = 2 \times 92.19 = 184.38 \text{ N/mm}^2$$

Actual expansion of the steel bar = Actual expansion of the copper bars

$$= \alpha_s\, lT + \frac{\sigma_{sT}}{E_s}\, l = 2$$

$$12 \times 10^{-6} \times 300 \times l + \frac{184.38}{2.1 \times 10^5} \times l = 2$$

$$756\, l + 184.38\, l = 420000$$

$$\therefore \quad l = (420000/940.38)$$

$$= 446.63 \text{ N/mm}^2$$

Example 7.49 A railway is laid so that there is no stress in the rails at 20°C. Determine (i) the stress in the rails at 60°C if there is no allowance for expansion, (ii) the stress in the rails if there is an expansion allowance of 8 mm per rail, (iii) the expansion allowance if the stress in the rail is to be zero when the temperature is 60°C, (iv) the maximum temperature to have no stress in the rails if the expansion allowance is 12 mm per rails. Take $\alpha = 12 \times 10^{-6}$ per °C, $E = 2.1 \times 10^5$ N/mm^2. This rails are 30 m long.

Solution Rise of temperature,

$$T = 60 - 20 = 40°\text{C}$$

(i) Thermal stress,

$$\sigma_T = \alpha\, TE = 12 \times 10^{-6} \times 40 \times 2.1 \times 10^5$$

$$= 100.8 \text{ N/mm}^2$$

(ii)

$$\frac{\sigma_T}{E}\, l = (\alpha\, lT - 8)$$

$$\sigma_T = (\alpha\, lT - 8) \times E/l$$

$$\therefore \quad \sigma_T = (12 \times 10^{-6} \times 30000 \times 40 - 8) \times \frac{2.1 \times 10^5}{30000}$$

$$= 44.8 \text{ N/mm}^2$$

(iii) If the stresses are to be zero, the expansion allowance

$$= \alpha\, lT = 12 \times 10^{-6} \times 300 \times 40$$

$$= 14.4 \text{ mm}$$

(iv) If the stresses are to be zero,

$$\sigma_T = (\alpha\, lT - 12) \times \frac{E}{l} = 0$$

$$\therefore \quad \alpha\, lT = 12$$

$$T = \frac{12}{\alpha L} = \frac{12}{12 \times 10^{-6} \times 30000}$$

$$= 33.33°\text{C}$$

Therefore, allowable temperature = 20 + 33.33 = 53.33°C

Example 7.50 A steel bar 25 mm diameter and 5 m long is connected to two grips one at each end at a temperature of 125°C. Determine the pull exerted when temperature falls to 25°C (i) if the ends do not yield, and (ii) if the ends yield by 1.25 mm. Take $\alpha = 12 \times 10^{-6}$ per °C, $E = 2.0 \times 10^5$ N/mm^2.

Solution Fall of temperature,

$$T = 125 - 25 = 100°\text{C}$$

(i) When the ends do not yield

Thermal stress, $\alpha_T = \alpha\, TE = 12 \times 10^{-6} \times 100 \times 2 \times 10^5$

$$= 240 \text{ N/mm}^2 \quad \text{(Tensile)}$$

Therefore, pull in the rod $= \sigma_T \times A = 240 \times (\pi/4) \times (25^2)$

$$= 117809.72 \text{ N}$$

(ii) When the ends yield by 1.25 mm

Thermal stress, $\sigma_T = (\alpha\, lT - 1.25) \times \dfrac{E}{l}$

$$\therefore \quad \sigma_T = (12 \times 10^{-6} \times 5000 \times 100 - 1.25) \times \frac{2 \times 10^5}{5000}$$

$$= 190 \text{ N/mm}^2$$

Therefore, pull in the rod $= \sigma_T \times A = 190 \times (\pi/4) \times (25^2)$

$$= 93266.03 \text{ N/mm}^2$$

Example 7.51 A steel tube 45 mm external diameter and 5 mm thick encloses centrally a solid copper bar of 30 mm diameter. The bar and tube are rigidly connected together at the ends at a temperature of 30°C. Determine the stress in each metal when heated to a temperature of 150°C. Also determine the increase in length if the original length of the unit is 300 mm. For steel, take $\alpha_s = 1.08 \times 10^{-5}$ per °C, $E_s = 2.1 \times 10^5$ N/mm^2 and for copper, $\alpha_c = 1.7 \times 10^{-5}$ per °C, $E_c = 1.1 \times 10^5$ N/mm^2.

Solution Rise of temperature,

$$T = 150 - 30 = 120°\text{C}$$

Area of cross section of steel tube,

$$A_s = (\pi/4) \times (45^2 - 35^2) = 628.2 \text{ mm}^2$$

Area of cross section of copper bar,

$$A_c = (\pi/4) \times 30^2 = 706.8 \text{ mm}^2$$

If the assembly is free to expand independently then, expansion of copper bar = $\alpha_c lT$ and expansion of steel tube = $\alpha_s\, lT$. But, in the composite bar, both the bars expand by the same amount δl, such that

$$\alpha_c\, lT > \delta l > \alpha_s\, lT.$$

Under equilibrium,

tensile force in steel tube = compressive force in copper bar

$$\sigma_{sT} \times A_s = \sigma_{cT} \times A_c$$

$$\therefore \quad \sigma_{sT} = \sigma_{cT} \times \frac{A_c}{A_s} = \frac{706.8}{628.2} \times \sigma_{cT}$$

$$= 1.125\, \sigma_{cT}$$

Now $$\frac{\sigma_{cT}}{E_c} + \frac{\sigma_{sT}}{E_s} = (\alpha_c - \alpha_s) \times T$$

$$\frac{\sigma_{cT}}{1.1 \times 10^5} + \frac{1.125\sigma_{cT}}{2.1 \times 10^5} = (1.7 \times 10^{-5} - 1.08 \times 10^{-5}) \times 120$$

$$3.3375\, \sigma_{cT} = 171.864$$

$$\therefore \quad \sigma_{cT} = \frac{171.864}{3.3375} = 51.49 \text{ N/mm}^2$$

$$\sigma_{sT} = 1.125 \times 51.49 = 57.93 \text{ N/mm}^2$$

Actual expansion of the steel tube = Actual expansion of the copper bar

Therefore, change in length of either component

$$= \alpha_s\, lT + \frac{\sigma_{sT}}{E_s}\, l$$

$$\therefore \quad \text{Change in length} = 1.08 \times 10^{-5} \times 300 \times 120 + \frac{57.93}{2.1 \times 10^5} \times 300$$

$$= 0.4715 \text{ mm}$$

Example 7.52 A copper bar 25 mm diameter is completely enclosed in a steel tube of 25 mm internal diameter and 40 mm external diameter. A pin 7.5 mm in diameter is fitted transversely to the axis of the bar near each end to secure the bar to the tube. Determine the intensity of shear stress induced in the pins when the temperature of the whole is raised by 60°C. Take for steel, $\alpha_s = 11 \times 10^{-6}$ per °C, $E_s = 2 \times 10^5$ N/mm^2 and for copper $\alpha_c = 17 \times 10^{-6}$ per °C, $E_c = 1 \times 10^5$ N/mm^2.

Solution Rise of temperature,

$$T = 60°\text{C}$$

Area of cross section of steel tube,

$$A_s = (\pi/4) \times (40^2 - 25^2) = 765.76 \text{ mm}^2$$

Area of cross section of copper bar,

$$A_c = (\pi/4) \times 25^2 = 490.87 \text{ mm}^2$$

Under equilibrium,

tensile force in steel tube = compressive force in copper bar

$$\sigma_{sT} \times A_s = \sigma_{cT} \times A_c$$

$$\therefore \quad \sigma_{sT} = \sigma_{cT} \times \frac{A_c}{A_s} = \frac{490.87}{765.76} \times \sigma_{cT}$$

$$= 0.641\ \sigma_{cT}$$

Now $$\frac{\sigma_{cT}}{E_c} + \frac{\sigma_{sT}}{E_s} = (\alpha_c - \alpha_s) \times T$$

$$\frac{\sigma_{cT}}{1 \times 10^5} + \frac{0.641\sigma_{cT}}{2 \times 10^5} = (17 \times 10^{-6} - 11 \times 10^{-6}) \times 60$$

$$2.641\ \sigma_{cT} = 72$$

$$\therefore \quad \sigma_{cT} = (72/2.641) = 27.26 \text{ N/mm}^2$$

Therefore, force between tube and bar,

$$\sigma_{cT} \times A_c = 27.26 \times 490.87 = 13381.2 \text{ N}$$

Area of cross section of pin,

$$A_p = (\pi/4) \times 7.5^2 = 44.18 \text{ mm}^2$$

Since the pin is fitted transversely and passes through the tube and the rod. So, it will be in double shear.

Therefore, shear stress in the pin = $(13381.2/2 \times 44.18) = 151.44$ N/mm^2

Example 7.53 A steel bolt of length l passes through a copper tube of the same length, and the nut at the end is turned up just snug at room temperature. Subsequently, the nut is tightened up by a number of turns and the entire assembly is raised by a temperature T. Calculate the stress in the bolt and tube if $l = 1000$ mm, $T = 60°$C, $n = 1/4$ turn, pitch of nut is 2 mm, area of copper tube = 600 mm^2, and area of steel bolt = 500 mm^2. Take $\alpha_s = 12 \times 10^{-6}$ per °C, $E_s = 2 \times 10^5$ N/mm^2 and for copper $\alpha_c = 17.5 \times 10^{-6}$ per °C, $E_c = 1 \times 10^5$ N/mm^2.

Solution As there is no external loading, therefore,

tensile force in steel bolt = compressive force in copper tube

$$P_s = P_c = P$$

Since final length of bolt and tube is same, therefore,

extension of bolt + compression of tube = movement of nut

$$\therefore \quad \delta l_s + \delta l_c = np$$

Now
$$\delta l_s = \frac{Pl}{A_s E_s} + \alpha_s Tl \quad \text{(Extension)}$$

$$\delta l_c = \frac{Pl}{A_c E_c} + \alpha_c Tl \quad \text{(Compression)}$$

$$\therefore \quad \left(\frac{Pl}{A_s E_s} + \alpha_s Tl\right) + \left(\frac{Pl}{A_c E_c} + \alpha_c Tl\right) = np$$

$$P\left(\frac{1}{A_s E_s} + \frac{1}{A_c E_c}\right) = (\alpha_c - \alpha_s)\, T + \frac{np}{l} \qquad \text{(i)}$$

Now substituting the values in equation (i), we have,

$$P\left(\frac{1}{500 \times 2 \times 10^5} + \frac{1}{600 \times 1 \times 10^5}\right)$$

$$= (17.5 \times 10^{-6} - 12 \times 10^{-6}) \times 60 + \frac{1}{4} \times \frac{2}{1000}$$

$$P = 28457.14 \text{ N}$$

$$\therefore \quad \sigma_s = \frac{28457.14}{500} = 56.91 \text{ N/mm}^2$$

$$\sigma_c = \frac{28457.14}{600} = 47.43 \text{ N/mm}^2$$

Example 7.54 A weight of 300 kN is supported by three short pillars, each 500 mm^2 in section. The central pillar is of steel and the outer ones are of copper. The pillars are so adjusted that at a temperature of 25°C each carries equal load. The temperature is then raised to 125°C. Calculate the stress in each pillar at 25°C and 125°C. For streel take $\alpha_s = 12 \times 10^{-6}$ per °C, $E_s = 2 \times 10^5$ N/mm^2 and for copper $\alpha_c = 18.5 \times 10^{-6}$ per °C, $E_c = 0.8 \times 10^5$ N/mm^2.

Solution Initial stresses: At 25°C each pillar carries weight = (300/3) = 100 kN

Stress in each pillar = $(100 \times 10^3)/500$ = 200 kN/mm^2

Thermal stresses: Let the stresses due to rise of temperature alone be σ_{sT} N/mm^2 (tensile) in steel and σ_{cT} N/mm^2 (compressive) in copper

Under equilibrium, tensile force in steel = compressive force in copper

$$\sigma_{sT} \times A_s = \sigma_{cT} \times A_c$$

$$\therefore \quad \sigma_{sT} = \sigma_{cT} \times \frac{A_c}{A_s} = \frac{2 \times 500}{500} \times \sigma_{cT}$$

$$= 2\,\sigma_{cT}$$

Now $$\frac{\sigma_{cT}}{E_c} + \frac{\sigma_{sT}}{E_s} = (\alpha_c - \alpha_s) \times T$$

$$\frac{\sigma_{cT}}{0.8 \times 10^5} + \frac{2\sigma_{cT}}{2 \times 10^5} = (18.5 \times 10^{-6} - 12 \times 10^{-6}) \times 100$$

$$3.6\ \sigma_{cT} = 104$$

$$\therefore \qquad \sigma_{cT} = (104/3.6) = 28.89 \text{ N/mm}^2$$

$$\sigma_{sT} = 2 \times 28.89 = 57.78 \text{ N/mm}^2$$

Therefore, final stress in steel = 200 – 57.78 = 142.22 N/mm^2

Final stress in copper = 200 + 28.89 = 228.89 N/mm^2

7.18 STRAIN ENERGY AND RESILIENCE

Stress and strain develop in an elastic material body due to application of an external load which may be applied gradually, or suddenly, or with impact. Work done on the body is converted into strain energy and this strain energy within the elastic limit is known as resilience. Whatever amount of energy is absorbed during loading, the same energy is recovered during unloading. Machine components such as helical, spiral, and leaf springs possess the property of resilience. For an example, in an internal combustion engine during the opening and closing of inlet or exhaust valve, the spring on the stem of the valve is compressed by which it stores energy and when it releases this energy the valve gets closed.

(i) Gradually Applied Load When a load on a body is gradually increased, the change in its length also increases gradually, such type of load is called gradual load. Within the elastic limit, stress is directly proportional to strain. The stress-strain diagram of a body under tensile test up to elastic limit is shown in Figure 7.42.

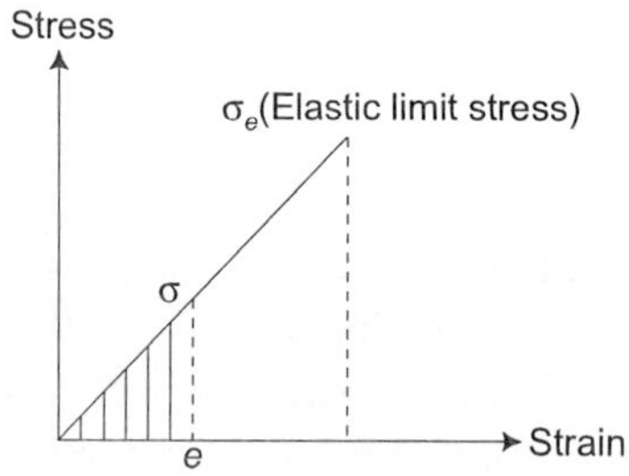

Fig. 7.42

Strain energy per unit volume at the stress σ is given by,

$$u = (1/2) \times \sigma \times e$$

But strain $$e = \frac{\sigma}{E}$$

$$\therefore \qquad u = \frac{1}{2} \times \sigma \times \left(\frac{\sigma}{E}\right) = \frac{\sigma^2}{2E}$$

Total strain energy absorbed by the body,

$$U = \frac{\sigma^2}{2E} \times \text{Volume of body} \tag{7.35}$$

The maximum strain energy absorbed by a body up to its elastic limit is called as proof resilience and this proof resilience per unit volume is called modulus of resilience.

$$\text{Proof resilience} = \frac{\sigma_e^2}{2E} \times \text{Volume} \tag{7.36}$$

$$\text{Modulus of resilience} = \frac{\sigma_e^2}{2E} \tag{7.37}$$

Similarly, the expressions for shear strain energy u_s in terms of shear stress τ, shear modulus G, and and volumetric strain energy u_v in terms of volumetric stress p and bulk modulus K can be given as,

Shear strain energy

$$u_s = \frac{\tau^2}{2G} \tag{7.38}$$

Volumetric strain energy

$$u_v = \frac{p^2}{2K} \tag{7.39}$$

(ii) Suddenly Applied Load When whole of the load is applied on the body at once there is a sudden change in its length, such type of load is called sudden load. For an example, when a load W being lowered on a platform and the load is very close to the platform suddenly the wire rope breaks and whole of the load W acts on the platform producing sudden change in its length by δl.

When the sudden load applied on a body, the internal resistance (R) of the body gradually develops from zero to the maximum value.

Work done on the body by sudden load = $W \times \delta l$

Strain energy stored in body,

$$U = \frac{1}{2} R \times \delta l$$

For equilibrium,

$$\frac{1}{2} R \times \delta l = W \times \delta l$$

or

$$R = 2W$$

If A is the area of cross section of the body, then

$$\sigma_{\text{sudden}} = \frac{R}{A} = \frac{2W}{A}$$

But

$$\frac{W}{A} = \sigma_{\text{gradual}}$$

$\therefore$

$$\sigma_{\text{sudden}} = 2 \times \sigma_{\text{gradual}} \tag{7.40}$$

(iii) Impact Loads When a load drops on a body from a certain height it possesses some kinetic energy which is utilized in deforming the body, such type of load is called impact load. For example, a blacksmith gives hammer blows to the hot iron pieces to give a desired shape.

Consider a rod of cross sectional area A and length l fixed at one end and has a collar at the other end as shown in Figure 7.43. A weight W slides freely on the rod which drops on the collar through a height h. The weight falling on the collar produces an instantaneous extension δl_i in the rod. Let at this instant the stress developed in the rod be σ_i and E is the Young's modulus of elasticity of the rod material.

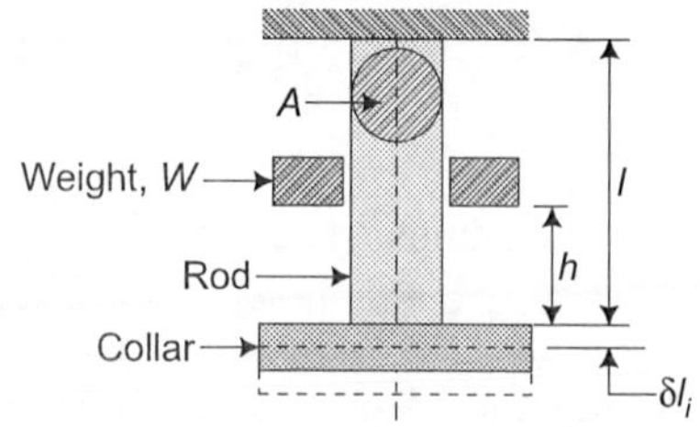

Fig. 7.43

External work done on the rod = Energy lost by the weight

$$= W \times (h + \delta l_i)$$

Strain energy stored by the rod $= \dfrac{\sigma_i^2}{2E} \times$ Volume of rod

$$= \frac{\sigma_i^2}{2E} \times A \times l$$

Under equilibrium,

$$W \times (h + \delta l_i) = \sigma_i^2 \times A \times l \qquad \text{(i)}$$

Substituting the instantaneous elongation,

$$\delta l_i = \frac{\sigma_i}{E} \times l$$

in equation (i), we have,

$$W \times \left(h + \frac{\sigma_i}{E} \times l\right) = \frac{\sigma_i^2}{2E} \times A \times l$$

$$Wh + \frac{W\sigma_i}{E} \times l = \frac{\sigma_i^2}{2E} \times A \times l$$

$$\sigma_i^2 - \frac{2W}{A} \times \sigma_i - \frac{2WhE}{Al} = 0$$

$$\sigma_i = \frac{W}{A} \pm \sqrt{\left(\frac{W^2}{A^2} + \frac{W}{A} \times \frac{2hE}{l}\right)}$$

Negative sign is not admissible as the stress cannot be compressive when the bar is elongated.

$$\therefore \qquad \sigma_i = \frac{W}{A}\left[1 + \sqrt{1 + \frac{2hAE}{Wl}}\right] \qquad (7.41)$$

Specific Cases

(i) If $h = 0$,

then $$\sigma_i = \frac{W}{A}\left[1 + \sqrt{1 + 0}\right]$$

$$= \frac{2W}{A} = \sigma_{\text{sudden}} \qquad (7.42)$$

$$\therefore \qquad \delta l_i = \frac{2Wl}{AE} \qquad (7.43)$$

(ii) If $\delta l_i <<< h$, then from equation (i), we have,

$$W \times h = \frac{\sigma_i^2}{2E} \times A \times l$$

$$\sigma_i = \sqrt{\frac{2WhE}{Al}} \qquad (7.44)$$

Example 7.55 A tensile load of 100 kN is applied suddenly to a circular bar of 50 mm diameter and 3 m long. If the value of Young's modulus is 2×10^5 N/mm^2, determine maximum instantaneous stress induced, instantaneous elongation in the rod, and strain energy absorbed in the rod.

Solution Area of cross section

$$A = (\pi/4) \times 50^2 = 1963.49 \text{ mm}^2$$

Using equation (7.40), maximum instantaneous stress induced

$$\sigma_{\text{sudden}} = 2 \times \frac{P}{A} = 2 \times \frac{100 \times 1000}{1963.49}$$

$$= 101.86 \text{ N/mm}^2$$

Instantaneous elongation in the rod,

$$\delta l_i = \frac{\sigma}{E} \times l = \frac{101.86 \times 3000}{2 \times 10^5}$$

$$= 1.53 \text{ mm}$$

Strain energy, $$U = \frac{\sigma^2}{2E} \times \text{Volume}$$

$$= \frac{101.86^2}{2 \times 2 \times 10^5} \times (1963.49 \times 3000)$$

$$= 152790.83 \text{ Nmm}$$

Example 7.56 A rectangular block of dimensions 100 mm × 80 mm × 60 mm fixed at the bottom edge subjected to a shear force of 1000 kN at the top surface as shown in Figure 7.44. Considering that shear stress is proportional to shear strain, determine shear angle, shear strain energy per unit volume, total shear strain energy for the block, proof resilience if elastic limit for the material is 1000 N/mm^2, and the modulus of resilience if shear modulus for material is 84 kN/mm^2.

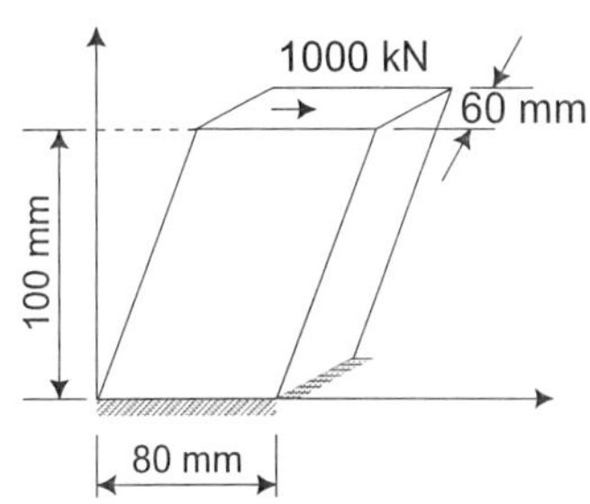

Fig. 7.44

Solution Refer Figure 7.44.

Area on which shear force is applied,

$$A = 80 \times 60 = 4800 \text{ mm}^2$$

Volume of the block, $V = 100 \times 80 \times 60$

$$= 480000 \text{ mm}^3$$

Shear stress, $\tau = \dfrac{1000 \times 1000}{4800}$

$$= 208.33 \text{ N/mm}^2$$

Shear angle, $\phi = \dfrac{\tau}{G} = \dfrac{208.33}{840 \times 1000}$

$$= 2.48 \text{ radian}$$

Shear strain energy per unit volume,

$$u_s = \frac{\tau^2}{2G} = \frac{(208.33)^2}{2 \times 84 \times 1000}$$

$$= 0.258 \text{ Nmm/mm}^3$$

Total strain energy, $U_s = u_s \times V = 0.258 \times (480000)$

$$= 123840 \text{ Nmm}$$

$$\text{Proof resilience} = \frac{\tau_e^2}{2G} \times \text{Volume} = \frac{(1000)^2}{2 \times 84 \times 1000} \times 480000$$

$$= 2857142.86 \text{ Nmm}$$

$$\text{Modulus of resilience} = \frac{\tau_e^2}{2G} = \frac{(1000)^2}{2 \times 84 \times 1000}$$

$$= 5.95 \text{ Nmm/mm}^3$$

Example 7.57 A prismatic metallic rod of rectangular section 500 mm × 200 mm and 2000 mm long is subjected to a load of 250 kN applied gradually on it. If the stress at the elastic limit of the rod material is 200 N/mm^2, and the modulus of elasticity is 200 kN/mm^2, determine strain energy at the given load, proof resilience, and modulus of resilience.

Solution Area on which shear force is applied,

$$A = 500 \times 200 = 10000 \text{ mm}^2$$

Volume of the rod, $V = 500 \times 200 \times 2000 = 2 \times 10^8 \text{ mm}^3$

Stress, $\sigma = (250 \times 10^3/100000) = 2.5 \text{ N/mm}^2$

Strain energy, $U = \frac{\sigma^2}{2E} \times V = \frac{(2.5)^2}{2 \times 200 \times 1000} \times 10^8$

$= 1562.5$ Nmm

Proof resilience $= \frac{\sigma_e^2}{2E} \times V = \frac{(200)^2}{2 \times 200 \times 1000} \times (2 \times 10^8)$

$= 20$ MNmm

Modulus of resilience $= \frac{\sigma_e^2}{2E} = \frac{(200)^2}{2 \times 2 \times 10^5}$

$= 0.1$ N/mm^2

Example 7.58 The water under a pressure of 7.5 N/mm^2 is suddenly admitted on a plunger of diameter 10 cm which is connected to a connecting rod 2.5 cm in diameter and 300 cm. Find the value of stress developed suddenly and deformation of the rod if the modulus of elasticity is 2×10^5 N/mm^2.

Solution Area of cross section of plunger,

$$A = (\pi/4) \times (100)^2 = 7853.98 \text{ mm}^2$$

Therefore, sudden load on plunger,

$$W = p \times A = 7.5 \times 7853.98 = 58904.85 \text{ N}$$

Area of cross section of connecting rod,

$$A' = (\pi/4) \times (25)^2 = 490.87 \text{ mm}^2$$

Stress developed in rod,

$$\sigma_{\text{sudden}} = \frac{2W}{A'} = \frac{2 \times 58904.85}{490.87}$$

$$= 240 \text{ N/mm}^2$$

Sudden deformation in rod

$$= \frac{\sigma_{\text{sudden}}}{E} \times l = \frac{240 \times 3000}{2 \times 10^5}$$

$$= 3.6 \text{ mm}$$

Example 7.59 A bar of 300 mm long and 20 mm in diameter is hanged vertically which has a collar attached at its lower end. Determine the instantaneous stress induced in the bar, its corresponding extension when a weight of 100 N falls on the collar from a height of 40 mm. Also determine the percentage error in the stress induced if the extension of the bar is neglected. Take the Young's modulus of elasticity = 2×10^5 N/mm^2.

Solution Area of cross section of bar,

$$A = (\pi/4) \times (20)^2 = 314.16 \text{ mm}^2$$

The instantaneous stress,

$$\sigma_i = \frac{W}{A}\left[1 + \sqrt{1 + \frac{2hAE}{Wl}}\right]$$

$$\therefore \qquad \sigma_i = \frac{100}{314.16} \times \left[1 + \sqrt{1 + \frac{2 \times 40 \times 314.16 \times 2 \times 10^5}{100 \times 300}}\right]$$

$$= 130.61 \text{ N/mm}^2$$

Instantaneous extension,

$$\delta l_i = \frac{\sigma_i}{E} \times l = \frac{130.61}{2 \times 10^5} \times 300$$

$$= 0.196 \text{ mm}$$

If the extension produced is neglected, the instantaneous stress is given by,

$$\sigma_i = \sqrt{\frac{2WhE}{Al}}$$

$$\therefore \qquad \sigma_i = \sqrt{\frac{2 \times 100 \times 40 \times 314.16 \times 2 \times 10^5}{314.16 \times 300}}$$

$$= 130.29 \text{ N/mm}^2$$

$$\text{Percentage error} = \left(\frac{130.61 - 130.29}{130.61}\right) \times 100$$

$$= 0.245\%$$

Example 7.60 A rod of 3000 mm long and 20 mm in diameter is stressed by a weight of 24 kg dropping freely from a height of 100 mm on a collar provided at the end of rod before commencing to stretch the rod. Determine the instantaneous stress induced and the instantaneous extension. Comment whether the rod will break under this impact load. Take the Young's modulus of elasticity = 2.1×10^5 N/mm^2 and ultimate strength of rod material = 450 N/mm^2.

Solution Area of cross section of rod,

$$A = (\pi/4) \times (20)^2 = 314.16 \text{ mm}^2$$

Falling load, $\quad W = 24 \times 9.8 = 235.2$ kg

The instantaneous stress,

$$\sigma_i = \frac{W}{A}\left[1 + \sqrt{1 + \frac{2hAE}{Wl}}\right]$$

$$\therefore \qquad \sigma_i = \frac{235.2 \times 9.8}{314.16} \times \left[1 + \sqrt{1 + \frac{2 \times 100 \times 314.16 \times 2 \times 10^5}{235.2 \times 3000}}\right]$$

$$= 100.66 \text{ N/mm}^2$$

As $\sigma_i < \sigma_u$, therefore, the rod will not break.

Instantaneous extension,

$$\delta l_i = \frac{\sigma_i}{E} \times l = \frac{100.66}{2 \times 10^5} \times 3000$$
$$= 1.51 \text{ mm}$$

7.19 STRESSES ON AN INCLINED PLANE (OR OBLIQUE SECTION)

Majority of engineering problems are subjected to tensile (or compressive) and shear stresses. Under such condition the resultant stress across any section will neither be normal nor tangential to the plane. In the following sections the stresses acting on an inclined plane (or oblique plane) is explained. To design a component of a machine member, or a structure a designer has to evaluate the maximum stress developed due to the applied loads. At every section of a member, there exists a set of three orthogonal planes, perpendicular to each other on which the stresses are wholly normal and shear stress does not exist. The planes which carry only normal stresses and shear stress is zero are called principal planes. The normal stresses acting on the principal planes are called principal stresses. Out of these three principal stresses, one has maximum value called major principal stress and the other one has the minimum value called minor principal stress and third one has some intermediate value.

(i) Stresses on a Plane Inclined to the Direction of Loading (Uniaxial Loading) Figure 7.45 shows a rectangular bar of uniform cross sectional area A and of unit thickness at right angles to the plane of the paper. The bar is subjected to a single tensile stress σ on the faces BC and AD.

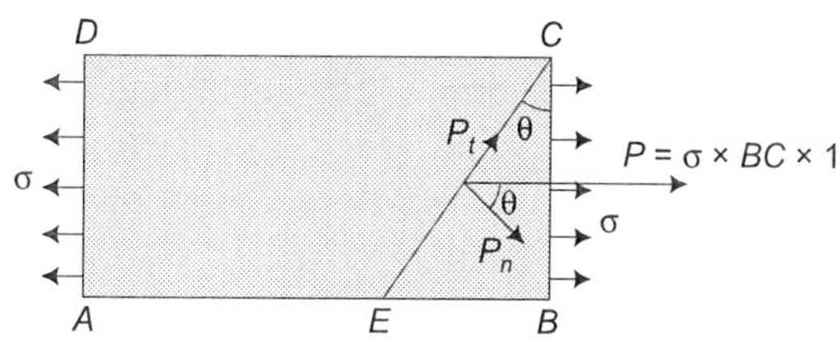

Fig. 7.45

The plane EC on which stresses are to be calculated is inclined at an angle θ with the normal cross section BC.

Tensile force on BC,

$$P = \text{stress} \times \text{area} = \sigma \times (BC \times 1)$$

The tensile force P acting on the inclined section EC in axial direction is resolved into two components, namely, P_n normal to the plane EC and P_t along the plane EC.

Let $\quad P_n = P \cos\theta = \sigma \times BC \times 1 \times \cos\theta,$

and $\quad P_t = P \sin\theta = \sigma \times BC \times 1 \times \sin\theta$

Normal stress on the section *EC*,

$$\sigma_n = \frac{\text{Force normal to section } EC}{\text{Area of section } EC}$$

$$= \frac{P_n}{EC \times 1}$$

$$\therefore \quad \sigma_n = \frac{\sigma \times BC \times \cos\theta}{EC}$$

$$= \sigma \times \cos\theta \times \cos\theta$$

$$= \sigma \cos^2\theta \quad \left(\because \frac{BC}{EC} = \cos\theta\right) \tag{7.45}$$

Similarly, tangential (or shear) stress,

$$\sigma_t = \frac{\text{Force along section } EC}{\text{Area of section } EC}$$

$$= \frac{P_t}{EC \times 1}$$

$$\sigma_t = \frac{\sigma \times BC \times \sin\theta}{EC}$$

$$= \sigma \times \cos\theta \times \sin\theta$$

$$\therefore \quad \sigma_t = \frac{\sigma}{2} \times 2 \times \cos\theta \times \sin\theta$$

$$= \frac{\sigma}{2} \times \sin 2\theta \quad (\because \sin 2\theta = 2\cos\theta\sin\theta) \tag{7.46}$$

From equation (7.45) it is observed that normal stress on oblique section *EC* will be maximum when $\cos^2\theta$ or $\cos\theta$ will have maximum value which is possible at $\theta = 0°$. But, when $\theta = 0°$, the section *EC* will coincide with section *BC* which is normal to the line of action of loading. Therefore, the plane normal to the axis of loading will carry maximum normal stress.

$$\therefore \quad \text{Maximum normal stress} = \sigma \cos^2 0°$$

$$= \sigma \tag{7.47}$$

From equation (7.46) it is observed that tangential stress (or shear stress) on oblique section *EC* will be maximum when $\sin 2\theta$ will have maximum value which is given by $\sin 2\theta = 1$ or $2\theta = 90°$ or $\theta = 45°$ or $135°$.

$$\therefore \quad \text{Maximum shear stress} = \frac{\sigma}{2}\sin 90° = \frac{\sigma}{2} \tag{7.48}$$

From equations (7.47) and (7.48) it is observed that maximum shear stress is half of the maximum normal stress value.

Example 7.61 A rectangular rod of cross sectional area 11000 mm^2 is subjected to an axial load of 44 kN. Determine the normal and shear stresses on a section which is inclined at an angle of 30° with normal cross section of the rod.

Solution Direct stress, $\sigma = (P/A) = (44 \times 1000)/11000$

$$= 4 \text{ N/mm}^2$$

The normal stress,

$$\sigma_n = \sigma \cos^2\theta = 4 \times \cos^2 30°$$

$$= 4 \times (0.866)^2 = 3 \text{ N/mm}^2$$

The shear stress,

$$\sigma_i = (\sigma/2) \times \sin 2\theta = (4/2) \times \sin (2 \times 30°)$$

$$= 1.732 \text{ N/mm}^2$$

Example 7.62 A rectangular rod of cross sectional area 10000 mm^2 is subjected to a tensile load P as shown in Figure 7.46. The permissible normal and shear stresses on the oblique plane AB are given as 14 N/mm^2 and 7 N/mm^2. Evaluate the safe value of P.

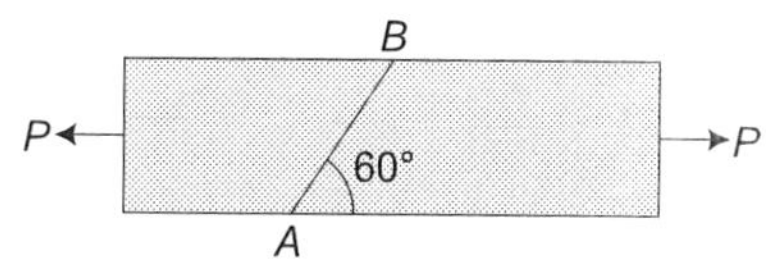

Fig. 7.46

Solution Refer Figure 7.46.

Angle of oblique plane AB with the normal cross section of the bar,

$$\theta = 90° - 60° = 30°$$

The normal stress, $\sigma_n = \sigma \cos^2\theta$

$$\therefore \quad \sigma = \frac{\sigma_n}{\cos^2\theta} = \frac{14}{\cos^2(30°)}$$

$$= 18.67 \text{ N/mm}^2$$

The shear stress, $\sigma_t = \frac{\sigma}{2} \times \sin 2\theta$

$$\therefore \quad \sigma = \frac{2\sigma_t}{\sin 2\theta} = \frac{2 \times 7}{\sin (2 \times 30°)}$$

$$= 16.17 \text{ N/mm}^2$$

The safe stress will be the least of two,

$$\therefore \quad \sigma_{safe} = 16.17 \text{ N/mm}^2$$

Therefore, safe value of pull,

$$P = \sigma_{safe} \times \text{area}$$

$$= 16.17 \times 10000 = 161700 \text{ N}$$

$$= 161.7 \text{ kN}$$

(ii) Stresses on Inclined Plane of a Member Subjected to Direct Stresses in Two Mutually Perpendicular Directions (Biaxial Loading) Figure 7.47 shows a rectangular bar of uniform cross sectional area A and of unit thickness at right angles to the plane of the paper. The bar is subjected to two direct tensile stresses σ_1 and σ_2 as shown in Figure 7.47. The plane EC on which stresses is to be calculated is inclined at an angle θ with the normal cross section BC.

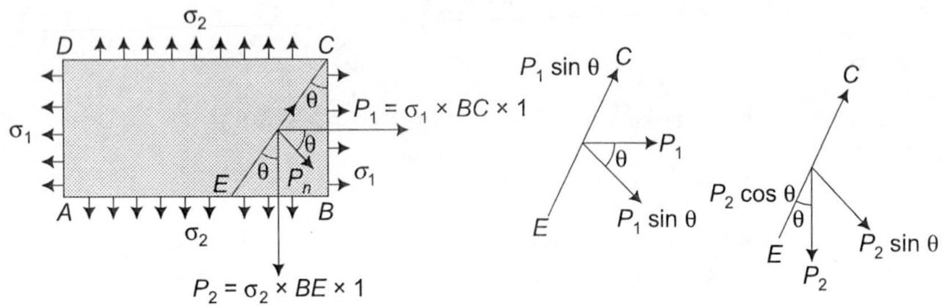

Fig. 7.47

Tensile force on BC,

$$P_1 = \text{stress on } BC \times \text{area of } BC$$

$$= \sigma_1 \times (BC \times 1),$$

and tensile force on EB,

$$P_2 = \text{stress on } EB \times \text{area of } EB$$

$$= \sigma_2 \times (EB \times 1)$$

The tensile forces P_1 and P_2 also act on the inclined section EC. The force P_1 acts in the axial direction and the force P_2 acts in downward direction. These forces P_1 and P_2 can be resolved into two components, namely, P_n normal to the plane EC and P_t along the plane EC. $P_1 \cos\theta$ and $P_2 \sin\theta$ are the components of P_1 and P_2 normal to plane EC, whereas $P_1 \sin\theta$ and $P_2 \cos\theta$ are along the plane EC.

Total force normal to plane EC,

$$P_n = P_1 \cos\theta + P_2 \sin\theta$$

$$\therefore \quad P_n = \sigma_1 \times BC \times 1 \times \cos\theta + \sigma_2 \times BE \times 1 \times \sin\theta$$

Total force along plane EC,

$$P_t = P_1 \sin\theta + (-P_2 \cos\theta)$$

$$= P_1 \sin\theta - P_2 \cos\theta$$

$$\therefore \quad P_t = \sigma_1 \times BC \times 1 \times \sin\theta - \sigma_2 \times BE \times 1 \times \cos\theta$$

Normal stress across oblique section EC,

$$\sigma_n = \frac{\text{Total force normal to } EC}{\text{Area of section } EC}$$

$$= \frac{P_n}{EC \times 1}$$

$$\sigma_n = \frac{\sigma_1 \times BC \times \cos\theta + \sigma_2 \times BE \times \sin\theta}{EC}$$

$$= \sigma_1 \times \frac{BC}{EC} \times \cos\theta + \sigma_2 + \frac{BE}{EC} \times \sin\theta$$

$$\sigma_n = \sigma_1 \times \cos\theta \times \cos\theta + \sigma_2 + \sin\theta \times \sin\theta$$

$$= \sigma_1 \cos^2\theta + \sigma_2 \sin^2\theta$$

$$\sigma_n = \sigma_1\left(\frac{1 + \cos 2\theta}{2}\right) + \sigma_2\left(\frac{1 - \cos 2\theta}{2}\right)$$

$$\therefore \quad \sigma_n = \frac{\sigma_1 + \sigma_2}{2} + \frac{\sigma_1 - \sigma_2}{2} \cos 2\theta \qquad (7.49)$$

Tangential (or shear) stress along section EC,

$$\sigma_t = \frac{\text{Total force along } EC}{\text{Area of section } EC}$$

$$= \frac{P_t}{EC \times 1}$$

$$\sigma_t = \frac{\sigma_1 \times BC \times \sin\theta - \sigma_2 \times BE \times \cos\theta}{EC}$$

$$= \sigma_1 \times \frac{BC}{EC} \times \sin\theta - \sigma_2 + \frac{BE}{EC} \times \cos\theta$$

$$\sigma_t = \sigma_1 \times \cos\theta \times \sin\theta - \sigma_2 + \sin\theta \times \cos\theta = (\sigma_1 - \sigma_2) \sin\theta \cos\theta$$

$$\sigma_t = \frac{(\sigma_1 - \sigma_2)}{2} 2 \sin\theta \cos\theta$$

$$\therefore \quad \sigma_t = \frac{(\sigma_1 - \sigma_2)}{2} \sin 2\theta \qquad (7.50)$$

The resultant stress,

$$\sigma_r = \sqrt{\sigma_n^2 + \sigma_t^2} \qquad (7.51)$$

The obliquity means the angle made by the resultant stress with the normal of the oblique plane is given by,

$$\tan\phi = (\sigma_t/\sigma_n) \qquad (7.52)$$

The shear stress is given by equation (7.50) which will be maximum when

$$\sin 2\theta = 1$$

or $$2\theta = 90°$$

or $$\theta = 45° \text{ or } 135°$$

Therefore, maximum shear stress

$$(\sigma_t)_{max} = \frac{(\sigma_1 - \sigma_2)}{2} \quad (7.53)$$

To locate the position of principal planes, the shear stress equation (7.50) should be equated to zero.

$$\therefore \quad \frac{(\sigma_1 - \sigma_2)}{2} \sin 2\theta = 0$$

or $$\sin 2\theta = 0^\circ$$

or $$2\theta = 180^\circ$$

$$\therefore \quad \theta = 0^\circ \text{ or } 90^\circ$$

When $\theta = 0^\circ$,

$$\sigma_n = \frac{(\sigma_1 + \sigma_2)}{2} + \frac{(\sigma_1 - \sigma_2)}{2} \times \cos 0^\circ$$

$$= \frac{(\sigma_1 + \sigma_2)}{2} + \frac{(\sigma_1 - \sigma_2)}{2} \times 1$$

$$= \sigma_1$$

When $\theta = 90^\circ$,

$$\sigma_n = \frac{(\sigma_1 + \sigma_2)}{2} + \frac{(\sigma_1 - \sigma_2)}{2} \times \cos 180^\circ$$

$$= \frac{(\sigma_1 + \sigma_2)}{2} + \frac{(\sigma_1 - \sigma_2)}{2} \times (-1)$$

$$= \sigma_2$$

(iii) A Member Subjected to Simple Stress (Shear Stresses) Figure 7.48 (a) shows a rectangular bar of uniform cross sectional area *A* and of unit thickness at right angles to the plane of the paper. The bar is subjected to a simple shear stress τ across the faces *BC* and *AD*. The plane *EC* on which stresses are to be calculated is inclined at an angle θ with the normal cross section *BC*. A shear stress is always accompanied by an equal shear stress at right angles to it, so the faces *AB* and *CD* will be subjected to a shear stress τ.

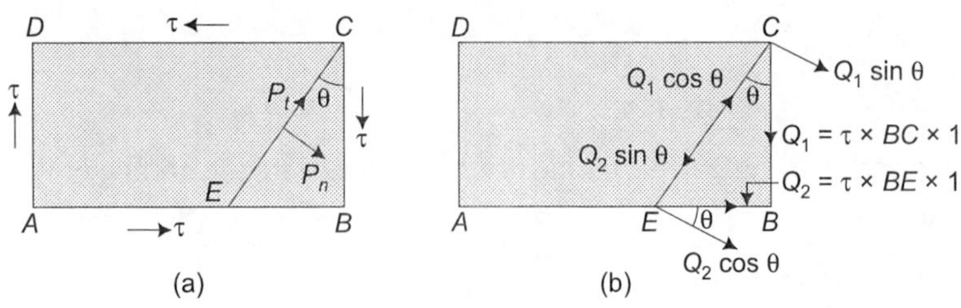

Fig. 7.48

The force Q_1 acts along *BC* and Q_2 acts along *EB* as shown in Figure 7.48 (b)

Shear force on *BC*, Q_1 = shear stress on *BC* × area of *BC*

$$= \tau \times (BC \times 1)$$

Shear force on EB, Q_2 = shear stress on $EB \times$ area of EB

$$= \tau \times (EB \times 1)$$

The force Q_1 is resolved into two components, namely, $Q_1 \cos\theta$ and $Q_1 \sin\theta$ along the plane CE and normal to plane CE respectively (Figure 7.47 b).

The force Q_2 is resolved into two components, namely, $Q_2 \sin\theta$ and $Q_2 \cos\theta$ along the plane CE and normal to plane CE respectively (Figure 7.48 b).

Total force normal to plane EC,

$$P_n = Q_1 \sin\theta + Q_2 \cos\theta$$

$$\therefore \quad P_n = \tau \times BC \times 1 \times \sin\theta + \tau \times BE \times 1 \times \cos\theta$$

Total force along plane EC,

$$P_t = Q_2 \sin\theta - Q_1 \cos\theta$$

$$\therefore \quad P_t = \tau \times EB \times 1 \times \sin\theta - \tau \times BC \times 1 \times \cos\theta$$

Normal stress across oblique section EC,

$$\sigma_n = \frac{\text{Total force normal to } EC}{\text{Area of section } EC}$$

$$= \frac{P_n}{EC \times 1}$$

$$\sigma_n = \frac{\tau \times BC \times \sin\theta + \tau \times BE \times \cos\theta}{EC}$$

$$= \tau \times \frac{BC}{EC} \times \sin\theta + \tau + \frac{BE}{EC} \times \cos\theta$$

$$\sigma_n = \tau \times \cos\theta \times \sin\theta + \tau + \sin\theta \times \cos\theta$$

$$= \tau \times (2\sin\theta \times \cos\theta)$$

$$\therefore \quad \sigma_n = \tau \sin 2\theta \quad (\because 2\sin\theta\cos\theta = \sin 2\theta) \qquad (7.54)$$

Tangential (or shear) stress along section EC,

$$\sigma_t = \frac{\text{Total force along } EC}{\text{Area of section } EC}$$

$$= \frac{P_t}{EC \times 1}$$

$$\sigma_t = \frac{\tau \times BE \times \sin\theta - \tau \times BC \times \cos\theta}{EC}$$

$$= \tau \times \frac{BE}{EC} \times \sin\theta - \tau + \frac{BC}{EC} \times \cos\theta$$

$$\sigma_t = \tau \times \sin\theta \times \sin\theta - \tau + \cos\theta \times \cos\theta$$

$$= \tau \times (\sin^2\theta - \cos^2\theta)$$

$\therefore$ $$\sigma_t = -\tau \cos 2\theta \quad (\because \sin^2\theta - \cos^2\theta = \cos 2\theta) \qquad (7.55)$$

The negative sign in equation (7.55) indicates that σ_t is acting in the downward direction on the plane EC.

Example 7.63 The tensile stresses at a point across two mutually perpendicular planes are 100 N/mm^2 and 50 N/mm^2. Evaluate the normal, tangential, and resultant stresses on a plane inclined at 30° to the axis of minor stress.

Solution Refer Figure 7.49.

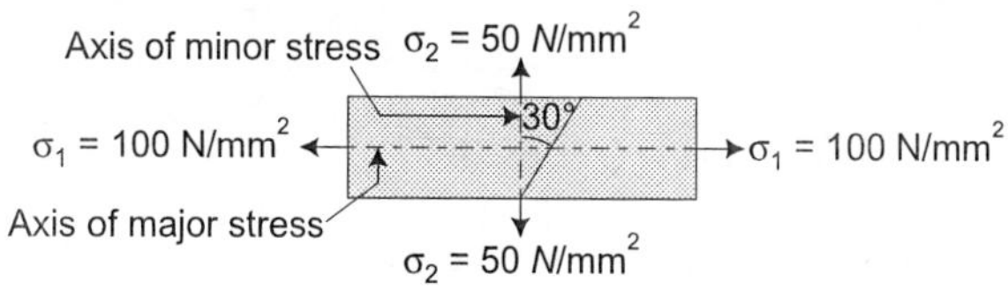

Fig. 7.49

Major tensile stress, $\sigma_1 = 100$ N/mm^2,

and minor tensile stress,

$$\sigma_2 = 50 \text{ N/mm}^2$$

Normal stress, $$\sigma_n = \frac{(\sigma_1 + \sigma_2)}{2} + \frac{(\sigma_1 - \sigma_2)}{2} \cos 2\theta$$

$\therefore$ $$\sigma_n = \frac{100 + 50}{2} + \frac{100 - 50}{2} \times \cos(2 \times 30)$$

$$= 75 + 25 \times 0.5 = 87.5 \text{ N/mm}^2$$

Tangential stress, $$\sigma_t = \frac{(\sigma_1 - \sigma_2)}{2} \sin 2\theta$$

$\therefore$ $$\sigma_t = \frac{(100 - 50)}{2} \times \sin(2 \times 30) = 25 \times \sin 60°$$

$$= 25 \times 0.866 = 21.65 \text{ N/mm}^2$$

Resultant stress, $$\sigma_r = \sqrt{\sigma_n^2 + \sigma_t^2}$$

$\therefore$ $$\sigma_r = \sqrt{(87.5)^2 + (21.65)^2} = \sqrt{9531.14}$$

$$= 90.14 \text{ N/mm}^2$$

Example 7.64 The stresses at a point in a rod are 120 N/mm^2 (tensile) and 60 N/mm^2 (compressive). Determine the resultant stress in magnitude and direction on a plane inclined at 60° to the axis of major stress. Also determine the maximum intensity of shear stress in the given rod material at the point.

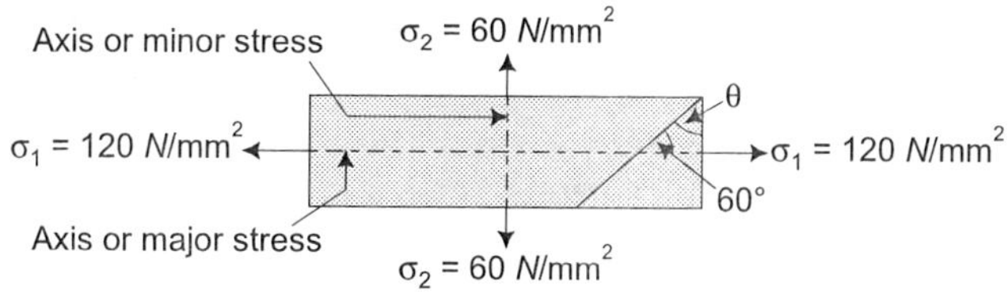

Fig. 7.50

Solution Refer Figure 7.50.

Major stress (tensile),

$$\sigma_1 = 120 \text{ N/mm}^2,$$

and minor stress (compressive),

$$\sigma_2 = 60 \text{ N/mm}^2$$

Angle of the plane made with major principal stress = 60°

$\therefore$ $$\theta = 90° - 60° = 30°$$

Normal stress, $$\sigma_n = \frac{(\sigma_1 + \sigma_2)}{2} + \frac{(\sigma_1 - \sigma_2)}{2} \cos 2\theta$$

$\therefore$ $$\sigma_n = \frac{120 + (-60)}{2} + \frac{120 - (-60)}{2} \times \cos (2 \times 30)$$

$$= 30 + 90 \times 0.5 = 75 \text{ N/mm}^2$$

Tangential stress, $$\sigma_t = \frac{(\sigma_1 - \sigma_2)}{2} \sin 2\theta$$

$\therefore$ $$\sigma_t = \frac{120 - (-60)}{2} \times \sin (2 \times 30)$$

$$= 90 \times \sin 60° = 90 \times 0.866$$

$$= 77.94 \text{ N/mm}^2$$

Resultant stress, $$\sigma_r = \sqrt{\sigma_n^2 + \sigma_t^2}$$

$\therefore$ $$\sigma_r = \sqrt{(75)^2 + (77.94)^2} = \sqrt{11699.64}$$

$$= 108.16 \text{ N/mm}^2$$

Maximum shear stress,

$$(\sigma_t)_{\max} = \frac{(\sigma_1 - \sigma_2)}{2}$$

$\therefore$ $$(\sigma_t)_{\max} = \frac{120 - (-60)}{2} = \frac{180}{2}$$

$$= 90 \text{ N/mm}^2$$

(iv) Stresses on Inclined Plane of a Member Subjected to Direct Stresses in Two Mutually Perpendicular Directions and Shear Stress at Right Angles to the Direct Stresses (Complex Stresses) Figure 7.51 (a) shows a rectangular bar of uniform cross sectional area A and of unit thickness at right angles to the plane of the paper. The bar is subjected to two direct tensile stresses (σ_1 and σ_2) and a simple shear stress (τ) as shown in Figure 7.51. The oblique plane EC on which stresses are to be calculated is inclined at an angle θ with the normal cross section BC.

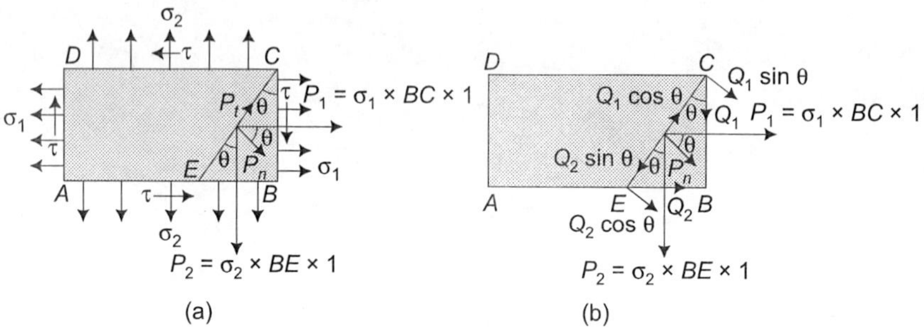

Fig. 7.51

The forces acting on the wedge EBC are shown in Figure 7.51 (b).

Tensile force on BC,

$$P_1 = \text{stress on } BC \times \text{area of } BC$$
$$= \sigma_1 \times (BC \times 1)$$

Tensile force on EB,

$$P_2 = \text{stress on } EB \times \text{area of } EB$$
$$= \sigma_2 \times (EB \times 1)$$

Shear force on BC,

$$Q_1 = \text{shear stress on } BC \times \text{area of } BC$$
$$= \tau \times (BC \times 1)$$

Shear force on EB,

$$Q_2 = \text{shear stress on } EB \times \text{area of } EB$$
$$= \tau \times (EB \times 1)$$

The force Q_1 is resolved into two components, namely, $Q_1 \cos\theta$ and $Q_1 \sin\theta$ along the plane CE and normal to plane CE respectively (Figure 7.51 b).

The force Q_2 is resolved into two components, namely, $Q_2 \sin\theta$ and $Q_2 \cos\theta$ along the plane CE and normal to plane CE respectively (Figure 7.51 b).

Total force normal to plane *EC*,

$$P_n = P_1 \cos\theta + P_2 \sin\theta + Q_1 \sin\theta + Q_2 \cos\theta$$

$$\therefore \quad P_n = \sigma_1 \times BC \times 1 \times \cos\theta + \sigma_2 \times BE \times 1 \times \sin\theta + \tau \times BC \times 1 \times \sin\theta + \tau \times BE \times 1 \times \cos\theta$$

Total force along plane *EC*,

$$P_t = P_1 \sin\theta - P_2 \cos\theta + Q_2 \sin\theta - Q_1 \cos\theta$$

$$\therefore \quad P_t = \sigma_1 \times BC \times 1 \times \sin\theta - \sigma_2 \times BE \times 1 \times \cos\theta + \tau \times EB \times 1 \times \sin\theta - \tau \times BC \times 1 \times \cos\theta$$

Normal stress across oblique section *EC*,

$$\sigma_n = \frac{\text{Total force normal to } EC}{\text{Area of section } EC} = \frac{P_n}{EC \times 1}$$

$$\sigma_n = \frac{\sigma_1 \times BC \times \cos\theta + \sigma_2 \times BE \times \sin\theta + \tau \times BC \times \sin\theta + \tau \times BE \times \cos\theta}{EC}$$

$$\sigma_n = \sigma_1 \times \frac{BC}{EC} \times \cos\theta + \sigma_2 \times \frac{BE}{EC} \times \sin\theta + \tau \times \frac{BC}{EC} \times \sin\theta + \tau \times \frac{BE}{EC} \times \cos\theta$$

$$\sigma_n = \sigma_1 \times \cos\theta \times \cos\theta + \sigma_2 \times \sin\theta \times \sin\theta + \tau \times \cos\theta \times \sin\theta + \tau \times \sin\theta \times \cos\theta$$

$$\sigma_n = \sigma_1 \times \cos^2\theta + \sigma_2 \times \sin^2\theta + 2\tau \sin\theta \cos\theta$$

$$\sigma_n = \sigma_1 \left(\frac{1 + \cos 2\theta}{2}\right) + \sigma_2 \left(\frac{1 - \cos 2\theta}{2}\right) + \tau \sin 2\theta$$

$$\left(\because \sin 2\theta = 2\sin\theta \cos\theta,\ \cos^2\theta = \frac{1 + \cos 2\theta}{2},\ \sin^2\theta = \frac{1 - \cos 2\theta}{2}\right)$$

$$\therefore \quad \sigma_n = \frac{\sigma_1 + \sigma_2}{2} + \frac{\sigma_1 - \sigma_2}{2} \cos 2\theta + \tau \sin 2\theta \qquad (7.56)$$

Tangential (or shear) stress along section *EC*,

$$\sigma_t = \frac{\text{Total force along } EC}{\text{Area of section EC}} = \frac{P_t}{EC \times 1}$$

$$\sigma_t = \frac{\sigma_1 \times BC \times \sin\theta - \sigma_2 \times BE \times \cos\theta + \tau \times BE \times \sin\theta - \tau \times BC \times \cos\theta}{EC}$$

$$\sigma_t = \sigma_1 \times \frac{BC}{EC} \times \sin\theta - \sigma_2 \times \frac{BE}{EC} \times \cos\theta + \tau \times \frac{BE}{EC} \times \sin\theta - \tau \times \frac{BC}{EC} \times \cos\theta$$

$$\sigma_t = \sigma_1 \times \cos\theta \times \sin\theta - \sigma_2 \times \sin\theta \times \cos\theta + \tau \times \sin\theta \times \sin\theta - \tau \times \cos\theta \times \cos\theta$$

$$\sigma_t = (\sigma_1 - \sigma_2) \times \cos\theta \sin\theta + \tau \times \sin^2\theta - \tau \times \cos^2\theta$$

$$\sigma_t = \left(\frac{\sigma_1 - \sigma_2}{2}\right) \times 2\cos\theta \sin\theta - \tau(\cos^2\theta - \sin^2\theta)$$

$$\therefore \quad \sigma_t = \frac{\sigma_1 - \sigma_2}{2} \sin 2\theta - \tau \cos 2\theta \quad (\because \cos^2\theta - \sin^2\theta = \cos 2\theta) \tag{7.57}$$

(v) Principal Stresses The planes on which shear stress or tangential stress is zero are called principal planes and the stresses acting on these planes are called principal stresses. In this section, we will determine two principal stresses taking the third principal stress (i.e., the stress on the plane of the paper) zero.

The shear stress on inclined plane *EC* is given by equation (7.57) as,

$$\sigma_t = \frac{\sigma_1 - \sigma_2}{2} \sin 2\theta - \tau \cos 2\theta$$

On principal planes, shear stress σ_t is zero,

$$\therefore \quad \frac{\sigma_1 - \sigma_2}{2} \sin 2\theta - \tau \cos 2\theta = 0$$

or

$$\tan 2\theta = \frac{2\tau}{\sigma_1 - \sigma_2} \tag{7.58}$$

Also

$$\tan 2\theta = \tan(180 + 2\theta),$$

therefore, with reference to the plane *EC*, there exist two planes at angles θ_1 and θ_2, on which the shear stress is zero. These two planes are called principal planes.

The angles are:

$$\theta_1 = \frac{1}{2} \tan^{-1} \frac{2\tau}{\sigma_1 - \sigma_2},$$

and

$$\theta_2 = 90° + \theta_1 \tag{7.59}$$

The normal stress on the inclined plane *EC* is given by equation (7.56) as,

$$\sigma_n = \frac{\sigma_1 + \sigma_2}{2} + \frac{\sigma_1 - \sigma_2}{2} \cos 2\theta + \tau \sin 2\theta$$

On the principal planes, normal stress is maximum and minimum,

$$\therefore \quad \frac{d}{d\theta}(\sigma_n) = 0$$

$$\frac{d}{d\theta}\left[\frac{\sigma_1 + \sigma_2}{2} + \frac{\sigma_1 - \sigma_2}{2} \cos 2\theta + \tau \sin 2\theta\right] = 0$$

$$\frac{\sigma_1 - \sigma_2}{2}(-2\sin 2\theta) + \tau(2\cos 2\theta) = 0$$

$$\frac{\sigma_1 - \sigma_2}{2} \sin 2\theta - \tau \cos 2\theta = 0,$$

which is the expression for shear stress and shows that shear stress is zero on planes on which normal stress is either maximum or minimum. Such planes are called principal planes.

For principal planes,

$$\tan 2\theta = \frac{2\tau}{\sigma_1 - \sigma_2}$$

2τ and $(\sigma_1 - \sigma_2)$ represents height and base respectively of a right angled triangle *XYZ* as shown in Figure 7.52.

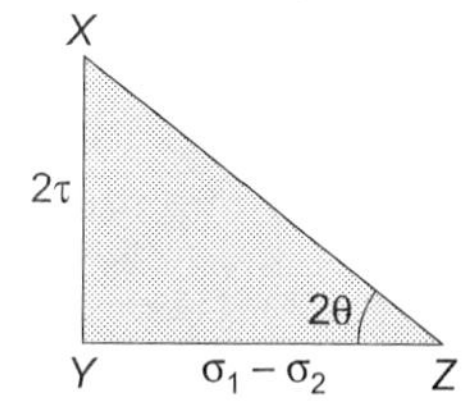

Fig. 7.52

So, diagonals of the right angle triangle

$$= \pm \sqrt{(\sigma_1 - \sigma_2)^2 + (2\tau)^2}$$

Taking positive value of diagonal

$$= \sqrt{(\sigma_1 - \sigma_2)^2 + (2\tau)^2} = \sqrt{(\sigma_1 - \sigma_2)^2 + 4\tau^2}$$

$$\therefore \qquad \sin 2\theta = \frac{2\tau}{\sqrt{(\sigma_1 - \sigma_2) + 4\tau^2}},$$

and

$$\cos 2\theta = \frac{\sigma_1 - \sigma_2}{\sqrt{(\sigma_1 - \sigma_2)^2 + 4\tau^2}}$$

The value of major principal stress (p_1) or maximum normal stress is obtained by substituting the values of sin 2θ and cos 2θ in equation (7.56)

$$\therefore \qquad p_1 = \frac{\sigma_1 + \sigma_2}{2} + \frac{\sigma_1 - \sigma_2}{2} \times \frac{\sigma_1 - \sigma_2}{\sqrt{(\sigma_1 - \sigma_2)^2 + 4\tau^2}} + \frac{\tau \times 2\tau}{\sqrt{(\sigma_1 - \sigma_2) + 4\tau^2}}$$

After simplification, we have,

$$p_1 = \frac{\sigma_1 + \sigma_2}{2} + \sqrt{\left(\frac{\sigma_1 - \sigma_2}{2}\right)^2 + \tau^2} \qquad (7.60)$$

Now taking negative value of diagonal

$$= -\sqrt{(\sigma_1 - \sigma_2)^2 + (2\tau)^2} = -\sqrt{(\sigma_1 - \sigma_2)^2 + 4\tau^2}$$

$$\therefore \qquad \sin 2\theta = \frac{2\tau}{-\sqrt{(\sigma_1 - \sigma_2) + 4\tau^2}},$$

and

$$\cos 2\theta = \frac{\sigma_1 - \sigma_2}{-\sqrt{(\sigma_1 - \sigma_2)^2 + 4\tau^2}}$$

The value of minor principal stress (p_2) or minimum normal stress is obtained by substituting the values of sin 2θ and cos 2θ in equation (7.56)

$$\therefore \qquad p_2 = \frac{\sigma_1 + \sigma_2}{2} + \frac{\sigma_1 - \sigma_2}{2} \times \frac{\sigma_1 - \sigma_2}{-\sqrt{(\sigma_1 - \sigma_2)^2 + 4\tau^2}} + \frac{\tau \times 2\tau}{-\sqrt{(\sigma_1 - \sigma_2) + 4\tau^2}}$$

After simplification, we have,

$$p_2 = \frac{\sigma_1 + \sigma_2}{2} - \sqrt{\left(\frac{\sigma_1 - \sigma_2}{2}\right)^2 + \tau^2} \qquad (7.61)$$

Equation (7.60) gives the maximum principal stress and equation (7.61) gives the minimum principal stress. The principal planes on which principal stresses act are at right angles.

Maximum Shear Stress The shear stress on inclined plane *EC* is given by equation (7.57).

For maximum shear stress,

$$\frac{d}{d\theta}(\sigma_t) = 0$$

$$\frac{d}{d\theta}\left(\frac{\sigma_1 - \sigma_2}{2}\sin 2\theta - \tau \cos 2\theta\right) = 0$$

$$\frac{\sigma_1 - \sigma_2}{2} \times (\cos 2\theta) \times 2 - \tau(-\sin 2\theta) \times 2 = 0$$

$$\tan 2\theta = -\frac{(\sigma_1 - \sigma_2)/2}{\tau} = \frac{(\sigma_2 - \sigma_1)/2}{\tau}$$

Also, $\tan 2\theta = \tan(180 - 2\theta)$,

therefore, there exist two planes at angles θ_3 and θ_4, on which the shear stress is maximum.

The angles of these planes are

$$\theta_3 = \frac{1}{2}\tan^{-1}\frac{(\sigma_2 - \sigma_1)/2}{\tau},$$

and $\theta_4 = 90° + \theta_3$

From the expression of tan 2θ, we can write,

$$\sin 2\theta = \frac{(\sigma_2 - \sigma_1)/2}{\sqrt{\left(\frac{\sigma_2 - \sigma_1}{2}\right)^2 + \tau^2}},$$

and

$$\cos 2\theta = \frac{\tau}{\sqrt{\left(\frac{\sigma_2 - \sigma_1}{2}\right)^2 + \tau^2}}$$

Substituting the values of sin 2θ and cos 2θ in equation (7.57) and solving, we get,

$$(\sigma_t)_{max} = \pm\sqrt{\left(\frac{\sigma_1 - \sigma_2}{2}\right)^2 + \tau^2} \qquad (7.62)$$

Example 7.65 At a point in a strained material, on plane *BC* there are normal and shear stresses 60 N/mm^2 and 15 N/mm^2 respectively. On plane *AC*, perpendicular to plane *BC*, there are normal and shear stresses of 300 N/mm^2 and 15 N/mm^2 as shown in Figure 7.53(a). Determine (i) principal stresses and location of principal planes, and (ii) maximum shear stress and the planes on which it acts.

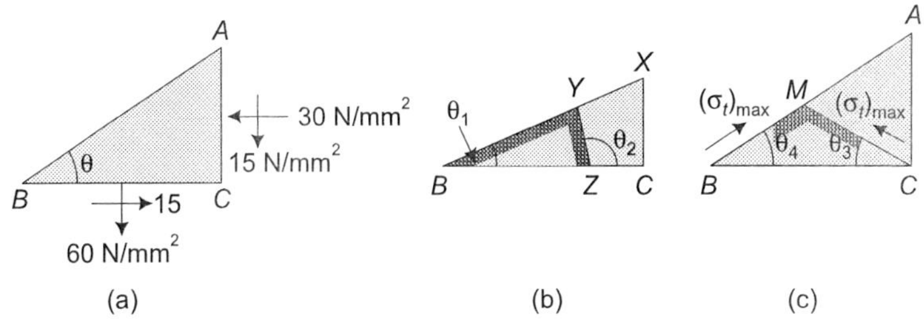

Fig. 7.53

Solution Refer Figure 7.53(a). Taking plane *BC* as the reference plane, stresses are

$$\sigma_1 = +60 \text{ N/mm}^2, \qquad \sigma_2 = -30 \text{ N/mm}^2,$$

$$\tau = -15 \text{ N/mm}^2 \quad \text{(on reference plane)}$$

(i) Principal stresses and location of principal planes

$$p_1 = \frac{\sigma_1 + \sigma_2}{2} + \sqrt{\left(\frac{\sigma_1 - \sigma_2}{2}\right)^2 + \tau^2}$$

$$\therefore \quad p_1 = \frac{60 - 30}{2} + \sqrt{\left(\frac{60 + 30}{2}\right)^2 + (15)^2}$$

$$= 62.43 \text{ N/mm}^2$$

$$p_2 = \frac{\sigma_1 + \sigma_2}{2} - \sqrt{\left(\frac{\sigma_1 - \sigma_2}{2}\right)^2 + \tau^2}$$

$$\therefore \quad p_1 = \frac{60 - 30}{2} - \sqrt{\left(\frac{60 + 30}{2}\right)^2 + (15)^2}$$

$$= -32.43 \text{ N/mm}^2$$

The angles are,

$$\theta_1 = \frac{1}{2}\tan^{-1}\frac{2\tau}{\sigma_1 - \sigma_2}$$

$$= \frac{1}{2}\tan^{-1}\frac{2 \times 15}{60 + 30}$$

$$= 9°13'$$

and

$$\theta_2 = 90° + \theta_1 = 90 + 9°13'$$

$$= 99°13'$$

Principal plane *BX* is inclined at an angle of θ_1 with *BC* and principal plane *YZ* is inclined at θ_2 with *BC* as shown in Figure 7.53(b).

(ii) Maximum shear stress and the planes of maximum shear

$$(\sigma_t)_{max} = \pm\sqrt{\left(\frac{\sigma_1 - \sigma_2}{2}\right)^2 + \tau^2}$$

$$= \pm\sqrt{\left(\frac{60 + 30}{2}\right)^2 + (15)^2}$$

$$= \pm\ 47.43\ \text{N/mm}^2$$

$$\theta_3 = \frac{1}{2}\tan^{-1}\frac{(\sigma_1 - \sigma_2)/2}{\tau}$$

$$= \frac{1}{2}\tan^{-1}\frac{(-30 - 60)/2}{15}$$

$$= -\ 35°47'$$

and

$$\theta_4 = 90° + \theta_3 = 90° - 35°47'$$

$$= 54°13'$$

The planes carrying maximum shear stress $(\sigma_t)_{max}$ and $-(\sigma_t)_{max}$ are shown in Figure 7.53(c) by *BM* and *CM* at angles of θ_4 and θ_3 respectively with respect to plane *BC*.

Example 7.66 A stressed element is subjected to unidirectional stress σ_1 = 600 N/mm^2. If maximum allowable tensile principal stress p_1 = 700 N/mm^2, determine the maximum value of τ and minor principal stress.

Solution The major principal stress is given by,

$$p_1 = \frac{\sigma_1 + \sigma_2}{2} + \sqrt{\left(\frac{\sigma_1 - \sigma_2}{2}\right)^2 + \tau^2}$$

or

$$700 = \frac{600 + 0}{2} + \sqrt{\left(\frac{600 - 0}{2}\right)^2 + \tau^2}$$

$$= 300 + \sqrt{90000 + \tau^2}$$

$$\tau^2 = (700 - 300)^2 - 90000 = 70000$$

$\therefore$

$$\tau = \sqrt{70000} = 264.57\ \text{N/mm}^2$$

The minor principal stress is given by,

$$p_2 = \frac{\sigma_1 + \sigma_2}{2} - \sqrt{\left(\frac{\sigma_1 - \sigma_2}{2}\right)^2 + \tau^2}$$

$\therefore$

$$p_2 = \frac{600 + 0}{2} - \sqrt{\left(\frac{600 - 0}{2}\right)^2 + (264.57)^2}$$

$$= 150 - 304.13 = -\ 154.13\ \text{N/mm}^2$$

Example 7.67 Direct stresses of 100 N/mm^2 (tensile) and 70 N/mm^2 (compressive) exist on two perpendicular planes at a certain point in a body which are also accompanied by shear stress τ. If maximum allowable tensile principal stress 130 N/mm^2, determine the maximum value of τ and the maximum shearing stress at the point.

Solution The major principal stress is given by,

$$p_1 = \frac{\sigma_1 + \sigma_2}{2} + \sqrt{\left(\frac{\sigma_1 - \sigma_2}{2}\right)^2 + \tau^2}$$

$$130 = \frac{100 - 70}{2} + \sqrt{\left(\frac{100 + 70}{2}\right)^2 + \tau^2}$$

$$= 15 + \sqrt{7225 + \tau^2}$$

$$\tau^2 = (130 - 15)^2 - 7225$$

$$= 13225 - 7225 = 6000$$

$$\therefore \quad \tau = \sqrt{6000} = 77.46 \text{ N/mm}^2$$

Maximum shear stress at the point,

$$(\sigma_t)_{max} = \pm\sqrt{\left(\frac{\sigma_1 - \sigma_2}{2}\right)^2 + \tau^2}$$

$$\therefore \quad (\sigma_t)_{max} = \pm\sqrt{\left(\frac{100 + 70}{2}\right)^2 + (77.46)^2}$$

$$= \pm\sqrt{13225.05} = \pm 115 \text{ N/mm}^2$$

Example 7.68 The stresses which act at a point on a body are σ_1 = 100 N/mm^2 (tensile), σ_2 = 60 N/mm^2 (compressive), and τ = 40 N/mm^2. The loading on the body has been increased to values which are k times the provided values. Determine the maximum value of k if the body material can withstand maximum normal and shear stresses of 300 N/mm^2 and 200 N/mm^2 respectively.

Solution The major principal stress is given by,

$$p_1 = \frac{\sigma_1 + \sigma_2}{2} + \sqrt{\left(\frac{\sigma_1 - \sigma_2}{2}\right)^2 + \tau^2}$$

$$300 = \frac{100k - 60k}{2} + \sqrt{\left(\frac{100k + 60k}{2}\right)^2 + (40k)^2}$$

$$= 20k + 89.44k = 109.44k$$

$$\therefore \quad k = (300/109.44) = 2.74 \text{ N/mm}^2$$

Maximum shear stress at the point,

$$(\sigma_t)_{max} = \pm\sqrt{\left(\frac{\sigma_1 - \sigma_2}{2}\right)^2 + \tau^2}$$

$$200 = \pm \sqrt{\left(\frac{100k + 60k}{2}\right)^2 + (40k)^2}$$

$$= 89.44k$$

$$\therefore \quad k = \frac{200}{89.44} = 22.36 \text{ N/mm}^2$$

Example 7.69 The intensity of resultant stress on a plane *PQ* at a point in a particular material under stress is 1000 kN/m^2 and it is inclined at 30° to the normal to that plane as shown in Figure 7.54. The normal component of stress on another plane *QR* at right angle to the plane *PQ* is 750 kN/m^2. Determine: (i) resultant stress on plane *QR* and its angle, (ii) the principal stresses and their directions, (iii) the maximum shear stress and their planes.

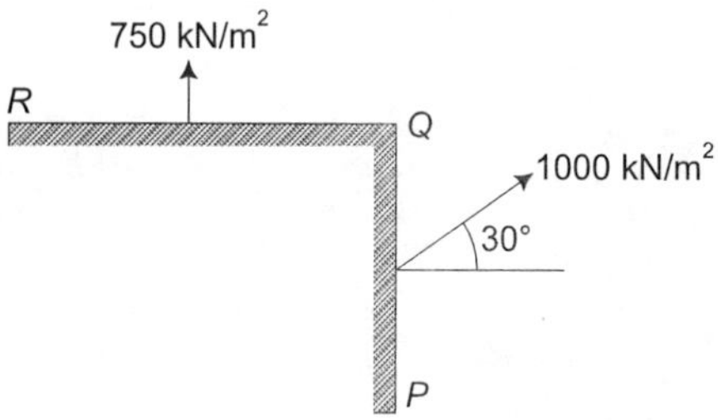

Fig. 7.54

Solution Refer Figure 7.54.

Resultant stree on plane

$$PQ = 1000 \text{ kN/m}^2$$

Normal stress on plane *QR*,

$$\sigma_2 = 750 \text{ kN/m}^2$$

The normal stress on the plane *PQ*,

$$\sigma_1 = 1000 \times \cos 30°$$

$$= 866.02 \text{ kN/m}^2$$

The shear (tangential) stress on the plane *PQ*,

$$\sigma_1 = 1000 \times \cos 30°$$

$$= 866.02 \text{ kN/m}^2$$

(i) Resultant stress on plane

$$QR = \sqrt{\sigma_2^2 + \tau^2} = \sqrt{750^2 + 500^2}$$

$$= 901.39 \text{ kN/m}^2$$

Angle of resultant with plane *QR*,

$$\tan\theta = \frac{\sigma_2}{\tau} = \frac{750}{500} = 1.5$$

$$\therefore \quad \theta = \tan^{-1}(1.5) = 56°19'$$

(ii) The principal stresses are given by,

$$p_1, p_2 = \frac{\sigma_1 + \sigma_2}{2} \pm \sqrt{\left(\frac{\sigma_1 - \sigma_2}{2}\right)^2 + \tau^2}$$

$$p_1, p_2 = \frac{866.02 + 750}{2} \pm \sqrt{\left(\frac{866.02 - 750}{2}\right)^2 + (500)^2}$$

$$= 808.01 \pm 503.35$$

$$\therefore \quad p_1 = 808.01 + 503.35 = 1311.36 \text{ kN/m}^2$$

$$p_2 = 808.01 - 503.35 = 304.66 \text{ kN/m}^2$$

The direction of principal stresses are given by,

Angles, $$\theta_1 = \frac{1}{2}\tan^{-1}\frac{2\tau}{\sigma_1 - \sigma_2}$$

$$= \frac{1}{2}\tan^{-1}\frac{2 \times 500}{866.025 + 750} = 41°41'$$

$$\theta_2 = 90° + \theta_1 = 90 + 41°41'$$

$$= 131°41'$$

(iii) Maximum stress is given by,

$$(\sigma_t)_{max} = \pm\sqrt{\left(\frac{\sigma_1 - \sigma_2}{2}\right)^2 + \tau^2} = \pm\sqrt{\left(\frac{866.02 - 750}{2}\right)^2 + (500)^2}$$

$$= \pm\ 503.35 \text{ kN/m}^2$$

The planes of maximum shear are given by,

$$\theta_3 = \frac{1}{2}\tan^{-1}\frac{(\sigma_2 - \sigma_1)/2}{\tau}$$

$$= \frac{1}{2}\tan^{-1}\frac{(-750 - 866.02)/2}{500}$$

$$= -6°37'$$

$$\theta_4 = 90° + \theta_3 = 90° - 6°37'$$

$$= 83°23'$$

7.20 MOHR'S STRESS CIRCLE

Mohr's circle is a graphical method to determine the normal, tangential (shear), and resultant stresses on an oblique plane when a body is subjected to biaxial or complex stresses.

7.20.1 Mohr's Circle When a Body is Subjected to Two Mutually Perpendicular and Unequal Principal Tensile Stresses

Considering a rectangular body subjected to two mutually perpendicular principal stresses of unequal intensities. It is required to determine the resultant stresses on an inclined plane.

Let σ_1 and σ_2 be the major and minor tensile stresses respectively and θ is the angle made by the inclined plane with the axis of minor tensile stress.

Steps to Draw Mohr's Circle

1. Draw the X-axis and Y-axis meeting at point A.
2. Take $AB = \sigma_1$ and $AC = \sigma_2$ on the same side (both are tensile) to some suitable scale.
3. Bisect BC at O. With O as centre and OB or OC as radius, draw a circle which is known as Mohr's circle.
4. Through O, draw a line OE making an angle 2θ with OB.
5. From E, drop a perpendicular ED on AB. Join AE

AD and ED respectively represent the normal and tangential stresses on the inclined plane and AE represents the resultant stress. The Mohr' circle is shown in Figure 7.55 in which angle ϕ represents the obliquity. The radius of Mohr's circle is given by, $(\sigma_1 - \sigma_2)/2$.

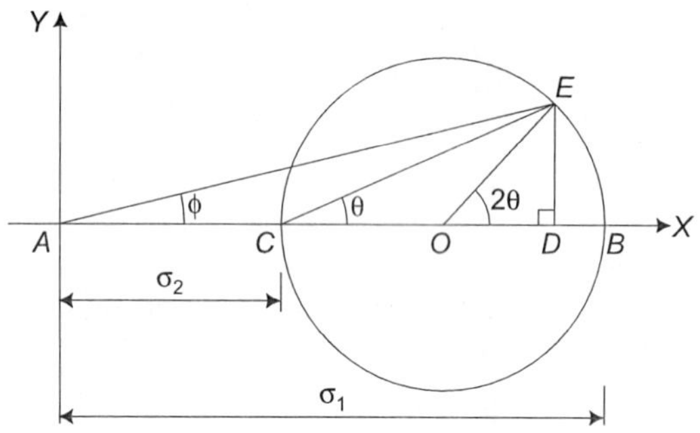

Fig. 7.55

Proof Refer Figure 7.55.

From the stress diagram,

$$CO = OB = OE = \text{radius of Mohr's circle}$$

$$= \frac{\sigma_1 - \sigma_2}{2}$$

$$AO = AC + CO = \sigma_2 + \frac{\sigma_1 - \sigma_2}{2}$$

$$= \frac{\sigma_1 + \sigma_2}{2}$$

and $$OD = OE \cos 2\theta$$

$$= \frac{\sigma_1 - \sigma_2}{2} \cos 2\theta$$

$\therefore$ $$AD = AO + OD$$

$$= \frac{\sigma_1 + \sigma_2}{2} + \frac{\sigma_1 - \sigma_2}{2} \cos 2\theta$$

$$= \sigma_n \quad \text{or} \quad \text{Normal stress}$$

And $$ED = OE \sin 2\theta$$

$$= \frac{\sigma_1 - \sigma_2}{2} \sin 2\theta$$

$$= \sigma_t \quad \text{or} \quad \text{Tangential stress}$$

Example 7.70 Solve Example 7.63 by using the Mohr's circle diagram method and compare the results obtained by the analytical method. The tensile stresses at a point across two mutually perpendicular planes are 100 N/mm^2 and 50 N/mm^2. Determine the normal, tangential, and resultant stresses on a plane inclined at 30° to the axis of the minor stress. Also determine the angle of obliquity.

Solution

$$\sigma_1 = 100 \text{ N/mm}^2 \quad \text{(Tensile)}$$

$$\sigma_2 = 50 \text{ N/mm}^2 \quad \text{(Tensile)}$$

$$\theta = 30°$$

Let $$1 \text{ cm} = 10 \text{ N/mm}^2$$

∴ $$\sigma_1 = (100/10) = 10 \text{ cm}$$

$$\sigma_2 = (50/10) = 5 \text{ cm}$$

Steps to Draw Mohr's Circle

1. Draw the X-axis and Y-axis intersecting at point A.
2. Take $AB = \sigma_1 = 10$ cm, and $AC = \sigma_2 = 5$ cm.
3. Bisect BC at O. With O as centre and OB or OC as radius, draw a circle which is known as Mohr's circle (Figure 7.56).
4. Through O, draw a line OE making an angle $2\theta = 60°$ with OB.
5. From E, drop a perpendicular ED on AB. Join AE and CE. Measure AD, ED, and AE.

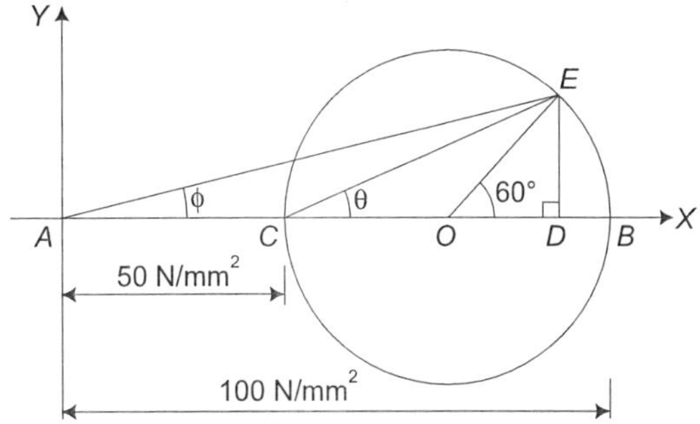

Fig. 7.56 (Not to scale)

By measurements: $AD = 8.8$ cm, $ED = 2.2$ cm, and $AE = 9.1$ cm

$$\therefore \quad \sigma_n = \text{length } AD \times \text{scale}$$

$$= 8.75 \times 10 = 87.5 \text{ N/mm}^2$$

(By analytical method, normal stress is 87.5 N/mm^2).

$$\sigma_t = \text{length } ED \times \text{scale}$$

$$= 2.15 \times 10 = 21.5 \text{ N/mm}^2$$

(By analytical method, tangential stress is 21.65 N/mm^2).

$$\sigma_r = \text{length } AE \times \text{scale}$$

$$= 9.05 \times 10 = 90.5 \text{ N/mm}^2,$$

(By analytical method, resultant stress is 90.14 N/mm^2).

The angle of obliquity,

$$\phi = 24.5°$$

7.20.2 Mohr's Circle When a Body is Subjected to Two Mutually Perpendicular Principal Stresses which are Unlike and Unequal

Considering a rectangular body subjected to two mutually perpendicular principal stresses which are not equal and one of them is tensile and the other is compressive. It is required to determine the resultant stresses on an inclined plane.

Let σ_1 be the major tensile stress, σ_2 be the minor compressive stress, and θ the angle made by the inclined plane with the axis of minor compressive stress.

Steps to Draw Mohr's Circle

1. Draw the X-axis and Y-axis intersecting at point A.
2. Take $AB = \sigma_1$ (tensile stress as positive) towards right of A and $AC = \sigma_2$ (compressive stress as negative) towards left of A to some suitable scale.
3. Bisect BC at O. With O as centre and OB or OC as radius, draw a circle which is known as Mohr's circle.
4. Through O, draw a line OE making an angle 2θ with OB.
5. From E, drop a perpendicular ED on AB. Join AE and CE.

AD and ED respectively represent the normal and tangential stresses (i.e., shear stress) on the inclined plane and AE represents the resultant stress. The Mohr' circle is shown in Figure 7.57 in which angle ϕ represents the obliquity. The radius of Mohr's circle is given by, $(\sigma_1 + \sigma_2)/2$.

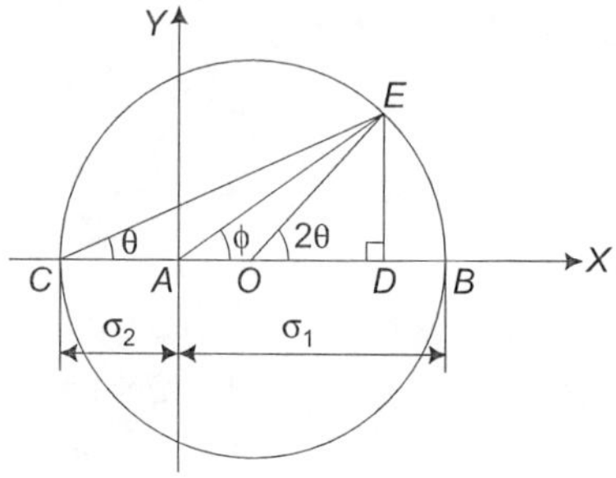

Fig. 7.57

Proof Refer Figure 7.57.

From the stress diagram,

$$CO = OB = OE = \text{radius of Mohr's circle}$$

$$= (\sigma_1 + \sigma_2)/2$$

$$AO = OC - AC = \frac{\sigma_1 + \sigma_2}{2} - \sigma_2 = \frac{\sigma_1 - \sigma_2}{2}$$

$$OD = OE \cos 2\theta = \frac{\sigma_1 + \sigma_2}{2} \cos 2\theta$$

$$\therefore \quad AD = AO + OD = \frac{\sigma_1 - \sigma_2}{2} + \frac{\sigma_1 + \sigma_2}{2} \cos 2\theta$$

$$= \sigma_n \quad \text{or} \quad \text{Normal stress}$$

$$ED = OE \sin 2\theta = \frac{\sigma_1 + \sigma_2}{2} \sin 2\theta$$

$$= \sigma_t \quad \text{or} \quad \text{Tangential stress}$$

Example 7.71 By using Mohr's circle diagram determine the normal, tangential, and resultant stresses on a plane inclined at 30° to the axis of the major stress when the stresses at a point in a bar are 100 N/mm^2 as tensile and 50 N/mm^2 as compressive. Also determine the maximum intensity of shear stress in the material at the point and the angle of obliquity.

Solution

$$\sigma_1 = 100 \text{ N/mm}^2 \quad \text{(Tensile)}$$

$$\sigma_2 = -50 \text{ N/mm}^2 \quad \text{(Compressive)}$$

$$\theta = 30^\circ$$

Let $1 \text{ cm} = 10 \text{ N/mm}^2$

$$\therefore \quad \sigma_1 = (100/10) = 10 \text{ cm}$$

$$\sigma_2 = (-50/10) = -5 \text{ cm}$$

Steps to Draw Mohr's Circle

1. Draw the X-axis and Y-axis intersecting at point A.
2. Take $AB = \sigma_1 = 10$ cm towards right of A and $AC = \sigma_2 = -5$ cm towards left of A.
3. Bisect BC at O. With O as centre and OB or OC as radius, draw a circle which is known as Mohr's circle (Figure 7.58).
4. Through O, draw a line OE making an angle $2\theta = 60°$ with OB.
5. From E, drop a perpendicular ED on AB. Join AE and CE. Measure AD, ED, and AE.

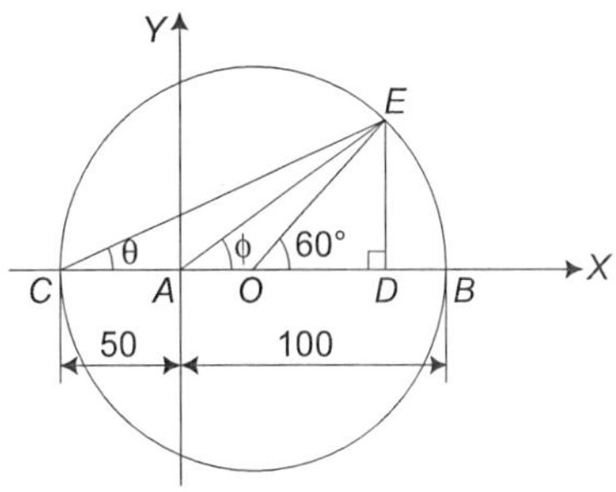

Fig. 7.58 (Not to scale)

By measurements: $AD = 8.8$ cm, $ED = 2.2$ cm, and $AE = 9.1$ cm

$$\therefore \quad \sigma_n = \text{length } AD \times \text{scale} = 6.25 \times 10$$

$$= 62.5 \text{ N/mm}^2$$

$$\sigma_t = \text{length } ED \times \text{scale}$$

$$= 6.5 \times 10 = 65 \text{ N/mm}^2$$

$$\sigma_r = \text{length } AE \times \text{scale}$$

$$= 9.1 \times 10 = 91 \text{ N/mm}^2$$

And the angle of obliquity,

$$\phi = 46°$$

7.20.3 Mohr's Circle When a Body is Subjected to Two Mutually Perpendicular Principal Tensile Stresses Accompanied by a Simple Shear

Considering a rectangular body subjected to two mutually perpendicular principal tensile stresses, namely, σ_1 (major tensile stress), σ_2 (minor tensile stress) accompanied by a simple shear stress τ as shown in Figure 7.59. θ is the angle made by oblique plane with the plane of major tensile stress.

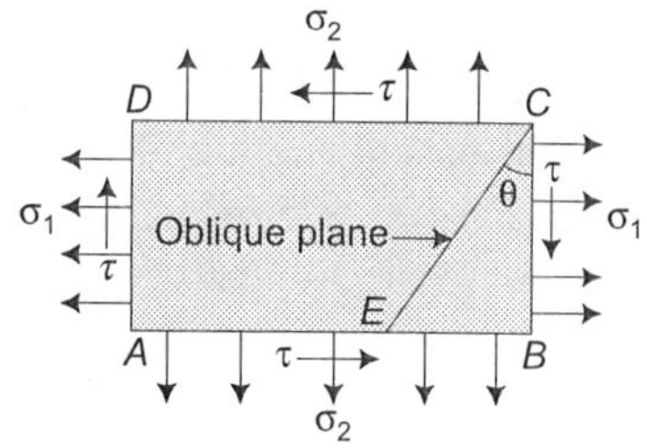

Fig. 7.59

The following sign conventions can be taken for the stresses: (1) Tensile stress as positive direct stress and compressive stress as negative direct stress to be drawn along the abscissa in a *X-Y* coordinate system. (2) Shear stress tending to rotate the body in the clockwise direction as positive shear stress and shear stress tending to rotate the body in the anticlockwise direction as negative shear stress to be drawn along the ordinate axis.

Steps to Draw Mohr's Circle

1. Draw the *X*-axis and *Y*-axis meeting at point *A*.
2. Take $AB = \sigma_1$ and $AC = \sigma_2$ on the same side (both are tensile) to some suitable scale.
3. Draw perpendiculars at *B* and *C* and cut off *BF* and *CG* equal to shear stress τ to the same scale. Bisect *BC* at *O*. With *O* as centre and radius equal to *OG* or *OF* draw a circle which is known as Mohr's circle.
4. Through *O*, draw a line *OE* making an angle 2θ with *OF* as shown in Figure 7.60.
5. From *E*, drop a perpendicular *ED* on *AB*. Join *AE*.

AD and *ED* respectively represents the normal and tangential stresses on the inclined plane and *AE* represents the resultant stress. The Mohr' circle is shown in Figure 7.60 in which angle ϕ represents the obliquity. The radius of Mohr's circle is given by,

$$\sqrt{\left(\frac{\sigma_1 - \sigma_2}{2}\right)^2 + \tau^2}$$

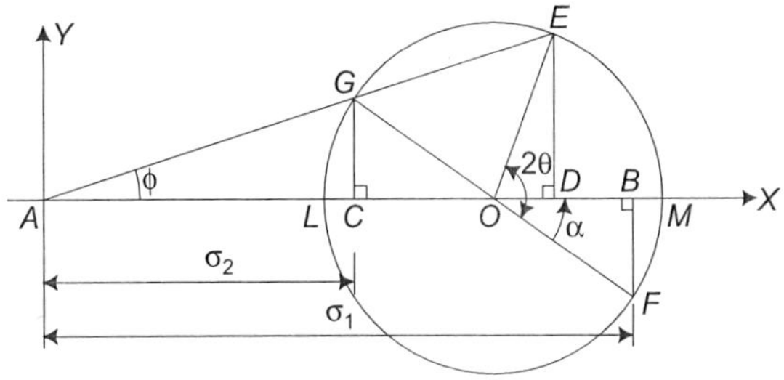

Fig. 7.60

Proof Refer Figure 7.60.

From the stress diagram,

$$CO = \frac{1}{2}CB = \frac{1}{2}[\sigma_1 - \sigma_2]$$

$$AO = AC + CO$$

$$= \sigma_2 + \frac{\sigma_1 - \sigma_2}{2}$$

$$= \frac{\sigma_1 + \sigma_2}{2}$$

$$OD = OE \cos(2\theta - \alpha)$$

$$\therefore \quad AD = AO + OD$$

$$= \frac{\sigma_1 + \sigma_2}{2} + OE \cos(2\theta - \alpha)$$

$$= \frac{\sigma_1 + \sigma_2}{2} + OE(\cos 2\theta \,.\, \cos\alpha + \sin 2\theta \,.\, \sin\alpha)$$

$$= \frac{\sigma_1 + \sigma_2}{2} + OE\cos 2\theta \,.\, \cos\alpha + OE\sin 2\theta \,.\, \sin\alpha$$

$$= \frac{\sigma_1 + \sigma_2}{2} + OF\cos\alpha \,.\, \cos 2\theta + OF\sin\alpha \,.\, \sin 2\theta$$

$$\therefore \quad AD = \frac{\sigma_1 + \sigma_2}{2} + OB\cos 2\theta + BF\sin 2\theta$$

$$= \frac{\sigma_1 + \sigma_2}{2} + \frac{\sigma_1 - \sigma_2}{2}\cos 2\theta + \tau\sin 2\theta$$

$$= \sigma_n$$

$$ED = OE\sin(2\theta - \alpha)$$

$$= OE(\sin 2\theta\cos\alpha - \cos 2\theta\sin\alpha)$$

$$= OE\cos\alpha\sin 2\theta - OE\sin\alpha\cos 2\theta$$

$$\therefore \qquad ED = OB \sin 2\theta - BF \cos 2\theta$$

$$= \frac{\sigma_1 - \sigma_2}{2} \sin 2\theta - \tau \cos 2\theta$$

$$= \sigma_t$$

Maximum and Minimum Value of Normal Stress The normal stress is given by *AD* in Figure 7.60. Maximum value of *AD* will be when *D* coincides with *M* and minimum value of *AD* will be when *D* coincides with *L*.

Therefore, maximum value of normal stress p_1 is given as,

$$p_1 = AM = AO + OM$$

$$= \frac{\sigma_1 + \sigma_2}{2} + OF$$

$$\therefore \qquad p_1 = \frac{\sigma_1 + \sigma_2}{2} + OF$$

$$= \frac{\sigma_1 + \sigma_2}{2} + \sqrt{OB^2 + BF^2}$$

$$= \frac{\sigma_1 + \sigma_2}{2} + \sqrt{\left(\frac{\sigma_1 - \sigma_2}{2}\right)^2 + \tau^2}$$

Minimum value of normal stress,

$$p_2 = AL = AO - LO$$

$$= \frac{\sigma_1 + \sigma_2}{2} - OF$$

$$\therefore \qquad p_2 = \frac{\sigma_1 + \sigma_2}{2} - \sqrt{\left(\frac{\sigma_1 - \sigma_2}{2}\right) + \tau^2}$$

For maximum normal stress, the point *D* coincides with *M*. But, when point *D* coincides with *M*, the point *E* also coincides with *M*. Therefore, for maximum value of normal stress angle becomes,

$$2\theta = \alpha$$

$$\therefore \qquad \theta = \frac{\alpha}{2}$$

also,

$$\tan 2\theta = \tan \alpha = \frac{BF}{OB}$$

$$= \frac{\tau}{\left(\frac{\sigma_1 - \sigma_2}{2}\right)} = \frac{2\tau}{(\sigma_1 - \sigma_2)}$$

For minimum normal stress, point *D* coincides with *L*. But, when point *D* coincides with *L*, point *E* also coincides with *L*. Therefore, for minimum value of normal stress angle becomes,

$$2\theta = \pi + \alpha$$

$$\therefore \quad \theta = \frac{\pi}{2} + \frac{\alpha}{2}$$

This shows that the plane of minimum normal stress is inclined at an angle of 90° to the plane of maximum normal stress.

Maximum Value of Shear Stress In Figure 7.60, shear stress is represented by *ED* which will become maximum when point *E* coincides with *G*, and *D* coincides with *O*.

Therefore, maximum shear stress

$$(\sigma_t)_{max} = OF = \sqrt{OB^2 + BF^2}$$

$$= \sqrt{\left(\frac{\sigma_1 - \sigma_2}{2}\right)^2 + \tau^2}$$

Example 7.72 Using Mohr's circle diagram method, determine the normal and tangential stresses across the inclined plane as shown in Figure 7.61. Also check your answer analytically.

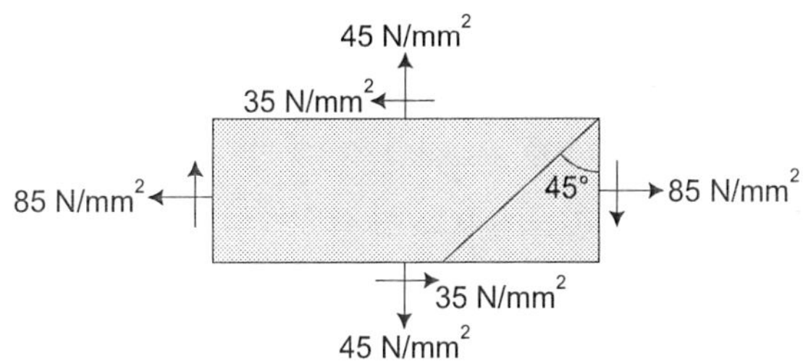

Fig. 7.61

Solution Refer Figure 7.61.

$$\sigma_1 = 85 \text{ N/mm}^2 \quad \text{(Tensile)}$$

$$\sigma_2 = 45 \text{ N/mm}^2 \quad \text{(Tensile)}$$

$$\theta = 45°$$

Let $\quad 1 \text{ cm} = 10 \text{ N/mm}^2$

$$\therefore \quad \sigma_1 = (85/10) = 8.5 \text{ cm}$$

$$\sigma_2 = (45/10) = 4.5 \text{ cm}$$

Steps to Draw Mohr's Circle

1. Draw the *X*-axis and *Y*-axis meeting at point *A*.
2. Take $AB = \sigma_1 = 8.5$ cm and $AC = \sigma_2 = 4.5$ cm on the same side.
3. Draw perpendiculars at *B* and *C* and cut off $BF = CG = \tau = 3.5$ cm. Bisect *BC* at *O*. With *O* as centre and radius equal to *OG* or *OF* draw a circle which is known as Mohr's circle.

4. Through O, draw a line OE making an angle $2\theta = 90°$ with OF as shown in Figure 7.62.
5. From E, drop a perpendicular ED on AB. Join AE.

AD and ED represent the normal and tangential stresses on the inclined plane respectively and AE represents the resultant stress. The Mohr' circle is shown in Figure 7.62 in which angle ϕ represents the obliquity.

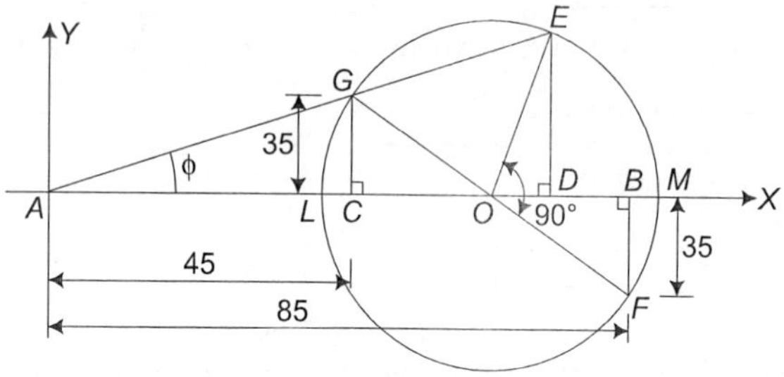

Fig. 7.62 (Not to scale)

By measurements: $AD = 10$ cm

$ED = 2$ cm, $\phi = 11°$

$\therefore$ σ_n = length AD × scale

$= 10 \times 10 = 100$ N/mm^2

σ_t = length ED × scale

$= 2 \times 10 = 20$ N/mm^2

Analytical Solution

$$\sigma_n = \frac{\sigma_1 + \sigma_2}{2} + \frac{\sigma_1 - \sigma_2}{2}\cos 2\theta + \tau \sin 2\theta$$

$$\therefore \quad \sigma_n = \frac{85 + 45}{2} + \frac{85 - 45}{2}\cos (2 \times 45) + 35 \sin (2 \times 45)$$

$$= 100 \text{ N/mm}^2$$

$$\sigma_t = \frac{\sigma_1 - \sigma_2}{2}\sin 2\theta - \tau \cos 2\theta$$

$$\therefore \quad \sigma_t = \frac{85 - 45}{2}\sin (2 \times 45) - 35 \cos (2 \times 45)$$

$$= 20 \text{ N/mm}^2$$

Example 7.73 The tensile stresses on two planes at right angles in a strained material at certain point are 50 N/mm^2 and 20 N/mm^2. These stresses are accompanied by a shear stress of 20 N/mm^2. Using Mohr's circle diagram method, determine principal stresses and the location of principal planes.

Solution

$$\sigma_1 = 50 \text{ N/mm}^2 \quad \text{(Tensile)}$$

$$\sigma_2 = 20 \text{ N/mm}^2 \quad \text{(Tensile)}$$

Let $1 \text{ cm} = 5 \text{ N/mm}^2$

$\therefore$ $\sigma_1 = (50/5) = 10 \text{ cm}$

$$\sigma_2 = \tau = (20/5) = 4 \text{ cm}$$

Steps to Draw Mohr's Circle

1. Draw the X-axis and Y-axis meeting at point A.
2. Take $AB = \sigma_1 = 10$ cm and $AC = \sigma_2 = 4$ cm on the same side.
3. Draw perpendiculars at B and C and cut off $BF = CG = \tau = 4$ cm. Bisect BC at O. With O as centre and radius equal to OG or OF draw a circle which is known as Mohr's circle as shown in Figure 7.63. AM and AL respectively represents the major principal (p_1) and minor principal stresses (p_2).

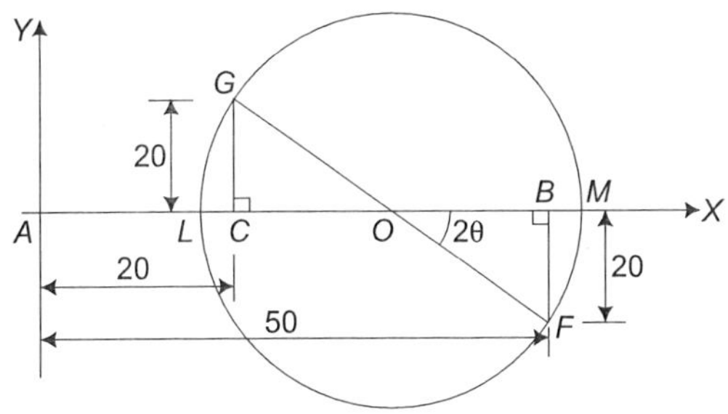

Fig. 7.63 (Not to scale)

By measurements: $AM = 12.1$ cm and $AL = 2$ cm

$$\angle FOB = 2\theta = 54^\circ$$

$\therefore$ $p_1 = \text{length } AM \times \text{scale} = 12.1 \times 5 = 60.5 \text{ N/mm}^2$

$\therefore$ $p_2 = \text{length } AL \times \text{scale} = 2 \times 5 = 10 \text{ N/mm}^2$

Location of principal planes:

$$2\theta = 54^\circ, \ \theta = (54^\circ/2) = 27^\circ$$

Angle for second plane = 27 + 90 = 117°

7.21 STRAIN ON AN INCLINED PLANE

To determine the strain on an inclined plane due to stresses σ_1, σ_2, and τ, the strains due to each stress have been determined separately and then combined to get the result.

(i) Strain on an Inclined Plane due to Stress σ_1 Consider a block *ABCD* subjected to a stress σ_1. Due to this stress it is deformed to $AB_1\ C_1\ D$, means *B* is displaced to B_1 and *C* is displaced to C_1, so there is an increase in length δl_x as shown in Figure 7.64. We have to determine the strain on the inclined plane *AC*.

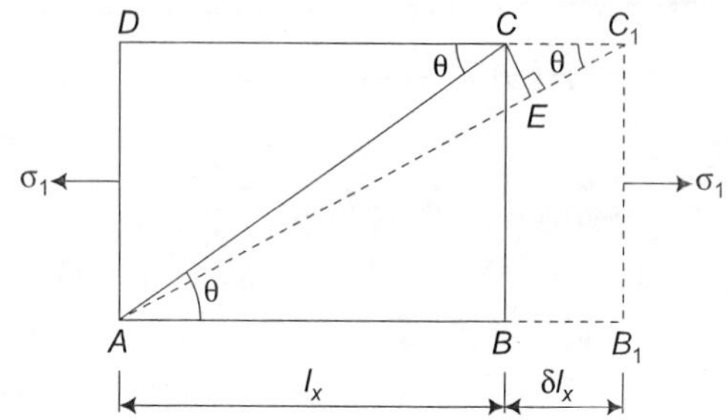

Fig. 7.64

Let e_1 be the strain corresponding to σ_1 in *X*-direction which is given by,

$$e_1 = \frac{\delta l_x}{l_x}$$

$\therefore$ $$\delta l_x = e_1 \times l_x$$

also $$\delta l_x = BB_1 = CC_1 = e_1 \times l_x$$

Let *e* be the strain in the diagonal *AC* due to stress σ_1, which is given by,

$$e = \frac{AC_1 - AC}{AC} = \frac{(AE + EC_1) - AC}{AC}$$

$$= \frac{EC_1}{AC} \quad (\because AE = AC)$$

$$e = \frac{CC_1 \cos\theta}{AC}$$

$$= \frac{e_1 \times l_x \times \cos\theta}{AC} \quad (\because EC_1 = CC_1 \cos\theta \ \&\ CC_1 = e_1 \times l_x)$$

$$e = \frac{e_1 \times (AC\cos\theta) \times \cos\theta}{AC} \quad (\because l_x = AC\cos\theta)$$

$\therefore$ $$e = e_1 \cos^2\theta \tag{7.63}$$

(ii) Strain on an Inclined Plane due to Stress σ_2 Refer Figure 7.65.

Let e_2 be the strain corresponding to σ_2 in *Y*-direction which is given by,

$$E_2 = \frac{\delta l_y}{l_y}$$

$\therefore$ $$\delta l_y = e_2 \times l_y$$

$$\delta l_y = CC_1 = DD_1 = e_2 \times l_y$$

Let e be the strain in the diagonal AC due to stress σ_2, which is given by,

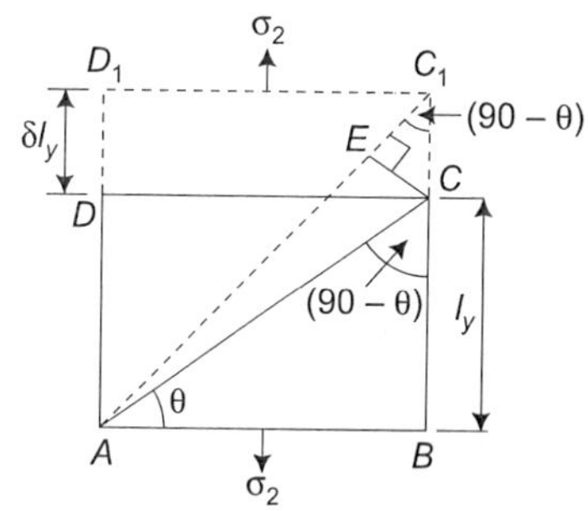

Fig. 7.65

$$e = \frac{AC_1 - AC}{AC} = \frac{(AC + EC_1) - AC}{AC}$$

$$= \frac{EC_1}{AC} \qquad (\because AE = AC \text{ \& } AC_1 = AC + EC_1)$$

$$e = \frac{CC_1 \cos(90 - \theta)}{AC}$$

$$= \frac{CC_1 \sin\theta}{AC} \qquad (\because EC_1 = CC_1 \cos(90-\theta) \text{ \& } \cos(90-\theta) = \sin\theta)$$

$$e = \frac{(e_2 \times l_y) \times \sin\theta}{AC}$$

$$= \frac{e_2 \times (AC \sin\theta) \times \sin\theta}{AC} \qquad (\because CC_1 = e_2 \times l_y \text{ \& } l_y = AC \times \sin\theta)$$

$$\therefore \qquad e = e_2 \sin^2\theta \qquad (7.64)$$

(iii) Strain on an Inclined Plane due to Stress τ Refer Figure 7.66.

Let ϕ be the shear strain produced due to shear stress τ, which is given by,

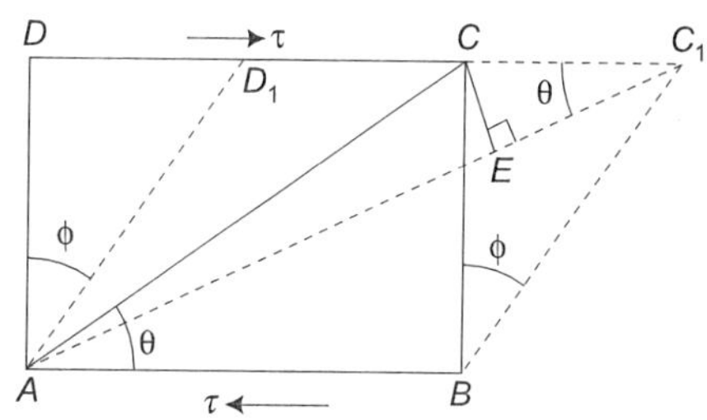

Fig. 7.66

$$\phi = \frac{CC_1}{CB}$$

$$CC_1 = \phi \times CB$$

$$= \phi \times AC \sin\theta \qquad (\because CB = AC \sin\theta)$$

Let e be the strain in the diagonal AC due to stress τ, which is given by,

$$e = \frac{AC_1 - AC}{AC}$$

$$= \frac{(AC + EC_1) - AC}{AC}$$

$$= \frac{EC_1}{AC} \qquad (\because AE = AC \text{ \& } AC_1 = AC + EC_1)$$

$$e = \frac{CC_1 \cos\theta}{AC}$$

$$= \frac{\phi \times AC \sin\theta \times \cos\theta}{AC}$$

$$(\therefore EC_1 = CC_1 \cos\theta \text{ \& } CC_1 = \phi \times AC \sin\theta)$$

$$e = \frac{\phi}{2} \times 2 \times \sin\theta \times \cos\theta$$

$$= \frac{\phi}{2} \times \sin 2\theta \qquad \text{(Multiplying and dividing by 2)}$$

$$\therefore \qquad e = \frac{\phi}{2} \times \sin 2\theta \tag{7.65}$$

(iv) Strain on an Inclined Plane due to Stresses σ_1, σ_2, and τ

The total strain (e) on an inclined plane is obtained by adding the right hand side of equations (7.63), (7.64), and (7.65) which is given by,

$$e = e_1 \cos^2\theta + e_2 \sin^2\theta + \frac{\phi}{2} \times \sin 2\theta$$

$$e = e_1 \frac{(1 + \cos 2\theta)}{2} + e_2 \frac{(1 - \cos 2\theta)}{2} + \frac{\phi}{2} \times \sin 2\theta$$

$$\therefore \qquad e = \frac{e_1 + e_2}{2} + \frac{e_1 - e_2}{2} \times \cos 2\theta + \frac{\phi}{2} \times \sin 2\theta \tag{7.66}$$

Principal Strains For the strain e to become maximum or minimum on an inclined plane

$$\frac{de}{d\theta} = 0$$

$$\therefore \qquad \frac{d}{d\theta}\left[\frac{e_1 + e_2}{2} + \frac{e_1 - e_2}{2} \times \cos 2\theta + \frac{\phi}{2} \times \sin 2\theta\right] = 0$$

On solving this equation, we get,

$$\tan 2\theta = \frac{\phi}{e_1 - e_2}$$

$$2\theta = \tan^{-1}\left(\frac{\phi}{e_1 - e_2}\right)$$

$$\theta = \frac{1}{2}\tan^{-1}\left(\frac{\phi}{e_1 - e_2}\right)$$

From the expression,

$$\tan 2\theta = \frac{\phi}{e_1 - e_2},$$

we have,

Diagonals of the right angled triangle $= \pm \sqrt{(e_1 - e_2)^2 + \phi^2}$

Taking positive value of diagonal $= \sqrt{(e_1 - e_2)^2 + \phi^2}$

$$\therefore \qquad \sin 2\theta = \frac{\phi}{\sqrt{(e_1 - e_2)^2 + \phi^2}}$$

$$\cos 2\theta = \frac{e_1 - e_2}{\sqrt{(e_1 - e_2)^2 + \phi^2}}$$

The value of major principal strain (e_{p1}) or maximum strain value is obtained by substituting the values of sin 2θ and cos 2θ in equation (7.66).

$$\therefore \qquad e_{p1} = \frac{e_1 + e_2}{2} + \frac{e_1 - e_2}{2} \times \frac{e_1 - e_2}{\sqrt{(e_1 - e_2)^2 + \phi^2}} + \frac{\phi}{2} \times \frac{\phi}{\sqrt{(e_1 - e_2)^2 + \phi^2}}$$

After simplification, we get,

$$e_{p1} = \frac{e_1 + e_2}{2} + \frac{1}{2}\sqrt{(e_1 - e_2)^2 + \phi^2}$$

$$= \frac{e_1 + e_2}{2} + \sqrt{\left(\frac{e_1 - e_2}{2}\right)^2 + \left(\frac{\phi}{2}\right)^2} \qquad (7.67)$$

Now, taking the negative value of diagonal

$$= -\sqrt{(e_1 - e_2)^2 + \phi^2}$$

$$\therefore \qquad \sin 2\theta = \frac{\phi}{-\sqrt{(e_1 - e_2)^2 + \phi^2}}$$

$$\cos 2\theta = \frac{e_1 - e_2}{-\sqrt{(e_1 - e_2)^2 + \phi^2}}$$

The value of minor principal strain (e_{p2}) or minimum strain value is obtained by substituting the values of sin 2θ and cos 2θ in equation (7.66).

$$\therefore \qquad e_{p2} = \frac{e_1 + e_2}{2} + \frac{e_1 - e_2}{2} \times \frac{(e_1 - e_2)}{-\sqrt{(e_1 - e_2)^2 + \phi^2}}$$

$$+ \frac{\phi}{2} \times \frac{\phi}{-\sqrt{(e_1 - e_2)^2 + \phi^2}}$$

After simplification, we get,

$$e_{p2} = \frac{e_1 + e_2}{2} - \frac{1}{2}\sqrt{(e_1 - e_2)^2 + \phi^2}$$

$$= \frac{e_1 + e_2}{2} - \sqrt{\left(\frac{e_1 - e_2}{2}\right)^2 + \left(\frac{\phi}{2}\right)^2} \qquad (7.68)$$

Equations (7.67) and (7.68) give the value of major principal strain and minor principal strain values respectively.

7.22 MOHR'S STRAIN CIRCLE

Mohr's strain circle is similar to Mohr's stress circle. The linear strains e_1 and e_2 are drawn along horizontal axis and half of shear strain is drawn along ordinate axis.

Let e_1 and e_2 be linear strains and ϕ the shear strain. To find the values of principal strains and the position of principal plane the following steps are to be followed:

1. Draw the X-axis and Y-axis meeting at point A.
2. Take $AB = e_1$ and $AC = e_2$ on the same side (if both are positive) to some suitable scale.
3. Draw perpendiculars at B and C and cut off BF and CG equal to half of shear stress (i.e., $\phi/2$) to the same scale. Join points G and F cutting the horizontal line at O. With O as centre and radius equal to OG or OF draw a circle which is known as Mohr's strain circle shown in Figure 7.67.
4. The circle cuts the horizontal line AX at points L and M where shear strain is zero. The AM represents the major principal strain e_{p1} and AL represents the minor principal strain e_{p2}.
5. OH or OP represents half of the maximum shear strain. Half of the angle MOF (i.e., θ_1) gives the position of principal plane and the second principal plane will be at an angle of $(90 + \theta_1)$.

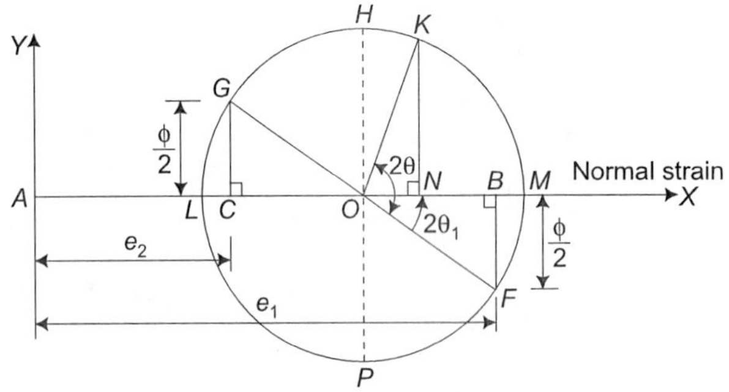

Fig. 7.67

Proof Refer Figure 7.67.

(i) Principal Strain.

$$e_1 = AM = AC + CO + OM$$
$$= AC + CO + R$$

where R is the radius of the circle which is given by,

$$R = \sqrt{OB^2 + BF^2} = \sqrt{\left(\frac{e_1 - e_2}{2}\right)^2 + \left(\frac{\phi}{2}\right)^2}$$

$$\therefore \quad e_{p1} = e_2 + \frac{e_1 + e_2}{2} + \sqrt{\left(\frac{e_1 - e_2}{2}\right)^2 + \left(\frac{\phi}{2}\right)^2}$$

$$= \frac{e_1 + e_2}{2} + \frac{1}{2}\sqrt{(e_1 - e_2)^2 + (\phi)^2}$$

$$e_2 = AL = AO + OL = AO - R$$

$$\therefore \quad e_{p2} = \frac{e_1 + e_2}{2} - \frac{1}{2}\sqrt{(e_1 - e_2)^2 + (\phi)^2}$$

Points H and P on the circle represent the maximum shear strain which is given by,

$$\phi_{max} = \pm \frac{1}{2}\sqrt{(e_1 - e_2)^2 + (\phi)^2}$$

(ii) Principal Angles Consider $\angle FOM$ to find the directions of principal planes carrying principal strains with respect to the reference plane BC.

$$\angle FOM = 2\theta_1 = \tan^{-1}\left(\frac{BF}{BO}\right) = \tan^{-1}\left(\frac{\phi}{e_1 - e_2}\right)$$

$$\therefore \quad \theta_1 = \frac{1}{2}\tan^{-1}\left(\frac{\phi}{e_1 - e_2}\right)$$

$$\angle FOL = 2\theta_2 = 2\theta_2 + 180°$$

$$\therefore \quad \theta_2 = \theta_1 + 90°$$

(iii) Strains on the Inclined Plane Plane BF is inclined at an angle θ to the reference plane BC. Draw $\angle FOK$ intersecting the Mohr's strain circle at K. Then coordinates of the point K determine the normal and shear strains on the inclined plane.

Normal strain on the inclined plane is given by,

$$e_\theta = AN = AO + ON = \frac{e_1 + e_2}{2} + R\cos(2\theta - 2\theta_1)$$

$$e_\theta = \frac{e_1 + e_2}{2} + R\cos 2\theta \cos 2\theta_1 + R\sin 2\theta \sin 2\theta_1$$

$$\therefore \quad e_\theta = \frac{e_1 + e_2}{2} + \frac{e_1 - e_2}{2}\cos 2\theta + \frac{\phi}{2}\cos 2\theta \quad (7.69)$$

Shear strain on the inclined plane is given by,

$$\phi_\theta = KN = R\sin(2\theta - 2\theta_1)$$

$$= R\sin 2\theta \frac{OB}{R} - R\cos 2\theta \frac{FB}{R}$$

$$\therefore \quad \phi_\theta = \frac{e_1 - e_2}{2}\sin 2\theta - \frac{\phi}{2}\cos 2\theta \quad (7.70)$$

7.23 PRINCIPAL STRAINS IN TERMS OF PRINCIPAL STRESSES

Consider three principal stresses p_1, p_2, and p_3 at a point on three principal planes $CEFD$, $BCEG$, and $ABCD$ respectively (Figure 7.68). Let E be the Young's modulus of elasticity and ν be the Poisson's ratio.

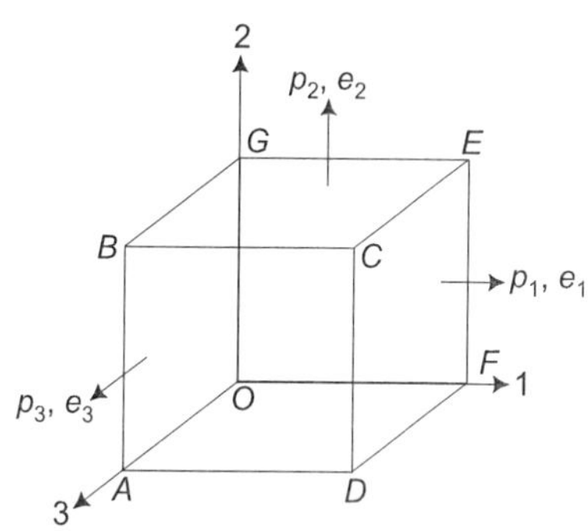

Fig. 7.68

Linear strain due to principal stress p_1 in direction

$$1 = (p_1/E)$$

Lateral strain due to principal stress p_1 in direction 2 = – ($\nu\, p_1/E$)

Lateral strain due to principal stress p_1 in direction 3 = – ($\nu\, p_1/E$)

Similarly, the linear strain and lateral strain due to principal stresses p_2, and p_3 can be determined.

Therefore, total strain or principal strain in direction 1,

$$e_{p1} = \frac{p_1}{E} - \frac{\nu\, p_2}{E} - \frac{\nu\, p_3}{E} \tag{7.71}$$

Total strain or principal strain in direction 2,

$$e_{p2} = \frac{p_2}{E} - \frac{\nu\, p_1}{E} - \frac{\nu\, p_3}{E} \tag{7.72}$$

Total strain or principal strain in direction 3,

$$e_{p3} = \frac{p_3}{E} - \frac{\nu\, p_1}{E} - \frac{\nu\, p_2}{E} \tag{7.73}$$

Example 7.74 The normal and shear strains acting at a point are e_1 = 750 μcm/cm, e_2 = – 300 μcm/cm, and $\phi/2$ = ± 225 μcm/cm. Determine (i) principal strains, (ii) principal angles, and (iii) normal and shear strain on a plane inclined at an angle of 30° to the plane e_1.

Solution

$$e_1 = 750 \times 10^{-6}\,\frac{\text{cm}}{\text{cm}} = 750 \times 10^{-6}$$

$$e_2 = -\,300 \times 10^{-6}\,\frac{\text{cm}}{\text{cm}} = -\,300 \times 10^{-6}$$

$$\frac{\phi}{2} = \pm\, 225 \times 10^{-6}\,\frac{\text{cm}}{\text{cm}} = \pm\, 225 \times 10^{-6}$$

$$\frac{e_1 + e_2}{2} = \frac{(750 - 300) \times 10^{-6}}{2} = 225 \times 10^{-6}$$

$$\frac{e_1 - e_2}{2} = \frac{(750 + 300) \times 10^{-6}}{2} = 525 \times 10^{-6}$$

(i) Principal strains are given by,

$$e_{p1} = \frac{e_1 + e_2}{2} - \sqrt{\left(\frac{e_1 - e_2}{2}\right)^2 + \left(\frac{\phi}{2}\right)^2}$$

and

$$e_{p2} = \frac{e_1 + e_2}{2} - \sqrt{\left(\frac{e_1 - e_2}{2}\right)^2 + \left(\frac{\phi}{2}\right)^2}$$

∴

$$e_{p1} = 225 \times 10^{-6} + \sqrt{(525 \times 10^{-6})^2 + (225 \times 10^{-6})^2}$$

$$= 7.96 \times 10^{-4}$$

∴

$$e_{p2} = 225 \times 10^{-6} - \sqrt{(525 \times 10^{-6})^2 + (225 \times 10^{-6})^2}$$

$$= -\,0.02193$$

(ii) Principal angles are given by,

$$\theta_1 = \frac{1}{2}\tan^{-1}\left(\frac{\phi}{e_1 - e_2}\right)$$

$$= \frac{1}{2}\tan^{-1}\left(\frac{450 \times 10^{-6}}{1050 \times 10^{-6}}\right)$$

$$= 11°36'$$

$$\phi_2 = 90 + \phi_1 = 90 + 11°36'$$

$$= 101°36'$$

(iii) Strains on inclined plane at 30° are given by,

$$e_\theta = \frac{e_1 + e_2}{2} + \frac{e_1 - e_2}{2}\cos 2\theta + \frac{\phi}{2}\cos 2\theta$$

$$\phi_\theta = \frac{e_1 - e_2}{2}\sin 2\theta - \frac{\phi}{2}\cos 2\theta$$

$\therefore$ $$e_\theta = 225 \times 10^{-6} + 525 \times 10^{-6}\cos 2(30) + 225 \times 10^{-6}\cos 2(30)$$

$$= 6 \times 10^{-4}$$

$\therefore$ $$\phi_\theta = 525 \times 10^{-6}\sin 2(30) - 225 \times 10^{-6}\cos 2(30)$$

$$= 3.422 \times 10^{-4}$$

Example 7.75 The principal strains acting at a point in a strained material subjected to principal stresses p_1 and p_2 are e_1 = 360 μcm/cm, e_2 = – 280 μcm/cm. Determine the magnitude of the principal stresses if Young's modulus of elasticity is 200 kN/mm^2 and Poisson's ratio is 0.30.

Solution Principal strains are given by,

$$e_{p1} = \frac{p_1}{E} - \frac{\nu\, p_2}{E}$$

$$e_{p2} = \frac{p_2}{E} - \frac{\nu\, p_1}{E}$$

(Neglecting principal stresses in third direction)

$$360 \times 10^{-6} \times 200 \times 10^3 = p_1 - 0.30 \times p_2$$

$$p_1 - 0.30 \times p_2 = 72 \qquad \text{(i)}$$

$$-280 \times 10^{-6} \times 200 \times 10^3 = p_2 - 0.30 \times p_1$$

$$p_2 - 0.30 \times p_1 = -56 \qquad \text{(ii)}$$

Solving equations (i) and (ii), we get,

$$p_1 = 66.66 \text{ kN/mm}^2$$

$$p_2 = -37.8 \text{ kN/mm}^2$$

EXERCISE

(i) Objective Type Questions

1. The ratio of lateral strain to linear strain is called:
   (a) modulus of elasticity (b) modulus of rigidity
   (c) bulk modulus (d) Poisson's ratio
2. Relation between E and K is given by:
   (a) $E = 3K (1 + 2\nu)$ (b) $E = 3K (1 - 2\nu)$
   (c) $E = 3K/(1 + 2\nu)$ (d) $E = K (1 + 2\nu)$
3. The ratio between stress and strain either tensile or compressive is called:
   (a) modulus of elasticity (b) modulus of rigidity
   (c) bulk modulus (d) Poisson's ratio
4. Rigidity modulus is related to:
   (a) tensile stress and strain (b) compressive stress and strain
   (c) shear stress and strain (d) none of the above
5. Unit of stress is:
   (a) N/mm^2 (b) Nmm (c) kg/mm (d) kg/mm^2
6. When a tensile or compressive load P acts on a body, the change in length is equal to:
   (a) Pl/AE (b) PA/lE (c) lE/PA (d) none
7. Hooke's law states that:
   (a) stress is directly proportional to strain
   (b) stress is inversely proportional to strain
   (c) stress is not directly proportional to strain
   (d) none of the above
8. For engineering material, Poisson's ratio lies in between:
   (a) 0 and 1 (b) –1 and +1 (c) 0 and 1/2 (d) – 1/2 and +1/2
9. The bulk modulus of elasticity of a material is twice its modulus of rigidity, the Poisson's ratio becomes:
   (a) 1/7 (b) 2/7 (c) 3/7 (d) 4/7
10. The bulk modulus of elasticity of a material is equal to its Young's modulus of elasticity, the Poisson's ratio becomes:
    (a) 4/3 (b) 3/4 (c) 3/7 (d) 1/3
11. If a material is restrained to deform and heated, it develops:
    (a) compressive stress (b) shear stress
    (c) tensile stress (d) no stress

12. If a material is not restrained to deform and heated, it develops:

(a) compressive stress (b) shear stress
(c) tensile stress (d) no stress

13. Thermal stress is given by:

(a) $E/\alpha T$ (b) $E\alpha T$ (c) $1/E\alpha T$ (d) none

14. Thermal strain is given by:

(a) $1/\alpha T$ (b) $E\alpha T$ (c) αT (d) none

15. There is a limiting value of load up to which the strain disappears on the removal of load, the corresponding stress is called:

(a) unit stress (b) yield stress
(c) compressive stress (d) elastic limit

16. Strain energy stored in a body of volume V subjected to uniform stress σ is given by:

(a) $\sigma^2/2E$ (b) $(\sigma^2/2E) \times V$ (c) $(\sigma^2/2E) \times (1/V)$ (d) none

17. The relation between modulus of elasticity E, modulus of rigidity G, and bulk modulus K is given by:

(a) $E = 9K/(G + 3K)$ (b) $E = K/(G + 3K)$
(c) $E = 9KG/(G + 3K)$ (d) none

18. In a uniaxial state of stress, the maximum shear stress occurs on a plane which is inclined to the load line at:

(a) 90° (b) 60° (c) 30° (d) 45°

19. The modulus ratio is given by:

(a) E_1/E_2 (b) E_1/G_2 (c) E_1/K_2 (d) none

20. The stress due to suddenly applied load as compared to the stress due to same load gradually applied is:

(a) equal (b) half (c) thrice (d) double

21. The Poisson's ratio for steel is:

(a) 0.1 (b) 0.2 (c) 0.3 (d) 0.4

22. Factor of safety is the ratio of:

(a) ultimate stress to working stress (b) working stress to ultimate stress
(c) normal stress to ultimate stress (d) none

23. The following material is more elastic:

(a) steel (b) fibre (c) leather (d) rubber

24. The normal stress on a plane inclined to the direction of loading (i.e., uniaxial loading) is given by:

(a) $\sigma \cos \theta$ (b) $\sigma \cos^2\theta$ (c) $\sigma/\cos^2\theta$ (d) none

25. The maximum shear stress on a plane inclined at an angle θ to the direction of loading (i.e., uniaxial loading) is:

(a) half of the maximum normal stress (b) double of the maximum normal stress

(c) equal to the maximum normal stress (d) none

26. Shear stress on the principal plane is:

(a) maximum (b) 1 (c) 0 (d) none

27. The resultant stress on a plane inclined at an angle θ to the direction of loading (i.e., uniaxial loading) is given by:

(a) $\sigma \cos \theta$ (b) $\sigma \cos^2\theta$ (c) $\sigma/\cos^2\theta$ (d) none

28. In a biaxial state of stress, the shear stress is equal to:

(a) half of the difference of stresses (b) double of the difference of stresses

(c) half of the sum of stresses (d) none of the above

29. In a biaxial state of stress, the maximum shear stress is maximum on a plane making an angle with the stresses equal to:

(a) 90° (b) 60° (c) 45° (d) 30°

30. Maximum shear stress in Mohr's circle is equal to:

(a) diameter of the circle (b) radius of the circle

(c) chord of the circle (d) none of the above

31. The deformation of bar under its own weight as compared to that when subjected to direct axial load equal to its own weight is:

(a) double (b) equal (c) half (d) zero

32. The Young' modulus for the perfectly rigid body is:

(a) 0 (b) 1 (c) 2 (d) infinite

33. The work done in producing strain on a material per unit volume is known as:

(a) elasticity (b) yield strength (c) toughness (d) resilience

34. Which one of the following is a brittle material:

(a) mild steel (b) brass (c) copper (d) cast iron

35. Ability of a material to absorb strain energy is called:

(a) toughness (b) malleability (c) brittleness (d) none

36. The ultimate tensile stress for mild steel lies in the range of:

(a) 100-200 N/mm^2 (b) 170-270 N/mm^2

(c) 270-370 N/mm^2 (d) 370-400 N/mm^2

37. The modulus of elasticity for mild steel lies in the range of:

(a) 100-200 GPa (b) 200-210 GPa (c) 210-310 GPa (d) 310-410 GPa

38. The modulus of elasticity is highest for

(a) aluminium (b) cast iron (c) mild steel (d) bronze

39. For dividing ductile and brittle material, generally an arbitrary percentage elongation considered is:

(a) 1% (b) 2% (c) 5% (d) none of the above

40. The Poisson's ratio of rubber lies in the range of:

(a) 0.1 to 0.2 (b) 0.2 to 0.3 (c) 0.3 to 0.4 (d) 0.45 to 0.5

41. In a tensile test of a specimen, the ratio of maximum load to the original cross sectional area of the test piece is termed:

(a) breaking stress (b) safe stress (c) ultimate stress (d) yield stress

ANSWERS

1. (d)	2. (b)	3. (a)	4. (c)	5. (a)	6. (a)	7. (a)	8. (c)	9. (b)	10. (d)
11. (a)	12. (d)	13. (b)	14. (c)	15. (d)	16. (b)	17. (c)	18. (d)	19. (a)	20. (d)
21. (c)	22. (a)	23. (a)	24. (b)	25. (a)	26. (c)	27. (a)	28. (a)	29. (c)	30. (b)
31. (c)	32. (d)	33. (d)	34. (d)	35. (a)	36. (d)	37. (b)	38. (c)	39. (c)	40. (d)
41. (c)									

(ii) Review Questions

1. Define stress and strain. Also discuss the different types of stresses and strains.

2. State Hooke's law. What do you mean by Young's modulus, and rigidity modulus?

3. Define the terms:

(i) elastic limit, (ii) limit of proportionality,
(iii) upper yield point, (iv) lower yield point,
(v) point of ultimate stress, (vi) breaking point.

4. What do you mean by working stress, and factor of safety?

5. Define (i) a composite bar, (ii) modular ratio, (iii) thermal stresses, (iv) thermal strain, and (v) Poisson's ratio.

6. Define principal stresses and principal strains.

7. Derive an expression for normal and shear stresses in a body subjected to biaxial loading.

8. Derive an expression for principal stresses in a body subjected to complex state of stress.

9. Explain the procedure for drawing Mohr's circle under complex stresses.

10. Derive the relationship between elastic constants.

11. What do you mean by

(i) resilience, (ii) proof resilience,
(iii) modulus of resilience?

12. Draw the sketch for stress versus strain diagram of mild steel and discuss the salient points.
13. Prove that stress due to sudden load is twice the stress due to gradual load.
14. Derive an expression for normal and shear stresses in a body subjected to uniaxial loading.

(iii) Numerical Problems

1. An elastic rod 30 mm in diameter, 250 mm long extends by 0.35 mm under a tensile load of 50 kN. Determine (i) the stress, (ii) the strain, and the elastic modulus for the material of rod. [**Ans.** 70.73 N/mm^2, 0.0014, 50521.43 N/mm^2]
2. A copper rod 5 mm in diameter when subjected to a pull of 600 N extends by 0.0875 mm over a gauge length of 150 mm. Determine the Young's modulus for copper. [**Ans.** 52.42 kN/mm^2]
3. A load of 5 kN has to be raised at the end of a steel wire. If the unit stress in the wire must not exceed 100 N/mm^2 then determine the minimum diameter required. Also determine the extension of 2.5 metre length of wire. Take the Young's modulus for elasticity as 0.2 MN/mm^2. [**Ans.** 7.979 mm, 1.25 mm]
4. A load of 6 kN has to be raised at the end of a steel wire. If the unit stress in the wire must not exceed 120 MN/m^2 then determine the minimum diameter required. Also determine the extension of 3 metre length of wire. Take the Young's modulus for elasticity as 200 GPa. [**Ans.** 7.98 mm, 1.8 mm]
5. A wooden tie is 50 mm wide, 100 mm deep, and 1200 mm long. It is subjected to an axial pull of 30 kN. The stretch of the member is found to be 0.5 mm. Determine the Young's Modulus for the material. [**Ans.** 11999.04 N/mm^2]
6. A steel rod 30 mm in diameter and 1.5 m long is subjected to an axial pull of 35 kN. Determine (i) the intensity of stress, (ii) the strain, and (iii) elongation. [**Ans.** 49.51 N/mm^2, 2.4755×10^{-4}, 0.3713 mm]
7. A steel bar 1500 mm long, 50 mm wide, and 20 mm thick is subjected to an axial tensile load of 120 kN. If the extension in the length of the bar is 0.9 mm, make calculations for (i) the intensity of stress, (ii) the strain, and (iii) modulus of elasticity of the bar material. [**Ans.** 120 N/mm^2, 0.0006, 2×10^5 N/mm^2]
8. The wire working on a railway signal is 8.5 mm in diameter and 350 m long. If the movement at the signal end is to be 17.5 cm, make calculations for the movement which should be given to the end of the wire at the signal box. A pull of 1650 N on the wire is applied. Take modulus of elasticity of wire material as 0.2 MN/mm^2. [**Ans.** 22.59 cm]
9. The following data are related to a tensile test conducted on a specimen: original diameter and gauge length = 16 mm and 62 mm respectively, minimum diameter at fracture = 11 mm, distance between gauge points at fracture = 76 mm, load at yield point and fracture

point = 42 kN, and 47 kN respectively. Maximum load that can be applied on specimen = 72 kN. Determine (i) the yield stress or strength, (ii) ultimate tensile stress, (iii) breaking stress, (iv) the percentage elongation and the percentage decrease in area after fracture, and (v) nominal and true stress at fracture.

[**Ans.** 208.89 N/mm^2, 358.10 N/mm^2, 233.76 N/mm^2, 22.58%, 52.73%, 233.76 N/mm^2, 494.58 N/mm^2]

10. The following data were observed for a steel bar 25 mm in diameter and 25 cm long tested for a tensile test: load at elastic limit = 65 kN, extension at elastic limit = 0.25 mm, load at upper yield point = 72 kN, maximum load = 135 kN, breaking load = 115 kN, diameter at the neck = 16 mm, final length = 30 cm. Find (i) modulus of elasticity, (ii) upper yield stress, (iii) ultimate stress, (iv) stress at breaking point, (v) percentage elongation, (vi) percentage reduction in cross sectional area.

[**Ans.** 134.4 kN/mm^2, 0.1467 kN/mm^2, 0.275 kN/mm^2, 0.2343 kN/mm^2, 20%, 59.04%]

11. The safe stress for a hollow steel column which carries an axial load of 2 MN is 120 MN/m^2. If the external diameter of the column is 25 cm, determine the internal diameter.

[**Ans.** 20.32 cm]

12. A steel plate is pulled with a force of 75 kN. The plate is 100 mm wide and 10 mm thick. A hole of diameter 10 mm is drilled at the centre of the plate. Find the maximum tensile stress set up in the plate. [**Ans.** 0.0833 kN/mm^2]

13. The bar shown in Figure 7.69 is subjected to an axial pull of 160 kN. If the stress in the middle portion is limited to 150 kN/mm^2 then determine the diameter of the middle portion. Proceed to determine the length of the middle portion if total elongation of the bar is 0.25 mm. The Young's modulus for the material of the bar is 200 kN/mm^2.

[**Ans.** 36.85 mm, 292.94 mm]

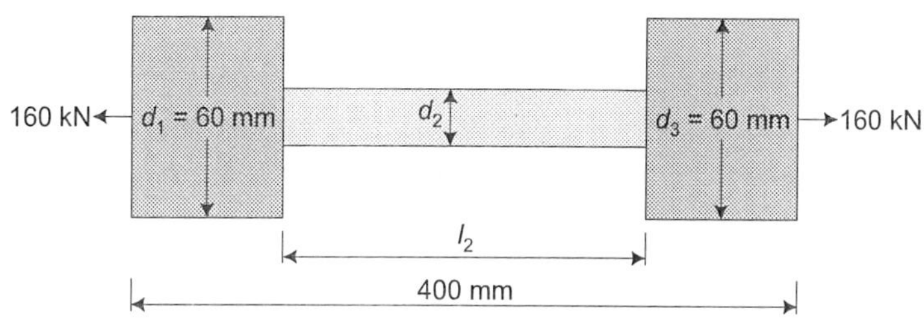

Fig. 7.69

14. A member formed by connecting a mild steel bar to an aluminium bar is shown in Figure 7.70.

Assume that there is no sideways buckling, determine the magnitude of force P that will cause the total length of the member to decrease 0.5 mm. The values of elastic modulus for mild steel and aluminium are 200 kN/mm^2 and 70 kN/mm^2 respectively.

[**Ans.** 1315.96 kN]

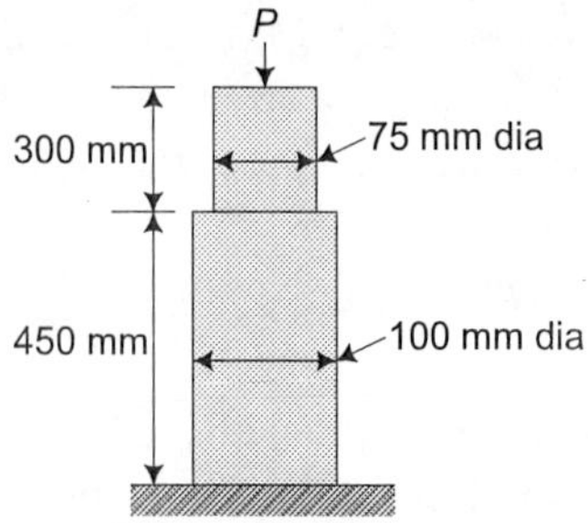

Fig. 7.70

15. A wrought iron bar is 5 m long and is 60 mm in diameter for 1 m of its length, 50 mm diameter for 2 m of its length, and 30 mm in diameter for its remaining length. The bar is in tension and the stress in the smallest section is 75 N/mm^2. Determine the total length of the bar if modulus of elasticity 2×10^5 N/mm^2. [**Ans.** 1.114 mm]

16. A mild steel bar is 9 m long and is 100 mm in diameter for 2 m of its length, 75 mm in diameter for 3 m of its length, and 50 mm in diameter for its remaining length. The bar is in compression under a compressive load of 5000 N. Find the total contraction of the bar if the modulus of elasticity is 2×10^5 N/mm^2. [**Ans.** 0.07428 mm]

17. A steel tie rod of 60 mm in diameter and 2.5 m long is subjected to an axial pull of 250 kN. Determine the percentage change in elongation produced under the same pull by boring the rod centrally for 1.1 m length of 30 mm diameter. The Young's modulus for the material of the bar is 200 kN/mm^2. [**Ans.** 14.69%]

18. A member *ABCD* is subjected to point loads P_1, P_2, P_3, and P_4 as shown in Figure 7.71. Calculate the force P_2 necessary for equilibrium if P_1 = 50 kN, P_3 = 455 kN, and P_4 = 135 kN. Determine the total elongation of the member if the Young's modulus for the material for the bar is 2×10^5 N/mm^2. [**Ans.** 370 kN, 0.445 mm]

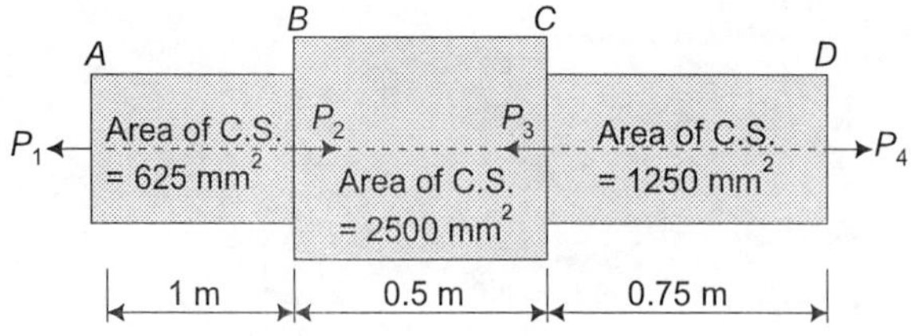

Fig. 7.71

19. A steel tie rod of 40 mm in diameter and 2 m long is subjected to an axial pull of 80 kN. To what length the bar should be bored centrally so that the total extension will increase by 20% under the same pull, the size of the bore is 20 mm in diameter. The Young's modulus for the material of the bar is 200 kN/mm^2. [**Ans.** 1.21 m]

20. A square bar of 30 mm side is held between two rigid plates and loaded by an axial pull equal to 500 kN as shown in Figure 7.72. Find the reactions at the end *A* and *C* and the elongation of the portion *AB*. Take the Young's modulus for the material of the bar as 2×10^5 N/mm^2. [**Ans.** 294.41 kN, 0.572 mm]

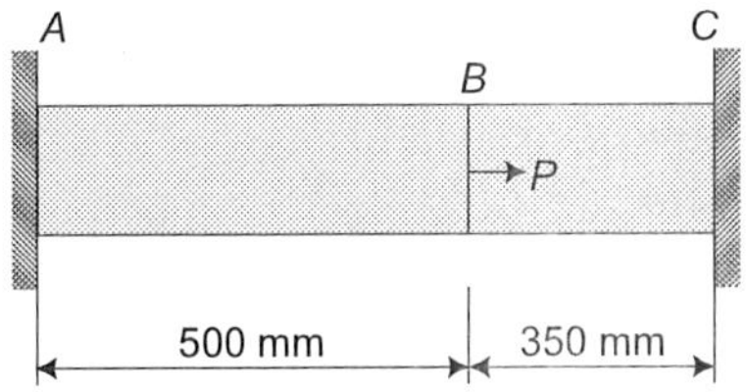

Fig. 7.72

21. A steel wire 2.5 m long and 3.5 mm in diameter is extended by 0.8 mm when a weight *W* is suspended from the wire. If the same weight is suspended from a brass wire 3 m long and 2.5 mm in diameter, it is elongated by 4.75 mm. Determine the modulus of elasticity of brass if that of steel is 2×10^5 N/mm^2. [**Ans.** 79224.98 N/mm^2]

22. A bar *ABCD* rigidly fixed at the ends *A* and *B* is subjected to two equal and opposite forces $P = 30$ kN at *C* and *D* as shown in Figure 7.73. Determine axial stresses in each section of the bar. [**Ans.** 33.3 kN, 44.4 kN]

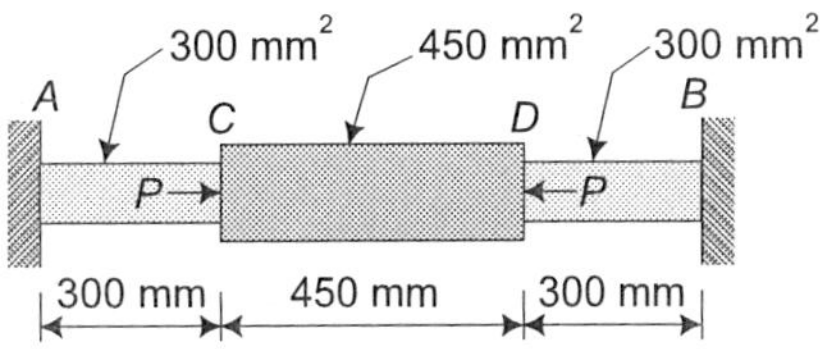

Fig. 7.73

23. A circular bar *ABCD* rigidly fixed at the ends *A* and *D* is subjected to axial loads as shown in Figure 7.74. Determine axial forces in each section of the bar and also determine the displacements of the ends *B* and *C* if Young's modulus for the material of the bar is 200 kN/mm^2.

[**Ans.** 133.33 kN, 33.33 kN, – 166.67 kN, 0.333 mm, 0.416 mm]

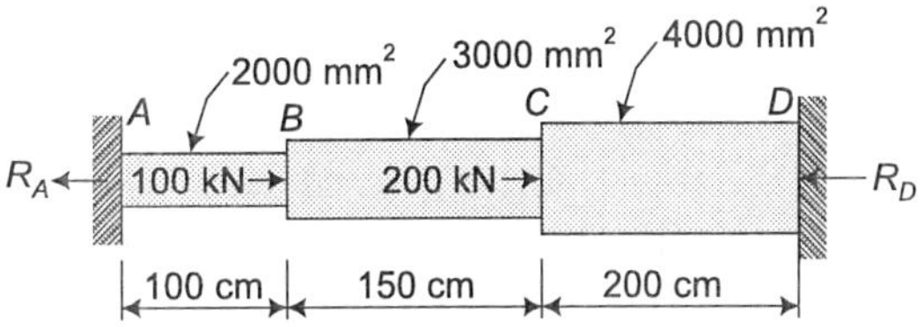

Fig. 7.74

24. Determine the reaction produced by the lower support on the bar as shown in Figure 7.75. Take Young's modulus for the material of the bar as 2×10^5 N/mm^2. Also determine the stresses in the bar.

[**Ans.** 463.69 kN, 136.31 kN, 0.059 kN/mm^2, 0.012052 kN/mm^2]

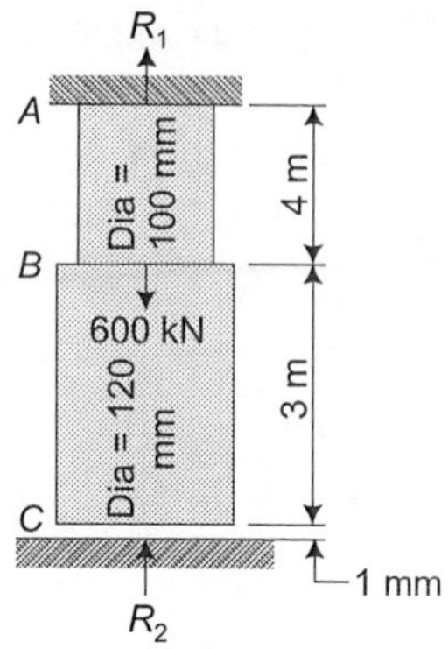

Fig. 7.75

25. A rod tapers uniformly from 20 mm to 10 mm diameter in a length of 200 mm. If the rod is subjected to an axial load of 5 kN, determine the extension of the rod. Take Young's modulus for the material of the bar is 2×10^5 N/mm^2. [**Ans.** 0.032 mm]

26. A straight bar of steel rectangular in section is 3 m long and is of uniform thickness 1.5 cm. The width of the rod uniformly varies from 10 cm to 4 cm. If the rod is subjected to an axial load of 30000 N, determine the extension of the rod. Take Young's modulus for the material of the bar is 2×10^5 N/mm^2.

[**Ans.** 0.458 mm]

27. For the bar shown in Figure 7.76, determine the reaction produced by the lower support on the bar. Take modulus of elasticity as 200 GN/m^2. Also determine the stresses in each section of the bar.

[**Ans.** 38.2 kN, 11.8 kN, 347.27 MN/m^2, 53.64 MN/m^2]

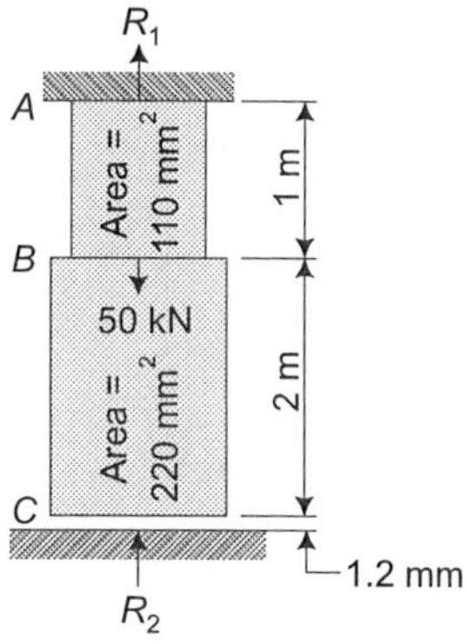

Fig. 7.76

28. A vertical circular bar 2 m long is subjected to a uniform stress of 0.3 N/mm^2 throughout its length. The diameter of the bottom edge of the bar is 60 mm. The weight density of the bar is 0.07644×10^{-3} N/mm^3. Determine its diameters at the middle section and at the top section if the bar is rigidly fixed in the ceiling.

[**Ans.** 68.15 mm, 77.4 mm]

29. A steel rod and two copper rods together support a load of 600 kN as shown in Figure 7.77. The cross sectional area of steel rod is 2500 mm^2 and of each copper rod is 1600 mm^2. Determine the stresses in the rods. Take E for steel = 2×10^5 N/mm^2 and for copper = 1×10^5 N/mm^2.

[**Ans.** 100 N/mm^2, 120 N/mm^2]

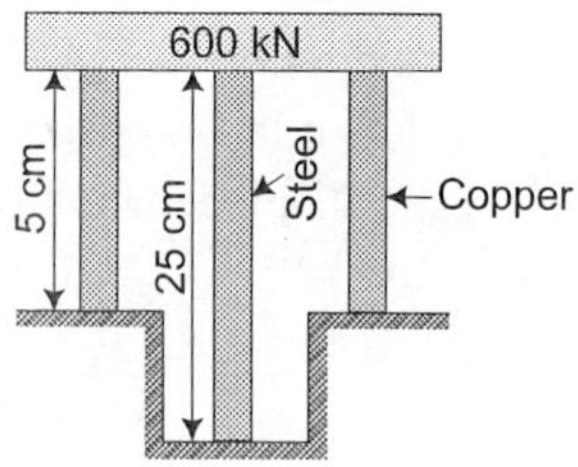

Fig. 7.77

30. A rigid beam weighing 1 kN is held in horizontal position by three wires of the same length. The outer wires are of brass of 1.5 mm diameter and attached to each one of the beam. The central wire is of steel of 0.7 mm diameter and attached to the middle of the horizontal beam. Determine the stresses induced in each of the wire if E for steel = 2×10^5 N/mm^2 and for brass = 0.8×10^5 N/mm^2.

[**Ans.** 222.42 N/mm^2, 556.05 N/mm^2]

31. A load of 420 kN is applied on a short concrete column 250 mm × 250 mm. The column is reinforced with steel bars of total area 2512 mm^2. If the modulus for steel is 18 times that of concrete, determine the stresses in concrete and steel. If the stress in concrete shall not exceed 4 N/mm^2, determine the area of steel required so that the column may support a load of 550 kN. [**Ans.** 3.99 N/mm^2, 71.82 N/mm^2, 4411.76 mm^2]

32. A load of 250 kN is applied on a short concrete column 250 mm × 250 mm. The column is reinforced with steel bars of total area 5600 mm^2. If the modulus for steel is 15 times that of concrete, determine the stresses in concrete and steel. If the stress in concrete shall not exceed 4 N/mm^2, determine the area of steel required so that the column may support a load of 550 kN. [**Ans.** 1.77 N/mm^2, 26.55 N/mm^2, 4464.8 mm^2]

33. A compound tube consists of a steel tube 150 mm internal diameter and 10 mm thickness and on outer brass tube 170 mm internal diameter and 10 mm thickness. The two tubes are of same length of 100 mm. The compound tube carries an axial load of 500 kN, determine the stresses and the load carried by each tube and the amount it shortens. Take E for steel = 2×10^5 N/mm^2 and for brass = 1×10^5 N/mm^2.

[**Ans.** 31.83 N/mm^2, 63.66 N/mm^2, 179.99 kN, 320.01 kN, 0.032 mm]

34. A steel rod 18 mm in diameter passes centrally through a steel tube 25 mm in internal diameter and 30 mm in external diameter. The tube is 750 mm long and is closed by rigid washers of negligible thickness which are fastened by nuts threaded on the rod. The nuts are tightened until the compressive load on the tube is 20 kN. Determine the stresses in the tube and the rod. Also determine the increase in these stresses when one nut is tightened by one quarter of a turn relative to the other. There are 4 threads per 10 mm. Young's modulus for steel is 200 kN/mm^2.

[**Ans.** 92.6 N/mm^2, 78.6 N/mm^2, 90.15 N/mm^2, 76.52 N/mm^2]

35. Determine the Poisson's ratio and bulk modulus of a material when Young's modulus and modulus of rigidity are 1.2×10^5 N/mm^2 and 0.45×10^5 N/mm^2 respectively.

[**Ans.** 0.33, 1.18×10^5 N/mm^2]

36. A 10 mm diameter rod of a material is subjected to an axial tensile force of 10 kN and the change in its diameter is observed to be 0.003. If the modulus of rigidity is 40 kN/mm^2, find the Poisson's ratio and modulus of elasticity. [**Ans.** 0.2323, 98.58 kN/mm^2]

37. A bar 30 mm × 30 mm × 250 mm long is subjected to an axial pull of 80 kN in the direction of its length. The extension of the bar was found to be 0.12 mm and the decrease in each lateral dimension is found to be 0.0037 mm. Evaluate Young's modulus, Poisson's ratio, modulus of rigidity, and bulk modulus for the material of the given bar.

[**Ans.** 1.85×10^5 N/mm^2, 0.26, 0.73×10^5 N/mm^2, 1.285×10^5 N/mm^2]

38. A bar of 25 mm diameter is subjected to a pull of 50 kN. The measured extension on a gauge length of 200 mm is 0.08 mm and the change in diameter is 0.0035 mm. Determine the Poisson's ratio and the values of the three moduli.

[**Ans.** 0.35, 254650 N/mm^2, 94314.81 N/mm^2, 282944.4 N/mm^2]

39. Two parallel walls 5 m apart are stayed together by a steel rod 30 mm in diameter passing through metal plates and nuts at each end. The nuts are screwed up to the plates while the bar is at a temperature of 423 K. Find the pull exerted by the bar after it has cooled to 293 K (i) if the ends do not yield, and (ii) if the total yielding at the two ends is 6 mm. Take $\alpha = 12 \times 10^{-6}$, $E = 2.1 \times 10^5$ N/mm^2. [**Ans.** 231.57 kN, 83.13 kN]

40. A steel rod 40 mm in diameter is completely enclosed in a copper tube of 40 mm internal diameter and 50 mm external diameter. A pin 25 mm in diameter is fitted transversely to the axis of the bar near each end to secure the bar to the tube. Determine the intensity of shear stress induced in the pins when the temperature of the whole is raised by 60°C. Also determine the stresses in the steel rod and the copper tube. Take coefficient of linear expansion for steel and copper as 1.2×10^{-5} per °C and 1.6×10^{-5} per °C respectively. Take the Young's modulus for steel and copper as 2×10^5 N/mm^2 and 1×10^5 N/mm^2 respectively.
[**Ans.** 18.286 N/mm^2, 10.286 N/mm^2, 13.17 N/mm^2]

41. A tensile load of 100 kN is gradually applied to a circular rod of 50 mm diameter and 3 m long. If the value of Young's modulus coefficient is 2.1×10^5 N/mm^2, determine stretch in the rod, stress in the rod, and strain energy absorbed by the rod.
[**Ans.** 0.764 mm, 50.93 N/mm^2, 38197.71 N mm]

42. A weight of 10 kN falls by 25 mm on a collar rigidly attached to a vertical bar 3 m long and 1000 mm^2 in section. Determine the instantaneous expansion of the bar if the value of Young's modulus coefficient is 2.1×10^5 N/mm^2. Also determine the strain energy stored in the bar. [**Ans.** 2.815 mm, 277432.33 N mm]

43. At a point in a strained material, the principal stresses are 400 N/mm^2 (tensile) and 300 N/mm^2 (compressive). Determine normal stress, shear stress, and resultant stress.
[**Ans.** 225 N/mm^2, 303.2 N/mm^2, 377.48 N/mm^2, 36.59°]

44. The stresses which act at a point on a body are $\sigma_1 = 120$ N/mm^2 (tensile), $\sigma_2 = 80$ N/mm^2 (compressive), and $\tau = 60$ N/mm^2. The loading on the body has been increased to values which are *k* times the provided values. Determine the maximum value of *k* if the body material can withstand maximum normal and shear stresses of 300 N/mm^2 and 200 N/mm^2 respectively. [**Ans.** 2.196, 7.715]

45. A rectangular block of material is subjected to a tensile stress of 108 N/mm^2 on one plane and a tensile stress of 47 N/mm^2 on the plane at right angles to the former. Each of the above stresses is accompanied by a shear stress of 61 N/mm^2 and that associated with the former tensile stress tends to rotate the block anticlockwise. Determine (i) the direction and magnitude of each of the principal stress, and (ii) magnitude of the greatest shear stress.
[**Ans.** 137.76 N/mm^2, 15.24 N/mm^2, 31°20′, 121°20′, 68.65 N/mm^2]

46. Using Mohr's circle diagram method, determine the normal and tangential stresses across the inclined plane as shown in Figure 7.78. Also check your answer analytically.
[**Ans.** 60 N/mm^2, 20 N/mm^2]

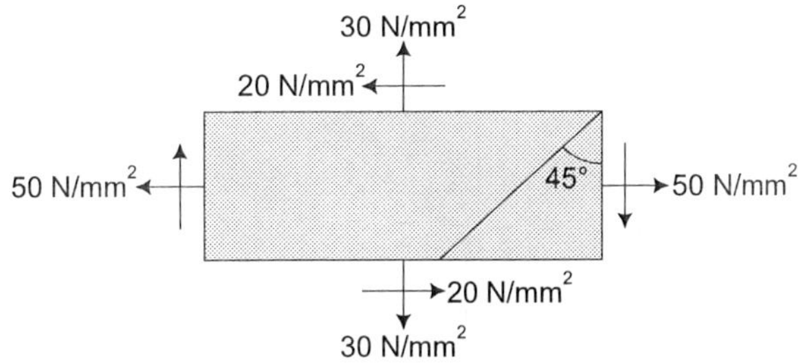

Fig. 7.78

47. The tensile stresses on two planes at right angles in a strained material at a certain point are 20 N/mm^2 and 10 N/mm^2. These stresses are accompanied by a shear stress of 10 N/mm^2. Using Mohr's circle diagram method, determine principal stresses and the location of principal planes. [**Ans.** 26.2 N/mm^2, 3.82 N/mm^2, 31.85°, 121.85°]

48. The intensity of resultant stress on a plane *PQ* at a point in a particular material under stress is 40 MN/m^2 and it is inclined at 30° to the normal to that plane as shown in Figure 7.54. The normal component of stress on another plane *QR* at right angle to the plane *PQ* is 25 MN/m^2. Determine (i) resultant stress on plane *QR* and its angle, (ii) the principal stresses and their directions, and (iii) the maximum shear stress and their planes.
[**Ans.** 32.02 MN/m^2, 50.39 MN/m^2, 9.25 MN/m^2, 38°13′, 128°13′, 20.57 MN/m^2]

8

Shear Force and Bending Moment

8.1 INTRODUCTION

A beam is a structural member which is sufficiently long as compared to its lateral dimensions and is subjected to loads normal to its longitudinal axis. Generally, the beams are horizontal which are supported along their length. Usually, the beams have three different types of support, namely, hinged (or pinned) support, roller support, and fixed support. A beam may be made of timber, timber reinforced with mild steel strips, and steel and reinforced concrete. The reinforced concrete beams are used in construction of bridges, flyovers, and buildings. Generally, the shape of most of the beams is of rectangular and I-sections. In any section of the beam, shear force and bending moment develop due to external loads and couples. In this unit, we will study the effect of forces or loads applied transverse to the axis of the member, producing bending in the member. The analysis of shear force and bending moment is required to design a beam.

8.2 TYPES OF BEAMS

The following are the types of beams which are used in general practice:

(i) **Cantilever Beam** A beam with one end fixed or built in and the other end free is called cantilever beam (Figure 8.1). There will be no deflection or rotation at the fixed end.

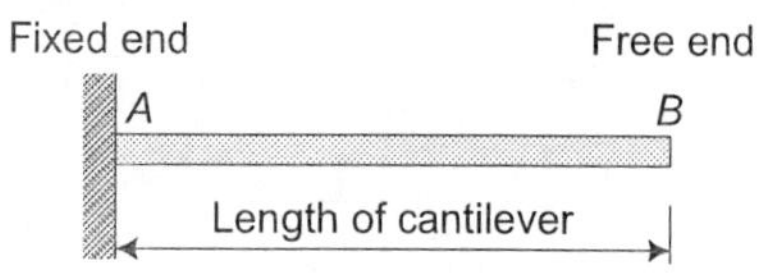

Fig. 8.1

(ii) **Simply Supported Beam** A beam which rests freely on two supports at its both ends is known as simply supported beam. Figure 8.2 shows a simply supported beam which is free to rotate at its ends when it bends.

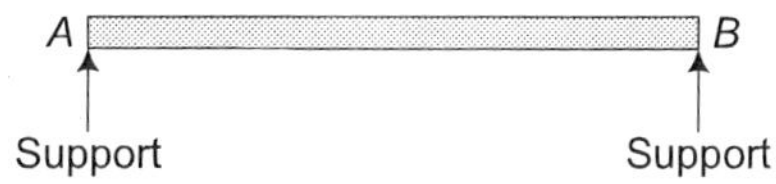

Fig. 8.2

(iii) **Overhanging Beam** A beam with one or both ends extended over the supports is called overhanging beam. Figure 8.3 shows an overhanging beam.

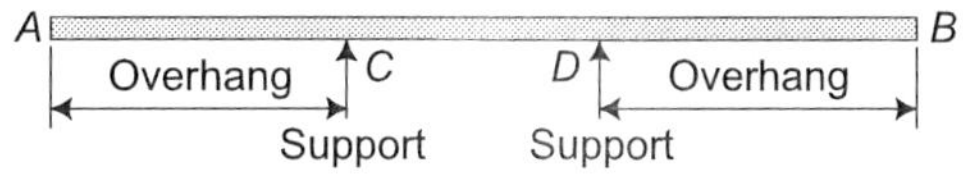

Fig. 8.3

(iv) **Continuous Beam** A beam having more than two supports is called a continuous beam (Figure 8.4). The supports at the ends are called end supports and the other supports are called intermediate supports.

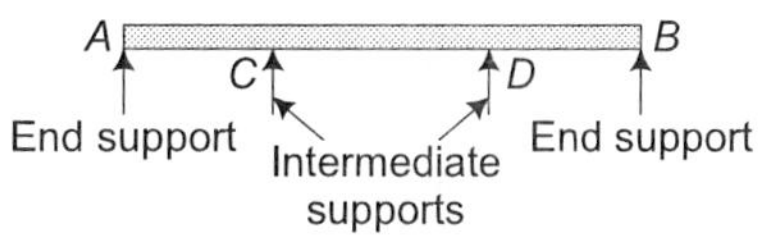

Fig. 8.4

(v) **Fixed Beam** A beam with its both ends fixed or built in walls is called fixed beam (Figure 8.5). A fixed beam is also called built in or encastre beam.

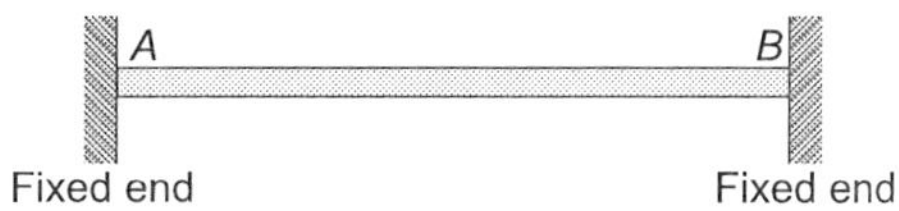

Fig. 8.5

8.3 TYPES OF LOADS

Generally, the beams are horizontal and the applied loads act in a vertical plane. The following are the important types of load which act on a beam:

(i) **Concentrated Load or Point Load** When the load is considered to act at a point on the beam, it is called concentrated or point load. In actual practice, it is not possible to apply a load at a point. However, it is not spread over any measurable part of the beam. The concentrated loads are shown in Figure 8.6.

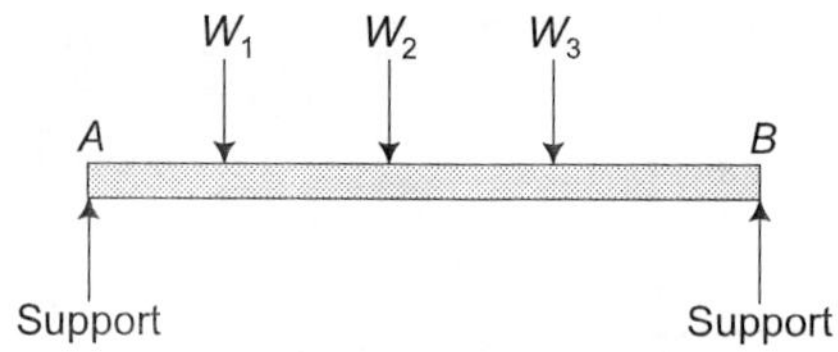

Fig. 8.6

(ii) Uniformly Distributed Load A uniformly distributed load is that which spreads evenly over a part or the entire length of a beam so that load per unit length of the beam remains same anywhere in it. A uniformly distributed load is shown in Figure 8.7. The rate of loading is denoted by u.d.l. and is expressed as w N/m run.

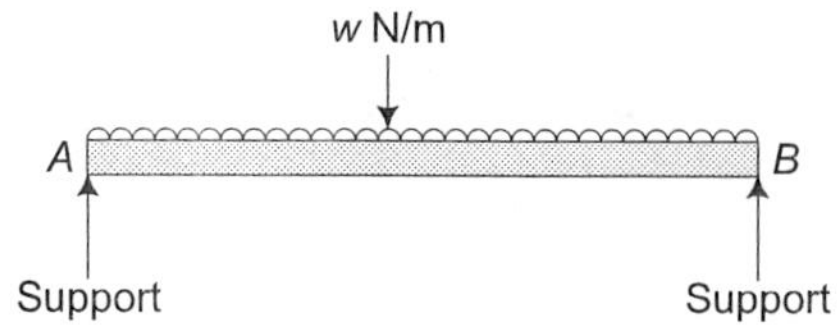

Fig. 8.7

(iii) Uniformly Varying Load The load which spreads over a beam in such a way that the load per unit length varies uniformly from a minimum value at one end to a maximum value at the other end. If the load varies uniformly from zero at one end to a maximum value at the other end, the load is called triangular load. In the case of a uniformly varying load, the area of the load diagram gives the total value of the load. A uniformly varying load or triangular load is shown in Figure 8.8.

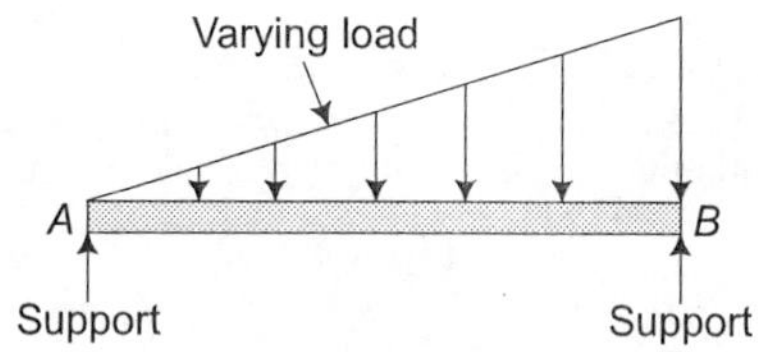

Fig. 8.8

(iv) Couple Load A beam may be loaded by a couple (Figure 8.9) whose magnitude is expressed in Nm.

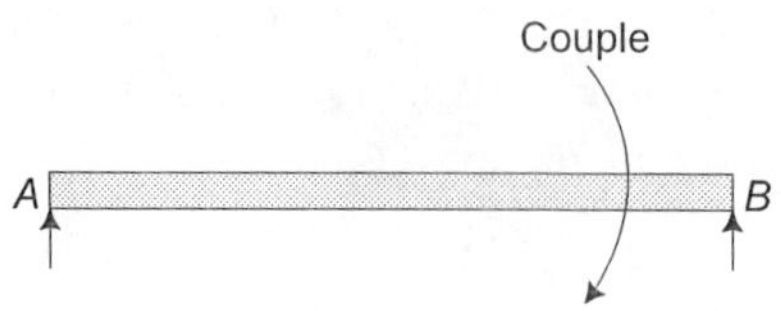

Fig. 8.9

A beam may carry any one of the above loading or a combination of them.

8.4 SHEAR FORCE AND BENDING MOMENT

8.4.1 Shear Force

The algebraic sum of all forces at any section of beam to the right or left side of the section is known as shear force. It is denoted by S.F.

Consider a simply supported beam *AB* carrying a load of 40 kN at its middle point as shown in Figure 8.10.

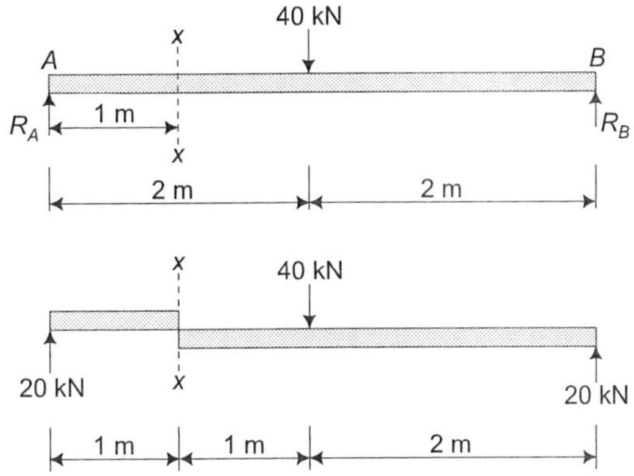

Fig. 8.10

Let R_A be the reaction at the support *A* and R_B the reaction at the support *B*. The load is transferred to supports and for equilibrium conditions,

$$R_A + R_B = 40 \text{ kN}$$

Also the sum of moment about *A* is equal to zero,

$$\Sigma M_A = 0$$

We have,

$$R_B \times 4 = 40 \times 2$$

$$\therefore \quad R_B = 20 \text{ kN},$$

$$R_A = 40 - R_B = 40 - 20 = 20 \text{ kN}$$

Now imagine the beam is divided into two parts by the section $x - x$.

Considering equilibrium of loads or forces on each portion of the beam, the resultant of the load and reaction to the left of section $x - x$ is given by,

$$F_{\text{left}} = R_A = 20 \text{ kN} \uparrow \text{(Vertically upwards)}$$

And the resultant of the load and reaction to the right of section $x - x$ is given by,

$$F_{\text{right}} = 40 - R_B = 40 \downarrow - 20 \uparrow$$
$$= 20 \text{ kN} \downarrow \quad \text{(Vertically downwards)}$$

It is observed that forces on the left and right sides of the section $x - x$ are equal in magnitudes but opposite in direction. The resultant force acting on any one of the parts normal to the axis of the beam is called shear force. Here the section $x - x$ is subjected to a shear force of 20 kN which is trying to shear off the beam.

Sign Convention The shear force at a section is considered positive when the resultant of the forces to the left of the section is upwards, or to the right of the section is downwards. Similarly, the shear force is taken as negative when the resultant of the forces to the left of the section is downwards, or to the right of the section is upwards.

8.4.2 Bending Moment

The algebraic sum of the moments of all the forces acting to the right or left side of the section is known as bending moment. It is denoted by B.M.

Considering equilibrium of moments of all the forces on each portion of the beam, the moments of the forces to the left of the section $x - x$ is given by,

$$M_{\text{left}} = R_A \times 1 = 20 \times 1 = 20 \text{ kNm} \quad \text{(Clockwise)}$$

And the moments of the forces to the right of the section $x - x$ is given by,

$$M_{\text{right}} = R_B \times 3 - 40 \times 1 = 20 \times 3 - 40 \times 1$$
$$= 20 \text{ kNm} \quad \text{(Anticlockwise)}$$

It is observed that moments on the left and right sides of the section $x - x$ are equal in magnitudes but opposite in direction. The section $x - x$ is acted upon by a moment of 20 kNm which is trying to bend the beam and thus it is called bending moment.

Sign Convention The bending moment at a section is considered positive when the moment of the forces on the left of the section is clockwise and to the right of the section is anticlockwise which causes concavity upwards and is called sagging bending moment as shown in Figure 8.11.

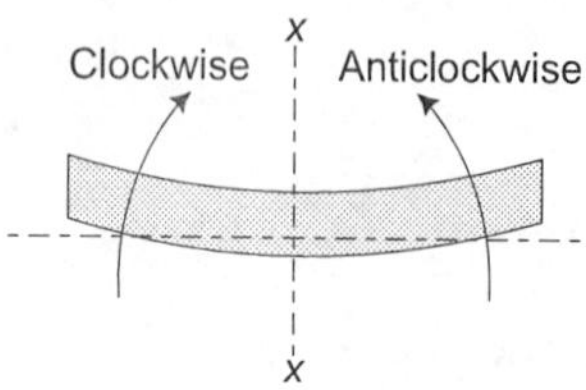

Positive B.M. (Sagging)

Fig. 8.11

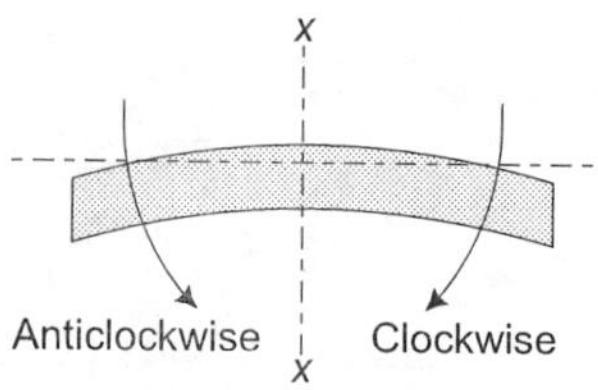

Negative B.M. (Hogging)

Fig. 8.12

Similarly, the bending moment is taken as negative when the moment of the forces on the left of the section is anticlockwise and to the right of the section is clockwise which causes convexity upwards and is called hogging bending moment as shown in Figure 8.12.

8.5 SHEAR FORCE AND BENDING MOMENT DIAGRAMS

The shear force and bending moment diagrams are the graphical representation of shear force and bending moment along the length of the beam. The shear force diagram is denoted by S.F.D. and the bending moment diagram is denoted by B.M.D.

In S.F. and B.M. diagrams, the abscissa represents the length of the beam and the ordinate represents the value of shear force and bending moment respectively. By these diagrams the maximum value of shear force and bending moments acting at the sections of beam can be known.

The salient points for drawing S.F.D. and B.M.D. are given below:

(i) Draw the beam along with loading to some suitable scale.

(ii) Choose the left or right portion of the section on the beam.

(iii) Add all the forces normal to the beam on the chosen portion by following the sign convention. Determine the shear force and bending moment at all the salient points.

(iv) Plot the S.F.D. and B.M.D. to a suitable scale under the loading diagram. Take the positive values of shear force and bending moments above the base line and negative values below the base line.

(v) The shear force changes by a vertical straight line at a section where there is a vertical concentrated load. The shear force does not change between two sections if there is no loading. Thus, the S.F.D. between two vertical loads will be horizontal and thus consists of rectangle for point loads. It consists of an inclined line for the portion on which u.d.l. acts and a parabolic curve for the uniformly varying load.

(vi) The bending moment at the free end of a cantilever and at the two supports of a simply supported beam is zero. The B.M.D. consists of inclined lines for the point loads acting on the beam. It consists of a parabolic curve for the portion on which u.d.l. acts and a cubic curve for the uniformly varying load.

(vii) The point in the B.M.D. where the bending moment becomes zero is called point of inflexion. The point in the B.M.D. where the bending moment changes its sign is called the point of contraflexure. Contra means opposite and the flexure means bending. So, at the point of contraflexure the beam flexes in opposite direction. Somewhere, the point of inflexion and the point of contraflexure are considered synonymous by different authors. The point of contraflexure can be located by equating the bending moment expression equal to zero for the portion of the beam where bending moment changes its sign.

The S.F.D. and B.M.D. for the various cases of beams and loads are plotted and analyzed in the next sections.

8.6 SHEAR FORCE AND BENDING MOMENT DIAGRAMS FOR A CANTILEVER

8.6.1 Cantilever Subjected to a Point Load at the Free End

Figure 8.13 (a) shows a cantilever *AB* of length *l* fixed at end *A* and free at end *B* which carries a point load *W*.

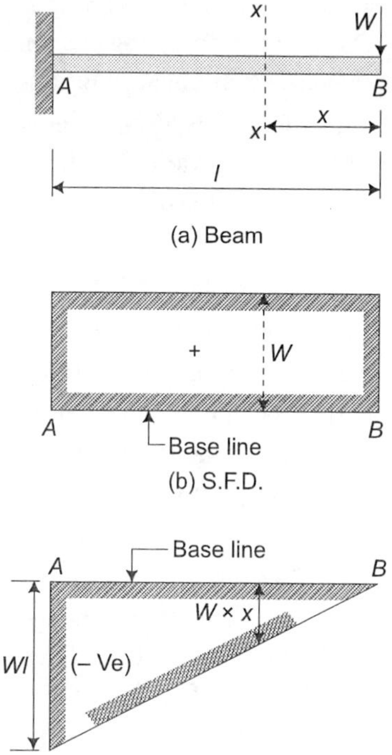

(a) Beam (b) S.F.D. (c) B. M. D.

Fig. 8.13

S.F.D. Consider a section $x - x$ at a distance x from the free end *B*. Consider the right portion of the section $x - x$. The resultant force on the right portion of the section $x - x$ is *W* which acts in the downward direction. As per the sign convention, the force on the right portion acting downwards is considered positive. Therefore, the shear force at the section $x - x$ is given by,

$$F_x = +W$$

As there is no other load between *A* and *B*, so the shear force will be constant at all the sections and it will be a rectangle of height *W* as shown in Figure 8.13 (b).

B.M.D. The bending moment at the section $x - x$ is given by,

$$M_x = - W \times x \qquad \text{(i)}$$

As per the sign convention, for the right portion of the section $x - x$, the clockwise moment will be taken as negative. This causes convexity upwards, so it is hogging bending moment.

From equation (i) it is observed that bending moment at any section of the cantilever is proportional to the distance of the section from the free end.

At $x = 0$, i.e., bending moment at *B*,

$$M_B = 0$$

At $x = l$, i.e., bending moment at *A*,

$$M_A = - W \times l$$

Thus bending moment follows the straight-line law. The bending moment diagram is shown in Figure 8.13 (c).

Example 8.1 A cantilever beam is subjected to several point loads as shown in Figure 8.14 (a). Draw the shear force and bending moment diagrams.

Solution

S.F.D. Consider a section $x - x$ between *C* and *D*. The shear force at the section $x - x$ will be equal to the resultant force acting on the right portion of the section. The resultant force is acting downwards, so as per the sign convention it is taken as positive.

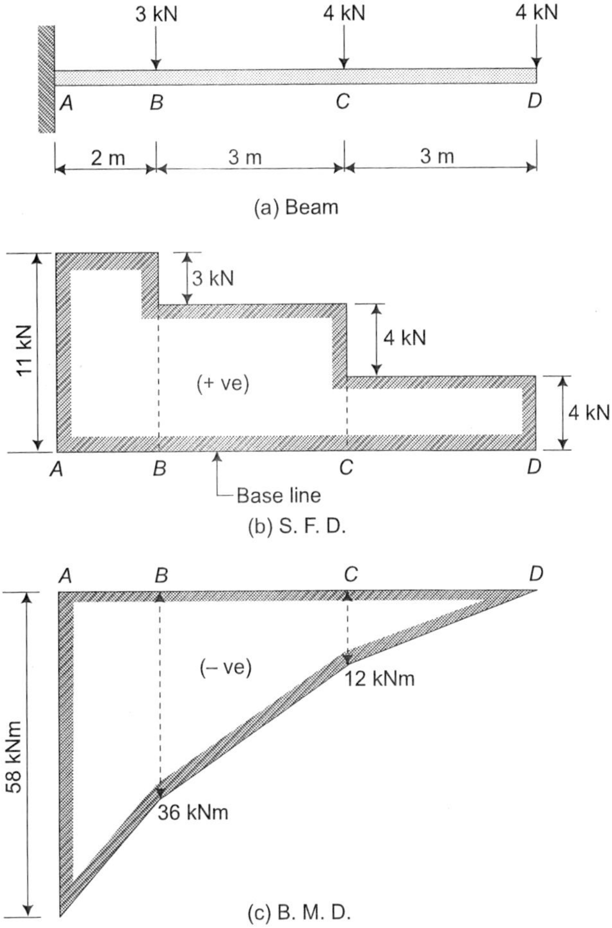

Fig. 8.14

$\therefore \qquad F_x = +4$ kN (Constant)

Similarly, consider section $x - x$ between B and C,

$$F_x = 4 + 4 = +8 \text{ kN} \quad \text{(Constant)}$$

Similarly, consider section $x - x$ between A and B,

$$F_x = 4 + 4 + 3 = +11 \text{ kN} \quad \text{(Constant)}$$

The S.F.D. is shown in Figure 8.14 (b) which consists of several rectangles of different ordinates.

B.M.D. Consider a section $x - x$ between C and D, the bending moment at the section $x - x$ is given by, $M_x = -4x$ kNm (Linear)

(For the right portion of the section $x - x$, the clockwise moment is taken negative)

At $x = 0$, $M_D = -4 \times 0 = 0$

At $x = 3$, $M_C = -4 \times 3 = -12$ kNm

Similarly, consider section $x - x$ between B and C, the bending moment at the section is given by,

$$M_x = -4x - 4(x - 3) \text{ kNm}$$

At $x = 3$, $M_C = -4 \times 3 = -12$ kNm (As before)

At $x = 6$, $M_B = -4 \times 6 - 4 \times (6 - 3) = -36$ kNm

Similarly, consider section $x - x$ between A and B, the bending moment at the section is given by,

$$M_x = -4x - 4(x - 3) - 3(x - 6) \text{ kNm}$$

At $x = 6$, $M_B = -4 \times 6 - 4 \times (6 - 3) - 3 \times (6 - 6)$

$= -36$ kNm (As before)

At $x = 8$, $M_B = -4 \times 8 - 4 \times (8 - 3) - 3 \times (8 - 6)$

$= -58$ kNm

The bending moment diagram is shown in Figure 8.14 (c).

8.6.2 Cantilever Subjected to Uniformly Distributed Load (u.d.l.) over its Whole Span

Consider a cantilever beam AB of length l carrying a u.d.l. of intensity w per unit length over the whole span of the cantilever as shown in Figure 8.15 (a). Take a section $x - x$ at a distance x from the free end B.

S.F.D. The resultant force on the right portion of the section $x - x$ is given by wx. As per the sign convention the downward force to the section is taken as positive.

$\therefore$ $F_x = + wx$ (Linear)

At $x = 0$, $F_B = w \times 0 = 0$

At $x = l$, $F_A = wl$

The shear force diagram will be a triangle as shown in Figure 8.15 (b).

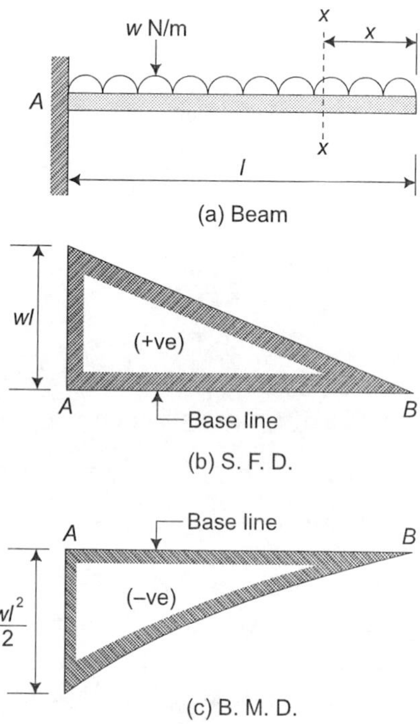

Fig. 8.15

B.M.D. The bending moment at the section $x - x$ is given by,

$$M_x = -\text{ (Total load on right side of section)}$$
$$\times \text{ (Distance of centre of gravity of right portion from section } x - x)$$

$\therefore$ $$M_x = -(wx) \times (x/2) = -w\,.\,(x^2/2) \quad \text{(Parabolic)}$$

(For the right portion of the section $x - x$, the clockwise moment is taken as negative)

At $x = 0$, i.e., bending moment at B,

$$M_B = 0$$

At $x = l$, i.e., bending moment at A,

$$M_A = -w\,.\,(l^2/2)$$

The B.M.D. is a parabola as shown in Figure 8.15 (c).

Example 8.2 A cantilever beam of length 4 m carries a uniformly distributed load of intensity 3 kN/m over a distance of 3 m from the support. Draw the shear force and bending moment diagrams.

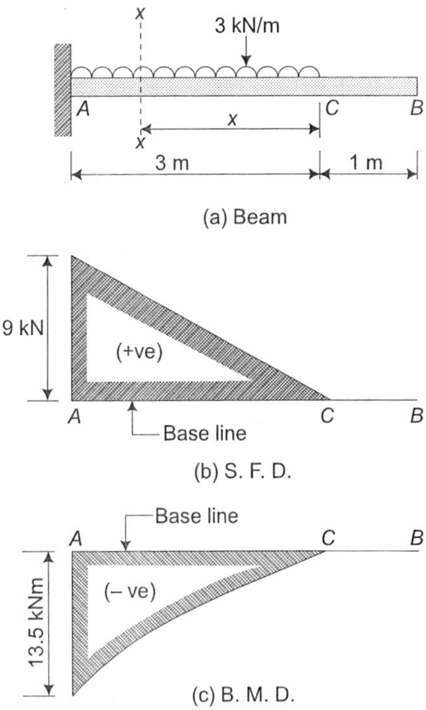

Fig. 8.16

Solution Refer Figure 8.16.

S.F.D. For BC, there is no loading from B to C, so, the shear force will be zero.

For AC, take a section $x - x$ at a distance x from the end C.

$$F_x = +\,3x \quad \text{(Linear)}$$

(The downward force to the right portion of the section $x - x$ is taken positive)

At $x = 0$, $\quad F_C = 3 \times 0 = 0$

At $x = 3$, $\quad F_A = 3 \times 3 = 9$ kN

The shear force diagram is shown in Figure 8.16 (b).

B.M.D. For BC, there is no loading from B to C, so, the BM will be zero.

For AC, $$M_x = -\,3x \times \frac{x}{2}$$
$$= -\frac{3x^2}{2} \text{ kNm} \quad \text{(Parabolic)}$$

(For the right portion of the section $x - x$, the clockwise moment is taken negative)

At $x = 0$, $\quad M_C = 3 \times 0 = 0$

At $x = 3$, $\qquad M_A = \dfrac{3 \times 3^2}{2} = -\ 13.5$ kNm

The B.M.D. is shown in Figure 8.16 (c).

Example 8.3 A cantilever beam of length 4 m carries a uniformly distributed load of intensity 3 kNm over a distance of 3 m from the free end. Draw the shear force and bending moment diagrams.

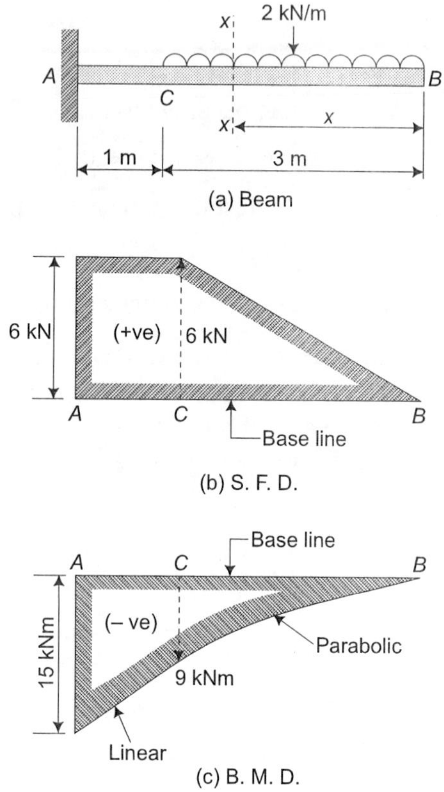

Fig. 8.17

Solution Refer Figure 8.17. Consider a section $x - x$ at a distance x from the free end.

S.F.D. For *BC*, $\qquad F_x = +\ 2x$ (Linear)

(The downward force to the right portion of the section $x - x$ is taken positive)

At $x = 0$, $\qquad F_B = 2 \times 0 = 0$

At $x = 3$, $\qquad F_C = 2 \times 3 = 6$ kN

For *CA*, $\qquad F_x = +\ (2 \times 3)$

$= +\ 6$ kN (Constant)

The shear force diagram is shown in Figure 8.17 (b).

B.M.D. For *BC*, $\qquad M_x = -\ 2x \times (x/2)$

$= x^2$ kNm (Parabolic)

(For the right portion of the section $x - x$, the clockwise moment is taken negative)

At $x = 0$, $\qquad M_B = 0$

At $x = 3$, $\qquad M_C = 3^2 = 9$ kNm

For *CA*, $\qquad M_x = -\ (2 \times 3) \times (x - 3/2) = -\ 6(x - 1.5)$ kNm

At $x = 3$, $\qquad M_C = -\ 6 \times (3 - 1.5) = -\ 9$ kNm (As before)

At $x = 4$, $\qquad M_A = -\ 6 \times (4 - 1.5) = -\ 15$ kNm (Linear)

The B.M.D. is shown in Figure 8.17 (c).

Example 8.4 A cantilever beam of length 4 m carries a uniformly distributed load of intensity 2 kN/m over a distance of 3 m as shown in Figure 8.18 (a). Draw the shear force and bending moment diagrams.

Solution Refer Figure 8.18.

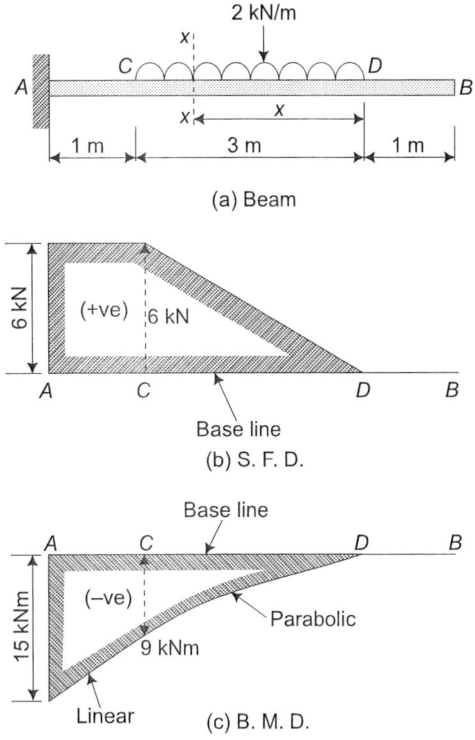

Fig. 8.18

S.F.D. For BD, there is no loading from B to D, so, the shear force will be zero.

For DC, consider a section $x - x$ at a distance x from the point D

$$F_x = +2x \quad \text{(Linear)}$$

(The downward force to the right portion of the section $x - x$ is taken positive)

At $x = 0$, $\quad F_D = 2 \times 0 = 0$

At $x = 3$, $\quad F_C = 2 \times 3 = 6$ kN

For CA, $\quad F_x = + (2 \times 3)$

$= + 6$ kN (Constant)

The shear force diagram is shown in Figure 8.18 (b).

B.M.D. For BD, there is no loading from B to D, so, the BM will be zero.

For DC, $\quad M_x = - 2x \times (x/2)$

$= x^2$ kNm (Parabolic)

(For the right portion of the section $x - x$, the clockwise moment is taken negative)

At $x = 0$, $\quad M_D = 0$

At $x = 3$, $\quad M_C = 3^2 = 9$ kNm

For CA, $\quad M_x = - (2 \times 3) \times (x - 3/2) = - 6 (x - 1.5)$ kNm

At $x = 3$, $\quad M_C = - 6 \times (3 - 1.5) = - 9$kNm (As before)

At $x = 4$, $\quad M_A = - 6 \times (4 - 1.5) = - 15$ kNm (Linear)

The B.M.D. is shown in Figure 8.18 (c).

Example 8.5 A cantilever beam of length 4 m carries a uniformly distributed load of intensity 4 kN/m over the whole span of the beam and a point load of 2 kN at the free end. Draw the shear force and bending moment diagrams.

Solution Refer Figure 8.19.

S.F.D. Consider a section $x - x$ at a distance x from the free end B.

The shear force at the section $x - x$ is given by,

$$F_x = 2 + 4 \times x \quad \text{(Linear)}$$

(The downward force to the right portion of the section $x - x$ is taken positive)

At $x = 0$, $F_B = 2 + 4 \times 0 = 2$ kN

At $x = 4$, $F_A = 2 + 4 \times 4 = 18$ kN

The shear force diagram is shown in Figure 8.19 (b).

B.M.D. The bending moment at any section $x - x$ at a distance of x from free end B is given by,

$$M_x = -2x - 4 \times x \times (x/2)$$

$$= (-2x - 2x^2) \text{ kNm} \quad \text{(Parabolic)}$$

(For the right portion of the section $x - x$, the clockwise moment is taken negative)

At $x = 0$, $M_B = 0$

At $x = 4$, $M_A = -2 \times 4 - 2 \times 4^2$

$= -40$ kNm

The B.M.D. is shown in Figure 8.19 (c).

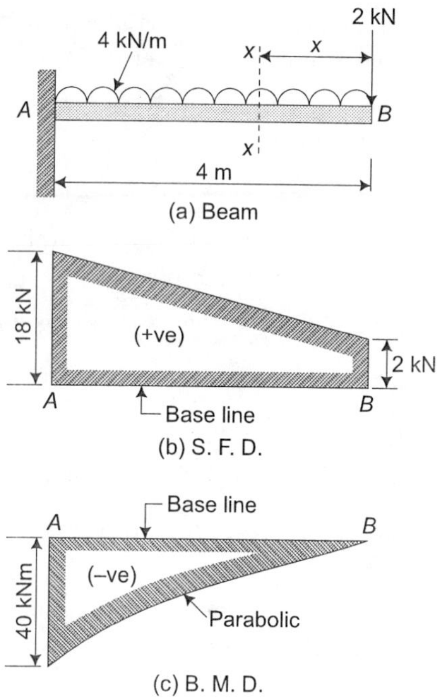

Fig. 8.19

Example 8.6 A cantilever beam of length 4.5 m carries a uniformly distributed load of intensity 6 kN/m over a length of 2.5 m from the free end. It also carries a point load of 7 kN at a distance of 0.5 m from the free end. Draw the shear force and bending moment diagrams.

Solution Refer Figure 8.20.

S.F.D. For BD, consider a section $x - x$ at a distance x from the free end B.

The shear force at the section $x - x$ is given by,

$$F_x = +6 \times x \quad \text{(Linear)}$$

(The downward force to the right portion of the section $x - x$ is taken positive)

At $x = 0$, $F_B = +6 \times 0 = 0$

At $x = 0.5$, $F_D = +6 \times 0.5 = 3$ kN

Similarly, for CD, consider a section $x - x$ at a distance x from the free end B.

The shear force at the section $x - x$ is given by,

$$F_x = +6 \times x + 7 \text{ kN} \quad \text{(Linear)}$$

At $x = 0.5$, $F_D = +6 \times 0.5 + 7 = 10$ kN

At $x = 2.5$, $F_C = +6 \times 2.5 + 7 = 22$ kN

For CA, $F_x = +(6 \times 2.5 + 7) = +22$ kN (Constant)

The shear force diagram is shown in Figure 8.20 (b).

B.M.D. For *BD*, the bending moment at any section *x* – *x* at a distance of *x* from free end *B* is given by,

$$M_x = -6 \times x \times x/2$$
$$= -3x^2 \text{ kNm} \quad \text{(Parabolic)}$$

(For the right portion of the section $x - x$, the clockwise moment is taken negative)

At $x = 0$, $\quad M_B = 0$

At $x = 0.5$, $\quad M_D = -(3 \times 0.5^2)$
$$= -0.75 \text{ kNm}$$

Similarly, for *CD*, the bending moment at any section $x - x$ at a distance of *x* from free end *B* is given by,

$$M_x = -6 \times x \times x/2 - 7 \times (x - 0.5)$$
$$= -3x^2 - 7 \times (x - 0.5) \text{ kNm} \quad \text{(Parabolic)}$$

At $x = 0.5$, $\quad M_D = -3 \times 0.5^2 - 7 \times (0.5 - 0.5)$
$$= -0.75 \text{ kNm} \quad \text{(As before)}$$

At $x = 2.5$, $\quad M_C = -3 \times 2.5^2 - 7 \times (2.5 - 0.5)$
$$= -32.75 \text{ kNm}$$

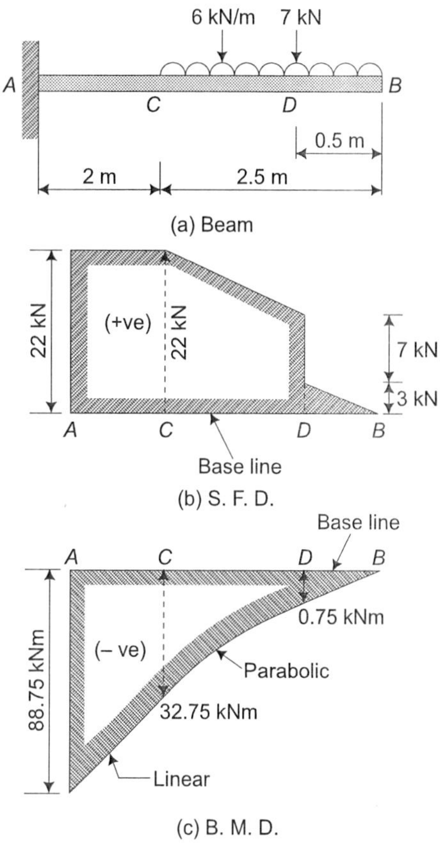

Fig. 8.20

Similarly, for *AC*, the bending moment at any section $x - x$ at a distance of *x* from free end *B* is given by,

$$M_x = -6 \times 2.5 \times (x - 2.5/2) - 7 \times (x - 0.5) \text{ kNm} \quad \text{(Linear)}$$

At $x = 2.5$, $\quad M_C = -6 \times 2.5 \times (2.5 - 1.25) - 7 \times (2.5 - 0.5)$
$$= -32.75 \text{ kNm} \quad \text{(As before)}$$

At $x = 4.5$, $\quad M_A = -6 \times 2.5 \times (4.5 - 1.25) - 7 \times (4.5 - 0.5)$
$$= -88.75 \text{ kNm}$$

The B.M.D. is shown in Figure 8.20 (c).

Example 8.7 A cantilever beam of length 5 m is loaded as shown in Figure 8.21 (a). Draw the shear force and bending moment diagrams.

Solution Refer Figure 8.21.

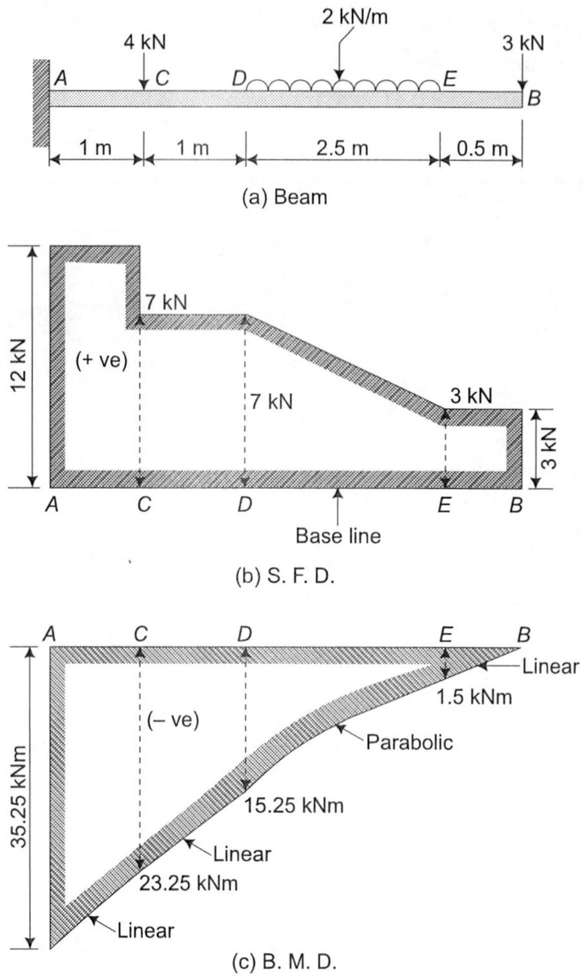

Fig. 8.21

S.F.D. Consider a section $x - x$ between B and E. The shear force at the section $x - x$ will be equal to the resultant force acting on the right portion of the section. The resultant force is acting downwards, so as per the sign convention it is taken positive.

$\therefore$ $$F_x = +3 \text{ kN} \quad \text{(Linear)}$$

This will remain constant for the section BE.

For DE, consider a section $x - x$ at a distance x from the free end B.

The shear force at the section $x - x$ is given by,

$$F_x = +2 \times (x - 0.5) + 3 \quad \text{(Linear)}$$

(The downward force to the right portion of the section $x - x$ is taken positive)

At $x = 0.5$, $\qquad F_E = + 2 \times (0.5 - 0.5) + 3 = + 3$ kN

At $x = 2.5$, $\qquad F_D = + 2 \times (2.5 - 0.5) + 3 = + 7$ kN

The shear force will remain constant between C and D, so,

$$F_C = + 7 \text{ kN}$$

Consider a section $x - x$ between A and C, the shear force at the section is given by,

$$F_x = + 3 + 2 \times 2.5 + 4 = + 12 \text{ kN} \quad \text{(Linear)}$$

This will remain constant for the section AC.

The shear force diagram is shown in Figure 8.21 (b).

B.M.D. For BE the bending moment at any section $x - x$ at a distance of x from free end B is given by,

$$M_x = - 3x \text{ kNm} \quad \text{(Linear)}$$

(For the right portion of the section $x - x$, the clockwise moment is taken negative)

At $x = 0$, $\qquad M_B = 0$

At $x = 0.5$, $\qquad M_E = - (3 \times 0.5) = - 1.5$ kNm

For ED, the bending moment at any section $x - x$ at a distance of x from free end B is given by,

$$M_x = - 3x - 2 \times (x - 0.5) \times (x - 0.5)/2 \text{ kNm} \quad \text{(Parabolic)}$$

(For the right portion of the section $x - x$, the clockwise moment is taken negative)

At $x = 0.5$, $\qquad M_E = - 3 \times 0.5 - 2 \times 0 = - 1.5$ kNm (As before)

At $x = 3$, $\qquad M_D = - 3 \times 4 - 2 \times (3 - 0.5) \times (3 - 0.5)/2$

$$= - 15.25 \text{ kNm}$$

Similarly, for CD, the bending moment at any section $x - x$ at a distance of x from free end B is given by,

$$M_x = - 3x - 2 \times 2.5 \times [x - (2.5/2 + 0.5)]$$

$$= - 3x - 5 \times (x - 1.75) \text{ kNm} \quad \text{(Linear)}$$

(For the right portion of the section $x - x$, the clockwise moment is taken negative)

At $x = 3$, $\qquad M_D = - 3 \times 3 - 5 \times (3 - 1.75)$

$$= - 15.25 \text{ kNm} \quad \text{(As before)}$$

At $x = 4$, $\qquad M_C = - 3 \times 4 - 5 \times (4 - 1.75) = - 23.25$ kNm

Similarly, for AC, the bending moment at any section $x - x$ at a distance of x from free end B is given by,

$$M_x = -3x - 2 \times 2.5 \times [x - (2.5/2 + 0.5)] - 4(x - 4)$$

$\therefore$ $$M_x = -3x - 5 \times (x - 1.75) - 4(x - 4) \text{ kNm} \quad \text{(Linear)}$$

(For the right portion of the section $x - x$, the clockwise moment is taken negative)

At $x = 4$, $$M_C = -3 \times 4 - 5 \times (4 - 1.75) - 4(4 - 4)$$
$$= -23.25 \text{ kNm} \quad \text{(As before)}$$

At $x = 5$, $$M_A = -3 \times 5 - 5 \times (5 - 1.75) - 4(5 - 4)$$
$$= -35.25 \text{ kNm}$$

The B.M.D. is shown in Figure 8.21 (c).

Example 8.8 A cantilever beam of length 5 m is loaded as shown in Figure 8.22 (a). Draw the shear force and bending moment diagrams.

Solution Refer Figure 8.22.

S.F.D. Consider a section $x - x$ between B and F. The shear force at the section $x - x$ will be equal to the resultant force acting on the right portion of the section. The resultant force is acting downwards, so as per the sign convention it is taken positive.

$\therefore$ $$F_x = +1.5 \text{ kN} \quad \text{(Linear)}$$

This will remain constant for the section BE.

For FE, consider a section $x - x$ at a distance x from the free end B.

The shear force at the section $x - x$ is given by,

$$F_x = +2 \times (x - 1) + 1.5 + 1.5 \quad \text{(Linear)}$$

(The downward force to the right portion of the section $x - x$ is taken positive)

At $x = 1$, $$F_F = +2 \times (1 - 1) + 1.5 + 1.5 = +3 \text{ kN}$$

At $x = 2$, $$F_E = +2 \times (2 - 1) + 1.5 + 1.5 = +5 \text{ kN}$$

For DE, consider a section $x - x$ at a distance x from the free end B. The shear force at the section $x - x$ is given by,

$$F_x = +2 \times (x - 1) + 1.5 + 1.5 + 2 \quad \text{(Linear)}$$

At $x = 2$, $$F_E = +2 \times (2 - 1) + 1.5 + 1.5 + 2 = +7 \text{ kN}$$

At $x = 3$, $$F_D = +2 \times (3 - 1) + 1.5 + 1.5 + 2 = +9 \text{ kN}$$

The shear force will remain constant between D and C, so,

$$F_C = +9 \text{ kN}$$

Consider a section $x - x$ between A and C, the shear force at the section is given by,

$$F_x = +1.5 + 1.5 + 2 \times 2 + 2 + 1 = 10 \text{ kNm} \quad \text{(Linear)}$$

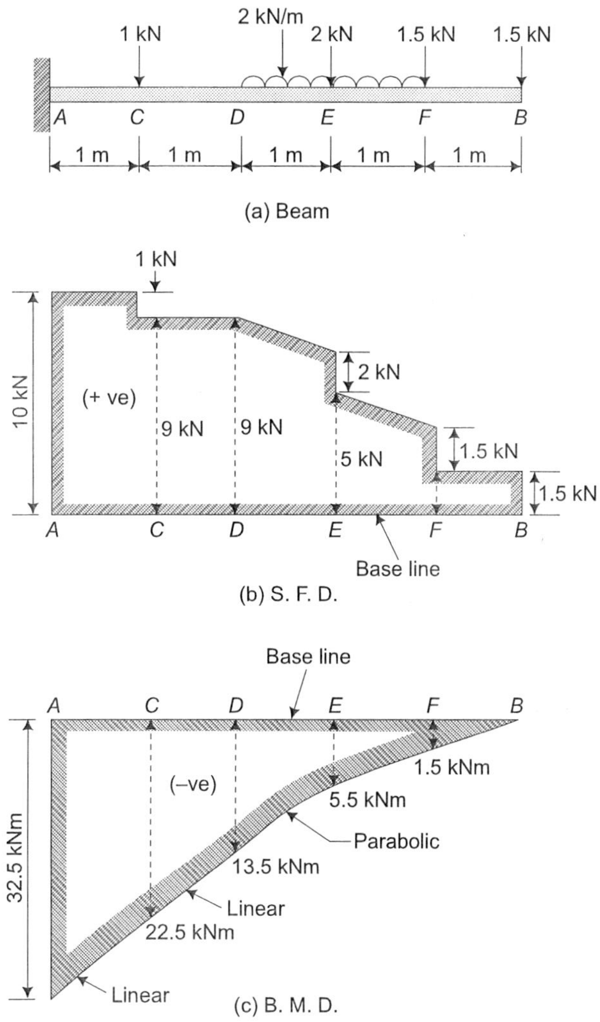

Fig. 8.22

This will remain constant for the section *AC*.

The shear force diagram is shown in Figure 8.22 (b).

B.M.D. For *BF* the bending moment at any section $x - x$ at a distance of x from free end *B* is given by,

$$M_x = -1.5x \text{ kNm} \quad \text{(Linear)}$$

(For the right portion of the section $x - x$, the clockwise moment is taken negative)

At $x = 0$, $\quad M_B = 0$

At $x = 1$, $\quad M_F = -(1.5 \times 1) = -1.5$ kNm

For *FE*, the bending moment at any section $x - x$ at a distance of x from free end B is given by,

$$M_x = -1.5x - 1.5 \times (x - 1) - 2 \times (x - 1) \times (x - 1)/2 \text{ kNm} \quad \text{(Parabolic)}$$

(For the right portion of the section $x - x$, the clockwise moment is taken negative)

At $x = 1$, $M_F = -1.5 \times 1 - 1.5 \times 0 - 2 \times 0$

$= -1.5$ kNm (As before)

At $x = 2$, $M_E = -1.5 \times 2 - 1.5 \times (2 - 1) - 2 \times (2 - 1) \times (2 - 1)/2$

$= -5.5$ kNm

Similarly, for *ED*, the bending moment at any section $x - x$ at a distance of x from free end B is given by,

$$M_x = -1.5x - 1.5 \times (x - 1) - 2 \times (x - 2) - 2 \times (x - 1)/2 \text{ kNm (Parabolic)}$$

(For the right portion of the section $x - x$, the clockwise moment is taken negative)

At $x = 2$, $M_E = -1.5 \times 2 - 1.5 \times (2 - 1) - 2 \times (2 - 2) - 2 \times (2 - 1) \times (2 - 1)/2$

$= -5.5$ kNm (As before)

At $x = 3$, $M_D = -1.5 \times 3 - 1.5 \times (3 - 1) - 2 \times (3 - 2) - 2 \times (3 - 1) \times (3 - 1)/2$

$= -13.5$ kNm

Similarly, for *CD*, the bending moment at any section $x - x$ at a distance of x from free end B is given by,

$$M_x = -1.5x - 1.5 \times (x - 1) - 2 \times (x - 2) - 2 \times 2 \times [x - (2/2 + 1)] \text{ kNm}$$

$\therefore$ $$M_x = -1.5x - 1.5 \times (x - 1) - 2 \times (x - 2) - 4 \times (x - 2) \text{ kNm} \quad \text{(Linear)}$$

(For the right portion of the section $x - x$, the clockwise moment is taken negative)

At $x = 3$, $M_D = -1.5 \times 3 - 1.5 \times (3 - 1) - 2 \times (3 - 2) - 4 \times (3 - 2)$

$= -13.5$ kNm (As before)

At $x = 4$, $M_C = -1.5 \times 4 - 1.5 \times (4 - 1) - 2 \times (4 - 2) - 4 \times (4 - 2)$

$= -22.5$ kNm

Similarly, for *AC*, the bending moment at any section $x - x$ at a distance of x from free end B is given by,

$$M_x = -1.5x - 1.5 \times (x - 1) - 2 \times (x - 2) - 2 \times 2 \times [x - (2/2 + 1)] - 1 \times (x - 4) \text{ kNm}$$

$\therefore$ $$M_x = -1.5x - 1.5 \times (x - 1) - 2 \times (x - 2) - 4 \times (x - 2) - (x - 4) \text{ kNm} \quad \text{(Linear)}$$

At $x = 4$, $M_C = -1.5 \times 4 - 1.5 \times (4-1) - 2 \times (4-2) - 4 \times (4-2) - (4-4)$

$= -22.5$ kNm (As before)

At $x = 5$, $M_A = -1.5 \times 5 - 1.5 \times (5-1) - 2 \times (5-2) - 4 \times (5-2) - (5-4)$

$= -32.5$ kNm

The B.M.D. is shown in Figure 8.22 (c).

Example 8.9 A cantilever beam of length 3 m is loaded as shown in Figure 8.23 (a). Draw the shear force and bending moment diagrams.

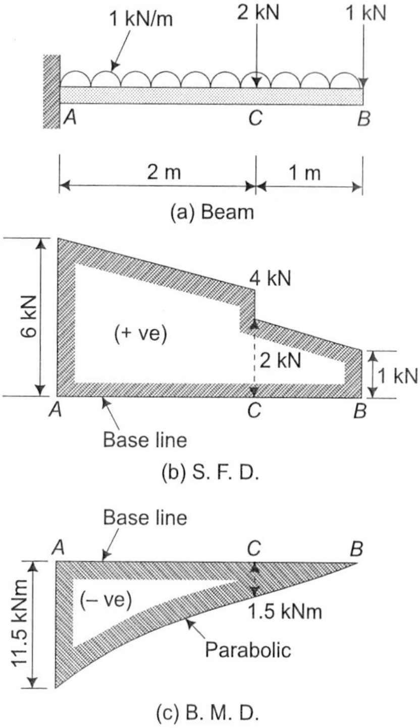

Fig. 8.23

Solution Refer Figure 8.23.

S.F.D. Consider a section $x - x$ between B and C. The shear force at the section $x - x$ will be equal to the resultant force acting on the right portion of the section. The resultant force is acting downwards, so as per the sign convention it is taken positive.

$\therefore$ $F_x = +[1 + 1 \times x]$

$= +[1 + x]$ kN (Linear)

At $x = 0$, $F_B = +1$ kN

At $x = 1$, $F_C = +[1 + 1] = +2$ kN

Similarly, consider section $x - x$ between A and C at a distance x from free end B. The shear force is given by,

$F_x = +[1 + 2 + 1 \times x]$

$= +[3 + x]$ kN (Linear)

(The downward force to the right portion of the section is taken positive)

At $x = 1$, $F_C = +[3 + 1] = +4$ kN

At $x = 3$, $F_A = +[3 + 3] = +6$ kN

The shear force diagram is shown in Figure 8.23 (b).

B.M.D. For BC the bending moment at any section $x - x$ at a distance of x from free end B is given by,

$$M_x = -1 \times x - 1 \times x \times x/2 = -x - x^2/2 \text{ kNm} \quad \text{(Parabolic)}$$

(For the right portion of the section $x - x$, the clockwise moment is taken negative)

At $x = 0$, $M_B = 0$

At $x = 1$, $M_C = -1 - (1)^2/2 = -1.5$ kNm (Parabolic)

Similarly, for the portion *AC*, the bending moment at any section is given by,

$$M_x = -1 \times x - 2 \times (x - 1) - 1 \times x \times x/2$$

$$= -x - 2(x - 1) - x^2/2 \text{ kNm} \quad \text{(Parabolic)}$$

At $x = 1$, $\quad M_C = -1 - 2(1 - 1) - 1^2/2 = -1.5$ kNm (As before)

At $x = 3$, $\quad M_A = -3 - 2(3 - 1) - 3^2/2 = -11.5$ kNm

The B.M.D. is shown in Figure 8.23 (c).

8.6.3 Cantilever Subjected to a Gradually Varying Load

Consider a cantilever beam *AB* of length *l* carrying a uniformly varying load of intensity zero at the free end to *w* at the fixed end as shown in Figure 8.24. Take a section $x - x$ at a distance *x* from the free end.

The rate of loading at a distance *x* is given by $(w/l) \times x$.

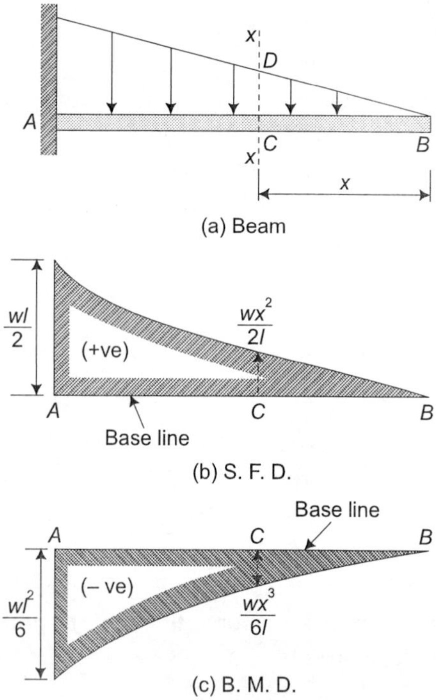

Fig. 8.24

S.F.D. The shear force at the section $x - x$ at a distance *x* from the free end is given by,

$$F_x = \text{Area of } \Delta BCD$$

$$= \frac{1}{2} \times \text{base} \times \text{height}$$

$$= \frac{1}{2} \times x \times \frac{wx}{l} = \frac{wx^2}{2l}$$

So, the shear force varies according to the parabolic law and it is a positive shear force.

At $x = 0$, $\quad F_B = + \dfrac{w(0)^2}{2l} = 0$

At $x = l$, $\quad F_A = + \dfrac{w(l)^2}{2l} = + \dfrac{wl}{2}$

B.M.D. The bending moment at any section $x - x$ at a distance of *x* from free end *B* is given by,

$M_x = -$ (Area of ΔBCD) $\times$ distance of centre of gravity of ΔBCD from section $x - x$

$$\therefore \quad M_x = -\frac{wx^2}{2l} \times \frac{x}{3} = -\frac{wx^3}{6l}$$

So, the bending moment varies according to the cubic law and it is a negative bending moment.

At $x = 0$, $$M_B = -\frac{w(0)x^3}{6l} = 0$$

At $x = l$, $$M_A = -\frac{w(l)^3}{6l} = -\frac{wl^2}{6}$$

Example 8.10 A cantilever beam of length 6 m carries a gradually varying load from zero at the free end to 2 kN/m at the fixed end. Draw the shear force and bending moment diagrams.

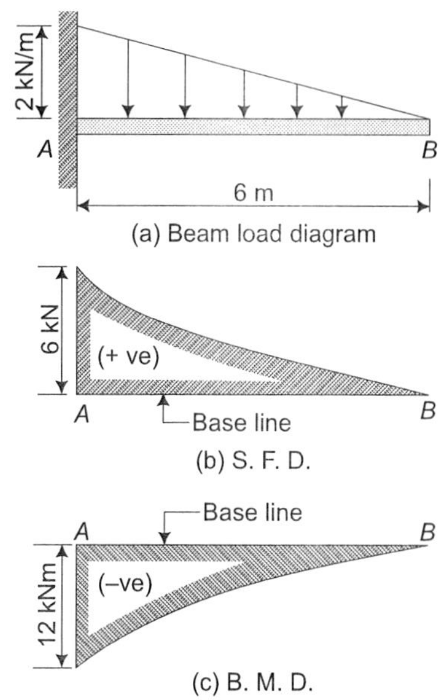

Fig. 8.25

Solution Refer Figure 8.25.

S.F.D. The shear force at the section $x - x$ at a distance x from the free end is given by,

$$F_x = \frac{wx^2}{2l},$$

this varies according to the parabolic law and is positive.

At $x = 0$, $$F_B = +\frac{w(0)^2}{2l} = 0$$

At $x = l$, $$F_A = +\frac{w(l)^2}{2l} = +\frac{wl}{2}$$

$$= +\frac{2 \times 6}{2} = +6 \text{ kN}$$

The shear force diagram is shown in Figure 8.25 (b).

B.M.D. The bending moment at any section $x - x$ at a distance of x from free end B is given by,

$$M_x = -\frac{wx^2}{2l} \times \frac{x}{3} = -\frac{wx^3}{6l},$$

this varies according to the cubic law and is negative.

At $x = 0$, $$M_B = -\frac{w(0)x^3}{6l} = 0$$

At $x = l$, $$M_A = -\frac{w(l)^3}{6l} = -\frac{wl^2}{6} = -\frac{2 \times (6)^2}{6} = -12 \text{ kNm}$$

The B.M.D. is shown in Figure 8.25 (c).

8.6.4 Cantilever Subjected to Couples

When a cantilever is subjected to a couple at a section the shear force does not change at that section as there is no change in load due to couple. But the bending moment changes suddenly in magnitude equal to that of couple at the section. Cantilever with couples have been explained in the Examples 8.11 and 8.12.

Example 8.11 Determine the reaction at the fixed end of a cantilever beam of length 8 m as shown in Figure 8.26 (a). Draw the shear force and bending moment diagrams.

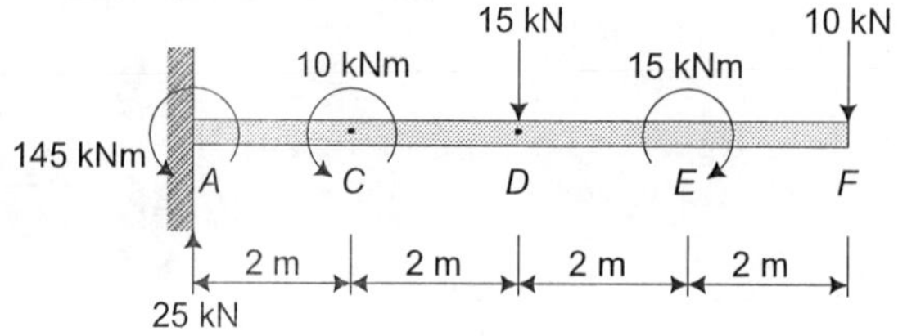

(a) Loaded beam

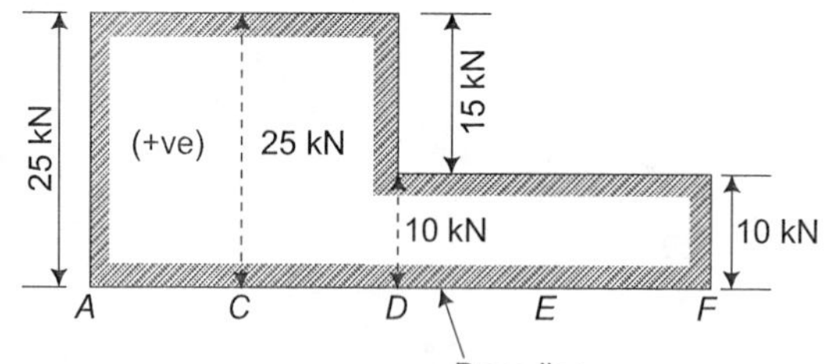

(b) S. F. D.

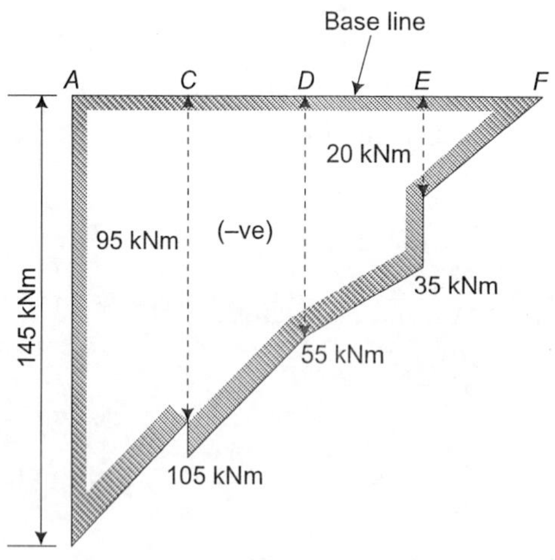

(c) B. M. D.

Fig. 8.26

Solution Total vertical load on the cantilever = 10 + 15 = 25 kN

∴ Vertical reaction at $A = 25$ kN ↑ (Upwards)

Taking moments about A, we have net moment

$$= + 10 \times 8 + 15 \text{ (couple at } E) + 15 \times 4 - 10 \text{ (couple at } C)$$

$$= 145 \text{ kNm} \quad \text{(Clockwise)}$$

S.F.D. Consider a section $x - x$ between E and F. The shear force at the section $x - x$ will be equal to the resultant force acting on the right portion of the section. The resultant force is acting downwards, so as per the sign convention it is taken positive.

$\therefore \quad F_x = + 10$ kN.

This is linear and will remain constant for the portion DF.

Similarly, consider section $x - x$ between A and D at a distance x from free end B. The shear force is given by,

$\therefore \quad F_x = + 10 + 15 = + 25$ kN.

This is linear and will remain constant for the portion AD.

The shear force diagram is shown in Figure 8.26 (b).

B.M.D. For FE the bending moment at any section $x - x$ at a distance of x from free end F is given by,

$$M_x = - 10x \text{ kNm} \quad \text{(Linear)}$$

(For the right portion of the section $x - x$, the clockwise moment is taken negative)

At $x = 0$, $\quad M_F = 0$

At $x = 2$, $\quad M_E = - (10 \times 2) = - 20$ kNm

For DE the bending moment at any section is given by,

$$M_x = - 10x - 15 \text{ kNm} \quad \text{(Linear)}$$

At $x = 2$, $\quad M_E = - 10 \times 2 - 15 = - 35$ kNm

At $x = 4$, $\quad M_D = - 10 \times 4 - 15 = - 55$ kNm

Similarly, for CD the bending moment at any section $x - x$ at a distance of x from free end F is given by,

$$M_x = - 10x - 15(x - 4) - 15 \text{ kNm} \quad \text{(Linear)}$$

At $x = 4$, $\quad M_D = - 10 \times 4 - 15(4 - 4) - 15$

$= - 55$ kNm (As before)

At $x = 6$, $\quad M_C = - 10 \times 6 - 15 \times (6 - 4) - 15$

$= - 105$ kNm

Similarly, for AC the bending moment at any section $x - x$ at a distance of x from free end F is given by,

$$M_x = -10x - 15(x-4) - 15 + 10 \text{ kNm} \quad \text{(Linear)}$$

At $x = 6$, $\quad M_C = -10 \times 6 - 15 \times (6-4) - 15 + 10$

$= -95$ kNm

At $x = 8$, $\quad M_A = -10 \times 8 - 15 \times (8-4) - 15 + 10$

$= -145$ kNm

The B.M.D. is shown in Figure 8.26 (c).

Example 8.12 A cantilever beam of length 4 m is loaded as shown in Figure 8.27 (a). Draw the shear force and bending moment diagrams and locate the position of the point of contraflexure if any.

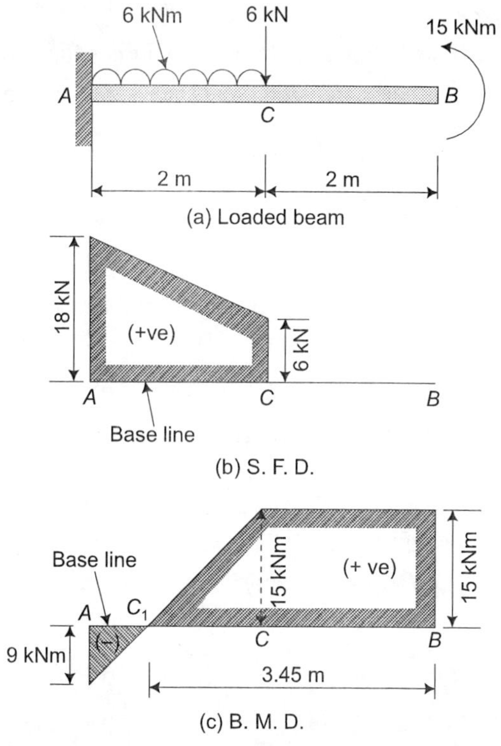

Fig. 8.27

Solution Refer Figure 8.27.

S.F.D. Consider a section $x - x$ at a distance x from the free end B. For the portion BC, as there is no loading so, the shear force is zero.

For the portion CA, consider a section $x - x$ at a distance x from the free end B. The shear force at the section $x - x$ is given by,

$$F_x = +6 + 6 \times (x - 2) \quad \text{(Linear)}$$

(The downward force to the right portion of the section $x - x$ is taken positive)

At $x = 2$, $\quad F_C = +6 + 6 \times (2 - 2)$

$= +6$ kN

At $x = 4$, $\quad F_A = +6 + 6 \times (4 - 2)$

$= +18$ kN

The shear force diagram is shown in Figure 8.27 (b).

B.M.D. For BC, the bending moment at any section $x - x$ at a distance of x from free end B is given by,

$$M_x = +15 \text{ kNm.}$$

This will remain constant for the portion BC.

(For the right portion of the section, the anticlockwise moment is taken positive)

For AC, the bending moment at any section $x - x$ at a distance of x from free end B is given by,

$$M_x = + 15 - 6 \times (x - 2) - 6 \times (x - 2) \times \frac{(x - 2)}{2}$$

$$= + 15 - 6(x - 2) - 3(x - 2)^2 \text{ kNm} \quad \text{(Parabolic)}$$

At $x = 2$, $\quad M_C = + 15$ kNm

At $x = 4$, $\quad M_A = + 15 - 6(4 - 2) - 3(4 - 2)^2$

$$= - 9 \text{ kNm}$$

The bending moment changes its sign in the portion AC. The point of contraflexure (C_1) occurs at a point where

$$M_x = 0.$$

That is, $\quad + 15 - 6(x - 2) - 2(x - 2)^2 = 0$

$$x^2 - 2x - 5 = 0,$$

solving this expression, we get,

$$x = \frac{-(-2) \pm \sqrt{(-2)^2 - 4 \times 1 \times (-5)}}{2 \times 1}$$

$$= \frac{2 \pm \sqrt{24}}{2} = 3.45 \text{ m} \quad \text{(Taking +ve sign only)}$$

So, the point of contraflexure (C_1) occurs at a distance of 3.45 m from the free end B. The B.M.D. is shown in Figure 8.27 (c).

8.7 SHEAR FORCE AND BENDING MOMENT DIAGRAMS FOR A SIMPLY SUPPORTED BEAM

8.7.1 Simply Supported Beam Subjected to a Point Load at the Centre

Consider a beam AB of length l carrying a concentrated load W at its midpoint C as shown in Figure 8.28 (a).

Let R_A be the reaction at the support A and R_B the reaction at the support B. The load is transferred to supports and for equilibrium conditions,

$$R_A + R_B = W$$

Also the sum of moment about A is equal to zero, i.e.,

$$\Sigma M_A = 0$$

We have, $\quad R_B \times l = W \times l/2$

$\therefore \quad R_B = W/2$

$$R_A = W - W/2 = W/2$$

S.F.D. Take section $x - x$ at a distance x from the end A. As per the sign convention the resultant force on the left side of the section acting upwards will be taken as positive.

$$\therefore \qquad F_x = + W/2,$$

this remains constant between A and C.

Now take section $x - x$ at a distance x from end A between C and B. The resultant force on the left side of the section is given by,

$$F_x = (W/2 - W)$$

$$= - W/2,$$

this will remain constant between C and B.

At the section C the shear force changes from $+ W/2$ to $- W/2$. The shear force diagram is shown in Figure 8.28 (b).

B.M.D. The bending moment at any section $x - x$ at a distance of x from end A is given by,

$$M_x = R_A \times x$$

$$= (W/2) \times x \qquad \text{(i)}$$

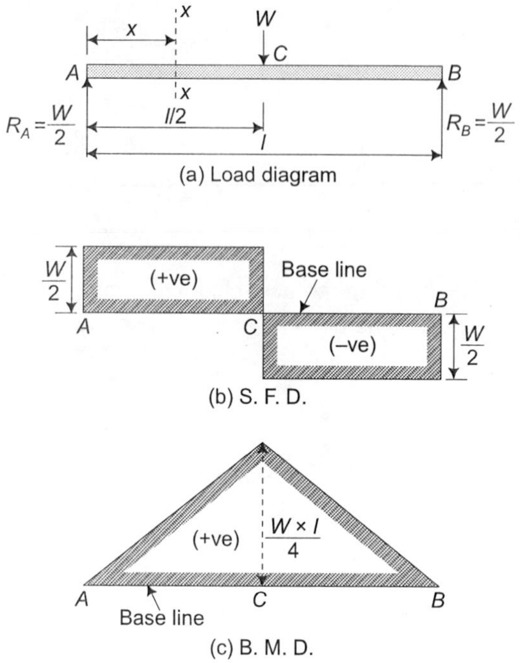

Fig. 8.28

As per the sign convention, for the left portion of the section $x - x$, the clockwise moment is taken positive. This causes concavity upwards, so it is sagging bending moment.

From equation (i) it is observed that bending moment at any section of the beam is proportional to the distance of the section from the end A.

At $x = 0$, i.e., bending moment at A,

$$M_A = 0$$

At $x = l/2$, i.e., bending moment at C,

$$M_C = (W/2) \times (l/2) = (Wl/4) \qquad \text{(ii)}$$

From equation (ii) it is observed that bending moment at any section of the beam is proportional to the distance of the section from the end A which increases from zero at A to $Wl/4$ at C.

The bending moment between C and B is given by,

$$M_x = R_A \times x - W \times \left(x - \frac{l}{2}\right)$$

$$= \frac{W}{2} \times x - W \times x + \frac{Wl}{2}$$

$$= \frac{Wl}{2} - \frac{Wx}{2}$$

At $x = l/2$, i.e., bending moment at C,

$$M_C = \frac{Wl}{2} - \frac{Wx}{2} \times \frac{l}{2} = \frac{Wl}{4} \quad \text{(As before)}$$

At $x = l$, i.e., bending moment at B,

$$M_B = \frac{Wl}{2} - \frac{Wl}{2} = 0$$

So, the bending moment at C is $Wl/4$ and it decreases to zero at B. The bending moment diagram is shown in Figure 8.28 (c).

8.7.2 Simply Supported Beam Subjected to a Point Load not at the Centre (Eccentric Point Load)

Consider a beam AB of length l carrying a concentrated load W at C at a distance of a from the support A as shown in Figure 8.29 (a).

Let R_A be the reaction at the support A and R_B is the reaction at the support B. The load is transferred to supports and for equilibrium conditions,

$$R_A + R_B = W$$

Also the sum of moment about A is equal to zero, i.e.,

$$\Sigma M_A = 0$$

We have,

$$R_B \times l = W \times a$$

$\therefore$

$$R_B = \frac{Wa}{l}$$

$$R_A = W - R_B = W - \frac{Wa}{l}$$

$$= W\left[1 - \frac{a}{l}\right] = W\left[\frac{l-a}{l}\right] = \frac{Wb}{l}$$

S.F.D. Take section $x - x$ at a distance x from the end A. As per the sign convention the resultant force on the left side of the section acting upwards is taken positive.

$\therefore$

$$F_x = +R_A = +Wb/l \quad \text{(Constant between } A \text{ and } C\text{)}$$

Now take section $x - x$ at a distance x between C and B. The resultant force on the left side of the section is given by,

$$F_x = \left(\frac{Wb}{l} - W\right) = W\left(\frac{b-l}{l}\right)$$

$$= -W\left(\frac{l-b}{l}\right) = -\frac{Wa}{l} \quad \text{(Remains constant between } C \text{ and } B\text{)}$$

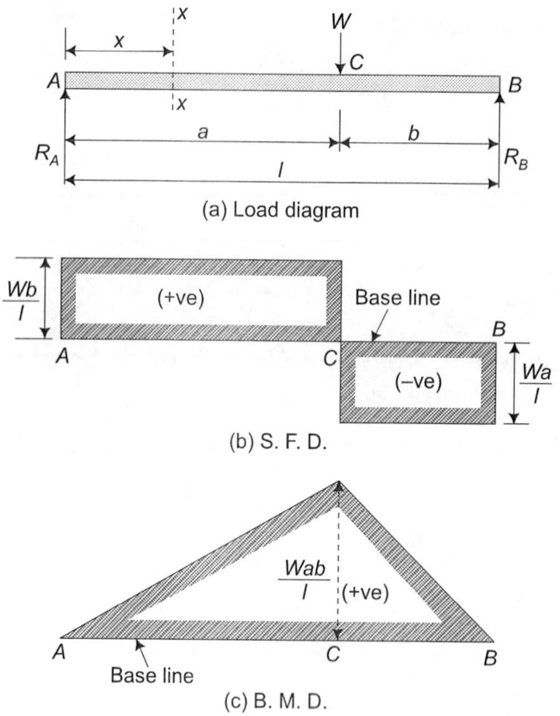

Fig. 8.29

At the section C the shear force changes from $+Wb/l$ to $-Wa/l$. The shear force diagram is shown in Figure 8.29 (b).

B.M.D. The bending moment at any section $x - x$ at a distance of x from end A is given by,

$$M_x = R_A \times x = (Wb/l) \times x$$

(For the left portion of the section, the clockwise moment is positive).

At $x = 0$, i.e., bending moment at A,

$$M_A = 0$$

At $x = a$, i.e., bending moment at C,

$$M_C = (Wb/l) \times a = Wab/l \quad \text{(Linear)}$$

The bending moment between C and B is given by,

$$M_x = R_A \times x - W \times (x - a)$$
$$= (Wb/l) \times x - W \times (x - a)$$

At $x = a$, i.e., bending moment at C,

$$M_C = \frac{Wb}{l} \times a - W \times (a - a) = \frac{Wab}{l} \quad \text{(As before)}$$

At $x = l$, i.e., bending moment at B,

$$M_B = \frac{Wb}{l} \times l - W \times (l - a)$$

$$= \frac{Wb}{l} \times l - W \times b = 0$$

The bending moment diagram is shown in Figure 8.29 (c).

Example 8.13 A simply supported beam of length 5 m carries concentrated load of 2 kN and 5 kN at distances of 1.5 m and 3.5 m from the left end. Draw the shear force and bending moment diagrams.

Solution Refer Figure 8.30. Let R_A be the reaction at the support A and R_B is the reaction at the support B. First calculate the reactions.

Taking moments of the forces about A,

$$R_B \times 5 = 5 \times 3.5 + 2 \times 1.5$$

$$= 20.5$$

$$\therefore \quad R_B = 20.5/5$$

$$= 4.1 \text{ kN}$$

Under equilibrium condition,

$$R_A + R_B = 2 + 5$$

$$\therefore \quad R_A = 7 - R_B$$

$$= 7 - 4.1$$

$$= 2.9 \text{ kN}$$

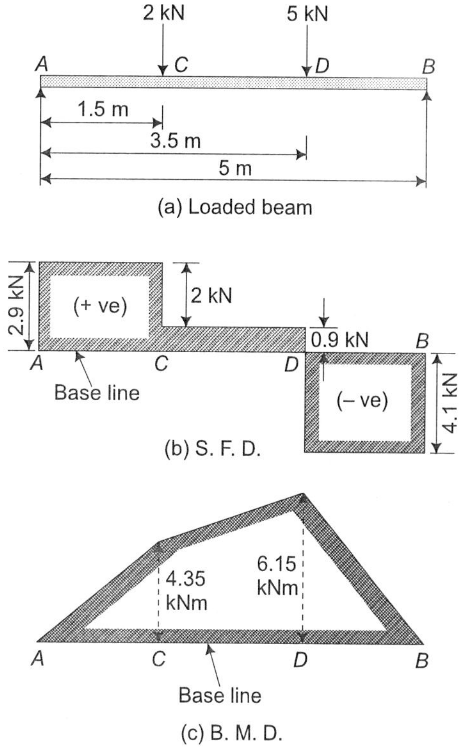

Fig. 8.30

S.F.D. Take section $x - x$ at a distance x from the end A.

$$F_x = + R_A = + 2.9 \text{ kN},$$

this remains constant between A and C.

(As per the sign convention the resultant force on the left side of the section acting upwards will be taken positive.)

Similarly, considering section between C and D, the shear force can be given by,

$$F_x = + 2.9 - 2 = + 0.9 \text{ kN},$$

this remains constant between C and D.

(The force on the left side of the section acting upwards is taken positive and the force acting downwards is taken negative.)

Similarly, considering section between D and B, the shear force can be given by,

$$F_x = +0.9 - 5 = -4.1 \text{ kN},$$

this remains constant between D and B.

The shear force diagram is shown in Figure 8.30 (b).

B.M.D. The bending moment at any section $x - x$ at a distance of x from end A is given by,

$$M_x = R_A \times x = 2.9x \quad \text{(Linear)}$$

(For the left portion of the section, the clockwise moment is taken positive.)

At $x = 0$, i.e., bending moment at A,

$$M_A = 0$$

At $x = 1.5$, i.e., bending moment at C,

$$M_C = 2.9 \times 1.5 = +4.35 \text{ kNm}$$

The bending moment at any section between C and D is given by,

$$M_x = 2.9 \times x - 2 \times (x - 1.5) \quad \text{(Linear)}$$

(For the left portion of the section, the clockwise moment is taken positive and the anticlockwise moment is taken negative.)

At $x = 1.5$, i.e., bending moment at C,

$$M_C = 2.9 \times 1.5 = +4.35 \text{ kNm} \quad \text{(As before)}$$

At $x = 3.5$, i.e., bending moment at D,

$$M_D = 2.9 \times 3.5 - 2 \times (3.5 - 1.5) = +6.15 \text{ kNm}$$

The bending moment at any section between D and B is given by,

$$M_x = 2.9 \times x - 2 \times (x - 1.5) - 5 \times (x - 3.5) \quad \text{(Linear)}$$

(For the left portion of the section, the clockwise moment is taken positive and the anticlockwise moment is taken negative.)

At $x = 3.5$, i.e., bending moment at D,

$$M_D = 2.9 \times 3.5 - 2 \times (3.5 - 1.5)$$
$$= +6.15 \text{ kNm} \quad \text{(As before)}$$

At $x = 5$, i.e., bending moment at B,

$$M_B = 2.9 \times 5 - 2 \times (5 - 1.5) - 5 \times (5 - 3.5) = 0$$

The bending moment diagram is shown in Figure 8.30 (c).

Example 8.14 A simply supported beam carries point loads as shown in Figures 8.31 (a). Draw the shear force and bending moment diagrams.

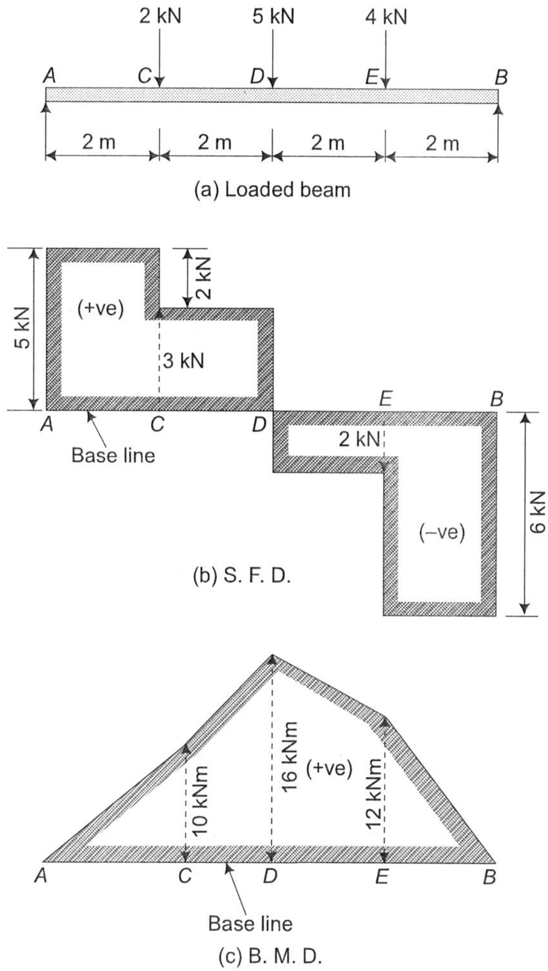

Fig. 8.31

Solution Refer Figure 8.31. Let R_A be the reaction at the support A and R_B is the reaction at the support B. In order to calculate reactions, take moments of the forces about A,

$$R_B \times 8 = 2 \times 2 + 5 \times 4 + 4 \times 6 = 48$$

$$\therefore \quad R_B = 48/8 = 6\text{kN}$$

Under equilibrium condition,

$$R_A + R_B = 2 + 5 + 4 = 11$$

$$\therefore \quad R_A = 11 - R_B = 11 - 6 = 5 \text{ kN}$$

S.F.D. Take section $x - x$ at a distance x from the end A.

$$F_x = + R_A = + 5 \text{ kN},$$

this remains constant between A and C.

(The resultant force on the left side of the section acting upwards is taken positive.)

Similarly, considering section between C and D, the shear force can be given by,

$$F_x = + 5 - 2 = + 3 \text{ kN},$$

this remains constant between C and D.

(The force on the left side of the section acting upwards is taken positive and the force acting downwards is taken negative.)

Similarly, considering section between D and E, the shear force can be given by,

$$F_x = + 3 - 5 = - 2\text{kN},$$

this remains constant between D and E.

Similarly, considering section between E and B, the shear force can be given by,

$$F_x = - 2 - 4 = - 6 \text{ kN},$$

this remains constant between E and B.

The shear force diagram is shown in Figure 8.31 (b).

B.M.D. The bending moment at any section $x - x$ at a distance of x from end A is given by,

$$M_x = R_A \times x = 5x \quad \text{(Linear)}$$

(For the left portion of the section, the clockwise moment is taken positive.)

At $x = 0$, i.e., bending moment at A,

$$M_A = 0$$

At $x = 2$, i.e., bending moment at C,

$$M_C = 5 \times 2 = + 10 \text{ kNm}$$

The bending moment at any section between C and D is given by,

$$M_x = 5x - 2\,(x - 2) \quad \text{(Linear)}$$

(For the left portion of the section, the clockwise moment is taken positive and the anticlockwise moment is taken negative.)

At $x = 2$, i.e., bending moment at C,

$$M_C = 5 \times 2 = + 10 \text{ kNm} \quad \text{(As before)}$$

At $x = 4$, i.e., bending moment at D,

$$M_D = 5 \times 4 - 2\,(4 - 2) = + 16 \text{ kNm}$$

The bending moment at any section between D and E is given by,

$$M_x = 5x - 2\,(x - 2) - 5\,(x - 4) \quad \text{(Linear)}$$

(For the left portion of the section, the clockwise moment is taken positive and the anticlockwise moment is taken negative.)

At $x = 4$, i.e., bending moment at D,

$$M_D = 5 \times 4 - 2 \times (4 - 2) - 5\,(4 - 4)$$

$$= +\,16 \text{ kNm} \quad \text{(As before)}$$

At $x = 6$, i.e., bending moment at E,

$$M_E = 5 \times 6 - 2 \times (6 - 2) - 5 \times (6 - 4) = +\,12 \text{ kNm}$$

The bending moment at any section between E and B is given by,

$$M_x = 5x - 2\,(x - 1) - 5\,(x - 4) - 4\,(x - 6) \quad \text{(Linear)}$$

(For the left portion of the section, the clockwise moment is taken positive and the anticlockwise moment is taken negative.)

At $x = 6$ i.e., bending moment at E,

$$M_E = 5 \times 6 - 2 \times (6 - 2) - 5 \times (6 - 4) - 4\,(6 - 6)$$

$$= +\,12 \text{ kNm} \quad \text{(As before)}$$

At $x = 8$ i.e., bending moment at B,

$$M_B = 5 \times 8 - 2 \times (8 - 2) - 5 \times (8 - 4) - 4\,(8 - 6) = 0$$

The bending moment diagram is shown in Figure 8.31 (c).

8.7.3 Simply Supported Beam Subjected to Uniformly Distributed Load (u.d.l.) over its Entire Span

Consider a beam AB of length l carrying a uniformly distributed load of intensity w per unit length over the whole span as shown in Figure 8.32 (a).

Due to symmetrical loading the reactions at the supports will be equal and half of the total load on the whole span and are given by,

$$R_A = R_B = wl/2.$$

S.F.D. Take section $x - x$ at a distance x from the end A. As per the sign convention the resultant force on the left side of the section acting upwards is taken as positive and the downward force is taken negative.

$$\therefore \quad F_x = +\,R_A - w \times x = +\,(wl/2) - wx \quad \text{(Linear)}$$

At $x = 0$, i.e., shear force at A,

$$F_A = +\,(wl/2) - w(0) = +\,wl/2$$

At $x = l$, i.e., shear force at B,

$$F_B = +\,(wl/2) - w(l)$$

$$= -\,wl/2$$

At $x = l/2$, i.e., shear force at C,

$$F_C = +\,(wl/2) - w \times (l/2)$$

$$= 0$$

So, the shear force changes uniformly from $+\,wl/2$ at point A to $-\,wl/2$ at point B and it becomes zero at $x = l/2$, i.e., at mid of the beam. The shear force diagram is shown in Figure 8.32 (b).

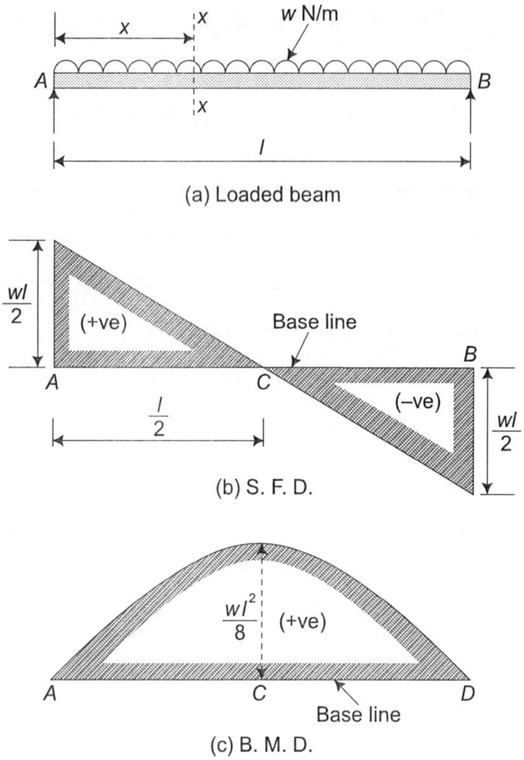

(a) Loaded beam

(b) S. F. D.

(c) B. M. D.

Fig. 8.32

B.M.D. The bending moment at any section $x - x$ at a distance of x from end A is given by,

$$M_x = +\,R_A \times x - w \times x \times \frac{x}{2}$$

$$= \frac{wlx}{2} - \frac{wx^2}{2} \quad \text{(Parabolic).}$$

At $x = 0$, i.e., bending moment at A,

$$M_A = 0$$

At $x = \frac{l}{2}$, i.e., bending moment at C,

$$M_C = \frac{wl}{2} \times \frac{l}{2} - \frac{w}{2} \times \left(\frac{l}{2}\right)^2 = +\,\frac{wl^2}{8}$$

At $x = l$, i.e., bending moment at B,

$$M_B = \frac{wl \times l}{l} - \frac{wl^2}{2} = 0$$

It is observed that bending moment is maximum at the mid of the beam where shear force is zero. The bending moment diagram is shown in Figure 8.32 (c).

Example 8.15 A simply supported beam of 5 m length carries uniformly distributed load of 5 kNm as shown in Figure 8.33 (a). Draw the shear force and bending moment diagrams. Also determine the maximum bending moment on the section.

Solution Refer Figure 8.33. Let R_A be the reaction at the support A and R_B be the reaction at the support B. In order to calculate reactions, take moments of the forces about A,

$$R_B \times 5 = 5 \times 4 \times 4/2 = 40$$

$$\therefore \quad R_B = 40/5 = 8 \text{ kN}$$

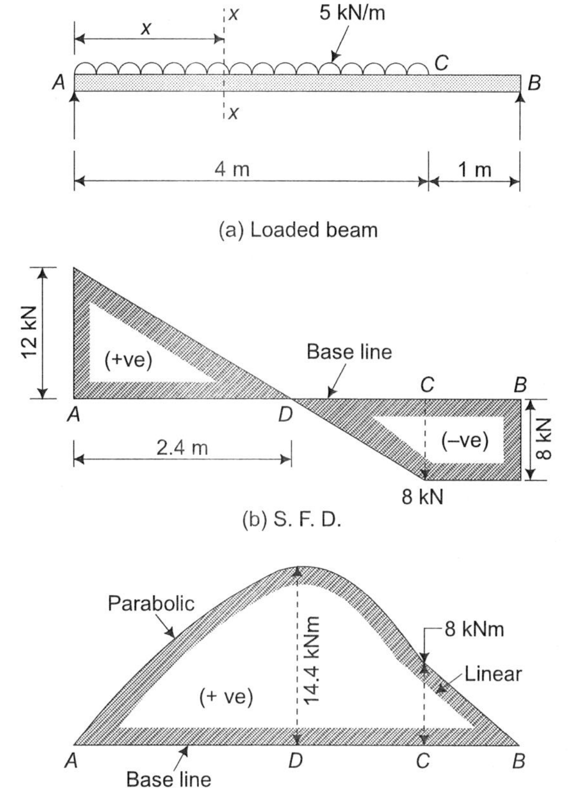

(a) Loaded beam

(b) S. F. D.

(c) B. M. D.

Fig. 8.33

Under equilibrium condition,

$$R_A + R_B = 5 \times 4 = 20$$

$$\therefore \quad R_A = 20 - R_B = 20 - 8 = 12 \text{ kN}$$

S.F.D. Take section $x - x$ at a distance x from the end A between A and C. The shear force at the section is given by,

$$F_x = + R_A - 5 \times x = + 12 - 5x \quad \text{(Linear)}$$

At $x = 0$, i.e., shear force at A,

$$F_A = + 12 - 5(0) = + 12 \text{ kN}$$

At $x = 4$, i.e., shear force at C,

$$F_C = + 12 - 5 \times 4 = - 8 \text{ kN} \quad \text{(Constant)}$$

The shear force changes from + 12 kN at A to – 8 kN at C. So, shear force will be zero somewhere between A and C. Let it be zero at a distance of x m from end A. Then equating F_x equal to zero, i.e.,

$$+12 - 5x = 0$$

$$\therefore \qquad x = 12/5 = 2.4 \text{ m.}$$

Thus shear force is zero at a distance of 2.4 m from end A. The shear force diagram is shown in Figure 8.33 (b).

B.M.D. The bending moment at any section $x - x$ between A and C at a distance of x from end A is given by,

$$M_x = + R_A \times x - 5 \times x \times x/2$$

$$= 12x - 5x^2/2 \quad \text{(Parabolic).}$$

(For the left portion of the section, the clockwise moment is taken positive and the anticlockwise moment is taken negative.)

At $x = 0$, i.e., bending moment at A,

$$M_A = 0$$

At $x = 4$, i.e., bending moment at C,

$$M_C = 12 \times 4 - 5 \times (4)^2/2 = +8 \text{ kNm}$$

At $x = 2.4$, i.e., bending moment at D,

$$M_D = 12 \times 2.4 - 5 \times (2.4)^2/2$$

$$= +14.4 \text{ kNm} \quad \text{(Maximum)}$$

The bending moment at any section between C and B is given by,

$$M_x = + R_A \times x - 5 \times 4 \times [x - 4/2]$$

$$= 12x - 20\,(x - 2) \quad \text{(Linear)}$$

At $x = 4$, i.e., bending moment at C,

$$M_C = 12 \times 4 - 20 \times (4 - 2) = +8 \text{ kNm} \quad \text{(As before)}$$

At $x = 5$, i.e., bending moment at B,

$$M_B = 12 \times 5 - 20 \times (5 - 2) = 0$$

It is observed that bending moment is maximum at the point where shear force is zero. The bending moment diagram is shown in Figure 8.33 (c). From the shear force diagram it can be seen that shear force is zero at point D. Thus bending moment is maximum at point D.

Example 8.16 A simply supported beam of 8 m span is loaded as shown in Figure 8.34 (a). Draw the shear force and bending moment diagrams indicating principal values.

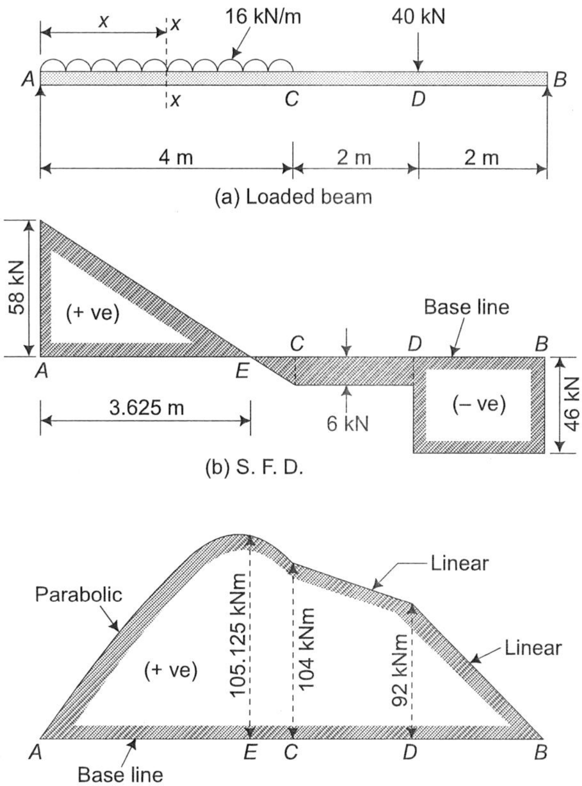

(a) Loaded beam

(b) S. F. D.

(c) B. M. D.

Fig. 8.34

Solution Refer Figure 8.34. Let R_A be the reaction at the support A and R_B is the reaction at the support B. In order to calculate reactions, take moments of the forces about A,

$$R_B \times 8 = 16 \times 4 \times 4/2 + 40 \times 6 = 368$$

$$\therefore \qquad R_B = 368/8 = 46 \text{ kN}$$

Under equilibrium condition,

$$R_A + R_B = 16 \times 4 + 40 = 104$$

$$\therefore \qquad R_A = 104 - 46 = 58 \text{ kN}$$

S.F.D. Take section $x - x$ at a distance x from the end A between A and C. The shear force at the section is given by,

$$F_x = + R_A - 16 \times x = + 58 - 16x \quad \text{(Linear)}$$

At $x = 0$, i.e., shear force at A,

$$F_A = + 58 - 16(0) = + 58 \text{ kN}$$

At $x = 4$, i.e., shear force at C,

$$F_C = +58 - 16 \times 4 = -6 \text{ kN} \quad \text{(Constant)}$$

The shear force changes from + 58 kN at A to – 6 kN at C. So, shear force will be zero somewhere between A and C. Let it be zero at a distance of x m from end A. Then equating F_x equal to zero, i.e.,

$$+58 - 16x = 0$$

$$\therefore \quad x = 58/16 = 3.625 \text{ m.}$$

Thus shear force is zero at a distance of 3.625 m from end A.

The shear force for the portion CD is given by,

$$F_x = +R_A - 16 \times x = +58 - 16 \times 4$$

$$= -6 \text{ kN} \quad \text{(Constant)}$$

The shear force for the portion DB is given by,

$$F_x = +R_A - 16 \times x - 40 = +58 - 16 \times 4 - 40$$

$$= -46 \text{ kN} \quad \text{(Constant)}$$

The shear force diagram is shown in Figure 8.34 (b).

B.M.D. The bending moment at any section $x - x$ between A and C at a distance of x from end A is given by,

$$M_x = +R_A \times x - 16 \times x \times x/2$$

$$= 58x - 8x^2 \quad \text{(Parabolic).}$$

(For the left portion of the section, the clockwise moment is taken positive and the anticlockwise moment is taken negative.)

At $x = 0$, i.e., bending moment at A,

$$M_A = 0$$

At $x = 4$, i.e., bending moment at C,

$$M_C = 58 \times 4 - 8 \times (4)^2 = +104 \text{ kNm}$$

At $x = 3.625$, i.e., bending moment at E,

$$M_E = 58 \times 3.625 - 8 \times (3.625)^2$$

$$= +105.125 \text{ kNm} \quad \text{(Maximum)}$$

The bending moment at any section between C and D is given by,

$$M_x = +R_A \times x - 16 \times 4 \times (x - 4/2)$$

$$= 58x - 64(x - 2) \quad \text{(Linear)}$$

At $x = 4$, i.e., bending moment at C,

$$M_C = 58 \times 4 - 64 \times (4 - 2)$$
$$= +\ 104 \text{ kNm} \quad \text{(As before)}$$

At $x = 6$, i.e., bending moment at D,

$$M_D = 58 \times 6 - 64 \times (6 - 2) = +\ 92 \text{ kNm}$$

The bending moment at any section between D and B is given by,

$$M_x = +\ R_A \times x - 16 \times 4 \times (x - 4/2) - 40 \times (x - 6)$$
$$= 58x - 64(x - 2) - 40\ (x - 6) \quad \text{(Linear)}$$

At $x = 6$, i.e., bending moment at D,

$$M_D = 58 \times 6 - 64 \times (6 - 2) - 40 \times (6 - 6)$$
$$= +\ 104 \text{ kNm} \quad \text{(As before)}$$

At $x = 8$, i.e., bending moment at B,

$$M_B = 58 \times 8 - 64 \times (8 - 2) - 40 \times (8 - 6) = 0$$

It is observed that bending moment is maximum at the point where shear force is zero. The bending moment diagram is shown in Figure 8.34 (c). From the shear force diagram it can be seen that shear force is zero at point E. Thus bending moment is maximum at point E.

Example 8.17 A simply supported beam of 8 m span is loaded as shown in Figure 8.35 (a). Draw the shear force and bending moment diagrams indicating principal values.

Solution Refer Figure 8.35. Let, R_A be the reaction at the support A and R_B the reaction at the support B. In order to calculate reactions, take moments of the forces about A,

$$R_B \times 8 = 40 \times 7 + 10 \times 4 \times (4/2 + 1) = 400 \text{ kN}$$

$$\therefore \qquad R_B = 400/8 = 50 \text{ kN}$$

Under equilibrium condition,

$$R_A + R_B = 10 \times 4 + 40 = 80$$

$$\therefore \qquad R_A = 80 - 50 = 30 \text{ kN}$$

S.F.D. Take section $x - x$ at a distance x from the end A between A and C. The shear force at the section is given by,

$$F_x = +\ R_A = +\ 30 \text{ KN} \quad \text{(Constant)}$$

The shear force at the section $x - x$ between C and D is given by,

$$F_x = +\ R_A - 10 \times (x - 1)$$
$$= +\ 30 - 10(x - 1) \quad \text{(Linear)}$$

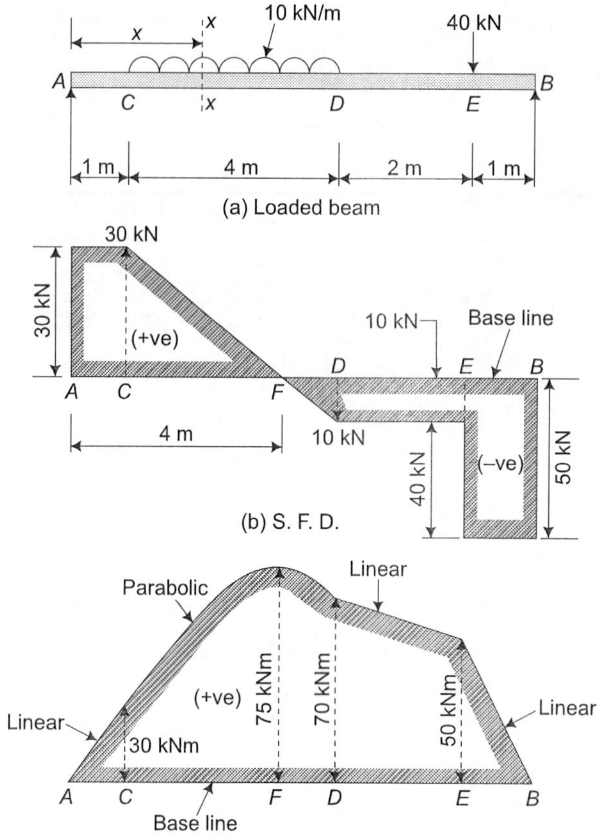

Fig. 8.35

At $x = 1$, i.e., shear force at C,

$$F_C = + 30 - 10(0) = + 30 \text{ kN}$$

At $x = 5$, i.e., shear force at D,

$$F_D = + 30 - 10 \times (5 - 1) = - 10 \text{ kN} \quad \text{(Constant)}$$

This will remain constant for the portion DE.

The shear force changes from + 30 kN at C to – 10 kN at D. So, shear force will be zero somewhere between C and D. Let it be zero at a distance of x m from end A. Then equating F_x equal to zero, i.e.,

$$+ 30 - 10 (x - 1) = 0$$

$$\therefore \quad x = 4 \text{ m.}$$

Thus shear force is zero at a distance of 4 m from end A.

The shear force for the portion *EB* is given by,

$$F_x = + R_A - 10 \times 4 - 40 = + 30 - 80$$

$$= - 50 \text{ kN} \quad \text{(Constant)}$$

The shear force diagram is shown in Figure 8.35 (b).

B.M.D. The bending moment at any section $x - x$ between *A* and *C* at a distance of *x* from end *A* is given by,

$$M_x = + R_A\, x = + 30\, x \text{ kNm} \quad \text{(Linear)}$$

(For the left portion of the section, the clockwise moment is taken positive)

At $x = 0$, i.e., bending moment at *A*,

$$M_A = 0$$

At $x = 1$, i.e., bending moment at *C*,

$$M_C = 30 \times 1 = + 30 \text{ kNm}$$

The bending moment at any section $x - x$ between *C* and *D* is given by,

$$M_x = + R_A \times x - 10 \times (x - 1) \times (x - 1)/2$$

$$= + 30x - 5(x - 1)^2 \quad \text{(Parabolic).}$$

(For the left portion of the section, the clockwise moment is taken positive and the anticlockwise moment is taken negative.)

At $x = 1$, i.e., bending moment at *C*,

$$M_C = + 30 \times 1 - 5 \times (0)$$

$$= + 30 \text{ kNm} \quad \text{(As before)}$$

At $x = 5$, i.e., bending moment at *D*,

$$M_D = 30 \times 5 - 5 \times (5 - 1)^2 = + 70 \text{ kNm}$$

At $x = 4$, i.e., bending moment at *F*,

$$M_F = 30 \times 4 - 5 \times (4 - 1)^2$$

$$= + 75 \text{ kNm} \quad \text{(Maximum)}$$

The bending moment at any section between *D* and *E* is given by,

$$M_x = + R_A \times x - 10 \times 4 \times [x - (4/2 + 1)]$$

$$= 30x - 40(x - 3) \quad \text{(Linear)}$$

At $x = 5$, i.e., bending moment at *D*,

$$M_D = 30 \times 5 - 40 \times (5 - 3)$$

$$= + 70 \text{ kNm} \quad \text{(As before)}$$

At $x = 7$, i.e., bending moment at E,

$$M_E = 30 \times 7 - 40 \times (7 - 3) = +50 \text{ kNm}$$

The bending moment at any section between E and B is given by,

$$M_x = +R_A \times x - 40 \times [x - (4/2 + 1)] - 40 \times (x - 7)$$

$$= 30x - 40(x - 3) - 40(x - 7) \quad \text{(Linear)}$$

At $x = 7$, i.e., bending moment at E,

$$M_E = 30 \times 7 - 40 \times (7 - 3) - 40 \times (7 - 7)$$

$$= +50 \text{ kNm} \quad \text{(as before)}$$

At $x = 8$, i.e., bending moment at B,

$$M_B = 30 \times 8 - 40 \times (8 - 3) - 40 \times (8 - 7) = 0$$

It is observed that bending moment is maximum at the point where shear force is zero. The bending moment diagram is shown in Figure 8.35 (c). From the shear force diagram it can be seen that shear force is zero at point F. Thus bending moment is maximum at point F.

Example 8.18 Draw the shear force and bending moment diagrams for a simply supported beam shown in Figure 8.36 (a).

Solution Refer Figure 8.36. Let R_A be the reaction at the support A and R_B is the reaction at the support B.

In order to calculate reactions, take moments of the forces about A,

$$R_B \times 10 = 20 \times 8 + 4 \times 2 \times (2/2 + 2 + 2) + 10 \times 2$$

$$= 220 \text{ kN}$$

$$\therefore \quad R_B = 220/10 = 22 \text{ kN}$$

Under equilibrium condition,

$$R_A + R_B = 10 + 4 \times 2 + 20 = 38$$

$$\therefore \quad R_A = 38 - 22 = 16 \text{ kN}$$

S.F.D. Take section $x - x$ at a distance x from the end A between A and C. The shear force at the section is given by,

$$F_x = +R_A = +16 \text{ kN} \quad \text{(Constant)}$$

The shear force at the section $x - x$ between C and D is given by,

$$F_x = +R_A - 10 = +16 - 10$$

$$= +6 \text{ kN} \quad \text{(Constant)}$$

This will remain constant for the portion CD.

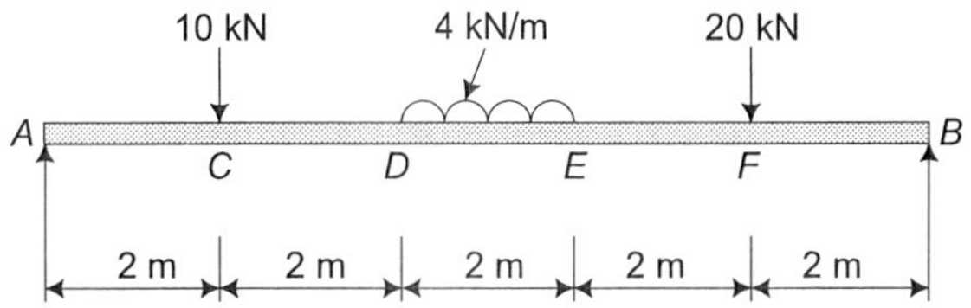

(a) Loaded beam

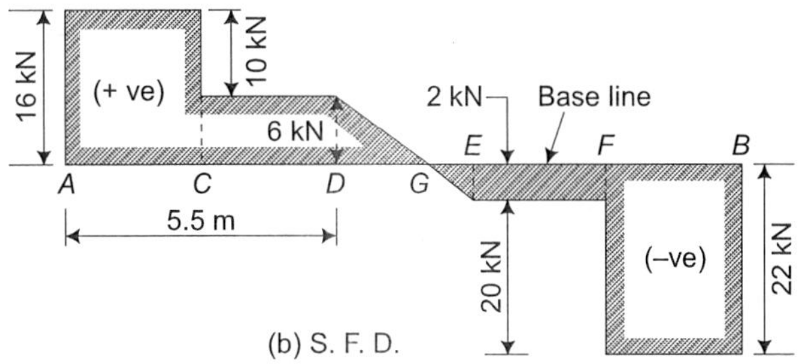

(b) S. F. D.

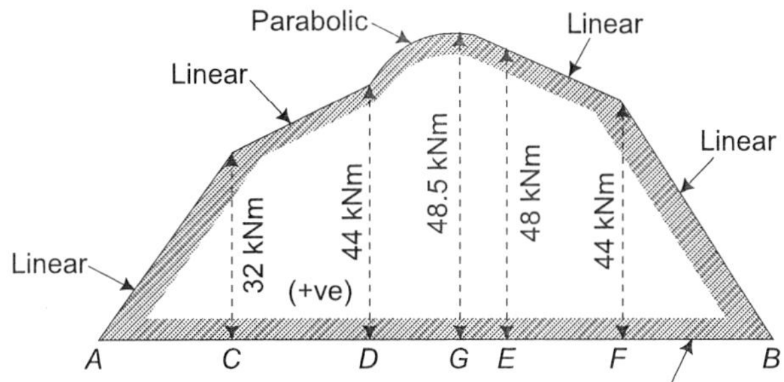

(c) B. M. D.

Fig. 8.36

The shear force at the section $x - x$ between D and E is given by,

$$F_x = + R_A - 10 - 4 \times (x - 4) \quad \text{(Constant)}$$

At $x = 4$, i.e., shear force at D,

$$F_D = + 16 - 10 - 4(0) = + 6 \text{ kN}$$

At $x = 6$, i.e., shear force at E,

$$F_E = + 16 - 10 - 4 \times (6 - 4) = - 2 \text{ kN}$$

This will remain constant for the portion EF.

The shear force changes from + 6 kN at D to – 2 kN at E. So, shear force will be zero somewhere between D and E. Let it be zero at a distance of x m from end A. Then equating F_x equal to zero, i.e.,

$$+ 16 - 10 - 4 \times (x - 4) = 0$$

$$\therefore \quad x = 5.5 \text{ m.}$$

Thus shear force is zero at a distance of 5.5 m from end A.

The shear force for the portion FB is given by,

$$F_x = + R_A - 10 - 4 \times 2 - 20$$
$$= + 16 - 10 - 8 - 20$$
$$= - 22 \text{ kN} \quad \text{(Constant)}$$

The shear force diagram is shown in Figure 8.36 (b).

B.M.D. The bending moment at any section $x - x$ between A and C at a distance of x from end A is given by,

$$M_x = + R_A x = + 16\,x \text{ kNm}$$

(For the left portion of the section, the clockwise moment is taken positive)

At $x = 0$, i.e., bending moment at A,

$$M_A = 0$$

At $x = 2$, i.e., bending moment at C,

$$M_C = + 16 \times 2 = + 32 \text{ kNm}$$

The bending moment at any section $x - x$ between C and D is given by,

$$M_x = + R_A \times x - 10 \times (x - 2)$$
$$= + 16x - 10(x - 2) \quad \text{(Linear)}$$

(For the left portion of the section, the clockwise moment is taken positive and the anticlockwise moment is taken negative)

At $x = 2$, i.e., bending moment at C,

$$M_C = + 16 \times 2 - 10 \times (2 - 2)$$
$$= + 32 \text{ kNm} \quad \text{(As before)}$$

At $x = 4$, i.e., bending moment at D,

$$M_D = + 16 \times 4 - 10 \times (4 - 2)$$
$$= + 44 \text{ kNm}$$

The bending moment at any section $x - x$ between D and E is given by,

$$M_x = + R_A \times x - 10 \times (x - 2) - 4 \times (x - 4) \times (x - 4)/2$$

$$\therefore \quad M_x = + 16x - 10(x - 2) - 2(x - 2)^2 \quad \text{(Parabolic)}$$

At $x = 4$, i.e., bending moment at D,

$$M_D = + 16 \times 4 - 10 \times (4 - 2) - 2(4 - 4)^2$$
$$= + 44 \text{ kNm} \quad \text{(As before)}$$

At $x = 6$, i.e., bending moment at E,

$$M_E = +16 \times 6 - 10 \times (6 - 2) - 2(6 - 4)^2$$
$$= +48 \text{ kNm}$$

At $x = 5.5$, i.e., bending moment at G,

$$M_G = +16 \times 5.5 - 10 \times (5.5 - 2) - 2(5.5 - 4)^2$$
$$= +48.5 \text{ kNm} \quad \text{(Maximum)}$$

The bending moment at any section between E and E is given by,

$$M_x = +R_A \times x - 10 \times (x - 2) - 4 \times 2 \times [x - (2/2 + 2 + 2)]$$

$\therefore$
$$M_x = +16x - 10(x - 2) - 8(x - 5) \quad \text{(Linear)}$$

At $x = 6$, i.e., bending moment at E,

$$M_E = +16 \times 6 - 10 \times (6 - 2) - 8 \times (6 - 5)$$
$$= +48 \text{ kNm} \quad \text{(As before)}$$

At $x = 8$, i.e., bending moment at F,

$$M_F = +16 \times 8 - 10 \times (8 - 2) - 8 \times (8 - 5)$$
$$= +44 \text{ kNm}$$

The bending moment at any section between F and B is given by,

$$M_x = +R_A \times x - 10 \times (x - 2) - 4 \times 2 \times [x - (2/2 + 2 + 2)] - 20 \times (x - 8)$$

$\therefore$
$$M_x = +16x - 10(x - 2) - 8(x - 5) - 20(x - 8) \quad \text{(Linear)}$$

At $x = 8$, i.e., bending moment at F,

$$M_F = +16 \times 8 - 10 \times (8 - 2) - 8 \times (8 - 5) - 20(8 - 8)$$
$$= +44 \text{ kNm} \quad \text{(As before)}$$

At $x = 10$, i.e., bending moment at B,

$$M_B = +16 \times 10 - 10 \times (10 - 2) - 8 \times (10 - 5) - 20(10 - 8) = 0$$

It is observed that bending moment is maximum at the point where shear force is zero. The bending moment diagram is shown in Figure 8.36 (c). From the shear force diagram it can be seen that shear force is zero at point G. Thus bending moment is maximum at point G.

Example 8.19 Draw the shear force and bending moment diagrams for a simply supported beam of 9 m span carrying uniformly distributed load as shown in Figure 8.37 (a).

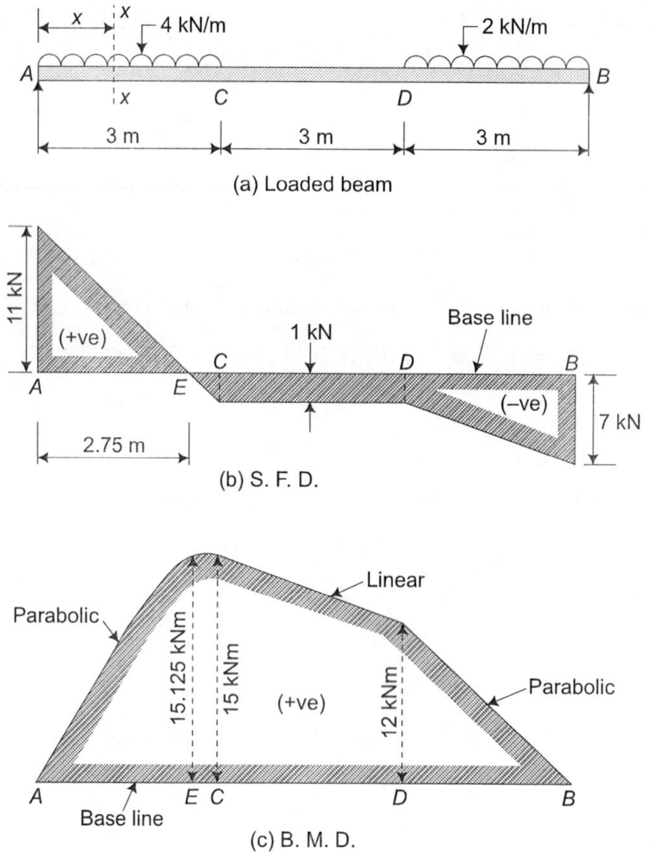

Fig. 8.37

Solution Refer Figure 8.37. Let R_A be the reaction at the support A and R_B be the reaction at the support B. In order to calculate reactions, take moments of the forces about A,

$$R_B \times 9 = 2 \times 3 \times (3/2 + 3 + 3) + 4 \times 3 \times 3/2 = 63 \text{ kN}$$

$$\therefore \qquad R_B = 63/9 = 7 \text{ kN}$$

Under equilibrium condition,

$$R_A + R_B = 4 \times 3 + 2 \times 3 = 18$$

$$\therefore \qquad R_A = 18 - 7 = 11 \text{ kN}$$

S.F.D. Take section $x - x$ at a distance x from the end A between A and C. The shear force at the section is given by,

$$F_x = + R_A - 4x = (+ 11 - 4x) \text{ kN} \quad \text{(Linear)}$$

At $x = 0$, i.e., shear force at A,

$$F_A = +11 - 4(0) = +11 \text{ kN}$$

At $x = 3$, i.e., shear force at C,

$$F_C = +11 - 4 \times 3 = -1 \text{ kN.}$$

This will remain constant between C and D.

The shear force changes from + 11 kN at A to – 1 kN at C. So, shear force will be zero somewhere between A and C. Let it be zero at a distance of x m from end A. Then equating F_x equal to zero, i.e.,

$$+11 - 4x = 0$$

$$\therefore \quad x = 2.75 \text{ m.}$$

Thus shear force is zero at a distance of 2.75 m from end A.

The shear force at the section $x - x$ between D and B is given by,

$$F_x = +R_A - 4 \times 3 - 2 \times (x - 6)$$

$$= +11 - 12 - 2(x - 6) \quad \text{(Linear)}$$

At $x = 6$, i.e., shear force at D,

$$F_D = +11 - 12 - 2(0) = -1 \text{ kN}$$

At $x = 9$, i.e., shear force at B,

$$F_B = +11 - 12 - 2 \times (9 - 6) = -7 \text{ kN}$$

The shear force diagram is shown in Figure 8.37 (b).

B.M.D. The bending moment at any section $x - x$ between A and C at a distance of x from end A is given by,

$$M_x = +R_A x - 4 \times x \times x/2$$

$$= (+11x - 2x^2) \text{ kNm} \quad \text{(Parabolic)}$$

(For the left portion of the section, the clockwise moment is taken positive and the anticlockwise moment is taken negative.)

At $x = 0$, i.e., bending moment at A,

$$M_A = 0$$

At $x = 3$, i.e., bending moment at C,

$$M_C = +11 \times 3 - 2 \times (3)^2 = +15 \text{ kNm}$$

At $x = 2.75$, i.e., bending moment at E,

$$M_E = +11 \times 2.75 - 2 \times (2.75)^2$$

$$= +15.125 \text{ kNm} \quad \text{(Maximum)}$$

The bending moment at any section $x - x$ between C and D is given by,

$$M_x = + R_A \times x - 4 \times 3 \times (x - 3/2)$$

$$= + 11x - 12(x - 3/2) \quad \text{(Linear)}$$

(For the left portion of the section, the clockwise moment is taken positive and the anticlockwise moment is taken negative.)

At $x = 3$, i.e., bending moment at C,

$$M_C = + 11 \times 3 - 12 \times (3 - 1.5)$$

$$= + 15 \text{ kNm} \quad \text{(As before)}$$

At $x = 6$, i.e., bending moment at D,

$$M_D = + 11 \times 6 - 12 \times (6 - 1.5)$$

$$= + 12 \text{ kNm}$$

The bending moment at any section between D and B is given by,

$$M_x = + R_A \times x - 12 \times (x - 3/2) - 2 \times (x - 6) \times (x - 6)/2 \text{ (Parabolic)}$$

At $x = 6$, i.e., bending moment at D,

$$M_D = + 11 \times 6 - 12 \times (6 - 3/2) - 2 \times (6 - 3) \times (6 - 3)/2$$

$$= + 12 \text{ kNm (As before)}$$

At $x = 9$, i.e., bending moment at B,

$$M_B = + 11 \times 9 - 12 \times (9 - 3/2) - 2 \times (9 - 6) \times (9 - 6)/2$$

$$= 0$$

It is observed that bending moment is maximum at the point where shear force is zero. The bending moment diagram is shown in Figure 8.37 (c). From the shear force diagram it can be seen that shear force is zero at point E. Thus bending moment is maximum at point E.

Example 8.20. Draw the shear force and bending moment diagrams for a simply supported beam shown in Figure 8.38 (a).

Solution. Refer Figure 8.38. Let R_A be the reaction at the support A and R_B be the reaction at the support B. In order to calculate reactions, take moments of the forces about A,

$$R_B \times 10 = 2 \times 8 + 4 \times 6 + 1 \times 4 \times (4/2 + 2) + 5 \times 2$$

$$= 66 \text{ kN}$$

$$\therefore \quad R_B = 66/10 = 6.6 \text{ kN}$$

Under equilibrium condition,

$$R_A + R_B = 5 + 1 \times 4 + 4 + 2 = 15$$

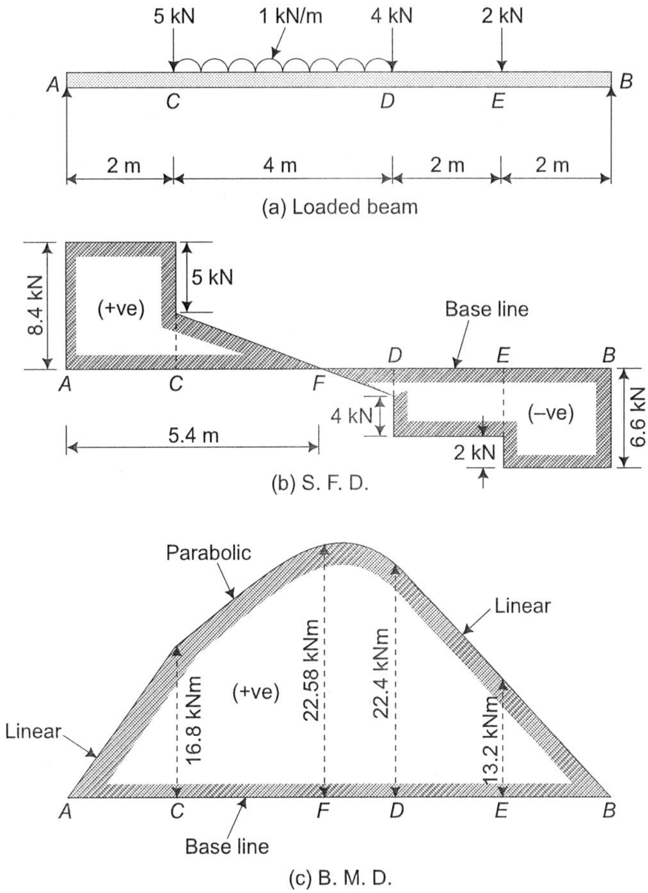

Fig. 8.38

$\therefore \qquad R_A = 15 - 6.6 = 8.4$ kN

S.F.D. Take section $x - x$ at a distance x from the end A between A and C. The shear force at the section is given by,

$$F_x = + R_A = + 8.4 \text{ kN} \quad \text{(Constant)}$$

The shear force at the section $x - x$ between C and D is given by,

$$\therefore \qquad F_x = + R_A - 5 - 1 \times (x - 2)$$

$$= + 8.4 - 5 - (x - 2) \quad \text{(Linear)}$$

At $x = 2$, i.e., shear force at C,

$$F_C = + 8.4 - 5 - (2 - 2) = + 3.4 \text{ kN}$$

At $x = 6$, i.e., shear force at D,

$$F_D = +8.4 - 5 - (6 - 2) = -0.6 \text{ kN}$$

The shear force changes from + 3.4 kN at C to – 0.6 kN at D. So, shear force will be zero somewhere between C and D. Let it be zero at a distance of x m from end A. Then equating F_x equal to zero, i.e.,

$$+8.4 - 5 - (x - 2) = 0$$

$$\therefore \quad x = 5.4 \text{ m.}$$

Thus shear force is zero at a distance of 5.4 m from end A.

The shear force for the portion DE is given by,

$$F_x = +R_A - 5 \times 1 \times 4 - 4$$

$$= +8.4 - 5 - 4 - 4 = -4.6 \text{ kN} \quad \text{(Constant)}$$

The shear force for the portion EB is given by,

$$F_x = +R_A - 5 - 1 \times 4 - 4 - 2$$

$$= +8.4 - 5 - 4 - 4 - 2 = -6.6 \text{ kN} \quad \text{(Constant)}$$

The shear force diagram is shown in Figure 8.38 (b).

B.M.D. The bending moment at any section $x - x$ between A and C at a distance of x from end A is given by,

$$M_x = +R_A x = +8.4x \text{ kNm} \quad \text{(Constant)}$$

(For the left portion of the section, the clockwise moment is taken positive)

At $x = 0$, i.e., bending moment at A,

$$M_A = 0$$

At $x = 2$, i.e., bending moment at C,

$$M_C = +8.4 \times 2 = +16.8 \text{ kNm}$$

The bending moment at any section $x - x$ between C and D is given by,

$$M_x = +R_A \times x - 5 \times (x - 2) - 1 \times (x - 2) \times (x - 2)/2 \quad \text{(Parabolic)}$$

(For the left portion of the section, the clockwise moment is taken positive and the anticlockwise moment is taken negative.)

At $x = 2$, i.e., bending moment at C,

$$M_C = +8.4 \times 2 - 5 \times (0) - 1 \times (0)$$

$$= +16.8 \text{ kNm} \quad \text{(As before)}$$

At $x = 6$, i.e., bending moment at D,

$$M_D = +\,8.4 \times 6 - 5 \times (6-2) - 1 \times (6-2) \times \frac{(6-2)}{2}$$

$$= +\,22.4 \text{ kNm}$$

At $x = 5.4$ i.e., bending moment at F,

$$M_F = 8.4 \times 5.4 - 5 \times (5.4-2) - 1 \times (5.4-2) \times \frac{(5.4-2)}{2}$$

$$= +\,22.58 \text{ kNm} \quad \text{(Maximum)}$$

The bending moment at any section between D and E is given by,

$$M_x = +\,R_A \times x - 8 \times (x-2) - 1 \times 4 \times [x - (4/2 + 2)] - 4(x-6) \quad \text{(Linear)}$$

At $x = 6$, i.e., bending moment at D,

$$M_D = 8.4 \times 6 - 5 \times (6-2) - 4 \times [6 - (2+2)] - 4 \times (0)$$

$$= +\,22.4 \text{ kNm (As before)}$$

At $x = 8$, i.e., bending moment at E,

$$M_E = 8.4 \times 8 - 5 \times (8-2) - 4 \times [8 - (2+2)] - 4 \times (8-6)$$

$$= +\,13.2 \text{ kNm}$$

The bending moment at any section between E and B is given by,

$$M_x = +\,R_A \times x - 5 \times (x-2) - 1 \times 4 \times [x - (4/2 + 2)] - 2(x-6) - 2 \times (x-8) \quad \text{(Linear)}$$

At $x = 8$, i.e., bending moment at E,

$$M_E = 8.4 \times 8 - 5 \times (8-2) - 4 \times [8-4] - 4(8-6) - 2 \times (8-8)$$

$$= +\,13.2 \text{ kNm} \quad \text{(As before)}$$

At $x = 10$, i.e., bending moment at B,

$$M_B = +\,8.4 \times 10 - 5 \times (10-2) - 4 \times [10-4] - 4(10-6) - 2 \times (10-8) = 0$$

It is observed that bending moment is maximum at the point where shear force is zero. The bending moment diagram is shown in Figure 8.38 (c). From the shear force diagram it can be seen that shear force is zero at point F. Thus bending moment is maximum at point F.

8.7.4 Simply Supported Beam Subjected to Couples

When a simply supported beam is subjected to a couple at a section the shear force does not change at that section as there is no change in load due to couple. But the bending moment changes suddenly in magnitude equal to that of couple at the section. Simply supported beam subjected to couples have been explained in the examples 8.21 to 8.23.

Example 8.21 A simply supported beam *AB* of length 4 m is subjected to a clockwise couple of 20 kNm and anticlockwise couple of 10 kNm at a distance of 1 m and 3 m respectively from the left end *A*. Draw the shear force and bending moment diagrams.

Solution Refer Figure 8.39. Let R_A be the reaction at the support *A* and R_B the reaction at the support *B*. In order to calculate reactions, take moments of the forces about *A*,

$$R_B \times 4 + 10 - 20 = 0$$

$$\therefore \quad R_B = 10/4 = 2.5 \text{ kN}$$

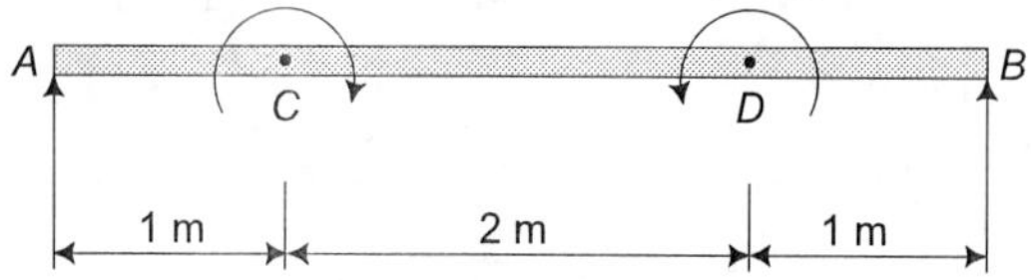

(a) Loaded beam

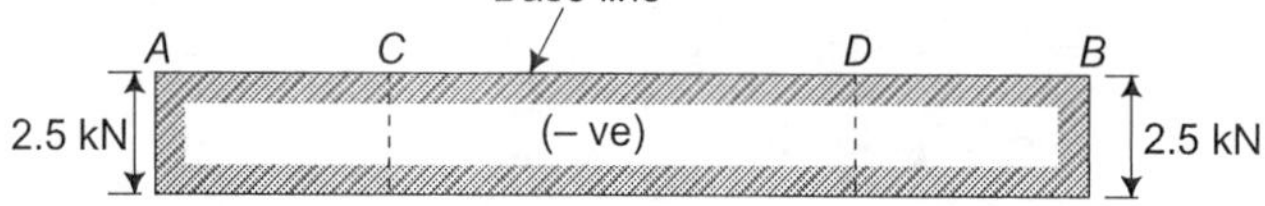

(b) S. F. D.

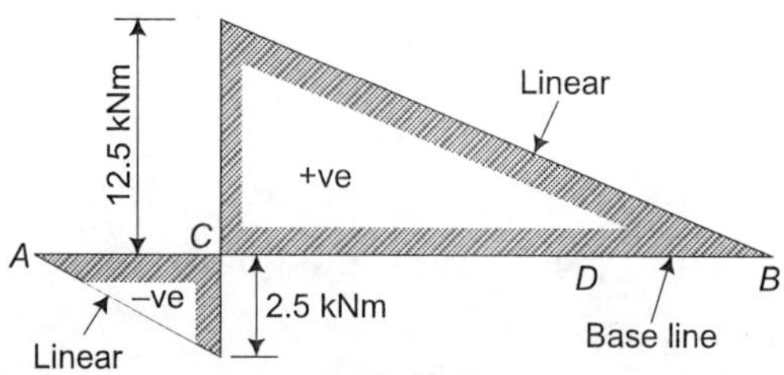

(c) B. M. D.

Fig. 8.39

Under equilibrium condition,

$$R_A + R_B = 0$$

$$\therefore \quad R_A = -2.5 \text{ kN}$$

S.F.D. Take section $x - x$ at a distance x from the end *A* between *A* and *B*. The shear force at the section is given by,

$$F_x = + R_A = + (-2.5) = -2.5 \text{ kN} \quad \text{(Constant)}$$

The shear force will remain constant between A and B. The shear force diagram is shown in Figure 8.39 (b).

B.M.D. Bending moment at A,

$$M_A = 0$$

Bending moment at C,

$$M_C = -2.5 \times 1 + 20 = +17.5 \text{ kNm}$$

(For the left portion of the section, the clockwise moment is taken positive)

Bending moment at D,

$$M_D = -2.5 \times 3 + 20 - 10 = +2.5 \text{ kNm}$$

(For the left portion of the section, the clockwise moment is taken positive and anticlockwise moment is taken negative)

Bending moment at B,

$$M_B = -2.5 \times 4 + 20 - 10 = 0$$

The bending moment diagram is shown in Figure 8.39 (c).

Example 8.22 Draw the shear force and bending moment diagrams for a simply supported beam shown in Figure 8.40 (a).

Solution Refer Figure 8.40. Let R_A be the reaction at the support A and R_B be the reaction at the support B. In order to calculate reactions, take moments of the forces about A,

$$R_B \times 10 = 10 \times 2 - 5 + 10 - 5 \times 8 = 65 \text{ kN}$$

$$\therefore \quad R_B = 65/10 = 6.5 \text{ kN}$$

Under equilibrium condition,

$$R_A + R_B = 10 + 5 = 15$$

$$\therefore \quad R_A = 15 - 6.5 = 8.5 \text{ kN}$$

S.F.D. Take section $x - x$ at a distance x from the end A between A and C. The shear force at the section is given by,

$$F_x = +R_A = +8.5 \text{ kN} \quad \text{(Constant)}$$

The shear force at the section $x - x$ between C and D is given by,

$$F_x = +R_A - 10 = +8.5 - 10 = -1.5 \text{ kN} \quad \text{(Constant)}$$

This will remain constant for the portion CF.

The shear force at the section $x - x$ between F and B is given by,

$$F_x = +R_A - 10 - 5 = 8.5 - 10 - 5$$

$$= -6.5 \text{ kN} \quad \text{(Constant)}$$

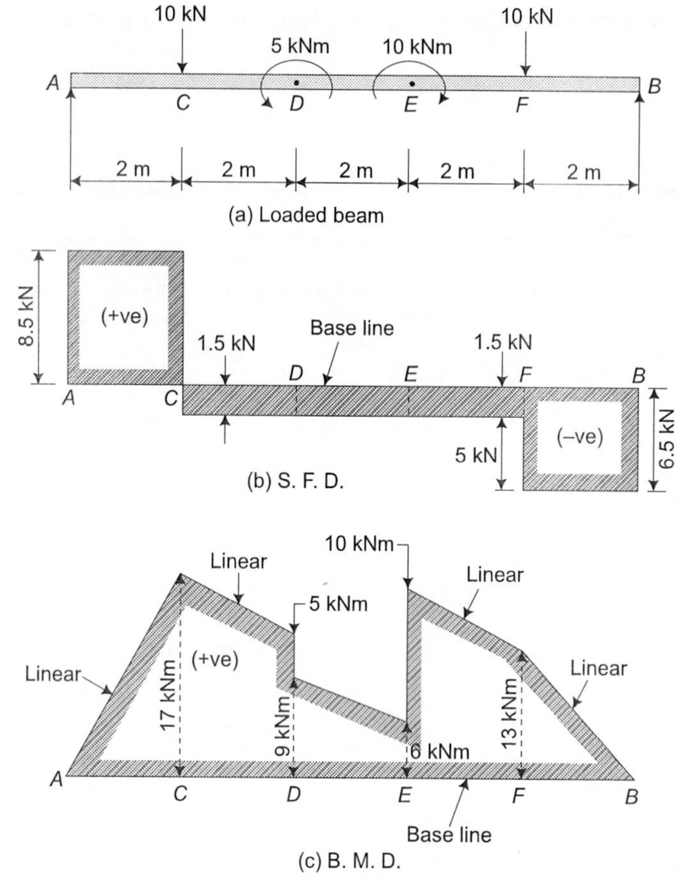

Fig. 8.40

The shear force diagram is shown in Figure 8.40 (b).

B.M.D. The bending moment at any section $x - x$ between A and C at a distance of x from end A is given by,

$$M_x = + R_A x = + 8.5x \text{ kNm}$$

(For the left portion of the section, the clockwise moment is taken positive)

At $x = 0$, i.e., bending moment at A,

$$M_A = 0$$

At $x = 2$, i.e., bending moment at C,

$$M_C = + 8.5 \times 2 = + 17 \text{ kNm}$$

The bending moment at any section $x - x$ between C and D is given by,

$$M_x = + R_A \times x - 10 \times (x - 2)$$

$$= + 8.5x - 10(x - 2) \quad \text{(Linear)}$$

(For the left portion of the section, the clockwise moment is taken positive and the anticlockwise moment is taken negative)

At $x = 2$, i.e., bending moment at C,

$$+ 8.5 \times 2 - 10 \times (2 - 2) = + 17 \text{ kNm} \quad \text{(As before)}$$

At $x = 4$, i.e., bending moment at D,

$$M_D = + 8.5 \times 4 - 10 \times (4 - 2) = + 14 \text{ kNm}$$

The bending moment at any section $x - x$ between D and E is given by,

$$M_x = + R_A \times x - 10 \times (x - 2) - 5$$

$$= 8.5x - 10 \times (x - 2) - 5 \quad \text{(Linear)}$$

At $x = 4$, i.e., bending moment at D,

$$M_D = + 8.5 \times 4 - 10 \times (4 - 2) - 5 = + 9 \text{ kNm}$$

At $x = 6$, i.e., bending moment at E,

$$M_E = + 8.5 \times 6 - 10 \times (6 - 2) - 5 = + 6 \text{ kNm}$$

The bending moment at any section between E and E is given by,

$$M_x = + R_A \times x - 10 \times (x - 2) - 5 + 10$$

$$= 8.5x - 10(x - 2) + 5 \quad \text{(Linear)}$$

At $x = 6$, i.e., bending moment at E,

$$M_E = + 8.5 \times 6 - 10 \times (6 - 2) + 5 = + 16 \text{ kNm}$$

At $x = 8$, i.e., bending moment at F,

$$M_F = + 8.5 \times 8 - 10 \times (8 - 2) + 5 = + 13 \text{ kNm}$$

The bending moment at any section between F and B is given by,

$$M_x = + R_A \times x - 10 \times (x - 2) - 5 + 10 - 5 \times (x - 8) \quad \text{(Linear)}$$

At $x = 8$, i.e., bending moment at F,

$$M_F = + 8.5 \times 8 - 10 \times (8 - 2) - 5 + 10 - 5 \times (8 - 8)$$

$$= + 13 \text{ kNm} \quad \text{(As before)}$$

At $x = 10$, i.e., bending moment at B,

$$M_B = + 8.5 \times 10 - 10 \times (10 - 2) - 5 + 10 - 5 \times (10 - 8) = 0$$

The bending moment diagram is shown in Figure 8.40 (c).

Example 8.23 Draw the shear force and bending moment diagrams for a simply supported beam shown in Figure 8.41 (a).

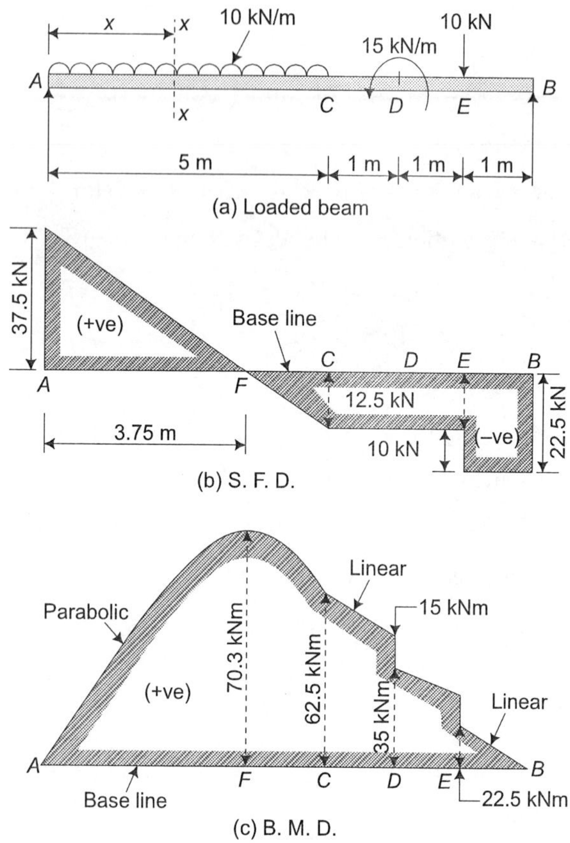

Fig. 8.41

Solution Refer Figure 8.41. Let R_A be the reaction at the support A and R_B be the reaction at the support B. In order to calculate reactions, take moments of the forces about A,

$$R_B \times 8 = 10 \times 5 \times 5/2 - 15 + 10 \times 7$$

$$= 180 \text{ kN}$$

$$\therefore \quad R_B = 180/8 = 22.5 \text{ kN}$$

Under equilibrium condition,

$$R_A + R_B = 10 \times 5 + 10 = 60$$

$$\therefore \quad R_A = 60 - 22.5 = 37.5 \text{ kN}$$

S.F.D. Take section $x - x$ at a distance x from the end A between A and C. The shear force at the section is given by,

$$F_x = + R_A - 10 \times x = + 37.5 - 10x \quad \text{(Linear)}$$

At $x = 0$, i.e., shear force at A,

$$F_A = + 37.5 - 10 \times (0) = + 37.5 \text{ kN}$$

At $x = 5$, i.e., shear force at C,

$$F_C = + 37.5 - 10 \times 5 = - 12.5 \text{ kN (Constant)}$$

For CD and DE this will remain constant.

For EB the shear force is given by,

$$F_x = + R_A - 10 \times x - 10 = + 37.5 - 10 \times 5 - 10$$
$$= - 22.5 \text{ kN}$$

This will remain constant for the section EB.

The shear force changes from + 37.5 kN at A to – 12.5 kN at C. So, shear force will be zero somewhere between A and C. Let it be zero at a distance of x m from end A. Then equating F_x equal to zero, i.e.,

$$+ 37.5 - 10x = 0$$

$$\therefore \quad x = 37.5/16 = 3.75 \text{ m.}$$

Thus, shear force is zero at a distance of 3.75 m from end A.

The shear force diagram is shown in Figure 8.41 (b).

B.M.D. The bending moment at any section $x - x$ between A and C at a distance of x from end A is given by,

$$M_x = + R_A \times x - 10 \times x \times x/2$$
$$= + 37.5x - 5x^2 \quad \text{(Parabolic).}$$

(For the left portion of the section, the clockwise moment is taken positive and the anticlockwise moment is taken negative.)

At $x = 0$, i.e., bending moment at A,

$$M_A = 0$$

At $x = 5$, i.e., bending moment at C,

$$M_C = 37.5 \times 5 - 5 \times (50)^2 = + 62.5 \text{ kNm}$$

At $x = 3.75$, i.e., bending moment at F,

$$M_F = 37.8 \times 3.75 - 5 \times (3.75)^2$$
$$= + 70.3 \text{ kNm} \quad \text{(Maximum)}$$

The bending moment at any section between C and D is given by,

$$M_x = + R_A \times x - 10 \times 5 \times (x - 5/2)$$

$$= 37.5x - 50 \times (x - 5/2) \quad \text{(Linear)}$$

At $x = 5$, i.e., bending moment at C,

$$M_C = 37.5 \times 5 - 50 \times (5 - 2.5)$$

$$= + 62.5 \text{ kNm} \quad \text{(As before)}$$

At $x = 6$, i.e., bending moment at D,

$$M_D = 37.5 \times 6 - 50 \times (6 - 2.5) = + 50 \text{ kNm}$$

The bending moment at any section between D and E is given by,

$$M_x = + R_A \times x - 10 \times 5 \times (x - 5/2) - 15$$

$$= 37.5x - 50 \times (x - 5/2) - 15 \quad \text{(Linear)}$$

At $x = 6$, i.e., bending moment at E,

$$M_D = 37.5 \times 6 - 50 \times (6 - 2.5) - 15 = + 35 \text{ kNm}$$

At $x = 7$, i.e., bending moment at B,

$$M_E = 37.5 \times 7 - 50 \times (7 - 2.5) - 15 = 22.5 \text{ kNm}$$

The bending moment at any section between E and B is given by,

$$M_x = + R_A \times x - 10 \times 5 \times (x - 5/2) - 15 - 10 \times (x - 7) \quad \text{(Linear)}$$

At $x = 7$, i.e., bending moment at E,

$$M_E = + 37.5 \times 7 - 10 \times 5 \times (7 - 2.5) - 15 - 10 \times (7 - 7)$$

$$= + 22.5 \text{ kNm} \quad \text{(As before)}$$

At $x = 8$, i.e., bending moment at B,

$$M_B = + 37.5 \times 8 - 10 \times 5 \times (8 - 2.5) - 15 - 10 \times (8 - 7) = 0$$

It is observed that bending moment is maximum at the point where shear force is zero. The bending moment diagram is shown in Figure 8.41 (c). From the shear force diagram it can be seen that shear force is zero at point F. Thus, bending moment is maximum at point F.

8.7.5 Simply Supported Beam Subjected to a Uniformly Varying Load form Zero at One End to *w* per Unit Length at the Other End

Consider a beam AB of length l carrying a uniformly varying load from zero at support A to w per unit length at support B linearly as shown in Figure 8.42 (a). Let R_A be the reaction at the support A and R_B be the reaction at the support B. In order to calculate reactions, take moments of the forces about A,

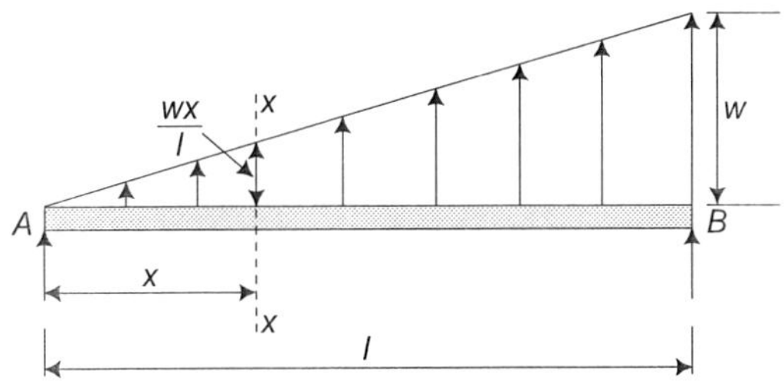

(a) Loaded beam

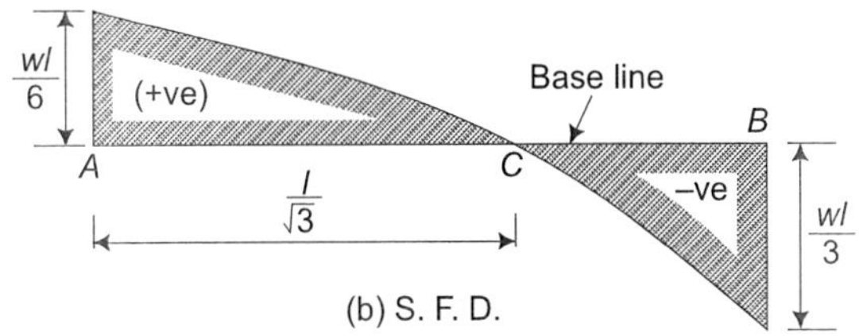

(b) S. F. D.

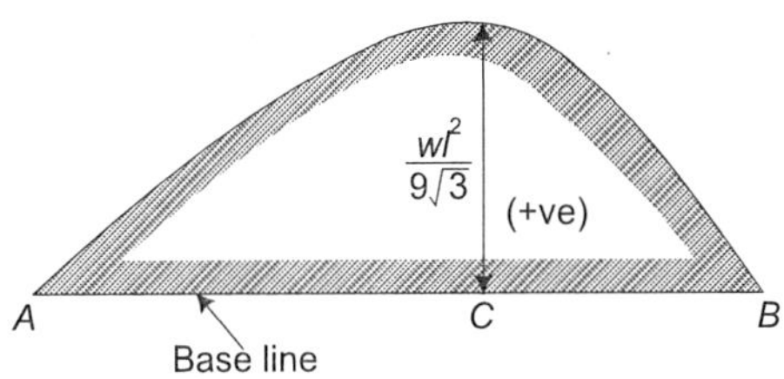

(c) B. M. D.

Fig. 8.42

$$R_B \times l = \frac{1}{2} \times w \times l \times \left(\frac{2}{3}\, l\right) = \frac{wl^2}{3}$$

$$\left(\because \text{Total load } \frac{w \times l}{2} \text{ is acting at } \frac{2}{3}\, l \text{ from end } A\right)$$

$\therefore$

$$R_B = \frac{wl}{3}$$

$$R_A = \text{Total load} - R_B$$

$$= \frac{wl}{2} - \frac{wl}{3} = \frac{wl}{6}$$

S.F.D. Take section $x - x$ at a distance x from the end A between A and B.

The load intensity at section = wx/l

Therefore, the total load at a distance x of the section

$$= \frac{1}{2} \times x \times \frac{wx}{l} = \frac{wx^2}{2l}$$

The shear force at the section is given by,

$$F_x = +R_A - \text{Load on length } x$$

$$= \frac{wl}{6} - \frac{wx^2}{2l} \quad \text{(Parabolic)}$$

At $x = 0$, i.e., shear force at A,

$$F_A = \frac{wl}{6} - \frac{wl}{2l} \times (0) = +\frac{wl}{6}$$

At $x = l$, i.e., shear force at B,

$$F_B = \frac{wl}{6} - \frac{w}{2l} \times (l)^2 = -\frac{wl}{3}$$

The shear force changes from + $wl/6$ at support A to – $wl/3$ at support B. So, somewhere between A and B shear force will be zero. Let it be zero at a distance of x from the support A. To determine x equating the shear force expression to zero, i.e.,

$$F_x = 0.$$

We have,

$$\frac{wl}{6} - \frac{wx^2}{2l} = 0$$

$$x^2 = \frac{l^2}{3}$$

$$\therefore \quad x = \frac{l}{\sqrt{3}} = 0.577\, l$$

The shear force diagram is shown in Figure 8.42 (b).

B.M.D. The bending moment at any section $x - x$ between A and B at a distance of x from end A is given by,

$$M_x = +R_A \times x - \frac{wx^2}{2l} \times \frac{x}{3}$$

$$= \frac{wlx}{6} - \frac{wx^3}{6l} \quad \text{(Cubic)}$$

(For the left portion of the section, the clockwise moment is taken positive and the anticlockwise moment is taken negative.)

At $x = 0$, i.e., bending moment at A,

$$M_A = 0$$

At $x = l$, i.e., bending moment at B,

$$M_B = \frac{wl \times l}{6} - \frac{wx^3}{6l} = \frac{wx^2}{6} - \frac{wx^2}{6} = 0$$

Maximum bending moment occurs at the point where shear force is zero. Therefore, it occurs at a distance of $x = l/\sqrt{3}$ from support A and is obtained by substituting the value of x in the bending moment expression.

$$\therefore \quad M_{max} = \frac{wl}{6} \times \frac{l}{\sqrt{3}} - \frac{w}{6l} \times \left(\frac{l}{\sqrt{3}}\right)^3$$

$$= \frac{wl^2}{6\sqrt{3}} - \frac{wl^2}{18\sqrt{3}} = \frac{wl^2}{9\sqrt{3}}$$

The bending moment diagram is shown in Figure 8.42 (c).

Example 8.24 Draw the shear force and bending moment diagrams for a simply supported beam shown in Figure 8.43 (a).

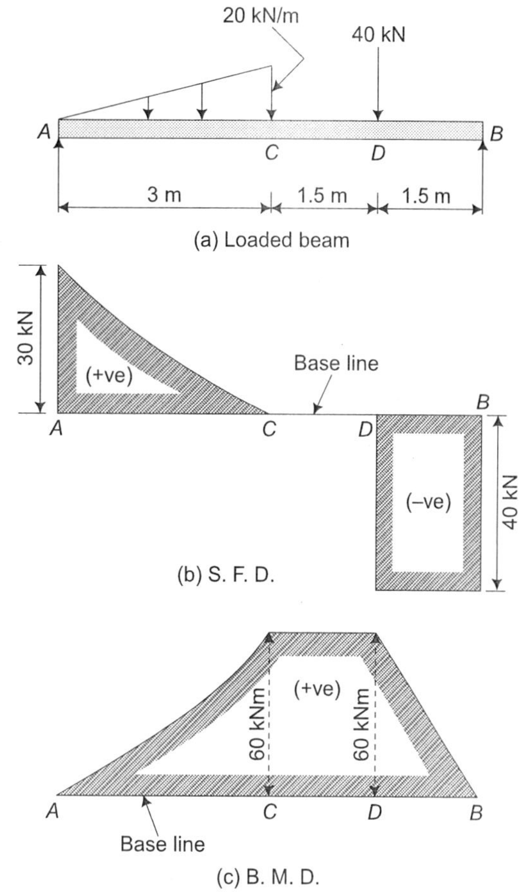

Fig. 8.43

Solution Refer Figure 8.43. Let R_A be the reaction at the support A and R_B be the reaction at the support B. In order to calculate reactions, take moments of the forces about A,

$$R_B \times 6 = 40 \times 4.5 + \frac{1}{2} \times 3 \times 20 \times \left(\frac{2}{3} \times 3\right)$$

$$= 240 \text{ kN}$$

$$\therefore \quad R_B = \frac{240}{6} = 40 \text{ kN}$$

Under equilibrium condition,

$$R_A + R_B = \frac{1}{2} \times 3 \times 20 + 40 = 70$$

$$\therefore \quad R_A = 70 - 40 = 30 \text{ kN}$$

S.F.D. Take section $x - x$ at a distance x from the end A between A and C.

The load intensity at section = $20x/3$

Therefore, the total load at a distance x of the section

$$= \frac{1}{2} \times x \times \frac{20x}{3}$$

$$= \frac{10x^2}{3}$$

The shear force at the section is given by,

$$F_x = +R_A - \text{Load on length } x$$

$$= 30 - \frac{10x^2}{3} \quad \text{(Parabolic)}$$

At $x = 0$, i.e., shear force at A,

$$F_A = 30 - (10/3) \times (0) = +30 \text{ kN}$$

At $x = 3$, i.e., shear force at C,

$$F_C = 30 - (10/3) \times (3)^2 = 0$$

For the portion CD the shear force is given by,

$$F_x = +R_A - (1/2) \times 3 \times 20$$

$$= 30 - 30 = 0 \quad \text{(Constant)}$$

Similarly, the shear force for the portion DB is given by,

$$F_x = +R_A - (1/2) \times 3 \times 20 - 40$$

$$= -40 \text{ kN} \quad \text{(Constant)}$$

The shear force diagram is shown in Figure 8.43 (b).

B.M.D. The bending moment at any section $x - x$ between A and C at a distance of x from end A is given by,

$$M_x = +R_A \times x - \left(\frac{1}{2} \times x \times \frac{20x}{3}\right) \times \frac{x}{3}$$

$$= 30x - \frac{10x^3}{9} \quad \text{(Cubic)}$$

(For the left portion of the section, the clockwise moment is taken positive and the anticlockwise moment is taken negative.)

At $x = 0$, i.e., bending moment at A,

$$M_A = 0$$

At $x = 3$, i.e., bending moment at C,

$$M_C = 30 \times 3 - \frac{10 \times (3)^3}{9} = +60 \text{ kNm}$$

For the portion CD the bending moment at any section $x - x$ is given by,

$$= +R_A \times x - \left(\frac{1}{2} \times 3 \times 20\right) \times \left(x - \frac{2}{3}l\right)$$

$$= 30x - 30 \times \left(x - \frac{2}{3} \times 3\right)$$

$$= 30x - 30(x - 2)$$

At $x = 3$, i.e., bending moment at C,

$$M_C = 30 \times 3 - 30 \times (3 - 2) = +\ 60 \text{ kNm} \quad \text{(As before)}$$

At $x = 4.5$, i.e., bending moment at D,

$$M_D = 30 \times 4.5 - 30 \times (4.5 - 2) = +\ 60 \text{ kNm}$$

For the portion DB the bending moment at any section $x - x$ is given by,

$$M_x = +\ R_A \times x - 30(x - 2) - 40 \times (x - 4.5)$$

At $x = 4.5$, i.e., bending moment at D,

$$M_D = 30 \times 4.5 - 30 \times (4.5 - 2) - 40 \times (0)$$

$$= +\ 60 \text{ kNm} \quad \text{(As before)}$$

At $x = 4.5$, i.e., bending moment at B,

$$M_B = 30 \times 6 - 30 \times (6 - 2) - 40 \times (6 - 4.5) = 0$$

The bending moment diagram is shown in Figure 8.43 (c).

Example 8.25 Draw the shear force and bending moment diagrams for a simply supported beam shown in Figure 8.44 (a).

Solution Refer Figure 8.44. Let R_A be the reaction at the support A and R_B be the reaction at the support B. In order to calculate reactions, take moments of the forces about A,

$$R_B \times 8 = 20 \times 4 + 20 + \frac{1}{2} \times 3 \times 10 \times \left(\frac{2}{3} \times 3\right)$$

$$= 130 \text{ kN}$$

$$\therefore \quad R_B = \frac{130}{8} = 16.25 \text{ kN}$$

Under equilibrium,

$$R_A + R_B = (1/2) \times 3 \times 10 + 20 = 35$$

$$\therefore \quad R_A = 35 - 16325 = 18.75 \text{ kN}$$

S.F.D. Take section $x - x$ at a distance x from the end A between A and C.

The load intensity at section = $(10x/3)$

Therefore, the total load at a distance x of the section

$$= (1/2) \times x \times (10x/3) = 5x^2/3$$

The shear force at the section is given by,

$$F_x = +\ R_A - \text{Load on length } x$$

$$= 18.75 - 5x^2/3 \quad \text{(Parabolic)}$$

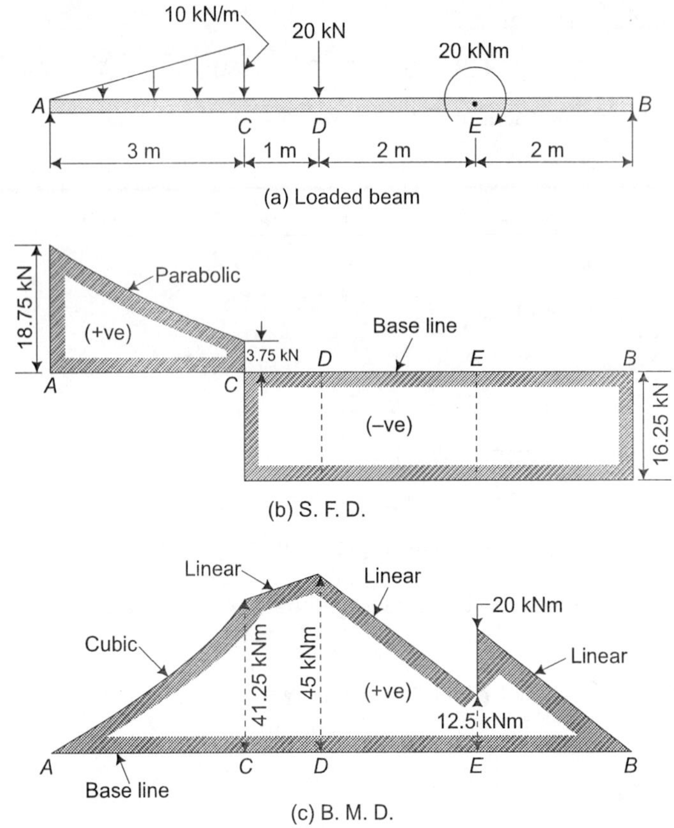

Fig. 8.44

At $x = 0$, i.e., shear force at A,

$$F_A = 18.75 - (5/3) \times (0) = + 18.75 \text{ kN}$$

At $x = 3$, i.e., shear force at C,

$$F_C = 18.75 - (5/3) \times (3)^2 = 3.75 \text{ kN}$$

For the portion CB the shear force is given by,

$$F_x = + R_A - (1/2) \times 3 \times 10 - 20$$

$$= 18.75 - 35 = - 16.25 \text{ kN}$$

This will remain constant for the portion CB.

The shear force diagram is shown in Figure 8.44 (b).

B.M.D. The bending moment at any section $x - x$ between A and C at a distance of x from end A is given by,

$$M_x = + R_A \times x - \left(\frac{1}{2} \times x \times \frac{10x}{3}\right) \times \frac{x}{3}$$

$$= 18.75x - \frac{5x^3}{9} \quad \text{(Cubic)}$$

(For the left portion of the section, the clockwise moment is taken positive and the anticlockwise moment is taken negative.)

At $x = 0$, i.e., bending moment at A,

$$M_A = 0$$

At $x = 3$, i.e., bending moment at C,

$$M_C = 18.75 \times 3 - \frac{5 \times (3)^3}{9}$$

$$= + 41.25 \text{ kNm}$$

For the portion CD the bending moment at any section $x - x$ is given by,

$$M_x = + R_A \times x - \left(\frac{1}{2} \times 3 \times 10\right) \times \left(x - \frac{2}{3}\, l\right)$$

$$= 18.75x - 15 \times \left(x - \frac{2}{3} \times 3\right)$$

$$= 18.75 - 15(x - 2)$$

At $x = 3$, i.e., bending moment at C,

$$M_C = 18.75 \times 3 - 15 \times (3 - 2)$$

$$= + 41.25 \text{ kNm} \quad \text{(As before)}$$

At $x = 4$, i.e., bending moment at D,

$$M_D = 18.75 \times 4 - 15 \times (4 - 2) = + 45 \text{ kNm}$$

For the portion DE the bending moment at any section $x - x$ is given by,

$$M_x = + R_A \times x - 15(x - 2) - 20 \times (x - 4)$$

At $x = 4$, i.e., bending moment at D,

$$M_D = 18.75 \times 4 - 15 \times (4 - 2) - 20 \times (0)$$

$$= + 45 \text{ kNm} \quad \text{(As before)}$$

At $x = 6$, i.e., bending moment at E,

$$M_E = 18.75 \times 6 - 15 \times (6 - 2) - 20 \times (6 - 4)$$

$$= + 12.5 \text{ kNm}$$

For the portion EB the bending moment at any section $x - x$ is given by,

$$M_x = + R_A \times x - 15(x - 2) - 20 \times (x - 4) + 20$$

At $x = 6$, i.e., bending moment at E,

$$M_E = 18.75 \times 6 - 15 \times (6 - 2) - 20 \times (6 - 4) + 20$$
$$= +\,32.5 \text{ kNm}$$

At $x = 8$, i.e., bending moment at B,

$$M_B = 18.75 \times 8 - 15 \times (8 - 2) - 20 \times (8 - 4) + 20 = 0$$

The bending moment diagram is shown in Figure 8.44 (c).

Example 8.26 A simply supported beam of length 6 m carries a gradually increasing load of 50 N per m run at one end to 100 N per m run at the other end. Draw the shear force and bending moment diagrams and also determine the position and magnitude of maximum bending moment.

Solution Refer Figure 8.45. The load is considered as a uniformly distributed load of 50 N/m over the whole span and a gradually varying load of zero at the support A to 50 N/m at the support B. Let R_A be the reaction at the support A and R_B the reaction at the support B. In order to calculate reactions, take moments of the forces about A,

$$R_B \times 6 = 50 \times 6 \times \frac{6}{2} + \frac{1}{2} \times 6 \times 50 \times \left(\frac{2}{3} \times 6\right)$$
$$= 1500 \text{ N}$$

$$\therefore \quad R_B = \frac{1500}{6} = 250 \text{ N}$$

Under equilibrium,

$$R_A + R_B = 50 \times 6 + (1/2) \times 6 \times 50$$
$$= 450 \text{ N}$$

$$\therefore \quad R_A = 450 - 250$$
$$= 200 \text{ N}$$

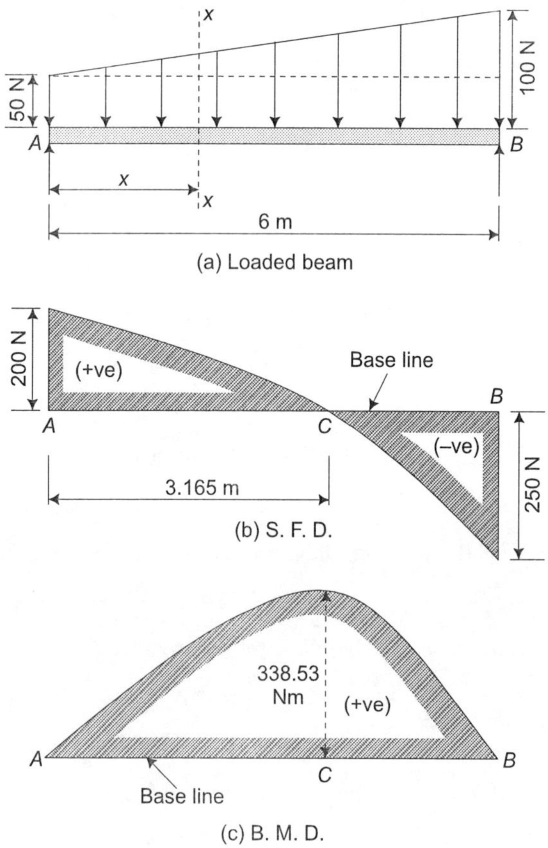

Fig. 8.45

S.F.D. Take section $x - x$ at a distance x from the end A between A and B.

The shear force at the section is given by,

$$F_x = +R_A - \text{Total load on length } x$$
$$= 200 - \left[50 \times x + \frac{1}{2} \times x \times \frac{50x}{6}\right]$$

At $x = 0$, i.e., shear force at A,

$$F_A = +\ 200 \text{ N}$$

At $x = 6$, i.e., shear force at B,

$$F_B = 200 - \left[50 \times 6 + \frac{1}{2} \times 6 \times \frac{50 \times 6}{6}\right]$$

$$= -\ 250 \text{ N}$$

The shear force changes from + 200 N at A to – 250 N at B. So, shear force will be zero somewhere between A and B. Let it be zero at a distance of x m from end A. Then equating F_x equal to zero, i.e.,

$$200 - \left[50 \times x + \frac{1}{2} \times x \times \frac{50x}{6}\right] = 0$$

$$x^2 + 12x - 48 = 0$$

$$\therefore \quad x = \frac{-\ 12 \pm \sqrt{(12)^2 + 4 \times 1 \times 48}}{2 \times 1}$$

$$= 3.165 \text{ m.}$$

Thus shear force is zero at a distance of 3.165 m from end A. The shear force diagram is shown in Figure 8.45 (b).

B.M.D. The bending moment at any section $x - x$ between A and B at a distance of x from end A is given by,

$$M_x = +\ R_A \times x - 50 \times x \times \frac{x}{2} - \left(\frac{1}{2} \times x \times \frac{50x}{6}\right) \times \frac{x}{3}$$

$$\therefore \quad M_x = +\ 200x - 25x^2 - (25/18)x^3 \quad \text{(Cubic)}$$

(For the left portion of the section, the clockwise moment is taken positive and the anticlockwise moment is taken negative.)

At $x = 0$, i.e., bending moment at A,

$$M_A = 0$$

At $x = 6$, i.e., bending moment at B,

$$M_B = 200 \times 6 - 25 \times (6)^2 - (25/18) \times (6)^3 = 0$$

Maximum bending moment occurs where shearing force is zero. Here shearing force is zero at a distance of 3.165 m. Therefore, maximum bending moment is obtained by substituting $x =$ 3.165 m in the bending moment expression.

At $x = 3.165$ m, i.e., bending moment at C,

$$M_B = 200 \times 3.165 - 25 \times (3.165)^2 - (25/18) \times (3.165)^3$$

$$= 338.53 \text{ Nm} \quad \text{(Maximum)}$$

The bending moment diagram is shown in Figure 8.45 (c).

Example 8.27 A simply supported beam of length l m is loaded with distributed load of intensity zero at the ends and w/m per unit length at the centre. Draw the shear force and bending moment diagrams.

Solution Figure 8.46 shows a simply supported beam AB of length l which carries a uniformly varying load from zero at each end to the w per unit length at the centre C.

$$\text{The total load} = \frac{1}{2}wl$$

As the load on the beam is symmetrical so, the reactions at the supports will be equal and will be half of the total load on the beam.

$$\therefore \qquad R_A = R_B = \frac{1}{2} \times \left(\frac{1}{2}wl\right) = \frac{wl}{4}$$

S.F.D. Take section $x - x$ at a distance x from the end A between A and C.

$$\text{The rate of loading at the section} = \frac{x}{(l/2)} w = \frac{2wx}{l}$$

The load on the beam at a distance

$$x = \frac{1}{2} \times x \times \frac{2wx}{l} = \frac{wx^2}{l}$$

The shear force at the section is given by,

$$F_x = +R_A - \text{Total load on length } x$$

$$= \frac{wl}{4} - \frac{wx^2}{l}$$

At $x = 0$, i.e., shear force at A,

$$F_A = \frac{wl}{4} - \frac{w(0)}{l} = \frac{wl}{4}$$

At $x = \frac{l}{2}$, i.e., shear force at C,

$$F_C = \frac{wl}{4} - \frac{w(l/2)^2}{l} = 0$$

As the beam is symmetrically loaded about its centre so, there is no need of determining shear force in the portion CB. The shear force diagram is shown in Figure 8.46 (b).

B.M.D. Since the beam is simply supported so, the bending moment at A and B will be zero.

The bending moment at any section $x - x$ between A and C at a distance of x from end A is given by,

$$M_x = +R_A \times x - \frac{wx^2}{l} \times \frac{x}{3}$$

$$= \frac{wlx}{4} - \frac{wx^3}{3l} \quad \text{(Cubic)}$$

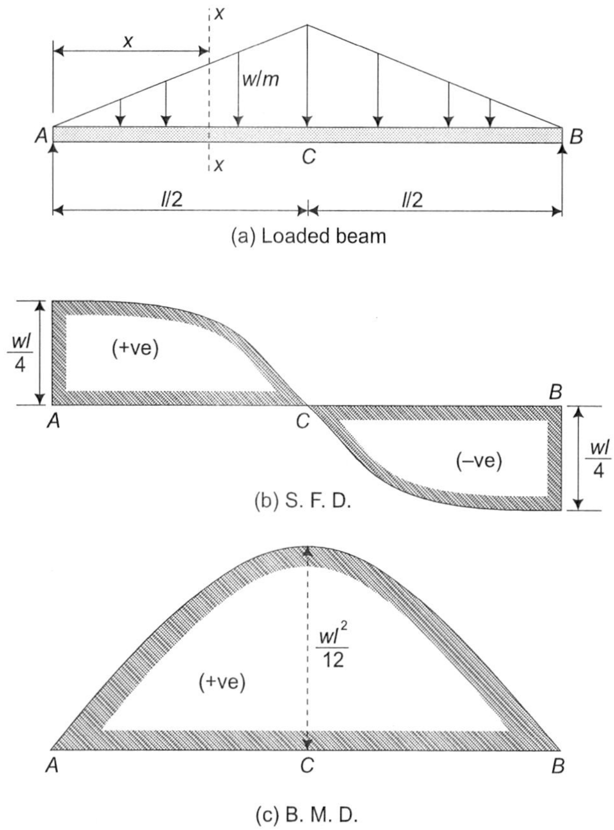

Fig. 8.46

(For the left portion of the section, the clockwise moment is taken positive and the anticlockwise moment is taken negative.)

At $x = 0$, i.e., bending moment at A,

$$M_A = 0$$

At $x = l/2$, i.e., bending moment at C,

$$M_C = \frac{wl}{4} \times \frac{l}{2} - \frac{w}{3l} \times \left(\frac{l}{2}\right)^3$$

$$= \frac{wl^2}{8} - \frac{wl^2}{24} = \frac{wl^2}{12}$$

Maximum bending moment occurs at the point where shear force is zero. Therefore, it occurs at a distance of $x = l/2$ form support A. Therefore, maximum bending moment occurs at C which is given by,

$$M_{max} = M_C = wl^2/12$$

Again there is no necessity of determining bending moments in the portion *CB*, as the beam is symmetrically loaded about its centre. The bending moment diagram is shown in Figure 8.46 (c).

Example 8.28 A horizontal beam 6 m long is hinged at *A* and supported on roller at *B* as shown in Figure 8.47 (a). It carries inclined loads of 4 kN, 2 kN, and 6 kN inclined at 45°, 45°, and 30° at the points *C*, *D*, and *E* respectively. Draw (i) thrust diagram, (ii) shear force diagram, and (iii) bending moment diagram.

Solution Refer Figure 8.47.

Resolving the inclined load into their vertical and horizontal components.

The vertical system consists of:

$$4 \sin 45° = 2\sqrt{2} = 2.828 \text{ kN}$$

$$2 \sin 45° = \sqrt{2} = 1.414 \text{ kN}$$

$$6 \sin 30° = 3 \text{ kN}$$

at *C*, *D*, and *E* respectively.

And horizontal system consists of:

$$4 \cos 45° = 4 \times 1/\sqrt{2} = 2\sqrt{2} = 2.828 \text{ kN}$$

$$2 \cos 45° = 2 \times 1/\sqrt{2} = \sqrt{2} = 1.414 \text{ kN}$$

$$6 \cos 30° = 6 \times \sqrt{3}/2 = 3\sqrt{3} = 5.196 \text{ kN}$$

at *C*, *D*, and *E* respectively.

The vertical and horizontal components are shown in Figure 8.47 (b).

The horizontal reaction at hinge $A = 5.196 + 1.414 - 2.828$

$= 3.782$ kN (2.828 kN acts in opposite direction).

The horizontal thrust diagram is shown in Figure 8.47 (c).

Let R_A be the reaction at the support *A* and R_B the reaction at the support *B*. In order to calculate reactions, take moments of the forces about *A*,

$$R_B \times 6 = 2.828 \times 1 + 1.414 \times 2.5 + 3 \times 4$$

$$= 18.363$$

$$\therefore \quad R_B = 18.363/6 = 3.06 \text{ kN}$$

Under equilibrium,

$$R_A + R_B = 2.828 + 1.414 + 3 = 7.242$$

$$\therefore \quad R_A = 7.242 - 3.06 = 4.182 \text{ kN}$$

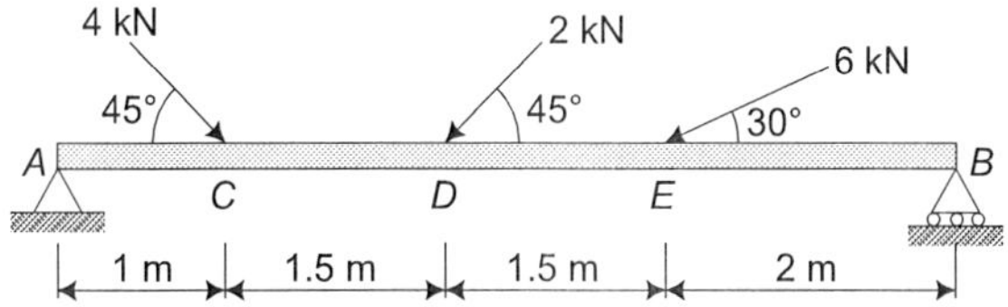

(a) Loaded beam

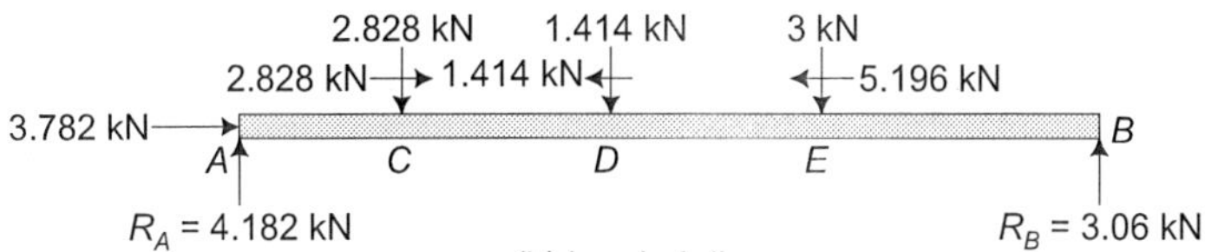

(b) Loaded diagram

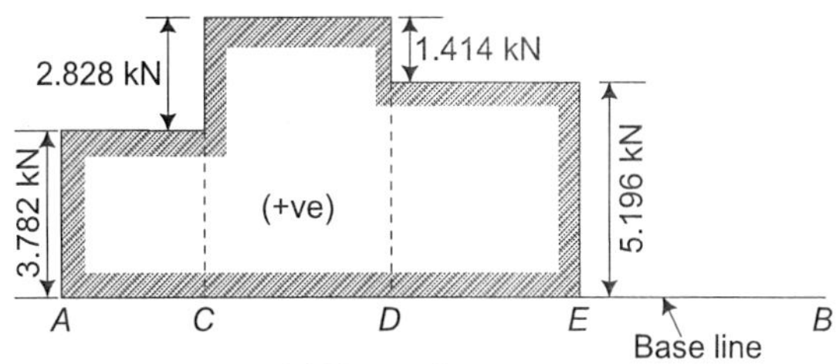

(c) Thrust diagram

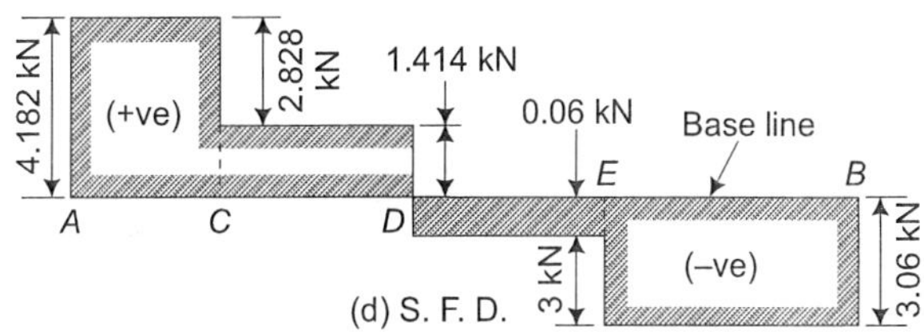

(d) S. F. D.

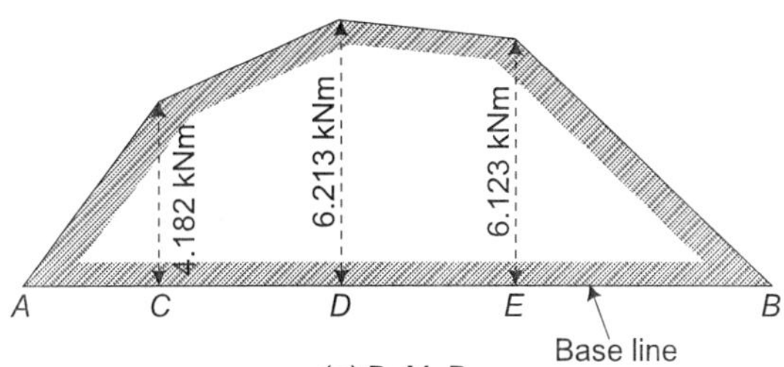

(e) B. M. D.

Fig. 8.47

S.F.D. Take section $x - x$ at a distance x from the end A between A and C. The shear force is given by,

$$F_x = + R_A = 4.182 \text{ kN} \quad \text{(Constant)}$$

Similarly, the shear force between C and D is given by,

$$F_x = + R_A - 2.828 = 4.182 - 2.828$$
$$= 1.354 \text{ kN} \quad \text{(Constant)}$$

The shear force between D and E is given by,

$$F_x = + R_A - 2.828 - 1.414 = 4.182 - 2.828 - 1.414$$
$$= - 0.06 \text{ kN} \quad \text{(Constant)}$$

The shear force between E and B is given by,

$$F_x = + R_A - 2.828 - 1.414 - 3 = 4.182 - 2.828 - 1.414 - 3$$
$$= - 3.06 \text{ kN} \quad \text{(Constant)}$$

The shear force diagram is shown in Figure 8.47 (d).

B.M.D. The bending moment at any section $x - x$ between A and C at a distance of x from end A is given by,

$$M_x = + R_A x = + 4.182x \text{ kNm}$$

(For the left portion of the section, the clockwise moment is taken positive)

At $x = 0$, i.e., bending moment at A,

$$M_A = 0$$

At $x = 1$, i.e., bending moment at C,

$$M_C = + 4.182 \times 1 = + 4.182 \text{ kNm}$$

The bending moment at any section $x - x$ between C and D at a distance of x from end A is given by,

$$M_x = + R_A x - 2.828 \times (x - 1)$$
$$= + 4.182x - 2.828 (x - 1) \text{ kNm}$$

(For the left portion of the section, the clockwise moment is taken positive and anticlockwise moment is taken negative)

At $x = 1$, i.e., bending moment at C,

$$M_C = + 4.182 \times 1 - 2.828 \times 0$$
$$= + 4.182 \text{ kNm} \quad \text{(As before)}$$

At $x = 2.5$, i.e., bending moment at D,

$$M_D = + 4.182 \times 2.5 - 2.828 \times (2.5 - 1)$$
$$= + 6.213 \text{ kNm} \quad \text{(Maximum)}$$

The bending moment at any section between D and E at a distance of x from end A is given by,

$$M_x = +R_A x - 2.828 \times (x - 1) - 1.414 \times (x - 2.5)$$

$$= +4.182x - 2.828(x - 1) - 1.414(x - 2.5) \text{ kNm.}$$

At $x = 2.5$, i.e., bending moment at D,

$$M_D = +4.182 \times 2.5 - 2.828 \times (2.5 - 1) - 1.414 \times 0$$

$$= +6.213 \text{ kNm} \quad \text{(As before)}$$

At $x = 4$, i.e., bending moment at E,

$$M_E = +4.182 \times 4 - 2.828 \times (4 - 1) - 1.414 \times (4 - 2.5)$$

$$= +6.123 \text{ kNm}$$

The bending moment at any section between E and B at a distance of x from end A is given by,

$$M_x = +R_A x - 2.828 \times (x - 1) - 1.414 \times (x - 2.5) - 3(x - 4)$$

$$\therefore \quad M_x = +4.182x - 2.828 \times (x - 1) - 1.414 \times (x - 2.5) - 3(x - 4)$$

At $x = 4$, i.e., bending moment at E,

$$M_E = +4.182 \times 4 - 2.828 \times (4 - 1) - 1.414 \times (4 - 2.5) - 3 \times 0$$

$$= +6.123 \text{ kNm} \quad \text{(As before)}$$

At $x = 6$, i.e., bending moment at B,

$$M_B = +4.182 \times 6 - 2.828 \times (6 - 1) - 1.414 \times (6 - 2.5) - 3(6 - 4) = 0$$

The bending moment diagram is shown in Figure 8.47 (e).

8.8 SHEAR FORCE AND BENDING MOMENTS FOR OVERHANGING BEAMS

A beam with one or both ends extended over the supports is called overhanging beam. In the case of overhanging beams the bending moment is negative for the overhanging portion, whereas it is positive between the supports. In the following examples the shear force and bending diagrams for the overhanging beams have been drawn.

Example 8.29 Draw the shear force and bending moment diagrams for the beam shown in Figure 8.48. Also determine the points of contraflexure.

Solution Refer Figure 8.48. Let R_A be the reaction at the support A and R_B the reaction at the support B. Taking moments of the forces about A,

$$R_B \times 4 + 400 \times 1.5 = 500 \times 5 + 1000 \times 2.5$$

$$= 5000 \text{ N}$$

$$\therefore \quad R_B = (5000 - 600)/4$$

$$= 1100 \text{ N}$$

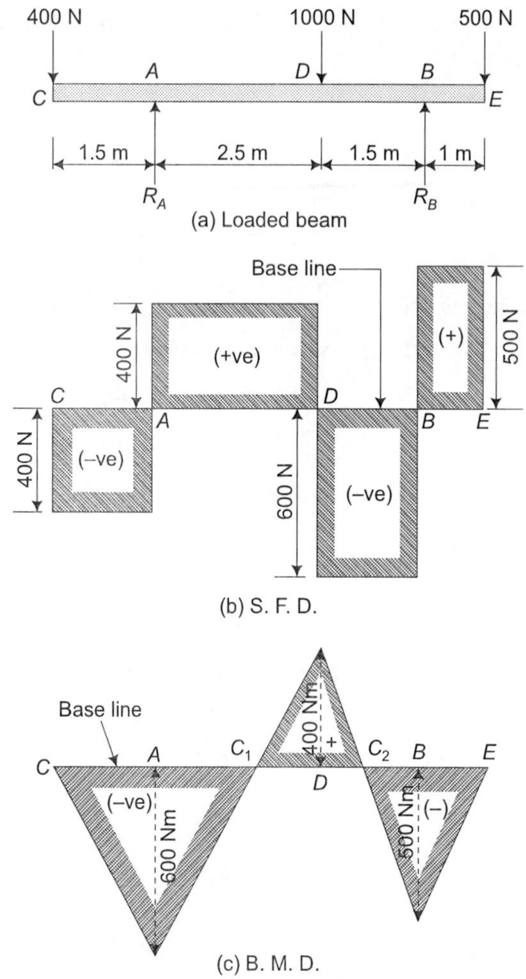

Fig. 8.48

Under equilibrium,

$$R_A + R_B = 400 + 1000 + 500$$

$$= 1900 \text{ N}$$

$$\therefore \quad R_A = 1900 - 1100$$

$$= 800 \text{ N}$$

S.F.D. Take section $x - x$ at a distance x from the end A between A and C. The shear force is given by,

$$F_x = -400 \text{ N} \quad \text{(Constant)}$$

Similarly, the shear force between A and D is given by,

$$F_x = -400 + R_A$$

$$= -400 + 800$$

$$= +400 \text{ N} \quad \text{(Constant)}$$

The shear force between D and B is given by,

$$F_x = +400 - 1000$$

$$= -600 \text{ N} \quad \text{(Constant)}$$

The shear force between B and E is given by,

$$F_x = -600 + R_B$$

$$= -600 + 1100$$

$$= +500 \text{ N} \quad \text{(Constant)}$$

The shear force diagram is shown in Figure 8.48 (b).

B.M.D. The bending moment at any section $x - x$ between C and A at a distance of x from end C is given by,

$$M_x = -400x \text{ Nm} \quad \text{(Linear)}$$

(For the left portion of the section, the anticlockwise moment is taken negative.)

At $x = 0$, i.e., bending moment at C,

$$M_C = 0$$

At $x = 1.5$, i.e., bending moment at A,

$$M_A = -400 \times 1.5 = -600 \text{ Nm}$$

The bending moment at any section between A and D at a distance of x from end A is given by,

$$M_x = -400\,(1.5 + x) + R_A x$$

$$= -400\,(1.5 + x) + 800x \text{ Nm} \quad \text{(Linear)}$$

(For the left portion of the section, the clockwise moment is taken positive and anticlockwise moment is taken negative)

At $x = 0$, i.e., bending moment at A,

$$M_A = -400 \times 1.5 + 800 \times 0$$

$$= -600 \text{ Nm} \quad \text{(As before)}$$

At $x = 2.5$, i.e., bending moment at D,

$$M_D = -400 \times (1.5 + 2.5) + 800 \times 2.5$$

$$= +400 \text{ Nm}$$

The bending moment changes from – 600 Nm at A to + 400 Nm at D. So, it will be zero somewhere at point C_1 (point of contraflexure) in between A and D. Let the point C_1 be at a distance of x m from A.

Then, $$-400\,(1.5 + x) + 800x = 0$$

$$-600 - 400x + 800x = 0$$

$\therefore$ $$x = (600/400) = 1.5 \text{ m}$$

The bending moment at any section between D and B at a distance of x from end A is given by,

$$M_x = -400\,(1.5 + x) + R_A x - 1000\,(x - 2.5) \text{ Nm} \quad \text{(Linear)}$$

At $x = 2.5$, i.e., bending moment at D,

$$M_D = -400 \times (1.5 + 2.5) + 800 \times 2.5 - 1000 \times 0$$

$$= +400 \text{ Nm} \quad \text{(As before)}$$

At $x = 4$, i.e., bending moment at B,

$$M_B = -400 \times (1.5 + 4) + 800 \times 4 - 1000 \times (4 - 2.5)$$

$$= -500 \text{ Nm}$$

The bending moment changes from + 400 Nm at D to – 500 Nm at B. So, it will be zero somewhere at point C_2 (point of contraflexure) in between D and B. Let the point C_2 be at a distance of x from A.

$$\therefore \quad -400(1.5 + x) + R_A x - 1000(x - 2.5) = 0$$

$$-400(1.5 + x) + 800x - 1000(x - 2.5) = 0$$

$$600 - 400x + 800x - 1000x + 2500 = 0$$

$$600x = 1900$$

$$\therefore \quad x = (1900/600) = 3.17 \text{ m}$$

The bending moment at any section between B and E at a distance of x from end A is given by,

$$M_x = -400(1.5 + x) + R_A x - 1000(x - 2.5) + R_B(x - 4) \text{ Nm} \quad \text{(Linear)}$$

At $x = 4$, i.e., bending moment at B,

$$M_B = -400 \times (1.5 + 4) + 800 \times 4 - 1000 \times (4 - 2.5) + 1100 \times 0$$

$$= -500 \text{ Nm} \quad \text{(As before)}$$

At $x = 5$, i.e., bending moment at E,

$$M_E = -400 \times (1.5 + 5) + 800 \times 5 - 1000 \times (5 - 2.5) + 1100 \times (5 - 4) = 0$$

The bending moment diagram is shown in Figure 8.48 (c).

Example 8.30 Draw the shear force and bending moment diagrams for the beam shown in Figure 8.49. Also locate the point of contraflexure.

Solution Refer Figure 8.49. Let R_A be the reaction at the support A and R_B is the reaction at the support B. Taking moments of the forces about A,

$$R_B \times 8 = 4 \times 8 \times (8/2) + 4 \times 12 = 176 \text{ kN}$$

$$\therefore \quad R_B = 176/8 = 22 \text{ kN}$$

Under equilibrium,

$$R_A + R_B = 4 \times 8 + 4 = 36 \text{ kN}$$

$$\therefore \quad R_A = 36 - 22 = 14 \text{ kN}$$

S.F.D. Take section $x - x$ at a distance x from the end A between A and B. The shear force is given by,

$$F_x = + R_A - 4x = + 14 - 4x \quad \text{(Linear)}$$

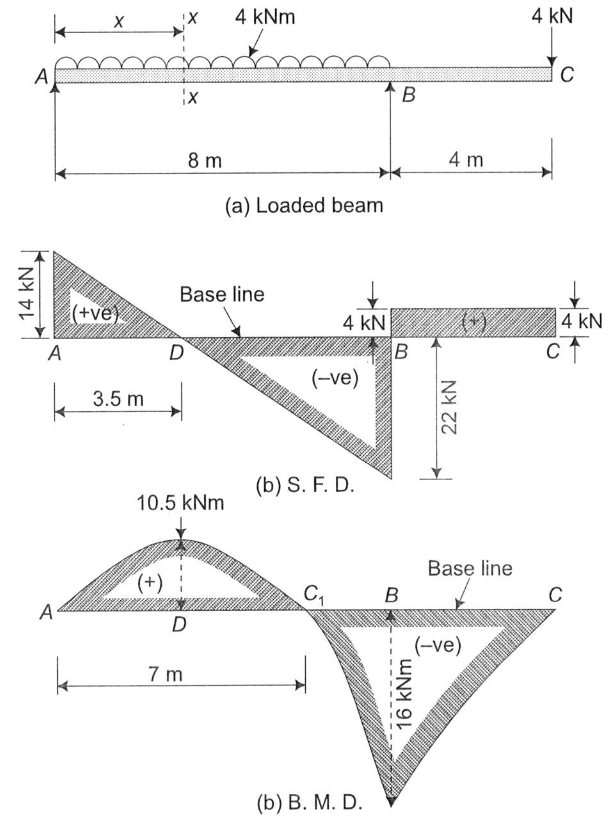

Fig. 8.49

At $x = 0$, i.e., shear force at A,

$$F_A = + 14 \text{ kN}$$

At $x = 8$, i.e., shear force at B,

$$F_B = + 14 - 4 \times 8 = - 18 \text{ kN}$$

The shear force is positive at A and negative at B, so it will be zero somewhere between A and B.

$$\therefore \quad + 14 - 4x = 0$$

$$\therefore \quad x = 14/4 = 3.5 \text{ m.}$$

The shear force is zero at point D which lies at a distance of 3.5 m from A.

The shear force between B and C is given by,

$$F_x = + 14 - 4 \times 8 + 22 = 4 \quad \text{(Constant)}$$

The shear force diagram is shown in Figure 8.49 (b).

B.M.D. The bending moment at any section $x - x$ between A and B at a distance of x from end A is given by,

$$M_x = + R_A x - 4 \times x \times (x/2)$$
$$= 14x - 2x^2 \quad \text{(Parabolic)}$$

(For the left portion of the section, the clockwise moment is taken positive and anticlockwise moment is taken negative)

At $x = 0$, i.e., bending moment at A,

$$M_A = 0$$

At $x = 8$, i.e., bending moment at B,

$$M_B = + 14 \times 8 - 2 \times (8^2) = - 16 \text{ kNm}$$

At $x = 3.5$, i.e., bending moment at D,

$$M_D = + 10 \times 3.5 - 2 \times 3.5^2 = + 10.5 \text{ kNm}$$

Point of contraflexure is located between A and B since the B.M. value is negative at B and positive at D. Let it be at a distance of x m from A.

Then, $$14x - 2x^2 = 0$$

$\therefore$ $$x = 7 \text{ m}$$

The bending moment at any section between B and C at a distance of x from end A is given by,

$$M_x = + R_A x - 4 \times 8 \times (x - 8/2) + R_B (x - 8)$$
$$= 14x - 32(x - 4) + 22 \times (x - 8) \quad \text{(Parabolic)}$$

(For the left portion of the section, the clockwise moment is taken positive and anticlockwise moment is taken negative)

At $x = 8$, i.e., bending moment at B,

$$M_B = + 14 \times 8 - 32 \times (8 - 4) + 22 \times (8 - 8)$$
$$= - 16 \text{ kNm} \quad \text{(As before)}$$

At $x = 12$, i.e., bending moment at C,

$$M_C = + 14 \times 12 - 32 \times (12 - 4) + 22 \times (12 - 8) = 0$$

The bending moment diagram is shown in Figure 8.49 (c).

Example 8.31 Draw the shear force and bending moment diagrams for the beam shown in Figure 8.50. Also clearly indicate the point of contraflexure.

Solution Refer Figure 8.50. Let R_A be the reaction at the support A and R_B be the reaction at the support B. Taking moments of the forces about A,

$$R_B \times 4 = 10 \times 5 + 20 \times 2 + 10 \times 2 \times (2/2)$$
$$= 110 \text{ kN}$$
$$\therefore \quad R_B = 110/4 = 27.5 \text{ kN}$$

Under equilibrium,

$$R_A + R_B = 10 \times 2 + 20 + 10 = 50 \text{ kN}$$
$$\therefore \quad R_A = 50 - 27.5 = 22.5 \text{ kN}$$

S.F.D. Take section $x - x$ at a distance x from the end A between A and C. The shear force is given by,

$$F_x = + R_A - 10x$$
$$= + 22.5 - 10x \quad \text{(Linear)}$$

At $x = 0$, i.e., shear force at A,

$$F_A = + 22.5 \text{ kN}$$

At $x = 2$, i.e., shear force at C,

$$F_C = + 22.5 - 10 \times 2$$
$$= 2.5 \text{ kN}$$

The shear force between C and B is given by,

$$F_x = + R_A - 10 \times 2 - 20$$
$$= + 22.5 - 20 - 20$$
$$= - 17.5 \text{ kN} \quad \text{(Constant)}$$

The shear force between B and D is given by,

$$F_x = + R_A - 10 \times 2 - 20 + R_B$$
$$= + 22.5 - 20 - 20 + 27.5$$
$$= + 10 \text{ kN} \quad \text{(Constant)}$$

The shear force diagram is shown in Figure 8.50 (b).

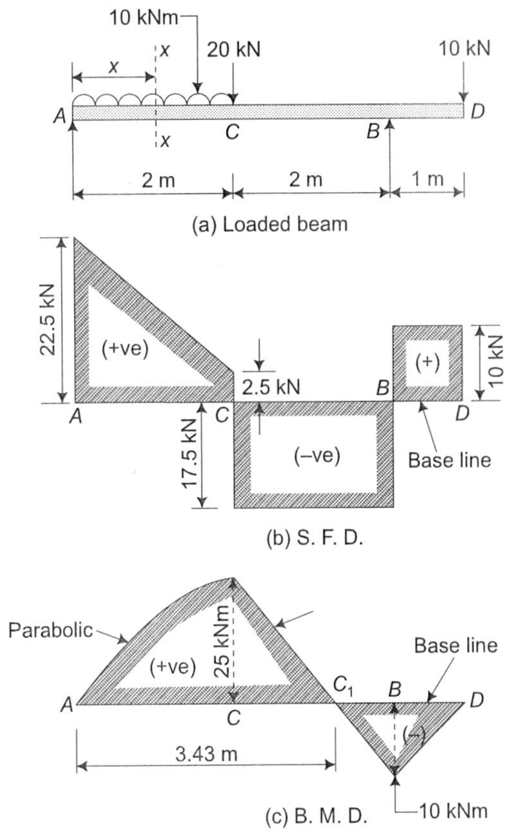

Fig. 8.50

B.M.D. The bending moment at any section $x - x$ between A and C at a distance of x from end A is given by,

$$M_x = + R_A x - 10 \times x \times (x/2)$$
$$= 22.5x - 5x^2 \quad \text{(Parabolic)}$$

(For the left portion of the section, the clockwise moment is taken positive and anticlockwise moment is taken negative)

At $x = 0$, i.e., bending moment at A,

$$M_A = 0$$

At $x = 2$, i.e., bending moment at C,

$$M_C = + 22.5 \times 2 - 5 \times 2^2 = + 25 \text{ kNm}$$

The bending moment at any section between C and B at a distance of x from end A is given by,

$$M_x = + R_A x - 10 \times 2 \times (x - 2/2) - 20(x - 2)$$

$$= 22.5x - 20(x - 1) - 20(x - 2) \quad \text{(Linear)}$$

At $x = 2$, i.e., bending moment at C,

$$M_C = + 22.5 \times 2 - 20(2 - 1) - 20 \times 0$$

$$= + 25 \text{ kNm} \quad \text{(As before)}$$

At $x = 4$, i.e., bending moment at B,

$$M_B = + 22.5 \times 4 - 20(4 - 1) - 20 \times (4 - 2)$$

$$= - 10 \text{ kNm}$$

Point of contraflexure is located between C and B since its value is positive at C and negative at B. Let it be located at a distance of x m from A.

Then, $22.5x - 20(x - 1) - 20(x - 2) = 0$

$\therefore$ $x = 3.43$ m

The bending moment at any section between B and D at a distance of x from end A is given by,

$$M_x = R_A x - 20(x - 1) - 20(x - 2) + R_B (x - 4)$$

$$= 22.5x - 20(x - 1) - 20(x - 2) + R_B (x - 4)$$

(For the left portion of the section, the clockwise moment is taken positive and anticlockwise moment is taken negative)

At $x = 4$, i.e., bending moment at B,

$$M_B = + 22.5 \times 4 - 20(4 - 1) - 20 \times (4 - 2) + 27.5 \times 0$$

$$= - 10 \text{ kNm} \quad \text{(As before)}$$

At $x = 5$, i.e., bending moment at D,

$$M_D = + 22.5 \times 5 - 20(5 - 1) - 20 \times (5 - 2) + 27.5 \times (5 - 4) = 0$$

The bending moment diagram is shown in Figure 8.50 (c).

Example 8.32 For the beam AD shown in Figure 8.51, determine the magnitude of the load P acting at D such that the reaction at supports A and B are equal. Draw the shear force and bending moment diagrams. Also clearly locate the point of contraflexure.

Solution Refer Figure 8.51. Let R_A be the reaction at the support A and R_B the reaction at the support B. Taking moments of the forces about A,

$$R_B \times 6 = P \times 7 + 15 + 25 \times 4 \times (4/2)$$

$$\therefore \quad 6R_B = 7P + 215 \qquad \text{(i)}$$

$$\text{And} \quad R_A + R_B = 25 \times 4 + P = 100 + P$$

$$\text{But} \quad R_A = R_B$$

$$\therefore \quad 2R_B = 100 + P \qquad \text{(ii)}$$

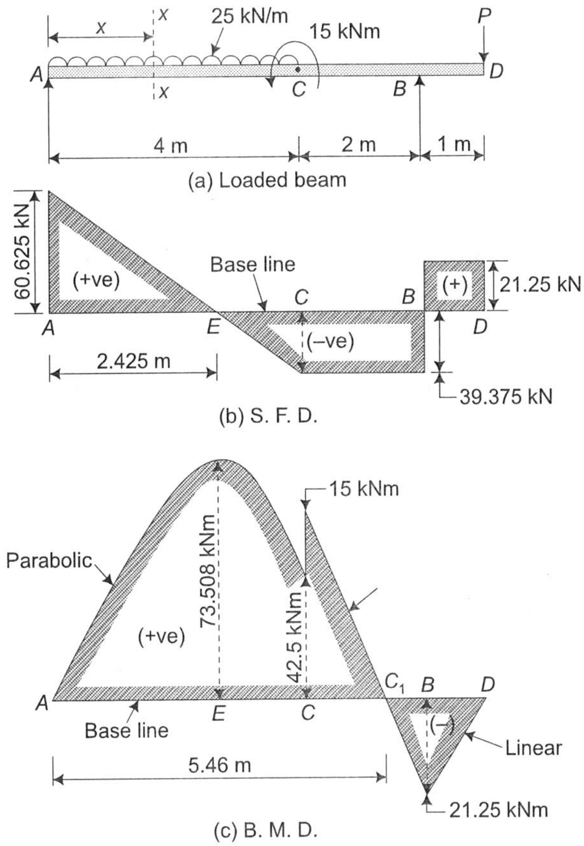

Fig. 8.51

Solving (i) and (ii),

$$(3(100 + P) = 7P + 215,\ 7P - 3P = 300 - 215$$

$$\therefore \qquad P = 85/4 = 21.25 \text{ kN}$$

$$R_A = R_B = (100 + 21.25)/2 = 121.25/2 = 60.625 \text{ kN}$$

S.F.D. Take section $x - x$ at a distance x from the end A between A and C. The shear force is given by,

$$F_x = +R_A - 25x = +60.625 - 25x \quad \text{(Linear)}$$

At $x = 0$, i.e., shear force at A,

$$F_A = +60.625 \text{ kN}$$

At $x = 4$, i.e., shear force at C,

$$F_C = +60.625 - 25 \times 4 = -39.375 \text{ kN}$$

The shear force is positive at A and negative at C, so it will be zero somewhere between A and B.

$$\therefore \qquad +60.625 - 25x = 0$$

$$\therefore \qquad x = 60.625/25 = 2.425 \text{ m.}$$

The shear force is zero at point E which lies at a distance of 2.425 m from A.

The shear force between C and B is given by,

$$F_x = +60.625 - 25 \times 4 = -39.375 \text{ kN} \quad \text{(Constant)}$$

The shear force between B and D is given by,

$$F_x = +60.625 - 25 \times 4 + 60.625 = +21.25 \text{ kN} \quad \text{(Constant)}$$

The shear force diagram is shown in Figure 8.51 (b).

B.M.D. The bending moment at any section $x - x$ between A and C at a distance of x from end A is given by,

$$M_x = +R_A x - 25 \times x \times (x/2)$$

$$= 60.625x - (25x^2/2) \quad \text{(Parabolic)}$$

(For the left portion of the section, the clockwise moment is taken positive and anticlockwise moment is taken negative)

At $x = 0$, i.e., bending moment at A,

$$M_A = 0$$

At $x = 4$, i.e., bending moment at C,

$$M_C = +60.625 \times 4 - 25 \times 4^2/2$$

$$= 42.5 \text{ kN/m}$$

At $x = 2.425$, i.e., bending moment at E,

$$M_E = + 60.625 \times 2.425 - 25 \times 2.425^2/2$$
$$= + 73.508 \text{ kNm} \quad \text{(Maximum)}$$

The bending moment at any section between C and B at a distance of x from end A is given by,

$$M_x = + R_A x - 25 \times 4 \times (x - 4/2) + 15$$
$$= 60.625x - 100(x - 2) + 15 \quad \text{(Linear)}$$

At $x = 4$, i.e., bending moment at C,

$$M_C = 60.625 \times 4 - 100(4 - 2) + 15 = + 57.5 \text{ kNm}$$

At $x = 6$, i.e., bending moment at B,

$$M_B = 60.625 \times 6 - 100(6 - 2) + 15 = - 21.25 \text{ kNm}$$

Point of contraflexure lies between C and B, since its value is positive at C and negative at B. Let the point of contraflexure (C_1) be located at a distance of x m from A.

Then, $$60.625x - 100(x - 2) + 15 = 0$$

$\therefore$ $$x = 5.46 \text{ m}$$

The bending moment at any section between B and D at a distance of x from end A is given by,

$$M_x = + R_A x - 25 \times 4 \times (x - 4/2) + 15 + R_B(x - 6)$$

$\therefore$ $$M_x = 60.625x - 100(x - 2) + 15 + 60.625\,(x - 6) \quad \text{(Linear)}$$

(For the left portion of the section, the clockwise moment is taken positive and anticlockwise moment is taken negative)

At $x = 6$, i.e., bending moment at B,

$$M_B = 60.625 \times 6 - 100\,(6 - 2) + 15 + 60.625 \times 0$$
$$= - 21.25 \text{ kNm} \quad \text{(As before)}$$

At $x = 7$, i.e., bending moment at D,

$$M_D = + 60.625 \times 7 - 100 \times (7 - 2) + 15 + 60.625 \times (7 - 6) = 0$$

The bending moment diagram is shown in Figure 8.51 (c).

Example 8.33 For the beam shown in Figure 8.52, draw the shear force and bending moment diagrams.

Solution Refer Figure 8.52. Let R_A be the reaction at the support A and R_B the reaction at the support B. Taking moments of the forces about A,

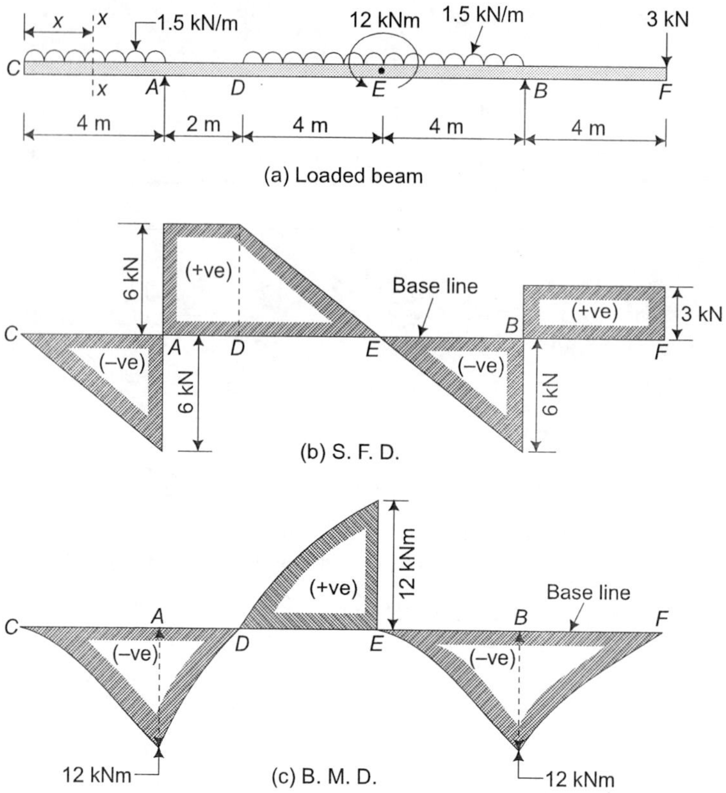

Fig. 8.52

$$R_B \times 10 + 1.5 \times 4 \times (4/2) = 3 \times 14 + 1.5 \times 8 \times [(8/2) + 2] - 12 = 102$$

$$10R_B = 102 - 12 = 90$$

$\therefore$ $$R_B = 90/10 = 9 \text{ kN}$$

And $$R_A + R_B = 1.5 \times 4 + 1.5 \times 8 + 3 = 21$$

$\therefore$ $$R_A = 21 - 9 = 12 \text{ kN}$$

S.F.D. Take section $x - x$ at a distance x from the end C between A and C. The shear force is given by, $F_x = -1.5x$ (Linear)

At $x = 0$, i.e., shear force at C,

$$F_C = 0$$

At $x = 4$, i.e., shear force at A,

$$F_A = -1.5 \times 4 = -6 \text{ kN}$$

The shear force between A and D is given by,

$$F_x = -1.5 \times 4 + R_A = -6 + 12$$

$$= +6 \text{ kN} \quad \text{(Constant)}$$

The shear force between D and B at any section from end D is given by,

$$F_x = -1.5 \times 4 + 12 - 1.5x$$

At $x = 4$, i.e., shear force at E,

$$F_E = -1.5 \times 4 + 12 - 1.5 \times 4 = 0$$

At $x = 8$, i.e., shear force at B,

$$F_B = -1.5 \times 4 + 12 - 1.5 \times 8 = -6 \text{ kN}$$

The shear force between B and F is given by,

$$F_x = -1.5 \times 4 + 12 - 1.5 \times 8 + 9$$

$$= 3 \text{ kN} \quad \text{(Constant)}$$

The shear force diagram is shown in Figure 8.52 (b).

B.M.D. The bending moment at any section $x - x$ between C and A at a distance of x from end C is given by,

$$M_x = -1.5 \times x \times (x/2) = -1.5x^2/2 \quad \text{(Parabolic)}$$

(For the left portion of the section, the anticlockwise moment is taken negative)

At $x = 0$, i.e., bending moment at C,

$$M_C = 0$$

At $x = 4$, i.e., bending moment at A,

$$M_A = -(1.5 \times 4^2)/2 = -12 \text{ kNm}$$

The bending moment at any section between A and D at a distance of x from end A is given by,

$$M_x = -1.5 \times 4 \times [(4/2) + x] + R_A x \quad \text{(Linear)}$$

At $x = 2$, i.e., bending moment at A,

$$M_A = -1.5 \times 4 \times (2 + 0) + 12 \times 0$$

$$= -12 \text{ kNm} \quad \text{(As before)}$$

At $x = 2$, i.e., bending moment at D,

$$M_D = -1.5 \times 4 \times (2 + 2) + 12 \times 2 = 0$$

The bending moment at any section between D and E at a distance of x from end A is given by,

$$M_x = -1.5 \times 4 \times (2 + x) + R_A x - 1.5 \times (x - 2) \times (x - 2)/2 \quad \text{(Parabolic)}$$

At $x = 2$, i.e., bending moment at D,

$$M_D = -1.5 \times 4 \times (2 + 2) + 12 \times 2 - 1.5 \times 0$$

$$= 0 \quad \text{(As before)}$$

At $x = 6$, i.e., bending moment at E,

$$M_E = -1.5 \times 4 \times (2 + 6) + 12 \times 6 - 1.5 \times 4 \times 2$$

$$= +12 \text{ kNm}$$

The bending moment at any section between E and B at a distance of x from end A is given by,

$$M_x = -1.5 \times 4 \times (2 + x) + R_A x - 1.5 \times (x - 2) \times (x - 2)/2 - 12 \quad \text{(Parabolic)}$$

At $x = 6$, i.e., bending moment at E,

$$M_E = -1.5 \times 4 \times (2 + 6) + 12 \times 6 - 1.5 \times 4 \times 2 - 12 = 0$$

At $x = 10$, i.e., bending moment at B,

$$M_B = -1.5 \times 4 \times (2 + 10) + 12 \times 10 - 1.5 \times 8 \times 4 - 12$$

$$= -12 \text{ kNm}$$

The bending moment at any section between B and F at a distance of x from end A is given by,

$$M_x = -1.5 \times 4 \times (2 + x) + R_A x - 1.5 \times 8 \times (x - 6) + R_B \times (x - 10) - 12$$

At $x = 10$ i.e., bending moment at B,

$$M_B = -1.5 \times 4 \times (2 + 10) + 12 \times 10 - 1.5 \times 8 \times 4 + 9 \times 0 - 12$$

$$= -12 \text{ kNm} \quad \text{(As before)}$$

At $x = 14$, i.e., bending moment at F,

$$M_F = -1.5 \times 4 \times (2 + 14) + 12 \times 14 - 1.5 \times 8 \times 8 + 9 \times 4 - 12 = 0$$

The bending moment diagram is shown in Figure 8.52 (c).

8.9 RELATIONS BETWEEN LOAD, SHEAR FORCE, AND BENDING MOMENT

Consider a simply supported beam carrying a uniformly distributed load of intensity *w/m* as shown in Figure 8.53. Take a small element *CD* of length *dx* at a distance of *x* from the left support. Let the element be subjected to shear force *F* and bending moment *M* on its left side (i.e., at *C*) and shear force *F* + *dF* and bending moment *M* + *dM* on its right side (i.e., at *D*).

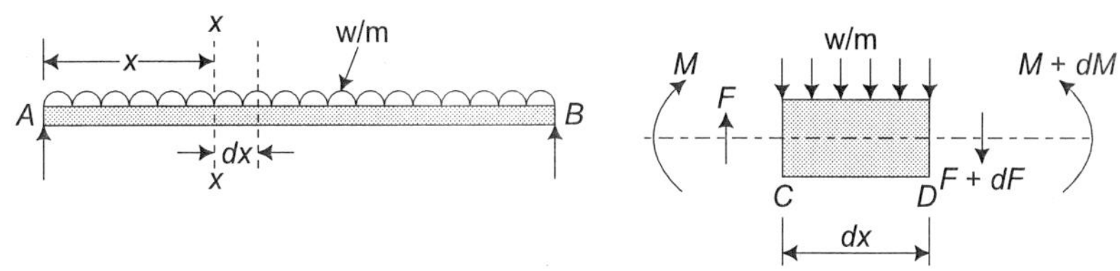

Fig. 8.53

The intensity of loading on the elemental length may be taken as constant. Consider the vertical equilibrium of forces, we have,

$$F - w.dx - (F + dF) = 0$$

$$- dF = w.dx$$

$$\therefore \qquad \frac{dF}{dx} = - w \qquad \text{(i)}$$

Therefore, the rate of change of shear force is equal to the rate of loading.

Now taking moments about the right side *D*, we have,

$$M - w.dx.\frac{dx}{2} + F.dx = M + dM$$

$$- w\frac{(dx)^2}{2} + F.dx = dM$$

Neglecting the term which is the product two differential, i.e.,

$$w\frac{(dx)^2}{2} = 0,$$

we have,

$$F.dx = dM$$

$$F = \frac{dM}{dx} \qquad \text{(ii)}$$

Thus, the shear force is equal to the rate of change of bending moment.

Taking the derivative of (ii), we have,

$$\frac{dF}{dx} = \frac{d^2M}{dx^2} \qquad \text{(iii)}$$

From (i) and (iii), we have,

$$\frac{d^2M}{dx^2} = \frac{dF}{dx} = -w \qquad \text{(iv)}$$

EXERCISE

(i) Objective Type Questions

1. A cantilever has:
   (a) both ends fixed
   (b) both ends simply supported
   (c) one end is fixed and other is free
   (d) both ends are free
2. In an overhanging beam:
   (a) supports are at the ends
   (b) both ends are simply supported
   (c) one end is fixed and other is free
   (d) the supports are not situated at the ends
3. A continuous beam has:
   (a) only two supports at the ends
   (b) one support only
   (c) more than two supports
   (d) none of the above
4. The following beams are the statically determinate beams:
   (a) simply supported beam
   (b) cantilever
   (c) overhanging beam
   (d) none of the above
   (e) all of the above
5. The following beams are the statically indeterminate beams:
   (a) simply supported
   (b) cantilever
   (c) fixed beams
   (d) continuous beams
   (e) (c) & (d)
6. The beams fixed at both of its ends are called:
   (a) simply supported
   (b) encastered beam
   (c) fixed beams
   (d) built-in beam
   (e) beams (b) (c) & (d)
7. The bending moment diagram for a simply supported beam carrying a point load at its centre will be:
   (a) triangle with maximum ordinate at the centre of the beam
   (b) parabola with maximum ordinate at the centre of the beam
   (c) both (a) and (b)
   (d) none of the above
8. The bending moment diagram for a beam carrying u.d.l. over its whole span will be:
   (a) triangle with maximum ordinate at the centre of the beam
   (b) parabola with maximum ordinate at the centre of the beam

(c) both (a) and (b)

(d) none of the above

9. In a cantilever the bending moment at the end section is equal to:

(a) area of the S.F.D. between the free end of the beam and that section

(b) area of the B.M.D. between the free end of the beam and that section

(c) area of the S.F.D. between the fixed end of the beam and that section

(d) none

10. For cantilever beam loaded by u.d.l. B.M. at the fixed end is:

(a) $-wl/2$ (b) $-wl^2$ (c) $-wl$ (d) $-wl^2/2$

11. For cantilever beam loaded by u.d.l. B.M. at the free end is:

(a) $-wl/2$ (b) $-wl^2$ (c) 0 (d) $-wl^2/2$

12. For cantilever beam loaded by gradually varying load, B.M. at the fixed end is:

(a) $-wl/2$ (b) $-wl^2/6$ (c) 0 (d) $-wl^2/2$

13. In a simply supported beam carrying a u.d.l. over the whole span the maximum B.M. is:

(a) $wl/2$ (b) $wl^2/6$ (c) $wl^2/8$ (d) $wl^2/2$

14. Bending moments at the supports in simply supported beam is:

(a) maximum (b) minimum (c) unity (d) zero

15. The maximum B.M. for a simply supported beam under varying distributed load is:

(a) $wl/9$ (b) $wl/9\sqrt{3}$ (c) $wl^2/8$ (d) $wl^2/9\sqrt{3}$

16. The shear force for a simply supported beam under varying distributed load at a distance of $l/\sqrt{3}$ is:

(a) maximum (b) positive (c) zero (d) negative

17. The maximum B.M. for a simply supported beam carrying a load whose intensity varies from zero at each end to the w per unit run at the mid span under varying distributed load is:

(a) $wl^2/12$ (b) $wl^2/9$ (c) $wl^2/8$ (d) $wl^2/9\sqrt{3}$

18. Which of the following statement is correct:

(a) shear force is the first derivative of B.M.

(b) shear force is the second derivative of B.M.

(c) shear force is the first derivative of intensity of load

(d) shear force is the second derivative of intensity of load

19. If the S.F.D. of a simply supported beam is parabolic, then load on the beam is:

(a) linearly varying distributed load (b) u.d.l.

(c) concentrated load (d) none

20. For a simply supported beam carrying a point load at its centre, S.F.D. is:

(a) a rectangle (b) a triangle

(c) two equal and opposite triangles (d) two equal and opposite rectangles

21. For a simply supported beam carrying a point load at its centre, B.M.D. diagram is:
    (a) a rectangle (b) a triangle
    (c) two equal and opposite triangles (d) two equal and opposite rectangles
22. For a simply supported beam carrying a u.d.l. over its whole span, S.F.D. diagram is:
    (a) a rectangle (b) a triangle
    (c) two equal and opposite triangles (d) two equal and opposite rectangles
23. The B.M. is maximum at a section where S.F. is:
    (a) changing sign (b) maximum (c) zero (d) minimum
    (e) none
24. The B.M. at a section where S.F. is zero is:
    (a) minimum (b) maximum
    (c) zero (d) either maximum or minimum
    (e) none
25. The point of contraflexure means:
    (a) the point of inflexion (b) the point of virtual hinge
    (c) both (a) & (b) (d) none
26. The point of contraflexure occurs in:
    (a) simply supported beam (b) continuous beam
    (c) both (a) & (b) (d) overhanging beam
27. The S.F. and B.M. for a cantilever at its free end when it carries u.d.l. over its whole span is:
    (a) 0 (b) 1 (c) maximum (d) none
28. Which one is correct:
    (a) at the point of contraflexure, B.M. is maximum
    (b) in case of beam, maximum B.M. always occurs at the mid span
    (c) in case of beam, maximum B.M. is zero at the two ends
    (d) none of the above
29. In a cantilever with u.d.l. the S.F. varies as per:
    (a) linear law (b) parabolic law (c) cubic law (d) none
30. Negative bending moment is also called:
    (a) sagging (b) hogging (c) both (a) & (b) (d) none of the above
31. Positive bending moment is also called:
    (a) sagging (b) hogging (c) both (a) & (b) (d) none of the above
32. For a simply supported beam carrying a u.d.l. of intensity w over its whole span, S.F. at the supported ends is:
    (a) $\pm wl$ (b) $\pm wl/8$ (c) $\pm wl/4$ (d) $\pm wl/2$

ANSWERS

1. (c)	2. (d)	3. (c)	4. (e)	5. (e)	6. (e)	7. (a)	8. (b)	9. (a)	10. (d)
11. (c)	12. (b)	13. (c)	14. (d)	15. (d)	16. (c)	17. (a)	18. (a)	19. (a)	20. (d)
21. (b)	22. (c)	23. (a)	24. (d)	25. (c)	26. (d)	27. (a)	28. (c)	29. (a)	30. (b)
31. (a)	32. (d).								

(ii) Review Questions

1. Define a beam. What are the different types of beams?
2. What are the various types of loads which act on a beam?
3. Define shear force and bending moment and give the sign conventions for drawing shear force and bending moment diagrams.
4. What is the point of contraflexure?
5. State the relation between load, shear force, and bending moment at any section of a beam.
6. Draw the S.FD. and B.M.D. for a cantilever of length l for the following cases:
   (i) when a point load acts on the free end,
   (ii) when a u.d.l. of intensity w acts over the entire length, and
   (iii) when a gradually varying load acts on it which vary from zero at the free end to w per unit length at the fixed end.
7. Draw the S.FD. and B.M.D. for a simply supported beam of length l for the following cases:
   (i) when a point load acts at its middle point,
   (ii) when a point load acts eccentrically, and
   (iii) when a u.d.l. of intensity w acts over the entire length.

(iii) Numerical Problems

1. A cantilever beam of length 8 m carries point load of 2 kN at the free end, 4 kN, and 3 kN at a distance of 3 m and 6 m from the free end. Draw the S.F.D. and B.M.D.
   [**Ans.** F_{max} = + 9 kN, M_{max} = – 42 kNm]
2. A cantilever beam of length 8 m carries a uniformly distributed load of intensity 6 kN/m over a distance of 5 m from the support. Draw the shear force and bending moment diagrams.
   [**Ans.** F_{max} = + 30 kN, M_{max} = – 75 kNm]
3. A cantilever beam of length 4 m carries a uniformly distributed load of intensity 5 kN/m over a distance of 3 m extending from the free end. Draw the shear force and bending moment diagrams.
   [**Ans.** F_{max} = + 15 kN, M_{max} = – 37.5 kNm]

4. A cantilever beam of length 6 m carries a uniformly distributed load of intensity 3 kN/m over a length of 3 m at a distance of 1.5 m from the fixed end. Draw the shear force and bending moment diagrams. [**Ans.** F_{max} = + 9 kN, M_{max} = – 27 kNm]
5. A cantilever beam of length 4 m carries a uniformly distributed load of intensity 8 kN/m over the whole span of the beam and a point load of 4 kN at the free end. Draw the shear force and bending moment diagrams. [**Ans.** F_{max} = + 36 kN, M_{max} = – 80 kNm]
6. A cantilever beam of length 6 m carries a uniformly distributed load of intensity 3 kN/m over a length of 3 m from the free end. It also carries a point load of 4 kN at a distance of 1 m from the free end. Draw the shear force and bending moment diagrams. [**Ans.** F_{max} = + 13 kN, M_{max} = – 60.5 kNm]
7. A cantilever beam of length 10 m carries two point loads of 6 kN and 8 kN at the free end and at a distance of 8 m from the free end respectively. A uniformly distributed load of intensity 3 kN/m also acts over a length of 5 m at a distance of 1 m from the free end. Draw the shear force and bending moment diagrams. [**Ans.** F_{max} = + 29 kN, M_{max} = – 173.5 kNm]
8. A cantilever beam of length 5 m carries four point loads of 3 kN, 4 kN, 3 kN, and 2 kN at the free end and at a distance of 1 m, 2 m, and 4 m from the free end respectively. A uniformly distributed load of intensity 4 kN/m also acts over a length of 2 m at a distance of 1 m from the free end. Draw the shear force and bending moment diagrams. [**Ans.** F_{max} = + 20 kN, M_{max} = – 66 kNm]
9. A cantilever beam of length 6 m carries two point loads of 2 kN and 4 kN at the free end and at a distance of 2 m from the free end respectively. In addition to this the cantilever beam also carries a uniformly distributed load of intensity 2 kN/m over its entire length. Draw the shear force and bending moment diagrams. [**Ans.** F_{max} = + 18 kN, M_{max} = – 64 kNm]
10. A cantilever beam of length 3 m carries a gradually varying load from zero at the free end to 1 kN/m at the fixed end. Draw the shear force and bending moment diagrams. [**Ans.** F_{max} = + 1.5 kN, M_{max} = – 1.5 kNm]
11. A cantilever beam of length 5 m carries two point loads of 10 kN and 20 kN at the free end and at a distance of 2 m from the free end respectively. In addition to this the cantilever beam also carries a uniformly distributed load of intensity 4 kN/m over a length of 2 m from the fixed support. Draw the shear force and bending moment diagrams. [**Ans.** F_{max} = + 38 kN, M_{max} = – 118 kNm]
12. A simply supported beam of length 5 m carries concentrated load of 3 kN and 6 kN at distances of 1.5 m and 3.5 m from the left end. Draw the shear force and bending moment diagrams. [**Ans.** M_{max} = + 7.65 kNm]
13. A simply supported beam of length 8 m carries concentrated load of 3 kN, 7 kN, and 5 kN at distances of 2 m, 4 m, and 6 m from the left end. Draw the shear force and bending moment diagrams. [**Ans.** M_{max} = + 29 kNm]

14. A simply supported beam of 5 m length carries uniformly distributed load of 2.5 kN/m over a length of 4 m from the left support. Draw the shear force and bending moment diagrams. Also determine the maximum bending moment on the section.

[**Ans.** M_{max} = + 7.2 kNm]

15. A simply supported beam of 8 m length carries uniformly distributed load of 20 kN/m over a length of 4 m at a distance of 1 m from the left support. This beam also carries a point load of 80 kN at a distance of 1 m from the right support. Draw the shear force and bending moment diagrams. Also determine the maximum bending moment on the section.

[**Ans.** M_{max} = + 150 kNm]

16. A simply supported beam of 10 m length carries uniformly distributed load of 5 kN/m over a length of 4 m at a distance of 2 m from the left support. In addition to this beam also carries point loads of 25 kN and 20 kN at a distance of 2 m and 6 m from the left support. Draw the shear force and bending moment diagrams. Also determine the maximum bending moment on the section. [**Ans.** M_{max} = + 102.5 kNm]

17. A simply supported beam of 8 m length carries a uniformly distributed load of 20 kN/m over a length of 3 m from the left support and another uniformly distributed load of 10 kN/m over a length of 2 m from the right support. Draw the shear force and bending moment diagrams. Also determine the maximum bending moment on the section.

[**Ans.** M_{max} = + 62.5 kNm]

18. A simply supported beam of 8 m length carries uniformly distributed load of 20 kN/m over a length of 4 m from the right support. In addition to this beam also carries point loads of 40 kN and 60 kN at a distance of 4 m and 6 m from the right support. Draw the shear force and bending moment diagrams. Also determine the maximum bending moment on the section. [**Ans.** M_{max} = + 220 kNm]

19. A simply supported beam of 10 m length carries uniformly distributed load of 4 kN/m over a length of 4 m at a distance of 2 m from the left support. In addition to this beam also carries point loads of 20 kN, 16 kN, and 8 kN at a distance of 2 m, 6 m, and 8 m from the left support. Draw the shear force and bending moment diagrams. Also determine the maximum bending moment on the section. [**Ans.** M_{max} = + 90.32 kNm]

20. Draw the shear force and bending moment diagrams for a simply supported beam shown in Figure 8.54. [**Ans.** M_{max} = + 41.25 kNm]

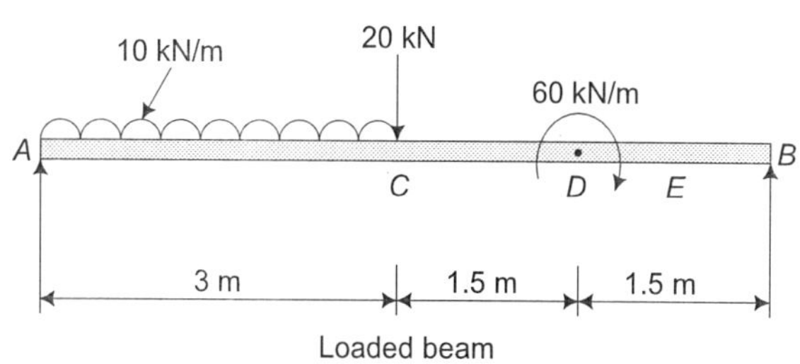

Fig. 8.54

21. Draw the shear force and bending moment diagrams for a simply supported beam shown in Figure 8.55. [**Ans.** M_{max} = + 37.05 kNm]

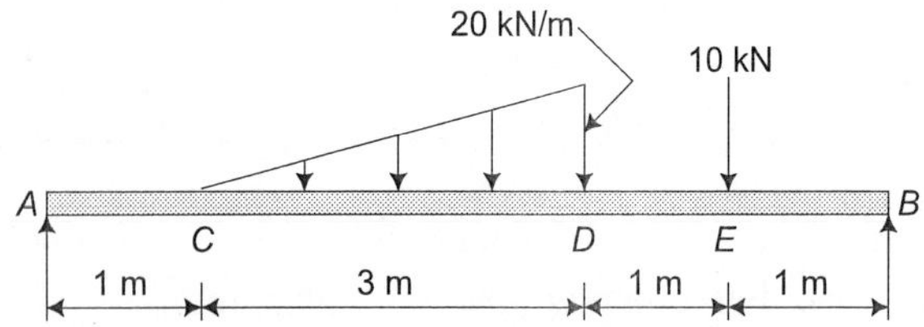

Loaded beam

Fig. 8.55

22. A simply supported beam of length 8 m rests on supports 6 m apart, the right hand end overhangs by 2 m. The beam carries a uniformly distributed load of 3 kN/m over the whole span. Draw the shear force and bending moment diagrams and also determine the point of contraflexure if any? [**Ans.** M_{max} = + 10.66 kNm, 5.33 m from left support]

23. Draw the shear force and bending moment diagrams for the overhanging beam shown in Figure 8.56. Also locate the point of contraflexure.

[**Ans.** M_{max} = + 17.698 kNm, 1.5 m from the right support]

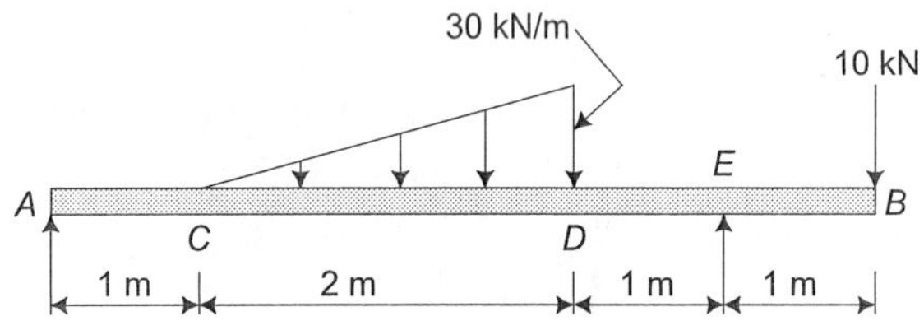

Loaded beam

Fig. 8.56

24. A simply supported beam carries inclined loads as shown in Figure 8.57. Draw the shear force and bending moment diagrams. [**Ans.** M_{max} = + 3.896 kNm]

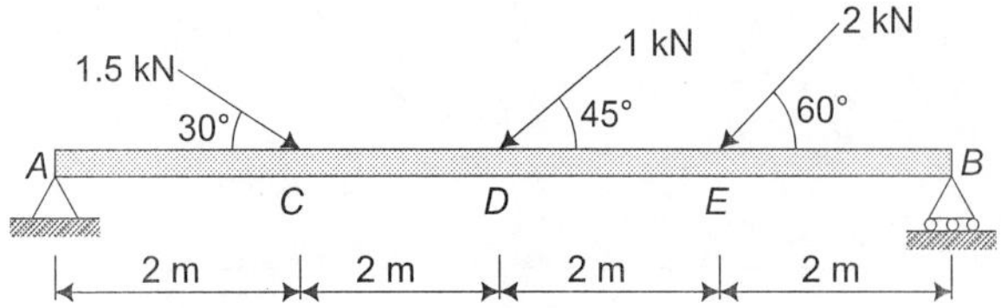

Loaded beam

Fig. 8.57

25. A simply supported beam of length 4 m carries a uniformly distributed load of 4 kN/m over a length of 0.5 m from the left end and it caries another uniformly distributed load of 4 kN/m over a length of 1 m from the right end. In addition to these the beam also carries concentrated loads of 8 kN and 20 kN at a distance of 0.5 m and 1.5 m respectively from the left end. Draw the shear force and bending moment diagrams.

[**Ans.** M_{max} = + 9.7 kNm]

26. A simply supported beam of length 7 m rests on supports 6 m apart, the right hand end overhangs by 1 m. The beam carries a uniformly distributed load of 45 kN/m over a length of 4 m extending the left support. A clockwise couple acts at a distance of 4 m from the left support and a concentrated load of 37.5 kN also acts at the extreme right of the beam. Draw the shear force and bending moment diagrams and also determine the point of contraflexure if any? [**Ans.** M_{max} = + 131.41 kNm, 1.526 m from the right support]

27. A simply supported beam of length 6 m rests on supports 4 m apart, both the right hand and left hand ends are overhanging by 1 m. The beam carries a uniformly distributed load of 55 kN/m over a length of 4 m extending the left support. Draw the shear force and bending moment diagrams. [**Ans.** M_{max} = + 17.5 kNm]

Appendix

Table A-1.1 Saturated water and steam properties based on temperature

Temp. T	Pressure p	Specific volume m³/kg		Specific enthalpy kJ/kg		Specific entropy kJ/kgK	
(°C)	(bar)	v_f	v_g	h_f	h_g	s_f	s_g
0.01	0.006113	0.001000	206.132	0.00	2501.3	0.0000	9.1562
05	0.008721	0.001000	106.377	41.99	2519.7	0.1510	8.9007
10	0.012276	0.001000	77.925	62.98	2528.9	0.2245	8.7813
15	0.017051	0.001001	77.925	62.98	2528.9	0.2245	8.7813
20	0.023385	0.001002	57.790	83.94	2538.1	0.2966	8.6671
25	0.031691	0.001003	43.359	104.87	2547.2	0.3673	8.5579
30	0.042461	0.001004	32.893	125.77	2556.2	0.4369	8.4533
35	0.056280	0.001006	25.216	146.66	2565.3	0.5052	8.3530
40	0.073837	0.001008	19.523	167.54	2574.3	0.5724	8.2569
45	0.095932	0.001010	15.258	188.42	2583.2	0.6386	8.1647
50	0.1235	0.001012	12.032	209.31	2592.1	0.7037	8.0762
55	0.15758	0.001015	9.568	230.20	2600.9	0.7679	7.9912
60	0.19941	0.001017	7.671	251.11	2609.6	0.8311	7.9095
65	0.25033	0.001020	6.197	272.03	2618.2	0.8934	7.8309
70	0.31188	0.001023	5.042	292.96	2626.8	0.9548	7.7552
75	0.38578	0.001026	4.131	313.91	2635.3	1.0154	7.6824
80	0.47390	0.001029	3.407	334.88	2643.7	1.0752	7.6121
85	0.57834	0.001032	2.828	355.88	2651.9	1.1342	7.5444
90	0.70139	0.001036	2.361	376.90	2660.1	1.1924	7.4790
95	0.84554	0.001040	1.982	397.94	2668.1	1.2500	7.4158
100	1.0135	0.001044	1.673	419.00	2676.0	1.3068	7.3548
105	1.2082	0.001047	1.4194	440.13	2683.8	1.3629	7.2958
110	1.4328	0.001052	1.2102	461.27	2691.5	1.4184	7.2386
115	1.6906	0.001056	1.0366	482.46	2699.0	1.4733	7.1832
120	1.9853	0.001060	0.8919	503.69	2706.3	1.5275	7.1295
125	2.3210	0.001065	0.7706	524.96	2713.5	1.5812	7.0774
130	2.7010	0.001070	0.6685	546.29	2720.5	1.6343	7.0269
135	3.1300	0.001075	0.5822	567.67	2727.3	1.6869	6.9777

Contd..

Contd..

140	3.6130	0.001080	0.5089	589.11	2733.9	1.7390	6.9298
145	4.1540	0.001085	0.4463	610.61	2740.3	1.7906	6.8832
150	4.7590	0.001090	0.3928	632.18	2746.4	1.8417	6.8378
155	5.4310	0.001096	0.3468	653.82	2752.4	1.8924	6.7934
160	6.1780	0.001102	0.3071	675.53	2758.1	1.9426	6.7501
165	7.0050	0.001108	0.2727	697.32	2763.5	1.9924	6.7078
170	7.9170	0.001114	0.2428	719.20	2768.7	2.0418	6.6663
175	8.9200	0.001121	0.2168	741.16	2773.6	2.0909	6.6256
180	10.022	0.001127	0.1941	763.21	2778.2	2.1395	6.5857
185	11.227	0.001134	0.1741	785.36	2782.4	2.1878	6.5464
190	12.544	0.001141	0.1565	807.61	2786.4	2.2358	6.5078
195	13.978	0.001149	0.1411	829.96	2790.0	2.2835	6.4697
200	15.538	0.001156	0.1274	852.43	2793.2	2.3308	6.4322
205	17.230	0.001164	0.1152	875.03	2796.0	2.3779	6.3951
210	19.063	0.001173	0.1044	897.75	2798.5	2.4247	6.3584
215	21.042	0.001181	0.0948	920.61	2800.5	2.4713	6.3221
220	23.178	0.001190	0.0862	943.61	2802.1	2.5177	6.2860
225	25.477	0.001199	0.0785	966.77	2803.3	2.5639	6.2502
230	27.949	0.001209	0.0716	990.10	2803.9	2.6099	6.2146
235	30.601	0.001219	0.0654	1013.61	2804.1	2.6557	6.1791
240	33.442	0.001229	0.0598	1037.31	2803.8	2.7015	6.1436
245	36.482	0.001240	0.0547	1061.21	2802.9	2.7471	6.1083
250	39.730	0.001251	0.0501	1085.34	2801.5	2.7927	6.0729
255	43.195	0.001263	0.0460	1109.72	2799.5	2.8382	6.0374
260	46.886	0.001276	0.0422	1134.35	2796.9	2.8837	6.0018
265	50.813	0.001289	0.0388	1159.27	2793.6	2.9293	5.9661
270	54.987	0.001302	0.0356	1184.49	2789.7	2.9750	5.9301
275	59.418	0.001317	0.0328	1210.05	2785.0	3.0208	5.8937
280	64.117	0.001332	0.0302	1235.97	2779.5	3.0667	5.8570
285	69.094	0.001348	0.0278	1262.29	2773.3	3.1129	5.8198
290	74.360	0.001366	0.0256	1289.04	2766.1	3.1593	5.7821
295	79.928	0.001384	0.0235	1316.27	2758.0	3.2061	5.7436
300	85.810	0.001404	0.0217	1344.01	2748.9	3.2533	5.7044
305	92.018	0.001425	0.0199	1372.33	2738.7	3.3009	5.6642
310	98.566	0.001447	0.0184	1401.29	2727.3	3.3492	5.6229
315	105.47	0.001472	0.0169	1430.97	2714.4	3.3981	5.5803

Contd..

Contd..

320	112.74	0.001499	0.0155	1461.45	2700.1	3.4479	5.5361
330	128.45	0.001561	0.0130	1525.29	2665.8	3.5506	5.4416
340	145.86	0.001638	0.0108	1594.15	2622.0	3.6593	5.3356
350	165.14	0.001740	0.0088	1670.54	2563.9	3.7776	5.2111
360	186.51	0.001892	0.0069	1760.48	2481.0	3.9146	5.0525
370	210.28	0.002213	0.0049	1890.37	2332.1	4.1104	4.7972
374.14	220.89	0.003155	0.003155	2099.26	2099.3	4.4297	4.4297

Table A-1.2 Saturated water and steam properties based on pressure

Pressure *P*	*Temp.* *T*	*Specific volume* m³/kg		*Specific enthalpy* kJ/kg		*Specific entropy* kJ/kgK	
(bar)	(°C)	v_f	v_g	h_f	h_g	s_f	s_g
0.006113	0.01	0.001000	206.132	0.00	2501.3	0.0000	9.1562
0.01	6.98	0.001000	129.208	29.29	2514.2	0.1059	8.9756
0.015	13.03	0.001001	87.98	54.7	2525.3	0.1956	8.8278
0.020	17.50	0.001001	67.004	73.47	2533.5	0.2607	8.7236
0.025	21.08	0.001002	54.254	88.47	2540.0	0.3120	8.6431
0.030	24.08	0.001003	45.665	104.03	2545.5	0.3545	8.5775
0.040	28.96	0.001004	34.80	121.44	2554.4	0.4226	8.4746
0.050	32.88	0.001005	28.193	137.79	2561.4	0.4763	8.3950
0.075	40.29	0.001008	19.238	168.77	2574.8	0.5763	8.2514
0.100	45.81	0.001010	14.674	191.81	2584.6	0.6492	8.1501
0.150	53.97	0.001014	10.022	225.91	2599.1	0.7548	8.0084
0.200	60.06	0.001017	7.649	251.38	2609.7	0.8319	7.9085
0.250	64.97	0.001020	6.204	271.90	2618.2	0.8930	7.8313
0.300	69.10	0.001022	5.229	289.21	2626.3	0.9439	7.7686
0.400	75.87	0.001026	3.993	317.55	2636.7	1.0258	7.6700
0.500	81.33	0.001030	3.240	340.47	2645.9	1.0910	7.5939
0.750	91.77	0.001037	2.217	384.36	2663.0	1.2129	7.4563
1.000	99.62	0.001043	1.6940	417.44	2675.5	1.3025	7.3593
1.01325	100.0	0.001043	1.6735	419.07	2675.7	1.3069	7.3545
1.25	105.99	0.001048	1.3749	443.30	2685.3	1.3739	7.2843
1.50	111.37	0.001053	1.1593	467.08	2693.5	1.4335	7.2232
1.75	116.06	0.001057	1.0036	486.97	2700.5	1.4848	7.1717
2.00	120.23	0.001061	0.8857	504.68	2706.6	1.5300	7.1371
2.25	124.0	0.001064	0.7933	520.69	2712.0	1.5705	7.0878

Contd..

Contd..

2.50	127.43	0.001067	0.7187	535.34	2716.9	1.6072	7.0526
2.75	130.60	0.001070	0.6573	548.87	2721.3	1.6417	7.0208
3.00	133.55	0.001073	0.6058	561.45	2725.3	1.6717	6.9918
3.25	136.30	0.001076	0.6520	573.26	2729.0	1.7005	6.9651
3.50	138.88	0.001079	0.5243	584.31	2732.4	1.7274	6.9404
3.75	141.32	0.001081	0.4914	594.79	2735.6	1.7527	6.9174
4.00	143.63	0.001084	0.4625	604.73	2738.5	1.7766	6.8958
4.50	147.93	0.001088	0.4140	623.24	2743.9	1.8206	6.8565
5.00	151.86	0.001093	0.3749	640.21	2748.7	1.8606	6.8212
5.50	155.48	0.001097	0.3427	655.91	2752.9	1.8972	6.7892
6.00	158.85	0.001101	0.3157	670.45	2756.8	1.9311	6.7600
6.50	162.01	0.001104	0.2927	684.26	2760.3	1.9627	6.7330
7.00	164.97	0.001108	0.2729	697.20	2763.5	1.9922	6.7080
7.50	170.43	0.001111	0.2556	709.45	2766.4	2.0199	6.6846
8.00	170.43	0.001115	0.2404	721.10	2769.1	2.0461	6.6627
8.50	172.96	0.001118	0.2270	732.20	2771.6	2.0709	6.6421
9.00	175.38	0.001121	0.2150	742.82	2773.9	2.0946	6.6225
9.50	177.69	0.001124	0.2042	753.00	2776.1	2.1171	6.6040
10.0	179.91	0.001127	0.1944	762.79	2778.1	2.1386	6.5864
11.0	184.09	0.001133	0.17751	781.21	2781.7	2.1791	6.5535
12.0	187.99	0.001139	0.1633	798.64	2784.8	2.2165	6.5233
13.0	181.64	0.001144	0.15125	814.91	2787.6	2.2514	6.4653
14.0	195.07	0.001149	0.14084	830.29	2790.0	2.2842	6.4692
15.0	198.32	0.001154	0.13177	844.87	2792.1	2.3150	6.4448
17.5	205.76	0.001166	0.11349	878.48	2796.4	2.3851	6.3895
20.0	212.42	0.001177	0.09963	908.77	2799.5	2.4473	6.3408
22.5	218.45	0.001187	0.08875	936.48	2801.7	2.5034	6.2971
25.0	223.99	0.001197	0.07998	962.09	2803.1	2.5546	6.2574
27.5	229.12	0.001307	0.07275	985.97	2803.9	2.6018	6.2208
30.0	233.90	0.001216	0.06668	1008.41	2804.1	2.6456	6.1869
32.0	238.38	0.001226	0.06152	1029.60	2804.0	2.6866	6.1551
35.0	242.60	0.001235	0.05707	1049.73	2803.4	2.7252	6.1252
40.0	250.40	0.001252	0.04978	1087.29	2801.4	2.7963	6.0700
50.0	263.99	0.001286	0.03944	1154.21	2794.3	2.9201	5.9733
60.0	275.64	0.001319	0.02144	1213.32	2784.3	3.0266	5.8891
70.0	285.88	0.001351	0.02737	1266.97	2772.1	3.1210	5.8132

Contd..

Contd..

80.0	295.06	0.001384	0.02351	1316.61	2757.9	3.2067	5.7431
90.0	303.40	0.001418	0.02048	1363.23	2742.1	3.2857	5.6771
100.0	311.06	0.001452	0.01803	1407.57	2724.7	3.3595	5.6140
110.0	318.15	0.001489	0.01599	1450.05	2705.6	3.4294	5.5527
120.0	324.75	0.001527	0.01426	1491.24	2684.8	3.4961	5.4923
130.0	330.93	0.001567	0.01278	1531.46	2662.2	3.5604	5.4323
140.0	663.75	0.001611	0.01149	1571.08	2637.5	3.6231	5.3716
150.0	342.24	0.001658	0.01034	1610.45	2610.5	3.6847	5.3097
160.0	347.43	0.001711	0.00931	1650.00	2580.6	3.7460	5.2454
170.0	352.37	0.001770	0.00836	1692.25	2547.2	3.8078	5.1776
180.0	357.06	0.001840	0.00749	1731.97	2509.1	3.8713	5.1044
190.0	361.54	0.001924	0.00666	1776.43	2464.5	3.9387	5.0227
200.0	365.81	0.002035	0.00583	1826.18	2409.7	4.0137	4.9269
210.0	369.89	0.002206	0.00495	1888.30	2334.7	4.1073	4.8015
220.0	373.80	0.002808	0.00353	2034.92	2159.0	4.3307	4.5224
220.9	374.14	0.003155	0.00315	0.00315	2099.3	4.4297	4.4297

Table A-1.3 Superheated steam properties

p = 0.1 bar				p = 0.5 bar			
T (°C)	v m^3/kg	h kJ/kg	s kJ/kgK	T (°C)	v m^3/kg	h kJ/kg	s kJ/kgK
45.81	14.674	2584.6	8.1501				
50	16.869	2592.6	8.1749	81.33	3.240	2645.9	7.5939
100	17.196	2687.5	8.4479	100	3.418	2682.5	7.6947
150	19.513	2783.0	8.6881	150	3.889	2780.1	7.9400
200	21.825	2879.5	8.9037	200	4.356	2877.6	8.1579
250	24.136	2977.3	9.1002	250	4.821	2976.0	8.2555
300	26.445	3076.5	9.2812	300	5.284	3075.5	8.5372
400	31.063	3279.5	9.6076	400	6.209	2178.9	8.8641
500	35.679	3489.0	9.8977	500	7.134	3488.6	9.1545
600	40.295	3705.4	10.1608	600	8.058	3705.1	9.4177
700	44.911	3928.7	10.4028	700	8.981	3928.5	9.6599
800	49.526	4159.1	10.6281	800	9.904	4158.9	9.8852
900	54.141	4396.4	10.8395	900	10.828	4396.3	10.0967
1000	58.757	4640.6	11.0392	1000	11.751	4640.5	10.2964
1100	63.382	4891.2	11.2287	1100	12.674	4891.1	10.4858
1200	67.987	5147.8	11.4090	1200	13.597	5147.7	10.6662
1300	72.603	5409.7	11.5810	1300	14.521	5409.6	10.8382

Contd..

Contd..

p = 1 bar				p = 2 bar			
T (°C)	v m³/kg	h kJ/kg	s kJ/kgK	T (°C)	v m³/kg	h kJ/kg	s kJ/kgK
99.62	1.6940	2675.5	7.3593				
100	1.6958	2676.2	7.3614	120.23	0.88573	2706.6	7.1271
150	1.9364	2776.4	7.6133	150	0.95964	2768.8	7.2795
200	2.1723	2875.3	7.8342	200	1.08034	2870.5	7.5066
250	2.4060	2974.3	8.0332	250	1.19880	2971.0	7.7085
300	2.6388	3074.3	8.2157	300	1.31616	3071.8	7.89226
400	3.1026	3278.1	8.5434	400	1.54930	3276.5	8.2217
500	3.5655	3488.1	8.8341	500	1.78139	3487.0	8.5132
600	4.0278	3704.7	0.0975	600	2.01297	3704.0	8.7769
700	4.4899	3928.2	9.3397	700	2.24426	3927.7	9.0194
800	4.9517	4158.7	9.5652	800	2.74539	4158.3	9.2450
900	5.4135	4396.1	9.7767	900	2.70643	4395.8	6.4565
1000	5.8753	4640.3	9.9764	1000	2.93740	4640.0	9.6563
1100	6.3370	4890.9	10.1658	1100	3.16834	4890.7	9.8458
1200	6.7986	5147.6	10.3462	1200	3.39927	5447.3	10.0262
1300	7.2603	5409.5	10.5182	1300	3.63018	5409.3	10.1982

p = 5 bar				p = 6 bar			
T (°C)	v m³/kg	h kJ/kg	s kJ/kgK	T (°C)	v m³/kg	h kJ/kg	s kJ/kgK
151.86	0.37489	2748.7	6.8212	158.55	0.31567	2756.8	6.7600
200	0.42492	2855.4	7.0592	200	0.35202	2850.1	6.9665
250	0.47436	2960.7	7.2708	250	0.39383	2957.2	7.1816
300	0.52256	3064.2	7.4598	300	0.43437	3061.6	7.3723
350	0.57012	3167.6	7.6328	350	0.47424	3165.7	7.5463
400	0.61725	3271.8	7.7937	400	0.51372	3270.2	7.7078
500	0.71093	3483.8	8.0872	500	0.59199	3482.7	8.0020
600	0.80406	3701.7	8.3521	600	0.66974	3700.9	8.2673
700	0.89691	3926.0	8.5952	700	0.74720	3925.4	8.5107
800	0.98959	4157.0	8.8211	800	0.82450	4156.5	8.7367
900	1.08217	4394.7	9.0329	900	0.90169	4394.4	8.9485
1000	1.17469	4639.1	9.2328	1000	0.97883	4638.8	9.1484
1100	1.26718	4889.9	9.4224	1100	1.05594	4889.6	9.3381
1200	1.35964	5146.6	9.6028	1200	1.13302	5146.3	9.5185
1300	1.45210	5408.6	9.7749	1300	1.21009	5408.3	9.6906

Contd..

Contd..

p = 8 bar				*p* = 10 bar			
T (°C)	*v* m^3/kg	*h* kJ/kg	*s* kJ/kgK	*T* (°C)	*v* m^3/kg	*h* kJ/kg	*s* kJ/kgK
170.43	0.24043	2769.1	6.6627	179.91	0.19444	2778.1	6.5864
200	0.26080	2839.2	6.8158	200	0.20596	2827.9	6.6939
250	0.29314	2950.0	7.0384	250	0.23268	2942.6	6.9246
300	0.32411	3056.4	7.22327	300	0.25794	3051.2	7.1228
350	0.35439	3161.7	7.1088	350	0.28247	3157.7	7.3010
400	0.38426	2167.1	7.5715	400	0.30659	3263.9	7.4650
500	0.44331	3480.6	7.8672	500	0.35411	3478.4	7.7621
600	0.50184	3699.4	8.1332	600	0.40109	3697.9	8.0289
700	0.56007	3924.3	8.3770	700	0.44779	3923.1	8.2731
800	0.61813	4155.7	8.6033	800	0.49432	4154.8	8.4996
900	0.67610	4393.6	8.8153	900	0.54075	4392.9	8.7118
1000	0.73401	4638.2	9.0153	1000	0.58712	4637.6	8.9119
1100	0.79188	4889.1	9.2049	1100	0.63345	4888.5	9.1016
1200	0.84974	5145.8	9.3854	1200	0.67977	5445.4	9.2821
1300	0.90758	5407.9	9.5575	1300	0.72608	5407.4	9.4542

p = 12 bar				*p* = 14 bar			
T (°C)	*v* m^3/kg	*h* kJ/kg	*s* kJ/kgK	*T* (°C)	*v* m^3/kg	*h* kJ/kg	*s* kJ/kgK
187.99	0.16333	2784.8	6.5233	195.07	0.14084	2790.0	6.4692
200	0.17930	2815.9	6.5898	200	0.14302	2803.3	6.4975
250	0.19235	2935.0	6.8293	250	0.16350	2927.2	6.7467
300	0.21382	3045.8	7.0316	300	0.18228	3040.4	6.9533
350	0.23452	3153.6	7.2120	350	0.20026	3149.5	7.1359
400	0.25480	3260.7	7.3773	400	0.21780	3257.4	7.3025
500	0.29463	3476.3	7.6758	500	0.25215	3474.1	7.6026
600	0.33393	3696.3	7.9434	600	0.28596	3694.8	7.8710
700	0.37294	3922.0	8.1881	700	0.31947	3920.9	8.1160
800	0.41177	4153.9	8.4149	800	0.35281	4153.0	8.3431
900	0.45051	4392.2	8.6272	900	0.38606	4391.5	8.5555
1000	0.48919	4637.0	8.8274	1000	0.41924	4636.4	8.7558
1100	0.52783	4888.0	9.0171	1100	0.45239	4887.5	8.9456
1200	0.56646	5144.9	9.1977	1200	0.48552	5144.4	9.1262
1300	0.60507	5406.9	9.3698	1300	0.51864	5406.5	9.2983

Contd..

Contd..

p = 16 bar				p = 18 bar			
T (°C)	v m³/kg	h kJ/kg	s kJ/kgK	T (°C)	v m³/kg	h kJ/kg	s kJ/kgK
201.4	0.12380	2794.0	6.4217	207.15	0.11042	2797.1	6.3793
225	0.13287	2857.2	6.5518	225	0.11673	2846.7	6.4807
250	0.14184	2919.2	6.6732	250	0.12497	2911.0	6.6066
300	0.15862	3034.8	6.8844	300	0.14021	3029.2	6.8226
350	0.17456	3145.4	7.0693	350	0.15457	3141.2	7.0099
400	0.19225	3254.2	7.2373	400	0.16847	3250.9	7.1793
500	0.22029	3471.9	7.5389	500	0.19550	3469.7	7.4824
600	0.24998	3693.2	7.8080	600	0.22199	3691.7	7.7523
700	0.27937	3919.7	8.0532	700	0.24818	3918.6	7.9983
800	0.30859	4152.1	8.2808	800	0.27420	4151.3	8.2258
900	0.33772	4390.8	8.4934	900	0.30012	4390.1	8.4386
1000	0.36678	4635.8	8.6938	1000	0.32598	4635.2	8.6390
1100	0.39581	4887.0	8.8837	1100	0.35180	4886.4	8.8290
1200	0.42482	5143.9	9.0642	1200	0.37761	5443.4	9.0096
1300	0.45382	5406.0	9.2364	1300	0.40340	5405.6	9.1817

p = 20 bar				p = 25 bar			
T (°C)	v m³/kg	h kJ/kg	s kJ/kgK	T (°C)	v m³/kg	h kJ/kg	s kJ/kgK
212.42	0.09963	2799.5	6.3408	223.99	0.07998	2803.1	6.2574
225	0.10377	2835.8	6.4146	225	0.08027	2806.3	6.2638
250	0.11144	2902.5	6.5452	250	0.08700	2880.1	6.4084
300	0.12547	3023.5	6.7663	300	0.09890	3008.8	6.6437
350	0.13587	3137.0	6.9562	350	0.10976	3126.2	6.8402
400	0.15120	3247.6	7.1270	400	0.12010	3239.3	7.0147
500	0.17568	3467.6	7.4316	500	0.13998	3462.0	7.3233
600	0.19960	3690.1	7.7023	600	0.15930	3686.2	7.5960
700	0.22323	3917.5	7.9487	700	0.17932	3914.6	7.8435
800	0.24668	4150.4	8.1766	800	0.19716	4148.2	8.0270
900	0.27004	4389.4	8.3895	900	0.21590	4387.6	8.2853
1000	0.29333	4634.6	8.5900	1000	0.23458	4633.1	8.4860
1100	0.31659	4885.9	8.7800	1100	0.25322	4887.6	8.6761
1200	0.33984	5142.9	8.9606	1200	0.27185	5141.7	9.8569
1300	0.36306	5405.1	9.1325	1300	0.29046	5404.4	9.0291

Contd..

Contd..

p = 30 bar				p = 35 bar			
T (°C)	v m^3/kg	h kJ/kg	s kJ/kgK	T (°C)	v m^3/kg	h kJ/kg	s kJ/kgK
233.90	0.06668	2804.1	6.1869	242.60	0.05707	2803.4	6.1252
250	0.07058	2855.8	6.2871	250	0.05873	2829.2	6.1748
300	0.08114	2993.5	6.5389	300	0.06842	2877.5	6.4460
350	0.09053	3115.3	6.7427	350	0.07678	3104.0	6.6578
400	0.09936	3230.8	6.9211	400	0.08453	3222.2	6.8404
500	0.11619	3456.5	7.2337	500	0.09918	3450.9	7.1571
600	0.13243	3682.3	7.5084	600	0.11324	3678.4	7.4338
700	0.14838	3911.7	7.7571	700	0.12699	3908.8	7.6837
800	0.16414	4146.0	7.9862	800	0.14056	4143.8	7.9135
900	0.17980	4385.9	8.1999	900	0.15402	4384.1	8.1275
1000	0.19541	4631.6	8.4009	1000	0.16743	4630.1	8.3288
1100	0.21098	4883.3	8.5911	1100	0.18080	4881.9	8.5191
1200	0.22652	5140.5	8.7719	1200	0.19415	5139.3	8.7000
1300	0.24206	5402.8	8.9442	1300	0.20749	5401.7	8.8723

p = 40 bar				p = 45 bar			
T (°C)	v m^3/kg	h kJ/kg	s kJ/kgK	T (°C)	v m^3/kg	h kJ/kg	s kJ/kgK
250.40	0.04978	2801.4	6.0700	257.48	0.04406	2798.3	6.0198
275	0.05457	2886.2	6.2284	275	0.04730	2863.1	6.1401
300	0.05884	2960.7	6.3614	300	0.05135	2943.1	6.2828
350	0.06645	3092.4	6.5820	350	0.05840	3080.6	6.5130
400	0.07341	3213.5	6.7689	400	0.06475	2104.7	6.7046
500	0.08643	3445.2	7.0900	500	0.07651	3439.5	7.0300
600	0.09885	3674.4	7.3688	600	0.08765	3670.5	7.3109
700	0.11095	3905.9	7.6198	700	0.09847	3903.0	7.5631
800	0.12287	4141.6	7.8502	800	0.10911	4139.4	7.7942
900	0.13469	4382.3	8.0647	900	0.11965	4380.6	8.0091
100	0.14645	4628.7	8.2661	1000	0.13013	4627.2	8.2108
1100	0.15817	4880.6	8.4566	1100	0.14056	4879.3	8.4014
1200	0.16987	5138.1	8.6376	1200	0.15098	5136.9	8.5824
1300	0.18156	5400.5	8.8099	1300	0.16139	5399.4	8.7548

Contd..

Contd..

p = 50 bar				p = 60 bar			
T (°C)	v m^3/kg	h kJ/kg	s kJ/kgK	T (°C)	v m^3/kg	h kJ/kg	s kJ/kgK
263.99	0.03944	2794.3	5.9733	275.64	0.03244	2784.3	5.8891
275	0.04141	2338.3	6.0543	300	0.03616	2884.2	6.0673
300	0.04532	2924.5	6.2083	350	0.04223	3043.0	6.3334
350	0.05194	3068.4	6.4492	400	0.04739	3177.2	6.5407
400	0.05781	3195.6	6.6458	450	0.05214	3301.8	6.7192
500	0.06857	3433.8	6.9758	500	0.05665	3422.1	6.8802
600	0.07869	3666.5	7.2588	600	0.06525	3658.4	7.1676
700	0.08849	3900.1	7.5122	700	0.07352	3894.3	7.4234
800	0.08711	4137.2	7.7440	800	0.08160	4132.7	7.6566
900	0.10762	4378.8	7.9593	900	0.08958	4375.3	7.8727
1000	0.11707	4625.7	8.1612	1000	0.09749	4622.7	8.0751
1100	0.12648	4878.0	8.3519	1100	0.10536	4875.4	8.2661
1200	0.13587	5135.7	8.5330	1200	0.11321	5133.3	8.4473
1300	0.14526	5398.2	8.7055	1300	0.12106	5396.0	8.6199

p = 70 bar				p = 80 bar			
T (°C)	v m^3/kg	h kJ/kg	s kJ/kgK	T (°C)	v m^3/kg	h kJ/kg	s kJ/kgK
285.88	0.02737	2772.1	5.8132	295.06	0.02352	2757.9	5.7431
300	0.02947	2838.4	5.9304	300	0.02426	2987.3	5.7905
350	0.03524	3016.0	6.2282	350	0.02995	2987.3	6.1300
400	0.03993	3158.1	6.4477	400	0.03432	3138.3	6.3633
450	0.04416	3287.0	6.6326	450	0.03817	3272.0	6.5550
500	0.04814	3410.3	6.7974	500	0.04175	3398.3	6.7239
600	0.05565	3650.3	7.0894	600	0.04845	3642.0	7.0205
700	0.06283	3888.4	7.3476	700	0.05481	3882.5	7.2812
800	0.06981	4128.3	7.8522	800	0.06097	4123.8	7.5173
900	0.07996	4371.8	7.7991	900	0.06702	4368.3	7.7350
1000	0.08350	4619.8	8.0020	1000	0.07301	4616.9	7.9384
1100	0.09027	4872.8	8.1933	1100	0.07896	4870.3	8.1299
1200	0.09703	5130.7	8.3747	1200	0.08489	5128.5	8.3115
1300	0.10377	5393.7	8.5472	1300	0.09080	5391.5	8.4842

Contd..

Contd..

p = 90 bar				p = 100 bar			
T (°C)	v m³/kg	h kJ/kg	s kJ/kgK	T (°C)	v m³/kg	h kJ/kg	s kJ/kgK
303.40	0.02048	2742.1	5.6771	311.06	0.101803	2724.7	5.6140
325	0.02327	2855.9	5.8711	325	0.01986	2809.0	5.7568
350	0.02580	2956.5	6.0361	350	0.02242	2923.4	5.9442
400	0.02993	3117.8	6.2853	400	0.02641	3096.5	6.2119
450	0.03350	3256.6	6.4843	450	0.02975	3240.8	6.4189
500	0.03677	3386.1	6.6585	500	0.03279	3373.6	6.5965
600	0.04285	3633.7	6.9588	60	0.03837	3625.3	6.9028
700	0.04857	3876.5	7.2221	700	0.04358	3870.5	7.1687
800	0.05409	4119.4	7.4597	800	0.04859	4114.9	7.4077
900	0.05950	4364.7	7.6782	900	0.05349	4361.2	7.6272
1000	0.06485	4613.9	7.8821	1000	0.05832	4611.0	7.8315
1100	0.07016	4867.7	8.0739	1100	0.06312	4865.1	8.0236
1200	0.07544	5126.2	8.2556	1200	0.06789	5123.8	8.2054
1300	0.08072	5389.2	8.4283	1300	0.07265	5387.0	8.3783

p = 125 bar				p = 150 bar			
T (°C)	v m³/kg	h kJ/kg	s kJ/kgK	T (°C)	v m³/kg	h kJ/kg	s kJ/kgK
327.89	0.01350	2673.8	5.4623	342.24	0.010338	2610.5	5.3097
350	0.01613	2826.2	5.7117	350	0.011470	2692.4	5.4420
400	0.02000	3039.3	6.0416	400	0.015649	2975.4	5.8810
450	0.02299	3199.8	6.2718	450	0.018446	3156.2	6.1403
500	0.02560	3341.7	6.4617	500	0.020800	3308.5	6.3442
550	0.02801	3475.1	6.6289	550	0.022927	3448.6	6.5198
600	0.03029	3604.0	6.7810	600	0.024911	3582.3	6.6775
700	0.03460	3855.4	7.0536	700	0.028612	3840.1	6.9572
800	0.03869	4103.7	7.2965	800	0.032096	4092.4	7.2040
900	0.04267	4352.5	7.5181	900	0.035457	4343.8	7.4279
1000	0.04658	4603.8	7.7237	1000	0.038748	4596.6	7.6347
1100	0.05045	4858.8	7.9165	1100	0.042001	4852.6	7.8282
1200	0.05430	5118.0	8.0987	1200	0.045233	5112.3	8.0108
1300	0.05813	5381.4	8.2717	1300	0.048455	5375.9	8.1839

Contd..

Contd..

p = 175 bar				p = 200 bar			
T (°C)	v m³/kg	h kJ/kg	s kJ/kgK	T (°C)	v m³/kg	h kJ/kg	s kJ/kgK
354.75	0.0079204	2528.8	5.1418	365.81	0.0058342	2409.7	4.9269
400	0.0124477	2902.8	5.7212	400	0.0099423	2818.1	5.5539
450	0.0151740	3109.7	6.0182	450	0.0126953	3060.1	5.9016
500	0.0173585	3274.0	6.2382	500	0.0147683	3238.2	6.1400
550	0.0192877	3421.4	6.4229	550	0.0165553	3393.5	6.3347
600	0.0210640	3560.1	6.5866	600	0.0181781	3537.6	6.5048
650	0.0227372	3693.9	6.7356	650	0.016929	3675.3	6.6582
700	0.0243365	3824.7	6.8736	700	0.0211311	3809.1	6.7993
800	0.0273849	4081.1	7.1245	800	0.0238532	4069.8	7.0544
900	0.0303071	4335.1	7.3507	900	0.0264463	4326.4	7.2830
1000	0.0331580	4589.5	7.5588	1000	0.0289666	4582.5	7.4925
1100	0.0359695	4846.4	7.7530	1100	0.0314471	4840.2	7.6874
1200	0.0387605	5106.6	7.9359	1200	0.0339071	5101.0	7.8706
1300	0.0415417	5370.5	8.1093	1300	0.0363574	5365.1	8.0441

Index